Molecular Evolution of Life

Chemica Scripta: Volume 26B, 1986

Molecular Evolution of Life

Proceedings of a conference held at Södergarn, Lidingö, Sweden, 8–12 September 1985

Edited by
Herrick Baltscheffsky, Hans Jörnvall and Rudolf Rigler

Published on behalf of The Royal Swedish Academy of Sciences
by Cambridge University Press

The right of the
University of Cambridge
to print and sell
all manner of books
was granted by
Henry VIII in 1534.
The University has printed
and published continuously
since 1584.

Cambridge University Press
Cambridge
London New York New Rochelle
Melbourne Sydney

Published by the Press Syndicate of the University of Cambridge
The Pitt Building, Trumpington Street, Cambridge CB2 1RP
32 East 57th Street, New York, NY 10022, USA
10 Stamford Road, Oakleigh, Melbourne 3166, Australia

First published 1986

Printed in Great Britain at the University Press, Cambridge

British Library CIP data available

Library of Congress CIP data available

ISSN 0004-2056 paperback
ISBN 0 521 33642 2 hard covers

Contents

Proteins and Enzymatic Functions

Complex Systems and Organization

Preface

In the last few years great advances have been made in our understanding of fundamental molecular aspects of the prebiological and biological evolution of life. Nucleic acids and proteins are investigated with rapidly increasing efficiency. Molecular dimensions and dynamics are studied with biochemical and biophysical methods at high resolution and sensitivity.This has made possible discoveries of new relationships between functional forms and structural patterns in molecular evolution.

There is a mutual cause and effect situation between scientists in the current fast pace of development in molecular evolution and the intensified collaboration between biochemistry, biophysics, genetics, medical chemistry, medical physics, molecular biology, and related areas. It was considered timely and urgent to try to integrate the expanding knowledge in an international conference focusing, over the borders of different academic subjects, on the molecular evolution of life. The response, both national and international, was overwhelming. The number of active participants increased from a planned 40–50 to 73. The presence of so many qualified scientists contributing to the area was made possible thanks to generous grants from the Royal Swedish Academy of Sciences and the Nobel Institute of Chemistry, the Swedish Medical Research Council, the Swedish Natural Science Research Council, the Swedish Cancer Society, the Ministry of Education, and from the following industries with an active research interest in this and neighbouring fields: Astra AB, Boehringer-Mannheim GmbH, KabiGen AB, KabiVitrum AB, LKB Products AB, Pharmacia AB and Skandigen AB.

The conference took place at Södergarns kursgård, Lidingö, near Stockholm, 8–12 September, 1985. The final afternoon session with four overview lectures, was held at the Royal Swedish Academy of Sciences. These lectures were given by Professors C. Ponnamperuma, USA and Sri Lanka, M. Eigen, FRG, E. Davie, USA, and G. Edelman, USA.

This volume contains nearly all of the papers presented at the conference. The contributions are grouped according to the main topics treated at the conference. Naturally, the divisions in the list of contents are somewhat arbitrary since overlaps between sub-fields are considerable.

Stockholm, November, 1985

Herrick Baltscheffsky Hans Jörnvall Rudolf Rigler

List of participants

J. Abelson, Pasadena, USA
R. Amils, Madrid, SPAIN
L. Arlinger, Stockholm, SWEDEN
E. Arnold, West Lafayette, USA
H. von Bahr-Lindström, Stockholm, SWEDEN
H. Baltscheffsky, Stockholm, SWEDEN
K. Beaucamp, Tutzing, FRG
C. Biebricher, Göttingen, FRG
G. Björk, Umeå, SWEDEN
T. L. Blundell, London, ENGLAND
C.-I. Brändén, Uppsala, SWEDEN
P. Clarke, London, ENGLAND
B. Daneholt, Stockholm, SWEDEN
E. W. Davie, Seattle, USA
L. Demetrius, Göttingen, FRG
R. E. Dickerson, Los Angeles, USA
G. M. Edelman, New York, USA
M. Eigen, Göttingen, FRG
L. Ernster, Stockholm, SWEDEN
S. Falkmer, Stockholm, SWEDEN
S. Forsén, Lund, SWEDEN
A. von Gabain, Umeå, SWEDEN
R. C. Gallo, Bethesda, USA
P. Gruss, Heidelberg, FRG
H. Haglund, Stockholm, SWEDEN
J. D. Hempel, Pittsburgh, USA
A. Holmgren, Stockholm, SWEDEN
L. Hood, Pasadena, USA
J.-O.Höög, Stockholm, SWEDEN
R. Huber, München, FRG
J. Jeffery, Aberdeen, SCOTLAND
H. Jörnvall, Stockholm, SWEDEN
U. Lagerkvist, Göteborg, SWEDEN
J. A. Lake, Los Angeles, USA
R. M. Lawn, San Fransisco, USA
A. Liljas, Uppsala, SWEDEN
F. Lindberg, Umeå, SWEDEN
S. Magnusson, Århus, DENMARK
B. Malmström, Göteborg, SWEDEN
B. Mannervik, Stockholm, SWEDEN
O. Markovič, Bratislava, CZECHOSLOVAKIA
E. Marmstål, Stockholm, SWEDEN
B. W. Matthews, Eugen, USA
S. Miller, La Jolla, USA
V. Mutt, Stockholm, SWEDEN
H. Neurath, Seattle, USA
S. Numa, Kyoto, JAPAN
L. Pereira da Silva, Campinas, BRASIL
S. Ohno, Duarte, USA
U. Pettersson, Uppsala, SWEDEN
C. Ponnamperuma, College Park, USA
R. Rigler, Stockholm, SWEDEN
P. Reichard, Stockholm, SWEDEN
J. M. Rosenberg, Pittsburgh, USA
M. Rossmann, West Lafayette, USA
R. T. Schimke, Stanford, USA
P. Schuster, Wien, AUSTRIA
J. Shepherd, Basel, SWITZERLAND
P. P. Slonimski, Gif-sur-Yvette, FRANCE
J. Stenflo, Malmö, SWEDEN
K. Stråby, Umeå, SWEDEN
J. R. Tata, London, ENGLAND
S. Tonegawa, Cambridge, USA
O. C. Uhlenbeck, Urbana, USA
G. Wadell, Umeå, SWEDEN
B. Wahren, Stockholm, SWEDEN
J. E. Walker, Cambridge, ENGLAND
B. L. Vallee, Boston, USA
H.-E. Wanntorp, Stockholm, SWEDEN
L. Wieslander, Stockholm, SWEDEN
H. Wigzell, Stockholm, SWEDEN
A. S. Wilkins, Cambridge, ENGLAND
R. Winkler, Göttingen, FRG

Prebiotic Systems and Evolutionary Pathways

Chemica Scripta 1986, **26B**, 5–11

Current Status of the Prebiotic Synthesis of Small Molecules

Stanley L. Miller

Department of Chemistry, B-017, University of California, San Diego, La Jolla, California 92093, USA

Paper presented at the Conference on 'Molecular Evolution of Life', Lidingö, Sweden, 8–12 September, 1985

Abstract

The prebiotic synthesis of small molecules has been accomplished using various simulated atmospheres with CH_4, N_2, NH_3, H_2O being the most effective, but H_2, CO, N_2, H_2O and H_2, CO_2, N_2, H_2O also give good yields of organic compounds provided $H_2/CO > 1$ and $H_2/CO_2 > 2$. The spark discharge is a very effective source of energy in such experiments, because it is a good source of HCN. Ultraviolet light would also have been important on the primitive earth. Almost all prebiotic amino acids are made by the hydrolysis of an amino nitrile formed from an aldehyde, NH_3 and HCN (Strecker synthesis). There are reasonable prebiotic syntheses worked out for the twenty amino acids that occur in proteins, with the exception of lysine, arginine and histidine. The purines are derived from the polymerization of HCN, and the precursor of the pyrimidines is cyanoacetylene. The sugars (including ribose), would have been formed from the base catalyzed polymerization of formaldehyde. There is no good prebiotic synthesis of straight chain fatty acids. Of the vitamin coenzymes, only nicotinic acid has been synthesized under prebiotic conditions.

Many of the molecules that are produced in these simulated primitive earth experiments are found in a group of meteorites that contain organic compounds, called the carbonaceous chondrites. Since such prebiotic syntheses took place on the parent body of the carbonaceous chondrites, generally thought to be an asteroid, it is plausible, but not proved, that such syntheses took place on the primitive earth, and that the first living organisms were formed out of these compounds.

Introduction

In the past three decades there has been a wide variety of experiments designed to simulate conditions on the primitive earth and to demonstrate how the organic compounds that made up the first living organisms were synthesized. This paper will review this work and indicate the status of such syntheses. There is too much material to review in detail, and the reader is directed to a number of more complete discussions [1–3].

1. The composition of the primitive atmosphere

There is no agreement on the constituents of the primitive atmosphere. It is to be noted that there is no geological evidence concerning the conditions on the earth from 4.5×10^9 years to 3.8×10^9 years since no rocks older than 3.8×10^9 years are known. Even the 3.8×10^9 years old Isua Rocks in Greenland are not sufficiently well preserved to infer details of the atmosphere at that time. Proposed atmospheres and the reasons given to favor them will not be discussed here. As shown in Sections 2–4, the more reducing atmospheres favor the synthesis of organic compounds both in terms of yields and the variety of compounds obtained. Some of the organic chemistry can give explicit predictions about atmospheric constituents. Such considerations cannot prove that the earth had a certain primitive atmosphere, but the prebiotic synthesis constraints should be a major consideration.

2. Energy sources

A wide variety of energy sources has been utilized with various gas mixtures since the first experiments using electric discharges. The importance of a given energy source is determined by the product of the energy available and its efficiency for organic compound synthesis. Even though both factors cannot be evaluated with precision, a qualitative assessment of the energy sources can be made. It should be emphasized that a single source of energy or a single process is unlikely to account for all the organic compounds on the primitive earth [4]. An estimate of the sources of energy on the earth at the present time is given in Table I.

The energy from the decay of radioactive elements was probably not an important energy source for the synthesis of organic compounds on the primitive earth since most of the ionization would have taken place in silicate rocks rather than in the reducing atmosphere. The shock wave energy from the impact of meteorites on the earth's atmosphere and surface as well as the larger amount of shock waves generated in lightning bolts have been proposed as energy sources for

Table I. *Present sources of energy averaged over the earth*

Source	Energy ($cal\ cm^{-2}\ yr^{-1}$)	Energy ($J\ cm^{-2}\ yr^{-1}$)
Total radiation from sun	260000	1090000
Ultraviolet light		
< 3000 Å	3400	14000
< 2500 Å	563	2360
< 2000 Å	41	170
< 1500 Å	1.7	7
Electric discharges	4[a]	17
Cosmic rays	0.0015	0.006
Radioactivity (to 1.0 km depth)	0.8	3.0
Volcanoes	0.13	0.5
Shock waves	1.1[b]	4.6

[a] 3 $cal\ cm^{-2}\ yr^{-1}$ of corona discharge + 1 $cal\ cm^{-2}\ yr^{-1}$ of lightning.
[b] 1 $cal\ cm^{-2}\ yr^{-1}$ of this is in the shock wave of lightning bolts and is also included under electric discharges.

primitive earth organic synthesis. Very high yields of amino acids have been reported in some experiments [5], but it is doubtful whether such yields would be obtained in natural shock waves. Cosmic rays are a minor source of energy on the earth at present, and it seems unlikely that any increase in the past could have been so great as to make them a major source of energy.

The energy in the lava emitted at the present time is a significant but not a major source of energy. It is generally supposed that there was a much greater amount of volcanic activity on the primitive earth, but there is no evidence to support this. Even if the volcanic activity was a factor of 10 greater than at present, it would not have been the dominant energy source. Nevertheless, molten lava may have been important in the pyrolytic synthesis of some organic compounds.

Ultraviolet light was probably the largest source of energy on the primitive earth. The wavelengths absorbed by the atmospheric constituents are all below 2000 Å except for ammonia (<2300 Å) and H_2S (<2600 Å). Whether it was the most effective source of organic compounds is not clear. Most of the photochemical reactions would occur in the upper atmosphere, and the products formed would, for the most part, absorb the longer wavelengths, and so be decomposed before they reached the protection of the oceans. The yield of amino acids from the photolysis of CH_4, NH_3, and H_2O at wavelengths of 1470 and 1294 Å is quite low [6], probably due to the low yields of hydrogen cyanide. The synthesis of amino acids by the photolysis of CH_4, C_2H_6, NH_3, H_2O, and H_2S mixtures by ultraviolet light of wavelengths greater than 2000 Å [7] is also a low yield synthesis, but the amount of energy is much greater in this region of the sun's spectrum. Only H_2S absorbs the ultraviolet light, but the photodissociation of H_2S results in a hydrogen atom having a high kinetic energy, which activates or dissociates the methane, ammonia, and water. This appears to be very attractive prebiotic synthesis. However, it is not clear whether a sufficient partial pressure of H_2S could be maintained in the atmosphere since H_2S is photolyzed rapidly to elemental sulfur and hydrogen.

The most widely used sources of energy for laboratory syntheses of prebiotic compounds are electric discharges. These include sparks, semicorona, arc, and silent discharges with the spark being the most frequently used type. The ease of handling and high efficiency of electric discharges are factors favoring its use, but the most important reason is that electric discharges are very efficient in synthesizing hydrogen cyanide, whereas ultraviolet light is not. Hydrogen cyanide is a central intermediate in prebiotic synthesis, being needed for amino acid synthesis by the Strecker reaction, or by selfpolymerization to amino acids, and most importantly for the prebiotic synthesis of adenine and guanine.

An important feature of all these energy sources is the activation of molecules in a local area followed by quenching of this activated mixture, and then protecting the organic compounds from further influence of the energy source. The quenching and protective steps are critical because the organic compounds will be destroyed if subjected continuously to the energy source.

3. Prebiotic synthesis of amino acids

Mixtures of CH_4, NH_3, and H_2O with or without added H_2 are considered strongly reducing atmospheres. The atmosphere of Jupiter contains these species with the H_2 in large excess over the CH_4. The first successful prebiotic amino acid synthesis was carried out using an electric discharge as an energy source [8]. The result was a large yield of amino acids (the yield of glycine alone was 2.1% based on the carbon), together with hydroxy acids, short aliphatic acids, and urea. One of the surprising results of this experiment was that the products were not a random mixture of organic compounds, but rather a relatively small number of compounds were produced in substantial yield. In addition the compounds were, with a few exceptions, of biological importance.

Table II. *Yields from sparking a mixture CH_4, NH_3, H_2O and H_2. The present yields are based on carbon. 59 mmoles (710 mg) of carbon was added as CH_4*

Compound	Yield (μmol)	Yield (%)
Glycine	630	2.1
Glycolic acid	560	1.9
Sarcosine	50	0.25
Alanine	340	1.7
Lactic acid	310	1.6
N-Methylalanine	10	0.07
α-Amino-*n*-butyric acid	50	0.34
α-Aminoisobutyric acid	1	0.007
α-Hydroxybutyric acid	50	0.34
β-Alanine	150	0.76
Succinic acid	40	0.27
Aspartic acid	4	0.024
Glutamic acid	6	0.051
Iminodiacetic acid	55	0.37
Iminoacetic-propionic acid	15	0.13
Formic acid	2330	4.0
Acetic acid	150	0.51
Propionic acid	130	0.66
Urea	20	0.034
N-Methyl urea	15	0.051
Total		15.2

The mechanism of synthesis of the amino and hydroxy acids was investigated [9]. It was shown that the amino acids were not formed directly in the electric discharge but were the result of solution reactions of smaller molecules produced in the discharge, in particular hydrogen cyanide and aldehydes. The reactions are shown in Scheme 1. These reactions were

$$RCHO + HCN + NH_3 \rightleftharpoons RCH(NH_2)CN \xrightarrow{H_2O} RCH(NH_2)\overset{O}{\overset{\|}{C}}\text{-}NH_2 \xrightarrow{H_2O} RCH(NH_2)COOH$$

$$RCHO + HCN \rightleftharpoons RCH(OH)CN \xrightarrow{H_2O} RCH(OH)\overset{O}{\overset{\|}{C}}\text{-}NH_2 \xrightarrow{H_2O} RCH(OH)COOH$$

Scheme 1

studied subsequently in detail, and the equilibrium and rate constants of these reactions were measured [10]. These results show that amino and hydroxy acids can be synthesized at high dilutions of HCN and aldehydes in a primitive ocean. It is also to be noted that the rates of these reactions were rather rapid. The half-lives for the hydrolysis of the amino and hydroxy nitriles are less than 10^3 years at 0 °C.

This synthesis of amino acids, called the Strecker synthesis, requires the presence of NH_4^+ (and NH_3) in the primitive ocean. On the basis of the experimental equilibrium and rate constants it can be shown [10] that equal amounts of amino and hydroxy acids are obtained when the NH_4^+ concentration is about 0.01 M at pH 8 and 25 °C with this NH_4^+ concentration being insensitive to temperature and pH. This translates into a p_{NH_3} in the atmosphere of 2×10^{-7} atm at 0 °C and 4×10^{-6} atm at 25 °C. This is a low partial pressure, but it would seem to be necessary for amino acid synthesis. A similar estimate of the NH_4^+ concentration in the primitive ocean can be obtained from the equilibrium decomposition of aspartic acid [11]. Ammonia is decomposed by ultraviolet light, but mechanisms for resynthesis are available. The details of the ammonia balance on the primitive earth remain to be worked out.

In a typical electric discharge experiment, the partial pressure of CH_4 is 0.1–0.2 atm. This pressure is used for convenience, and it is likely, but never demonstrated, that organic compound synthesis would work at much lower partial pressures of methane. There are no estimates available for p_{CH_4} on the primitive earth but 10^{-5} to 10^{-3} atm seems plausible. Higher pressures are not reasonable because the sources of energy would convert the CH_4 to organic compounds in the oceans too rapidly for higher pressures of CH_4 to build up.

As discussed above, ultraviolet light acting on this mixture of gases is not effective in producing amino acids except at very short wavelengths or in the presence of an absorber such as H_2S.

Pyrolysis of CH_4 and NH_3 gives very low yields of amino acids. The pyrolysis conditions are from 800 to 1200 °C with contact times of a second or less [12]. However, the pyrolysis of CH_4 and other hydrocarbons gives good yields of benzene, phenylacetylene, and many other hydrocarbons. It can be shown that phenylacetylene would be converted to phenylalanine and tyrosine in the primitive ocean [13]. Pyrolysis of the hydrocarbons in the presence of NH_3 gives substantial yields of indole, which can be converted to tryptophan in the primitive ocean [14].

A mixture of CH_4, N_2, and traces of NH_3, and H_2O is a more realistic atmosphere for the primitive earth because large amounts of NH_3 would not have accumulated in the atmosphere since the NH_3 would dissolve in the ocean. It is still, however, a strongly reducing atmosphere.

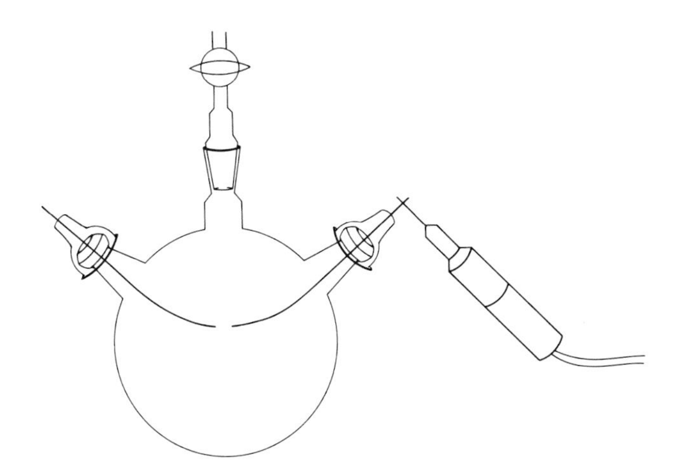

Fig. 1. Spark discharge apparatus. The 3 l flask is shown with the two tungsten electrodes and a spark generator. The second electrode is usually not grounded. In the experiments described in Table III, the flask contained 100 ml of 0.05 M-NH_4Cl brought to pH 8.7 giving p_{NH_3} of 0.1 torr. The p_{CH_4} was 200 torr and p_{N_2} was 80 torr. Since the temperature was about 30 °C during the sparking, p_{H_2O} was 32 torr.

This mixture of gases is quite effective with an electric discharge in producing amino acids [15]. The yields are somewhat lower than with higher partial pressures of NH_3, but the products are more diverse. Hydroxy acids, short aliphatic acids, and dicarboxylic acids are produced along with the amino acids. Ten of the 20 amino acids that occur in proteins are produced directly in this experiment. Counting asparagine and glutamine, which are formed but hydrolyzed before analysis, and methionine, which is formed when H_2S is added [16], one can say that 13 of the 20 amino acids in proteins can be formed in this single experiment. Cysteine was found in the photolysis of CH_4, NH_3, H_2O, and H_2S [7]. The pyrolysis of hydrocarbons, as discussed above, leads to phenylalanine, tyrosine, and tryptophan [13, 14]. This leaves only the basic amino acids: lysine, arginine, and histidine. There are so far no established prebiotic syntheses of these amino acids. There is no fundamental reason that the basic amino acids cannot be synthesized, and this problem may be solved before too long.

4. Mildly reducing and non-reducing atmospheres

There has been less experimental work with gas mixtures containing CO and CO_2 as carbon sources instead of CH_4.

Table III. *Yields from sparking CH_4 (336 mmoles); N_2, and H_2O with traces of NH_3*

	μmol
Glycine	440
Alanine	790
α-Amino-*n*-butyric acid	270
α-Aminoisobutyric acid	~30
Valine	19.5
Norvaline	61
Isovaline	~5
Leucine	11.3
Isoleucine	4.8
Alloisoleucine	5.1
Norleucine	6.0
tert-Leucine	<0.02
Proline	1.5
Aspartic acid	34
Glutamic acid	7.7
Serine	5.0
Threonine	~0.8
Allothreonine	~0.8
α,γ-Diaminobutyric acid	33
α-Hydroxy-γ-aminobutyric acid	74
α,β-Diaminopropionic	6.4
Isoserine	5.5
Sarcosine	55
N-Ethylglycine	30
N-Propylglycine	~2
N-Isopropylglycine	~2
N-Methylalanine	~15
N-Ethylalanine	<0.2
β-Alanine	18.8
β-Amino-*n*-butyric acid	~0.3
β-Amino-isobutyric acid	~0.3
γ-Aminobutyric acid	2.4
N-Methyl-β-alanine	~5
N-Ethyl-β-alanine	~2
Pipecolic acid	~0.05

Yield based on the carbon added as CH_4. Glycine = 0.26%, Alanine = 0.71%, total yield of amino acids in the table = 1.90%.

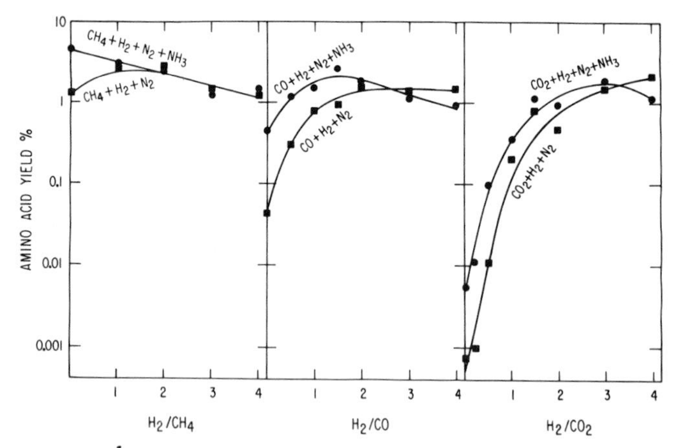

Fig. 2. Amino acid yields based on initial carbon. The apparatus of Fig. 1 was used with a sparking time of 48 h. In all experiments $p_{N_2} = 100$ torr, p_{CH_4} or p_{CO} or $p_{CO_2} = 100$ torr. For experiments containing NH_3, the spark discharge flask contained 100 ml of 0.05 M-NH_4Cl brought to ~pH 8.7 so that p_{NH_3} was 0.1 torr. For experiments not containing NH_3, the flask had 100 ml of H_2O.

Spark discharges have been the source of energy most extensively investigated [17. 18]. Figure 2 compares amino acid yields using CH_4, CO, and CO_2 as a carbon source with various amounts of H_2 [18]. Separate experiments were performed with and without added NH_3. In the case of CH_4 without added NH_3, the yield of amino acids is 4.7% at $H_2/CH_4 = 0$ and drops to 1.4% at $H_2/CH_4 = 4$. With CO and no added NH_3, the amino acid yield is 0.05% at $H_2/CO = 0$ and rises to a maximum of 2.7% at $H_2/CO = 3$. With CO_2 and no added NH_3, the amino acid yield is 7×10^{-4} at $H_2/CO_2 = 0$. This is close to the level of reagent contamination and is so low that this could not be considered as a significant source of amino acids on the primitive earth. At higher H_2/CO_2 ratios, however, the yield rises to about 2%. With both CO and CO_2, the presence of added NH_3 increases the yield of amino acids by a factor of ~10 at low H_2/CO and H_2/CO_2 ratios. The amino acids produced in the CH_4 experiments were similar to those shown in Table III. With CO and CO_2, glycine was the predominant amino acid with only a small amount of alanine being produced.

The hydrogen cyanide and formaldehyde yields were also measured in these experiments. In both cases the yields paralleled those of the amino acids but were a factor 5 to 10 higher. Considerable amounts of NH_3 were also produced in these experiments.

A mixture of $CO+H_2$ is used in the Fischer–Tropsch reaction to make hydrocarbons in high yields. The reaction requires a catalyst, usually Fe or Ni supported on silica, a temperature of 200–400 °C and a short contact time. Depending on the conditions, aliphatic hydrocarbons, aromatic hydrocarbons, alcohols, and acids can be produced. If NH_3 is added to the $CO+H_2$, then amino acids, purines, and pyrimidines can be formed [19]. The intermediates in these reactions are not known, but it is likely that HCN is involved together with some of the intermediates postulated for the electric discharge processes.

A mixture of $CO+H_2O$ with electric discharges is not particularly effective in organic compound synthesis, but ultraviolet light that is absorbed by the water (<1849 Å) results in the production of formaldehyde and other aldehydes, alcohols, and acids in fair yields [20, 21]. The mechanism seems to involve splitting the H_2O to H+OH with the OH converting CO to CO_2 and the H reducing another molecule of CO.

Electric discharges and ultraviolet light do not give substantial amounts of organic compounds with a mixture of CO_2+H_2O. Ionizing radiation (e.g. 40 MeV helium ions) gives small yields of formic acid and formaldehyde [24].

Calculations using one dimensional photochemical models of CO_2+H_2O atmospheres show that substantial amounts of H_2CO can be produced in these atmospheres by solar ultraviolet light [22, 23].

The action of gamma-rays on an aqueous solution of CO_2 and ferrous ion gives fair yields of formic acid, oxalic acid, and other simple products [25]. Ultraviolet light gives similar results. In these reactions, the Fe^{2+} is a stoichiometric reducing agent rather than a catalyst. Nitrogen in the form of N_2 does not react, and experiments with NH_3 have not been tried.

The implications of these results in considering the composition of the primitive earth is that CH_4 is the best carbon source for prebiotic synthesis, especially for amino acid synthesis. Although glycine was essentially the only amino acid synthesized in the spark discharge experiments with CO and CO_2, other amino acids (e.g. serine, aspartic acid, alanine) would probably have been formed from this glycine with the H_2CO and HCN in the primitive ocean over longer periods of time. Since we do not know which amino acids were required for the origin of the first living organism, atmospheres containing CO and CO_2 can not be excluded, but a CH_4 containing atmosphere is favored if a wide variety of amino acids is needed. The synthesis of purines and sugars described below would not be greatly different with CH_4, CO or CO_2 with adequate H_2. Although the spark discharge yields of amino acids, HCN and H_2CO are about the same with CH_4 and with $H_2/CO > 1$ and $H_2/CO_2 > 2$, it is not clear how such high H_2/carbon ratios could have been maintained in the primitive atmosphere since H_2 escapes from the earth's atmosphere into outer space. These problems are poorly understood and beyond the scope of this outline.

5. Purine and pyrimidine synthesis

Hydrogen cyanide is used in the synthesis of purines as well as amino acids. This is illustrated in a remarkable synthesis of adenine. If concentrated solutions of ammonium cyanide are refluxed for a few days, adenine is obtained in up to 0.5% yield along with 4-aminoimidazole-5-carboxamide and the usual cyanide polymer [26].

The mechanism of adenine synthesis in these experiments is probably as shown in Scheme 2.

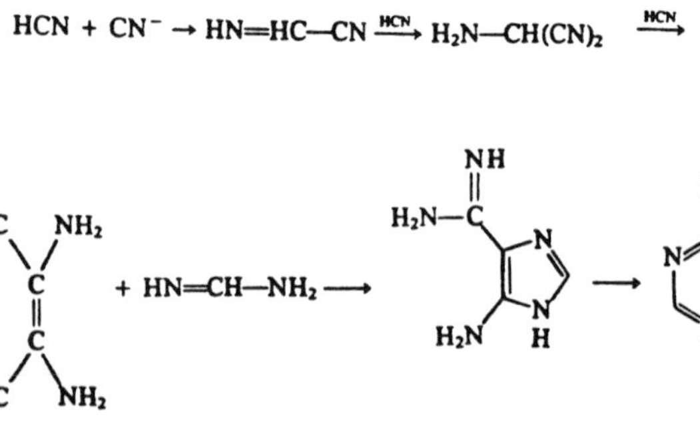

$$HCN + CN^- \rightarrow HN{=}HC{-}CN \xrightarrow{HCN} H_2N{-}CH(CN)_2 \xrightarrow{HCN}$$

$$(NC)(H_2N)C{=}C(NC)(NH_2) + HN{=}CH{-}NH_2 \rightarrow \text{(4-aminoimidazole-5-carboxamidine)} \rightarrow \text{(adenine)}$$

Scheme 2

The difficult step in the synthesis of adenine just described is the reaction of tetramer with formamidine. This step may be bypassed by the photochemical rearrangement of tetramer to aminoimidazole nitrile, a reaction that proceeds readily in contemporary sunlight [27] (Scheme 3).

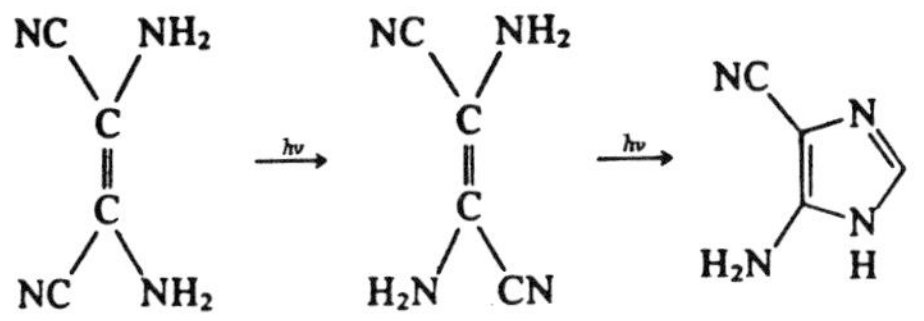

Scheme 3

A further possibility is that tetramer formation may have occurred in a eutectic solution. High yield of tetramer (> 10%) can be obtained by cooling dilute cyanide solutions to between −10 and −30 °C for a few months [27].

The prebiotic synthesis of the pyrimidine cytosine involves cyanoacetylene, which is synthesized in good yield by sparking mixtures of $CH_4 + N_2$. Cyanoacetylene reacts with cyanate to give cytosine [28] (Scheme 4), and the cytosine can be

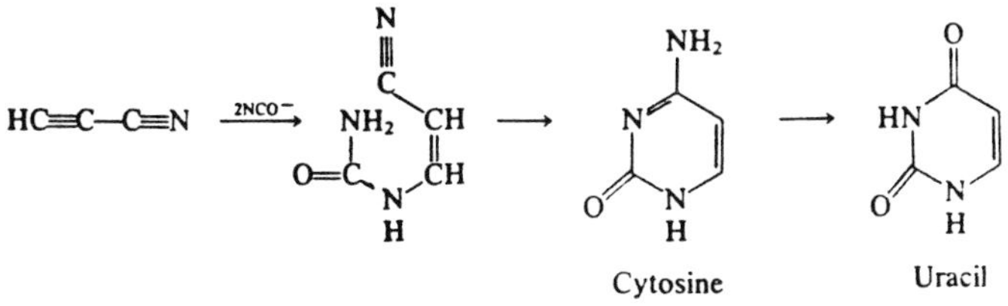

Cytosine Uracil

Scheme 4

converted to uracil. Cyanate can come from cyanogen or by the decomposition of urea.

A related synthesis starts with cyanoacetaldehyde, from the hydration of cyanoacetylene, which reacts with guanidine to give diaminopyrimidine. This is then hydrolyzed to cytosine and uracil [29] (Scheme 5).

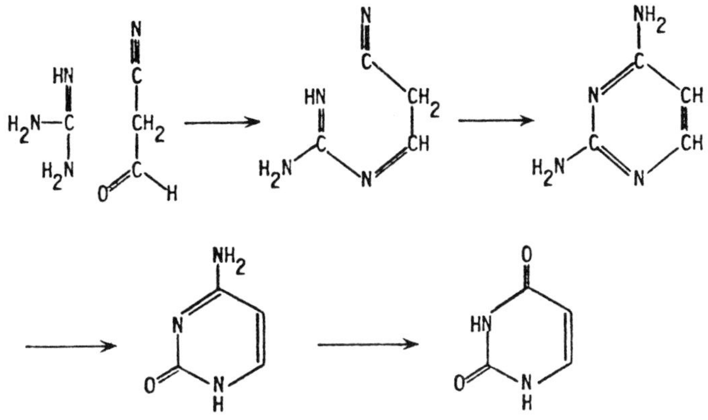

Scheme 5

Another prebiotic synthesis of uracil starts from β-alanine and cyanate and ultraviolet light [30] (Scheme 6).

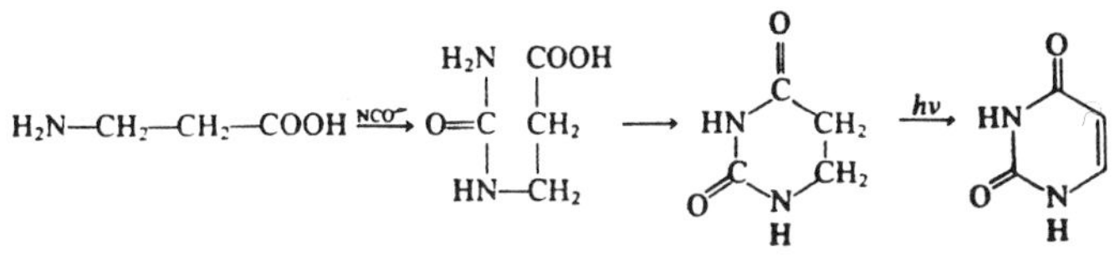

Scheme 6

6. Sugars

The synthesis of reducing sugars from formaldehyde under alkaline conditions was discovered long ago. However, the process is very complex and incompletely understood. It depends on the presence of a basic catalyst, with calcium hydroxide and calcium carbonate being frequently used. In the absence of catalysts, little or no sugar is obtained. Particularly attractive is the finding that at 100 °C, clays such as kaolin serve to catalyze formation of monosaccharides, including ribose, in good yield from dilute (0.01 M) solutions of formaldehyde [31, 32].

The reaction is autocatalytic and proceeds in stages through glycolaldehyde, glyceraldehyde, and dihydroxyacetone, tetroses, and pentoses to give finally hexoses including glucose and fructose. One proposed reaction sequence is as shown in Scheme 7.

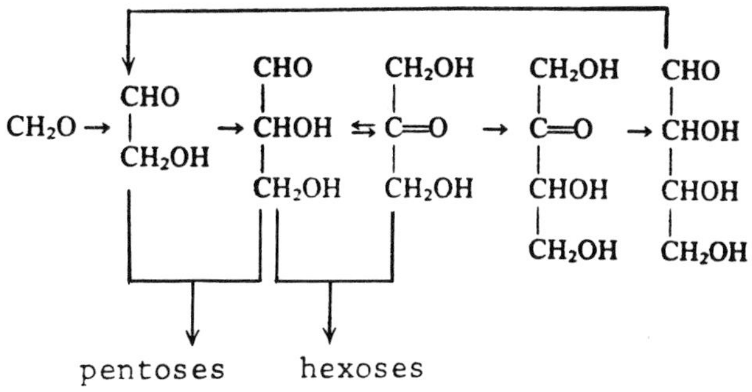

pentoses hexoses

Scheme 7

The problem with sugars on the primitive earth is not their synthesis, but rather their stability. They decompose in a few hundred years at most at 25 °C. There are a number of possible ways to stabilize sugars, the most interesting being to convert the sugar to a glycoside of a purine or pyrimidine.

7. Other prebiotic compounds

There are a number of compounds that have been synthesized under primitive earth conditions, but space does not permit an adequate discussion. These include:

dicarboxylic acids, tricarboxylic acids, fatty acids (C_2–C_{10}, branched and straight), fatty alcohols (straight chain via Fischer–Tropsch reaction), Porphin, nicotinonitrile and nicotinamide, triazines, imidazoles.

Other prebiotic compounds that may have been involved in polymerization reactions include:

cyanate [NCO^-], cyanamide [H_2NCN], cyanamide dimer [$H_2NC(NH)NH$—CN], dicyanamide [NC—NH—CN], cyanogen [NC—CN], HCN tetramer, diimino succinonitrile, acylthioesters, phosphate polymers.

8. Compounds that have not been synthesized prebiotically

It is a matter of opinion as to what constitutes a prebiotic synthesis. In some cases the conditions are so forced (e.g. the use of anhydrous solvents) or the concentrations so high (e.g. 10 M formaldehyde) that such conditions could not have occurred extensively on the primitive earth. Reactions under these and other extreme conditions can not be considered prebiotic.

There have been many claimed prebiotic synthesis in which the compound has not been properly identified. The best method for unequivocal identification these days is gas chromatography–mass spectrometry of a suitable derivative, although melting points and mixed melting points can sometimes be used. The amino acid analyzer alone or chromatography in multiple solvent systems does not prove the identification of a compound.

Some of the compounds which do not yet have adequate prebiotic syntheses are: arginine, lysine, histidine, straight-chain fatty acids, porphyrins, pyridoxal, thiamine, riboflavin, folic acid, lipoic acid, biotin. It is probable that prebiotic syntheses will be available before too long for some of these compounds. In other cases the compounds may not have been synthesized prebioticly, so their occurrence in living systems started after the origin of life.

9. Organic compounds in carbonaceous chondrites

On 28 September 1969 a type II carbonaceous chondrite fell in Murchison, Australia. Surprisingly large amounts of amino acids were found by Kvenvolden *et al.* [33, 34]. The first report identified seven amino acids (glycine, alanine, valine, proline, glutamic acid, sarcosine, and α-aminoisobutyric acid), of which all but valine and proline had been found in the original electric discharge experiments [8, 9]. The most striking are sarcosine and α-aminoisobutyric acid. The second report identified 18 amino acids of which nine had previously been identified in the original electric discharge experiment, but the remaining nine had not.

At that time we had identified the hydrophobic amino acids from the low temperature electric discharge experiments described above, and therefore we examined the products for the non-protein amino acids found in Murchison. We were able to find all of them [15].

There is a striking similarity between the products and relative abundances of the amino acids produced by electric discharge and the meteorite amino acids. Table IV compares the results. The most notable difference between the meteorite and the electric discharge amino acids is the pipecolic acid, the yield being extremely low in the electric discharge.

Table IV. *Relative abundances of amino acids in the Murchison meteorite and in an electric discharge synthesis*

Amino acid	Murchison meteorite	Electric discharge
Glycine	****	****
Alanine	****	****
α-Amino-*n*-butyric acid	***	****
α-Aminoisobutyric acid	****	**
Valine	***	**
Norvaline	***	***
Isovaline	**	**
Proline	***	*
Pipecolic acid	*	< *
Aspartic acid	***	***
Glutamic acid	***	**
β-Alanine	**	**
β-Amino-*n*-butyric acid	*	*
β-Aminoisobutyric acid	*	*
γ-Aminobutyric acid	*	**
Sarcosine	**	***
N-Ethylglycine	**	***
N-Methylalanine	**	**

Mol. ratio to glycine (= 100): * 0.05–0.5; ** 0.5–5; *** 5–50; **** < 50.

Proline is also present in relatively low yield from the electric discharge. The amount of α-aminoisobutyric acid is greater than α-amino-*n*-butyric acid in the meteorite, but the reverse is the case in the electric discharge. We do not believe that reasonable differences in ratios of amino acids detract from the overall picture. Indeed, the ratio of α-aminoisobutyric acid to glycine is quite different in two meteorites of the same type, being 0.4 in Murchison and 3.8 in Murray [35]. A similar comparison has been made between the dicarboxylic acids in Murchison [36] and those produced by an electric discharge [37], and the product ratios are quite similar.

The close correspondence between the amino acids found in the Murchison meteorite and those produced by an electric discharge synthesis, both as to the amino acids produced and their relative ratios, suggests that the amino acids in the meteorite were synthesized on the parent body by means of an electric discharge or analogous processes. A quantitative comparison of the amino acid and hydroxy acid abundances [38] shows that these compounds can be accounted for by a Strecker–Cyanohydrin synthesis on the parent body [39]. Electric discharges appear to be the most favored source of energy but sufficient data are not available to make realistic comparison with other energy sources.

Our ideas on the prebiotic synthesis of organic compounds are based largely on the results of experiments in model systems. So it is extremely gratifying to see that such synthesis really did take place on the parent body of the meteorite, and so it becomes quite plausible that they took place on the primitive earth.

10. Interstellar molecules

In the past 15 years a large number of organic molecules have been found in interstellar dust clouds mostly by emission lines in the microwave region of the spectrum (for a summary see ref. 40). The concentration of these molecules is very low (a few molecules per cm^3 at the most) but the total amount in a dust cloud is large. The molecules found include formaldehyde, hydrogen cyanide, acetaldehyde, and cyanoacetylene. These are important prebiotic molecules, and this immediately raises the question of whether the interstellar molecules played a role in the origin of life on the earth. In order for this to have taken place it would have been necessary for the molecules to have been greatly concentrated in the solar nebula and to have arrived on the earth without being destroyed by ultraviolet light or pyrolysis. This appears to be difficult to do. In addition, it is necessary for some molecules to be continuously synthesized (unless life started very quickly) because of their instability, and an interstellar source could not be responsible for these.

For these reasons, it is generally felt that the interstellar molecules played at most a minor role in the origin of life. However, the presence of so many molecules of prebiotic importance in interstellar space, combined with the fact that their synthesis must differ from that on the primitive earth where the conditions were very different, indicates that some molecules are particularly easily synthesized when radicals and ions recombine. Another way of saying this is that there appears to be a universal organic chemistry, which shows up in interstellar space, in the atmospheres of the major planets, and in the reducing atmosphere of the primitive earth.

Acknowledgement

This work was supported by NASA Grant NAGW-20.

References

1. Miller, S. L. and Orgel, L. E., *The Origins of Life on the Earth*. Prentice Hall, Englewood Cliffs, New Jersey (1974).
2. Kenyon, D. H. and Steinman, G., *Biochemical Predestination*. McGraw-Hill, New York (1969).
3. Lemmon, R. M., *Chem. Rev.* **70**, 95–109 (1970).
4. Miller, S. L., Urey, H. C. and Oro, J., *J. Mol. Evol.* **9**, 59–72 (1976).
5. Bar-Nun, A., Bar-Nun, N., Bauer, S. H. and Sagan, C., *Science* **168**, 470–473 (1970).
6. Groth, W. and Weyssenhoff, H. von, *Planet. Space Sci.* **2**, 79–85 (1960).
7. Sagan, C. and Khare, B. N., *Science* **173**, 417–420 (1971); *Nature (London)* **232**, 577–578 (1971).
8. Miller, S. L., *Science* **117**, 528–529 (1953); *J. Am. Chem. Soc.* **77**, 2351–2361 (1955).
9. Miller, S. L., *Biochim. Biophys. Acta* **23**, 480–489 (1957); *Ann. N.Y. Acad. Sci.* **69**, 260–274; also in *The Origin of Life on the Earth* (ed. A. Oparin). Pergamon Press, Oxford (1959), pp. 123–135.
10. Miller, S. L. and Van Trump, J. E., in *Origin of Life* (ed. Y. Wolman). Reidel, Dordrecht, Holland (1981), pp. 135–141.
11. Bada, J. L. and Miller, S. L., *Science* **159**, 423–425 (1968).
12. Lawless, J. G. and Boynton, C. D., *Nature (London)* **243**, 405–407 (1973).
13. Friedmann, N. and Miller, S. L., *Science* **166**, 766–767 (1969).
14. Friedmann, N., Haverland, W. J. and Miller, S. L., in *Chemical Evolution and the Origin of Life* (ed. R. Buvet and C. Ponnamperuma), pp. 123–135. North Holland, Amsterdam (1971).
15. Ring, D., Wolman, Y., Friedmann, N. and Miller, S. L., *Proc. Natl. Acad. Sci., USA* **69**, 765–768 (1972); Wolman, Y., Haverland, W. J. and Miller, S. L., *Proc. Natl. Acad. Sci., USA* **69**, 809–811 (1972).
16. Van Trump, J. E. and Miller, S. L., *Science* **178**, 859–860 (1972).
17. Abelson, P. H., *Proc. Natl. Acad. Sci., USA* **54**, 1490–1494 (1965).
18. Schlesinger, G. and Miller, S. L., *J. Mol. Evol.* **19**, 376–382, 383–390 (1983).
19. Hayatsu, R., *et al.*, *Geochim. Cosmochim. Acta* **36**, 555–571 (1972); Yoshino, D., Hayatsu, R. and Anders, E., *Geochim. Cosmochim. Acta* **35**, 927–938 (1971).
20. Bar-Nun, A. and Hartman, H., *Orig. Life* **9**, 93–101 (1978).
21. Bar-Nun, A. and Chang, S., *J. Geophys. Res.* **88**, 6662–6672 (1983).
22. Pinto, J. P., Gladstone, C. R. and Yung, Y. L., *Science* **210**, 183–185 (1980).
23. Kasting, J. F., Pollack, J. B. and Crisp, D., *J. Atm. Chem.* **1**, 403–428 (1984).
24. Garrison, W. M., Morrison, D. C., Hamilton, J. G., Benson, A. A. and Calvin, M., *Science* **114**, 416–418 (1951).
25. Getoff, N., *Z. Naturforsch* **17b**, 87–90, 751–757 (1962).
26. Oró, J. and Kimball, A. P., *Arch. Biochem. Biophys.* **94**, 221–227 (1961); *ibid.* **96**, 293–313 (1962).
27. Sanchez, R. A., Ferris, J. P. and Orgel, L. E., *J. Mol. Biol.* **30**, 223–253 (1967); *ibid.* **38**, 121–128 (1968).
28. Sanchez, R. A., Ferris, J. P. and Orgel, L. E., *Science* **154**, 784–785 (1966); Ferris, J. P., Sanchez, R. A. and Orgel, L. E., *J. Mol. Biol.* **33**, 693–704 (1968).
29. Ferris, J. P., Zamek, O. S., Altbuch, A. M. and Freiman, H., *J. Mol. Evol.* **3**, 301–309 (1974).
30. Schwartz, A. W. and Chittenden, G. J. F., *Biosystems* **9**, 87–92 (1977).
31. Gabel, N. W. and Ponnamperuma, C., *Nature (London)* **216**, 453–455 (1967).
32. Reid C. and Orgel, L. E., *Nature (London)* **216**, 455 (1967).
33. Kvenvolden, K., Lawless, J. G., Pering, K., Peterson, E., Flores, J., Ponnamperuma, C., Kaplan, I. R. and Moore, C., *Nature (London)* **228**, 923–926 (1970).
34. Kvenvolden, K. A., Lawless, J. G. and Ponnamperuma, C., *Proc. Natl. Acad. Sci., USA* **68**, 486–490 (1971).
35. Cronin, J. R. and Moore, C. B., *Science* **172**, 1327–1329 (1971).
36. Lawless, J. G., Zeitman, B., Pereira, W. E., Summons, R. E. and Duffield, A. M., *Nature (London)* **251**, 40–42 (1974).
37. Zeitman, B., Chang, S. and Lawless, J. G., *Nature (London)* **251**, 42–43 (1974).
38. Peltzer, E. T. and Bada, J. L., *Nature (London)* **272**, 443–444 (1978).
39. Peltzer, E. T., Bada, J. L., Schlesinger, G. and Miller, S. L., *Adv. Space Res.* **4** (no. 12), 69–74 (1984).
40. Mann, A. P. C. and Williams, D. A., *Nature (London)* **283**, 721–725 (1980).

Chemica Scripta 1986, **26B**, 13–26

The Physics of Molecular Evolution

Manfred Eigen

Max-Planck-Institut für Biophysikalische Chemie, Am Fassberg, D-3400 Göttingen, Federal Republic of Germany

Paper presented at the Conference on 'Molecular Evolution of Life', Lidingö, Sweden, 8–12 September 1985

Abstract

The Darwinian concept of evolution through natural selection has been revised and put on a solid physical basis, in a form which applies to self-replicable macromolecules. Two new concepts are introduced: 'sequence space' and 'quasi-species'. Evolutionary change in the DNA- or RNA-sequence of a gene can be mapped as a trajectory in a sequence space of dimension ν, where ν corresponds to the number of changeable positions in the genomic sequence. Emphasis, however, is shifted from the single surviving wildtype, a single point in the sequence space, to the complex structure of the mutant distribution that constitutes the quasi-species. Selection is equivalent to an establishment of the quasi-species in a localized region of sequence space, subject to threshold conditions for the error rate and sequence length. Arrival of a new mutant may violate the local threshold condition and thereby lead to a displacement of the quasi-species into a different region of sequence space. This transformation is similar to a phase transition; the dynamical equations that describe the quasi-species have been shown to be analogous to those of the two-dimensional Ising model of ferromagnetism. The occurrence of a selectively advantageous mutant is biased by the particulars of the quasi-species distribution, whose mutants are populated according to their fitness relative to that of the wildtype. Inasmuch as fitness regions are connected (like mountain ridges) the evolutionary trajectory is guided to regions of optimal fitness. Experimental evidence for this modification of the simple 'chance and law' nature of the Darwinian concept is presented. The results of the theory can be applied to the construction of a machine that provides optimal conditions for a rapid evolution of functionally active macromolecules.

I. Introduction: the physical and the historical aspect of evolution

Life is the unique outcome of biological evolution. The historical process is not accessible to physics because physics doesn't deal with processes as such [1]. Physics only deals with regularities among events. Our question therefore is: Can life also be considered the manifestation of a physical regularity corresponding to certain initial and boundary conditions? The laws behind such a regularity might be identified, while the evolutionary process itself still escapes a complete reconstruction. In order to identify historical boundary conditions we need records, witnesses or remnants remaining in the present organisms.

As far as the regularity 'life' is concerned we can be sure that there is no simple universal formula that covers all its aspects. Understanding *E. coli*, for instance, is of little help in understanding the human brain. On the other hand there are theories which explain certain physical aspects of life. The theory of selection will tell us how to master complexity and how to achieve optimal adaptation. The theory of functional organization will tell us how ensembles of molecules can be integrated to form supramolecular entities such as the living cell. Network theory might tell us how a supracellular organization can achieve functions that are typical of higher organisms.

In this paper I shall review some theories relevant to the major steps in the origin and evolution of life.

The procedure will be:

- to start with the given facts and to identify the physical problems which are involved;
- to develop theories that solve the problems;
- to derive quantitative consequences of the theories;
- to subject these consequences to experimental tests.

Typical problems involved are:

- how to originate genetic information;
- how to achieve optimal performance of translation products;
- how to organize and control supramolecular complexity, and possibly
- how to recreate such systems in the test tube.

II. Selection and evolution at the molecular level

II.1. *Reducing complexity*

The major obstacle to a physical understanding of life is the tremendous complexity of its molecular organization. As demonstrated in Table I this complexity is encountered with even the smallest unit of life, the single gene. It is almost trivial to emphasize that neither random choice nor any type of equilibrium model could solve this riddle of complexity. Any two states i and k, that have established equilibrium, are populated relative to one another according to a Boltzmann exponential: $x_i/x_k = \exp(-\Delta G_{ik}/RT)$, where x_i and x_k are fractional population numbers and ΔG_{ik} is the difference of

Table I
A DNA sequence with ν positions has 4^ν states:

ν	Coding for	Number of states
30	Deca peptide	$\sim 10^{18}$
300	Small Protein	$\sim 10^{180}$
3000	Small virus	$\sim 10^{1800}$

Equivalents of complexity in our universe:

10^{22} DNA molecules ($\nu = 30$) $\sim$ One ton of 1-millimolar NTP solution

10^{42} DNA molecules ($\nu = 300$) $\sim$ A viscous soup made of all oceans

10^{102} DNA molecules ($\nu = 3000$) $\sim$ Closest packing of DNA filling the universe

10^{180} DNA molecules are beyond our imagination

free energies between both states. The ΔG-values of different polynucleotide chains do not exceed the thermal energy by factors large enough to allow a selective accumulation of any particular sequence among the huge manifold of alternative sequences. As Table I shows the total number of possible sequences by far exceeds any realistic value of population numbers meaning that true equilibrium can never be established and consequently the system must remain under random drift conditions.

In this situation, natural selection is the only tool to reduce the complexity in a deterministic manner. Hence an entirely reasonable answer to one of the questions posted above reads: *Information, in particular genetic information, originates through natural selection.* The problem next to solve is: *Who* selects, or better, *how* is *natural* selection as a process of self-organization *related to molecular properties*?

Natural selection is the consequence of a 'naturally' inherited property of certain states of matter. It simply follows from replication (including selfcomplementary as well as recombinative reproduction) under non-equilibrium conditions. All living beings – as Spiegelman [2] once very appropriately expressed – are subject to the biblical verdict: Reproduce! By the same reason, molecules that are able to reproduce show – just as living beings do – the phenomenon of 'natural selection'. Among biological macromolecules only nucleic acids are inherently endowed with the ability to reproduce. Proteins, though functionally more versatile, are devoid of this property.

We have studied in detail kinetics and mechanisms of replication of RNA, the most likely candidate for early evolution, and their implications on natural selection [3, 4]. Biebricher has reported on this work during this conference. Under experimental conditions relevant for early phases of evolution the replication rate terms assume the simple form

$$\dot{x}_i = (E_i - \bar{E}(t))\, x_i \qquad (1)$$

where x_i is a relative population number and $\dot{x}_i = \mathrm{d}x_i/\mathrm{d}t$. E_i is the rate coefficient for excess production of species i through self-copying and $\bar{E}(t)$ the corresponding average for all competitors present which enters the equation due to the fact that the time derivative of the relative population variables is considered. This equation is similar to Verhulst's equation of nineteenth century population dynamics. Formally such a system can be shown to involve a general potential-function, linking the time derivative with its gradient (with respect to the x_i's). Selection then is characterized by a straightforward extremum principle:

$$\bar{E}(t) \rightarrow E_{\max} \qquad (2)$$

where the maximum (excess) reproduction rate coefficient defines the 'fittest'. This property is lost if the system approaches true chemical equilibrium, if, for instance, detailed balance would be established in the reaction system:

$$\nu_i\,\mathrm{NTP} \xrightarrow[\text{template}]{\text{enzyme}} (\mathrm{NMP})_{\nu_i} + \nu_i\,\mathrm{pp} \rightarrow \nu_i(\mathrm{NMP}+\mathrm{pp}),$$

(NTP = nucleoside triphosphates of the bases A, U, G and C, NMP = nucleoside monophosphates or nucleotide monomers in the RNA chain, ν_i = number of monomeric units in polymeric chain i and pp = pyrophosphate). Under conditions far from equilibrium competition of all species for the (buffered) substrate enforces 'natural' selection, which means reduction of the complex variety of many sequences to only one form, the fittest among all initially present sequences. This formalism does not yet resolve any of the problems of evolutionary adaptation. However, it demonstrates clearly what 'natural selection' means, namely: *Natural selection is not some* a priori *property of states of matter, but rather a physical consequence of self-reproduction under conditions far from chemical equilibrium.*

We may call any state that can be characterized by such a behaviour a replicator. Replicators include various forms of quite differing complexity, e.g.

- RNA molecules (showing complementary replication of their plus and minus strands);
- DNA molecules (showing semiconservative self-replication of their double strands);
- viruses (both DNA and RNA);
- cells (that multiply vegetatively);
- organismus (with certain restrictions caused by their recombinative heredity).

These biological objects share the property of replication with typically non-biological forms, e.g. with laser modes. Let me briefly entertain a 'what if' example from physics which sheds more light on the causal relationship between self-reproduction and natural selection.

Suppose that materilization of γ-rays through e^+e^- pair production were an inherently autocatalytic process reproducing exactly the physical properties, such as charge and mass, of the catalytic particle. Then a maximum cross-section of pair production for a certain combination of mass and charge would inevitably lead to an irrevocable selection of that very combination of physical properties. It so happens that the Bethe–Heitler approximation for the relative cross-section of pair production is maximal for the elementary charge if it is combined with the muonium mass. This relation, first emphasized by Queisser [5], is shown in Fig. 1. Since the electron is the decay product of the muon an evolutionist's interpretation would be that elementary properties such as charge and mass belong to a secondary category, different from the universal invariants such as Planck's constant and vacuum light velocity; this secondary category were just a consequence of an extremely sharp selection caused by the fundamental properties of the physical universe. However, at the present state of knowledge such an interpretation is (as Gell-Mann [6] commented) mere science fiction.

For RNA or DNA, the 'elementary particles' of biology, such behaviour is a well established fact, if conditions for extremely precise replication are chosen. This was verified by experiments using *de novo* minivariants of phage $Q\beta$ as templates and $Q\beta$ replicase as enzyme [3, 4]. In biology natural selection quite generally *is* the tool for reducing complexity.

II.2. *The concept of sequence space*

As illustrative as the preceding treatment is with respect to a physical understanding of the phenomenon of natural selection, it contributes little to the solution of the riddles of life. In fact, because of its mathematical simplicity and elegance it may obscure the true problems.

One of the most important prerequisites of evolution, variability, is missing in the preceding treatment. Without

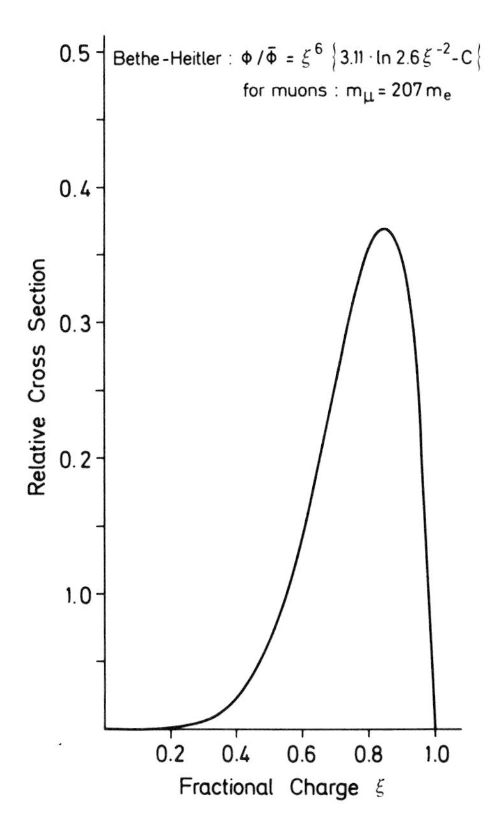

Fig. 1. Normalized cross-section for pair production, ϕ, by photons in screening muon matter, plotted against an assumed fraction ξ of elementary charge. The case $\xi = 1$ corresponds to the conventional charge of the electron. The only input parameter is the mass of the muon. The value of the constant C is not precisely known. It is quoted in literature as '3 or so'. For $C = 1.93$ the sharp maximum would exactly coincide with $\xi = 1$ (according to Queisser [5]).

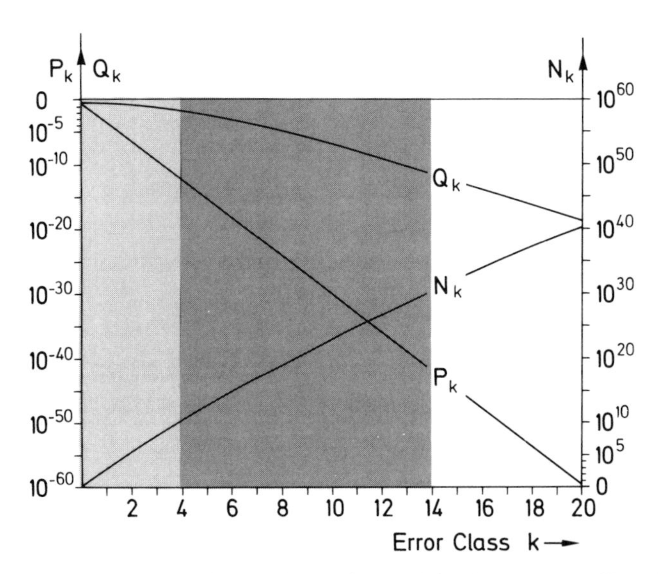

Fig. 2. Probability Q_k of producing any k-error defect in a sequence of length ν, number N_k of sequences belonging to error class k, and probability $P_k = Q_k/N_k$ of producing a given k-error mutant as functions of the mutation distance k (log plots). The shadowed regions correspond to $P_k = 10^{-12}$ (light shadow) and $Q_k = 10^{-12}$ (dark shadow) indicating that in a conventional laboratory population of 10^{12} individuals (RNA molecules or viruses) given mutants with $k > 4$ are becoming increasingly rare (expectation < 1) and that on the average one may still expect one mutant with $k = 14$.

mutations a system could never evolve, or adapt to optimal performance. Mutations (including single base changes, insertions, deletions or more sophisticated recombinative aberrations) provide the variability which, in combination with selection, makes evolution possible.

In classical theory mutations are introduced as purely random fluctuations superimposed upon population growth. Being stochastic in nature they render the evolutionary process indeterminate rather than target-directed, at least in a primary sense. Goal directedness is only introduced in a secondary way, namely via natural selection of the fittest. This, in fact, is the common interpretation given nowadays: an advantageous mutant, whenever it appears, may grow up and eventually dominate the population in a deterministic fashion. However, the initial phase of its appearance is an inherently stochastic event. Unfavourable mutants – usually not seen in population dynamics – must necessarily greatly outnumber the favourable cases. If this interpretation, in which the emphasis is entirely on the appearance of a mutant that shows a selective advantage, were true we would not exist. Such an interpretation could explain the evolution of life only if it were a simple hill-climbing process with a monotonic increase of selective values. Such a value distribution, however, is far from reality. Rather we should assume that the landscape is anything but monotone. Then, with a simple hill-climbing mechanism only some foot-hills in the value landscape could be reached, representing only minor degrees of fitness. The formidable problem which confronts Darwinian theory here will immediately become apparent if we do some quantitative estimates.

Figure 2 shows, for a sequence length of $\nu = 300$, a plot of

$$Q_k = \binom{\nu}{k} q^{\nu-k}(1-q)^k$$

the fraction of sequences showing any k-error defect,

$$N_k = \binom{\nu}{k} 3^k$$

the number of sequences belonging to error class k, and

$$P_k = Q_k/N_k = q^\nu \left(\frac{q^{-1}-1}{3}\right)^k$$

the probability of producing a given k-error mutant, as a function of the mutation distance k (or Hamming distance). (Here again, ν is the number of monomers in the sequence, q is the probability of correctly reproducing a monomer and $1-q$ the corresponding single digit error rate; the factor 3 in N_k and P_k results from the fact that any given nucleotide A, U, G, C can be replaced by 3 alternative ones.) Furthermore in Fig. 3, Q_k and P_k are plotted for sequences of varying length. For simplicity it is assumed that the single base mutation rate $(1-q)$ is uniform for all positions so that the distribution Q_k is exactly binomial or (with good approximation) Poissonian.

The numbers are striking if one relates them to the claims of Darwinian theory. They show that in typical populations (e.g. Petri dish with typically 10^{12} replicators) *given* mutants appear reproducibly only for extremely small mutation distances. This in turn means that the evolutionary process gets stuck on fairly minor value hills unless the landscape is so monotonous that by small jumps (e.g. $\nu = 300$: $k \ll 10$) a hill-climbing route can always be found. Any comparison with typical landscapes on earth discloses a hopeless situation. In fact, the problem of complexity which seemed to be solved by natural selection comes in again through the back door.

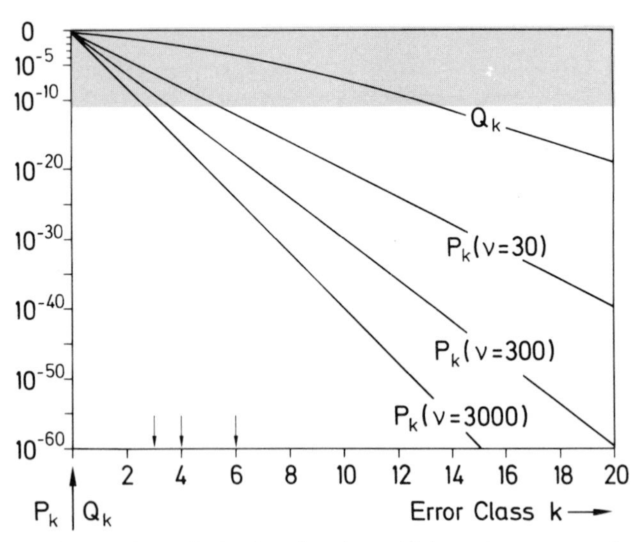

Fig. 3. Q_k and P_k (cf. Fig. 2) as functions of k for various sequence lengths ν. The shadowed region again corresponds to P_k-expectation values $> 10^{-12}$ again referring to a population of 10^{12} individuals. For $\nu \geqslant 30$, Q_k is almost independent of ν; whereas, P_k shows a strong dependence on ν. However, the deterministic appearance of mutants for $\nu = 3000$, 300 and 30 is ensured for upper limits of k that are quite similar, namely 3, 4 and 6, respectively.

However, this interpretation of Darwinian behaviour is not correct for three reasons.

(1) Mutants are not just perturbations of wildtype reproduction, but rather constitute the major fraction of the population.

(2) Due to a selective population of different mutant states there is a bias on mutant production, or to say it in a more provocative way: mutants already appear in a target-directed manner.

(3) The landscape of mutant space does not resemble anything like the landscape as we know it from earth. It is a landscape in ν-dimensional space which entirely escapes our imagination.

Figure 4 suggests how to construct a mutant space through extrapolation. We call this ν-dimensional point space the sequence space, a concept which was introduced first by Rechenberg [7] and later used by us in constructing the quasi-species model [8, 9]. It has particularly interesting applications in comparative sequence analysis and precursor reconstruction which were worked out in co-operation with Dress [10] of the University of Bielefeld and which were

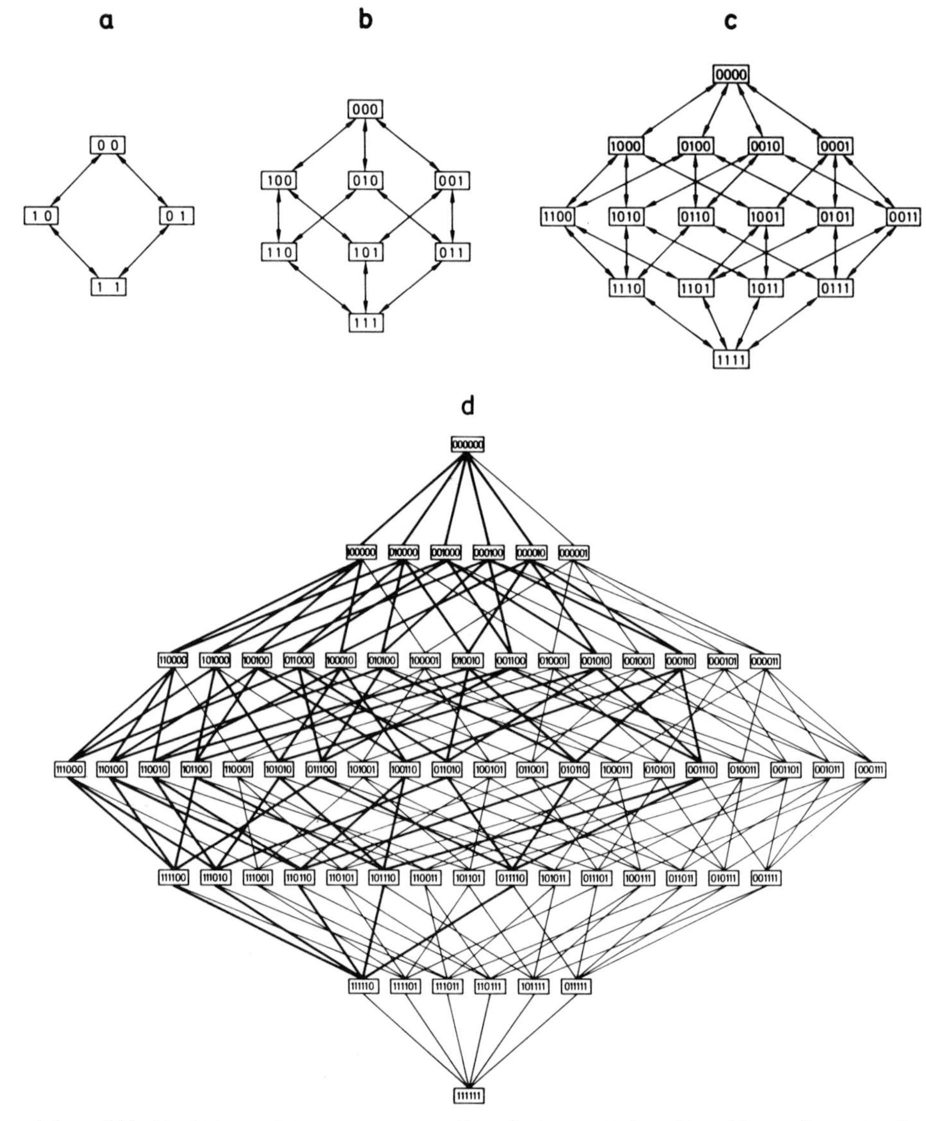

Fig. 4. The correct representation of kinship distances between sequences of length ν can only be achieved in a ν-dimensional point space as exemplified for binary sequences with $\nu = 2$ (*a*), $\nu = 3$ (*b*), $\nu = 4$ (*c*) and $\nu = 5$ (*d*, bold lines) or $\nu = 6$ (*d*, all lines). If 4 digit classes are involved each axis has 4 instead of 2 equivalent points. The enormous increase of connectivity with increasing ν is clearly seen. The number of states increases with 2^ν (or 4^ν), and the number of shortest alternative routes between two points separated by a distance k increases with k!

introduced at this meeting in the lecture of Winkler-Oswatitsch.

What is different with such spaces if we compare them with the lower-dimensional spaces more accessible to our imagination? The most important properties of high-dimensional spaces [11] are:

(*a*) there enormous storage capacity, the 'volume' being k^{ν} where k is the number of digit classes and ν the sequence length;

(*b*) the relatively short mutual distances between any two points which – despite the huge volume – remain $\leqslant \nu$, supposing detour-free connection;

(*c*) the tremendous connectivity providing $\binom{\nu}{k}$ alternative routes between any two points spaced by a distance k and hence facilitating an escape from possible selection traps.

Sure enough, there is not much advantage if we would have to orient ourselves in this space by totally unbiased random walk. We would get lost just as if we would wander about in the vast depth of the universe. However the situation would be quite different if the distribution of selective values would provide some guidance to the highest peaks of value landscape. The value distribution in this space is neither monotonically changing nor is it entirely random. Height distribution on the earth is neither close to any of these two extremes; points of high altitude occur in expanded mountain regions and not in the midst of a plane. Except for very few places there are not even any abrupt large changes in altitude. Mandelbrot [12] has suggested that altitude distribution be simulated by Brownian reliefs yielding self-similar contours if viewed on different distance scales. What is meant by a Brownian relief? Think of a steady shift along the abscissa while motion along the ordinate is Brownian, i.e. a random walk (Fig. 5). The fractal dimension D of the line to line function, obtained asymptotically, is $\frac{3}{2}$.

In a similar way Brownian spacial reliefs may be obtained with fractal surface dimensions > 2. Figure 6 shows two examples computed by Mandelbrot, the upper being of fractal surface dimension $D = 2.5$ where every vertical cut is a Brownian line to line function as in Fig. 5. The lower picture in Fig. 6 has a fractal surface dimension close to 2 ($D \sim 2.1$) and represents a more realistic model of landscapes on earth. In his book Mandelbrot asks the question, how a rain drop falling onto an island having such a surface structure would have to move in order to reach the ocean. Obviously, raindrops (or other material) would have to fill up all the 'cups' (as Mandelbrot calls the hollow areas surrounded by ridges). If a situation as depicted in Fig. 7 is reached, motion along a one-dimensional profile would follow a kind of 'devil's terrace' as represented by the function $B_H^*(x)$ which is related to the fractional Brown function $B_H(x)$ as $B_H^*(x) = \max_{x' \leqslant u \leqslant x} B_H(x)$. Mandelbrot then shows that an extension to higher dimensions $M = (x_1, \ldots, x_M)$ lowers the value of $B_H^*(x)$ relative to $B_H(x)$, because the cups can empty with increasing probability in a dimension other than depicted by the vertical cut in Fig. 7, until finally for $M = \infty$ the identity $B_H^* - B_H = 0$ must result (as was shown earlier by Levy [13]). In this case every drop falling on the landscape would end up in the ocean.

An analogous situation is found for the evolutionary uphill motion, where the role of the cup outlets, i.e. the lowest points, is now taken over by the gendarmes on the ridge, i.e. the highest points. The problem here is not to get stuck on an isolated gendarme. One cannot directly use the procedure outlined above. Raindrops, because of gravity try to reach the locally lowest points and thereby follow the path of a river bed. Evolutionary 'motion' heads for the locally highest points and therefore proceeds along ridges. As is seen from Fig. 7 a second type of a devil's terrace may be obtained by grinding off the peaks between two successive wells, defining a function $B_H^{**}(x_1, \ldots, x_M)$. This function in relation to $B_H(x_1, \ldots, x_M)$ changes upon increasing the number of dimensions in a similar way as $B_H^*(x_1, \ldots, x_M)$ in the above example

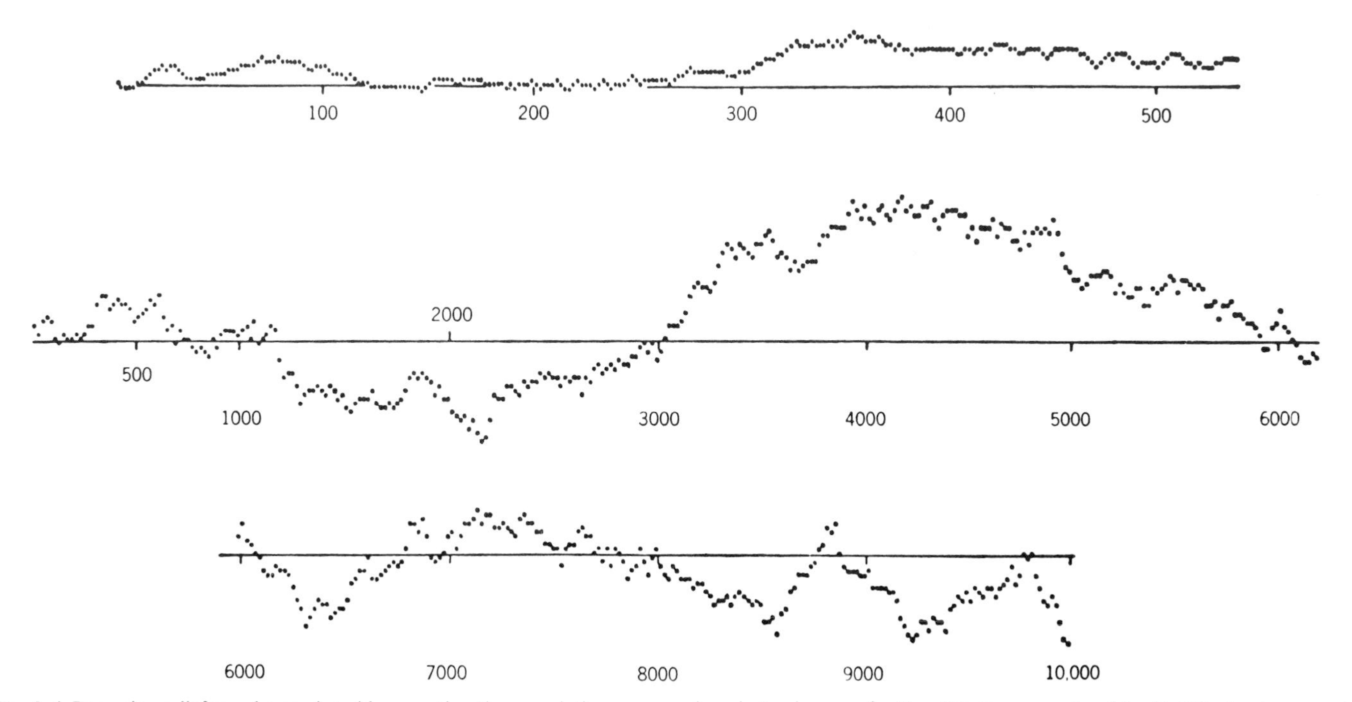

Fig. 5. A Brownian relief may be produced by counting the cumulative successes in coin tossing, e.g. for 'head' in the upward and for 'tail' in the downward direction as a function of the number of trials. Mandelbrot realized that the Brownian line to line function is a good representation for contour lines of mountainous landscapes on earth. The self-similarity is seen by viewing the Brownian function on different scales. The fractal dimension of this line is $D = 1.5$ (after Mandelbrot [12]).

(a)

(b)

Fig. 6. Brownian landscape modelled on top of a plane using different fractal dimensions yields the reliefs shown in this picture (*a*) for $D = 2.5$ and (*b*) for $D = 2.1$. The lower picture represents more realistic landscapes as found on the earth. A vertical cut through the upper picture yields the type of contour lines as shown in Fig. 5 (after Mandelbrot [12]).

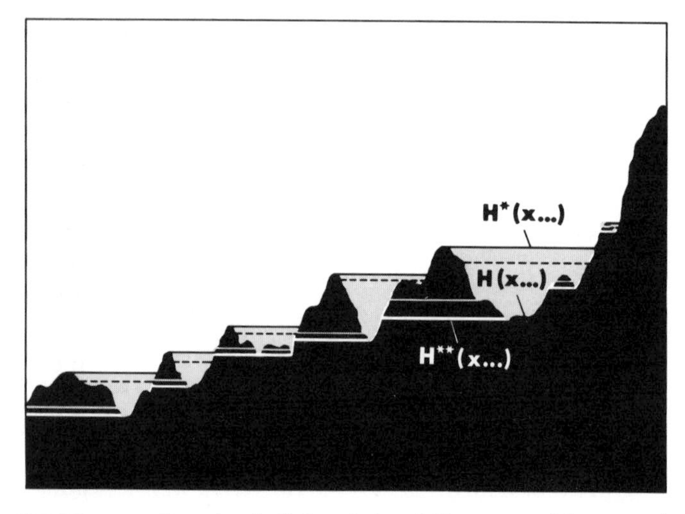

Fig. 7. In a one-dimensional relief, a raindrop falling on top of the mountain at the right-hand edge of the picture can only reach the ocean (assumed to be somewhere left of the picture) by filling the cups and then moving along the 'devilish terrace' (black solid line) denoted by $H^*(x\ldots)$. The larger the dimension M, the more outlets exist which do not require a complete filling of the cups (broken lines); finally, for $M \to \infty$, the functions $H^*(x_1, \ldots, x_M)$ and $H(x_1, \ldots, x_M)$ become identical, meaning that the raindrop will always find a way to the ocean along a natural path (i.e. without filling cups). As shown in the text, evolutionary adaptation is a complementary problem in which, however, the process proceeds along the white lines denoted by the function $H^{**}(x_1, \ldots, x_M)$.

does. The exact analogy of both cases can be shown by turning Fig. 7 upside down. In the inverse landscape obtained the wells or cups now represent peaks or gendarmes and vice versa and in this way the above argumentation applies. Mandelbrot's conjecture that there might be a $M_{\text{crit}} < \infty$ for which the escape problem is solved must depend on how fine-grained the landscape is and on the size of the allowed jumps. We shall come back to this problem in the next paragraph, but we mention now that any landscape in value space is at least connected as much and certainly more regularly than the Brownian landscape considered in these examples. This is suggested by several experimental facts.

First, experiments on site directed mutagenesis using modern molecular genetic techniques indicate that most mutants, which are close relatives of a given wildtype, show similar functional fitness as the wildtype molecule except for a few critical positions where abrupt changes may be observed.

Secondly, the existence of phylogenetic trees in which extreme positions may have appreciably less than 50% homology indicate a rather expanded and continuous distribution of functionally equivalent sequence states, ranging over quite large distances in sequence space. In other words, the sequence space is pervaded by a broad ridge of near-optimal value height from which protrusions descend into many different directions.

Finally, the known existence of relatively unrelated sequences with similar functional destination and efficiency (e.g. chymotrypsin and subtilisin) suggests that a number of quite independent such mountain regions exist in sequence space, each having its own more or less defined basin of attraction in which the evolutionary process may start.

The concept of sequence space has opened a new view of the problems of macromolecular evolution, and we are now facing new problems. Selection apparently now means localization of a distribution in sequence space. How are the different mutants of a wildtype distribution populated in sequence space? This appears to be a cardinal question for evolutionary optimization.

II.3. *The concept quasi-species*

Asking for the population distribution in sequence space is equivalent to solving the dynamical equations which regulate the population numbers. For this purpose we change to a new phase space in which each co-ordinate is assigned the population number (or concentration) of one of the 4^ν possible states. Hence this concentration space is of a much higher dimension than that of sequence space, i.e. 4^ν as compared to ν. Mutations are now introduced in a rational way considering their kinship relations with the wildtype as expressed by the distances in sequence space. This is quite different from the classical procedure where mutants are subsumed as an unspecified fluctuation term. Here each mutant is specified individually and this will turn out to be instrumental for understanding their role as bridges between successive wildtypes in the evolutionary process.

The system of differential equations of the 4^ν mutant states (again suggested by experiments on the kinetics of replication by Biebricher and discussed in more detail in the lecture of Schuster) reads:

$$\dot{x}_i(t) = (W_{ii} - \bar{E}(t))\, x_i(t) + \sum_{k \neq i} W_{ik}\, x_k(t) \tag{3}$$

where again $x_i(t)$ is a fractional population variable, W_{ii} the rate coefficient for excess production of an error-free copy of species i, $\bar{E}(t)$ the average excess production and W_{ik} a rate coefficient for producing the mutant i by erroneously copying the mutant k. It is obvious that the magnitude of these non-diagonal matrix coefficients depends on the kinship (or Hamming distance) between mutants i and k. This system of nonlinear equations has been shown to be equivalent to a system of differential equations having the form

$$\dot{y}_i(t) = (\lambda_i - \bar{\lambda}(t))\, y_i(t) \quad (4)$$

where the population variable $y_i(t)$ now refers to a whole dynamically ordered mutant clan having a defined (possibly degenerate) master sequence rather than to an individual replicator species. We call these newly arranged ensembles of replicators 'quasi-species'. The quasi-species is the target of selection, as is implied by the analogy between equations (4) and (1). Only those distributions for which $\lambda_i > \bar{\lambda}(t)$ will grow up thereby continuously raising the threshold of selection $\bar{\lambda}(t)$. Finally there is only one quasi-species distribution left which can match the above condition, yielding again an extremum principle for selection in analogy to (2)

$$\bar{\lambda}(t) = \bar{E}(t) \to \lambda_{\max} \quad (5)$$

The population variable y_m which refers to the maximum eigenvalue $\lambda_{\max}$ includes contributions from the various replicators that are selectively evaluated according to their diagonal coefficients W_{ii} and that are coupled with one another through their non-diagonal coefficients W_{ik}. One consequence that might be mentioned right away is that quasi-species selection does not necessarily favour the fittest individual (i.e. the individual defined by the highest diagonal coefficient W_{mm}), nor does equivalent appearance of two mutants (so-called neutral mutants) necessarily mean that they have identical diagonal coefficients: $W_{ii} = W_{kk}$. Selection may prefer the individual that is surrounded by a larger number of mutants with almost matching fitness, even if its absolute fitness is lower than that of an individual representing an isolated peak in the value landscape of sequence space. Or, in other words, evolution prefers massive mountain regions and is guided along ridges represented by mutants that are populated according to their relative fitness. Thus we arrive at a new view about the role of mutants signifying some fundamental shift of emphasis in the principle of natural selection, as summarized in Table II.

The classical view was: Only the advantageous mutants count and they appear on an entirely stochastic basis. This view has been modified through the works of Kimura [14], Jukes [15] and others as to show, that also the neutral mutants are of importance for evolution and that the 'fittest' may drift along neutral pathways – still entirely stochastically. We now learn: Also the disadvantageous mutants are of importance as long as they form part of a populated quasi-species, and as long as they are able to reproduce with some efficiency not too different from the fittest individual in the population. Then they may guide the evolution along ridges in the value landscape. Neutral or nearly neutral mutants are of eminent importance. If they are not too far distant from the wildtype (defined by the master sequence) they will appear deterministically in the quasi-species population.

Table II. *Evolution of formal Darwinian theory*

Model	Main emphasis	Target of selection	Mode of change
Neo-Darwinian	Advantageous mutant	Defined wildtype	Upgrowth
Neutralist	Neutral mutant	Degenerate wildtype	Random drift
Quasi-species	Less advantageous mutants	Widely dispersed mutant distribution	Phase transition of mutant distribution

Table III. *The quasispecies model*

1971–78	Ansatz+approximate solution Threshold relation Stochastic limitations	Eigen/Schuster [8, 9, 16]
1974–76	Transformation to vector space	Thompson/McBride [17]
	General solutions	Jones *et al.* [18, 19, 20]
1978–85	Specific cases, numerical solutions Phase transition behaviour Stochastic models	Schuster/Sigmund *et al.* [21–23]
1983	Renormalization General localization condition Stochastic theory	McCaskill [24, 25]
1985	Recursive calculation of eigenvectors and eigenvalues	Rumschitzki [26]
	Equivalence to two-dimensional Ising model	Leuthäusser [27]

Much could be said about the physical nature of the quasi-species model. I would rather highlight, in a few tables and figures, some of the features of theory and their impact on the possibility of studying molecular evolution phenomena under controlled conditions in the laboratory. The essential steps of developing the concept of the quasi-species are compiled in Table III, while the quantitative consequences of the theory are summarized in Table IV. They especially refer to the structure of the quasi-species and the error threshold.

The best-adapted sequence (wildtype) usually constitutes only a minor fraction of the quasi-species distribution. This is shown quantitatively in Fig. 8, where the fractional population number of the wildtype sequence is plotted versus the average symbol quality $\bar{q}$. (If $\bar{q}$ is the average symbol quality, $1-\bar{q}$ is the average error rate per symbol and $\nu(1-\bar{q}) \equiv \epsilon$ the error expectation value of the sequence having ν positions). It is seen that near the error threshold, where $1-\bar{q} \approx 1/\nu$, the fraction of wildtype generally becomes quite small, even if the superiority, σ, of the wildtype approaches infinity (for a definition of σ cf. refs [8] and [9]; in the simplest case σ is the ratio of wildtype and average mutant replication rate, i.e. $\sigma = W_{mm}/\bar{W}_{kk \neq mm}$). Usually σ is below ten and sometimes not far from one. In these cases the population fraction of the wildtype becomes quite small if the error rate surpasses $1/\nu$. The curves in Fig. 8, except for the case $\sigma = \infty$, intersect the abscissa. Those points represent the error threshold, where the quasi-species becomes unstable. The error threshold plays the role of a melting temperature. Violating the error threshold causes a kind of 'melting' of the distribution, or a delocalization in the sequence space. In fact, as Leuthäusser [27] has shown, there is an exact equivalence between the quasi-species and the two-dimensional Ising model of ferromagnetism. The order–disorder transition, however, does not refer to the geometrical, but rather to the sequence space. Figure 7 in Schuster's paper demonstrates how sharp such phase transitions are.

Table IV. *Characteristic properties of quasi-species*

Unique sequence:	Master sequence, consensus = wildtype.
Error threshold:	$\nu(1-\bar{q}) \equiv \epsilon \leqslant \ln \sigma$; (cf. Fig. 8). Means length limitation (ν_{max}). If either $(1-\bar{q})$ or ν variable, adaptation to $(1-\bar{q})_{max}$ or ν_{max} yields greatest evolutionary power [16]. Is extremely sharp (phase transition) (cf. Fig. 7 in the paper of Schuster, this volume). May be compared with Curie temperature [27]. Does not require singular fittest state (follows from renormalization work [24]).
At error threshold:	Very small fraction of master individuals [16] (cf. Fig. 10). Poissonian expectation of less-adapted mutants [8]. Overpopulation of well-adapted mutants [28]. Degeneracy of neutral mutants (value landscape, distance) [28]. Protruded mutant populations (ridges) [28].
Below error threshold:	Stable distribution [8, 9]. Localization in sequence space [24]. Increasing fraction of master individuals (cf. Fig. 9). Decreasing fraction of mutants (cf. Fig. 9).
Above error threshold:	Instability or delocalization (cf. Fig. 9) [8, 9, 24]. Yielding either: random population [22], evolution [8], or annealing [28].
Selection:	Localization in sequence space [8, 24] like phase transition [22, 27]; $\bar{\lambda}(t) \rightarrow \lambda_{max}$.
Evolution:	Arrival of advantageous mutant causes instability of former wildtype distribution and phase transition-like change to a new distribution. Sequence of such events is stochastic, but value biased.
Optimization:	Through value-guided evolution.

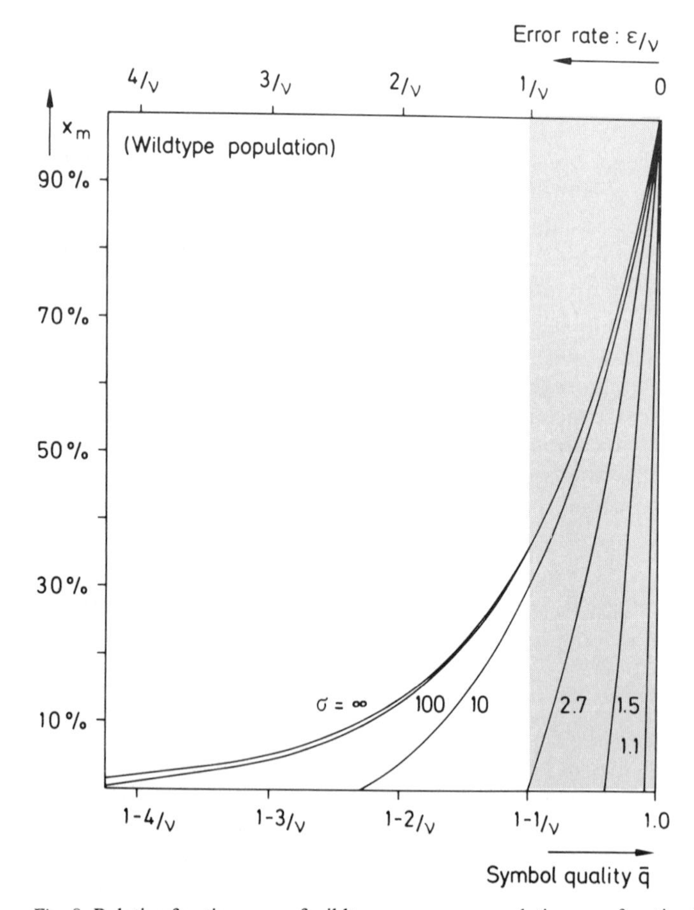

Fig. 8. Relative fraction, x_m, of wildtype sequence population as a function of the average symbol quality, $\bar{q}$, or error rate $(1-\bar{q}) \equiv \epsilon/\nu$. ($\epsilon$ is the expectation value of errors in the total sequence of length ν.) The parameter, σ, is the superiority of the wildtype (most simply measured as the ratio of wildtype to the average mutant reproduction rates). The curves intercept the abscissa at the error threshold, except for $\sigma = \infty$.

The error threshold relation is quantitatively represented in Fig. 9, where the maximum number, ν_{max}, of symbols of a (stable) wildtype sequence is plotted as a function of the average symbol quality $\bar{q}$ (or of the average symbol error rate $(1-\bar{q})$). These curves impressively demonstrate how restrictive the error threshold relation is with respect to the allowed number of symbols that are apt to be selected in a sequence.

I should emphasize the following fact. Just because only a small fraction of wildtype is present does not mean that the sequence that is obtained by sequence analysis is indeterminate, unless real degeneracies, i.e. neutral wildtypes, co-exist. The superposition of the sequences which constitute the mutant spectrum, usually yields a defined master sequence which either resembles, or is closely related to, the wildtype sequence.

Experimental evidence for the quasi-species structure has been obtained for RNA-viruses as well as for *de novo*–RNA products produced by Qβ-replicase. Weissmann [29] and his co-workers have cloned single copies of the *E. coli* phage Qβ and shown that true wildtype sequences are present to an extent of less than 5%. The experiment is described in the legend of Fig. 10. Similar findings have been reported by Domingo [30] for foot-and-mouth-disease virus. The presence of a large variety of mutants according to Palese [31] also seems to be typical for influenza virus. The error threshold relation has been tested for phage Qβ. The single-chain RNA genome is comprised of 4200 nucleotides. Weissmann and his co-workers [32] have produced site directed mutants and measured the appearance of revertants. They obtained an error rate $(1-\bar{q})$ of 3×10^{-4}, and from their data one can deduce a σ-value of about 4. The data show that phage Qβ operates exactly at its error threshold, and this is compatible with the presence of a very small fraction of the wildtype. That

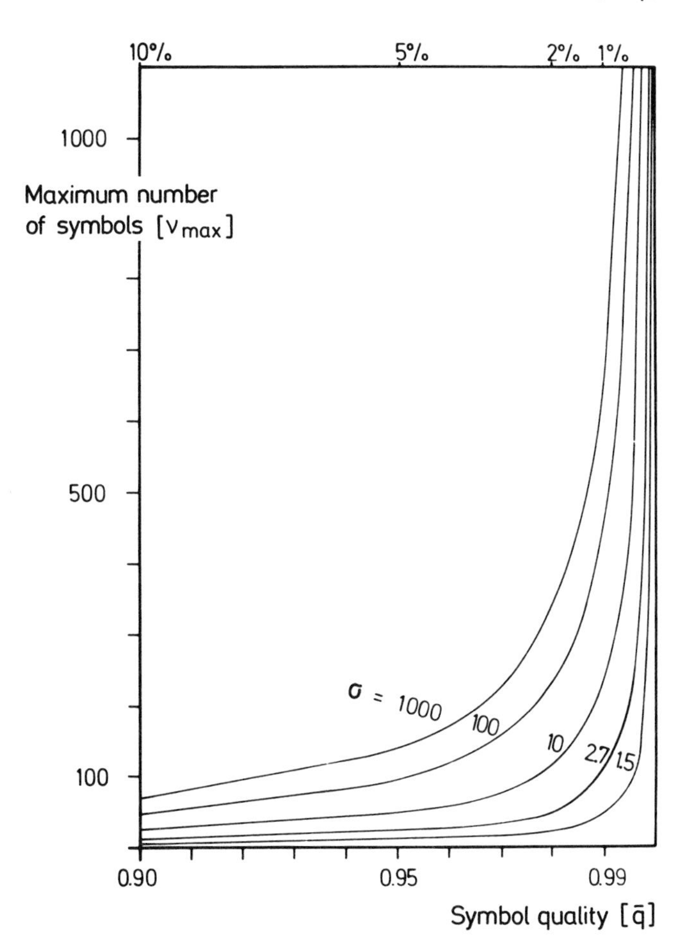

Fig. 9. Graphic representation of the error threshold relation for various values of σ as parameter. The maximum number, ν_{max}, of symbols (or monomers) that is allowed for a selected sequence to remain stable (i.e. not to accumulate errors) is plotted *vs.* the average symbol quality, $\bar{q}$, or error rate $(1-\bar{q})$. If ν is given, the curves indicate the maximum error rates $(1-\bar{q})$, i.e. the threshold values which must not be surpassed for systems with stable selection.

nature generally adheres to the threshold relation is seen from the fact that single stranded RNA viruses generally achieve, but never exceed, genomic sizes of about 10^4 nucleotides.

Typical selection and evolution experiments are described in Section III. They are also the subject of Biebricher's lecture which can be found in this volume.

There are important conclusions with respect to the procedure of optimization resulting from a combination of the sequence-space and quasi-species concepts [28]. As was shown in Figs 2 and 3 the probability $P(k)$ to produce a particular k-error mutant sharply drops with increasing k restricting unbiased mutational jumps to the very close vicinity of the master sequence. Mutants, however, are not only produced through erroneous copying of the master sequence, but also through correct self-replication as well as through erroneous replication of other closely related mutants. In fact, mutants far distant from the master copy owe their existence almost solely to their production from precursors which are positioned along the shortest connecting line to the master sequence. A selectively advantageous copy usually is to be expected among far distant mutants. Their probability of appearance therefore depends strongly on how the intermediates along the connecting line to the wildtype are populated.

This very fact of concomitant selective evaluation of all mutants in the quasi-species population drastically modifies the unbiased probability distribution P_k. According to second-order perturbation theory applied to equations (4) one obtains the fraction of population numbers of mutant i in error class k $(x_i(k))$ relative to the master copy m (x_m)

$$x_i(k)/x_m = P_k \frac{W_{mm}}{W_{ii}} f_i(k)$$

where $P(k)$ is the function introduced in Figs 2 and 3, k the Hamming distance between master copy and mutant i, W_{mm} and W_{ii} the diagonal coefficients introduced with equations (4) and $f_i(k)$ a function that can be calculated by recursion

$$f_i(k) = \frac{W_{ii}}{W_{mm}-W_{ii}} \left\{ 1 + \sum_{j=1}^{\binom{k}{1}} f_j(1) + \ldots + \sum_{j=1}^{\binom{k}{k-1}} f_j(k-1) \right\} \tag{6}$$

The first terms explicitly read

$$\left.\begin{aligned} f_i(1) &= \frac{W_{ii}}{W_{mm}-W_{ii}} \\ f_i(2) &= \frac{W_{ii}}{W_{mm}-W_{ii}} \left\{ 1 + \sum_{j=1}^{\binom{2}{1}} f_j(1) \right\} \\ f_i(3) &= \frac{W_{ii}}{W_{mm}-W_{ii}} \left\{ 1 + \sum_{j=1}^{\binom{3}{1}} f_j(1) + \sum_{j=1}^{\binom{3}{2}} f_j(2) \right\} \end{aligned}\right\} \tag{7}$$

etc....

As is seen the function $f_i(k)$ contains sums of products in which the last term is a k-fold product of hyperbolic terms like

$$\frac{W_{ii}}{W_{mm}-W_{ii}}$$

comprising all intermediates that appear along the connecting line between master copy and mutant i in the error class k. The last term of the sum contains $k!$ such products corresponding to $k!$ different shortest connecting lines between the master copy and the mutant. If mutants are nearly neutral, i.e. $W_{ii} \approx W_{mm}$, the hyperbolic terms

$$\frac{W_{ii}}{W_{mm}-W_{ii}}$$

may become quite large. (The second order perturbation approximation breaks down for $W_{ii} = W_{mm}$, but otherwise holds as long as $W_{ii}/(W_{mm}-W_{ii})$ is smaller than ν). A neutral area of this kind (e.g. the one depicted in Fig. 4*b*) indeed may reach, through multiplicative amplification, tremendously high population numbers as compared to mutant areas that represent low degrees of fitness. An example is demonstrated in Fig. 11. The value profile along the connecting line between master copy and mutant (that may lead to a selective advantage) is assumed here to pass through a minimum, i.e. $\frac{1}{2}W_{mm}$. The sequence length ν in this example is only 100. Curve *a* shows the decrease of P_k with increasing k that resembles the decrease of relative population numbers for mutants of low degree of fitness (as compared to the master), while curve *b* shows the modification due to the value profile given. Looking at the numbers one may say a given 12-error mutant in a low fitness area will 'never' be obtained under any realistic laboratory conditions, while a mutant of this error class situated in a high fitness area (cf. profile) is destined to appear in any typical Petri-dish population.

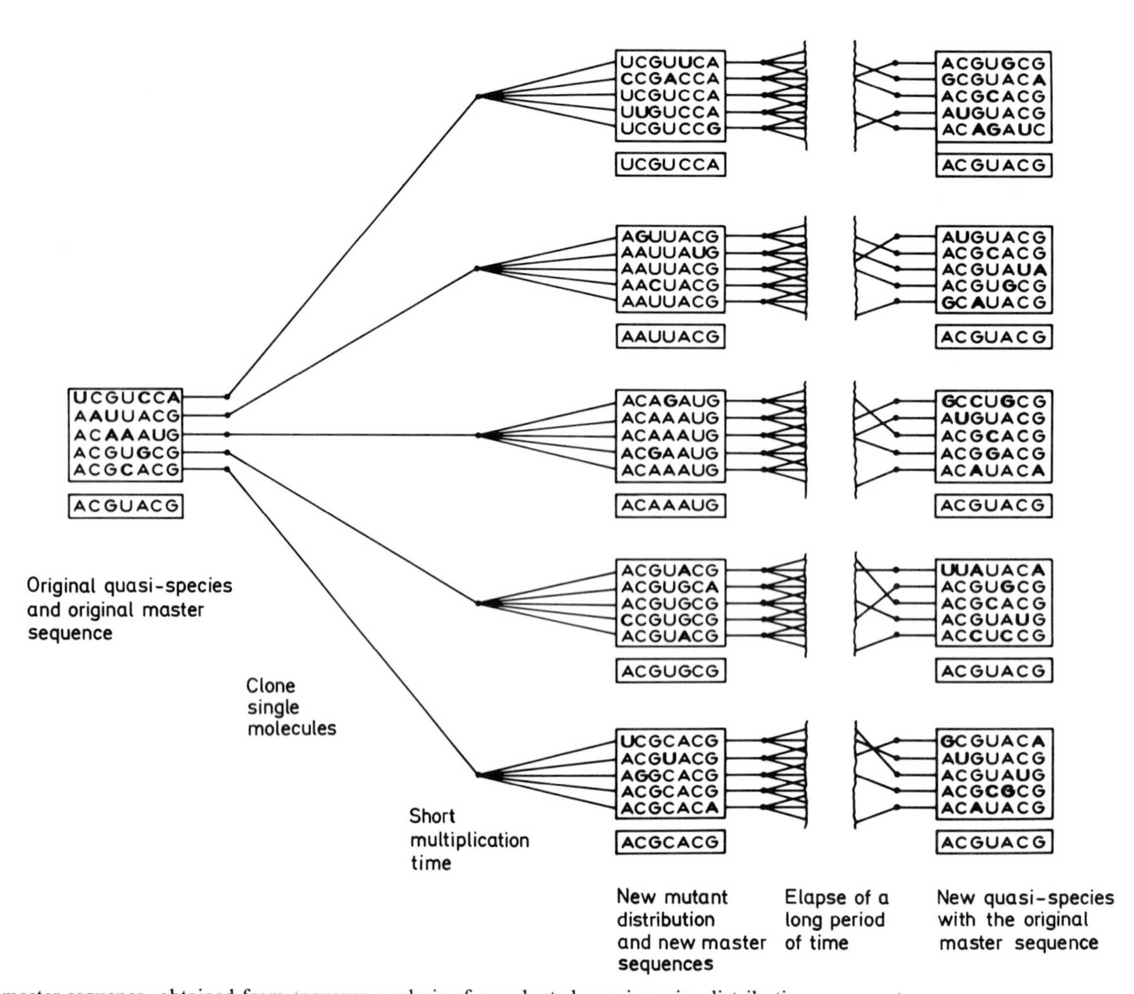

Fig. 10. The master sequence, obtained from sequence analysis of an adapted quasi-species distribution, represents a consensus sequence; each position represents the most frequently appearing nucleotide. This master sequence itself (or wildtype sequence) may be present to only a very small extent. This was shown experimentally, by Weissmann and his co-workers [29]. The experiment is represented schematically in this picture. The fact that amplification of clones is possible without restoring immediately the original master sequence is due to the abundance of those mutants whose selective values W_{kk} are close to the wildtype value W_{mm}. Those abundant mutants are essentially 'fished out' of the solution. They grow up as $\exp(W_{kk}\,t)$; their replacement by the original wildtype requires a lag time of revertant formation. The wildtype then slowly outgrows the mutant with a time course, $\exp\{(W_{mm}-W_{kk})t\}$, with $(W_{mm}-W_{kk}) \ll W_{kk}$.

We see that this selective bias caused by mutants that are below (but not far below) the level of wildtype fitness produces guidance which is decisive for the choice of the evolutionary route. Or expressed in a more provocative way: The appearance of a selective advantage is not a random event, it is rather biased by the value distribution of the mutants. Almost all large-error mutants that are present in such a distribution appear in high value areas, i.e. in areas where the next advantageous copy is to be expected. In fact, this guidance may be equivalent to testing through populations of something like 10^{30} mutants, although only 10^{12} replicators are actually present in the system. It is a true guidance effect which prefers advantageous over disadvantageous mutants, not changing the stochastic nature of the elementary process of mutation. It is a sort of mass action caused by selectively active mutants, preselecting the next advantageous wildtype. Upon appearance of such a copy a kind of phase transition will take place. The former mutant distribution will melt away and a new quasi-species will build up. Annealing, i.e. local or temporal violation of the error threshold, may largely favour such phase transitions.

II.4. *Networks of replicators*

The evolution of molecular replicators under primordial conditions on earth could not have produced very long sequences. One of the obstacles is the error threshold corresponding to the fidelity of primitive copying mechanisms. Using well adapted replicases even present day single stranded viruses were not able to extend their length to appreciably more than 5×10^3 to 10^4 nucleotides, although they were in need of information storage capacity and therefore compromised to overlapping encoding of information. Another obstacle is the obtainable mutation jump length which strongly decreases with sequence length. For instance, with $\nu = 300$ a ten error mutant, if close to wildtype fitness, might easily occur among the natural population. With $\nu = 3000$ and a correspondingly adapted mutation rate the appearance of such an error-copy is 10^{10} times less likely. Increasing the dimension offers advantages only for co-operative units such as single genes. A mutation must influence the function to be optimized, which means that only the corresponding dimensions of sequence space offer the chance of escape from traps such as local value peaks. The build up of a more sophisticated function requires the co-operation between several genes. Such a co-operation had to be organized since originally unrelated sequences are competitive. The most likely way to enlarge information capacity was gene doubling and there is some experimental evidence that this was nature's choice. We heard at this meeting, how larger protein molecules could have developed from smaller domaines through gene doubling. The same is most likely true for the evolution of

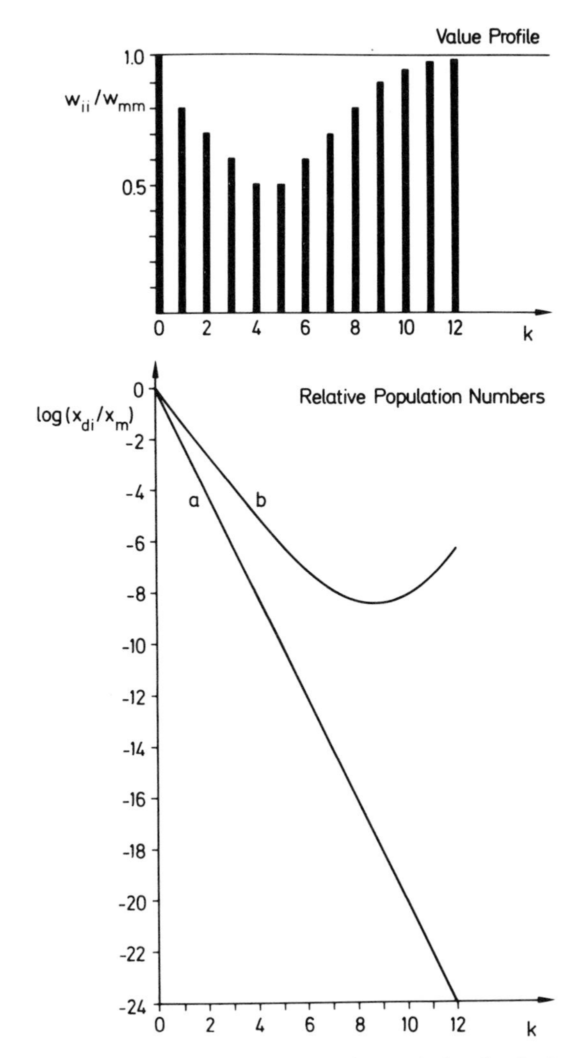

Fig. 11. Example for the modification of a population distribution of mutants because their selective values, W_{kk}, are close to W_{mm}, the selective value of the wildtype. In the upper part of the picture a particular value profile is shown for mutants with $k = 1$–12. In the lower part the probability, P_k, for producing a mutant with k errors (curve *a*) is compared to the fractional population number x_k/x_m (curve *b*) that the mutant achieves according to the given value profile. Curve *a* would be representative for the population numbers of any series of mutants for which $W_{kk} \ll W_{mm}$. The particular parameters assumed in these calculations are $\nu = 100$ and $(1-\bar{q}) = 1/\nu$.

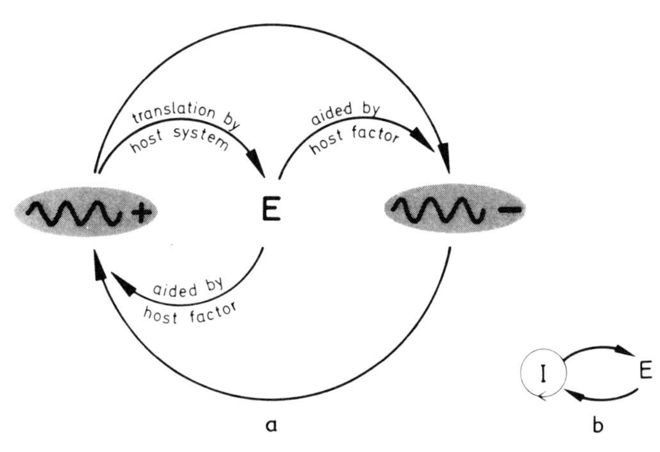

Fig. 12. RNA virus infection represents nature's simplest realization of a hypercycle. Replication of the viral genome is not only instructed through template action of the plus and the minus strand, but it is also effected by a factor that is encoded in the viral genome, namely a subunit of the replication enzyme. Therefore, the rate of replication depends not only on template concentrations, as in equations 1 and 3, but also on the concentration of active enzyme, which in turn is a function of template concentration (in equation (3) a constant environment containing the replication machinery – as materialized in test-tube experiments – is assumed). The feedback loop provided by the enzyme is superimposed upon the replication cycle. In more complex systems it may include more than one replication cycles.

divergent functions. Mutants of a quasi-species distribution may have diverged and adapted to individual, co-operative functions. We have heard from Winkler-Oswatitisch about such an example, namely the family of *t*-RNAs [33, 34]. There must be a mechanism which switches off competition, regulates growth to a mutually acceptable level and hence controls a stable level of relative population numbers. Schuster has reported on those co-operative systems. The functional link which is superimposed upon several replicators must include a feed back loop, by very similar reasons that require selective systems to be inherently autocatalytic. We termed such superimposed loops, which integrate several replication loops, hypercycles [8]. Compartmentation of the ensemble alone could not substitute for hypercyclic organization. Indeed competition would be even more fierce within the compartment and thus would require mutual self-control. Hypercycles are a large class of different systems involving various degrees of reaction order. They originate most easily within a quasi-species whose genotypes acquired some ability of feed back on their genotype, either through simple binding or through catalytic enhancement. Since all mutants and their phenotypes are similar this will immediately lead to looped networks from which the hypercycle may emerge through selective advantages. In combination with compartmentation it may most efficiently utilize phenotypically all inventions made by and encoded in the genotypes. Hypercycles, as was shown in a little book in 1978 [9], are systems with most intriguing properties. At the time translation was invented they became truely advantageous because they allowed for a specific feed back of the translation products to their genes. In fact, RNA virus infection is nature's simplest realization of a hypercycle as is illustrated in Fig. 12. Nature may have conserved other more complex hypercycles. Influenza viruses are compartments which hold together eight independently replicating RNA molecules, three of which encode replicases [35]. The organization which led to an evolution of such a reaction network has not yet been elucidated.

In this lecture I cannot enlarge further on hypercycles which represent a most interesting stage of evolution with many consequences such as unifying the genetic code and the chirality of biological macromolecules. I mentioned it merely to show that evolution had to go through several phases each having its own characteristic regularities of events and physical laws behind them. Table V will be my only speculation about a possible historical sequence of events allowing a molecular replicator to evolve to a cellular replicator and finally build up multicellular organisms.

This Table emphasizes the changing nature of selection which sometimes produces large varieties of coexisting species, while in other phases the distribution is extremely sharp. Evolution starts with prebiotic chemistry which – as we have heard at this meeting – provides an environment with all sorts of functionally active states, possibly including primitive enzyme catalysis. Evolutionary optimization, however, requires replication for two reasons. The information accumulated thus far must not get lost. Replication conserves information, but replication also provides the basis for

Table V*a*

Quasi-species states	Hypercyclic states
← Prebiotic chemistry	→
Molecular replicators: RNA	
Primitive translation; feedback	
Gene divergence →	Functional links
	Compartments, Gene doubling; Integration
Cellular replicators ←	Synchronized reproduction: DNA
Ur-Cell, recombination	
Various cell types →	Sexual reproduction
Vegetative multiplication	Eukaryotic cell, cell assemblies
→ Parasitic replicators	Multicellular organisms
Bacteriophages, plant and animal viruses	Central nervous system: Man
	Human society

Table V*b*

Years	Information content	Features
$>4\times10^9$	$\nu<10$	Proteinoids
		GNC→RNY code
$>3.5\times10^9$	$\nu\sim10^2$–10^3	Enzymic function
		Genetic code
$>3\times10^9$	$\nu>10^5$	Fermentation
		Photosynthesis
		Gene processing
$>2\times10^9$	$\nu>10^7$	Respiration
		Cell organelles
		Diversification
$>1\times10^9$		Differentiation
		Sensory/motor organs
5×10^8	$\nu\sim10^9$–10^{10}	Perception

natural selection, as was shown in this paper. The first molecular replicators which were able to store and conserve information were of quite limited length because of the error threshold relation. We could think of many localized quasi-species distributions.

For the build-up of translation machinery the co-operation between different replicators had to be organized. Co-operation among otherwise competing species, feedback of the phenotypes on their genotypes, and selective power of the whole ensemble requires both hypercyclic organization and compartmentation. The once and for all character of hypercyclic selection may have established a universal code and the universal chiralities in initially racemic structures.

Increased sophistication expands the threshold length to the size of the present single stranded RNA viruses or operon units. Optimization of enzymic function is to be attributed to this phase where fidelities and gene length were in optimal correspondence. Longer chains could not have been achieved with single stranded RNA. However, double stranded DNA provides a new means of error correction, in which the parental strand acts as reference in semi-conservative reproduction. This phase ends by synchronizing the DNA reproduction with the division of the compartment; only then does the whole unit represent a replicator, which, again, can form quasi-species distributions.

Diversity can be restored with the cellular replicator. Selection is of Darwinian nature, not as fierce as in the hypercyclic states. Different cell types can originate. Whether introns entered at this stage, or in an earlier phase when genes elongated through self-doubling, is uncertain. The error threshold suggests that introns had to establish a functional advantage, e.g. in the process of recombination, in order to be conserved.

The further development leads the system again to network or hypercyclic states with complex forms of organization. Sexual recombination is of a hypercyclic nature and so are the processes that led to cell differentiation and supra-genetic information storage. These phases are beyond the scope of this paper and, therefore, are merely mentioned.

III. An evolution machine?

We now come to our final question which addresses a practical aspect, the utilization of our knowledges about the regularities of evolution. Can we re-create simple forms of life or typical life products in the test tube?

Test tube experiments on molecular evolution have been carried out first by the late Spiegelman [36, 37] and his co-workers. He used the technique of serial transfer (Fig. 13) to adapt small variants of $Q\beta$-RNA to unfavourable environmental conditions such as the presence of ethidium bromide which inhibits replication. He obtained successively one-, two- and three-error mutants, which were slightly better adapted to the new environment than was the wildtype. At that time theories of RNA replication and models of molecular evolution were not available to the experimenters. On the basis of the present knowledge it turns out that experiments have not been done under optimal conditions. As Fig. 14(*a*) shows the mutant obtained in these experiments is only slightly better adapted to the new environment than the former wildtype. Our experiments with RNA strands, that were synthesized *de novo* by $Q\beta$ replicase, show that much higher efficiencies indeed can be obtained (cf. Fig. 14*b*). Biebricher has reported this work in his lecture [38, 39].

On the basis of the new insight gained by theory we have constructed an automatically controlled serial dilution machine which can work under optimal conditions (which often are not easily handled in test tube experiments) [40]. This machine tries to simulate the natural process of evolution and thereby is limited to systems which are able to express a selective advantage in terms of the reproduction rates. With the help of such a machine one may also adapt micro-organisms so that they synthesize certain gene products and adapt them to new environmental conditions. However, there are severe limits caused by the high fidelity of organismic replication. Single genes then cannot easily be adapted under any optimal or near-optimal conditions.

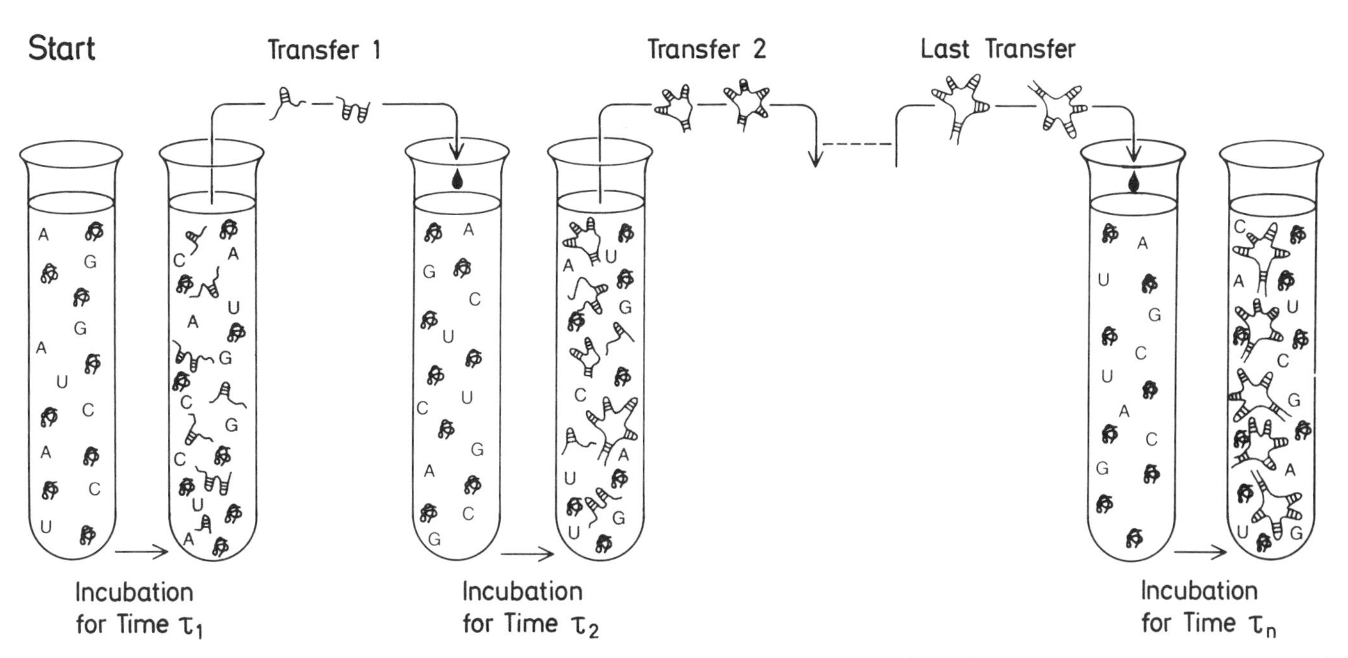

Fig. 13. Serial transfer was used by Spiegelman [36] and co-workers to prolong growth of $Q\beta$ RNA indefinitely. A constant reaction mixture, containing the four energy rich monomers of RNA in the form of nucleoside triphosphates and the active enzyme $Q\beta$ replicase, is first incubated with a given number of RNA templates. After a certain time interval, during which the RNA is amplified n-fold, an aliquot of the reaction mixture is transferred to the next test-tube and thereby diluted by the factor n. The subsequent generation of RNA templates favours those mutants that are best adapted to the particular environment. The procedure may be iterated indefinitely. It provides an effective way of 'evolution in the test-tube'.

What are our main disadvantages? First, we do not have the space of a planetary laboratory, nor do we have the time allotted for natural evolution. Moreover, we cannot easily build up complex cellular organisms that are better adapted to conditions other than through their natural genes. What we want is just single genes, and for their evolution we need error rates adapted to *their* threshold ν_{max}, which, on the other hand, are lethal for the whole organism. Furthermore, in order to test phenotypes we have to clone their genotypes. Again there is a restriction: the number of clones we could handle is very small compared to nature's resources of mutants.

And yet we have learned something from our theories. The value landscape is not random, neither is it completely regular. In any case, mountain or plane regions are clustered. Let us therefore first try to reconstruct a value landscape by a sampling procedure. In order to draw a map, e.g. of Europe, which shows where the highest mountains are, one does not have to determine contour lines or altitudes point by point. It is sufficient to have some peak density or average altitude of any given region, and for that it suffices to sample a relatively small number of points well distributed among the landscape. A thousand dots, the colour of which is an indication of their altitude, may already yield quite an informative map, showing at least the major mountain regions. Concentrating on those regions the procedure could be iterated several times so that finally a peak may be identified with quite high resolution. One important prerequisite of employing such a technique to value the landscape is the ability to localize the positions of the sampling points, at least on a relative scale. In other words, in order to draw a map I have to know at least the relative positions. Mutants to be sampled therefore must not be produced or selected on a purely random basis. One should rather know at least average mutation distances and one must be sure that they are well distributed in order to cover a sufficiently large area.

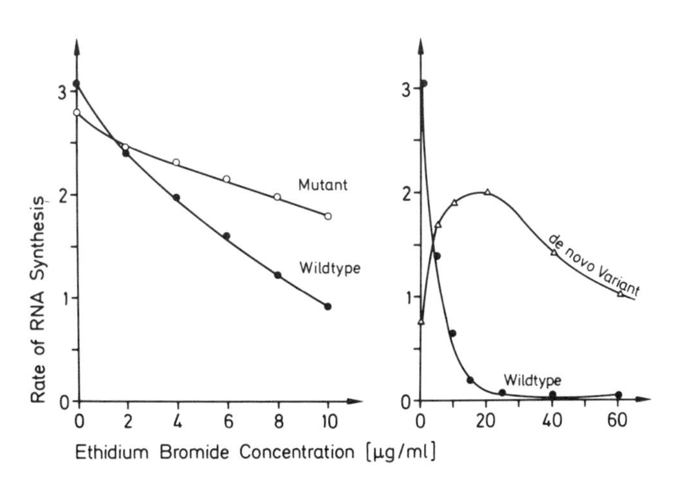

Fig. 14. Evolution experiments in the test-tube. A three-error mutant was obtained by Spiegelman [37] and co-workers through serial transfer that is slightly better adapted to an environment containing the replication inhibitor, ethidium bromide, than wildtype midivariant, which otherwise is an efficient template for $Q\beta$ replicase (left side). A more efficient mutant was obtained by Sumper and Luce [38] through *de novo* synthesis of templates by $Q\beta$ replicase in the presence of the inhibitor. As is seen from the profile of the curve, this *de novo* mutant is even addicted to ethidium bromide.

Here we can again benefit from the new insights we have gained through theory. A quasi-species becomes unstable at the error threshold which somehow plays the role of a melting point. Through annealing procedures we may obtain mutant spectra with variable distances but defined averages. In this way we could produce hierarchical mutant spectra and apply a fluorescence sampling technique as outlines in Fig. 15. We are then only one step apart from a true evolutionary bio-technology, in which restriction fragments replace our model systems.

In fact, we have started to construct such a machine [41] and presently aim at a parallel handling of about one thousand clones. The project presents many technical chal-

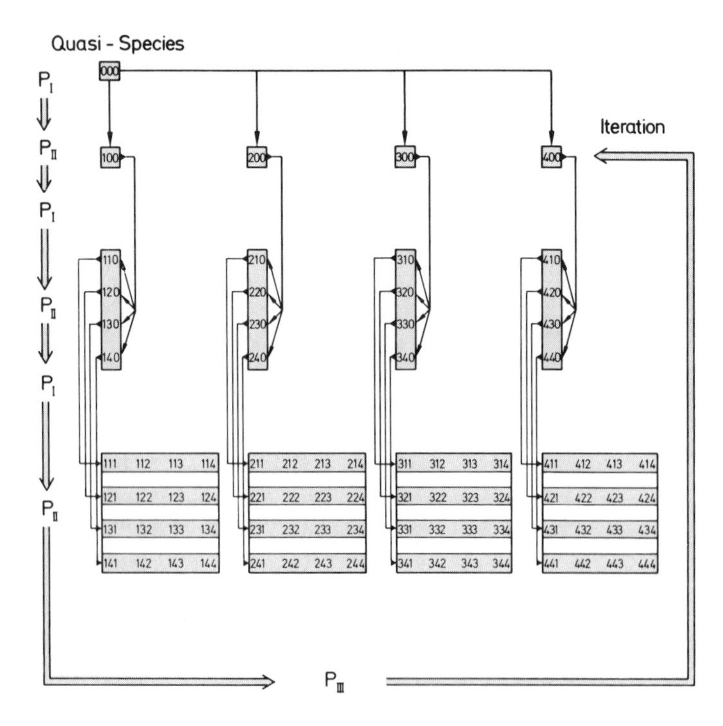

Fig. 15. The Evolution Reactor (which is under construction) tries to materialize the principles of optimization used in natural evolution. In each generation it produces a hierarchy of mutants with known average mutual distances (each step denoted by P_{II} produces a mutant spectrum with an average known distance from the wildtype, yielding three distance categories k_1, k_1+k_2 and $k_1+k_2+k_3$ denoted by the three letter code). The final distribution of 64 (or 10^3) clones is sampled by a multi-channel quartz fibre fluorimeter and a functional (or value) map is constructed; from this the 10 most suitable clones are selected according to an optimization algorithm. The procedure is iterated indefinitely. The method utilizes the cohesiveness of value landscapes to locate the highest value mountains.

lenges, such as the construction of a thousand channel quartz fibre fluorimeter, the processing of large numbers of clones (pipetting, serial dilution, pulsed incubation and growth), the production of hierarchically ordered mutant spectra, the development of evolutionary optimization algorithms and their automation. A solution of these problems may add a new flavour to biotechnology.

Acknowledgement

I wish to thank Dr Robert Clegg for carefully reading the manuscript and reviewing the English.

References

1. Wigner, E. P., *Proc. 11th Conf. Robert A. Welch Found.* Houston, Texas (1967).
2. Mills, D. R., Peterson, R. L. and Spiegelman, S., *Proc. Natl. Acad. Sci. USA* **58**, 217 (1967).
3. Biebricher, C. K., Eigen, M. and Gardiner, W. C., *Biochemistry* **22**, 2544 (1983).
4. Biebricher, C. K., Eigen, M. and Gardiner, W. C., Jr., *Biochemistry* **23**, 3186 (1984); **24** (1985).
5. Queisser, H. J., personal communication.
6. Gell-Mann, M., personal communication.
7. Rechenberg, I., *Evolutionsstrategie*. Problemata Frommann–Holzboog, Stuttgart–Bad Cannstatt (1973).
8. Eigen, M., *Naturwissenschaften* **58**, 465 (1971).
9. Eigen, M. and Schuster, P., *Naturwissenschaften* **64**, 541 (1977); **65**, 7 (1978); **65**, 341 (1978).
10. Dress, A., Eigen, M. and Winkler-Oswatitsch, R. (in preparation).
11. Hamming, R. W., *Coding and Information Theory*. Prentice Hall Inc., Englewood Cliffs, N.J. (1980).
12. Mandelbrot, B. B., *The Fractal Geometry of Nature*. W. H. Freeman, New York (1983).
13. Levy, P., *Rend. Mat., Roma* **22**, 24 (1963); cf. also ref. 12, p. 434.
14. Kimura, M. and Ohta, T., *Theoretical Aspects of Population Genetics*. Princeton University Press, N.J. (1971).
15. King, J. L. and Jukes, T. H., *Science* **164**, 788 (1969).
16. Eigen, M., *Adv. Chem. Phys.* **33**, 211 (1978).
17. Thompson, C. J. and McBride, J. L., *Math. Biosci.* **21**, 127 (1974).
18. Jones, B. L., Enns, R. H. and Rangnekar, S. S., *Bull. Math. Biol.* **38**, 15 (1976).
19. Jones, B. L., *J. Math. Biol.* **6**, 169 (1978).
20. Schuster, P. and Sigmund, K., *Ber. Bunsenges. Phys. Chem.* **89**, 668 (1985).
21. Swetina, J. and Schuster, P., *Biophys. Chem.* **16**, 329 (1982).
22. Hofbauer, J. and Sigmund, K., *Evolutionstheorie und dynamische Systeme*. Paul Parey, Berlin and Hamburg (1984).
23. Feistel, R. and Ebeling, W., *BioSystems* **15**, 291 (1982); and Ebeling, W., Engel, A., Esser, B. and Feistel, R., *J. Statist. Phys.* **37**, 314, 369 (1984).
24. McCaskill, J. S., *J. Chem. Phys.* **80**(10), 5194 (1984).
25. McCaskill, J. S., *Biol. Cybernet.* **50**, 63 (1984).
26. Rumschitzki, D., *J. Chem. Phys.* (in the press).
27. Leuthäusser, I., *J. Chem. Phys.* (in the press).
28. Eigen, M., *Ber. Bunsenges. Phys. Chem.* **89**, 658 (1985).
29. Domingo, E., Flavell, R. A. and Weissmann, C., *Gene* **1**, 3 (1976); and Domingo, E., Davilla, M. and Ortin, J., *Gene* **11**, 333 (1980).
30. Domingo, E., personal communication.
31. Palese, P., personal communication.
32. Domingo, E., Sabo, D., Taniguchi, T. and Weissmann, C., *Cell* **13**, 735 (1978).
33. Eigen, M. and Winkler-Oswatitsch, R., *Naturwissenschaften* **68**, 217 (1981).
34. Eigen, M. and Winkler-Oswatitsch, R., *Naturwissenschaften* **68**, 282 (1981).
35. Palese, P. and Kingsbury, D. W., *Genetics of Influenza Viruses*. Springer Verlag, Wien, New York (1983).
36. Spiegelman, S., *The Harvey Lectures*, Series 64. Academic Press, New York and London (1970).
37. Kramer, F. R., Mills, D. R., Cole, P. E., Nishihara, T. and Spiegelman, S., *J. Mol. Biol.* **89**, 719 (1974).
38. Sumper, M. and Luce, R., *Proc. Natl. Acad. Sci. USA* **72**, 162 (1975).
39. Biebricher, C. K., Eigen, M. and Luce, R., *J. Mol. Biol.* **148**, 369 (1981); **148**, 391 (1981).
40. Bauer, G., Otten, H. and Eigen, M., unpublished work.
41. Otten, H. and Eigen, M., unpublished work.

Chemica Scripta 1986, **26B**, 27–41

The Physical Basis of Molecular Evolution

Peter Schuster

Institut fuer Theoretische Chemie und Strahlenchemie, Universitaet Wien, Waehringerstr. 17, A-1090 Wien, Austria

Paper presented at the Conference on 'Molecular Evolution of Life', Lidingö, Sweden, 8–12 September 1985

Abstract

Molecular evolution is visualized as a dynamical process which is described by kinetic equations of polynucleotide replication. These equations are consistent with the results of extensive experimental studies on *in vitro* replication of virus specific RNA. Replication and mutation lead to stationary mutant distributions, called 'quasi-species'. The internal structure of quasi-species is responsible for the efficiency of adaptation. Among the many possible rare mutants only those which have precursors of high selective value in the population are formed with sufficiently high probability. Evolution proceeds from one local fitness optimum to the next via a series of slightly deleterious mutants.

The dynamics of systems with autocatalysis of second or higher order is extremely rich. Regular and chaotic oscillations of concentrations may occur in spatially homogeneous solutions. Dissipative structures may form in reaction–diffusion systems. Selection and local optimization of mean fitness occur only in special systems with strong constraints on the rate constants. One important special case of this type is Fisher's selection equation of population genetics. 'Catalytic hypercycles' represent another class of special systems with simple dynamical properties. They do not allow optimization of mean fitness. On the contrary, selection is suppressed. Autocatalytic reaction networks of this class show co-operative behavior or 'permanence': no member of the network is eliminated through selection.

The relevance of this dynamical model of molecular evolution for chemical and early biological evolution is discussed.

1. A dynamical view of molecular evolution

Conventional molecular evolution is static in the sense that it is based on the comparison of molecular structures. 'Structure' means here physical structure in a wide sense ranging from primary sequences of biopolymers to tertiary and quaternary structures which constitute the three-dimensional architecture of biomolecules. Sequence comparisons of homologous biopolymers in present-day organisms revealed relations between species and allowed the construction of phylogenetic trees independently of comparative morphology. These results also provided additional, indirect insight into molecular evolution. Selectively neutral mutations play a non-negligible role. They appear in all populations at an approximately constant rate and thus make the pace of an 'evolutionary clock'. This clock is independent of individual species and hence also independent of the phenotypically manifest changes. Comparison of structures of related protein molecules allowed the reconstruction of evolutionary trees of enzyme families which document the early history of life. Despite these and other spectacular results the 'static' approach to molecular evolution is unable to reveal the mechanism of biological evolution. To give an example, the data collected on selectively neutral mutants do not allow conclusions on their relevance for the evolutionary process as a whole. Mechanistic aspects are essentially dynamic. Their investigation requires a dynamical theory.

Population genetics is such a dynamical theory. It provides insight into some basic aspects of natural selection. Population genetics describes how genes spread in populations, what the stationary gene distributions are and how such distributions respond to changes in the environment. The present-day view of biological evolution is mainly based on the incorporation of the results of molecular biology into the theoretical frame of population genetics. To give an example, Kimura's extension of classical population genetics [1] combined with data from protein and DNA sequencing provided the presently accepted mechanism of the evolutionary clock.

Population genetics is incomplete in the sense that it treats the act of mutation as an event outside the theory. The mutant appears like the 'deus ex machina' of the baroque theater. Mutations are considered as rare events and, mutation rates are treated as empirical input data. Molecular biology provided, and is continuously providing a true wealth of information on the molecular details of mutations which calls for an appropriate incorporation into a dynamical theory of molecular evolution. This, however, turns out to be conceptually difficult within the formalism of population genetics.

Biochemical kinetics provides an alternative approach towards molecular evolution. Here, mutations need not be incorporated into the theory: the molecular mechanism of nucleic acid replication includes, in principle, all the various steps leading to mutants. The basic problem of biochemical kinetics lies in the enormous complexity of the reaction networks in real cells. The biochemistry of the metabolic pathways which are involved in the cellular synthesis of nucleic acids easily fills a whole monograph. The detailed dynamics of such a network is so complex that it seems completely hopeless to choose it as a basis of a theory of molecular evolution. Eigen [2] proposed an approach which parallels with the common strategy in physics: the essential unit of replication is selected out of unnecessary accessories in order to study the principles of molecular evolution. This unit of replication consists of a template and the molecular machinery of template-induced polynucleotide synthesis. We take this unit out of its natural context in the cell and incorporate it into an artificial but sufficiently simple environment. This is achieved by efficient external control which keeps most variables of the biochemical reaction network essentially constant. Then, the kinetic equations become simple and allow straightforward theoretical analysis.

Meanwhile, the kinetic theory of molecular evolution has

been worked out in detail [3–7]. New theoretical concepts like those of sequence space, quasi-species, error threshold and hypercycles appeared and turned out to be useful in molecular evolution. We shall describe these concepts in the forthcoming sections of this contribution.

An experimental set-up was derived from a simple RNA bacteriophagew, $Q\beta$, and its host the bacterium *Escherichia coli*. It resembles closely the idealized environmental conditions mentioned above (for details see Biebricher, this volume). The kinetic studies on this model system resulted in a molecular mechanism of template-induced RNA synthesis [8–10]. This mechanism allows to specify the conditions to which the kinetic theory of evolution applies.

The *in vitro* studies on RNA replication revealed another important fact. The experimental system fulfils the requirements of Darwinian evolution: it provides variability through replication errors and it shows selection. As a consequence of these two conditions, the system in the test tube adapts to the environment. Thus, Darwinian evolution is not a privilege of living organisms; molecules replicating outside living cells evolve by the same principles. Considering the history of life we may conclude that, almost certainly, Darwinian evolution has preceded biological evolution.

2. Quasi-species and evolutionary adaptation

At first we consider simple template-induced replication of polynucleotides. Additional catalytic action of polynucleotides like the kind of activity discovered in RNA self-splicing reactions is excluded. RNA catalysis will be discussed briefly in Section 3. Accordingly, polynucleotides replicate and mutate independently. In the language of reaction kinetics template induced, population independent replication is an example of first order autocatalysis.

RNA replication as observed with most RNA bacteriophages is an example of 'complementary replication'. A given single strand is replicated via an intermediate negative copy. Commonly, one distinguishes plus- and minus-strands: the plus-strand is the one which is translated into protein in the host cell. The analysis of complementary replication is straightforward: there is a fast process which leads to equilibrium of plus- and minus-strands and then, the plus–minus ensemble grows like a single replicating entity. We shall neglect the transient phase and always refer to the plus–minus ensemble in the forthcoming discussions.

2.1. *Replication and mutation*

RNA replication in an open system (Fig. 1) can be described properly by a series of single, reaction steps which are autocatalytic in first order

$$(A)+I_k \xrightarrow{A_k Q_{kk}} 2I_k \quad (k = 1,2,\ldots,n) \tag{1}$$

and the accompanying network of mutations

$$(A)+I_k \xrightarrow{A_k Q_{jk}} I_j+I_k \quad (j,k = 1,2,\ldots,n) \tag{2}$$

Herein, we denote the individual polynucleotide sequences by I_k. Accordingly, n such sequences form a set whose members are related by mutations ($k = 1,2,\ldots,n$). This set is completely connected in the sense that every mutant can be synthesized from every template (Fig. 2). The symbol (A) stands for the material required for RNA synthesis. In experimental systems this includes energy-rich monomers, the nucleoside triphosphates GTP, ATP, CTP and UTP, in the proper stoichiometric amounts. Laboratory studies utilize also an RNA polymerizing enzyme, e.g. $Q\beta$-replicase which is a virus specific RNA polymerase.

The rate constants A_k are a measure of the total rate of RNA synthesis, including correct copies and mutations, on the template I_k. The frequencies at which correct copies and individual mutations are obtained are represented by the elements of the mutation matrix Q: Q_{jk} is the frequency to obtain I_j as an error copy of I_k. Accordingly, we have a conservation relation:

$$\sum_{k=1}^{n} Q_{jk} = 1 \tag{3}$$

Every copy has to be either correct or erroneous.

The diagonal elements of the mutation matrix, Q_{kk}, were called the quality factors of the replication process [2, 3]. The case $Q_{kk} = 1$ means ultimate accuracy of replication, or no mutations.

2.2. *The basis of structural variability*

It is worth considering the numbers of polynucleotides which are involved in such a replication–mutation system. Even if we restrict ourselves to point mutations – that means we consider exclusively polymers of the same chain length ν – the numbers of possible sequences are 'hyperastonomically' large: we have 4^ν different sequences with ν digits! In a typical experiment we are dealing with 10^{12} polynucleotides or less. This number is approximately the number of oligomers of chain length $\nu = 20$.

In nature we are dealing with much longer polynucleotide sequences: the smallest RNA molecules, the transfer RNA's are about $\nu = 70$ bases long. The genetic information of the smallest viruses amounts to a few thousand bases. The DNA of bacteria is some million base pairs long, those of eukaryotic organisms even about 10^9 bp long.

Point mutations are not the only class of deviations from regular template induced polynucleotide synthesis. Several types of replication errors have been identified. They are distinguished as different classes of mutations. We mention the most important of them here.

(1) Point mutations are single base exchanges. The chain length (ν) is conserved.

(2) Deletions are mutations which lead to polynucleotides of smaller chain length. One base or a sequence of two or more bases is lost during replication.

(3) Insertions lead to longer polynucleotides. A base sequence is inserted between two neighboring bases of the template. Commonly, the sequence inserted is a copy of some part of the template.

(4) Rearrangements of the polynucleotide sequence during replication lead to new sequences of approximately the same length in which the parts appear in different order.

Thus, a natural ensemble of polynucleotide sequences which are related through replication and mutation is not confined to a constant chain length ν.

Any population size, *in vivo* or *in vitro*, is by far smaller than

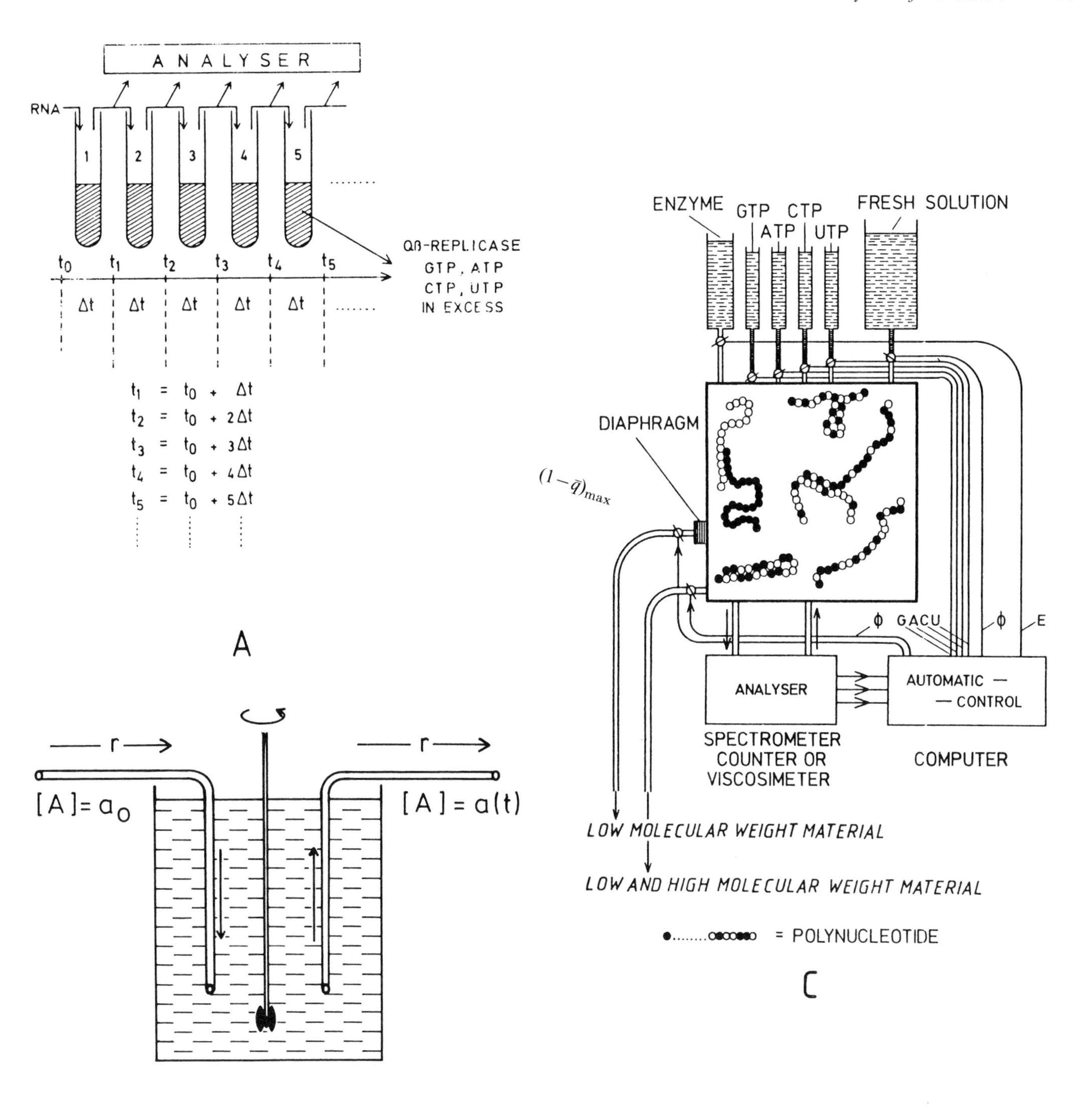

Fig. 1. Three types of open systems to keep RNA replication away from equilibrium. (*A*) Serial transfer. These experiments are an example of a simple experimental technique which is used frequently in test tube evolution studies. The material source is provided by fresh solution in the test tubes. This solution contains the energy rich nucleoside triphosphates, GTP, ATP, CTP and UTP, as well as an enzyme for RNA replication. A small sample of RNA suitable for replication is transferred after equal time intervals (Δt) from one test tube to the next. Thereby, the energy-rich material consumed by the reaction is renewed. The evolutionary constraint is provided by the discontinuous dilution of the RNA at the transfer steps. Most of the replicating molecules are discarded. (*B*) The continuously stirred flow reactor (CSTR). In this reactor which is known also as 'chemostat' in microbiology, the material source is provided by the continuous influx of a solution containing the material which is necessary for replication. We summarize all the compounds required under the symbol *A*. The constraint is provided by the continuous outflux of solution from the reactor. Replicating molecules are injected into the reactor at $t = t_0$. Then, their concentrations may increase and eventually reach a stationary value, or they may be diluted out of the reactor depending on the input solution (a_0) and the flow rate (r) which is commonly measured in terms of the reciprocal residence time (τ_R^{-1}) of the solution in the tank reactor. (*C*) The evolution reactor. This kind of flow reactor consists of a reaction vessel which has walls which are impermeable to polynucleotides. Energy-rich material is poured from the environment into the reactor. The degradation products are removed steadily. Material transport is adjusted in such a way that the concentrations of the energy rich monomers, GTP, ATP, CTP and UTP, are constant in the reactor. A dilution flux ϕ is installed in order to remove the excess of polynucleotides produced by replication. Thus, the sum of the concentrations, $[I_1]+[I_2]+\ldots+[I_n] = c$, may be controlled by the flux ϕ. Under 'constant organization' ϕ is adjusted such that $c = c_0$ is constant. The regulation of ϕ requires internal control, which may be achieved by analysis of the solution and data processing by a computer as indicated above.

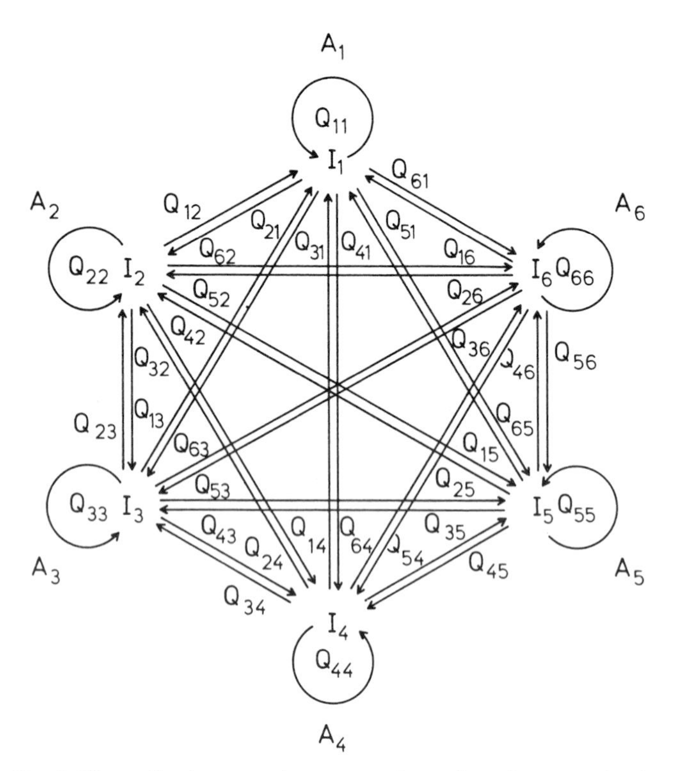

Fig. 2. The replication–mutation system. It consists of a network of n^2 template-induced polymerization reactions. The rate constants are of the form $A_j Q_{kj}$, where A_j is the rate constant of RNA synthesis on the template I_j and Q_{kj} the frequency of mutations from I_j to I_k. In the example shown here we chose $n = 6$.

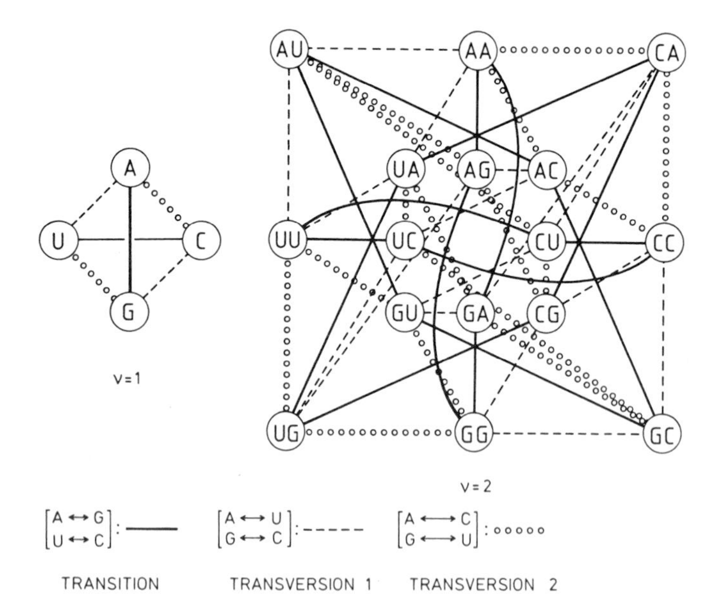

Fig. 3. The sequence space of polynucleotides based on the four-letter alphabet G, A, C, U. We show here only the cases $\nu = 1$ and $\nu = 2$. There are three classes of base exchanges which occur in point mutations: transitions, which are purine–purine and pyrimidine–pyrimidine exchanges, and two classes of transversions known as 'intrapair' and 'interpair' transversions.

the number of polynucleotide sequences which are related through mutations. Thus, we observe a striking difference to ordinary chemical reaction networks which involve but a few molecular species each of which is present in a very large number of copies. The converse situation is the rule here; the numbers of different molecular species that may be interconverted through replication and mutation, exceed by far the total number of molecules present in any experiment, or even the number of molecules available on the Earth. The applicability of conventional chemical kinetics, which, in principle, is a theory of infinite population sizes, to problems of molecular evolution is a subtle question which has to be considered carefully for each particular case. We cannot discuss this problem in full consequence here, but we refer to a forthcoming paper on molecular evolution described as a stochastic process. Here, only those aspects are presented for which the use of the kinetic equations can be well justified.

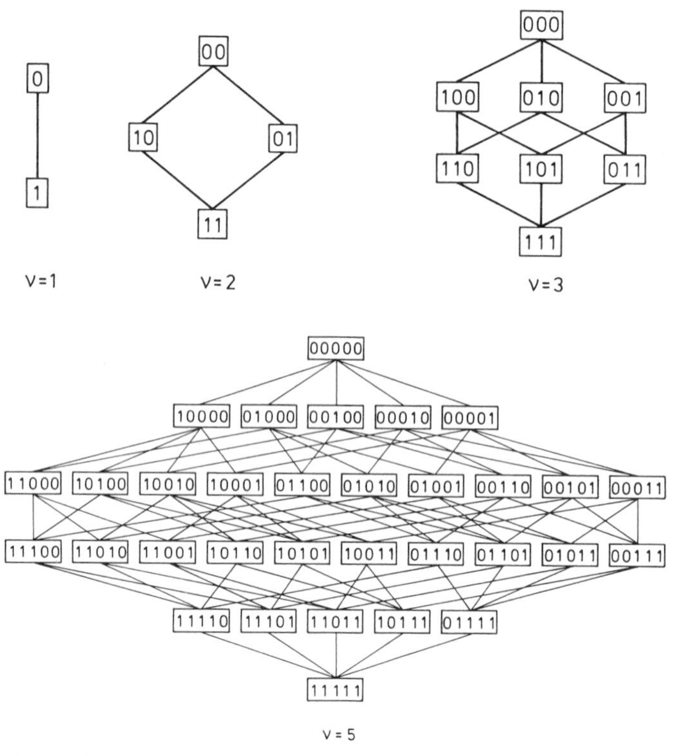

Fig. 4. The sequence space of binary sequences based on the two-letter alphabet 0, 1. We present the cases $\nu = 1$, 2, 3 and 5. All pairs of sequences, I_i and I_k, with Hamming distance $D(i, k) = 1$ are connected by a straight line. The graphs obtained are hypercubes of dimension ν.

2.3. *The concept 'sequence space'*

In order to illustrate a replicating system with many more possible sequences than molecules in the population we may use an abstract point space. We assign a point to every sequence. The points are arranged such that the sequence relations are reflected by neighborhoods. The appropriate measure of the relationship of two sequences I_i and I_k is the Hamming distance $D(i, k)$. It is defined as the number of positions in which the two properly aligned sequences differ. In order to illustrate the structure of such a sequence space we connect all pairs of sequences with Hamming distance $D = 1$ by a straight line. We call the diagram obtained the graph of sequence relations. In the case of a four-letter alphabet like the G, A, C, $U(T)$ alphabet of natural polynucleotides the graphs of sequence relations are quite complicated (Fig. 3, see also [11]). For the purpose of illustration, however, a two-letter alphabet, e.g. (0, 1) serves equally well (Fig. 4). Then, the graphs of sequence relations are simply the 'ν-cubes'. These are hypercubes of dimension ν. The number ν is the chain length of the polynucleotide sequence.

In any real system the number of points of the sequence space is gigantic. A population corresponds to some small section of connected points on the hypercube. Any high

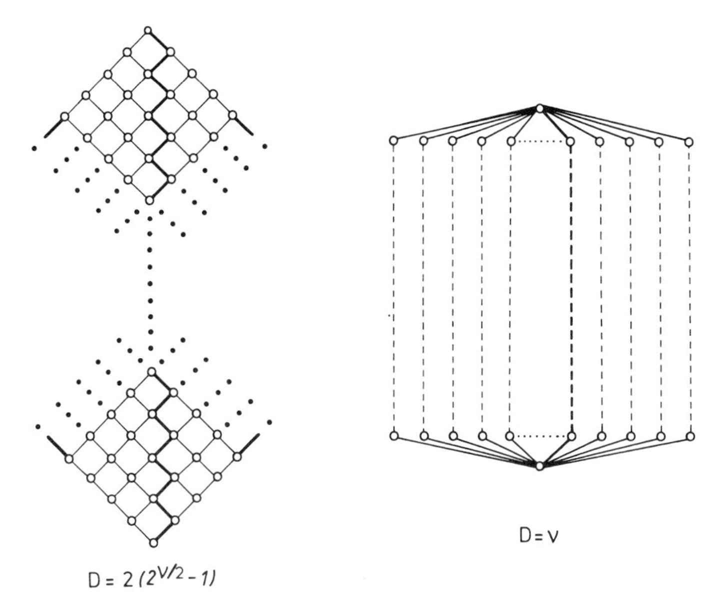

Fig. 5. Maximum distances in low and high dimensional sequence spaces. We compare two graphs, a two-dimensional array of points and the hypercube of dimension ν. In the first case the maximum Hamming distance is $2(2^{\nu/2}-1)$ whereas it amounts ν (only!) in the second graph.

dimensional space differs substantially from the two or three dimensional spaces of our experience. Mutation distances between related members of an ensemble become much shorter in high dimensions. For example let us arrange all 2^{50} binary sequences of chain length $\nu = 50$ on a two-dimensional lattice (Fig. 5). Then we have an array of $2^{25} \times 2^{25}$ points. The largest distance amounts $(2^{25}-1) \times \sqrt{2}$ times the unit length. If we count 'distance' by the numbers of edges we have to pass along the shortest path from one point to another, just as we would count Hamming distances. This largest distance consists of $2 \times (2^{25}-1)$ edges. Now consider the hypercube of dimension $\nu = 50$: the largest distance in this case is only 50 edges! In other words, it needs only 50 mutation steps (ordered in sequence!) to reach the most distant sequence which, of course, is the complementary sequence. This property of the high dimensional sequence space represents one fundamental factor of the efficiency of evolutionary optimization.

Every polynucleotide sequence is characterized by a selective value. We can imagine a 'value landscape' which is erected on the sequence space by assigning selective values to the individual points. Evolution can now be visualized as a 'hill climbing' or optimization process on the value landscape. In the forthcoming sections we shall work out a quantitative expression for the selective value and we shall consider the optimization process in more detail.

2.4. *Selection and constraints*

One basic concept of Darwin's theory is natural selection. In order to be able to provide a coherent picture of the evolutionary process we have to make the notion of selection more precise. Let us assume a population of several types of polynucleotides I_k $(k = 1, \ldots, n)$ at time $t = 0$. We denote their concentrations by $[I_k] = c_k$. We say that the molecular species I_m has been selected if all other concentrations except $c_m(t)$ vanish in the limit of infinite time

$$\lim_{t\to\infty} c_m(t) \neq 0 \quad \text{and} \quad \lim_{t\to\infty} c_k(t) = 0 \quad \text{for all } k \neq m \tag{4}$$

Selection occurs only in systems with some constraint concerning the total number of molecules. The basic effect of the constraint is to introduce finite lifetimes for the individual molecules. Constraints in nature are provided by the finiteness of reservoirs and by degradation processes. The simplest degradation process we can think of is first-order decay

$$I_k \xrightarrow{D_k} (B) \quad (k = 1, 2, \ldots, n) \tag{5}$$

Hydrolytic cleavage of RNA or DNA leads to energy poor monomers. These are the nucleoside monophosphates GMP, AMP, CMP and UMP or TMP, respectively. The symbol (B) stands here for the stoichiometric combination of monomers as obtained by degradation of a given polynucleotide.

One might well ask why we assume irreversibility of the reactions (1), (2) and (5). This question has been studied in some detail in previous papers (see, for example, [7, 12]). It turned out that irreversibility for practical purposes of these reaction steps is a necessary prerequisite of global selection. Only then the system is kept away from thermodynamical equilibrium for sufficiently long time. Moreover, we observe that the conditions of practical irreversibility of the formation and of the degradation proceses are well fulfilled in all real biological systems

$$\Delta G(\nu_G\,\mathrm{GTP} + \nu_A\,\mathrm{ATP} + \nu_C\,\mathrm{CTP} + \nu_U\,\mathrm{UTP}) \gg \Delta G(\mathrm{I} + \nu\mathrm{PP})$$

and

$$\Delta G(I) \gg \Delta G(\nu_G\,\mathrm{GMP} + \nu_A\,\mathrm{AMP} + \nu_C\,\mathrm{CMP} + \nu_U\,\mathrm{UMP})$$

We denote the stoichiometric coefficients by ν_G, ν_A, ν_C and ν_U with $\nu = \nu_G + \nu_A + \nu_C + \nu_U$ and use the symbol PP for pyrophosphate anion. Replication and degradation proceed under conditions of strongly negative ΔG.

In a continuously stirred tank reactor (CSTR, see, e.g. Fig. 1) the constraint is provided by the flow which is commonly measured in terms of the reciprocal residence time of the solution in the tank: $r = \tau_R^{-1}$. Then, the dilution flux introduces a constraint also in absence of degradation. The term in the kinetic equations is formally identical with that of first order decay. We just have

$$D_1 = D_2 = \ldots = D_n = r.$$

Another constraint called constant organization or constant population size is frequently used in molecular evolution and population genetics because it simplifies the mathematical analysis substantially. The total number of molecules, the population size, is kept constant by means of an unspecific dilution flux $\phi(t)$. This flux is adjusted instantaneously to the rate of replication in order to compensate for the excess production of the whole population (see Fig. 1 and [2, 4]).

2.5. *Rate equations and solutions*

In mass action kinetics and normalized concentrations, $x_k = c_k/\Sigma_j c_j$ with $\Sigma_k x_k = 1$ $(j, k = 1, 2, \ldots, n)$, the rate equations corresponding to the network of reactions (1), (2) and (5) under the constraint of constant population size are of the form

$$\dot{x}_k = x_k(A_k Q_{kk} - D_k - \phi) + \sum_{j \neq k} A_j Q_{kj} x_j \qquad (k, j = 1, 2, \ldots, n) \tag{6}$$

The unspecific dilution flux ϕ in this equation can be expressed by the mean excess production $\bar{E}(t)$ of the population:

$$\phi = \bar{E} = \sum_k E_k x_k = \sum_k (A_k - D_k) x_k$$

For convenience we introduced here an excess production of species I_k: $E_k = A_k - D_k$.

In case of equation (6) it is possible to remove the nonlinearity introduced by the flux term by means of a transformation of variables [13, 14]

$$y_k = x_k \exp\left[\int_0^t \bar{E}(\tau)\,d\tau\right] \quad (k = 1, 2, \ldots, n) \tag{7}$$

Then we are dealing with a linear differential equation

$$\dot{y}_k = \sum_{j=1}^{n} w_{kj} y_j \quad (k = 1, 2, \ldots, n) \tag{8}$$

with $w_{kj} = A_j Q_{kj} - D_k \delta_{kj}$ being an element of the value matrix W. Accordingly, the diagonal elements of W are the selective values, e.g. $w_{kk} = A_k Q_{kk} - D_k$ is the selective value of the polynucleotide sequence I_k.

Let us consider briefly error-free replication. Then, the matrix of mutation frequencies Q is diagonal: $Q_{ik} = Q_{kk}\,\delta_{ik}$. Accordingly, we have $w_{kk} = A_k - D_k = E_k$, the selective value of a polynucleotide sequence is identical with its excess production. The corresponding kinetic equations can be solved easily. By straightforward calculus we obtain

$$x_k(t) = x_k(0) \exp\left[E_k t - \int_0^t \bar{E}(\tau)\,d\tau\right] \quad (k = 1, 2, \ldots, n) \tag{9}$$

Clearly, the population converges asymptotically towards a homogeneous state in which the most efficiently replicating species is present exclusively:

$$\lim_{t\to\infty} x_m(t) = 1; \quad \lim_{t\to\infty} x_k(t) = 0 \quad \text{for} \quad k \neq m$$
$$\text{and} \quad E_m = \max[E_j \quad (j = 1, \ldots, n)] \tag{10}$$

We observe selection of the polynucleotide I_m in this case. The process obeys an optimization principle: the mean excess production $\bar{E}(t)$ is a non-decreasing function. It represents the quantity which is optimized during selection.

In the general case, replication with errors, the linear differential equation can be analysed by standard techniques. Let us assume that the matrix W is diagonalizable. Then the solutions of equation (6) can be expressed in terms of left-hand and right-hand eigenvectors, $\boldsymbol{u}_i = [u_{ki}(k = 1, \ldots, n)]$ and $\boldsymbol{v}_i = [v_{ik}(k = 1, \ldots, n)]$, and eigenvalues, $\lambda_i\,(i = 1, \ldots, n)$ of the matrix $W = [w_{ij}]$

$$y_k(t) = \sum_{i=1}^{n} u_{ki} \exp(\lambda_i t) \sum_{j=1}^{n} v_{ij} y_j(0) \quad (k = 1, \ldots, n) \tag{11}$$

In case we choose sufficiently small rate constants for the degradation reactions ($D_k < A_k Q_{kk}(k = 1, \ldots, n)$) the matrix W is a positive matrix which admits, by the theorem of Frobenius, an eigenvalue λ_1 which is dominant in the sense that $\lambda_1 > |\lambda_k|$ for all other eigenvalues λ_k of the matrix W. The eigenvalue λ_1 is simple and hence, standard linear theory shows that one has for long times t

$$y_k(t) \simeq u_{k1} \exp(\lambda_1 t) \sum_{j=1}^{n} v_{1j} y_j(0) \quad (k = 1, \ldots, n)$$

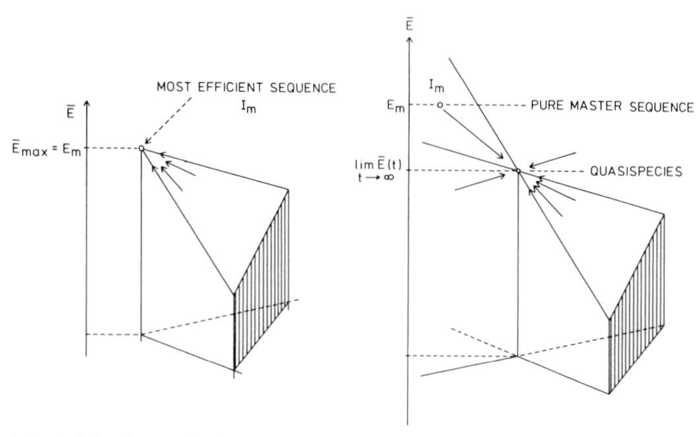

Fig. 6. Selection and optimization of mean excess production $\bar{E}$. In the case of error free replication we observe optimization of $\bar{E}$ for all initial conditions (left-hand sketch). The mean excess production in the replication–mutation system may increase or decrease depending on the choice of initial conditions (right-hand sketch).

In addition, the relative variables converge asymptotically to values which are independent of initial conditions

$$y_k(t)/\sum_j y_j(t) = x_k(t) \to u_{k1}/\sum_j u_{j1} \quad (k, j = 1, \ldots, n)$$

We called this stationary mutant distribution the 'quasi-species' of the system in order to indicate the analogy to the notion of species in biology. The properties of the quasi-species are determined completely by the dominant eigenvector, $\boldsymbol{u}_1$, of the matrix W, This dominant eigenvector depends on the distribution of replication rate constants, $A_k(k = 1, \ldots, n)$, as well as on the mutation matrix Q. We shall present one model for the Q matrix based on point mutations [11] in Section 2.6. It will be used to illustrate the error propagation problem.

In the case of the replication–mutation system we do not observe selection in the ordinary sense. Instead, the stationary solution consists of a master sequence and its mutants. The master sequence (I_m) is the sequence which is present at the highest concentration in the stationary distribution of polynucleotides. With the exception of some pathological cases with especially high mutation rates, the master sequence is the sequence with the maximum selective value:

$$I_m: w_{mm} = \max[w_{jj} \quad (j = 1, \ldots, n)]$$

As we showed above, this stationary mutant distribution, the quasi-species, is determined by the dominant eigenvector of W. In order to retain the concept of selection we formulate the replication–mutation problem in an abstract space whose coordinate axes are spanned by the eigenvectors of the matrix W. Then the vector of normalized concentrations can be written as

$$\boldsymbol{x}(t) = [x_1(t), \ldots, x_n(t)] = \sum_{k=1}^{n} x_k(t)\,\boldsymbol{x}_k = \sum_{k=1}^{n} u_k(t)\,\boldsymbol{u}_k.$$

The vectors $\boldsymbol{x}_k$ are unit eigenvectors of the Cartesian coordinate system and $\boldsymbol{u}_k$ the right-hand eigenvectors of the value matrix W. The variables $u_k(t)$ are simple exponential functions of time

$$u_k(t) = u_k(0) \exp(\lambda_k t) \exp\left[-\int_0^t \bar{E}(\tau)\,d\tau\right] \quad (k = 1, 2, \ldots, n)$$

After sufficiently long time only the variable corresponding to the dominant eigenvector does not approach zero:

$u_1(t) \to 1$, $u_k(t) \to 0$ $(k = 2, 3, \ldots, n)$. Hence, we do observe selection in the replication–mutation system if we use the transformed variables $u_k(t)$. Do we also observe optimization of the mean excess production $\bar{E}(t)$? No! We can easily construct a counterexample. Assume that we start from a homogeneous population consisting of the master sequence I_m exclusively. It has the largest selective value, w_{mm}. Less efficient mutants are produced as time proceeds and $\bar{E}(t)$ decreases until it reaches the stationary value which is characteristic of the quasi-species (Fig. 6). In the course of evolution in nature we do not start from homogeneous or nearly homogeneous populations already containing mainly master sequences and then optimization of the mean excess production commonly occurs.

2.6. *Error thresholds and 'random' replication*

The accuracy of replication sets a limit to the amount of genetic information which can be transferred stably from generation to generation. In order to explore the accuracy limit of replication we shall use a model for the mutation matrix, Q [11]. This model is restricted to point mutations and therefore, all polynucleotide sequences to be considered have the same chain length ν. This model may seem rather restrictive, but it turned out that the results obtained are of general validity.

We may factorize the quality factors of replication, the diagonal elements of the mutation matrix Q, into the contributions of single digits:

$$Q_{kk} = q_1^{(k)} q_2^{(k)} q_3^{(k)} \ldots q_\nu^{(k)} \tag{12}$$

Herein $q_1^{(k)}$ is the accuracy of the incorporation of the first base into the new polynucleotide sequence which is synthesized on the template I_k. The factor $q_2^{(k)}$ represents the accuracy of incorporation of the second base, $q_3^{(k)}$ that of the third base, etc. In analogy to the definition of quality factors (Q_{kk}) the limit $q = 1$ means no errors or ultimate precision of replication. In the model which we present here, we assume uniform single digit accuracies:

$$q_1 = q_2 = \ldots = q_\nu = q \tag{13}$$

This assumption implies also that the accuracy of replication does not depend on the base sequence of the polynucleotide. Then, the quality factors are of the simple form $Q_{kk} = q^\nu$.

In addition we restrict the model to 'binary sequences', these are sequences built from two digits (0, 1) only. Then, the frequency of the incorporation of the wrong digit is simply $1-q$. Off-diagonal elements of the mutation matrix can be calculated now directly from the Hamming distance of the two sequences to be interconverted by mutation

$$Q_{jk} = q^{\nu - D(j,k)} (1-q)^{D(j,k)} \tag{14}$$

Note, that the mutation matrix Q is symmetric in this case. Then, the off-diagonal elements of the value matrix W can be written as

$$w_{jk} = A_k q^{\nu - D(j,k)} (1-q)^{D(j,k)} \tag{15}$$

and hence, W is not symmetric. Nevertheless, Rumschitzky [15] presented a proof that the value matrix W has exclusively real eigenvalues within this model.

Complex conjugate pairs of eigenvalues, however, may well occur for non-symmetric Q matrices $(Q_{ik} \neq Q_{ki})$. Highly unsymmetric mutation frequencies are characteristic for polynucleotide sequences with so called 'hot spots'. These are positions in DNA or RNA at which one observes extraordinarily high mutation rates. In general, the frequencies of reverse mutations at these positions are much lower. The existence of complex conjugate pairs of eigenvalues of the value matrix W has an interesting consequence: transient oscillations of concentrations may occur. In the limit of long times the system, nevertheless, converges towards the dominant eigenvector. The corresponding eigenvalue is real and positive and hence, all oscillations in the concentrations fade out for sufficiently long times.

Let us now consider the dominant eigenvector of W as a function of q. As we showed in the previous section it represents the stationary mutant distribution or the quasi-species of the replication–mutation system. There is a precisely defined limit: highly accurate replication, $q \simeq 1$, converges towards error-free replication in the limit $q \to 1$.

Another special case of particular interest is the point $q = 0.5$. There incorporation of correct and wrong digits occurs with the same probability. Replication errors are so frequent that the selective values w_{kk} have no influence on the stationary mutant distributions. Inheritance breaks down and all polynucleotides are synthesized with equal probabilities. At the stationary state the sequences are present according to their statistical weights. The quasi-species degenerates to a uniform distribution of polynucleotides and the concept of a master sequence becomes obscure. In order to point at the lack of sequence correlations between template and copy we called this process 'random replication' [11]. We shall discuss the implications of random replication for the applicability of the kinetic equations to describe low accuracy replication later on.

Now we consider the results of a numerical computation which are shown graphically in Fig. 7. At the value $q = 1$ we observe selection of the master sequence (I_m) as expected. Mutants appear in the stationary distribution for $q < 1$ and increase in frequency with decreasing q-values. In the figure we show the sum of the relative concentrations of all one-error mutants, all two-error mutants, etc. Mutants with more errors become important at lower values q. The system shows a sharp transition from faithful to random replication at a critical minimum accuracy: $q = q_{\min}$. The logarithmic plot shown in Fig. 7 provides an even better insight into the sharpness of the transition.

The minimum accuracy of replication can be derived from the dominant eigenvalue of the value matrix W by means of second order perturbation theory [3, 11]

$$Q_{\min} = (q_{\min})^\nu = \sigma_m^{-1} \tag{16}$$

Herein, σ_m represents the 'superiority' of the master sequence, which is expressed easily in terms of the rate constants A_k and D_k:

$$\sigma_m = \frac{A_m}{D_m + E_{-m}} \tag{17}$$

with

$$\bar{E}_{-m} = \sum_{k \neq m} E_k x_k \Big/ \sum_{k \neq m} x_k \tag{18}$$

In terms of our model system we have $\sigma_m = A_m / \bar{A}_{-m}$, since

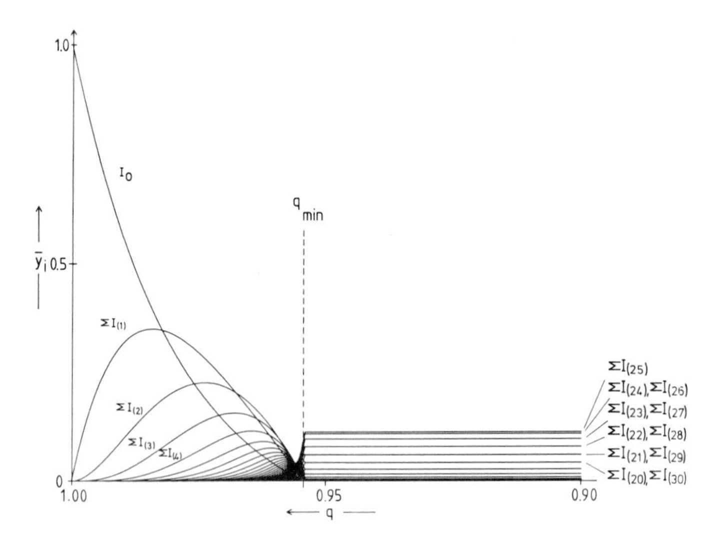

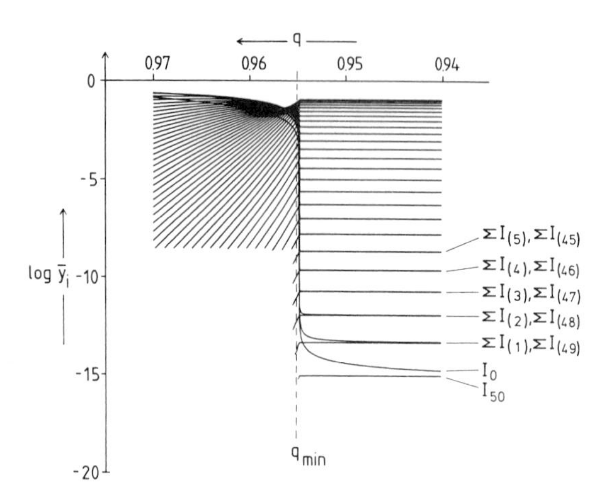

Fig. 7. The error threshold in the replication–mutation system. We show the stationary mutant distribution for a replicating ensemble of polynucleotides, the 'quasi-species'. The curves represent the relative concentrations of the master sequence (I_0), the sum of the relative concentrations of all one error mutants ($\Sigma I_{(1)}$), of all two error mutants ($\Sigma I_{(2)}$), etc., as functions of the mean single digit accuracy of replication ($\bar{q}$). In the numerical example presented here the chain length of the polynucleotide is $\nu = 50$. The replication rate constant for the master sequence is chosen to be $A_0 = 10$, for all other sequences $A_1 = A_2 = \ldots = 1$ in arbitrary time and concentration units. From these parameters we calculate a minimum accuracy $Q_{\min} = 0.1$ (for details see [11]). Thus, the critical single digit accuracy is $\bar{q}_{\min} = (Q_{\min})^{1/\nu} = 0.945$ in this case. Starting from pure master sequence at error-free replication we observe the expected decrease in the relative concentration of the master sequence with decreasing replication accuracy $\bar{q}$. In sequence $(1, 2, 3, \ldots, n)$, error mutants dominate the quasi-species. At the critical value of the single-digit accuracy, $\bar{q}_{\min}$, we observe a sharp transition in the structure of the quasi-species. The sharpness of the transition is even more pronounced in the logarithmic plot. Below the critical value the concentrations are determined exclusively by the statistical weights of the corresponding error mutant classes. All single polynucleotide sequences are equally frequent. Hence, the sum of the concentrations of 25 error mutants ($EI_{(25)}$) is largest, followed by 24 and 26 error mutants ($EI_{(24)}$, $EI_{(26)}$), etc.

we assumed equal rate constants of degradation. The calculated critical accuracy of replication falls right into the center of the transition computed numerically (Fig. 7).

The transitions become exceedingly sharp for still longer chains. It suggests the conjecture that we are dealing here with an analogue to higher order phase transitions in the limit of infinitely long chains $\nu \to \infty$. Recently, Leuthaeuser [16] was able to show the equivalence of this transition with the phase transition at the Curie point of a two dimensional Ising model.

Replication of polynucleotides with different chain lengths (ν_k $(k = 1, 2, \ldots, n)$) by the same replication machinery yields an interesting generalization of the accuracy limit. Then, q is a constant and the accuracy of replication, Q_{mm} is determined by the chain length ν:

$$Q_{mm} = q^{\nu_m} > Q_{\min} = q^{\nu_{\max}} \tag{19}$$

or

$$\nu_{\max} = -\frac{\ln \sigma}{\ln q} \simeq \frac{\ln \sigma}{1-q} \tag{20}$$

since $q \simeq 1$ in most real systems. The error threshold sets a limit to the lengths of polynucleotide chains which can be replicated with sufficient accuracy to allow stable transfer of genetic information.

Although we used a rather special model system for the demonstration of the existence of a sharply defined minimum accuracy of replication the results are valid in great generality. This has been shown, for example, by application to real systems as they were discussed in [2, 3, 17]. Empirical data from viruses and bacteria suggest [3] that at least these organisms operate close to the error threshold.

In case of the RNA bacteriophage *Qβ* the single digit accuracy of replication was determined experimentally by Weissmann and co-workers [18, 19]. Their results combined with an estimate of the superiority (for details see [3]) lead to a maximum chain length of $\nu_{\max} \simeq 4500$ bases. Natural *Qβ* RNA is 4220 bases long.

Let us now consider the transition from direct to random replication in more detail. The critical accuracy of replication ($q = q_{\min}$) defines an error threshold below which the concepts of quasi-species and master sequence become obscure. There is a fundamental conceptual problem with the deterministic error threshold: below error threshold the deterministic model predicts a uniform stationary distribution of sequences. Such a distribution can never be realized in any population, neither in nature nor in the laboratory. As we pointed out in Section 2.2 we are always dealing with many orders of magnitude more possible polynucleotide sequences than we have molecules in the population. A uniform distribution of polynucleotide sequences, thus, is far away from reality.

A stochastic analysis of the replication–mutation system by means of multitype branching processes [20], however, allows a physically meaningful interpretation of the error propagation problem in finite populations. Instead of characterizing the details of the stationary mutant distribution in the replication–mutation system we ask for the probability of survival of the master sequence (I_m) to infinite time. A non-zero probability of survival to infinite time is the stochastic equivalent to a stable quasi-species. The stochastic treatment shows that zero probability of survival of the master sequence implies zero probability of survival for all mutants. Below error threshold we are thus dealing with a steadily changing ensemble of polynucleotide sequences. Interestingly, the quantitative expressions for this stochastic error threshold become identical with those obtained from the analysis of differential equations provided both are derived from the same value matrix W. This interpretation of the error threshold by means of stochastic processes shows some analogy to the 'localization threshold' of mutant distributions which has been recently derived by McCaskill [21].

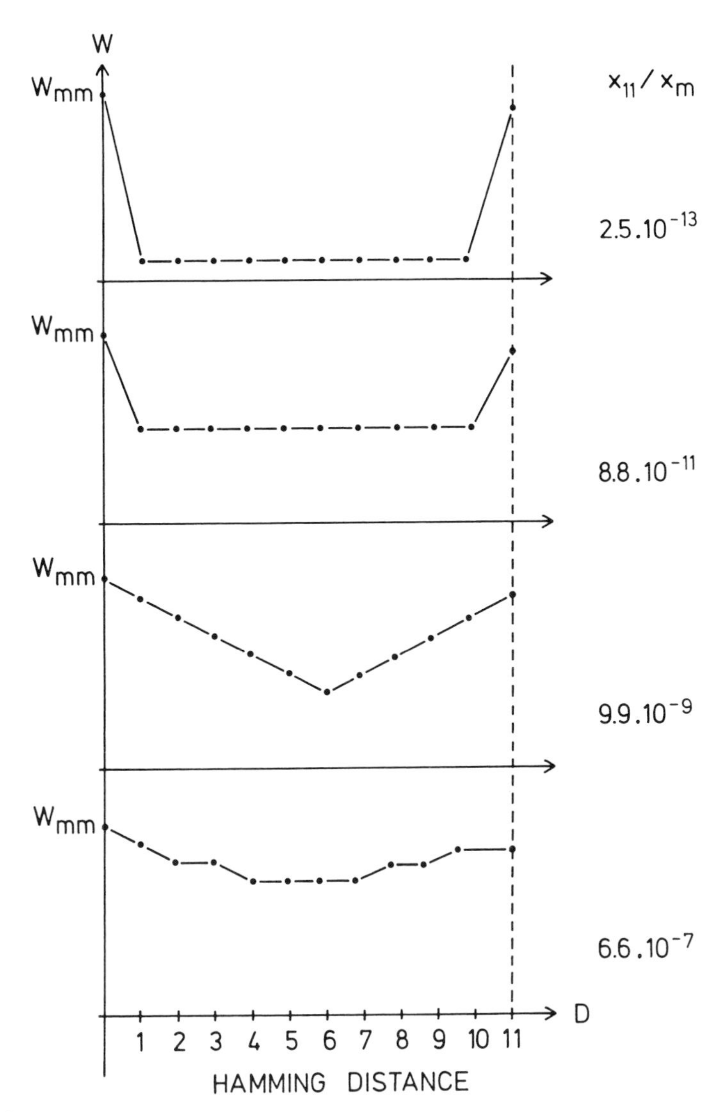

Fig. 8. Structure of the quasi-species. We show the frequency of the mutant I_{11} in a chain of sequences $I_m, I_1, \ldots, I_{11}$, as a function of the distribution of selective values of the intermediate mutants. Although the selective values w_{mm} and $w_{11,11}$ are the same in all cases, the ratio x_{11}/x_m varies over many orders of magnitude.

2.7. *The structure of the quasi-species and the pace of evolution*

The error threshold is of direct relevance for adaptation in evolution. Fast adaptation to a changing environment requires sufficiently high rates of mutation. Consider a system which is very precise in replication. It forms few mutants only and adaptation is a slow process. Soon, such a system will be driven out by its more flexible competitors. Too many replication errors, on the other hand, bring the system below error threshold, inheritance breaks down and deterioration is inevitable. Efficient evolution thus requires a compromise between the two extremes, an error frequency which is adjusted to the requirements dictated by the environment.

The internal structure of the quasi-species is of fundamental importance for the evolutionary process. Analytical expressions for the components of the dominant eigenvector of the value matrix W were derived by perturbation theory [6]. We present some illustrative examples in Fig. 8. The frequency at which a mutant I_k is present in the stationary distribution depends essentially on three quantities.

(1) The selective value of the mutant, w_{kk}.

(2) The Hamming distance between the master sequence I_m and the mutant, $D(m, k)$.

(3) The selective values of the intermediates along the mutation paths connecting I_m and I_k.

The mutant frequencies in the quasi-species are strongly biased by the structure of the value landscape. We find high relative concentrations of mutants which lie along 'ridges' of high selective value. Isolated high peaks in this landscape, i.e. points of high selective value which are surrounded by neighbors of low replication efficiency, are scarcely populated only.

Let us visualize the evolutionary process as 'hill climbing' in the value landscape. The optimization principle poses one important restriction: the mean excess productivity, $\bar{E}$, must not decrease. (This statement is not strictly true, but it holds under almost all relevant conditions. For further details see Section 2.5 and [7].) We assume that evolution proceeds through stages of 'quasi-stationarity'. By this we mean that the frequent mutants are present in the population in concentrations as predicted by the quasi-species.

The question whether the assumption of quasi-stationary is well justified or not, cannot be answered in general. It depends on the error rate, on the size of the popuation and on the structure of the value landscape. Recently we conceived a stochastic model system which allows to test quasi-stationarity by means of computer simulation [22]. Binary sequences are subjected to replication and mutation. The population size is controlled by a dilution flux. In a first test case thermodynamic stabilities of sequences are used as selective values. In Fig. 9 we show the results of one typical computer experiment. For high single-digit accuracies the condition of quasi-stationarity is fulfilled very well but the system is soon trapped in a local maximum of the value landscape. At lower accuracies but still above the error threshold, the quasi-species distribution is distorted and the system approaches the global optimum for sufficiently long times. Single-digit accuracies below error threshold are prohibitive for optimization. The systems drift randomly through the value landscape.

In Section 2.2 we showed that the majority of mutants is not present at all. Now, we distinguish two classes of rare mutants: rare mutants which have rare precursors and rare mutants which have frequent precursors in the population. Clearly, mutants of the first class are very unlikely to be formed at all. Thus, evolution proceeds from one local maximum to the next via a series of mutants of the second class. The structure of the quasi-species tells us where these mutants are: they are situated along the ridges of the value landscape.

Progress in evolution is made from one optimum to another through a series of intermediate mutants which are slightly deleterious compared to their master sequences. High dimensionality of sequence space turns out to be extremely important now: then the distances are short and the numbers of steps which are required to 'jump' from one hill to the next are fairly small. Therefore, real populations have a chance to proceed from one local optimum to another one in the vicinity. During the 'jump' those mutants are populated whose selective values form a ridge between the two peaks in the value landscape. The pace of molecular evolution is made by the rates at which the chains of mutants along the ridges of the value landscape are populated. Mutation rates, sufficiently

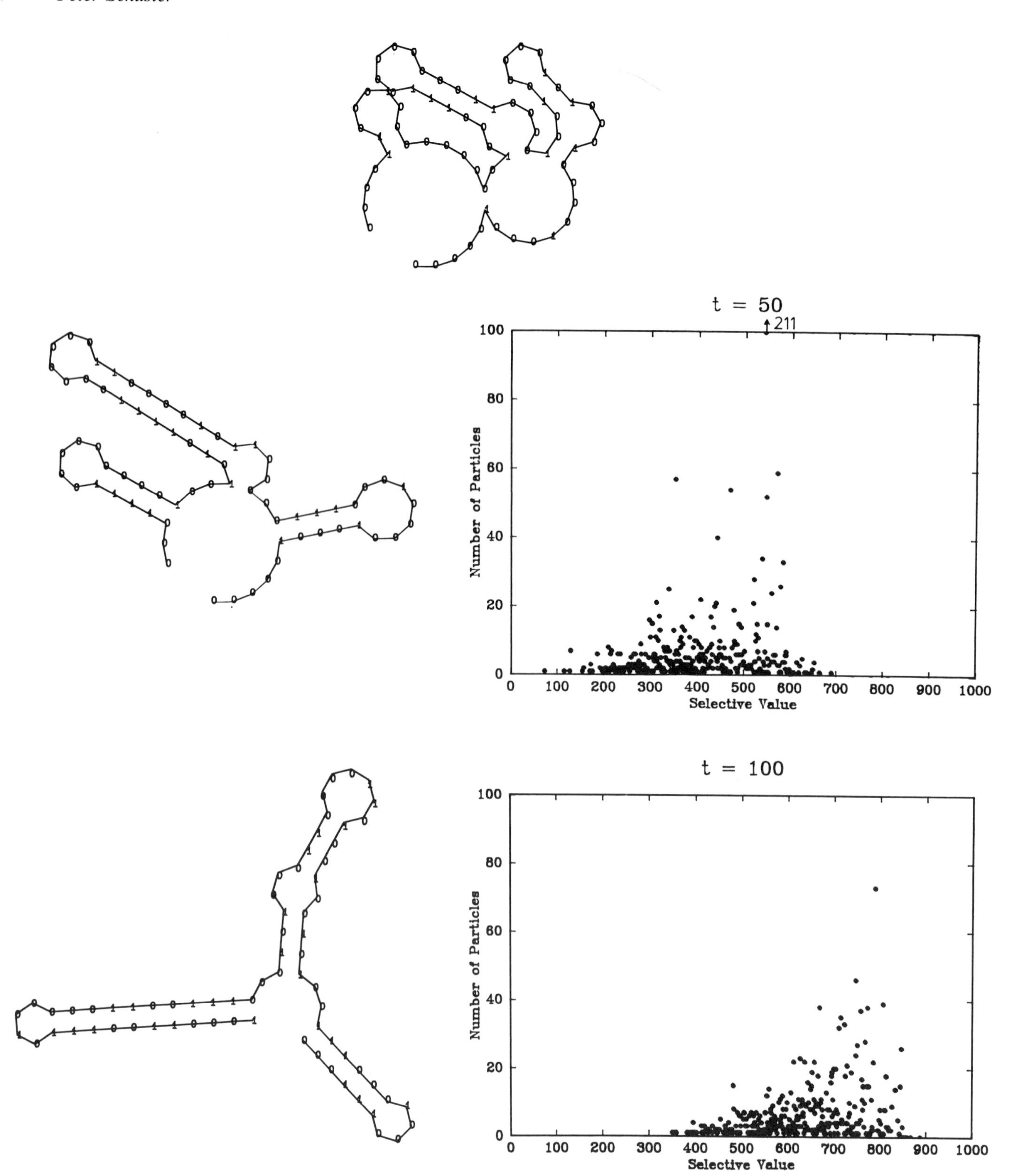

Fig. 9. The course of a computer-simulation experiment of adaptation in the replication–mutation system [22]. A population of $n = 2000$ binary sequences of chain length $\nu = 70$ is subjected to replication and mutation. The accuracy of replication is chosen to be $q = 0.997$. The selective value of a sequence is assumed to be equal to its thermodynamic stability. The excess production of sequences is removed by a dilution flux. We show the secondary structures of the initial sequence at times $t = 0$ and of the most frequent sequences at times $t = 50$, 100, 150 and 200. On the right hand side we present the distribution of sequences as a function of the selective values. At time $t = 200$ we observe a quasi-species-like distribution: the sequences with the highest selective values dominate. The ensemble reached 'quasi-stationarity'. The earlier plots show distortions of quasi-species because newly formed, more efficient mutant sequences are just outgrowing the previous master sequence. Comparison of the secondary structures documents the course of adaptation. The number of base pairs which, in essence, determines the thermodynamic stability increases steadily during evolutionary adaptation. Clearly, the assumption that the selective value is dominated by thermodynamic stability is unrealistic and has to be replaced by more sophisticated model assumptions [22]. The basic features of adaptation, nevertheless, are reflected very well by this simple model system.

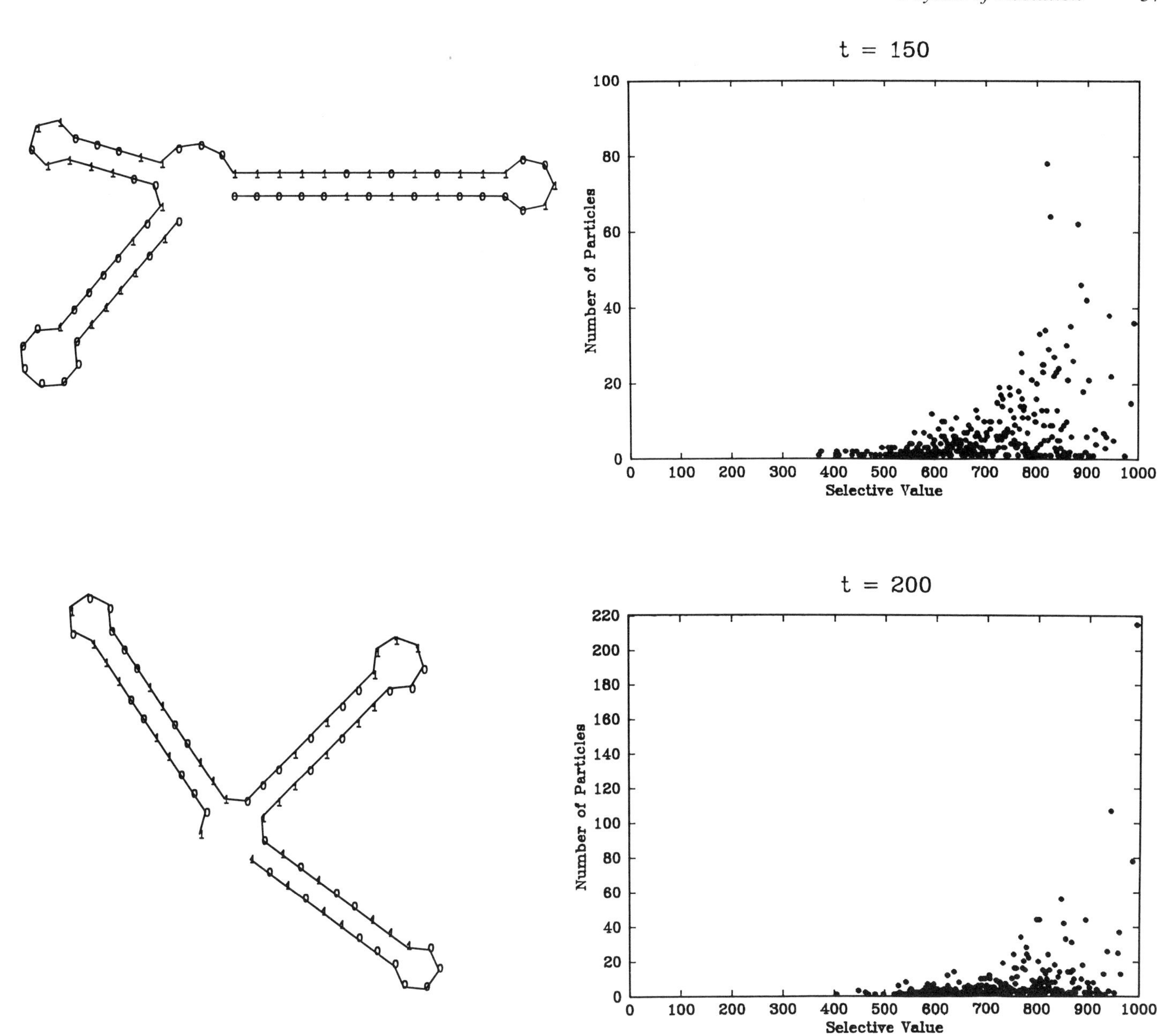

high but well above the error threshold are required for efficient progress of the evolving population.

Does evolution come close to or eventually reach the global maximum of the value landscape? Optimization of small entities and substructures occurs readily. Molecular biology tells us that most proteins have structures which are fully optimized for their tasks in the cell. Most viruses appear to be optimally adapted within the limits which are imposed on the amount of genetic information by the error threshold. Still larger entities like bacteria seem to be at the limits of full adaptation. From this size on, the sequence space becomes too large or, from another point of view, populations are too small to allow evolution to proceed near global optima.

3. Higher-order autocatalysis

Let us now consider examples of higher-order autocatalysis. In these cases polynucleotides do not only act as templates in replication. In addition, they have catalytic activities which are relevant for the replication process.

3.1. *RNA catalysis*

Is it reasonable to assign catalytic activity to an RNA molecule? Some years ago the answer would have to be 'no', since there was no evidence for catalytic activity of polynucleotides apart from ribosomal protein synthesis. Recent research, however, revealed such a catalytic activity of RNA in natural RNA processing. In particular, specific autocatalytic action of some RNA molecules on its own production has been reported [23]. RNA catalysis in the specific cleavage of some precursor RNA molecules to yield the biologically active forms was found as well [24]. In these examples RNA catalyses two classes of reactions: ligation and cleavage of RNA molecules. Both activities may be important for enzyme free RNA replication systems. The present stage of knowledge in this very new field of research, however, does not yet allow such far reaching extrapolations.

3.2. *Indirect catalysis via translation*

Catalytic activity, however, need not be exerted by the polynucleotide molecule itself. It may also be the result of the

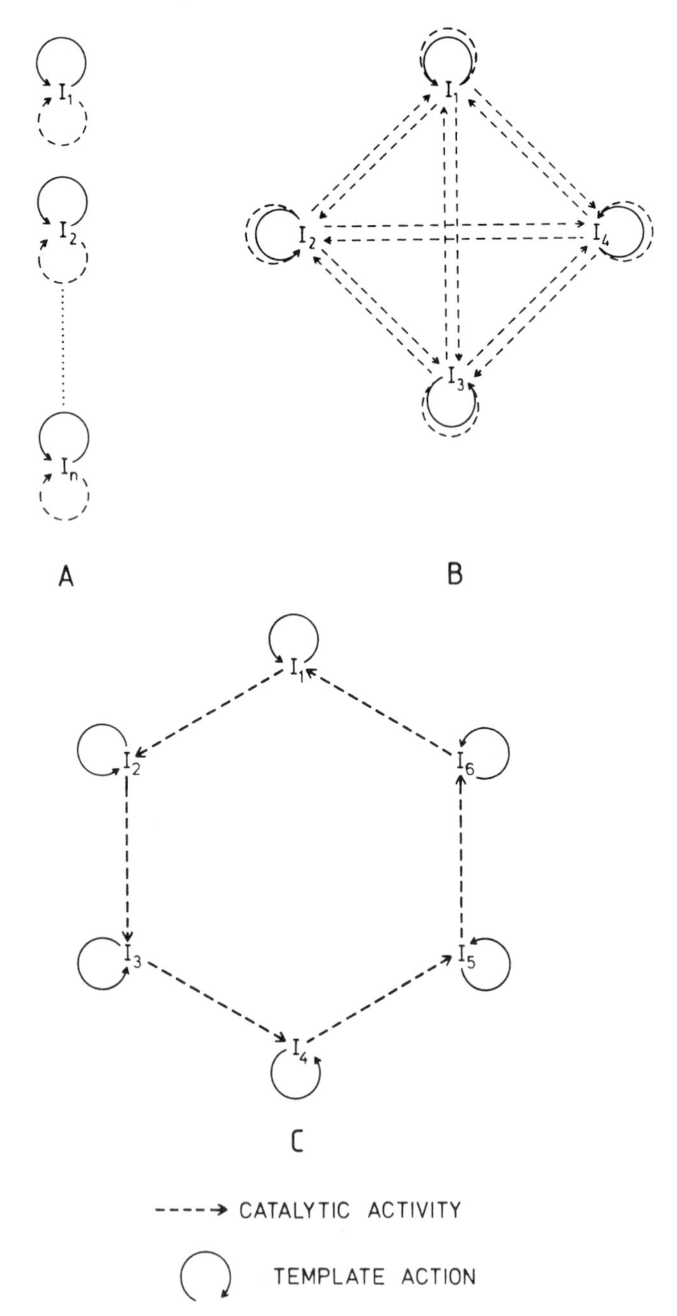

Fig. 10. Three special cases of reaction networks with second-order autocatalysis. (*A*) The multidimensional Schloegl model, (*B*) Fisher's selection equation (we choose $n = 4$) and (*C*) the catalytic hypercycle (we choose $n = 6$).

action of an intermediate. In the biochemistry of the cell, proteins are the usual carriers of catalytic activity. Proteins are synthesized from polynucleotides by template induced translation. Hence they are intermediates of the type mentioned above. This indirect catalytic action of polynucleotides on polynucleotides via proteins represents the basic mechanism of interaction between genes. The biochemistry of the cell, indeed, can be visualized as an enormously complicated network of catalytic activities. Some of them are autocatalytic in higher order. As we indicated in the introduction the cellular reaction network is so complicated that it seems to be hopeless at present to study its dynamics in detail. Nevertheless, it is important to know whether, when and how the general properties of higher order autocatalytic systems might differ from those of first-order replication–mutation systems. These questions can be studied by means of fairly simple models of reaction networks.

3.3. *Second-order autocatalysis*

In order to illustrate the effect of higher-order autocatalysis we shall analyse the simplest case: second order autocatalysis without mutations,

$$(A)+I_k+I_j \xrightarrow{A_{kj}} 2I_k+I_j \quad (k,j = 1,2,\ldots,n) \tag{21}$$

Here we have chosen I_k to act as template and I_j to represent the catalyst for replication. We restrict our considerations here to error-free replication and neglect the effect of mutations. Second and higher order autocatalysis give rise to problems in mass action kinetics. Cubic or higher order terms appear in the rate equations, if we take the concentration of A properly into account (see equation 21). These terms are highly improbable in elementary step reaction kinetics since they require simultaneous encounters of more than two molecules. Biochemical kinetics, however, is different in this respect. Very often mass action kinetics is inadequate to describe properly the dynamics of the many steps in enzyme catalysed reaction networks. Then it is advantageous to use simplifying assumptions and to apply a kind of overall kinetics. One of the best established approximations describes substrate binding together with the enzymic reaction. It is known as Michaelis–Menten kinetics. Allosteric enzymes are of particular importance because they can lead to co-operativity or self-enhancement which is equivalent to higher-order autocatalysis within certain ranges of substrate concentrations. One well investigated example of this type is the glycolytic chain. There are two allosteric enzymes giving rise to higher order autocatalysis; phosphofructokinase and pyruvate kinase [25].

3.4. *The rate equations*

In this contribution we shall deal exclusively with the simplest, but nevertheless, general kinetic equation of second-order autocatalysis. We apply mass action kinetics and find for normalized concentrations and under the constraint of constant organization

$$\dot{x}_k = x_k\left(\sum_j A_{kj}x_j - \phi\right) \quad (k,j = 1,2,\ldots,n) \tag{22}$$

Again we make use of an unspecific dilution flux ϕ which removes the excess production of the system

$$\phi = \sum_k \sum_j A_{kj}x_j x_k$$

The kinetic equations with second-order autocatalysis are more difficult to investigate than those of independent replication, because they contain genuine non-linearities which cannot be removed by simple transformation of variables. It was possible, nevertheless, to perform complete qualitative analysis for some important special cases [26–30].

In the general case all rate constants A_{kj} are different from zero and do not fulfil any special condition. This reaction system is characterized by very rich dynamics including multiple stationary states, stable limit cycles and strange attractors. Transitions occur by a whole variety of bifurcations including saddle-node bifurcations, super- and subcritical Hopf-bifurcations, 'blue sky' bifurcations and others.

It is interesting to point out in this context that the differential equation (22) with dimension n is equivalent to the generalized Lotka–Volterra equation of dimension $n-1$

$$\dot{y}_k = y_k\left(\alpha_k + \sum_{j=1}^{n-1} \beta_{kj} y_j\right) \quad (k = 1, 2, \ldots, n-1) \tag{23}$$

Hofbauer [31] presented a proof that both equations are related by a non-linear transformation. The n^2 rate constants A_{kj} are uniquely converted into a corresponding set of constants α_k *and* β_{kj}. This fact makes equation (22) particularly attractive because it demonstrates equivalence of classes of dynamical systems applied in different fields from molecular evolution to theoretical ecology [32].

3.5. *Special cases of second-order autocatalysis*

In Fig. 10 we show schematic diagrams of three special cases of reaction networks with second order autocatalysis. These systems have been studied thoroughly by the techniques of qualitative analysis. Commonly the constraint of constant organization was applied.

(1) The multidimensional generalization of the Schloegl model [33] which is characterized by a matrix of rate constants which has diagonal form, $A_{jk} = A_{kk}\,\delta_{jk}$ $(j, k = 1, \ldots, n)$.

(2) Fisher's selection equation of population genetics in which the matrix of rate constants is symmetric, $A_{jk} = A_{kj}$.

(3) The hypercycle equation whose rate coefficients fulfil the relation $A_{kj} = F_k\,\delta_{k,j+1}$ $(j, k \bmod n$, and $k = 1, \ldots, n)$ or, in terms of the kinetic equations:

$$\dot{x} = x_k(F_k x_j - \phi) \quad (j = k-1 \bmod n \text{ and } k = 1, \ldots, n) \tag{24}$$

with $\phi = \Sigma_k F_k x_j x_k$.

Fisher's selection equation is fundamental to population genetics since it combines Mendelian genetics with the concept of optimization in the sense of Darwin's theory. The equation describes the spreading of n genes, more precisely n 'alleles' $(A_1, A_2, \ldots, A_n)$ at a single locus, in a large, in principle infinitely large population. The diploid genotypes are either homozygous $(A_k A_k)$ or heterozygous $(A_k A_j)$. The rate constants A_{kk} or A_{kj} are a measure of the fitness of the phenotypes corresponding to $A_k A_k$ or $A_k A_j$, respectively. Clearly, the fitness of $A_k A_j$ is identical to that of $A_j A_k$ since it does not matter on which of the two equivalent chromosomes the gene A_k or A_j sits, and consequently, the matrix A is always symmetric. The multidimensional Schloegl model actually can be visualized as a pathological case of Fisher's selection equation, in particular as the case of vanishing fitness of all heterozygotes.

Let us now consider the relevance of Fisher's selection equation for replicating molecular systems. The condition of matrix A being symmetric implies equality of the catalytic coefficients: $A_{jk} = A_{kj}$, the catalytic activity of I_k on the replication of I_j has to be the same as that of I_j on I_k. Such a symmetry in catalytic action of molecules is very unlikely and presumably very few molecular systems will meet the prerequisites of the validity of Fisher's selection equation. The multidimensional Schloegl model is somewhat different in this respect. In order to obtain overall rate equations of this type we need for example coincidence of template induced RNA replication and autocatalytic self-splicing activity for a set of related RNA molecules. An experimental system with these properties seems to be constructable by systematic design of RNA sequences. Another possibility for the experimental realization of the multidimensional Schloegl model consists of a set of genes which code for their own specific replicases.

Hypercycles are model systems for symbiosis of molecules, of organisms or of species. They provide insight into a general mechanism of dynamical stabilization in autocatalytic reaction networks. On the molecular level hypercyclic organization seems to occur with some classes of RNA viruses which have split genomes and code for several replicases. To give an example the human and animal virus influenza A is a possible candidate for hypercycle dynamics.

The two special cases (1) and (2) lead to selection and allow optimization of mean excess productivities. In contrast to independent replication the optimization process is not of global nature; it leads only into local fitness maxima in which the system is trapped. In general it will be very hard to leave such a local trap by means of microscopic fluctuations [4]

The third special case, the hypercycle equation has been investigated in great detail. Its dynamics is now well understood [4, 25, 28–30]. There are no asymptotically stable stationary states for hypercycle equations of dimension $n \geqslant 5$. Instead, the systems approach asymptotically stable limit cycles and all concentrations oscillate with wave like patterns [34]. The dynamics of catalytic hypercycles according to equation (24) is nevertheless fairly simple; no cases of period doubling, quasi-periodicity or chaotic dynamics have been reported so far. Another interesting property of the hypercycle equation concerns the long time behavior of the solutions; no concentration variable approaches zero and hence no member of the hypercycle will disappear. This property is quite the opposite of selection and we called it 'co-operation' [4] or 'permanence' [35].

Reaction–diffusion equations with reaction terms closely related to the hypercycle equation were studied in the context of morphogenesis and biological pattern formation [36]. They yield dissipative structures in systems with suitable spatial boundaries and under the necessary chemical conditions to keep the system far away from thermodynamic equilibrium.

4. Generalizations

An important question which comes immediately to our mind concerns the legitimacy of generalizations of the results derived from the dynamical model of molecular evolution. The model is consistent with experimental studies on RNA replication *in vitro*. Does this fact justify extrapolation to molecular evolution in general? Can we draw conclusions to 'chemical evolution'? Are results obtained by means of an abstract dynamical model of relevance for biological evolution proper, the evolution of higher organisms?

At first we shall answer the easy question concerning formal generalizations of the model. Population independent replication and mutation is particularly insensitive to changes in the constraints applied. This is a characteristic property of first order autocatalysis in contrast to second and higher order autocatalytic systems. Recently, we were able to present a rigorous proof showing that general time dependent supplies of input concentrations have no influence on the outcome of the selection process, nor do they effect the structure of the quasi-species [7, 37]. There is only one restriction: the concentrations of the input materials must not vanish in the reactor during the period of observation. What actually enters the equations for generalized conditions of selection are the time averages of these input concentrations.

Similar generalizations are not possible for systems with second and higher order autocatalysis. Most of these systems are much more sensitive to time dependent constraints. The degree of sensitivity depends on the particular system, on the choice of rate constants and on the choice of initial conditions. Clearly, it is of crucial importance whether the system has one, two or many stable stationary states. Shape and size of the basins of attraction of these stationary states determine the behavior of the system. To give another example, it matters whether the system approaches fairly insensitive, simple limit-cycle oscillations or it undergoes highly sensitive multiphasic oscillations or has even quasi-periodic or chaotic dynamics. In the latter cases small changes in the input concentrations may drive the system from the domain of one type of behavior into that of another dynamics. Hence, no general predictions on the sensitivity of second- and higher-order autocatalytic systems can be made. The responses of the systems to the time program of the input parameters are highly variable.

Random fluctuations in the environment can be incorporated into the kinetic equations. They lead to fluctuating rate constants. Attempts to study such systems by means of stochastic differential equations [38] or population statistical methods [39] were made but no systematic investigations have been performed so far.

Systematic changes in the environment were observed with RNA replication in the $Q\beta$ system. If the net RNA production is not compensated by dilution the concentration of RNA increases and the binding of RNA to enzyme molecules reaches saturation (see Biebricher, this volume). This saturation of enzyme causes a change in the mechanism of selection which can be considered as a kind of environmental change. It is possible, nevertheless, to incorporate this change into an extended mechanism of replication which leads to different selection equations. Thereby, environmental changes of this class become part of the evolving system.

An experimental set-up for enzyme-free RNA replication has been worked out by Orgel and co-workers [40]. So far they were successful in finding conditions for template-induced RNA synthesis. Under conditions which are appropriate for polymerization the double strands formed do not separate into single strands and, hence, one essential step in enzyme-free RNA replication is not yet solved. Let us assume, nevertheless, that conditions will be found one day under which the reaction network leading to RNA replication can be completed *in vitro*. Then, the process will follow the same general overall kinetics as observed with the $Q\beta$ system and as incorporated into the dynamical model of molecular evolution. In principle, the results presented here apply also to prebiotic evolution.

Replication and evolution of RNA viruses *in vivo* follows closely the kinetic model discussed here. Several examples of quasi-species-like mutant distributions were observed with RNA viruses *in vivo*, in particular with $Q\beta$ and related bacteriophages, with foot-and-mouth-disease virus and with the virus influenza A.

The cellular metabolism of bacteria and other prokaryotes is very complex and highly organized. In particular, DNA replication is a complicated reaction network consisting of many individual steps which are catalysed by ten or more enzymes. The appropriate unit of replication, however, is the whole compartment. Theory becomes applicable on this higher hierarchical level; replication of bacterial cells, including mutations, follows the same overall laws as replication of RNA molecules. Unfortunately, precise experimental data on mutant distributions in bacterial populations are not available at present. The appropriate basis of comparison with the theory is still missing.

Generalization of the concepts derived from molecular evolution to evolutionary phenomena in higher organisms encounters substantial difficulties. The relation between the genotype, the polynucleotide sequence of the DNA in the germline, and the associated phenotype is all but transparent. There are two major factors in the replication of eukaryotes which make this relation extremely sophisticated.

(1) The genetic information of the two replicating individuals is 'chopped' into pieces and combined anew during sexual replication and genetic recombination.

(2) The process of somatic morphogenesis leading from the fertilized egg to the adult organism is of enormous dynamical complexity.

Despite the substantial progress made in the molecular biology of eukaryotes the molecular mechanisms of both processes are still largely unknown. It is unclear, therefore, what the unit of selection in higher organisms really is. Evolution of eukaryotes is such a complicated process, and any extrapolation of results derived from molecular evolution requires therefore special care.

Acknowledgements

Financial support of the work reported here has been provided by the Austrian 'Fonds zur Foerderung der wissenschaftlichen Forschung' (Project No. 5286), by the 'Stiftung Volkswagenwerk', BRD and the 'Hochschuljubilaeumsstiftung der Stadt Wien'. The preparation of computer plots by Mag. Walter Fontana and Dr Joerg Swetina as well as drawings by Mr Johann Koenig is gratefully acknowledged.

References

1. Kimura, M., *The Neutral Theory of Molecular Evolution*. Cambridge University Press, Cambridge, UK (1983).
2. Eigen, M., *Naturwissenschaften* **58**, 465 (1971).
3. Eigen, M. and Schuster, P., *Naturwissenschaften* **64**, 541 (1977).
4. Eigen, M. and Schuster, P., *Naturwissenschaften* **65**, 7 (1978).
5. Eigen, M. and Schuster, P., *Naturwissenschaften* **65**, 341 (1978).
6. Eigen, M., *Ber. Bunsenges. Phys. Chem.* **89**, 658 (1985).
7. Schuster, P. and Sigmund, K., *Ber. Bunsenges. Phys. Chem.* **89**, 668 (1985).
8. Biebricher, C. K., *Evolutionary Biology* **16**, 1 (1983).
9. Biebricher, C. K., Eigen, M. and Gardiner, W. C., Jr., *Biochemistry* **22**, 2544 (1983).
10. Biebricher, C. K., Eigen, M. and Gardiner, W. C., Jr., *Biochemistry* **23**, 3186 (1984).
11. Swetina, J. and Schuster, P., *Biophys. Chem.* **16**, 329 (1982).
12. Hofbauer, J. and Schuster, P., Dynamics of Linear and Nonlinear Autocatalysis and Competition. In *Stochastic Phenomena and Chaotic Behavior in Complex Systems* (ed. P. Schuster), p. 160. Springer-Verlag, Berlin (1984).
13. Thompson, C. J. and McBride, J. L., *Math. Biosc.* **21**, 127 (1974).
14. Jones, B. L., Enns, R. H. and Rangnekar, S. S., *Bull. Math. Biol.* **38**, 15 (1976).
15. Rumschitzky, D., *J. Chem. Phys.* (in the press).
16. Leuthaeuser, I., *J. Chem. Phys.* (in the press).
17. Eigen, M. and Schuster, P., *J. Mol. Evol.* **19**, 47 (1982).
18. Domingo, E., Flavell, R. A. and Weissmann, C., *Gene* **1**, 3 (1976).
19. Batschelet, E., Domingo, E. and Weissmann, C., *Gene* **1**, 27 (1976).

20. Demetrius, L., Schuster, P. and Sigmund, K., *Bull. Math. Biol.* **47**, 239 (1985).
21. McCaskill, J. S., *J. Chem. Phys.* **80**, 5194 (1984).
22. Fontana, W. and Schuster, P., Evolutionary Adaptation in Populations of RNA Molecules – A Stochastic Model. Preprint (1985).
23. Kruger, K., Grabowski, P. J., Zaug, A. J., Sands, J., Gottschling, D. E. and Cech, T. R., *Cell* **31**, 147 (1982).
24. Guerrier-Takada, C., Gardiner, K., Marsh, T., Pace, N. and Altmann, S., *Cell* **35**, 849 (1983).
25. Hess, B. and Markus, M., *Ber. Bunsenges. Phys. Chem.* **89**, 642 (1985).
26. Schuster, P., Sigmund, K. and Wolff, R., *Bull. Math. Biol.* **40**, 743 (1978).
27. Hofbauer, J., Schuster, P., Sigmund, K. and Wolff, R., *SIAM J. Appl. Math.* **38**, 282 (1980).
28. Schuster, P., Sigmund, K. and Wolff, R., *J. Math. Anal. Appl.* **78**, 88 (1980).
29. Hofbauer, J., Schuster, P. and Sigmund, K., *J. Math. Biol.* **11**, 155 (1981).
30. Phillipson, P. E. and Schuster, P., *J. Chem. Phys.* **79**, 3807 (1983).
31. Hofbauer, J., *Nonlinear Analysis, Theory, Methods and Appl.* **5**, 1003 (1981).
32. Schuster, P. and Sigmund, K., *J. theor. Biol.* **100**, 553 (1983).
33. Schloegl, F., *Z. Phys.* **253**, 147 (1972).
34. Phillipson, P. E., Schuster, P. and Kemler, F., *Bull. Math. Biol.* **46**, 339 (1984).
35. Sigmund, K. and Schuster, P., Permanence and Uninvadability for Deterministic Population Models. In *Stochastic Phenomena and Chaotic Behavior in Complex Systems* (ed. P. Schuster), p. 173. Springer-Verlag, Berlin (1984).
36. Meinhardt, H., *Ber. Bunsenges. Phys. Chem.* **89**, 691 (1985).
37. Schuster, P. and Sigmund, K., Selection and Evolution in General Open Systems. Preprint (1985).
38. Iganaki, H., *Bull. Math. Biol.* **44**, 17 (1982).
39. Demetrius, L., *J. theor. Biol.* **103**, 619 (1983).
40. Inoue, T. and Orgel, L. E., *Science* **219**, 859 (1983).

Chemica Scripta 1986, **26B**, 43–49

Repeats of Base Oligomers (*N = 3n ± 1 or 2*) as Immortal Coding Sequences of the Primeval World: Construction of Coding Sequences is Based Upon the Principle of Musical Composition

Susumu Ohno and Marty Jabara*

Beckman Research Institute of the City of Hope, 1450 East Duarte Road, Duarte, CA 91010, USA

Paper presented by Susumu Ohno at the Conference on 'Molecular Evolution of Life', Lidingö, Sweden, 8–12 September 1985
Supported by a Bixby Foundation Grant.

Abstract

There are three compelling reasons that suggest that the first set of prebiotic coding sequences to be translated were repeats of base oligomers, the numbers of bases in oligomeric units not being multiples of three. In subsequent evolution, new genes, as a rule, arose from redundant copies of the pre-existed genes, thus, new and old belonging to the same family or superfamily. Occasionally, however, there appeared to have been *de novo* recruitments of truly new coding sequences from the noncoding base sequences of the genome. In every instance, such newborn coding sequences have shown themselves to be near exact repeats of base oligomers thus recapitulating the prebiotic event.

Although most coding sequences of today are of considerable antiquity, they are nevertheless not unique. Instead each is comprised of a number of recurring base oligomers that are related to each other and their derivatives. Thus, they are constructed along the principle that governs musical composition. Musical transformation under the set rule of the entire mouse anti-NP^b I_gV^H coding sequence 98-codon-long has been accomplished and shown.

Introduction

Since replication of nucleic acids is based upon the inherent complementarity that exists between two purine-pyrimidine pairs, prebiotic propagation of nucleic acids has almost certainly taken place [1]. Nevertheless, the very fact that proteins are far more versatile structures than nucleic acids suggests the prebiotic establishment of a translation machinery that linked the base sequence of nucleic acids with the amino acid sequence of polypeptide chains as the necessary prelude of the emergence of life on this earth eons ago [1, 2]. In the past, it has often been assumed that such a prebiotic translation machinery was promiscuous to the extreme, translating any and all available base sequences to amino acid sequences. Assuming the average size of the first set of translated polypeptide chains to be 100-amino-acid-residue long, Hoyle thus argued that they could have been any of the astronomical 20^{100} variety and that the earth's history of mere several billion years was simply not long enough to sort out a miniscule fraction of useful from the sea of useless among this astronomical variety. Hence, the extraterrestrial origin of life on this earth [3]. The problem with the above argument is that time was no ally in the prebiotic world where natural selection as we understand it was yet to operate. Accordingly, what could not be accomplished in several billion years would not have been accomplished in one hundred billion years.

It is granted that the chain-initiating codon as such was likely to have been nonexistent in the prebiotic translation machinery. The invariable presence of chain-terminating base triplets, on the other hand, must have been the bane of its existence. Even if the redundancy of third bases was complete in the triplet coding system of the prebiotic machinery, the simultaneous emergence of all 16 different kinds of primitive *transfer* RNAs was not likely to have occurred.

Furthermore, there might not have been 16 different kinds of amino acids in equal abundance in the prebiotic world; some of them being so scarce to be virtually absent. In addition to the complete redundancy of third bases, let us assume that the prebiotic translation machinery could effectively utilize only 14 kinds of amino acid-specific *transfer* RNAs, the probability of randomly generated nucleic acid base sequences containing 300-base-long open reading frames then becomes $(14/16)^{100}$ which comes to roughly 1.58×10^{-5}. It follows then that were the first set of base sequences to be translated by the prebiotic machinery randomly generated unique sequences, the emergence of life on this earth could never have taken place.

Three virtues of oligomeric repeats as primordial coding sequences

What if they were repeats of base nanomers, the probability of those having open reading frames of indefinite length becomes $(14/16)^3$ which is respectably 67%. This then is the first virtue of oligomeric repeats that qualifies them as primordial coding sequences of the prebiotic world [4]. Their second virtue is that periodical polypeptide chains encoded by them were likely to have assumed either α-helical or β-sheet secondary structures [5]. It would be recalled that modern collagen that assumes α-helical structure along almost its entire length apparently had the original tripeptidic periodicity of Gly-Pro-X, while serum albumin that is comprised of 28 loops of β-sheet structures also demonstrates easily recognizable periodicity [6], the original periodicity likely to have been

* Musician: 828 North Beverly Glen, Los Angeles, CA 90077.

nanopeptidic [7]. The third virtue is conferred to only those oligomeric repeats where the number of bases in the primordial building block oligomer is not a multiple of three.

Immortality conferred to repeats of $N = 3n \pm 1$ or 2 base oligomers

There would be no way of knowing the exact number of polypeptide chain variety whose simultaneous presence in a particular prebiotic niche caused the emergence of the first cell on this earth eons ago. Nevertheless, one hundred different kinds appears not too far off the mark. For the accumulation of this magnitude to have taken place, earlier emerged base sequences that originally encoded potentially useful polypeptide chains should have continued to do so in spite of subsequently sustained base substitutions, deletions and insertions. In short, *conditio sine qua non* of the emergence of the first cells was a measure of immortality inherently embodied within a prebiotic set of coding sequences.

The concept of mutation load dictates that there should be an inverse relationship between the number of gene loci in the genome and the exactitude of DNA replication. The number of gene loci in the mammalian genome approaches 10^5. In order to conserve so large a number of genes in functional state, their copying as well as editing mechanism has been refined to the extreme; the inherent replication error rate becoming of the order of 10^{-6}/base pair/year [8]. At the other extreme, retroviruses being endowed with but a few gene loci in the genome can afford to be venturesome, thus, their reverse transcriptases are extremely error prone; the inherent replication error rate being of the order of 10^{-3}/base pair/year [9]; a thousand-fold difference between the conservative rich and the venturesome poor. One would not expect the error rate inherent in the nonenzymatic nucleic acid replication of the prebiotic world to have been less than the error rate of reverse transcriptase. At the error rate of 10^{-3}/base pair/years, it should be noted that a given polypeptide chain would have already undergone 100% amino acid sequence change every one thousand years. Indeed, the emergence of life on this earth had to depend upon the property of immortality inherent in the prebiotic set of coding sequences.

The nature of the coding system is such that the most deleterious of base changes sustainable by coding sequences are premature chain terminations and reading-frame shifts. A base substitution that changes an amino acid specifying codon to a chain terminator is obviously deleterious, unless it occurs very near to the 3′ end of a coding sequence, for a shorter polypeptide chain thus produced would have lost its originally assigned function. Deletions or insertions of bases that are not multiples of three in numbers cause shifts in the reading frame. Since unused reading frames are normally full of chain terminators, they also cause premature chain terminations thus depriving shorter polypeptide chains of their assigned functions.

Provided that the number of bases in the primordial building block oligomer is not a multiple of three, coding sequences that are oligomeric repeats acquire the inherent resistance to the above noted most deleterious of base changes [4]. This is illustrated below on the heptameric repeats

```
 Leu //Gln    Pro    Ala    Ala   Cys    Ser   Leu // Gln
C T G//CA  G C C T G/C A G C C T G/C A G C C T G//C A G C
 //Cys    Ser    Leu    Gln   Pro    Ala    Ala// Cys    Ser
  //Ala    Ala    Cys    Ser   Leu    Gln    Pro// Ala    Ala
```

Since 7 is not a multiple of 3, three consecutive copies of 7, the 21-base-long sequence becomes the unit periodicity thus encoding the heptapeptidic periodical polypeptide chain. Since within the 21-base-long unit, three copies of the heptamer are translated in all three reading frames, if one reading frame is free of a chain terminator, the other two reading frames are also free of this nemesis; all three reading frames encoding polypeptide chains of the identical heptapeptidic periodicity. Therefore, a base change that created an internal chain terminator (e.g. from cysteine codon T G C to a terminator T G A) would have silenced only one of the three open reading frames. Similarly, reading-frame shifts are of no consequence, since all three reading frames encode polypeptide chains of the identical periodicity. Hence a good measure of immortality is inherent in those coding sequences that are repeats of $N = 3n \pm 1$ or 2 base oligomers. There are additional benefits in that this type of oligomeric repeats invariably give longer periodicities to polypeptide chains they encode. For example, nanomeric repeats can give only the tripeptidic periodicity to their polypeptide chains, while heptameric repeats give the heptapeptidic periodicity. There is one drawback, however, it would be recalled that under one specific assumption of the prebiotic condition, the probability of nanomeric repeats having open reading frames of indefinite length was $(14/16)^3$ which is roughly 67%, this probability for hexameric repeats, on the other hand is $(14/16)^7$ which amounts to roughly 39% which nevertheless, is still tolerable. All in all, it would appear that the prebiotic set of coding sequences that presaged the emergence of the first cell on this earth almost had to be repeats of $N = 3n \pm 1$ or 2 base oligomers.

The exact periodicity of modern coding sequences that arose *de novo*

In evolution, new genes, as a rule, arise from redundant copies of the pre-existed genes, thus, new and old belong to the same family or superfamily [10]. Occasionally, however, there appeared to have been *de novo* recruitments of truly new coding sequences from the noncoding base sequences of the genome. In every instance, such newborn coding sequences have shown themselves to be near exact repeats thus encoding polypeptide chains of the exact periodicity.

Members of the cod order *Gadiformes* and those of the flounder order *Heterosomata* inhabiting Arctic and Antarctic oceans have recently evolved antifreeze proteins quite independent of each other. The simplest of the *Gadiformes* antifreeze proteins maintain the exact tripeptidic periodicity of Ala-Thr-Ala, with a galactose-galactosamine dissacharide attached to each threonine residue [11]. While the simplest of the *Heterosomata* antifreeze proteins has the nearly exact monodecapeptidic periodicity of Ala-Ala-Ala-Ala-Ala-Ala-Leu-Thr-Ala-Ala-Asp [11].

Similarly, two species of malarial parasites belonging to the same genus *Plasmodium* have apparently mobilized different parts of their genomic repetitious sequences to generate new independent coding sequences for their circumsporozoite antigens. Thus, this antigen of *Plasmodium falciparum* has the exact tetrapeptidic periodicity of Asn-Ala-Asn-Pro [12], whereas the corresponding antigen of *P. knowlesi* has the dodecapeptidic periodicity of Gly-Gln-Pro-Gln-Ala-Gln-Gly-Asp-Gly-Ala-Asn-Ala [13].

The same thing can be said of *de novo*-generated independent coding sequences of dipteran insects that are transcribed from giant polytene chromosomes of larval salivary glands and translated to yield secretory proteins. All in all, it would thus appear that truly new coding sequences occasionally generated by modern organisms recapitulate the first set of prebiotic coding sequences of eons ago that presaged the emergence of life on this earth.

Residual periodicity maintained by modern coding sequences of ancient origins render them melodious qualities

If coding sequences are unique *sensu stricto*, a given base hexamer is expected to recur once every 1000 (4^6) bases, whereas a base decamer should recur once every one million (4^{10}) bases. For a more exact calculation, however, AT/GC ratio of a base oligomer in question as well as that of a stretch of DNA has to be taken into account. Thus, in reasonably long, unique DNA sequences of variable AT/GC ratios, the chance-expected occurrence of a given base oligomer per number of DNA base pairs (N) can be calculated by the following formula [14]

$$1 = \frac{(g)^{n1}}{2}\left(\frac{1-g}{2}\right)^{n2} \times N$$

in which g is the A+T content of a unique DNA sequence and $n1$ and $n2$ are the numbers of A+T and G+C in the base oligomer. Curiously, the average restriction fragment length yielded by the nonrepetitious fraction of the genomic DNA corresponds well to the prediction made by the above formula [14]. It would be recalled that each restriction enzyme recognizes a specific base oligomer which is more often than not palindromic. The above, however, merely means that the type of base oligomers recognized by restriction enzymes were never members of the primordial building blocks. The fact is that most, if not all, of the apparently unique coding sequences are comprised of recurring base oligomers that are related to each other and their derivatives [4].

For example, the only 98-codon-long germline coding sequence for immunoglobulin heavy-chain variable (I_gV_H) region that is unique to C57BLACK6 inbred strain of the mouse encodes anti-NP^b (4-hydroxy-3-nitrophenyl-acetyl) antigen-binding pocket which when combined with $V_{\lambda L}$ that is essentially invariant in the mouse gives a peculiar heteroclitic to an antibody in that the antibody thus formed shows a higher binding affinity to a related hapten (NIP(5-iodo-NP) than NP itself used as an immunogen [15]. As already noted, so short a coding sequence contains recurring base decamers and monodecamers that are related to each other. 5th to 7th Gln-Gln-Pro are encoded by base decamer C A G C A G C C T G, while the same decamer translated in a different reading frame encodes Ser-Ser-Leu of 82A to 82C positions. The 3′ 6/10th of the former is also a part of monodecamer A G C C T G G G G C T encoding 6th to 9th Gln-Pro-Gly-Ala which recurs almost immediately downstream encoding 13th to 16th Lys-Pro-Gly-Ala [16].

One thus sees that coding sequences in general and mouse B1–8 anti-NP^b germline I_gV_H coding sequence in particular are constructed along the principle that governs musical composition. The traditional musical composition embodied in sonata form consists of: (1) the *exposition*, in which the principal and secondary subjects are presented; (2) the *development*, in which one or both subjects are developed or worked out; (3) the *recapitulation*, in which both subjects are repeated followed by a coda. The decamer C A G C A G C C T G that appears at the beginning and recapitulated near the end can be considered as the principal subject of B1–8 anti-NP^b I_gV_H coding sequence, whereas the tandemly recurring A G C C T G G G G C T at the beginning may not quite be considered as the secondary subject for it is not recapitulated near the end. Nevertheless, the principal subject C A G C A G C C T G is developed in a number of variations between the beginning and the end. For example, a two-base-deviant of the principal subject C A G A G G C C T G
(* *)
encodes Gln-Arg-Pro of 39th to 41st positions; asterisks denoting deviant bases, while another two-base-deviant C A — C A G C C T A encodes 77th to 79th Thr-Ala-Tyr, this time,
(* *)
in the third reading frame [16]. The hypervariable CDR-1 region contains a three-base-deviant of the principal subject C A G C T A — C T G for Ser-Tyr-Trp of 31st to 33rd, while
(* * *)
FR-3 region contains another 4-base-deviant C A A — AC C C T C for 73rd to 75th Lys-Pro-Ala. Truncated deviants of the
(* * * *)
principal subject also recur in tandem, e.g. A A G A G C/A A G G G C for 64th to 67th Lys-Ser-Lys-Gly as well as C A G C A/C A G C C and C A G C T for 76th to 78th Ser-Thr-Ala and then 81st, 82nd Gln-Leu.

Musical transformation *inter alia* of the entire anti-NP^b I_gV_H coding sequence

The above indeed indicated that coding sequences in general and I_gV_H in particular are quite amenable to transformation to musical scores. The shortness of I_gV_H was especially attractive, for when transformed, long coding sequences are bound to sound redundant and repetitious. Thus, we have established one invariant rule to abide by. This rule simply assigned two alternative positions each in the treble clef stave to A, G, T and C in the ascending order as shown in Fig. 1. Accordingly, the treble clef musical score of Fig. 2 can be transformed back to the coding base sequence with no ambiguity whatsoever. It seemed natural that purines being heavier than pyrimidines should occupy the lower half of the octave. With the rule shown in Fig. 1, the entire coding sequence of the mouse germline anti-NP^b B1–8 coding sequence has been transformed into the musical score as shown in four segments in Fig. 2. Since the decamer C A G C A G C C T G was the principal subject of this I_gV^H coding sequence, the 10/8 time signature should have been most appropriate. However, we have settled on the more conventional 9/8 time signature at the beginning and C Major. It would be noted, however, that the time signature

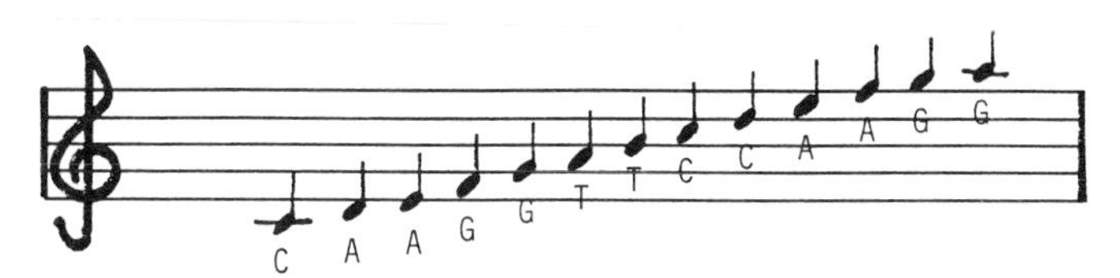

Fig. 1. Assignment of two alternative positions to four bases, A, G, T, and C in the ascending order, in the treble clef stave. This invariant rule permits the treble clef musical score to be transformed back to the base sequence with no ambiguity whatsoever.

Fig. 2–1

Fig. 2–1, 2–2 and 2–3. Musical transformation of the entire coding sequence of the mouse germline anti-NP^b I_g V_H peculiar to C57BLACK6 inbred strain [15]. Bases of the coding sequence accompanied by corresponding amino acid residues are shown immediately below each row of the treble clef musical score. As far as possible, bases are placed directly below corresponding musical notes. Hypervariable amino acid residues of CDR-1 and CDR-2 regions are shown in smaller capital letters.

ANTI-NP^B VERSION II (B)

Fig. 2–2

ANTI-NP^B IG VH VERSION II (C)
TYR
T A
60
ASN GLU LYS
C A A T G A G A A G
PHE LYS
T T C A A G
SER
A G C A
LYS
A
ALA THR
G G C C A C
LEU
A C T G
70
THR
A C T
VAL
G T A G
ASP
A C
LYS* PRO
A A A C C C
SER
T C C A
SER THR
G C A C A G
ALA
C
TYR
C T A
C
80
MET
A T G C
GLN
A G
LEU
C T C A G C
82A
SER
82B
SER
A G C C
82C
LEU
THR
T G A C A T
SER
C
T
GLU ASP
G A G G A C
CRESC.
SER
T C T
ALA
G C G
VAL
G T C
90
TYR
T A T
TYR
T A T
CYS
T G T G
ALA
C A A
ARG
G A
/ C

Fig. 2–3

changed to 3/4 with Fr-2 (2nd row of Fig. 2–2 and to F Major shortly thereafter (4th row of Fig. 2–2). Within CDR-2 region, it changed further to A Major (2nd row of Fig. 2–3). However, soon it returned to C Major, but changed to the time signature of 6/8 (4th row of Fig. 2–3), and ended thus. Needless to say, the musical score of the base clef of Fig. 2 was designed to complement the treble clef musical score, therefore, it had only indirect relationship with the coding sequence *per se*.

References

1. Miller, S. L. and Orgel, L. E., *Origin of Life on the Earth*. Prentice-Hall, New York (1974).
2. Eigen, M. and Schuster, P., *Naturwissenschaften* **64**, 541 (1977).
3. Hoyle, F., *Ten Faces of the Universe*. Freeman Press, London (1979).
4. Ohno, S., *J. Mol. Evol.* **20**, 313 (1984).
5. Ycas, M., *J. Mol. Evol.* **2**, 17 (1972).
6. Ohno, S., *Proc. Natl. Acad. Sci. USA* **78**, 7657 (1981).
7. Alexander, F., Young, P. R. and Tilghman, S. M., *J. Mol. Biol.* **173**, 159 (1984).
8. Ohno, S., *Trends Genetics* **1**, 160 (1985).
9. Gojobori, T. and Yokoyama, S., *Proc. Natl. Acad. Sci. USA* **82**, 4198 (1985).
10. Ohno, S., *Evolution by Gene Duplication*. Springer-Verlag, Heidelberg, Berlin, New York (1970).
11. DeVries, A. L., *Comp. Biochem. Biophysiol.* **73**, 627 (1982).
12. Zavala, F., Tam, J. P., Cochrane, A. H., Quakyi, I., Nussenzweig, R. S. and Nussenzweig, V., *Science* **228**, 1436 (1985).
13. Ozaki, L. S., Svec, P., Nussenzweig, R. S., Nussenzweig, V. and Godson, G. N., *Cell* **34**, 815 (1983).
14. Nei, M. and Li, W., *Proc. Natl. Acad. Sci. USA* **76**, 5269 (1979).
15. Bothwell, A. L. M., Paskind, M., Reth, M., Imanishi-Kari, T., Rajewsky, K. and Baltimore, D., *Cell* **24**, 625 (1981).
16. Ohno, S., Kato, K., Hozumi, T. and Matsunaga, T., *Proc. Natl. Acad. Sci. USA* **79**, 132 (1982).

Chemica Scripta 1986, **26B**, 51–57

Darwinian Evolution of Self-Replicating RNA

Christof K. Biebricher

Max-Planck-Institut für Biophysikalische Chemie, Am Fassberg, D-3400 Göttingen, Federal Republic of Germany

Paper presented at the Conference on 'Molecular Evolution of Life', Lidingö, Sweden, 8–12 September 1985

Abstract

We used as model system for the study of evolution and selection at the molecular level the *in vitro* replication of short-chained self-replicating RNA by Qβ replicase. A detailed kinetic analysis of the replication process can be achieved by splitting the total mechanism into its elementary steps. Analytical treatment leads to compact mathematical equations describing the experimental observations. The incorporation profiles simulated in the computer show excellent agreement with the measured profiles. The complicated selection behavior of two or more different RNA species can be quantitatively deduced from the rate values of replication steps crucial for selection. When enzyme is in excess over RNA, the overall replication rate constant values suffice for calculating the selection value. When RNA is in excess, there is no correlation between replication rate constants and selection values, but the selection values depend on the rate constant values of the enzyme-binding step and the rate constants of double-strand formation of the single-stranded templates.

In the absence of extraneously added template, Qβ replicase is able, after a long lag time, to synthesize self-replicating RNA *de novo*. When amplification of RNA is suppressed by omission of a pyrimidine nucleoside triphosphate, a slow accumulation of a mixture of oligonucleotides is observed. Qβ replicase also is able to modify a template RNA by condensing nucleotides at its 3′-end. These slow side reactions may have an intimate relationship to the *de novo* RNA synthesis.

Introduction

The expression of genetic information is an enormously complicated process. Quantitative description of the selection of mutants has thus not been possible. Spiegelman and collaborators [1–3] have developed a system derived from the replication system of RNA coliphage Qβ that has a particularly simple phenotypic expression of the genotype, not involving production of diffusible products (reviewed in [4]). Purified replicase and nucleoside triphosphate precursors are added to the incubation mixture as environmental factors. The efficiency of being replicated by the enzyme is the phenotypic expression. This system was used in our laboratory as model system where we tried to reduce Darwinian selection to clearly defined and measurable physical parameters.

In vivo, Qβ replicase replicates the viral RNA specifically while ignoring the vast excess of cellular RNA. *In vitro*, conditions have been found where other RNA is also accepted by Qβ replicase [5, 6]. However, there is a fundamental difference in stoichiometry. While Qβ RNA itself can be autocatalytically amplified by replicase to large concentrations, the RNA synthesis reaction with 'unphysiological' RNA templates generally stops after production of the complementary sequences. For evolution experiments, this reaction is of no significance. Qβ RNA itself, having a chain length of about 4000 nucleotides, is not suitable for evolution experiments. However, a number of self-replicating RNA species with nucleotide chain lengths of about 100 are available, and most evolution experiments have been performed with these short-chained self-replicating RNA species, some of which were sequenced [7–9]. The still somewhat contested origin of these molecules will be discussed later.

The replicase preparations used in our own experiments were purified to homogeneity. Two different forms were obtained: The holoenzyme, containing all 4 subunits S1, β, EFTu, EFTs [10, 11], and a core enzyme that lacks the subunit S1 [12] (Fig. 1). Both preparations can catalyze the replication of the short-chained self-replicating RNA species.

Kinetics of RNA production

Comparison of the genotypes, i.e. the nucleotide sequences of self-replicating RNA species [7–9] reveals surprisingly little information about recognition signals for Qβ replicase. The only unambiguous similarity is a C cluster at the 3′-ends

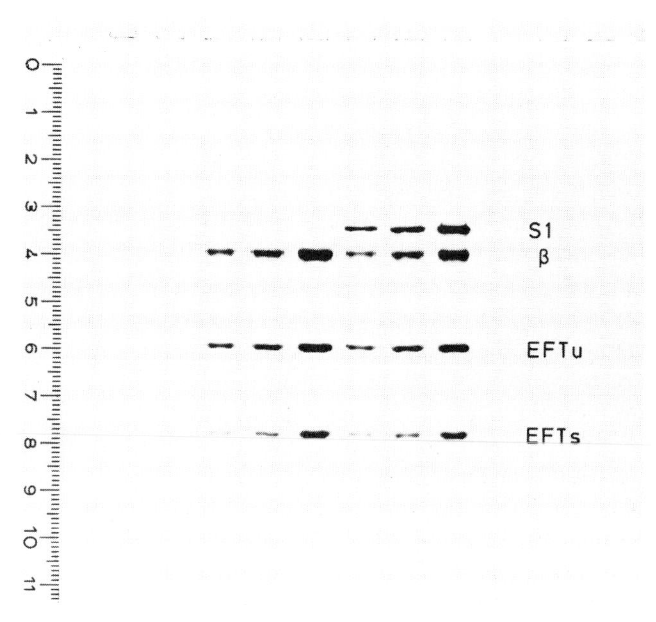

Fig. 1. Subunits of replicase. Sodium dodecyl sulfate gelelectrophoresis of a purified Qβ replicase preparation. Lane 1–3: core enzyme lacking subunit S1; lane 4–6: holoenzyme preparation. From [4].

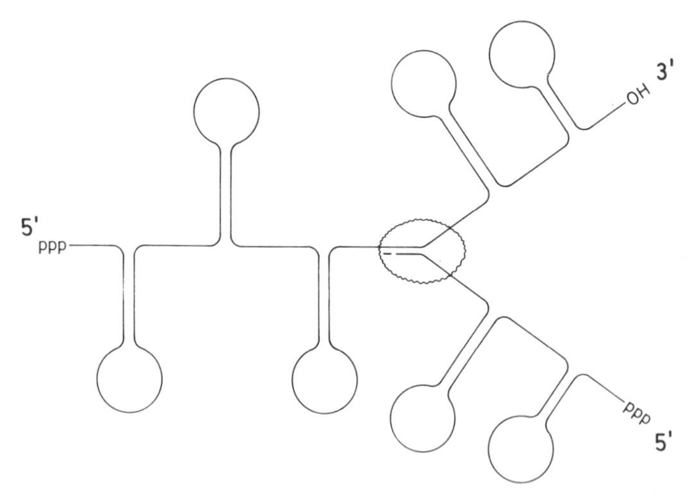

Fig. 2. Model of elongation complex. At the replication fork the growing replica forms a short double-helical region with the template. The replica and template double helix is opened by the enzyme and released by the proceeding enzyme from different sites. The free ends of the single-stranded RNA are protected against annealing by intra-molecular base-pairing. From [4].

and – as a consequence – a G cluster at the 5′-ends of the sequence [9]. This condition, however, is clearly not sufficient for recognition. Some crucial information is hidden in the tertiary structure of the RNA; it has been shown that some of the sequences can be folded to different tertiary structures, only one of which is accepted by Qβ replicase [13].

The kinetic behavior of RNA replication by Qβ replicase is now well understood. Replication requires single-stranded RNA templates and results in the synthesis of a single-stranded replica and the liberation of template and enzyme [14]. In a first step, replicase binds to template RNA; geminal association of two GTP molecules at the 3′-end of the template is required before the new replica chain is started by the priming phosphodiester bond formation [15]. The specificity of the replicase for its choice of templates is kinetically controlled by the lifetimes of the initiation complexes. Chain elongation proceeds by sequential nucleoside triphosphate complexation and phosphodiester formation accompanied by release of inorganic pyrophosphate. An additional feature, still not understood, is the progressive melting of replica and template, apparently catalyzed by the replicase itself (Fig. 2).

Once the replica chain is finished, it is adenylated at its 3′-end and released. The remaining inactive template–enzyme complex dissociates rather slowly and requires the participation of a triphosphate. Addition of all these steps to a complete mechanism gives the minimal mechanism [15]. Many elementary step rate constants can be measured or estimated with satisfactory accuracy, and thus the kinetic processes can be simulated in the computer by numerical integration of the pertaining differential equations. Excellent agreement between measured incorporation profiles and computer-simulated profiles has been found. Simulated profiles are used here to illustrate the main features of the replication. The agreement with measured profiles is readily seen by comparing the simulations shown here with published experimental profiles [13, 16].

Two main growth phases can be distinguished:

(*a*) an exponential growth phase, where enzyme is in excess over template, and

(*b*) a linear growth phase, where enzyme is saturated with enzyme.

We define an overall replication rate in the linear phase

$$\rho = \mathrm{d}[\mathrm{pp}]/(n-1)[E_c]\,\mathrm{d}t$$

where n is the nucleotide chain length of the template, [pp] the inorganic pyrophosphate liberated by the reaction and $[E_c]$ the enzyme bound to template. ρ was measured experimentally by determining the incorporation of a labelled triphosphate into RNA and calculating from it the RNA concentration with the help of the measured nucleotide chain length and base composition. $[E_c]$ was calculated from the protein concentration assuming 100% of the enzyme to be active and bound to template. The overall rate in the exponential growth phase κ is defined as $\mathrm{d}[I_0]/[I_0]\,\mathrm{d}t$, where $[I_0]$ is the total bound and unbound RNA concentration. Experimentally, κ was measured by measuring the incorporation profiles of incubations from serial dilutions of template by a factor F_{dil}. The resulting displacement of the profile on the time axis t_{dif} was determined. Since $F_{\mathrm{dil}} = \exp(\kappa t_{\mathrm{dif}})$, $\kappa = (\log F_{\mathrm{dil}})/t_{\mathrm{dif}}$ (Fig. 3).

The overall rates κ and ρ in the growth phases differ, ρ being typically only one-third of the κ values. Transition between the exponential and linear phase occurs sharply when the molar ratio of RNA to enzyme reaches values between 1 and 2 (Figs. 3, 4). In the linear phase, most of the enzyme is inactive since it is awaiting template release (Fig. 4). In the exponential phase the fraction of the complexed enzyme actively engaged in synthesis is larger.

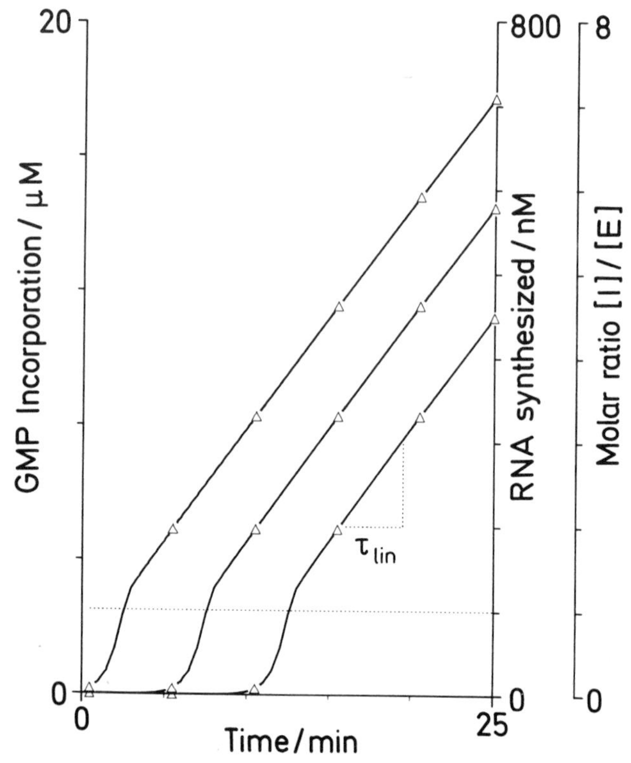

Fig. 3. Measurement of overall replication rates in the linear and exponential growth phases. Template was added in 10^{-1}, 10^{-4} and 10^{-7} dilutions and GMP incorporation determined by measuring the acid-insoluble radioactivity. The concentration of synthesized RNA strands can be calculated from the chain length and the nucleoside composition (right scale). Division by the enzyme concentration gives the moles RNA strands synthesized per enzyme molecule present. The slope of the linear increase gives the replication rate ρ in the linear phase. From the displacements t_{dif} of the profiles by dilutions by the factor 10^3 one can calculate the overall replication rate κ in the exponential phase. A simulation of the usual experimental procedure is shown; in practice, the curve has to be reconstructed from the measured points.

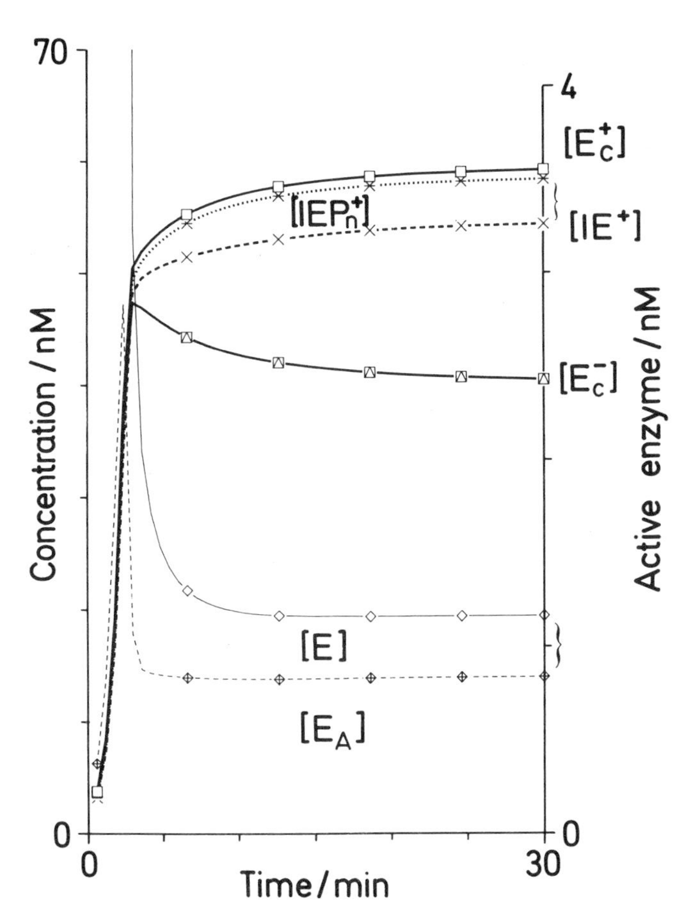

Fig. 4. Complexed and free enzyme. From a computer simulation of the replication process with standard rate constant values [15], except $k_{ds} = 5 \times 10^4\ \mathrm{M}^{-1}\ \mathrm{s}^{-1}$, $k^{+}_{D5'} = 4 \times 10^{-3}\ \mathrm{s}^{-1}$, $k^{-}_{D5'} = 6 \times 10^{-3}$. In the linear phase, complexed enzyme $[E_c^+]$ is mainly present in inactive forms waiting for release of template $[IE^+]$ or the completed replica $[IE_n^+]$ (left scale), while free enzyme $[E]$ and enzyme actively engaged in synthesis $[E_A]$ are present in low concentrations (right scale). In the exponential phase, the relative fraction of actively engaged complexed enzyme is higher. □—□, $[E_c^+]$; ⧄—⧄, $[E_c^-]$; ×---×, $[IE^+]$; ⋇....⋇, $[IE^+]+[IEP_n^+]$; ⊕-⊕, $[E_A]$; ◇—◇, $[E_A]+[E]$.

For satisfactory quantitative description (see Appendix), it suffices to consider the rates for the main steps, initiation, elongation and termination. It is of particular value to separate all steps involving free enzyme from those that do not. A simplified three-step mechanism (Fig. 5) is obtained where only a few rates have to be considered. Furthermore, even though all reactions are in principle reversible, the rates of the back reactions for these steps prove to be so much slower that the error made by assuming them to be irreversible is small. For enzyme concentrations between 40 and 250 nM the ρ values are found to be independent of enzyme concentration, and the κ values are found to be independent of enzyme concentration above 100 nM. Apparently the enzyme-binding step is so rapid in both growth phases that it does not contribute to the overall rate. The overall replication rates are thus functions only of the elongation and the termination rates; conversely, one can calculate the elongation and enzyme reactivation rates from measurements of the overall replication rates in both phases. As seen in Table I, both rates depend upon triphosphate concentration, upon the template and upon other conditions as well; the replication rates with holoenzyme are usually higher than with core enzyme.

Since replication always results in synthesizing the complementary strand, both single strands are present and may react with each other to form a double strand. Indeed, quite

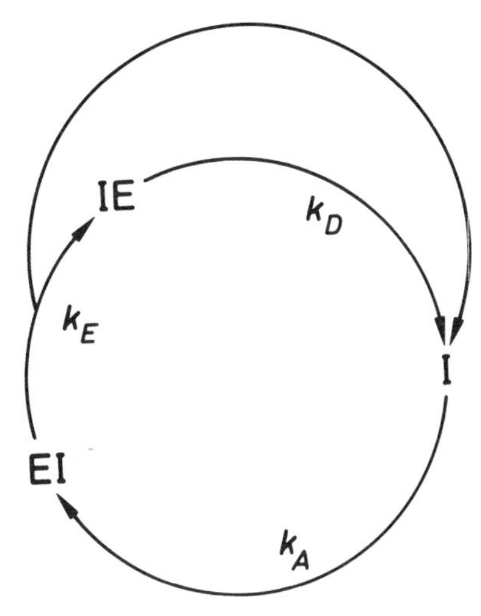

Fig. 5. Simplified three-step mechanism. The elementary steps of enzyme–template binding and the dissociation of the inactive template–enzyme complex (reactivation step) are considered. All back reactions are neglected, and all other steps from chain initiation to product release are combined to an elongation step. Uncomplexed complementary RNA strands are assumed to be able to form a double strand.

Table I. *Replication rates and retention times in the exponential and linear growth phase*

	κ (10^{-2} s^{-1})	ρ (10^{-3} s^{-1})	$\tau_{\exp}$ (s)	τ_{lin} (s)	τ_D (s)	τ_E (s)
(*a*)						
200 μM [NTP]	1.02	2.9	98	345	322	23
300 μM [NTP]	1.16	3.2	86	312	293	20
500 μM [NTP]	1.38	3.8	72	263	247	17
(*b*)						
80 nM [E]	0.95	3.25	106	308	279	29
100 nM [E]	1.00	3.25	100	309	282	23
140 nM [E]	1.00	3.37	100	297	270	27
(*c*)						
MNV-11	1.92	6.37	52	157	143	14
MDV-1	1.42	3.60	70	278	263	15

Conditions:
(*a*) Template MNV-12 RNA, 190 nM core Qβ replicase.
(*b*) Template MNV-12 RNA, 500 μM [NTP] (each), core Qβ replicase.
(*c*) Template as indicated, 500 μM [NTP] (each), 120 nM holo Qβ replicase. The τ-values are retention times (e.g. $\tau_E = 1/k_E$).

high rates of double-strand formation can be expected, since the sequence complexity is low due to the short chain. We always found remarkably strong secondary structures for the self-replicating species [13], which reduces their rates of double-strand formation. Probably sequences with double-strand formation rates that are normal for their chain length are not self-replicating because the replica irreversibly forms a double strand with the template before both are released from the enzyme. Despite the strong secondary structure, double strands appear to be produced exclusively in the late linear phase. There, single-stranded template has reached a steady state concentration (see Appendix) where the amount synthesized is balanced by the amount lost by double strand formation [17] (Fig. 6). The steady state template concentrations of different species vary considerably, depending on their rates of double strand formation, and decrease with increasing ionic strength because the rates of double-strand formation increase with ionic strength.

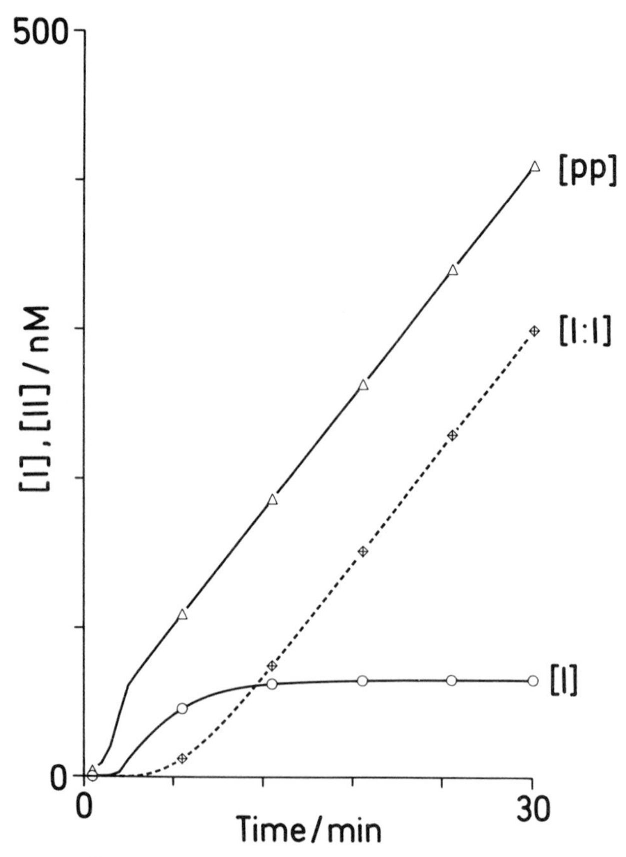

Fig. 6. Double strand formation. The conditions for simulation were as in Fig. 4. In the exponential growth phase (first 5 min) net synthesis (incorporation [pp]) produces predominantly enzyme-bound RNA, in the early linear phase (5–10 min) free single-stranded RNA, and in the late linear phase (> 10 min) double-stranded RNA is the main product.

Accumulation of double-stranded RNA leads to a marked suppression of the incorporation rates, because double-stranded RNA was found to compete for free enzyme (Fig. 7). The resulting inactive complex redissociates with a rate value about five orders of magnitude higher than dissociation of an active template-enzyme complex.

Competition of different species

In the exponential growth phase where enzyme and substrate is not limited the growth of one species is not influenced by the presence of another. Therefore, the species with the fastest replication rate is selected (see Appendix). When the differences between the overall rates are small, many replication rounds are necessary for selection of one species. Experimentally, care has to be taken to maintain exponential growth conditions by transferring into fresh incubation mixture before enzyme saturation is approached.

While selection in the exponential phase observes theoretical laws which are well understood, the selection behavior of self-replicating species in the linear phase is at first surprising: the concentration ratios between species can be altered by many orders of magnitude while the total RNA concentration increase is comparatively small. A comparison of the replication rates gives no explanation for that behavior, since it is often the species with the lower replication rate which wins the competition. However, simple biochemical reasoning, simulation of the results by numerical integration of the rate equations, and analytical treatment of the simplified mechanism agree in readily explaining the experimental results [18]. In the linear phase, the replication rate of one species is affected by the presence of other species, since they all must compete for enzyme. The overall replication rates ρ are governed by the reactivation rates of enzyme, a step unimportant for selection. For competition, the enzyme-binding rates of different species and their relative concentrations are of crucial importance for selection. Double-strand formation also affects the competition by limiting the number of free templates. Dramatic selections are observed because advantageous species grow up exponentially in the steady state of other species. Frequently species with a shorter chain length were found to have a selective advantage in the exponential phase, while in the linear phase longer chain lengths are preferred. The shorter chain length was correlated to a faster rate of enzyme reactivation, while the longer chain length

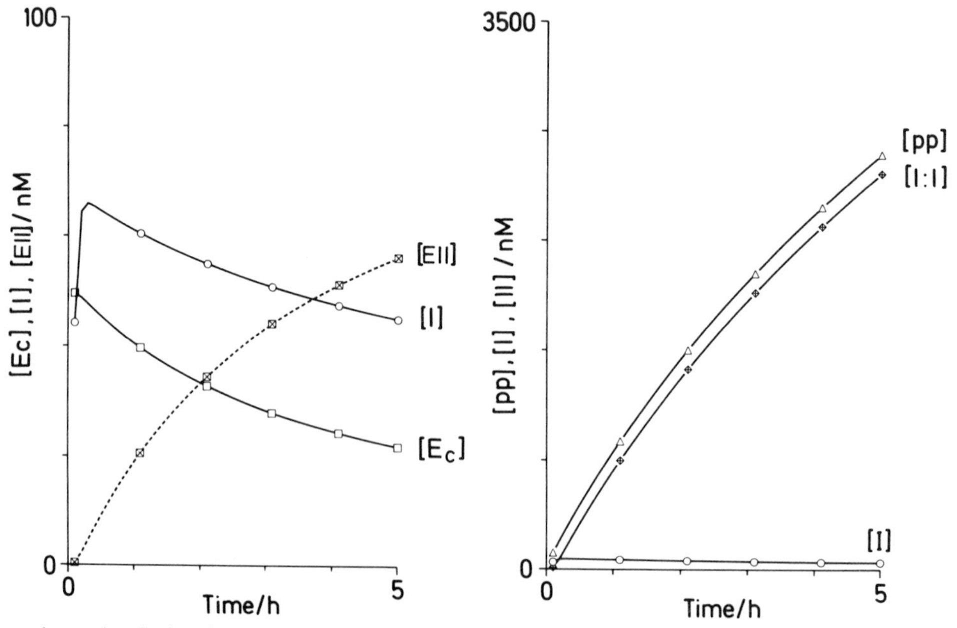

Fig. 7. Product inhibition at longer incubation times. In experimentally measured profiles, incorporation is linear only for a short period; at longer times, the incorporation rate drops due to product inhibition. With increasing double-stranded RNA concentration ($[I:I]$, ⊕), increasing amounts of enzyme bind to it ($[EII]$, ⊠), at the expense of enzyme-bound template ($[E_c^+]$, □). Incorporation, and with it the template concentration ($[I]$, ○), drops with time. Other inhibition terms (not considered in this simulation) include consumption of substrate and accumulation of pyrophosphate.

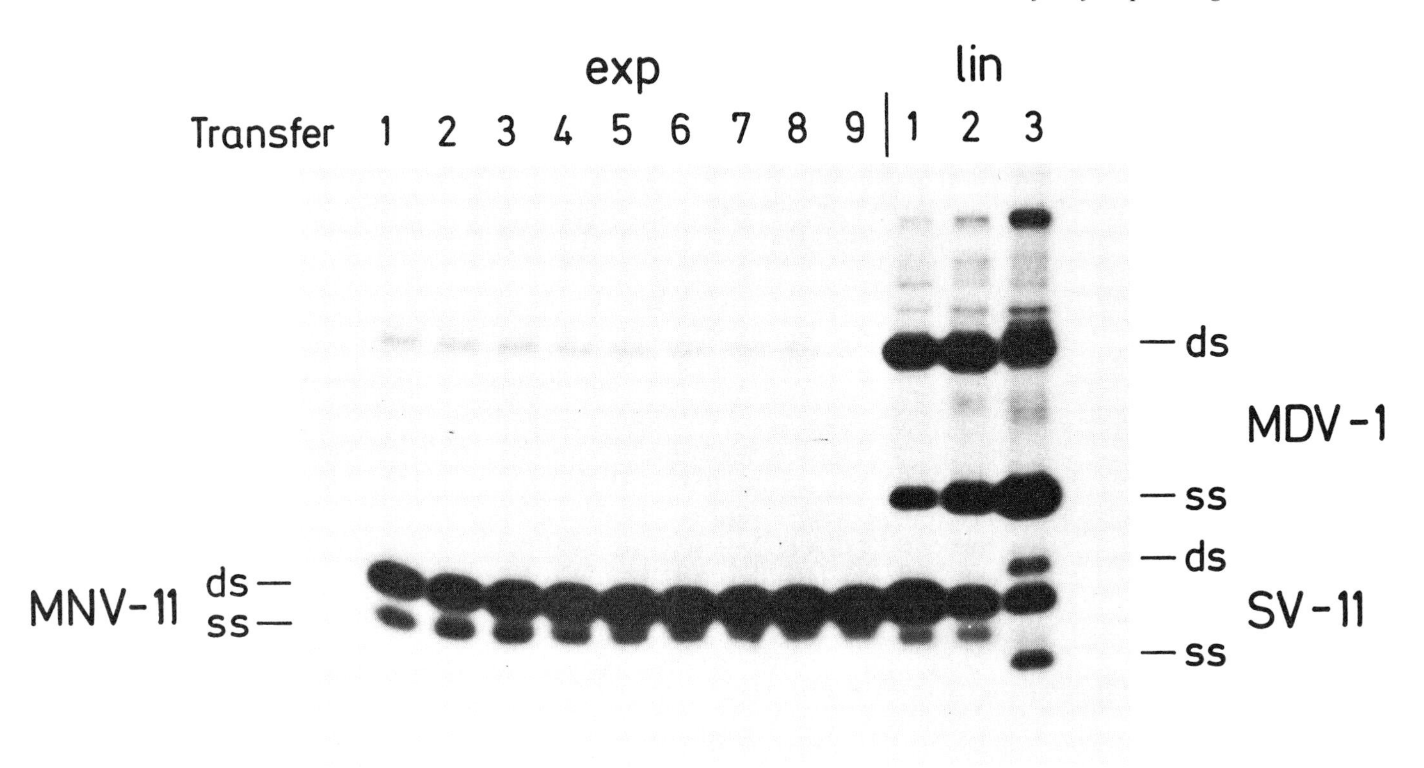

Fig. 8. Competition between MNV-11 and MDV-1 in the exponential and linear phase. Transfers in the exponential phase were done by incubating for 7 min at 30 °C and diluting it 1000-fold into fresh incubation mixture. Under these conditions, enzyme saturation was reached. Transfers in the linear phase were incubated for 2, 2, and 15 h, respectively, and diluted 20-fold. Exponential transfers with incubation times reduced to 6 min (not shown) outdiluted the MDV-1 completely while the MNV-11 concentration remained constant.

reduced particularly the rate of double-strand formation. Elongation rates are controlled by the presence of pause sites and show generally little correlation to chain lengths. This is illustrated by Fig. 8 and Table I where it is shown that MNV-11 is selected in the exponential phase while MDV-1 rapidly takes over when enzyme becomes limiting.

An additional reaction has to be considered for sequentially related species, e.g. for members of a quasi-species distribution. Not only the rate of production of homologous, but also those of heterologous double strands must be considered. When the 'hybridization' rate can be neglected, selection leads finally to a steady state where both species coexist at a concentration ratio independent of the initial conditions. With substantial hybridization, one species usually outcompetes the other completely.

While the phenotypic properties of our species can not be derived from the sequences since neither the tertiary structures nor their influence on the replication can be calculated, we succeeded in deducing the selection behavior from rate values that can be readily determined for each species in the absence of its competitors. Compact analytical equations (see Appendix) describing the selection in the exponential phase and in the steady state of template formation have been obtained. The selection in the growth phase transitions which is too complicated to be quantitized with analytical equations can be readily investigated by computer simulations.

Non-template-instructed RNA synthesis catalysed by Qβ replicase

In the absence of a suitable template or under conditions where normal replication cannot take place, Qβ replicase catalyses phosphodiester formation between nucleotide triphosphates without template instruction at greatly reduced rates. In the absence of template, the reaction requires both purine triphosphates but incorporates also pyrimidine triphosphates. The oligonucleotides are found to have the sequence pppGpR(pN)$_x$, where x is typically 15–20, but may reach values of 60 and more, R is purine and N any nucleoside [19]. Accumulation of substantial oligonucleotide amounts requires long incubation times, high triphosphate and enzyme concentrations and the absence of one of the pyrimidine triphosphates. In the presence of all 4 triphosphates, incubation leads after long lag times to a sudden outgrowth of self-replicating RNA species [20] with chain lengths of about 60–200 nucleotides [21]. Even when synthesized under completely identical conditions, their sequences are unrelated, as are their replication rates and chain lengths [21]. Sumper and Luce [20] called this reaction *de novo* synthesis in analogy to similar phenomena observed with DNA polymerase I and RNA polymerase [22, 23].

The kinetics of the *de novo* reaction is quite different from the kinetics of replication [16] excluding an amplification of pre-existing RNA impurities. The lag times observed are substantially longer than the amplification of one strand to macroscopic appearance and are inversely proportional to the 2.5th power of the triphosphate concentration and to the 2.5th power of the enzyme concentration. At low enzyme or substrate concentrations or at higher ionic strength no *de novo* reaction occurs and the RNA species can be cloned under these conditions. Template-instructed amplification is much more rapid and usually no interference with the *de novo* reaction is observed in the usual experimental time-spans [1–3, 5–9], provided the enzyme is not already contaminated with self-replicating RNA. Recently, it was reported that specially purified enzyme is unable to perform *de novo* synthesis [24]. However, the enzyme concentrations used in these experiments were so low that no *de novo* synthesis was to be expected. We repeated the purification procedure of this report and found that this enzyme has the same capability of *de novo* synthesis as observed with our preparations.

When Qβ replicase is incubated with RNA, but initiation of a new chain is inhibited by replacing GTP with ITP, a sequentially biased but heterogenous sequence of discrete length is attached slowly to RNA species that bind to replicase.

Is there a biological role for these slow reactions? It seems likely that the reactions also occur *in vivo* since they represent a plausible source for the heterogeneous 6S RNA observed *in vivo* late in the infection process. However, all enzymes also catalyze, in addition to their main reaction, chemically closely related reactions at a low rate. The special feature of Qβ replicase is that the side reaction may become readily detectable by a subsequent autocatalytic amplification of the rare self-replicating RNA products.

Acknowledgements

I thank Professor M. Eigen for his support and his many contributions to the work, Professor W. C. Gardiner for introducing me to the field of reaction mechanism modeling with computer simulations and many suggestions. I am indebted to Mr R. Luce and Ms M. Druminski for expert technical assistance.

Appendix

Analytical treatment of replication kinetics

Equilibrated exponential growth phase: $[I_0] \ll [E_0]$
Coherent exponential growth at all intermediates

$$d[I_i]/[I_i]\,dt = \kappa$$

Equilibrated linear growth phase: $[I_0] > [E_0]$
Steady state for replication intermediates

$$d[I_i]/dt = 0$$

Explanation of symbols

$[E_0]$ = total concentration of enzyme (complexed and free);
$[I_0]$ = total concentration of RNA (complexed and free);
$[I_i]$ = any replication intermediate.
Simplified palindromic 3-step mechanism (Fig. 5).

Definitions

$$[E_c] = [EI]+[IE], \quad [I_0] = [I]+[E_c], \quad [E_0] = [E]+[E_c].$$

Differential equations

$$d[I]/dt = -k_A[E][I]+k_D[IE]-k_E[EI];$$
$$d[EI]/dt = k_A[E][I]-k_E[EI];$$
$$d[IE]/dt = k_E[EI]-k_D[IE].$$

Experimentally found for $[E] > 100$ nM: $k_A[E] > k_E > k_D$ ($500, 60, 6\times 10^{-3}\ s^{-1}$).

Linear growth phase

$$\rho = v_{max}/[E_0] = k_E k_D/(k_E+k_D).$$

Exponential growth phase

$$\kappa^3+(k_A[E]+k_E+k_D)\kappa^2+k_D(k_A[E]+k_E)\kappa-k_A[E]k_Ek_D = 0$$
$$\kappa = k_D/2\{[1+4k_E/k_D]^{\frac{1}{2}}-1\} \approx (k_Ek_D)^{\frac{1}{2}}$$
$$k_D = \kappa/2\{[\kappa+3\rho)/(\kappa-\rho)]^{\frac{1}{2}}-1\}$$
$$k_E = \kappa/2\rho\{\kappa-\rho+(\kappa^2-2\kappa\rho-3\rho^2)^{\frac{1}{2}}\}$$

Double-strand formation

In the linear phase, the loss term for template by double-strand formation $d[II]/dt = k_{ds}[I^+][I^-] = 1/4k_{ds}[I]^2$ (for the palindromic case) must be considered, where $[II]$ is the concentration of double strand. A steady state for the free template is reached where

$$[I] = (2v/k_{ds})^{\frac{1}{2}}$$

where

$$v = \rho[E_c]$$

Only for double-stranded RNA is a net increase observed with

$$d[II]/dt = v/2$$

Competition between different species

Case A. $[^1I], [^2I] \ll [E_0]$. Double-strand formation can be neglected.

$$\frac{[^1I]}{[^2I]^t} = \frac{[^1I]}{[^2I]^{t=0}} e^{(^1\kappa-^2\kappa)t}$$

Case B. $[^1I] \ll [E_0] < [^2I]$, $k_{hy} \ll k_{ds}$.

Species 1 grows exponentially with the rate $^1\kappa$ as it would in the absence of the other species at the remaining steady-state free enzyme concentration. Hybridization considered: exponential growth with the rate $(^1\kappa-k_{hy}[^2I])$.

Case C. $[^1I], [^2I] > [E_0]$.

Steady-state conditions in the linear growth phase

$$\frac{[^1I]}{[^2I]} = \frac{^1k_A{}^2k_{ds}-{}^2k_Ak_{hy}}{^2k_A{}^1k_{ds}-{}^1k_Ak_{hy}}$$

Both, numerator and denominator positive: both species coexist.
Numerator negative, denominator positive: Species 1 dies out.
Numerator positive, denominator negative: Species 2 dies out.
Numerator negative, denominator negative: One of the species dies out, depending on initial conditions.

$k_{hy} \ll k_{ds}$ favors coexistence.

$k_{hy} \approx k_{ds}$ favors extinction.

References

1. Mills, D. R., Peterson, R. L. and Spiegelman, S., *Proc. Natl. Acad. Sci. USA* **58**, 217 (1967).
2. Levisohn, R. and Spiegelman, S., *Proc. Natl. Acad. Sci. USA* **63**, 807 (1968).
3. Kramer, F. R., Mills, D. R., Cole, P. E., Nishihara, T. and Spiegelman, S., *J. Mol. Biol.* **89**, 719 (1974).

4. Biebricher, C. K., in *Evolutionary Biology*, vol. 16, p. 1. Plenum, New York (1983).
5. Palmenberg, A. and Kaesberg, P., *Proc. Natl. Acad. Sci. USA* **71**, 1371 (1974).
6. Feix, G., *Nature (London)* **259**, 593 (1976).
7. Mills, D. R., Kramer, F. R. and Spiegelman, S., *Science* **180**, 916 (1973).
8. Mills, D. R., Kramer, F. R., Nishihara, T. and Spiegelman, S., *Proc. Natl. Acad. Sci. USA* **72**, 4252 (1975).
9. Schaffner, W., Ruegg, K. J. and Weissmann, C., *J. Mol. Biol.* **117**, 877 (1977).
10. Blumenthal, T., Landers, T. A. and Weber, K., *Proc. Natl. Acad. Sci. USA* **69**, 1313 (1972).
11. Wahba, A. J., Miller, M. J., Niveleau, A., Landers, T. A., Carmichael, G. G., Weber, K., Hawley, D. A. and Slobin, L. I., *J. Biol. Chem.* **249**, 3314 (1974).
12. Kamen, R., Kondo, M., Römer, W. and Weissmann, C., *Eur. J. Biochem.* **31**, 44 (1972).
13. Biebricher, C. K., Diekmann, S. and Luce, R., *J. Mol. Biol.* **154**, 629 (1982).
14. Weissmann, C., *FEBS Lett. (Suppl.)* **40**, S10 (1974).
15. Biebricher, C. K., Eigen, M. and Gardiner, W. C., *Biochemistry* **22**, 2544 (1983).
16. Biebricher, C. K., Eigen, M. and Luce, R., *J. Mol. Biol.* **148**, 391 (1981).
17. Biebricher, C. K., Eigen, M. and Gardiner, W. C., *Biochemistry* **23**, 3186 (1984).
18. Biebricher, C. K., Eigen, M. and Gardiner, W. C., *Biochemistry* **24**, 6550 (1985).
19. Biebricher, C. K., Eigen, M. and Luce, R., *Nature (London)*, in press.
20. Sumper, M. and Luce, R., *Proc. Natl. Acad. Sci. USA* **72**, 162 (1975).
21. Biebricher, C. K., Eigen, M. and Luce, R., *J. Mol. Biol.* **148**, 369 (1981).
22. Kornberg, A., Bertsch, L. L., Jackson, J. F. and Khorana, H. G., *Proc. Natl. Acad. Sci. USA* **51**, 315 (1964).
23. Krakow, J. S. and Karstadt, M., *Proc. Natl. Acad. Sci. USA* **58**, 2094 (1967).
24. Hill, D. and Blumenthal, T., *Nature (London)* **301**, 350 (1983).

Chemica Scripta 1986, **26B**, 59–66

Comparative Sequence Analysis

Exemplified with tRNA and 5S rRNA

Ruthild Winkler-Oswatitsch, Andreas Dress* and Manfred Eigen

Max-Planck-Institute for Biophysical Chemistry, D-3400 Göttingen, Federal Republic of Germany

Paper presented by Ruthild Winkler-Oswatitsch at the Conference on 'Molecular Evolution of Life', Lidingö, Sweden, 8–12 September 1985

Abstract

The advent of new sequencing techniques has brought a sudden increase in the data available for the study of evolutionary history on a quantitative basis. Criteria are put forward and methods are developed that allow an optimal alignment of sequences, a determination of the topology of their kinship relations, a reconstitution of precursors and a reliable establishment of their randomization. The criteria developed are tested by comparison to a large bulk of data from both tRNA and ribosomal 5S RNA sequences. Ancestral features such as base compositions and periodic sequence patterns could be restored. The data suggest that tRNA and 5S rRNA evolved concomitantly with an early genetic code favoring codons of the form RNY.

I. Introduction

Evolution takes place and is mapped in the sequence space of nucleic acids in which all sequences are represented by correct kinship distances. The concept of sequence space [1] is illustrated in Fig. 1. In order to reconstruct the evolutionary past from the information which is linearly arranged and stored in the DNA or RNA sequences a number of unrelated problems as listed below must be solved. (Literature on sequence analysis and phylogeny is reviewed in ref. 2.)

(1) Any comparative analysis of nucleotide composition and pattern critically depends on proper alignment. An unambiguous alignment can be achieved only if the degree of homology and invariance of the sequences is substantial.

(2) The sequences to be compared may have evolved in different fashions. If the genealogy of the sequences shows successive divergence it is represented by a tree-like topology (Fig. 2b). Whether the phylogenetic tree can be established and how well it represents the true temporal development depends primarily on the rate of mutations. Other topologies to be encountered are bundles and nets with internal loops. For a family of tRNAs within a given species an early common node can be established; from this node all the sequences have simultaneously diverged, independently acquiring their characteristic differences (Fig. 2b). A mutant distribution of a wildtype sequence, though little diverged, shows a random distribution of all low error variants including many loops and shortcuts (Fig. 2c).

(3) If the topology of the kinship relation among the sequences is known, one may try to establish the common ancestor. In a tree-like divergence this means to retrogressively reconstruct nodal sequences down to the earliest nodes. The true root, however, cannot be assigned unequivocally from distance data alone. In a bundle-like and looped net divergence the master sequence usually provides reliable ancestral information. The master sequence is obtained by superposing the aligned sequences and recording the most frequent nucleotide at each position.

* Department of Mathematics, University of Bielefeld, FRG.

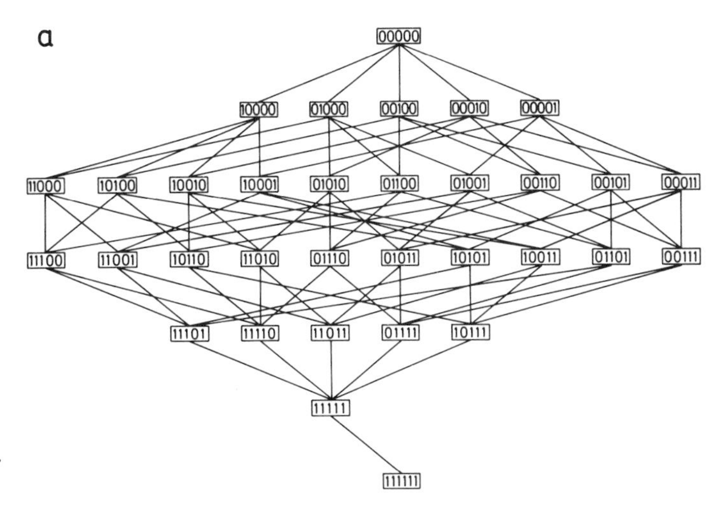

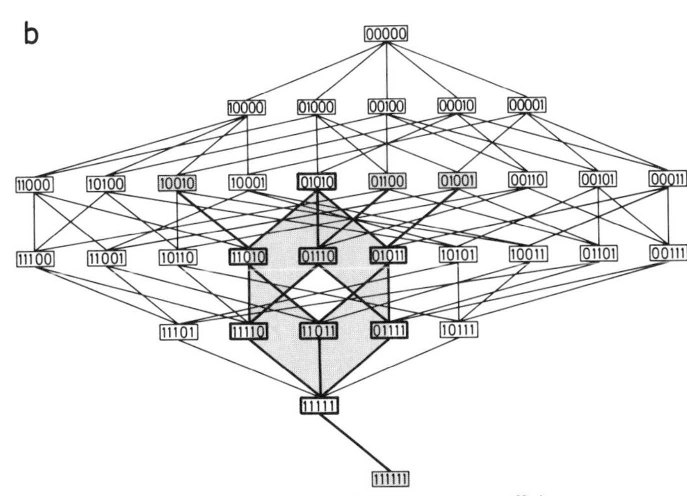

Fig. 1. (a) Example of a five dimensional sequence space sufficient to represent binary sequences comprising five positions. (b) Projection of a three dimensional subspace that correlates four sequences via their distances defined in sequence space.

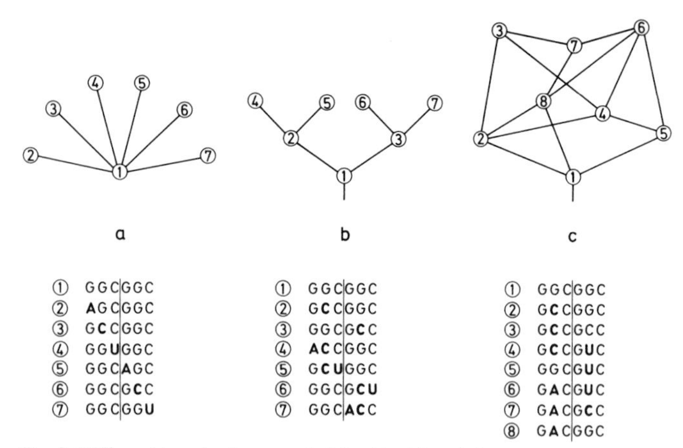

Fig. 2. Different topologies expected for kinship relations among sequences: (a) bundle, (b) (symmetric) tree, (c) looped net.

(4) In a bundle-like or looped net dendrogram, divergent and convergent evolution cannot be distinguished on the basis of distances, as is possible in a tree-like topology. A position which is uniformly occupied by only one type of nucleotide in all aligned sequences is called constant. Such positions indicate biases in an otherwise appreciably diverged set. Therefore information about possible precursors can only be obtained from those positions which show sufficient variability in their nucleotide composition and are not yet completely randomized.

(5) If ancestral sequences can be reconstructed one may search for special characteristics such as periodic nucleotide patterns. To assign them as ancestral properties rather than as biases for present requirements, one has to introduce temporal arguments in order to prove that the biased features depend upon time.

Sequence data were taken from various sources. tRNA data are compiled and continually brought up to date by M. Sprinzl and co-workers in *Nucleic Acid Research* [3]. This ensemble of sequences can be used to study the phylogeny of individual sequences, and to investigate the divergence within the family of tRNAs belonging to a given species.

Ribosomal 5S RNAs, of course, can only be studied in terms of their phylogeny. A wealth of data has been recently compiled. They are continually edited by V. A. Erdmann and co-workers and can also be found in *Nucleic Acid Research* [4]. In addition, 17 archaebacterial sequences were kindly provided by G. E. Fox, C. R. Woese and K. R. Luehrson [5]. With such copious information on eubacteria, archaebacteria, cytoplasmic sequences of eucaryotes, chloroplasts and mitochondria statistically significant conclusions can be drawn.

II. How to achieve an unambiguous alignment?

For tRNA it was easy to assign 76 common positions with sufficient homology to make an unequivocal alignment. Although various tRNAs differ in their length, the positioning of the anticodon and the largely symmetrical secondary structure allowed a clear distinction of insertions and deletions. This is not the case for ribosomal 5S RNAs even though these sequences also appear to be highly conserved down to the early branchings of eubacteria and archaebacteria. The difficulty to align 5S rRNA sequences is the assignment of those positions that are not uniformly occupied. Fig. 3 shows the alignment of 35 eubacterial sequences stretched into a 'procrustean bed' of 149 positions. Some positions are entirely unoccupied because archaebacteria, eucaryotes and, in particular, mitochondria are occupied at those places [6]. The degree of homology obtained from the distances between master and individual sequences for 5S rRNA varies between 73 and 79%. Therefore, alignment as such does not cause any serious problem; although there is no objective method to align the sequences. We can only achieve the most probable alignment characterized by a minimum number of deviations at each position.

tRNA and ribosomal 5S RNA sequences represent highly conserved groups well localized in a defined area of sequence space; only those axis of the sequence space referring to variably occupied positions may be considered (cf. Fig. 1b).

III. How to analyse the alignment for constant and variable positions?

To decide upon the topology of divergence and the degree of randomization one first has to determine, by comparative sequence analysis, which of the highly conserved positions are constant, and hence not subject to any change, and which of the positions – regardless of their conservation status – were allowed to mutate during the course of evolution.

It is common to analyze alignments according to pair distances. Two kinds of pair distances are differentiated: individual *i v.* individual *k* and individual *i v.* master sequence. Usually we begin with a specific alignment and add up, in the horizontal direction, those positions that are differently occupied in the two sequences under consideration. This number is called the Hamming distance. Then the degree of randomization follows from the average pair distance. An average difference of 75% in case of nucleic acids (four bases) would indicate complete randomization. However, this standard method may lead to erroneous results.

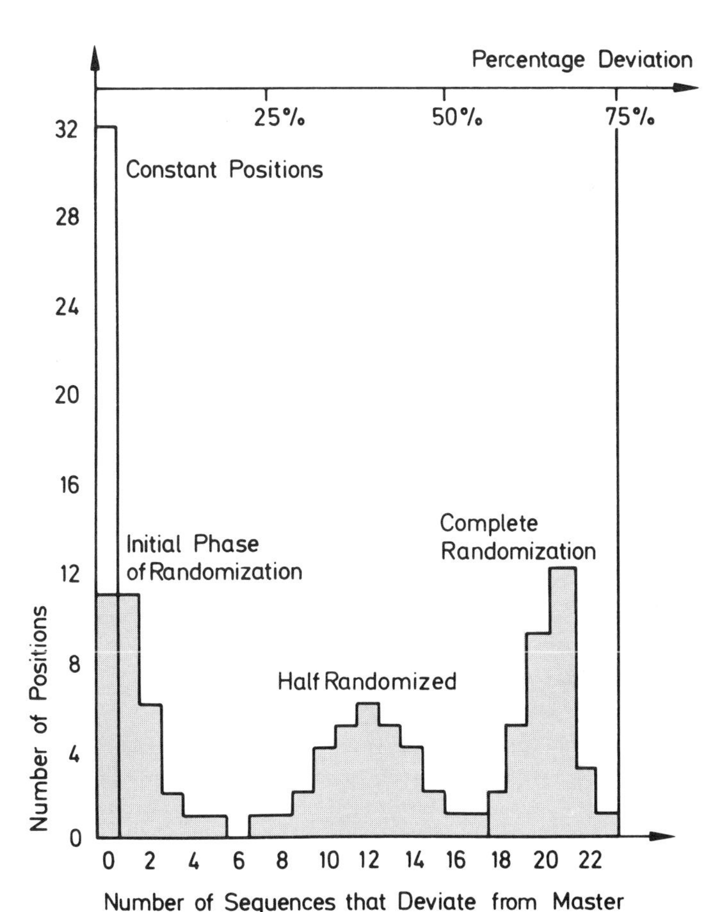

Fig. 4. Analysis of vertical positions in an alignment of a set of 30 sequences relative to the master sequence. Three possible situations of randomization are exemplified.

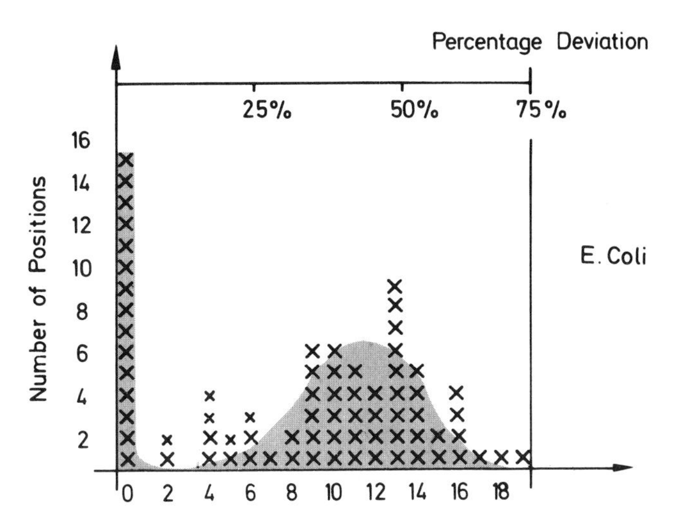

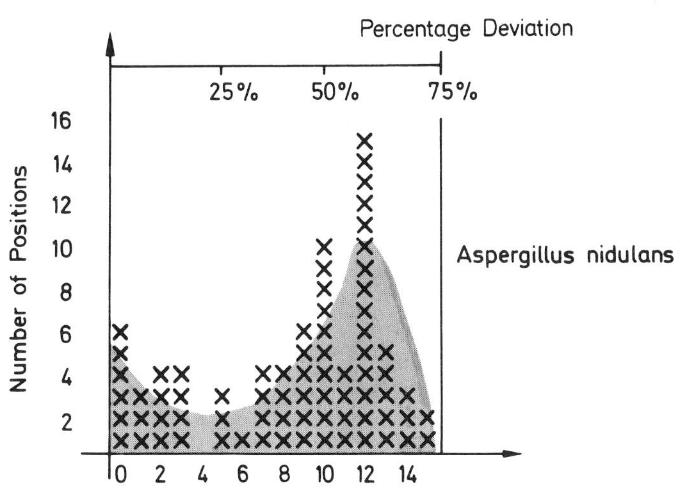

Fig. 5. Analysis of vertical positions relative to the master sequence applied to the tRNA families of two species (*E. coli* and *Aspergillus nidulans*).

To decide upon the true topology of divergence it is necessary to know how far randomization has advanced. For this purpose we introduce a procedure of analysis which intrinsically depends upon the individual positions. This procedure refers to the true sequence space rather than to mere distance space. Instead of summing up differences in the horizontal direction, we sum up vertically, that is position by position [7], and compare with a reference sequence, most suitably the master. The histogram in Fig. 4 describes the outcome of such an analysis [8]. On the abscissa we plot the absolute number of nucleotides that do not agree with the nucleotide appearing in the master sequence at corresponding positions starting with zero and running up to that number which relates to 75% of all the sequences under consideration. Zero means all nucleotides in a vertical row of the alignment are identical, one means all but one are the same, and so on. The largest number corresponds to the limit of randomization including all those positions where 75% of the sequences analyzed differ from the nucleotide in the corresponding positions of the master.

The ordinate tells us how many of the positions qualify for the condition at the value of the abscissa. For instance, if there are 10 positions where all sequences of the set show identical nucleotide occupation the ordinate 10 is assigned to the abscissa-point zero. Fig. 5 demonstrates the evaluation for two tRNA ensembles. In the *E. coli* histogram there is a sharply defined group of constant positions which is significantly separated from a broad distribution of mutated positions, yet far from the limit of randomization. Otherwise the maximum of this broad peak would be close to 75% deviation. The mitochondria data indicate that the tRNA sequences of *Aspergillus nidulans* have moved towards complete randomization. Such a treatment enables us to subdivide tRNA sequences into constant and variable positions [8].

IV. How to decide upon the correct topology?

Only the variable positions in the sequences can be used to reconstruct their genealogy provided that they have not randomized completely.

To achieve a quantitative measure of evolutionary distance we must distinguish two types of substitutions: those in which the nucleotide class R or Y is conserved, that means transitions are between A and G or U and C; and those in which the class is changed, that means transversion from G or A to C or U. Separate counting of transitions and transversions is quantitatively more correct than an undifferentiated sum of both, although for each of the procedures one discards some of the information available. Similarly one has to take into account whether the base under consideration is paired or unpaired, if true temporal distances are to be presented. Considering solely transversions means an analysis of binary (R and Y) sequences yielding a considerable simplification of the procedure.

Is there an objective, quantitative way to decide upon the topology? There are good approximate methods leading to quantitative results. However, ambiguities remain. For instance, an optimally adapted dendrogram, based on present sequences, may not necessarily reflect the correct chronological order of branching. It is technically difficult to handle a set of 100 sequences which includes 4950 pair distances. The smallest dendrogram correlation involving three sequences is based on 161 700 different three-sequence combinations. Moreover, assignment of tree-likeness requires four sequence correlations, which implies about four million different combinations. The optimal dendrogram would be the one based on all available correlations. It is quite obvious that there are technical limitations, even if the adjustment is carried out with a computer. If the topology is known, then the construction of the optimal dendrogram, founded on a suitable approximation, can always be improved by skill and experience.

The correct topology of branching can be tested in the following way [7]. Consider two binary sequences A and B in RY notation. The number of positions differently occupied defines a Hamming distance. Any related third sequence C can then be fitted to AB in such a way that the two distances AC and BC are exactly matched. In this way three binary sequences can always be represented by a tripod. Then think of three sequences in A, U, G, C notation and try to fit them in an analogous way. In the left example of Fig. 6 the distance

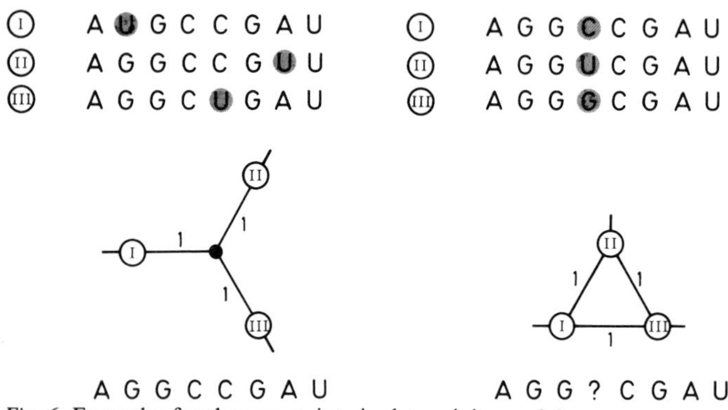

Fig. 6. Examples for the uncertainty in determining nodal sequences.

correlation again yields a tripod. In the right example, however, there is no evidence for any common precursor as suggested by the equilateral triangle in which any of the three corners could describe the precursor.

Whether or not a dendrogram of three sequences is correct can be checked by linking it with a fourth sequence D. If D can be connected with the ternary dendrogram matching all distances (i.e. DA, DB and DC) then the correlation is tree-like. In other words, in case *all* possible four sequence combinations in a set of n sequences to be analyzed fit exactly a dendrogram, all sequences are related in a tree-like fashion. This conjecture has been proven mathematically [7].

The procedure is exemplified in the graph in Fig. 7. Combine any four sequences A, B, C, D by determining their mutual distances AB, AC, AD, BC, BD and CD and form the three possible sums including all four distances: AB+CD, AC+BD, AD+BC. These distance sums can also be written as expressed in Fig. 7, where x stands for the deviation from tree-likeness and y for a branching distance. If all three sums of distances are identical, x and y are zero. Then the resulting topology resembles an ideal bundle. If only the two larger sums are identical, x is zero, and y is finite, the topology is

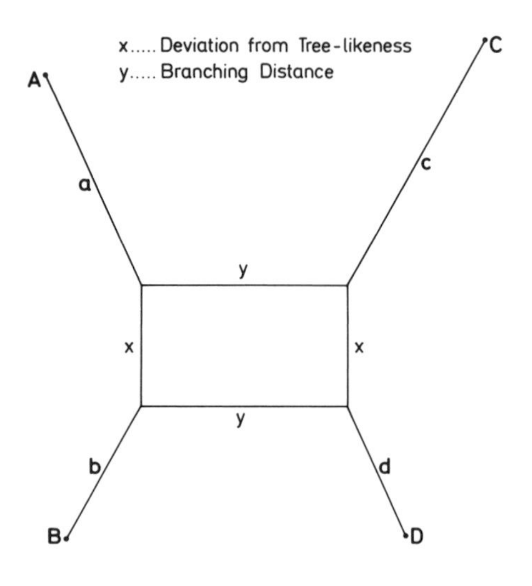

Fig. 7. Analysis of tree-likeness with a four sequence correlation (for a thorough investigation of n-sequence-correlations see [12]).

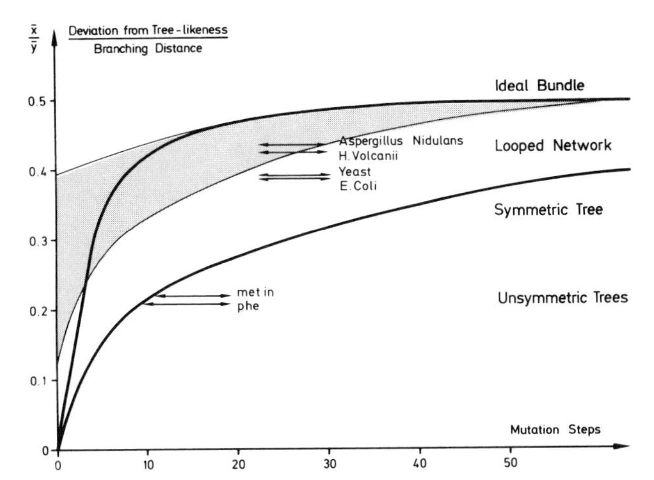

Fig. 8. Tree-likeness analysis applied to a determination of topology of tRNA kinship relations. Experimental values include two phylogenies of tRNAs specified by their anticodon and four tRNA families referring to four different species.

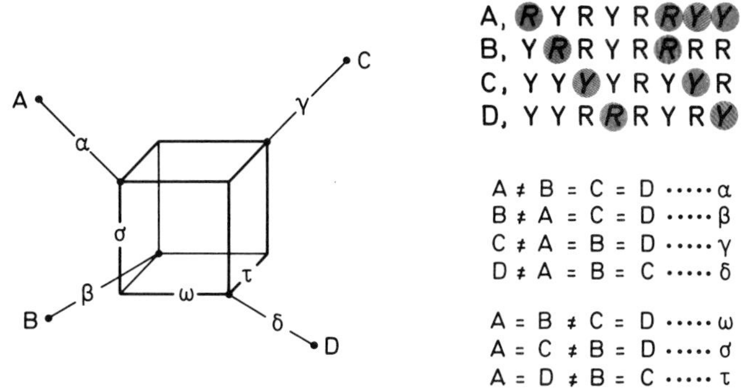

Fig. 9. Topology of distance relations of four sequences (A, B, C, D), in RY notation, represented in sequence space.

a true dendrogram with a branching distance y. If, finally all three sums are different and x and y are finite, the topology is neither bundle- nor tree-like but rather looks like a looped net. In case divergence occurred consecutively starting from a common ancestor, parallel or reverse mutations must have happened in one or several positions.

A determination of the parameters x and y for *all* four-sequence combinations of a given set of tRNA sequences results in a topology with a quantitative representation in terms of its degree of tree- or bundle-likeness.

In Fig. 8 the average ratio of deviations from tree-likeness, $\bar{x}$, and branching distance, $\bar{y}$, is shown for two categories of tRNA sequences. One set (Phe and Met-in) belongs to phylogenetic divergences of a given tRNA molecule (identified by its anticodon). The other is based on tRNA-families comprised of one eubacterium, one eucaryote, one mitochondrion and one archaebacterium. The phylogenetic data manifest themselves as tree-like dendrograms with large branching distances. The data for the tRNA-families of different species, on the other hand, specify a topology which is typical for partly randomized mutant spectra.

The method of four-sequence combinations – if a tree-like topology is confirmed – may be used as a basis for tree construction according to any of the known algorithms applying to distances. Those algorithms can be improved by taking into account not only the cumulative distances but also sequence dependent distances. In fact, one can do better by directly studying the sequence space rather than the mere distance space. For four symbols (A, U, G, C) there is however no simple geometric representation for such a treatment. For the sake of simplicity the method will be demonstrated in Fig. 9 with binary sequences, where a geometric picture can be given as in the case of distance space.

In an alignment of four binary sequences seven different distances of two different kinds may be recorded [7]. First, four distances may be registered, applying to positions in which one of the four sequences differs from the three others, which themselves are uniformly occupied. These positions mark the four peripherical distances α, β, γ and δ. Second, three distances may be registered, applying to positions where two of the four sequences mutually agree. These positions define the three dimensions of the box σ, ω and τ. With two symbols R and Y no other situations – except complete agreement – can exist. The correct topology of kinship relation among four different species is geometrically depicted by a three-dimensional box. It designates the range of uncertainty of node assignment.

If a correct assignment of nodal sequences is required – as is necessary for tracing out the common ancestor – then this method is superior to a simple distance space analysis. This statement is exemplified with four 5S rRNA master sequences in Fig. 10 [6]. The comparison of both diagrams reveals the following.

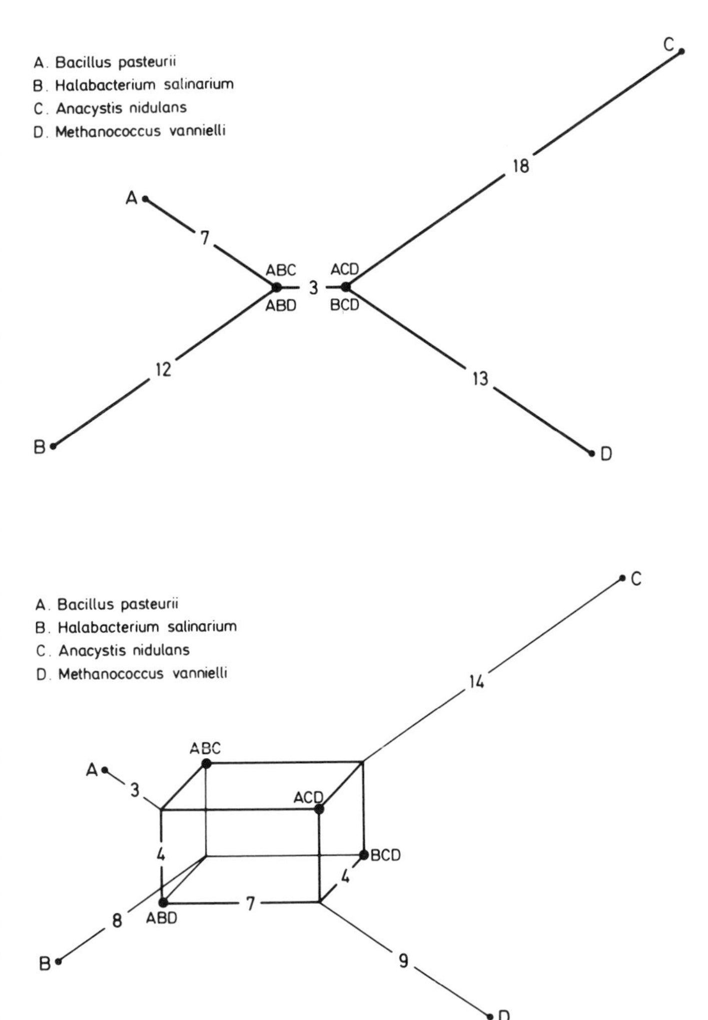

Fig. 10. Comparison of topology analyses, applied to four 5S rRNA sequences, in distance and in sequence space.

(1) The distance space analysis suggests that the master sequence ABC is identical to that of ABD. The same equivalence holds for the master sequences ACD and BCD. However, sequence space shows that those identities do not exist. They are artifacts caused by balancing two effects. Each of the two pairs of master sequences differ at eight positions.

(2) The distance space analysis pretends that the topology of kinship of the four master sequences is almost bundle-like, although the investigated sequences are fairly unrelated. The sequence space analysis illustrates the uncertainty of the early node. Any point within the box could be the root; the center of the box is more likely than any of the four corners, which are assigned to the four master sequences.

V. How far has ancient information randomized?

There are several quantitative procedures to measure the degree of randomization. One possible way is exhibited in Fig. 8. Another approach is to base this analysis on the changes of the various parameters as for instance $(a+b+c+d)$ *v.* x or y, and x *v.* y, or correspondingly $(\alpha+\beta+\gamma+\delta)$ *v.* σ, τ or ω, and σ *v.* τ *v.* ω. Since all these parameters are characteristic for the degree of randomization, such a plot of one parameter *v.* another has some internal calibration. In Fig. 11 a very simple but equivalent method is visualized by a computer experiment [8]. Identical copies of a given initial sequence are allowed to mutate independently. We then record, as functions of time:

(*a*) the average pair distance of individual copies;

(*b*) the average distance of individual copies from the given initial sequence;

(*c*) the distance between initial and the master sequence.

The results are given in the diagram [8] in Fig. 12. The average pair-difference deviates from zero twice as fast as the average distance of individual sequences from the initial sequence. This is because mutations in different sequences initially appear with higher probability at different positions. When two individual sequences mutate at different positions their mutual distance corresponds to the sum of their distances from the initial sequence. Later, when complete randomization is approached, parallel and reverse mutations dominate, and all curves level off at 75%.

The interesting point is that the distance between master and initial sequence – which eventually also will randomize to the same level – departs much more slowly from zero than the other two distances. Therefore, it is justified to measure the deviations from the (unknown) initial sequence as representative of the deviations from the (known) master sequence.

The experimental points refer to four different tRNA families for which we have determined the average pair distances and the average distances between individual sequences and master. Both are directly accessible numbers. The

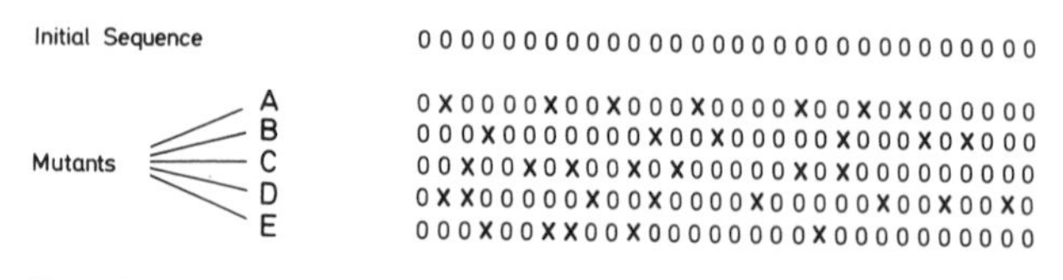

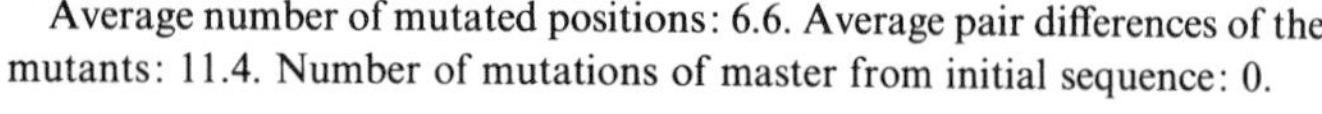

Fig. 11. Simulation of error accumulation and the concomitant evolution of distances.

Average number of mutated positions: 6.6. Average pair differences of the mutants: 11.4. Number of mutations of master from initial sequence: 0.

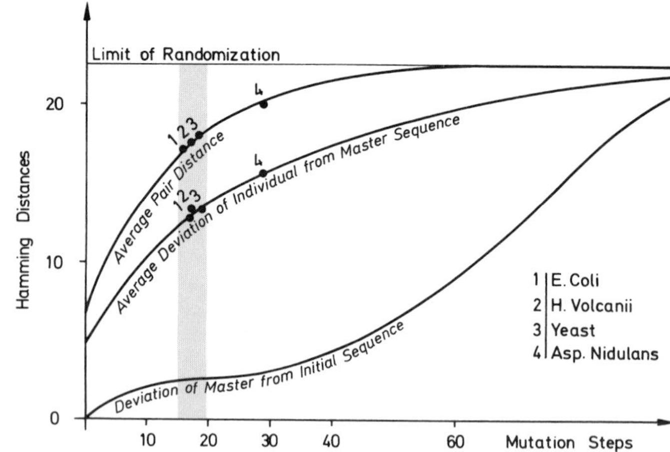

Fig. 12. The dependence of distances on time (as a function of the number of mutations) resulting from the simulation experiment in Fig. 11. The numbered points belong to experimental values obtained for tRNA families of different species.

distance criteria tell us that for *E. coli*, yeast and *H. volcanii* the master sequence has not departed very far from its ancestral sequence, whereas the divergence for mitochondria has clearly proceeded much further. In other words, as long as the average pair distance is appreciably larger than the average distance between the individual and the master sequence one expects the master to faithfully represent the ancient features. The two upper curves calibrate the time axis for any set of experimental data. It is surprising how well the master sequence conserves ancient properties. From this we conclude that the superposition of sequences is meaningful for the ancestral information which can be recorded.

This procedure is restricted to those tRNA sequences which have evolved according to a bundle-like topology. It does not work for a tree-like divergence. Loops such as those which appear in mutant distributions, may cause the master to depart rapidly from the initial sequence. The method leads us back into the time of evolution of the 64 code adaptors, namely the time of the origin of the genetic code. The data suggest that tRNAs diverged from a single precursor (or its mutant spectrum) which we can reconstitute at least for the major part of its variable positions.

VI. Is there a common ancestor?

The master sequence is obtained from the superposition of all sequences under consideration counting the most frequent nucleotide at each position. The meaning of the master sequence is obscured by the various mutations in each single tRNA copy. Superposition of all the tRNA sequences yields the master sequence and simultaneously restores the meaning of the ancestor sequence. In this manner we may restore the meaning of an ancestor from those present sequences that have mutated so far that the original features are largely disguised. However, the method applies well only to the sequence relationships with either a bundle-like topology, typical for independent parallel divergence, or to a network topology with internal loops typical for mutant distributions, whose variable positions have not yet randomized completely. It is obvious that this procedure does not detect early nodes in tree-like topologies. Consider two branches that have drifted extensively. If one of these branches is accidentally populated by a larger number of sequences than the other it will determine the master sequence. In this case each nucleotide in the master sequence might have been insufficiently determined due to the small selection of sequences that happened to be analyzed.

VII. Can ancestral features be identified?

Two striking and surprising ancestral features in both tRNA and ribosomal 5S RNA became evident from these procedures just described. Both RNAs have an abundance of RNY triplet patterns where R stands for purine, N for any of the four nucleotides, and Y for pyrimidine. This finding is surprising because tRNA and ribosomal 5S RNA are non-coding sequences. In other words, they are not expected to carry genetic information, as genes or messengers do. We reported this fact first for tRNA [9]. John Shepherd of the University of Basle then found analogous pattern frequencies in all kinds of coding sequences – to an extent that allowed him to use this method to distinguish introns from extrons [10]. Now we can report an even stronger RNY-bias in ribosomal 5S RNAs [6]. If we analyze these sequences in terms of short patterns we find a pronounced triplet bias. This is the only type of bias present. It appears in one reading frame as a strong RNY preference while, in the other two reading frames, it shows up as a weaker YNR bias. The histograms in Fig. 13 comprise the complete frequency distribution of all possible triplet patterns for eubacteria. It demonstrates that we are dealing with well established experimental facts, rather than with some arbitrary fluctuations. There is a significant prevalence of RNY triplet patterns. The histogram also exhibits some additional RNR bias. Such a behavior is exactly what one would expect for a partly randomized sequence which originally was composed of RNY triplets.

Another interesting observation is that the RNY-bias usually is stronger in master sequences than it is in the average sequences. This is particularly pronounced for the tRNA families where the master sequences more clearly resemble the ancestors. The RNY-bias in the master sequence obtained for a phylogenetic set of both tRNA and 5S rRNA is weaker since these master sequences are not identical with the earliest node-sequences.

For this purpose it was necessary to reconstruct early node sequences. We used the method based on sequence space analysis to reconstruct the topology. The dendrogram [6] obtained in Fig. 14 shows the experimental outcome. The four correlated sequences are chosen such that they include the earliest node somewhere in their branching. The numbers refer to the RNY/YNR ratios which, on the average, are higher inside than outside the box. This diagram demonstrates

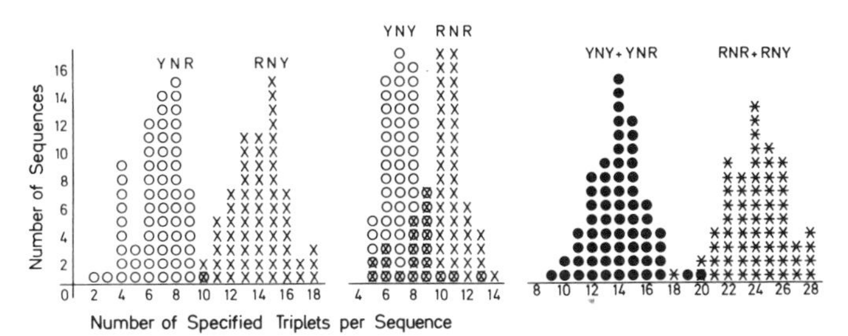

Fig. 13. Histograms of the frequencies of triplet patterns determined for 5S rRNA of 35 eubacteria.

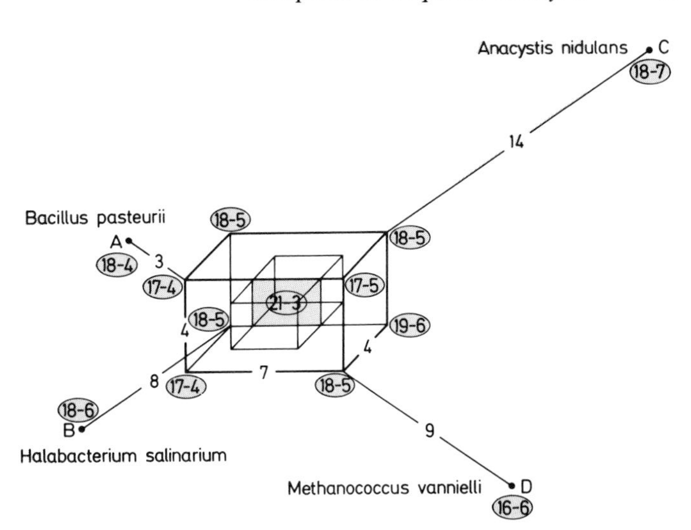

Fig. 14. Phylogenetic distribution of RNY *v.* YNR frequencies for four 5S rRNAs represented in sequence space.

that the RNY bias is an ancient feature of ribosomal 5S RNA sequences which has been partly randomized during evolution.

The specific nucleotide composition of tRNA and 5S rRNA sequences is an additional characteristic quality. For tRNAs we have demonstrated previously [9] that – except for mitochondria – the sequences are generally rich in G and C. This finding is depicted in Fig. 15. The biased nature of the effect is obvious. The master sequences exhibit the effect much more pronounced than the averages. The same is true for the AU-richness of mitochondrial sequences, probably a constraint of evolution related to their ATP production.

The phylogenetic set of ribosomal 5S RNAs, whose master sequences do not trace back as far as the tRNA-families – which originated along with the genetic code – should show a similar but weaker bias. This expectation is corroborated by the evidence in Fig. 16. Only two distinct biases are detectable. One is a definite preference for R in the first and Y in the third position, while the middle position is neutral with respect to R and Y. Second, all positions in the sequence reveal a general preference of G and C over A and U. The order of triplet frequencies RNY > RNR > YNY > YNR is, according to Shepherd [10], a general trait of coding sequences. It may reflect the evolution of the genetic code from an RNY structure, providing a comma-free readout via wobble-intermediates to the present form. The GC richness signifies that the first codons were of the form GNC with the assignments glycine, alanine, aspartic acid and valine. These four amino acids, in fact, occur with highest abundance in Miller's [11] experiments simulating prebiotic synthesis. The next step in the evolution of the code would then be the transition from GNC to RNY. Such a code specifies eight amino acids, again commonly found in simulation experiments. Degenerate replacements might have introduced RNN assignments which finally led to the present form NNN, a code that cannot be read in a comma-free manner without the help of machinery.

The consideration of the order within the frequency data, more than the mere existence of a triplet pattern as such, led us to believe that tRNA and ribosomal 5S RNA have descended from coding sequences used at the time the genetic code originated.

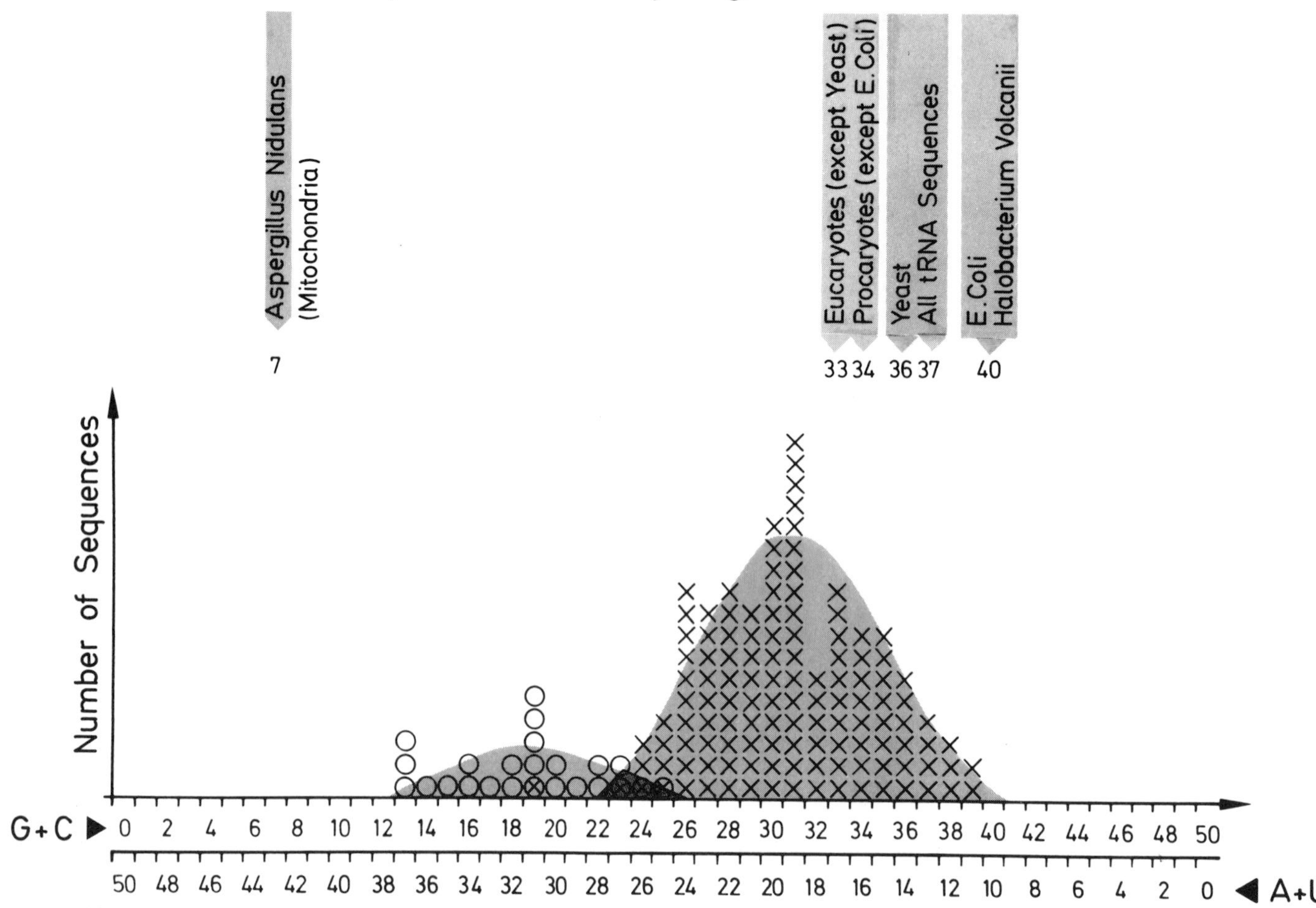

Fig. 15. Abundance of G+C or A+U in the variable positions of the tRNAs. (Each X or 0 stands for one of the 144 sequences analyzed, the 0 indicating the mitochondrial sequences.) The numbers in the upper part refer to the master sequences of the mentioned species or kingdoms.

Table I. *Base occupation of the three triplet positions in the main reading frame for eubacteria, archaebacteria and eucaryotes. All the data indicate a bias for G+C over A+U and a specific bias for R in the first and Y in the third positions (leaving the middle position completely neutral)*

Triplet position Occupied by	1	2	3
Eubacteria			
R	28	20	14
Y	11	19	25
A+U	15	16	15
G+C	24	23	24
Archaebacteria			
R	24.5*	19	15.5*
Y	14.5	20	23.5
A+U	12.5	15	15.5
G+C	26.5	24	23.5
Eucaryotes			
R	21	19	19
Y	17	18	19
A+U	17	15	13
G+C	21	22	25

* One triplet in the master sequence is related to equally frequent appearance of two nucleotides.

Acknowledgement

We wish to thank Dr Robert Clegg for carefully reading the manuscript and reviewing the English.

References

1. Eigen, M., *Ber. Bunsenges. Phys. Chem.* **89**, 658 (1985) and cf. paper this volume.
2. Wiley, E. O., *Phylogenetics*. John Wiley, New York (1981).
3. Gauss, D. H. and Sprinzl, M., *Nucleic Acids Res.* **6**, r1 (1981), including data of R. Gupta, Ph.D. Thesis, University of Illinois, Urbana (1981).
4. Erdmann, V. A., Wolters, J., Huysmans, E., Vandenberghe, A. and de Wachter, R., *Nucleic Acids Res.* **12**, r133 (1984).
5. Personal communication.
6. Eigen, M., Lindemann, B., Winkler-Oswatitsch, R. and Clarke, C. H., *Proc. Natl. Acad. Sci. USA* **82**, 2437 (1985).
7. Dress, A. and Eigen, M. *et al.*, to be published, cf. also Simoes-Pereira, J. M. S., *J. Comb. Theory* **6**, 303 (1969).
8. Eigen, M. and Winkler-Oswatitsch, R., *Naturwissenschaften* **68**, 217 (1981).
9. Eigen, M. and Winkler-Oswatitsch, R., *Naturwissenschaften* **68**, 282 (1981).
10. Shepherd, J. C. W., *J. Mol. Evol.* **17**, 94 (1981); and *Proc. Natl. Acad. Sci. USA* **78**, 1596 (1981); and cf. paper this volume.
11. Miller, S., cf. paper this volume.
12. Dress, A., *Adv. Math.* **53**, 321 (1984).

Chemica Scripta 1986, **26B**, 67–72

The Meaning of Selective Advantage in Macromolecular Evolution

Lloyd Demetrius

Max-Planck-Institut für Biophysikalische Chemie, D-3400 Göttingen, Federal Republic of Germany

Paper presented at the Conference on 'Molecular Evolution of Life', Lidingö, Sweden, 8–12 September 1985

Abstract

A population of macromolecular sequences is considered to be described by a dominant sequence together with an array of related sequences generated by mutation. Selective advantage of the population describes the ability of the population to perpetuate itself from generation to generation. This property is determined by two factors:

(*a*) The rate at which population numbers increase – the intrinsic rate of increase or the Malthusian parameter;

(*b*) The rate at which fluctuations in population numbers are damped – the population entropy.

During conditions of exponential growth, the Malthusian parameter determines selective advantage. When growth is stationary, the population entropy predicts the gain in Malthusian parameter due to small perturbations in the replication rates of the individual sequences. Hence during the stationary growth phase, population entropy determines selective advantage.

These two notions of selective advantage are invoked in the study of two phenomena in macromolecular populations.

(*a*) The effect of environmental variability on the sequence length of primers in the primed synthesis of DNA homopolymers. A study of these effects is relevant in understanding the transition from a random assembly of self-replicating macromolecules to a stable replicating population – a central problem in models of prebiotic evolution.

(*b*) The relative effect of selection within the genome and selection at the level of the organism in determining genome size. This study is relevant in understanding the evolution of size of multigene families such as satellite DNA and informational DNA.

Introduction

Genome size in viruses and DNA content in cells of organisms show considerable variability. RNA viruses have an average genome size of $3\text{–}4 \times 10^6$ Da. These sizes range from 1.1×10^6 Da in the case of $Q\beta$ phage to 20×10^6 Da for the Reoviruses. Moreover in many groups of RNA viruses, the genetic information is dispersed among separate molecules. These components also exhibit considerable size asymmetry. In Reoviruses, consisting of 10 modules, the size of the modules range from 0.61×10^6 to 2.5×10^6 Da [1].

The DNA content in groups of free living organisms also span a large range. The average bacterium has a DNA content of 4.3×10^9 Da. Eukaryotes have approximately 300 times more DNA per haploid cell than a bacterium. Within both prokaryotes and eukaryotes the range is large. In eukaryotes DNA content ranges from 5.4×10^{10} Da for fungi to 6×10^{13} Da for urodels [2].

There is consensus among biologists that genome size plays a major role in the evolutionary process and that the diversity observed among viruses and free living organisms is the result of the principle of natural selection. The concept of selective advantage is one of the central notions underlying this principle. The term selective advantage in its broadest sense refers to the ability of an organism or any self-replicating entity to perpetuate itself.

This article is concerned with the quantitative characterization of this notion and its applicability in studying the evolution of genome size at two levels of genomic organization.

(*a*) The transition from a random assembly of oligomers to a stable replicating population, as observed in unprimed synthesis of DNA homopolymers [3], and RNA polynucleotides [4, 5]. We observe that there is a threshold given in terms of genome size at which this transition occurs.

(*b*) The evolution of multigene families such as ribosomal RNA and histone genes. Genome size in multigene families is due to the relative effect of intragenomic selection and organismal selection. We observe that under certain conditions, the variance in genome size within individuals can be expressed in terms of the individual selective values. These selective values are shown to depend on demographic parameters such as generation time and iteroparity.

1. Selective advantage: individuals and populations

Within a population consisting of a group of macromolecules, cells or organisms, selective advantage of an individual describes the ability of the individual relative to other individuals to contribute to successive generations.

In a population of macromolecules the selective advantage of an *individual* can be simply described as its replication rate. This rate will be determined by certain kinetic properties of the macromolecule such as the rate at which synthesis is initiated and the rate at which the replicase binds to the template.

Errors in replication occur in autocatalytic systems. This implies that a macromolecular population is not a homogeneous unit but consists of a dominant sequence together with an array of related sequences generated by mutations [6, 7]. A phage clone, for example, which has been produced by the infection of a single phage on a bacterial cell, consists of the wild type sequence together with variant sequences derived by mutations. These mutants will in general have replication rates distinct from the wild type. The selective advantage of the *population* can be defined as the ability of the population relative to others to perpetuate itself. This property involves two components:

(i) *The ability of the population to establish itself when resources are unlimited.* When resources are unlimited, growth is exponential. In this case, the ability of the population to perpetuate itself depends on the magnitude of this exponential increase, that is, the Malthusian parameter.

(ii) *The ability of the population to persist when resources are limited.* Under conditions of limited resources, growth is stationary. The ability to persist will depend on the response of the Malthusian parameter to small perturbations in the replication rates of the polynucleotides in the population. The magnitude of this response is determined by the rate at which fluctuations in population numbers are damped, that is, the population entropy [8, 9].

The significance of the Malthusian parameter as a measure of selective advantage in macromolecular populations has been considerably analysed and developed in the theory described by Eigen [7] and elaborated in Eigen and Schuster [10]. The role of the entropy concept in understanding the persistence of macromolecular populations under constant and randomly varying environmental conditions was discussed in [8] and subsequently developed in [9, 11, 12]. In the next two sections we review the properties of these two concepts and their significance as measures of selective advantage under constant and randomly varying environmental conditions.

2. Population dynamics in constant environments

Experimental studies of macromolecular replication have shown that the population dynamics can be described by a few kinetic parameters:

(1) The rate of replication of the individual sequences, that is, the rate at which the polynucleotide is synthesized.

(2) The rate at which the polynucleotide is degraded.

(3) The mutation rate or the probability that replication results in an exact copy of the polynucleotide.

When environmental conditions are constant, the rates will be fixed numbers. A population which consists initially of wildtype sequences will generate new types on account of the errors in replication. The proportion of wild type to mutants will vary owing to the differences in replication rates of the two types of sequences. Since the rates are constant, a final state will be reached when the total numbers increase at a uniform rate with the relative proportions of the wild type and mutants remaining invariant. The dynamical equations describing this process in discrete time have been studied in [8] and are given by

$$\bar{x}(t+1) = A\bar{x}(t) \tag{2.0}$$

Here $\bar{x}(t)$ is a vector whose elements $\{x_i(t)\}$ represent the concentration of the polynucleotide I_i at time t. The matrix A has elements w_{ij}, where

$$w_{ii} = A_i Q_{ii} - D_i \tag{2.1}$$

$$w_{ij} = A_j Q_{ij} \tag{2.2}$$

Here A_i is the rate of polymer synthesis, D_i the decomposition rate and $Q = (Q_{ij})$ the matrix determining the mutation rates.

The ultimate growth rate and the entropy can be derived by studying the asymptotic properties of the model described by (2.0). The growth rate and the entropy parameters are determined by the replication and mutation rates of the individual sequences [8, 12].

Growth rate as selective advantage

The growth rate r is given implicitly as the unique real root of the characteristic equation of the matrix A. An explicit expression for r can be obtained if we assume that the only mutations that occur are point mutations. We will let q denote the mean single-digit accuracy. Let us assume that correct and erroneous digits are incorporated with equal frequency ($q = \frac{1}{2}$) and furthermore let us suppose that all mutants have the same replication rate. When these conditions hold, the growth rate r becomes [9, 12]

$$r = \log\left(\frac{\sigma - 1 + 2^{\nu}}{2^{\nu}}\right) \tag{2.3}$$

Here σ, called the superiority parameter, represents the ratio of the replication rate of the wild-type sequence to that of the mutants, and ν the number of nucleotides in a given sequence.

The graph given by Fig. 1 shows how r varies with the sequence length ν for different values of σ. Although this graph has been derived for the case $q = \frac{1}{2}$, its general property – a decrease of r with ν – is known to be valid for all values of the accuracy rate q.

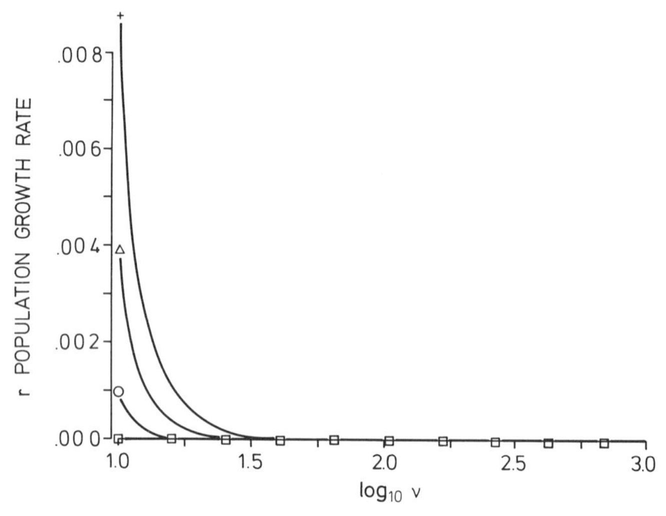

Fig. 1. The relation between growth rate r and the logarithm of the sequence length ν for different values of the superiority parameter σ. +, $\sigma = 10$; △, $\sigma = 5$; ○, $\sigma = 2$; □, $\sigma = 1$.

This model indicates that growth rate decreases with sequence length. This implies that under conditions in which selective advantage is determined by the growth rate, shorter sequences which confer a higher r will have a competitive advantage over longer sequences which have a lower r.

Population entropy as selective advantage

The entropy denoted by H measures the rate at which the fluctuations in the population numbers are damped. For populations in the stationary growth phase, the gain in the Malthusian parameter due to small perturbations in the replication and mutation rates is determined by H [9].

The entropy parameter H classifies the population according to two properties.

(1) The frequency with which mutations back to the wildtype or dominant sequence occurs.

(2) The variability in replication rates of the wildtype and mutant.

A simple expression for H, when the mean single-digit accuracy $q = \frac{1}{2}$, is given by

$$H = -\left(\frac{\sigma}{\sigma - 1 + 2^{\nu}}\right)\log\sigma + \log(\sigma - 1 + 2^{\nu}) \qquad (2.4)$$

In this case the maximum value of H is achieved when $\sigma = 1$, that is, when the replication rates of the wild type and mutants coincide. In this case, H is given by $H = \nu \log 2$.

The graph given by Fig. 2 shows how H varies with ν for different values of σ. This graph, as in the graph for the growth rate r, was derived for the value $q = \frac{1}{2}$. However, its general property – an increase of H with ν – is known valid for all values of the mutation rate q.

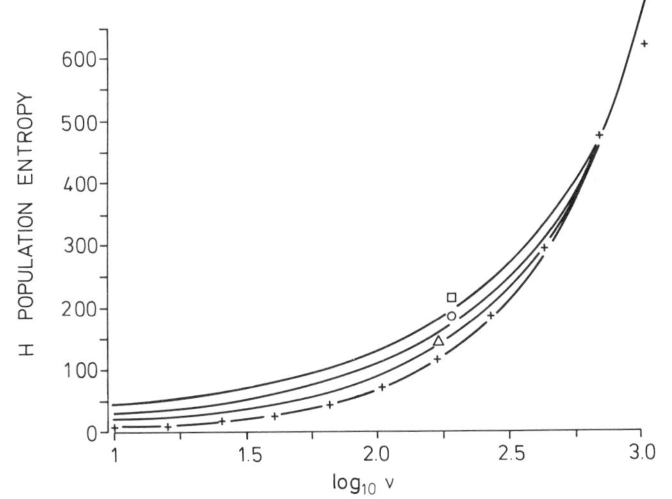

Fig. 2. The relation between the population entropy H and the logarithm of the sequence length ν for different values of the superiority parameter σ. See Fig. (1) for interpretation of the symbols +, ○, △, □.

This model shows that entropy increases with sequence length. This implies that under conditions in which selective advantage is determined by population entropy, the longer sequences which confer a higher H will have a competitive advantage over the shorter sequences which have a lower H.

3. Population dynamics in random environments

In randomly varying environments, the kinetic parameters describing replication and mutation rates of the polynucleotide sequences will change randomly over time. On account of these changes, the growth rate is now a random quantity which depends in general on the initial distribution of the wild type and mutant sequences [11, 13]. The dynamics of population in constant environments is quite distinct from that in randomly varying environments.

Under constant environmental conditions, repetition of an experiment using different distributions of wild type and mutants will always result in identical growth rates and the same final distribution of wild-type and mutant sequences.

Under randomly varying environmental conditions, repetition of an experiment with different initial distributions of wildtype and mutant sequences will in general result in different cumulative growth rates characterized by different distributions.

A dynamical model describing evolution under randomly varying environmental conditions has been studied in [13] and is described by

$$\bar{x}(t+1) = A(t)A(t-1)\ldots A(1)\bar{x}(0) \qquad (3.0)$$

In this model, in contrast to (2.0), the matrices $A(t)$ are random and the elements $\{w_{ij}(t)\}$ describing the replication and mutation rates are random variables.

The growth rate r, a random variable is given by

$$r = \lim_{t\to\infty} \frac{1}{t} \log \|\bar{x}(t)\| \qquad (3.1)$$

There exists in general no explicit expression for the mean growth rate given by (3.1).

The entropy parameter for the model (3.0), involves both properties of the population of macromolecules and properties of the environment. The number H represents the correlation between the variability in the environment and the variability in replication and mutation rates of the polynucleotides. There exists no general explicit expression for H. However, as shown in [13], the predictive value of H described for the constant environment model, is also valid for the random environment case: under conditions of mean stationary growth, H determines the gain in mean growth rate due to small perturbations in the replication and mutation rates of the polynucleotides in the population. This property implies as in the constant environment case, that whereas r determines selective advantage in the exponential growth phase, the entropy H determines selective advantage in the stationary growth phase.

4. The consequences of natural selection

Two modes of selection can be distinguished. The first mode is characterized by the case when growth rate determines selective advantage (r-selection), the second mode is characterized by the period in which entropy determines selective advantage (H-selection). The occurrence of these modes will be determined by the amount of resources which are available to the population.

The following table summarizes the effect of the resources on the mode of selection.

MODES OF SELECTION

Resources	Constant environments	Random environments
Unlimited	r-selection	r-selection
Limited	H-selection	H-selection

In the case of viruses and free-living organisms, the first phase in the evolutionary process will be described by the case where resources are unlimited. For example, during the initial phase of bacterial infection by phages, the resources for viral replication are in unlimited supply. Replication will be exponential and clones with higher growth rates will outcompete those with lower growth rates. The mode of selection in this period is r-selection. As the number of phage particles increases, the metabolites which are necessary for replication become rare. Resources can now be considered limited. Success in replication will be determined by populations with the higher entropy H. The process described will operate in both constant and randomly varying environments. Environmental factors, both in terms of the availability of the resources and in terms of the randomness of its variation thus exert a critical role in the evolutionary process.

5. Macromolecular self-organization

One of the central issues concerning the problem of the origin of life revolves around understanding the transition from a random assembly of self-replicating oligomers to a stable replicating population. Certain experimental studies [3–5, 14], although involving enzymes which are much too complex to be considered prebiotic elements, do reflect certain essential features of this transition. Kornberg *et al.* [3] have shown that in the presence of dATP, dTTP and DNA polymerase, long sequences of homopolymers are generated after a certain lag period. Krakow and Karstadt [14] have described the production of homopolymers in unprimed reactions with RNA polymerase. The series of experiments described by Biebricher, Eigen and Luce [5] using $Q\beta$ replicase, give rise to complex polynucleotides in template-free synthesis. In the experiments described in [3] and [14] it was also noted that when reactions are primed, the transitions to stable polynucleotide populations occurred in the absence of a lag period, only when sequences of a certain minimum length was used.

Mathematical models describing the transition from a random assembly of oligomers to a stable set of replicating polymers have been developed in [11]. These models which were introduced in order to understand certain aspects of prebiotic evolution can be described in terms of the formalism of the random environment models given in (3.0). These models of self-organization postulate that the transition to a stable replicating population is a cooperative effect determined by the interaction between the individual oligomers. We consider oligomers of length ν. This implies that there are 4^ν possible sequences. The interaction $w_{ij}(t)$ at time t between the sequences I_i and I_j is characterized by a random variable with mean zero and variance s^2. This randomness is assumed to be due to the small number of reactants in the population. The cooperative behavior is represented in terms of products of random matrices of dimension 4^ν. Stability of the population is characterized by the largest Liapunov exponent given by the number r as in (3.1). The number r can be expressed in terms of the parameters ν and s^2. Stability of the population and hence the transition to a stable self-replicating population is described by the condition $r > 0$. In [15], we invoke this condition, to show that the minimal sequence length for which the transition to a stable self-replicating population occurs is inversely related to the variance s^2.

Lower bounds on the size of primers have been observed in experiments involving primed synthesis of DNA homopolymers [3]. This minimum length was determined by temperature. For example at 10 °C, the sequence is $(AT)_4$, at 20 °C $(AT)_5$, at 37 °C $(AT)_6$. The connection between these experimental results and the theoretical predictions is discussed in [15].

6. The complication of phenotype

In order to understand the effect of selective forces on genome size and frequency in eukaryotic organisms, we distinguish between two phenotypic characteristics.

(1) *The intragenomic phenotype*. This determines the efficiency with which the sequence spreads within the genome. The mechanisms by which this occurs are transposition and duplication of sequences. These mechanisms are known to amplify and disperse segments of the DNA throughout the chromosome. Selective advantage of a population of sequences will be determined by the replication and mutation rates of the sequences that constitute the genome:

(2) *The organismal phenotype*. This determines the reproduction and survivorship of individuals in the population who possess repeated copies of the nucleotide sequence.

At the level of the organism, we distinguish also between selection at the level of the individual and selection at the level of the population.

At the individual level selective advantage can be simply measured by the number of offspring the individual produces. At the population level, we need to consider the age-dependence in reproduction and mortality of the individuals in the population. Let $m(x)$ denote the age-specific fecundity and $l(x)$, the probability that an individual born at age zero survives to age x. Under constant environmental conditions, the growth rate and entropy can be explicitly expressed in terms of these parameters [16]. The growth rate r is the unique real root of the equation

$$\int_0^\infty \exp(-rx)\, l(x)\, m(x)\, \mathrm{d}x = 1 \tag{6.0}$$

This parameter describes the rate of increase of population numbers when the stable age-distribution is attained. This parameter determines selective advantage during the exponential phase of growth.

The entropy H is given by

$$H = -\frac{\int_0^\infty p(x) \log p(x)\, \mathrm{d}x}{\int_0^\infty x p(x)\, \mathrm{d}x} \tag{6.1}$$

Here $p(x) = \exp(-rx)\, l(x)\, m(x)$.

The entropy H describes the degree of iteroparity of the population. H is zero in the case of annual plants which reproduce once in their life-time then die. H is positive for perennial plants whose reproductive activity is spread over several years.

When growth rate is stationary, H measures the gain in the Malthusian parameter due to small perturbations in the survivorship $l(x)$ and the fecundity $m(x)$ [9]. Thus H is a measure of selective advantage during the stationary growth phase.

The relationship between the measures of selective advantage at the level of the genome and at the level of the population are crucial in understanding the evolution of genome size in multigene families.

The multigene family

The multigene family encompasses a large family of tandemly repeated nucleotides that have arisen by gene duplication. This family whose evolution has been extensively studied in [17, 18] can be characterized by the properties of sequence homology and overlapping phenotypic functions.

The notions of selective advantage can be invoked to give an account of the evolution of genome size and variation in size in these families. We distinguish between satellite DNA families and the informational and multiplicational families.

Satellite DNA

Satellite DNA can be considered as a tandem array consisting of a predominant sequence together with the mutants which have been derived from it by substitution, deletions and insertions.

As this family does not seem code for proteins, the sequences can be considered to be mainly under selective constraints defined by the genome. Thus evolutionary change is determined primarily by the intragenomic phenotype.

The selective advantage of satellite DNA can be computed using a model analogous to that which was used to describe selective advantage of the macromolecular populations in Section 2. Certain simplifications are imposed by the properties of the satellite DNA. For example, as the satellite DNA appears to serve whatever function it has by its presence instead of by virtue of its exact sequence, we can assume that the mutant sequences have the same replication rate. The growth rate of the population which measures the overall rate at which a family consisting of the tandem array of sequences increases in size will be as in (2.3) inversely related to the sequence length.

The entropy H measures the heterogeneity in the family. This, under certain conditions, is given by (2.4). If there is no mutation and the family consists of identical short sequences repeated in tandem, $H = 0$. If the predominant short sequence accounts for a small minority of the copies and the other short sequences are related to the predominant sequence by mutation, then H is positive. Arthropods can be considered as belonging to the class $H = 0$, and Mammals to the class $H > 0$.

We may consider as in Section 2, the case where the only mutations are point mutations, given by $q = \frac{1}{2}$, and the mutants sequence have the same duplication rate as the dominant sequence. In this case, for sequences of length ν, the adaptive value H is given by

$$H = \nu \log 2 \tag{6.2}$$

These models predict that under conditions in which perpetuation to the next generation is determined by growth rate (r-selection), short sequences will be favored; whereas, when persistence is determined by the entropy of the genome (H-selection), longer sequences will be favored. These models assume that duplication and replication are the only mechanisms causing changes in genome size.

A more complete analysis of the change of genome size in satellite DNA would involve integrating these notions of selective advantage with the effect of mechanisms such as interchromosomal crossing over and random drift studied in [19, 20].

Informational and multiplicational gene family

The genes belonging to this family have important phenotypic properties. Hence, evolution of these genes will be determined primarily by the organismal phenotype. A model which integrates the notion of selective advantage measured by the demographic parameters given by (6.0) and (6.1), with a model based on evolution of gene size by stabilizing selection [19, 20] enables one to derive an expression giving the variance in the distribution of genome size. Stabilizing selection refers to the case where the selective advantage is a maximum for an intermediate family size and decreases with the square of the deviation from the optimum value. In this model the optimum selective advantage is assumed to be given by the maximum value of H. When the generation time T is fixed, this maximum denoted H^* is

$$H^* = \frac{1+\log T}{T} \tag{6.3}$$

Crossing over between types generates heterogeneity in genome size. The force of selection is assumed to act by eliminating the more extreme sequences.

The distribution in family size is known to be normally distributed when the difference of the family size between the new recombinant family and the parental one is small [21]. The variance V in genome size is then given by [21]

$$V = \sqrt{\frac{\gamma m^2}{2s}} \tag{6.4}$$

Here m is the mean number of genes expanded or deleted and γ the rate of unequal intrachromosomal crossing over and s the selective coefficient.

When the selective advantage is determined by the adaptive value H, a simple analysis yields

$$s = \frac{1}{H^*(H^* - H)} \tag{6.5}$$

where H^* is the maximum adaptive value and H the adaptive value of the population of individuals.

Thus V can be expressed in terms of the generation time and the degree of iteroparity H. This model makes explicit two demographic factors that determine size heterogeneity – the mean generation time and the degree of iteroparity or variability in age at which individuals are produced.

Conclusion

The perpetuation of a population from one generation to the next requires both the properties of growth and adaptation. These two processes are necessary prerequisites for evolution at all levels of organization. This article has discussed how these properties of growth and adaptation can be measured in the context of macromolecular evolution. The capacity of growth is measured by the Malthusian parameter. The capacity of adaptation is described by a parameter, the population entropy, which classifies populations of macromolecules according to the complexity of the population. In constant environments, this complexity parameter describes the variability in replication and mutation rates of the polynucleotide sequences in the population. In variable environments, this complexity parameter measures the correlation between the variability in the environment and variability in the replication and mutation rates.

The Malthusian parameter determines the outcome of selection during the exponential growth phase, that is, when resources are unlimited. The entropy determines the outcome of selection during the stationary phase of population growth, for example when resources are limited.

Acknowledgements

I would like to thank Christof Biebricher, Manfred Eigen and Peter Schuster for valuable discussions.

References

1. Matthews, R. E. F., *Intervirology* **12**, 129 (1979).
2. Hinegardner, R., in *Molecular Evolution* (ed. F. Ayala). Sinauer Associates, Sunderland, Massachusetts (1976).
3. Kornberg, A., Bertsch, L., Jackson, J. and Khorana, H. G., *Biochemistry* **51**, 351 (1964).
4. Sumper, M. and Luce, R., *Proc. Natl. Acad. Sci. USA* **72**, 162 (1975).
5. Biebricher, C., Eigen, M. and Luce, R., *J. Mol. Biol.* **148**, 369 (1981).
6. Domingo, E., Flavell, R. A. and Weismann, C., *Gene* **1**, 3 (1976).
7. Eigen, M., *Naturwissenschaften* **58**, 465 (1971).
8. Demetrius, L., *J. Theor. Biol.* **103**, 619 (1983).
9. Demetrius, L., *Proc. R. Soc. Lond.* B**225**, 147 (1985).
10. Eigen, M. and Schuster, P., *The Hypercycle*. Springer Verlag, Berlin, Heidelberg, New York (1979).
11. Demetrius, L., *Proc. Natl. Acad. Sci. USA* **81**, 6068 (1984).
12. Demetrius, L., Schuster, P. and Sigmund, K., *Bull. Math. Biology* **47**, no. 2, 239 (1985).
13. Demetrius, L., *J. Chem. Phys.* (submitted) (1986).
14. Krakow, J. S. and Karstadt, M., *Proc. Natl. Acad. Sci. USA* **75**, 5334 (1967).
15. Demetrius, L., *Proc. Natl. Acad. Sci. USA* (submitted) (1986).
16. Demetrius, L., *Proc. Natl. Acad. Sci. USA* **71**, 4745 (1974).
17. Hood, L., Campbell, J. H., Elgin, S. C . R., *Ann. Rev. Genetics* **9**, 305 (1975).
18. Hood, J. M., Huang, H. V. and Hood, L., *J. Mol. Evol.* **15**, 181 (1980).
19. Ohta, T. and Kimura, M., *Proc. Natl. Acad. Sci. USA* **78**, no. 2, 1129 (1981).
20. Ohta, T., *Nature* **292**, 648 (1981).
21. Crow, J. F. and Kimura, M., *An Introduction to Population Genetics Theory*. Harper and Row, New York, Evanston, London (1970).

Nucleic Acids and Informational Systems

Chemica Scripta 1986, 26B, 75–83

Origins of Life and Molecular Evolution of Present-day Genes

John C. W. Shepherd

Biocentre of the University of Basel, Klingelbergstrasse 70, CH-4056 Basel, Switzerland

Paper presented at the Conference on 'Molecular Evolution of Life', Lidingö, Sweden, 8–12 September 1985

Abstract

Purine–pyrimidine correlations from present-day DNA sequences support the hypothesis that a comma-less coding system (i.e. only readable in one frame) was used in the earliest self-replicating systems as life originated on earth. The indications are that the primeval coding triplets had the form RNY (R = purine, Y = pyrimidine and N = purine or pyrimidine), providing codons for a maximum of eight amino acids on the basis of the present genetic code. Corroborative evidence for this hypothesis is obtained by examining the mutated relics of such messages in present genomes. These relics are found almost universally, but most clearly seen in genes which are known to have been stable over long periods of time, and in other genes, particularly those coding for plentiful proteins, which have been long subject to restrictions preserving some features of their primeval codon usage. With the start point provided by the hypothesis, indicative simulations can be made of possible paths for subsequent mutation over the whole period of evolutionary time. Such simulations suggest a period of considerable functional improvement in which the majority of the mutations occurring are accepted, followed by one in which most mutations are rejected and only minor further improvements can be made.

Introduction

A study of present life forms and of the fossil record enabled Darwin to formulate his theory of evolution by natural selection. Lacking all the present knowledge of molecular biology, however, he could not have any definite ideas of how the first self-replicating systems and later the first cellular organisms could have originated. No theories had yet been formulated for the driving force and interactions within such processes and he spoke of 'life with its several powers, having been originally breathed into a few forms or into one' [1].

Today, in relation to these problems, we are in a much stronger position in all respects. The physical and chemical conditions on the primitive earth at this time, some 4000 million years ago, can now be better pictured [2]. We have considerable knowledge of the molecular mechanisms involved in present life processes, and, by extrapolation into the past, hypotheses can be made for the varied development along the branches of the evolutionary tree. For times prior to the earliest known points of divergence between present life forms, principles for life's origin and early organization have also been proposed [3].

Much of the earlier molecular information about the past came from a study of protein sequences [4], but in recent years a rapidly increasing treasure house of information is the large number of DNA sequences which have now been determined (and some RNA genomes). Besides yielding a maximum of information about present function and the not-too-distant evolutionary past, it is now suggested that DNA may yet provide our best source of information about the long distant past, dating back to the early development of the genetic code.

Primeval code

As the first longer DNA sequences were published some seven years ago, an analysis showed the presence of strong periodical positional correlations of purines (R) and pyrimidines (Y) with a period of three [5]. The phases and amplitudes, when examined in detail, strongly suggested that these Y,R patterns were remnants of primeval messages having codons of the form RNY (N is any base). They could be detected mainly in the coding regions of present genes but sometimes in non-coding regions.

A method was then developed [6] for examining present-day sequences to see where the relics of the old messages could best be found and in which reading frame. The DNA was divided into short successive lengths and for each length the reading frame was determined that gave the least mutation ($Y \rightleftharpoons R$) away from such an RNY message. This 'RNY search' method showed that the majority of genes examined, when considered over their whole length, are read in the RNY frame.

It was then realized that RNY was a purine–pyrimidine form of comma-less code, a concept first introduced by Crick, Griffith and Orgel in 1957 [7]. In this type of code, the codons in the reading frame ('sense' codons) cannot be found anywhere in the other two frames ('non-sense' codons). No start signal would then be required to indicate the correct reading frame and the 'sense' anticodons could not block translation by attaching themselves out of frame to the message. Coding for a maximum of twenty amino acids was shown possible if each position in the message can be assigned to any of the four bases. The actual genetic code did not prove to be of this form, but in 1976 Crick, Brenner, Klug and Pieczenik [8] gave grounds for the belief that the present code could have been derived from the former use of a simpler comma-less code, depending on a regular purine–pyrimidine pattern in the messages. They considered a central problem in the emergence of life on the primeval earth, namely the way in which the synthesis of polypeptides directed by a nucleic acid template could have started. Basing their arguments on a proposed simple transfer RNA with two possible conformations [9, 10] and the sequence regularities in the anticodon loops of present tRNA's, they deduced a likely primeval message of the form RRY with RYY as a less-likely alternative. Eigen and Schuster in 1978 [11] further considered the problem and favoured an RNY message on account of better balance between purines and pyrimidines and more symmetry between plus and minus strands. Assuming that the present genetic code then applied, it was also noted as a great advantage for this message that the majority of the eight amino acids for which it coded were likely to have been abundant in prebiotic synthesis [12] and are frequently found in meteorites [13]. Thus, on quite independent grounds, an RNY primeval message has also been proposed.

Subsequent to the above DNA analysis, Eigen and Winkler-Oswatitsch [14] applied a similar purine–pyrimidine analysis to an ancestral tRNA sequence derived by phylogenetic analysis from about 200 known tRNA sequences [15]. Although the statistical significance was less than in the longer lengths of DNA, their results indicated here also the presence of a primeval RNY message. Thus, the early translation process could have been helped by the primitive tRNA having a dual role as adaptor and message, assisted by the complementarity between plus and minus strands.

DNA evidence

Many reasons for believing that the DNA purine–pyrimidine patterns are relics of the remote evolutionary past, rather than the results of some process of natural selection, have been given elsewhere [5, 6, 16, 17]. Some of the main points are further discussed or summarized here.

Displaced patterns

Shorter regions of displaced pattern can be detected within genes from the sharp fall of rhythmic correlation amplitude as the separation of the correlated bases increases (relatively short separations considered). These regions can also be seen clearly by means of the RNY message search.

Simple tests confirm this conclusion. Randomly mixing the codons in genes containing regions of displaced pattern reduces the amplitude of correlations at short separations. Many 'out-of-phase' codons from the displaced region have now been mixed amongst the 'in-phase' codons, and vice versa.

Amino acid proportions

As a further test, an average mixture of amino acids (as given by Dayhoff [4]) for a typical protein can be taken and reverse translated into DNA, with random choice of the degenerate codons. The above effects due to displaced regions are not seen in this DNA. The conclusion is reached that the Dayhoff proportions of amino acids in present-day proteins are a natural consequence of a mutated RNY message. Simulations of mutations away from typical RNY genes confirm this conclusion and, moreover, reproduce all the features of the present Y,R patterns in the genes, allowing for some displacements by insertion or deletion within the gene.

Observations on the mutation away from RNY in degenerate codons (see below) also preclude the RNY patterns being simply due to the amino acid proportions of the protein.

Date of last use of the comma-less code

Although at first sight it might seem unreasonable to expect that traces of such primeval messages can still be seen, yet mutation simulations show that it is indeed possible if known mutation rates averaged over long periods are used. An estimate of the time of last use of the comma-less code has also been made from the time of divergence of related genes and very roughly appears to be some 3000 million years ago. Accepted mutations ($Y \rightleftharpoons R$) were estimated in various gene pairs since their time of divergence and since their derivation from RNY messages, and compared [16].

Proportions of codon types

A further simulation (to be described below) shows that if random mutation is made forward from an RNY message, the proportions of RNY codons, of the once mutated (transversions considered) forms YNY and RNR, and of the twice mutated YNR, fall for many present-day genes on the simulated curves. This phenomenon is difficult to explain other than by the RNY hypothesis.

RNY message search

Now that so many gene sequences are known, further convincing evidence comes from observing where the best preserved relics of the RNY message are found. If the hypothesis is believable, they should be found in genes which are known to have stabilized evolutionarily early near their present sequence and to have had low mutation rates since then, and this is generally the case. Genes which it is reasonable to suppose have first come into use much later, for example, the homeotic genes (such as those in *Drosophila*), controlling multi-cellular development, do not show such clear RNY messages and give the impression of more mutation away from RNY and more displacements from the RNY reading frame.

A few examples to illustrate the RNY message, as found in some genes thought likely to have had an ancient origin, are shown in Figs. 1–8. With the previously described method [6, 16], the DNA is examined in lengths, L, of 120 bases, moving forward in steps, S, of 15 or 30 bases. In the upper part of the figures, the percentage of transversions $m\%$ away from an RNY message in first and third positions is plotted for each reading frame against the base number b. Below these plots the reading frame which has the least mutations away from RNY is shown, recorded for each length at its mid-point. The horizontal lines in the upper plot give the limits of one standard deviation above and below the mean mutations, when the same method is applied to random sequences. Mutation counts near or within these lines indicate non-coding DNA regions or very mutated messages. 16S and 23S rRNA of *E. coli* have previously been tested and found to have this random nature [16].

Figures 1–8 show RNY search plots for: (1) yeast (*S. pombe*), alcohol dehydrogenase ADH [18]; (2) *E. coli*, major outer membrane protein OMPF [19]; (3) the nuclear gene for yeast (*S. cerevisiae*), mitochondrial cytochrome C oxidase sub-unit 4 [20]; (4) yeast (*S. pombe*), cytochrome C and flanks [21]; (5) the rat cytochrome C gene showing the two exons [22]; (6) rat alpha tubulin gene showing exons 2, 3 and 4 [23]; (7) amoeba (*A. castellani*), actin gene I with two exons [24]; (8) yeast (*S. cerevisiae*), actin gene with two exons [25, 26].

Considering the continuously coded genes in Figs. 1–4, the reading frame is correctly predicted for each gene and its extent shows up as a region of low mutation away from RNY against a background of sequences not carrying the message. Unexpected detail is often shown in such plots. For example, in the case of cytochrome C oxidase in Fig. 3, there is a good

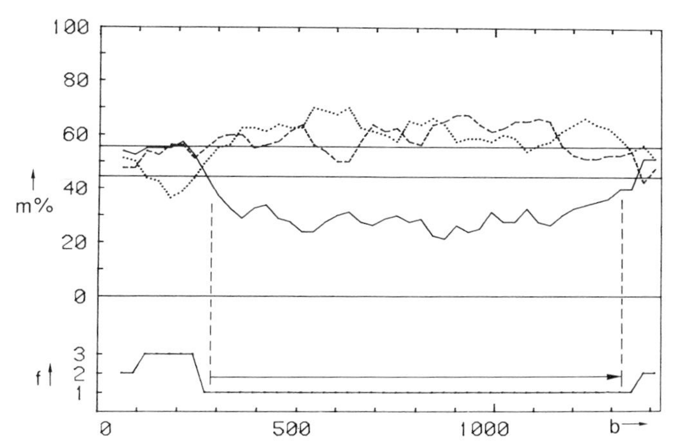

Fig. 1. Search for the primeval RNY message in yeast ADH. The transversion counts ($m\%$) away from an RNY message in first and third positions of the codons is plotted against b, the sequence base number of the mid-point of the test length L. Frame 1 values: full-line graph; frame 2 values: dashed graph; frame 3 values: dotted graph. (The horizontal lines show the limits of one standard deviation above and below the mean mutations when random sequences are tested by this method.) In the lower part of the figure, the reading frame f, having the least mutations $m\%$ away from RNY, is plotted at the mid-point of each length L, and the extent of the actual gene is indicated by an arrowed line alongside, showing that its reading frame agrees with that of the RNY message. For this gene, L (120 bases) is moved forwards in steps, S (30 bases).

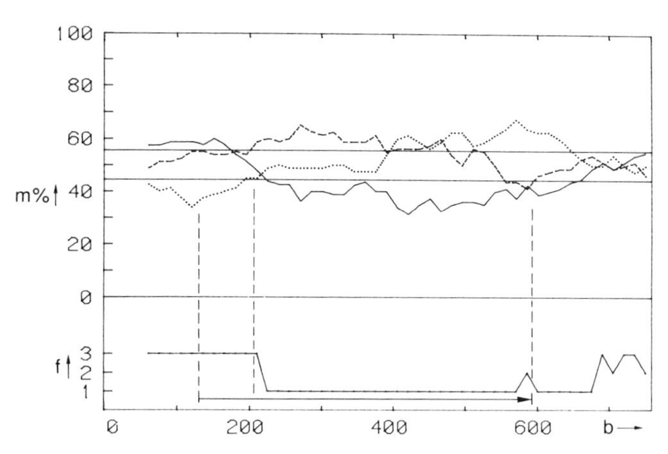

Fig. 3. RNY search plot for cytochrome C oxidase sub-unit 4 ($L = 120$, $S = 15$). The vertical dashed line at base 204 corresponds to the end of the signal peptide. (The change in reading frame f nearest to RNY occurring here may indicate an earlier frame shift in the pre-sequence or the addition of an RNY message from a different origin.)

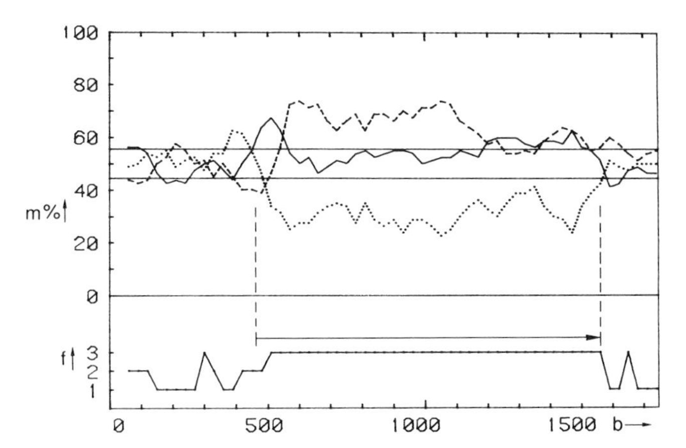

Fig. 2. RNY search plot for *E. coli* OMPF ($L = 120$, $S = 30$).

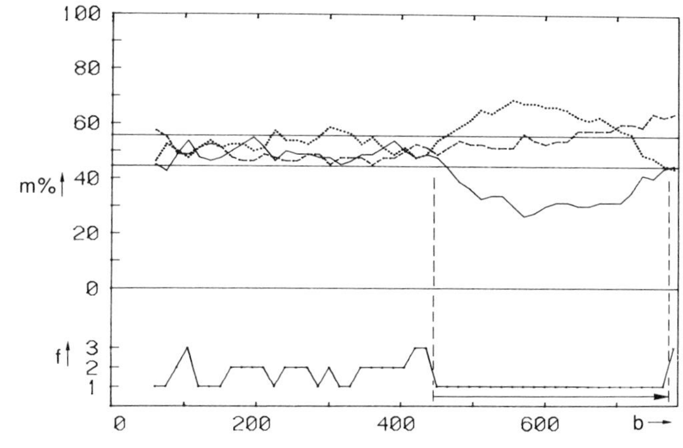

Fig. 4. RNY search plot for yeast cytochrome C ($L = 120$, $S = 15$).

indication of the extent of the signal peptide extending from base 130–204, and for *E. coli* OMPF there is some indication of a signal peptide, but not agreeing so well with its actual extent 456–521. Other such details, for example the detection of overlapping genes or the shortening of a gene by a stop signal in fairly recent times, are convincing evidence that events in the past are often being seen when tracing the RNY message.

In the four examples of interrupted genes in Figs. 4–8, the reading frame for each exon is correctly predicted (apart from the 10 base first exon of yeast actin in Fig. 8), and the exons show up as regions of low mutation from RNY. These regions, however, sometimes extend beyond the known exon limits (also previously noted for other exons and discussed [16]). For example, the region covering the first exon extends into the intron in Fig. 5. The method is capable of showing up a gene coded within an intron, for example the maturase in one intron of the yeast mitochondrial gene for cytochrome b [27] can be seen.

Many other genes are more mutated away from RNY than these eight examples. Many also have within them the above mentioned regions of displaced patterns and, in some of these, it appears that selection may have taken advantage of a mutational shift by a single base deletion to obtain a hydrophobic region in the protein, since NYR codes for more hydrophobic residues than RNY. Well expressed genes generally show the message more clearly than those with less expression, but there are exceptions. A varied selection of genes demonstrating these points has been previously analyzed and discussed [6, 16, 17]. Taken over their whole length most genes are still read in the frame nearest an RNY message, and this applies in all types of genomes at present investigated.

RNY in degenerate codons

A striking effect in many genes is that the RNY message is less mutated in the degenerate codons than elsewhere. Two main reasons for this effect can be conceived:

(1) Conservation of RNY in the whole of evolutionary time before the gene becomes specialized for its function;

(2) Selection pressure exerted by the populations of tRNA's.

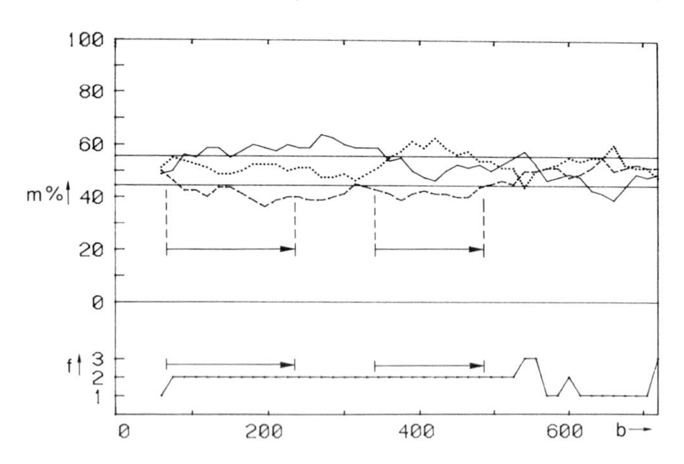

Fig. 5. RNY search plot for rat cytochrome C ($L = 120$, $S = 15$). The extent of the 2 exons is shown and the actual exon reading frames agree with the predicted frames *f*.

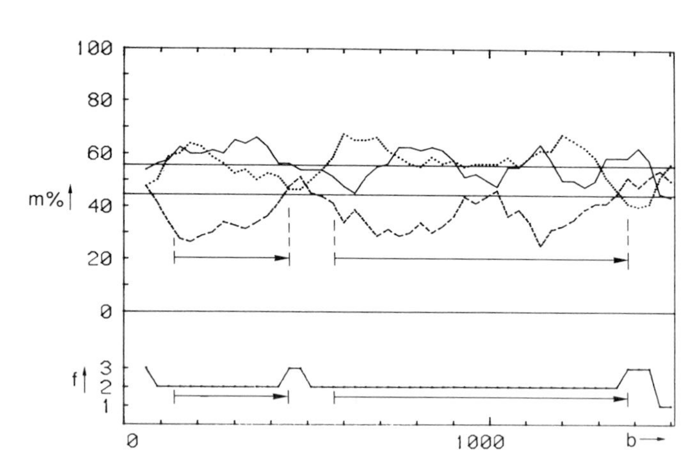

Fig. 7. RNY search plot for amoeba actin ($L = 120$, $S = 30$), showing the two exons.

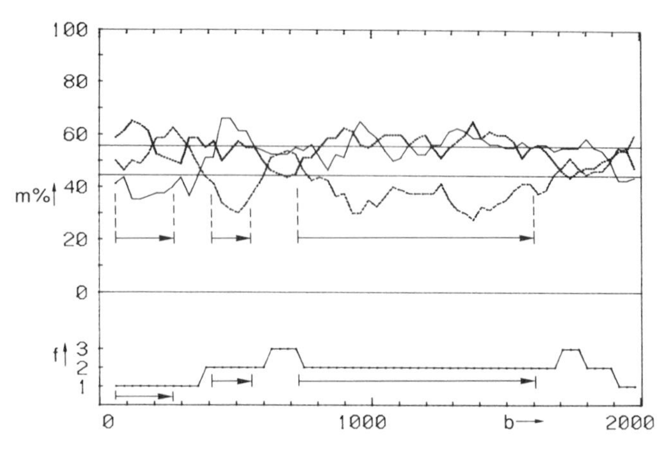

Fig. 6. RNY search plot for rat alpha tubulin ($L = 120$, $S = 30$), showing exons 2, 3 and 4.

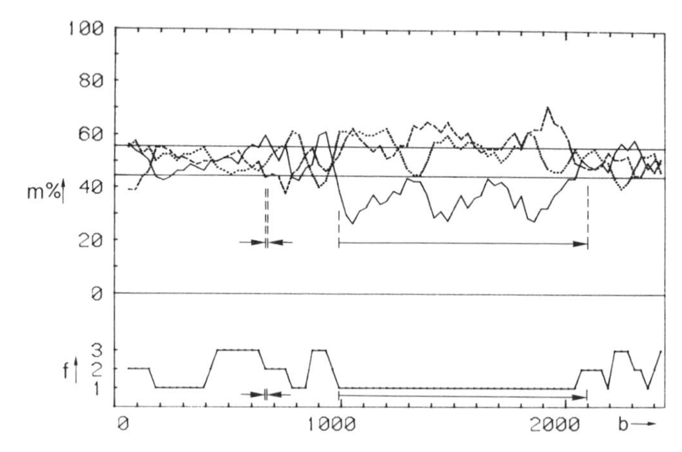

Fig. 8. RNY search plot for yeast actin ($L = 120$, $S = 30$), showing the two exons.

The arguments for (1) have been given previously [5] and the effects seen in the whole ϕX174 viral genome taken as an example. In primeval times, when the coded protein was improving by many changes of amino acids, those mutated phenotypes having much more efficient proteins would dominate in the population and be accepted. Changes from RNY to RNR giving no change of amino acid would have little chance of surviving. However, once the gene is almost specialized for its purpose, most mutations giving a new amino acid are rejected (particularly in those areas which correspond to vitally stable parts of the protein). There is now a large proportion of degenerate changes observed in the third codon positions, giving little, if any, selective advantage (only transversions need concern us here), and these, in the absence of any more efficient phenotypes, will be accepted (or a distribution of almost neutral alleles may be found). Considered over the whole of evolutionary time since the commaless code was abandoned, this period of stability may be relatively short, and consequently the higher RNY/RNR ratio in degenerate codons remains.

Additional to this selective effect will be superimposed item (2), another selection due to the available populations of tRNA's, especially in genes which are well expressed, and this selection may have been operative during the whole time up to the present-day. Sets of abundant tRNA's may have been always used for the translation of such genes, and any change to a less abundant tRNA species strongly selected against. Such a change would only have been allowed if it produced an amino acid change which increased the efficiency of the protein. Very good examples are found in yeast, one being the yeast alcohol dehydrogenase gene [18]. (Other yeast and *E. coli* examples have already been discussed [16].)

In yeast ADH when the change from RNY to RNR gives no new amino acid, as in the degenerate codons of glycine, alanine, valine and threonine, the RNY count is 125 whereas that of RNR is zero. The isoleucine count for the RNY form is also 21 whereas the single RNR (ATA) codon has zero counts. When RNY and RNR code for different amino acids, the ratio RNY/RNR is 44/57. In this gene a similar effect is seen for the degenerate changes from the once mutated form YNY to the twice mutated form YNR. When such a mutation yields no new amino acid, as for the degenerate codons of serine, leucine, proline and arginine, the YNY/YNR score is 60/0, whereas if a different amino acid is given it is 43/21.

Molecular evolution of genes

The comma-less code hypothesis gives for the first time a starting point from which simulations of gene development can be attempted to show the possible mutational pathways a gene can follow when more refined translational mechanisms allow the message to be read in any frame and more amino acids can be used than the eight for RNY messages.

Random mutation

The simplest possible assumption of random Y⇌R mutation away from an RNY message has been previously made [16] and a typical gene length of 900 considered with equal numbers of RRY and RYY randomly mixed in the primeval gene. All mutations were assumed accepted and the proportions of RNY, the mean of YNY and RNR, and YNR were plotted against the mutations applied. As noted above, the proportions of these types of codons in present-day genes fell on the plotted curves and particularly those genes with known low mutation rates and thought to be of ancient origin were found nearer the start point.

Such a simulation has now been repeated in Fig. 9, using the same methods as before [16], but plotting a different miscellaneous set of genes (only β-globin is a common feature). In order, however, that the simulation could be extended by applying various types of selection pressure, a typical primeval RNY gene expressed in T, C, A, G form was taken of similar length (912 bases), containing equal numbers (randomly mixed) of all the sixteen RNY type codons coding for eight amino acids. These have been assumed to be all used at the time the comma-less code was abandoned. Mutations were randomly applied and all mutations occurring between T, C, A and G recorded. The dotted curves in Fig. 9 were then deduced showing the proportions of RNY, YNY and RNR (mean), and YNR type codons from one such mutational simulation, plotted as before [16]. The full line curves represent the average of many such experiments, as calculated from rate equations for the mutational changes between the four codon types. The actual Y⇌R mutations of the new set of genes away from an RNY message (in first and third codon positions) were noted, corrections for multiple or back mutations applied to deduce the accepted Y⇌R mutations $t\%$, and their positions on the decay curves recorded as shown by the vertical lines. It is found that their codon type proportions again fall near the calculated and simulated curves and that a well preserved protein such as cytochrome C is near the start point.

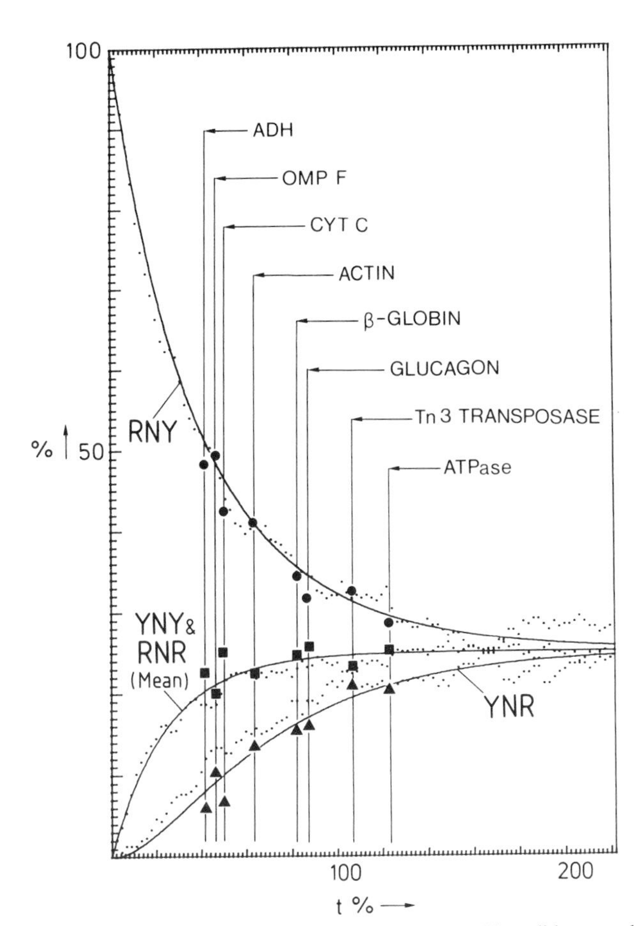

Fig. 9. Random mutation away from an RNY message. The solid curves show the expected percentages (averaged over many simulations) of the codons having the forms RNY, YNY and RNR (mean), and YNR, after $t\%$ transversions (all assumed accepted) have been applied to a primeval message (912 bases long), formed by randomly mixing 19 of each type of the 16 possible RNY codons. The dotted curves represent one simulated run. From their present sequences, the genes are located on the horizontal axis and their codon proportions (RNY, ●; YNY and RNR (mean), ■; YNR, ▲) are compared with the expected values. The genes are yeast ADH [18]; *E. coli* OMPF [19]; yeast (*S. pombe*), cytochrome C [21]; actin from the ciliate *Oxytricha fallax* [28]; rabbit β-globin [29]; anglerfish preproglucagon I [30]; *E. coli* Tn3 transposase [31]; tobacco chloroplast ATPase gene, β and ϵ subunits [32].

Avoidance of stop signals

A simple type of selection pressure was first applied, namely the avoidance of the translational stop signals TAA, TAG and TGA. This was found to make very little difference to the curves at lower accepted mutation ($t\%$) values since there are relatively few YNR type codons of any type here. As t increases, the differences between simulations, allowing and not allowing stops, usually fall within the statistical variations to be expected within a gene of this length. In the further simulations, avoidance of stops will be included. Being a small correction, however, it is not important for the simulation to know exactly when these stops came into use. (For some time after the comma-less code was abandoned any unrecognized codon may have functioned as a stop signal until all the present amino acids and the full present genetic code applied.)

Selection for a particular function

The results shown in Fig. 9 and previously [16] can be partly understood even though no simulation of selection pressure has yet been made. Mutations have been randomly applied along the whole length of the gene, but in different regions selection may have accepted or rejected quite different codons according to the local structural requirements of the protein or other functional necessities. Averaged over the whole gene, the effect might still appear random. This, however, does not seem to be a satisfactory explanation for selection pressure acting throughout the gene to use some types of codons and avoid others (for example as already described in yeast ADH and elsewhere [16]).

An attempt to see the effect of such a pressure was therefore made, with the full realization that any such effort must be highly speculative, since so many factors are unknown. A distinction will now be made between the total applied mutations $T\%$, the number $t\%$ spreading through and accepted by the population, and the actual mutations $m\%$, the number of base differences recorded between a gene at any given time

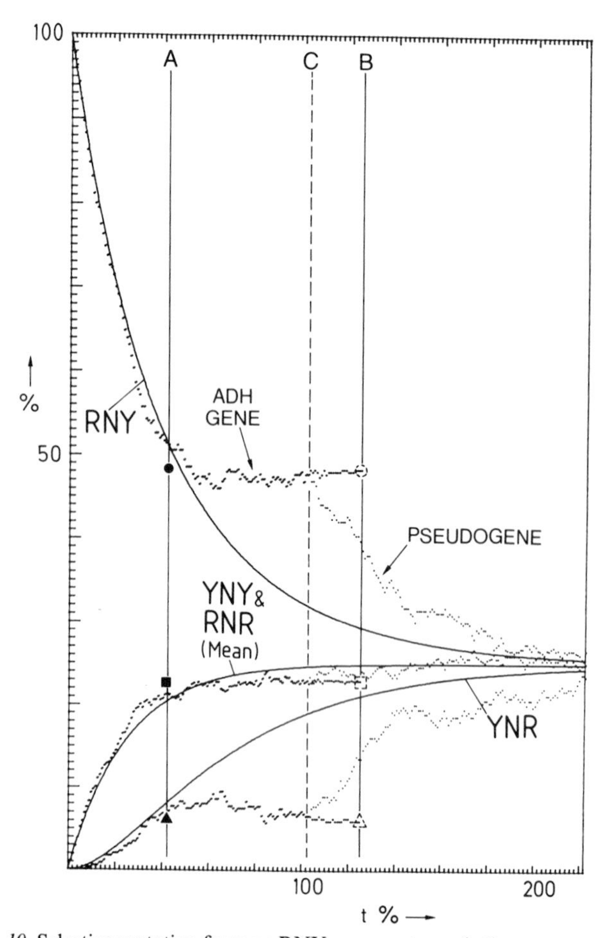

Fig. 10. Selective mutation from an RNY message towards the present yeast ADH gene. The percentages of RNY, YNY and RNR (mean), and YNR codons are plotted against $t\%$, the transversions accepted (heavily dotted curves). When the vertical line B is reached, the mutated gene has the same number of each of the 61 codon types as the ADH gene (stops are avoided). The actual ADH percentage of RNY is shown as ◌ on this line, of YNY and RNR (mean) as ⬚, and of YNR as △. The vertical line A corresponds to the t value estimated for ADH in Fig. 9 (on the assumption of random mutation), and here the actual ADH percentages are shown as ●, ■ and ▲, respectively. The vertical line C shows the point at which the selection is switched off (as for a pseudo-gene) and the lightly dotted curves show the subsequent decay of RNY, YNY and RNR (mean) and YNR towards 25%, as expected for a random sequence. The solid lines show mean random mutation away from an RNY message as in Fig. 9.

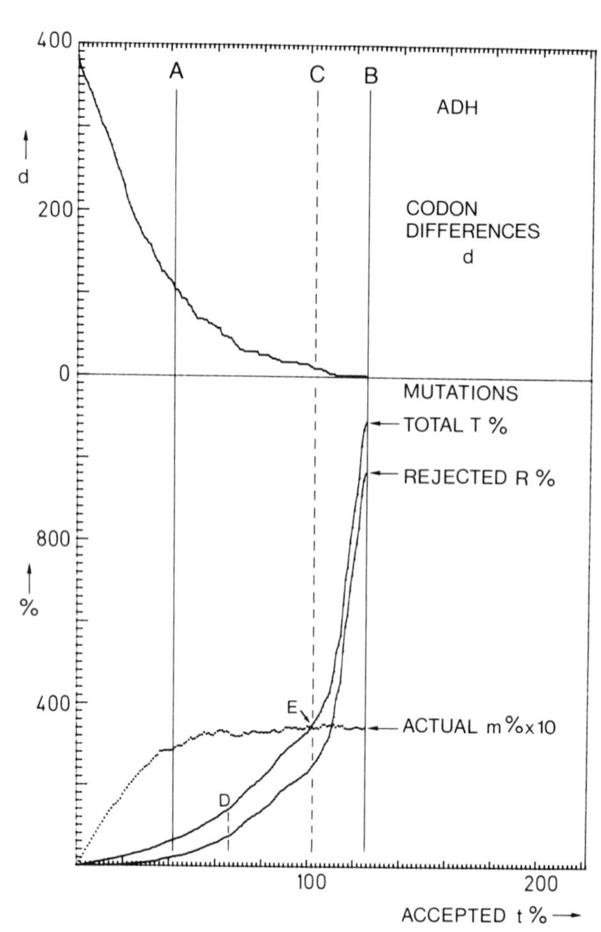

Fig. 11. Mutation and codon difference counts corresponding to the ADH selective simulation shown in Fig. 10. E is the point in the total mutation plot where selection is switched off. (The subsequent mutation counts in the pseudo-gene are not shown.)

and the primeval message ($m\%$ becomes much smaller than $t\%$ after long time periods because of multiple mutations at individual positions).

For the first example, an RNY primeval message has been taken of the same length (1050 bases) as yeast alcohol dehydrogenase [18], with approximately equal numbers of the sixteen RNY codons randomly mixed in it. Random mutations have then been applied but only accepted if the change made the proportions of the 61 codons nearer to those of the ADH gene (no stops were accepted). This was achieved by rejecting all mutations which increase the sum d of the differences (taken as positive numbers) between the number of codons for ADH and the primeval test gene for each codon type.

The result of one such simulation is shown in Fig. 10. The proportions of RNY, etc., follow the same random decay curves as before until the point A and then leave these curves to reach the exact codon counts of each codon in ADH at B, although not in the same codon order along the gene. At A the final proportions of RNY and the other types have almost been achieved and a minority of further codon changes is necessary between A and B.

Figure 11 gives more details of the $Y \rightleftharpoons R$ mutations occurring in this simulation, classified as total T, accepted t, rejected R, and actual m away from the primeval message, and again expressed as percentages of the total bases in the gene. The contrast between the situation at the beginning, when almost all of the randomly applied mutations T are accepted, and that at the end, when almost all the T mutations are rejected, is now seen. The decrease in the sum d of codon differences at the top of the figure indicates the gene's improvement towards ADH. Other simulations (not shown) followed the same curves with small statistical variations up to approximately the point E. After this the curves from the individual simulations were very variable. This can be expected, for the time taken before a random mutation removes one of the last few imperfections is very dependent indeed on the particular random number set used (or on the position and type of random mutation occurring in reality).

The great uncertainties in such a simulation are obvious. It is, for example, not at all clear that better proportions of the required codons and thus of the corresponding amino acids will result in an improvement in functional efficiency. Mutations giving the same value of d have also been accepted, a type of pseudoneutral mutations, which, however, in some cases may have a selective advantage. (A change in codon type at one position has occurred although d has stayed the same.)

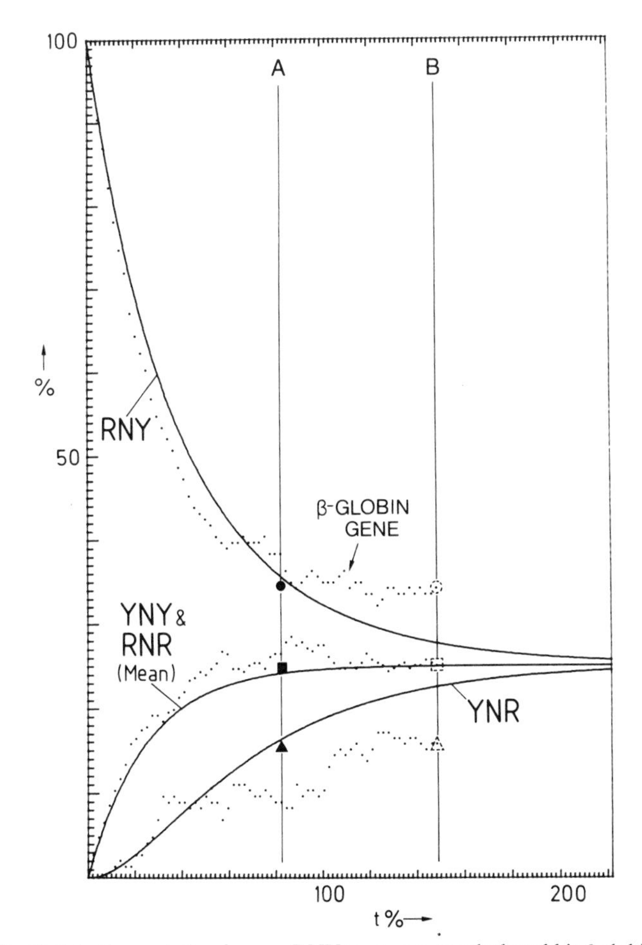

Fig. 12. Selective mutation from an RNY message towards the rabbit β-globin gene, plotted as in Fig. 10.

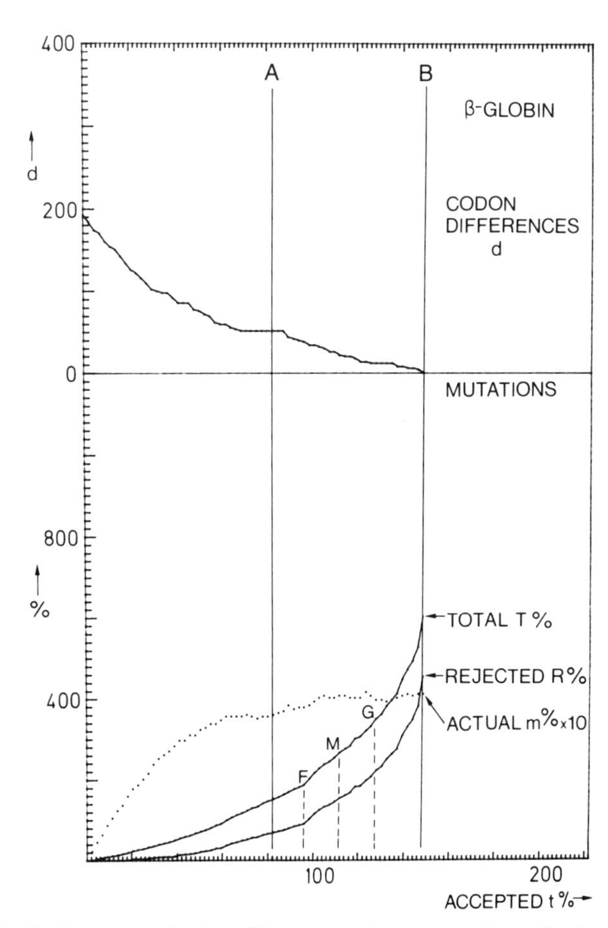

Fig. 13. Mutation and codon difference counts corresponding to the β-globin simulation shown in Fig. 12.

There is also no attempt to account for the many factors involved in the acceptance or rejection of a given mutant by the population. In spite of these undefined features, it is felt that such mutation of a gene towards an optimum may reveal by analogy some general concepts. Since there are no known criteria for functional optimization, the minimization of the codon differences has been taken as a reasonable substitute, since the final proportions of amino acids used and the exact method of coding for these is steadily approached. Other simulations, based on different principles, have not seemed so realistic, for example selecting for the right base at the correct position. In this case, the impression from the roughly known time scale for accepted $Y \rightleftharpoons R$ mutations [16] is that the gene attains its final form too quickly and the period of relative stability, known to be a fact for many ancient genes, is no longer seen.

A further similar simulation is recorded in Figs. 12 and 13, using rabbit β-globin [29], a gene further away from the RNY message in Fig. 9. The statistical variations between each individual simulation are more in this case, since the β-globin gene has only 438 bases, whereas that of ADH has 1050. The same tendencies are seen as for ADH. Particularly the RNY decline is a good indication of the mutational events, the other once mutated YNY and RNR, and the twice mutated YNR curves having more variations between individual simulations. The RNY proportions, as before, decrease along the random decay curve to the point A and then little further change is observed until the exact codon composition of β-globin is reached at B.

The simulations may give some hints as to the real course of events in the past history of some modern genes:

(1) Some further explanation is found of the fact that so many genes have codon proportions corresponding to the random decay curves, as shown in Fig. 9. A present-day gene, after it has completed a period of rapid improvement in function and is nearing optimization, may well find itself for a long time period on nearly horizontal curves such as those between A and B in Fig. 10, where the proportions of RNY, etc., stay almost constant and little further improvement is possible. The earlier plotting of the gene at A in Fig. 9 has been due to the fact that the accepted mutation value $t\%$ was obtained from the actual mutations away from RNY by a method assuming no selection [16]. This point is further discussed in Section 4 below.

(2) A good demonstration that the gene is following such a near constancy line as between A and B comes from the increased mutation rate in pseudogenes. In this case the gene may be for a time released from all selective restraints. This can be simulated as shown in Fig. 10, where the selection is switched off for a gene at C and stops are also no longer avoided. The proportions of RNY, etc., immediately start to alter more quickly, assuming new random decay curves towards 25% of each type. The switch-on of such a pseudo-gene could also be simulated, if it is later used for another purpose. Some genes may well have had a very chequered history with selective pressures released and re-applied.

The increase in mutation rate when a gene becomes a pseudo-gene can be estimated from the slope (total/accepted) of the $T\%$ curve at a given point. At E in Fig. 11 for ADH (where a switch-off of selection has been simulated in Fig. 10) it is approximately 7.5 and between D and E it is roughly 5.5; at M in Fig. 13 for β-globin, it is approximately 5. It is interesting to note that an increase by a factor of 3.5 has in fact been estimated [33], if the mean mutation rate (4.6×10^{-9} per site per year between T, C, A and G) in three globin pseudo-genes is compared with the mean of the rates in each codon position of the normal genes (0.71, 0.62 and 2.64×10^{-9} per site per year).

(3) A very speculative point concerns a region such as that between D and E on the $T\%$ curve in Fig. 11 and between F and G in Fig. 13 covering periods when the gene is nearly optimized. Here the $T\%$ curves are very approximately linear and yet they may well cover a period of up to 1000 million years or more on the figure's timescale (for example in the case of β-globin the $Y \rightleftharpoons R$ accepted rate $t\%/10^9$ years as been estimated at about 27 since the point of divergence of α- and β-globins, taken to be 500 million years ago [16]). This would signify that any changes in accepted mutations $t\%$ within these time ranges would correspond proportionally with corresponding changes in $T\%$ the total mutations. Two types of molecular clocks, based on T and t, respectively, should therefore agree for these evolutionary relatively recent times. The accepted rate of mutations has been found to be a reasonably accurate clock ar ⁻ᵈ to apply with the same rate for a given gene (say hemoglobin) in the branches of related lineages in such time ranges. The question of why this should be the case has been widely discussed. For example, Kimura has proposed a correlation of this fact with a neutral theory of evolution [34]. The present indication is that for genes which are nearing optimization for their function, the reason could be connected with this linear relation between t and T. Since T depends on such mechanisms as accuracy of replication and efficiency of repair, these processes might be much the same for the related organisms in the lineage group.

(4) When the curves relating accepted and actual mutations in Figs. 11 and 13 are compared with earlier correction curves derived on the assumption of no selection [16], it is evident that they disagree for regions beyond about 40% accepted mutations t. When the mutational simulations approach optimization of the gene, a small proportion of total mutations are still accepted but almost all of these are those which we have termed pseudo-neutral. These are, however, large in number in relation to the very small number of actual mutations still required for perfecting the gene. (This is very clearly seen for t values of more than 90–100% in Fig. 11.) Such a situation could well be imagined in a nearly optimized gene. Many pseudo-neutral changes are accepted, distributed along its whole length, for example at third codon positions of degenerate codons, and yet the losses and gains in actual mutations away from the primeval message balance out in total.

Conclusion

The attempt to delve back to the origins of life and draw conclusions about subsequent development is handicapped by the enormous number of unknown factors. Caution is needed to distinguish between the facts, such as positional correlations between bases in DNA sequences, and hypotheses proposed to explain these features (particularly speculative simulations of selection as just shown). Modern genes are thought to be a patchwork derived from shorter primeval genes and the evolution of genomes is now known to be subject to many events other than point mutations, for example recombination, duplication, transposition, insertion or deletion. The possibilities for sub-unit assembly offered by an exon–intron system will also have been utilized by selection, and a reverse transcriptase function gives additional evolutionary pathways. Apart from derivation from RNY codons, some genes (or parts of genes) may have had a completely different origin than the one just imagined, perhaps primevally, but more likely when a more sophisticated enzyme system has been developed (for example as conceived by Ohno [35]). The present DNA evidence, however, points to relics of RNY messages in all types of genes and from a wide variety of genomes (of viruses, plasmids, prokaryotes, and eukaryotes together with their mitochondria and chloroplasts), and suggests this comma-less code as the forerunner of today's genetic code.

Maintenance of some of these purine–pyrimidine patterns seems to have been helped by natural selection in some cases, and remnants of primitive beginnings may well be hidden in present-day more efficient molecular mechanisms, for example in the present translational system. Careful extrapolation back from our ever increasing molecular knowledge of life processes seems to offer the best chance of discovering more about its origins.

Acknowledgements

Particular thanks are due to Dr T. Bickle for many useful discussions and to Dr W. Arber for his constant support and encouragement. The supply of DNA library sequence tapes from EMBL and the National Biomedical Research Foundation is also gratefully acknowledged, as is the support given by the Swiss National Science Foundation.

Note added in proof:

A recent analysis of some two hundred 5S rRNA sequences, from eubacteria, chloroplasts, mitochondria, archaebacteria and eukaryotes suggest that these may also be derived from RNY coding sequences used at the time the genetic code originated [Eigen, M., Lindemann, B., Winkler-Oswatitsch, R. and Clarke, C. H., *Proc. Natl. Acad. Sci. USA* **82**, 2437 (1985)].

References

1. Darwin, C., *Origin of Species*. Murray, London (1859).
2. Miller, S. L. and Orgel, L. E., *The Origins of Life on the Earth*. Prentice-Hall, New Jersey (1974).
3. Eigen, M. and Schuster, P., *J. Mol. Evol.* **19**, 47 (1982).
4. Dayhoff, M. O., *Atlas of Protein Sequence and Structure*, vol. 5 (and supplements 1, 2 and 3). National Biomedical Research Foundation, Washington, D.C. (1972).
5. Shepherd, J. C. W., *J. Mol. Evol.* **17**, 94 (1981).
6. Shepherd, J. C. W., *Proc. Natl. Acad. Sci. USA* **78**, 1596 (1981).
7. Crick, F. H. C., Griffith, J. S. and Orgel, L. E., *Proc. Natl. Acad. Sci. USA* **43**, 416 (1957).
8. Crick, F. H. C., Brenner, S., Klug, A. and Pieczenik, G., *Origins Life* **7**, 389 (1976).
9. Fuller, W. and Hodgson, A., *Nature (London)* **215**, 817 (1967).
10. Woese, C., *Nature (London)* **226**, 817 (1970).
11. Eigen, M. and Schuster, P., *Naturwissenschaften* **65**, 341 (1978).

12. Miller, S. L., *Science* **117**, 528 (1953).
13. Cromin, J. R. and Moore, C. B., *Science* **172**, 1327 (1971).
14. Eigen, M. and Winkler-Oswatitsch, R., *Naturwissenschaften* **68**, 282 (1981).
15. Eigen, M. and Winkler-Oswatitsch, R., *Naturwissenschaften* **68**, 217 (1981).
16. Shepherd, J. C. W., *Cold Spring Harbor Symp. Quant. Biol.* **47**, 1099 (1982).
17. Shepherd, J. C. W., *TIBS* **9**, 8 (1984).
18. Russell, P. R. and Hall, B. D., *J. Biol. Chem.* **258**, 143 (1983).
19. Inokuchi, K., Mutoh, N., Matsuyama, S.-I. and Mizushima, S., *Nucl. Acids Res.* **10**, 6957 (1982).
20. Maarse, A. C., Van Loon, A. P. G. M., Riezman, H., Gregor, I., Schatz, G. and Grivell, L. A., *EMBO J.* **3**, 2831 (1984).
21. Russell, P. R. and Hall, B. D., *Mol. Cell. Biol.* **2**, 106 (1982).
22. Scarpulla, R. C., Agne, K. M. and Wu, R., *J. Biol. Chem.* **256**, 6480 (1981).
23. Lemischka, I. and Sharp, P. A., *Nature (London)* **300**, 330 (1982).
24. Nellen, W. and Gallwitz, D., *J. Mol. Biol.* **159**, 1 (1982).
25. Nellen, W., Donath, C., Moos, M. and Gallwitz, D., *J. Mol. Appl. Genet.* **1**, 239 (1981).
26. Gallwitz, D., Perrin, F. and Seidel, R., *Nucl. Acids Res.* **9**, 6339 (1981).
27. Lazowska, J., Jacq, C., Slonimski, P. P., *Cell* **22**, 333 (1980).
28. Kaine, B. P. and Spear, B. B., *Nature (London)* **295**, 430 (1982).
29. Efstratiadis, A., Kafatos, F. C. and Maniatis, T., *Cell* **10**, 571 (1977).
30. Lund, P. K., Goodman, R. H., Dee, P. C. and Habener, J. F., *Proc. Natl. Acad. Sci. USA* **79**, 345 (1982).
31. Heffron, F., McCarthy, B. J., Ohtsubo, H. and Ohtsubo, E., *Cell* **18**, 1153 (1979).
32. Shinozaki, K., Deno, H., Kato, A. and Sugiura, M., *Gene* **24**, 147 (1983).
33. Li, W.-H., Gojobori, T. and Nei, M., *Nature* **292**, 237 (1981).
34. Kimura, M., *The Neutral Theory of Molecular Evolution.* Cambridge University Press (1983).
35. Ohno, S., *J. Mol. Evol.* **20**, 313 (1984).

Chemica Scripta 1986, **26B**, 85–89

Evolutionary Aspects of Unconventional Codon Reading

Ulf Lagerkvist

Department of Medical Biochemistry, University of Gothenburg, Box 33031, S-400 33 Gothenburg, Sweden

Paper presented at the Conference on 'Molecular Evolution of Life', Lidingö, Sweden, 8–12 September 1985

Abstract

There is by now evidence that codon readings in protein synthesis both *in vitro* and *in vivo* may involve practically all the base combinations forbidden by the wobble rules in the reading of the third codon nucleotide. Tentative mechanisms of such unconventional codon readings are discussed as well as their role in the cells' normal translational repertoire and their possible evolutionary implications.

The genetic code comes in sixteen boxes with four codons in each box. All four codons in a box have the same nucleotides in the first two positions, the variation is in the third position. In half of the boxes all four codons signify the same amino acid and in what follows we will refer to such boxes as family boxes (Fig. 1) and the codons in such boxes as family codons. Obviously, from the point of view of translational fidelity, it makes no difference how the third position is read in a family codon since the first two positions are enough to specify the amino acid. These considerations led us to speculate that perhaps in the reading of such codons the wobble restrictions, that presumably govern the pairing between the wobble nucleotide of the anticodon and the third nucleotide of the codon, might be less strictly observed than in the reading of non-family codons. To test this idea we have used an *in vitro* protein synthesizing system from *E. coli* programmed with the phage message MS2-RNA and have observed a type of unconventional reading whereby a codon may be read by an anticodon that cannot, according to the wobble rules [1], form a stable base pair with the third codon nucleotide [2–6]. Furthermore it was found that, as a group, the family codons (in this investigation represented by the alanine, glycine and valine codons) were considerably more amenable to this type of unconventional reading, that seemingly ignored the wobble restrictions, than were the non-family codons, represented by the lysine and glutamine codons.

UUU	Phe	UCU	Ser	UAU	Tyr	UGU	Cys
UUC	"	UCC	"	UAC	"	UGC	"
UUA	Leu	UCA	"	UAA	Och	UGA	Umb
UUG	"	UCG	"	UAG	Amb	UGG	Trp
CUU	Leu	CCU	Pro	CAU	His	CGU	Arg
CUC	"	CCC	"	CAC	"	CGC	"
CUA	"	CCA	"	CAA	Gln	CGA	"
CUG	"	CCG	"	CAG	"	CGG	"
AUU	Ile	ACU	Thr	AAU	Asn	AGU	Ser
AUC	"	ACC	"	AAC	"	AGC	"
AUA	"	ACA	"	AAA	Lys	AGA	Arg
AUG	Met	ACG	"	AAG	"	AGG	"
GUU	Val	GCU	Ala	GAU	Asp	GGU	Gly
GUC	"	GCC	"	GAC	"	GGC	"
GUA	"	GCA	"	GAA	Glu	GGA	"
GUG	"	GCG	"	GAG	"	GGG	"

Fig. 1. The genetic code. Family boxes are indicated by heavy lines.

What, then, might be the mechanism of this unconventional codon reading? To answer this question we must consider the two basic principles of the classic codon-reading scheme. The first principle is embodied in the wobble rules while the second principle demands that an anticodon must form a stable base pair with all three nucleotides of a codon in order to read it. To explain our results one could assume that the wobble restrictions are not valid in the reading of family codons, meaning that base pairs involving the third codon nucleotide may be formed, that violate the wobble rules. Nevertheless, these pairs would be as stable as regular base pairs. Although this might be the case in some instances to be discussed below, as a general principle it appears unlikely both because of the seemingly intractable structural problems it presents [7] and because it fails to explain how, on this assumption, the translational apparatus could discriminate against codon–anticodon interactions that we know are not compatible with translational fidelity. Furthermore, when two tRNAs competed for the same family codon, our results indicated that the tRNA that read according to the wobble rules was typically an order of magnitude more efficient than the tRNA that read unconventionally. We have therefore suggested an alternative explanation which assumes that the wobble rules are basically correct [8].

The two-out-of-three reading hypothesis

According to this hypothesis a codon may be read by relying mainly on the Watson–Crick base pairs formed with the first two codon positions, while the mispaired nucleotides in the third codon and anticodon wobble positions make a comparatively small contribution to the total stability of the reading interaction. On the other hand, it is not implied that the nature of the mispair is without importance for the efficiency of the unconventional reading. In fact, our results indicate that some mispairs, for instance C·A, have a certain amount of stability and facilitate unconventional codon reading while others like C·U and C·C have very low stability and make no contribution, or perhaps even a negative one, to the overall stability of the codon–anticodon interaction [6]. This partial stability of C·A base pairs is consistent with a number of observations in different laboratories both *in vitro* and *in vivo* [9–16].

The hypothesis furthermore assumes that the probability of two-out-of-three reading is a function of the strength of the interaction between the anticodon and the first two codon nucleotides and that a G·C interaction is stronger than an A·U interaction. Therefore, the probability of such reading would be greatest for codons making only G·C interactions in these positions (G–C codons), intermediate for codons which make one G·C and one A·U interaction ('mixed' codons), and minimal for codons making only A·U interactions (A–U codons).

When the genetic code was examined for the distribution of the three codon categories it was immediately apparent that G–C codons were without exceptions restricted to the family boxes while the A–U codons were always outside the families (Fig. 1). This is the distribution to be expected as the result of an evolutionary process which selected for a code where the probability that reading by two-out-of-three should cause translational errors was at a minimum. On the other hand, this tentative structural explanation leaves the mixed codons unaccounted for. It was noted, however, that the distribution of the mixed codons with respect to the codon families is not random. In the left half of the codon square all mixed codons appear in families while in the right half they are all outside the families [8]. This non-random distribution is not likely to be fortuitous and instead suggests the possibility that mixed codons in the left half are confined to the family boxes because they stand a greater risk of being read by two-out-of-three than mixed codons in the right half.

Table I. *Efficiency of two-out-of-three reading as a function of the ratio of G·C to A·U pairs in the interaction between the anticodon and the first two codon nucleotides*

Codons	Mispair involved	Competing anticodons	Relative efficiency of unconventional reading (% of normal reading: mean value ± S.D.)
GGA	C·A	CCC / UCC	15±5
GGA	C·A	CCC / NCC	11±1
CAA	C·A	CUG / s²UUG	2.8±0.5
AAA	C·A	CUU / s²UUU	0.5±0.25

The table has been taken from ref. [6]. N is an unknown derivative of uridine; s²U is 2-thio 5-methylcarboxylmethyluridine (s²UUU) or an unknown derivative of 2-thiouridine (s²UUG).

Predictions of the hypothesis

The two-out-of-three hypothesis makes several predictions that can be tested experimentally. However, in this context it might be appropriate to consider also some predictions that have been mistakenly ascribed to the hypothesis. For instance, the hypothesis does not imply that family codons *in vivo* are normally read by two-out-of-three even in situations where all the prerequisites for unconventional codon reading are at hand. It only requires that in such circumstances the family codons should be read by two-out-of-three with a frequency that is not negligible, i.e. we assume it to be higher than the translational error frequency. Even on this limited assumption such reading would become a potential source of translational error and, therefore, also an important restriction in the evolution of the genetic code as pointed out above. Furthermore, the hypothesis does not necessarily require that two-out-of-three reading *in vivo* should be efficient enough to sustain protein synthesis on a level compatible with life in cases where the cell has lost an isoacceptor tRNA by mutation [17–19]. On the other hand, the hypothesis clearly demands that family codons should be easier to read by two-out-of-three than non-family codons, and our results with the alanine, glycine, and valine codons as compared to the glutamine and lysine codons show that this is indeed the case. At the same time, these results might be influenced by the nature of the mispairs formed in the different unconventional readings. We have therefore tried to assess the importance of the Watson–Crick base pairs between the anticodon and the first two codon nucleotides by comparing the efficiencies of readings that involve the same mispair between the wobble position and the third codon position, in this case the mispair C·A. It is immediately apparent from the results in Table I that the efficiency of the unconventional reading is a function of the ratio of G·C to A·U pairs in the interaction between the anticodon and the first two codon nucleotides as required by the two-out-of-three hypothesis. It is also obvious that an unconventional reading may involve the mispair C·A and still be very inefficient as in the reading of the lysine codon AAA by the $tRNA_1^{Lys}$ (anticodon CUU).

Two-out-of-three as an error mechanism *in vivo*

In its original version the two-out-of-three hypothesis regards such reading mainly as a potential source of translational errors and attempts to explain how it could have acted as a restriction in the evolution of the genetic code. One may then ask if there is any experimental evidence that two-out-of-three reading could cause translational errors *in vivo*. In this context it is important to emphasize that the most easily observed unconventional codon readings *in vivo* are those that actually lead to translational errors. It is possible to detect translational errors in the synthesized protein by two-dimensional gel electrophoresis provided that the translational error leads to an amino acid substitution that produces a difference in charge. For instance, substituting lysine for asparagine would result in a new protein spot on the gel (so-called stuttering). Using stuttering as an indication of translational error, Parker and co-workers [20, 21] have shown that an asparagine auxotroph of *E. coli*, containing a *rel* mutation, when starved for asparagine will incorporate lysine in response to the asparagine codons using what appears to be a two-out-of-three reading mechanism. It would therefore seem that, at least in this particular genetic background, amino acid starvation can increase two-out-of-three reading to a point where it will produce a high-error frequency *in vivo* even in the reading of codons with the lowest probability of such reading. We may thus assume with some confidence that this alternative codon reading method can function as a potential error-producing mechanism also *in vivo*.

Unconventional methods in the cell's normal repertoire of codon readings

The wobble rules prohibit codon readings that involve any of the base pairs U·U, U·C, C·U, C·C, C·A, I·G, G·A and G·G between the first position of the anticodon and the third codon position. However, over the years there has been a steady trickle of observations indicating that certain cells or cell organelles routinely use such readings as part of their normal reading repertoire. In fact, by now there is evidence for almost all the prohibited base pairs being used in protein synthesis [9–15, 22–28]. Some of this evidence is merely circumstantial but in the case of the mammalian and yeast mitochondria it is unrefutable and admits of no doubt as far as the reading of the family codons is concerned.

The analysis of the mitochondrial genome and its gene products from mammalian cells as well as from yeast and Neurospora has shown that in these organelles codon families are each read by only one tRNA. Such tRNAs have U in the wobble position, with the exception of the yeast tRNA reading the arginine family CGN which has A in this position. Outside the families, codons of the type NN^{A}_{G} are read by tRNAs with U (or substituted U) in the wobble position, while NN^{U}_{C} codons are read by tRNAs with G in this position [9, 22–24]. In all cases where we have information on the primary structure, tRNAs that read family codons have U in the wobble position unsubstituted while the tRNAs that read non-family codons in yeast and Neurospora have this U substituted. The initiator tRNAs are special in that they always contain C in the wobble position, as do the methionine tRNAs from yeast and *Neurospora*. These findings might indicate that a two-out-of-three reading mechanism operates in the mitochondrion [9, 22]. However, it is equally possible, as pointed out by Heckman *et al.* [23], that in the mitochondrion U in the wobble position of the anticodon can form stable base pairs with both U and C in the third position of the codon [29]. Thus, we may be certain that these mitochondria use a kind of unconventional codon reading that is operationally similar to two-out-of-three reading, i.e. it produces the same result – the reading of all four codons in a family box by a single tRNA – but the actual mechanism employed is still an open question.

It has often been suggested that the mitochondrion is of procaryotic origin, and even if this organelle has obviously had a very special evolutionary history it is, nevertheless, pertinent to ask if there are procaryotes today that use the same kind of unconventional codon reading as the mitochondria. Kilpatrick and Walker [30] have recently reported that *M. mycoides* sp. *capri* contains only one glycine tRNA and that this tRNA has an unsubstituted U in the wobble position. Nothing is known about codon usage in this *Mycoplasma*, but in the very closely related *Mycoplasma capricolum* the glycine codon GGU is used frequently, and unless there is an absolute bias against GGU and GGC in *M. mycoides* its single glycine tRNA must be able to read these codons. We have tested this tRNA in our *in vitro* protein-synthesizing system and have found that it was almost as efficient in the unorthodox reading of the codons GGU and GGC as it was

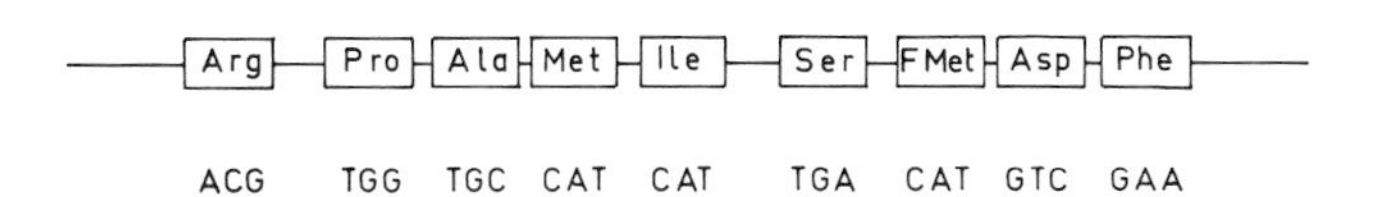

Fig. 2. A cluster of tRNA genes in *Mycoplasma mycoides*. The anticodon sequence is indicated under each gene.

in conventional reading. This is, of course, the result to be expected for a tRNA that had been specifically designed to read all four codons in a family. Nor can the enhancement of unconventional codon reading displayed by the mycoplasma $tRNA^{Gly}$ possibly be explained by some property of the *E. coli* ribosome. It would therefore seem that the structural context provided by the mycoplasma tRNA molecule enhances the ability of the anticodon to make unconventional codon readings. This is reminiscent of the report by Yarus and co-workers [31, 32] that the ability of an anticodon to suppress a nonsense mutation is influenced by its structural context in the tRNA molecule.

In order to elucidate further the possibility of unconventional codon reading in *M. mycoides* we have initiated an inventory of the tRNA genes in this organism. So far this has resulted in the cloning and sequencing of ten genes, i.e. one isolated gene for arginine tRNA and a cluster of nine genes for the arginine, proline, alanine, methionine, isoleucine, serine, formyl methionine, aspartate and phenylalanine tRNAs [33] (Fig. 2). For each family box represented in the cluster only one tRNA gene has been found and these genes have T in the position corresponding to the first anticodon nucleotide, with the exception of the gene for arginine tRNA, which has A in this position. Although the data are still very far from complete, taken together with what we know about the glycine tRNA they are at least not inconsistent with the notion that *M. mycoides* will prove to be a promising hunting ground for anyone interested in unconventional codon reading.

Some evolutionary considerations

The unorthodox and greatly simplified codon reading scheme used in mitochondria poses the question as to whether this represents the remnants of a very primitive reading method. However, one could equally well think of it as the result of a fairly recent evolutionary process by which the mitochondrion has been compelled to jettison tRNA genes in order to save coding space. These alternative explanations have a common denominator, in that the evolutionary pressure, which led to extreme economy with regard to the DNA available to the organelle, could have caused a regression to a translational apparatus similar to that existing in the very primitive cell.

Assuming that unconventional methods are used in the reading of at least some family codons in *M. mycoides* a similar line of argument might apply. This organism belongs to a group of mycoplasmas that have the smallest genomes of any micro-organisms capable of autonomous replication, i.e. that are able to grow outside a host cell. Their small genome and the fact that they lack a cell wall and might, like the mitochondria, be using a simplified codon reading scheme, raises the possibility that they could be relicts of a primitive ancestor of all procaryotes, and this has in fact been suggested by Morowitz [34]. On the other hand, based on their ribosomal RNA sequences they are related to the Gram-positive bacteria and particularly to the clostridia [35]. They would then be just another group of procaryotes, although for some unknown reason they have been obliged to reduce their genomes by shedding genes, among others ribosomal RNA genes and tRNA genes. In this context it should perhaps be pointed out that they have achieved nothing like the extreme genome economy that one finds in mammalian mitochondria; for instance, in the tRNA gene cluster shown in Fig. 2 the spacer regions between the genes are as long or longer than the spacers in a similar gene cluster in *B. subtilis* [36, 37], a micro-organism with a genome of normal size.

From the point of view of using the translational machinery of mitochondria (and possibly also of mycoplasmas) as a model for a primitive 'ur-translation' it does not necessarily matter very much if they are true vestiges of a primitive ancestor or if they represent more recent developments. This is illustrated by the following model for some aspects of the evolution of the genetic code [38, 39]. We will assume that, at some stage of its evolution, the code had reached a degree of sophistication where four distinct bases were employed to code for an assortment of amino acids more limited than that found in proteins today. However, the reading anticodons could not distinguish among the nucleotides occupying the third codon position: although the code was structurally a three-letter code, it was operationally a two-letter code, in terms of the number of letters per codon actually read with precision. This situation is reminiscent of the reading of codon families in the mitochondrion today. Thus all codons in the primitive code must have belonged to such families. Setting aside one of the 16 families for the stop words, this leaves a maximum of 15 amino acids that could be unambiguously coded for. At a more advanced stage of development, the cell would have perfected its translational machinery so that it could read the third codon nucleotide without ambiguity. Consequently, new amino acids could be accommodated in the code. It then became necessary to make a decision as to which groups of four codons could be divided up between two different amino acids without compromising translational fidelity. As a result of this selection, codons with a high probability of being read by 'two-out-of-three' would remain as family codons, and only those with a low probability would be used as non-family codons. Thus it is possible to understand some of the structural features of the present code in terms of an evolution aimed at minimizing translational errors resulting from the reading of codons by the two-out-of-three method.

Concluding remarks

The greatly simplified codon reading apparatus of the mitochondrion must have been obtained at a cost, either in the rate of translation of in translational fidelity. Indeed, if a simple set-up could work as efficiently and with as great precision as a more complicated one, it is difficult to see why the cell should have bothered to evolve the very sophisticated translational machinery of today from its presumably rather humble origin in the primitive ancestor of cells. Unfortunately we know nothing about rate and fidelity of translation in the mitochondrion and attempts to develop an efficient mitochondrial *in vitro* protein synthesizing system have so far met with little success. However, if a procaryote with a simplified translational system similar to that of the mitochondrion could be found, it might turn out to be more amenable to studies of protein synthesis. With this in mind an effort should be made to look for such organisms and the mycoplasmas would seem to be the prime candidates. A complete inventory of the tRNA genes and their gene products in a mycoplasma should provide an answer to this question. This search is important not only because unconventional codon reading is an interesting problem in its own right, but also because it is really the mirror image of the translational fidelity problem. Therefore, once a suitable system for the study of such reading as part of a cell's normal translational repertoire has been found, it might be expected to yield crucial information about the molecular basis for the selection of the correct codon–anticodon interaction in the reading of the messenger RNA.

References

1. Crick, F. H. C., *J. Mol. Biol.* **19**, 548–555 (1966).
2. Mitra, S. K., Lustig, F., Åkesson, B., Lagerkvist, U. and Strid, L., *J. Biol. Chem.* **252**, 471–478 (1977).
3. Mitra, S. K., Lustig, F., Åkesson, B., Axberg, T., Elias, P. and Lagerkvist, U., *J. Biol. Chem.* **254**, 6397–6401 (1979).
4. Samuelsson, T., Elias, P., Lustig, F., Axberg, T., Fölsch, G., Åkesson, B. and Lagerkvist, U., *J. Biol. Chem.* **255**, 4583–4588 (1980).
5. Lustig, F., Elias, P., Axberg, T., Samuelsson, T., Tittawella, I. and Lagerkvist, U., *J. Biol. Chem.* **256**, 2635–2643 (1981).
6. Samuelsson, T., Axberg, T., Borén, T. and Lagerkvist, U., *J. Biol. Chem.* **258**, 13178–13184 (1983).
7. Topal, M. P. and Fresco, J. R., *Nature* (*London*) **263**, 289–293 (1976).
8. Lagerkvist, U., *Proc. Natl. Acad. Sci. USA* **75**, 1759–1762 (1978).
9. Barrell, B. G., Anderson, S., Bankier, A. T., de Bruijn, M. H. L., Chen, E., Coulson, A. R., Drouin, J., Eperon, I. C., Nierlich, D. P., Roe, B. A., Sanger, F., Schreier, P. H., Smith, A. J. H., Staden, R. and Young, I. G., *Proc. Natl. Acad. Sci. USA* **77**, 3164–3166 (1980).
10. Hsu Chen, C.-C., Cleaves, G. R. and Dubin, D. T., *Nucl. Acids Res.* **11**, 8659–8662 (1983).
11. Gupta, R., *J. Biol. Chem.* **259**, 9461–9471 (1984).
12. Fukuda, K. and Abelson, J., *J. Mol. Biol.* **139**, 377–391 (1980).
13. Kashdan, M. A. and Dudock, B. S., *J. Biol. Chem.* **257**, 11191–11194 (1982).
14. Kuchino, Y., Watanabe, S., Harada, F. and Nishimura, S., *Biochemistry* **19**, 2085–2089 (1980).
15. Hirsch, D., *J. Mol. Biol.* **58**, 439–458 (1971).
16. Stern, L. and Schulman, L. H., *J. Biol. Chem.* **253**, 6132–6139 (1978).
17. Murgola, E. J. and Pagel, F. T., *J. Mol. Biol.* **138**, 833–844 (1980).
18. Piper, P. W., *J. Mol. Biol.* **122**, 217–235 (1978).
19. Munz, P., Leupold, U., Agris, P. and Kohli, J., *Nature* (*London*) **294**, 187–188 (1981).
20. Parker, J. and Friesen, J. D., *Mol. Gen. Genet.* **177**, 439–445 (1980).
21. Parker, J., Johnston, T. C. and Boyia, P. T., *Mol. Gen. Genet.* **180**, 275–281 (1980).
22. Bonitz, S. G., Berlani, R., Coruzzi, G., Li, M., Nobrega, F. G., Nobrega, M. P., Thalenfeld, B. E. and Tzagoloff, A., *Proc. Natl. Acad. Sci. USA* **77**, 3167–3170 (1980).
23. Heckman, J. E., Sarnoff, J., Alzner-Deweerd, B., Yin, S. and Raj-Bhandary, U. L., *Proc. Natl. Acad. Sci. USA* **77**, 3159–3163 (1980).
24. Anderson, S., Bankier, A. T., Barrell, B. G., de Bruijn, M. H. L., Coulson, A. R., Drouin, J., Eperon, I. C., Nierlich, D. P., Roe, B. A., Sanger, F., Schreier, P. H., Smith, A. J. H., Staden, R. and Young, I. G., *Nature* (*London*) **290**, 457–465 (1981).
25. Clary, D. O. and Wolstenholm, D. R., *Nucl. Acids Res.* **12**, 2367–2379 (1984).
26. Bibb, M. J., van Etten, R. A., Wright, C. T., Walberg, M. W. and Clayton, D. A., *Cell* **26**, 167–180 (1981).
27. Beier, H., Barciszewska, M., Krupp, G., Mitnacht, R. and Gross, H. J., *EMBO J.* **3**, 351–356 (1984).
28. Bienz, M. and Kubli, E., *Nature* (*London*) **294**, 188–190 (1981).
29. Grosjean, H. J., de Henan, S. and Crothers, D. M., *Proc. Natl. Acad. Sci. USA* **75**, 610–614 (1978).
30. Kilpatrick, M. W. and Walker, R. T., *Nucleic Acids Res.* **8**, 2783–2786 (1980).

31. Yarus, M., McMillan, C., III, Cline, S. W., Bradley, D. and Snyder, M., *Proc. Natl. Acad. Sci. USA* **77**, 5092–5096 (1980).
32. Yarus, M., *Science* **218**, 646–652 (1982).
33. Samuelsson, T., Elias, P., Lustig, F. and Guindy, Y. S. *Biochem. J.* **232**, 223–228 (1985).
34. Morowitz, H. J., *Israel J. Med. Sci.* **20**, 750–753 (1984).
35. Woese, C. R., Maniloff, J. and Zablen, L. B., *Proc. Natl. Acad. Sci. USA* **77**, 494–498 (1980).
36. Green, C. J. and Vold, B. S., *Nucl. Acids Res.* **11**, 5763–5774 (1983).
37. Wawrousek, E. F., Narasimhan, N. and Hansen, J. N., *J. Biol. Chem.* **259**, 3694–3702 (1984).
38. Lagerkvist, U., *Am. Sci.* **68**, 192–198 (1980).
39. Lagerkvist, U., *Cell* **23**, 305–306 (1981).

Chemica Scripta 1986, **26B**, 91–95

Transfer RNA Modification in Different Organisms

Glenn R. Björk

Department of Microbiology, University of Umeå, S-901 87 Umeå, Sweden

Paper presented at the Conference on 'Molecular Evolution of Life', Lidingö, Sweden, 8–12 September 1985

Abstract

Transfer RNAs from all organisms so far investigated contain modified nucleosides which are derivatives of the four major nucleosides adenosine (A), guanosine (G), cytosine (C) and uridine (U). Different enzymes can catalyze the formation of the same modified nucleoside in different positions of tRNA. The modification pattern in each position of the tRNA from the three kingdoms eukaryotes, eubacteria and archaebacteria was compared. It is revealed that tRNA from each kingdom contains group-specific modifications. Furthermore, some modified nucleosides are unique to eukaryotes, eubacteria or archaebacteria, respectively. However, some (Ψ13, Cm32, m^1G37, t^6A37, Ψ38, Ψ39, Ψ55, m^1A58) are present at comparative positions in tRNAs from all three kingdoms, suggesting either that convergent evolution of the formation of these modified nucleosides has occurred or that these modified nucleosides were present in the tRNA of the common ancestor to the three kingdoms. Interestingly, N^6-threonyladenosine (t^6A) itself and its derivatives in position 37 (3′-side of the anticodon) and 1-methylguanosine (m^1G) in the same position are present in tRNAs from all three kingdoms that read codons starting with A or C/G, respectively. However, tRNAs reading codons starting with U contain a great variety of modified nucleosides in position 37. Functional aspects of some modified nucleosides are discussed.

Introduction

Transfer RNA is fundamental to protein synthesis and interacts during the translation process with many different proteins and nucleic acids. It contains modified nucleosides which are derivatives of the four major nucleosides adenosine (A), guanosine (G), cytidine (C) and uridine (U) (Fig. 1).

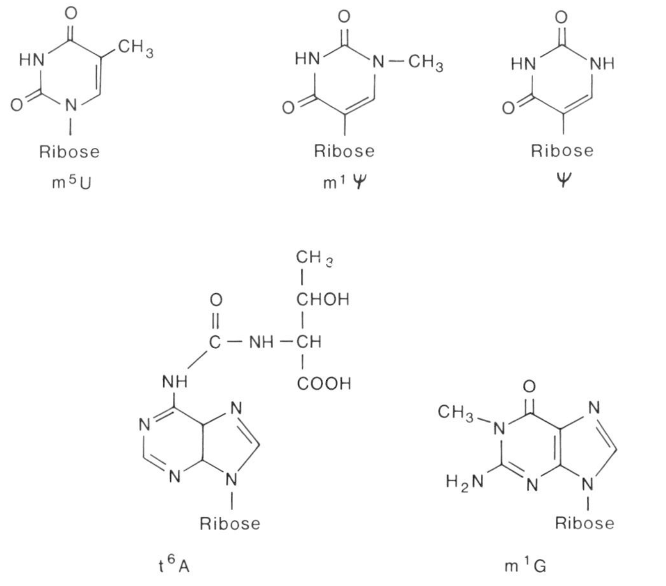

Fig. 1. The structure of some modified nucleosides and their abbreviations.

Transfer RNA from all organisms so far investigated contains modified nucleosides. The formation of the modified nucleosides is catalyzed by highly specific enzymes which operate during the maturation of the tRNA after the primary transcript is made. (For review, see [1].) There are different enzymes in yeast catalyzing the formation of the same nucleoside (1-methylguanosine, m^1G) in position 9 and position 37 [2]. (See Fig. 2 for designation of the positions.) Furthermore, a mutation in the *hisT* gene of *Salmonella typhimurium* affects the synthesis of pseudouridine (Ψ) at position 38, 39 and 40 but not in other positions, like position 55 [3]. Therefore there are reasons to believe that each position in tRNA prone to be modified is recognized by different enzymes even if the enzymes catalyze the formation of the same modified nucleoside. Some modified nucleosides like, N^6-(4-hydroxyisopentenyl)-2-methylthioadenosine (ms^2io^6A) and 5-methylaminomethyl-2-thiouridine (mnm^5s^2U) have complicated structures and more than one enzyme is therefore involved in their synthesis. In tRNA from *S. typhimurium*, about 30 different modified nucleosides have been observed, of which 26 have been identified [4]. Since some modified nucleosides, like Ψ, are present in many positions and some have complex structures, it has been estimated that at least 45 different tRNA-modifying enzymes may be present in *Escherichia coli*/*S. typhimurium*. One enzyme needs about 1 kb of genetic material, while 1% of the total genetic information present in bacteria is devoted to tRNA modification [1].

While tRNA(m^5U)methyltransferase, i.e. the enzyme catalyzing the formation of m^5U54, is regulated like stable RNA, the tRNAm(m^1G)- and tRNA(mnm^5s^2U)methyltransferase are not [5]. The structural gene (*trmA*) for the tRNA(m^5U)methyltransferase is a monocistronic operon, while the tRNA(m^1G)methyltransferase (the *trmD* gene product) and tRNA(Ψ)synthetase (the *hisT* gene product) are part of differentially expressed multicistronic operons [6–8]. Thus, the synthesis of tRNA modifying enzymes are precisely but not coordinately regulated and the organization of the corresponding genes is complex. Little is known about the regulation of these enzymes in other organisms than *E. coli*.

Excluding phage and organelle tRNAs, the sequence of about 250 tRNAs species from different organisms are known [9]. This paper will discuss the pattern of modification in tRNAs from the three kingdoms eukaryotes, eubacteria and archaebacteria. It should be noted, however, that of the 51 tRNA species sequenced from archaebacteria, 42 are from one species, *Halobacterium vulcanii* [9, 10]. Therefore, too-far-reaching generalizations cannot be made at present, since some specific characteristics may be confined to the extreme halophiles, of which *H. volcanii* is a member. The current state of knowledge of tRNA modification in archaebacteria has recently been reviewed [11].

Results and discussion

Presence of modified nucleosides in tRNA from different organisms

Figure 2 shows that certain positions are prone to be modified. This is true for tRNAs from all three kingdoms. The level of modified nucleosides as well as the variety of them are greater in eukaryotic tRNAs compared to eubacterial/archaebacterial tRNAs. Thus, many modified nucleosides and positions are specific for eukaryotic tRNAs. However, the three modified nucleosides $m^1\Psi$, m^2_2Gm and a structurally unknown modified guanosine (G*) are present only in tRNA from archaebacteria (Table IA, B, C and Table II). However, the absence of tRNA from these organisms of m^7G46 and m^5U54 (rT, ribothymidine), which are abundant in eukaryotic and eubacterial tRNAs, should be noted [12]. Furthermore, the mnm^5s^2U, $cmnm^5s^2U$, mo^5U, cmo^5U, $mcmo^5U$ are specific for tRNAs from eubacteria, while mcm^5U, mcm^5s^2U, m^5Um, m^3C, yW, O^2yW and i^6A are only found in eukaryotic tRNAs. Therefore, the corresponding enzymes must be specific for each group of organism. Furthermore, it is likely that there are different enzymes catalyzing the formation of the same modified nucleoside in different positions of the tRNA. Therefore, it is more relevant to compare the presence of modified nucleosides in each position of the tRNA (Table I A–C) than to only compare the presence of a modified nucleoside as such.

In the amino acid stem, positions 8 and 9 are modified in tRNA from all three kingdoms but the modified nucleosides are not the same (Table I A). In the D-loop (positions 10–26), Ψ13 is present in tRNA from all kingdoms. Position 15 is never modified in eubacterial and eukaryotic tRNAs, in contrast to tRNAs from archaebacteria, which have in this position the archaebacterial specific modified guanosine (G*). Furthermore, G*15 occurs in 55% of the sequenced archaebacterial tRNAs. In both eukaryotic and eubacterial tRNAs Gm18 is frequent (25%) but absent in archaebacterial tRNAs. No eubacterial specific modification is present in the D-loop region.

In the anticodon stem and loop (except position 34 and 37, see below) Gm29 is specific for archaebacterial tRNA (Table I B). Transfer RNAs from all three kingdoms have frequently Ψ in positions 38 and 39. While the variable loop is not modified in tRNAs from archaebacteria, the TΨC-loop shows some interesting features (Table I C): (i) in tRNAs from all kingdoms Ψ55 is conserved; (ii) tRNAs from all three kingdoms contain a modified U54, but the archaebacterial tRNA possess $m^1\Psi54$ compared to the frequently occurring m^5U (T54, ribothymidine) in eukaryotic and eubacterial elongator tRNAs; (iii) *all* tRNAs sequenced from archaebacteria have Cm56, while both eubacterial and eukaryotic tRNAs always have an unmodified C in this position.

Considering the presence of different modified nucleosides, archaebacterial tRNAs is more similar to eukaryotic than eubacterial tRNAs [12]. This is also true even if position

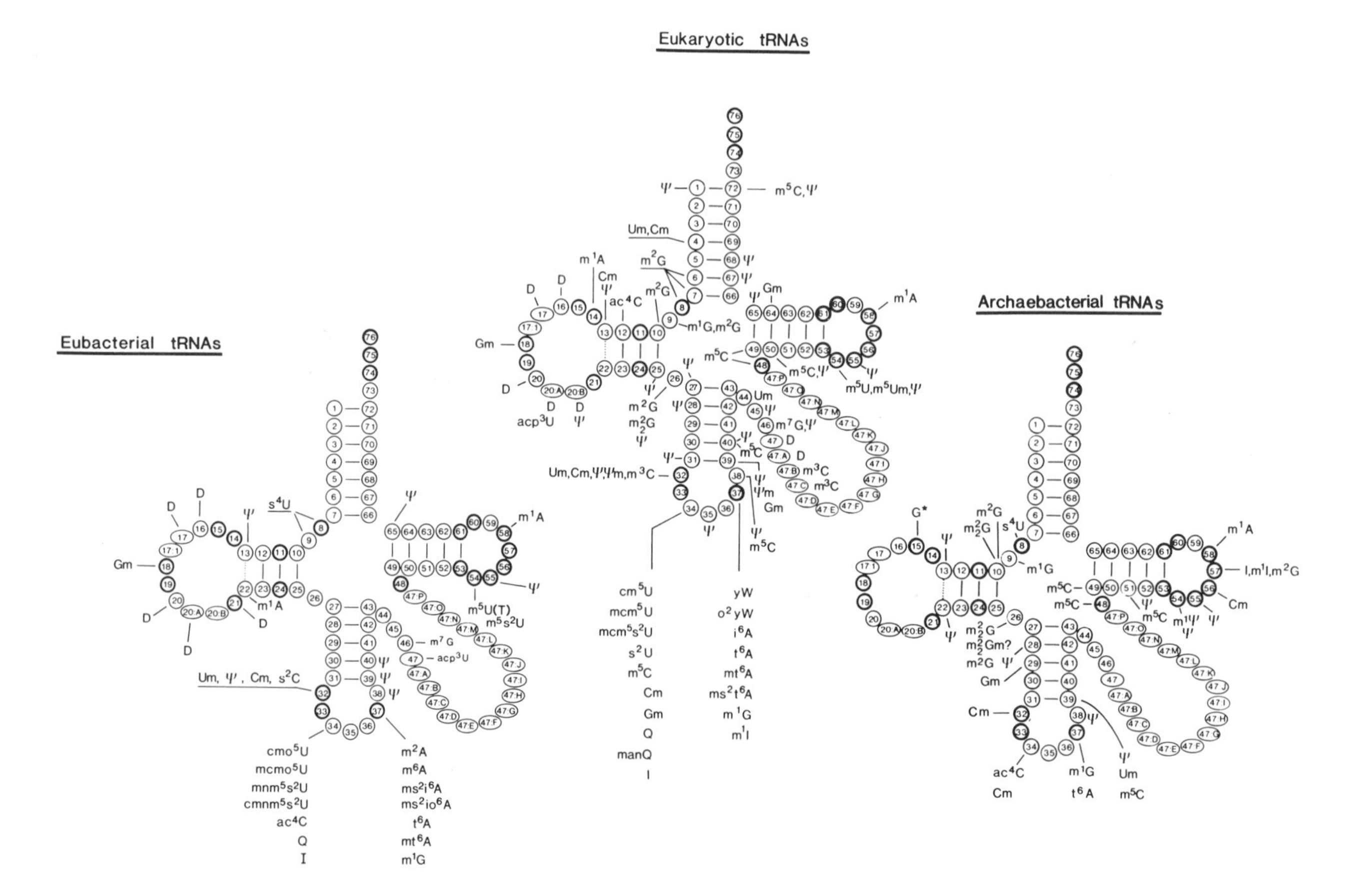

Fig. 2. Nucleoside modification patterns in eubacterial, eukaryotic and archaebacterial tRNAs. Abbreviations of the different modified nucleosides are given in Sprinzl *et al.* [9], from which also the data are compiled (about 250 sequenced tRNAs, excluding phage and organelle specific tRNAs). A methyl group is abbreviated m; a carboxyl group c; a thiol group s; an amino group n; an isopentenyl group i; an oxy- or hydroxyl group o. The position on the base is indicated by the numbers.

Table IA. *Pattern of modifications in the amino acid stem and D-stem and loop of tRNAs from different organisms*[a]

Position	Transfer RNA from:		
	Eukaryotes	Eubacteria	Archaebacteria
Amino acid stem			
1	Ψ	—	—
4	Um, Cm	—	—
6	m^2G	—	—
7	m^2G	—	—
8	m^2G	s^4U	s^4U
9	m^1G; m^2G	s^4U	m^1G
67	Ψ	—	—
68	Ψ	—	—
72	m^5C, Ψ	—	—
D-loop and stem			
10	m^2G	—	m^2G; m^2_2G
12	ac^4C	—	—
13	Ψ, Cm	Ψ	Ψ
14	m^1A	—	—
15	—	—	G*
16	D	D	—
17	D	D	—
18	Gm	Gm	—
20	D	—	—
20: A	D; acp^3U	D	—
20: B	D, Ψ	—	—
21	—	D	—
22	—	m^1A	Ψ
25	Ψ	—	—
26	m^2G; m^2_2G; Ψ	—	m^2G; m^2_2G; (m^2_2Gm)

[a] The tRNAs have been grouped according to Sprinzl *et al.* [9] although recently halobacteria has been suggested to be part of a photosynthetic group, the 'photocyta' [31]. The m^2_2Gm is placed in parentheses since its identification is tentative [32].

Table IB. *Pattern of modifications in the anticodon stem and loop, except positions 34 and 37 and the variable loop of tRNAs from different organisms*

Position	Transfer RNA from:		
	Eukaryotes	Eubacteria	Archaebacteria
Anticodon stem and loop, except positions 34 and 37			
27	Ψ	—	—
28	Ψ	—	Ψ
29	—	—	Gm
31	Ψ	—	—
32	Um; Cm; Ψ; Ψm; m^3C	Um; Cm; Ψ; s^2C	Cm
35	Ψ	—	—
38	Ψ; m^5C	Ψ	Ψ
39	Ψ, Ψm, Gm	Ψ	Ψ; Um; m^5C
40	Ψ; m^5C	Ψ	—
Variable loop			
44	Um	—	—
45	Ψ	—	—
46	m^7G; Ψ	m^7G	—
47	D	acp^3U	—
47: A	D	—	—
47: B	m^3C	—	—
47: C	m^3C	—	—
48	m^5C	—	m^5C

Table IC. *Pattern of modifications in the TΨC-loop of tRNAs from different organisms*

Position	Transfer RNA from:		
	Eukaryotes	Eubacteria	Archaebacteria
49	m^5C	—	m^5C
50	m^5C, Ψ	—	—
51	—	—	m^5C
52	—	—	Ψ
54	m^5U; m^5Um; Ψ	m^5U; s^2m^5U	$m^1\Psi$; Ψ; Um
55	Ψ	Ψ	Ψ
56	—	—	Cm(100%)[a]
57	—	—	(I)[b], m^1I; m^2G
58	m^1A	m^1A	m^1A
64	Gm	—	—
65	Ψ	Ψ	—

[a] All archaebacterial tRNAs so far sequenced have Cm in position 56.
[b] Tentative.

specificity is considered, since many positions (m^1G9, m^2G10, m^2_2G10, m^2G26, m^2_2G26, m^5C48 and m^5C49) are shared between archaebacterial tRNAs and eukaryotic tRNAs. Formation of s^4U8 ($tRNA^{Met}$ from *Thermoplasma acidophilus*) occurs both in archaebacteria and eubacteria. Some positions are specific for tRNAs from archaebacteria (Table II). Since most of the sequenced archaebacterial tRNAs originate from *H. vulcanii* it should be noted that G*15, Cm56 and m^1I57 are also present in tRNAs from *Methanobacteria thermophilus* but not in eubacterial tRNAs. Furthermore, $tRNA^{Met}$ from *Sulfolobulus acidocaldarius* and *Thermoplasma acidophilum* both contain m^2G10, m^2G26, Gm29, m^5C48 and Cm56, which are also present in tRNAs from *H. vulcanii* but not in eubacterial tRNAs. Therefore tRNA modification in *H. vulcanii* share many typical features with other archaebacterial tRNAs but not with eubacterial tRNAs.

Some modified nucleosides occur at the same positions in

Table II. *Position specific modification in tRNA from different kingdoms*[a]

Nucleoside modifications specific for tRNAs from:		
Archaebacteria	Eubacteria	All three kingdoms
m^2_2G10	—	—
_G*15_	s^4U9	$\Psi13$
$\Psi22$	D21	Cm32
m^2_2Gm26	m^1_2A22	m^1G37
Gm29	*s^2C32*	t^6A37
Um39	*mnm^5s^2U34*	$\Psi38$
m^5C39	—	$\Psi39$
m^5C51	*$cmnm^5s^2U34$*	$\Psi55$
$\Psi52$	*mo^5U34*	m^1A58
$m^1\Psi54$	*cmo^5U34*	—
Um54	*$mcmo^5U34$*	—
Cm56	*m^2A347*	—
m^1I57	*m^6A37*	—
m^2G57	—	—
—	*ms^2i^6A37*	—
—	ms^2io^6A37	—
—	acp^3U47	—
—	*m^5s^2U54*	—

[a] Nucleosides in italics are only present in tRNA from the indicated kingdom. Modified nucleosides only present in eukaryotes are: mcm^5U, mcm^5s^2U, m^5Um, m^3C, yW, o^2yW and i^6A. Many positions are eukaryotic-specific (cf. Table I A, B, C) and were not included in this table.

tRNAs from all three kingdoms (Table II). The corresponding enzymes might therefore have been present in the ancestor to the three kingdoms or, alternatively, a convergent evolution has occurred.

Modification at position 37 (3′-side of the anticodon) in different tRNAs and their encoding capacities

Positions 34 and 37 often contain modified nucleosides and a certain pattern between modification and encoding capacities of tRNAs has been pointed out earlier [13]. This is especially true for position 37. Codons starting with U or A are mostly read by tRNAs having a hydrophobic modification like i^6A/yW and their derivatives, and hydrophilic like t^6A and its derivatives, respectively [13]. It was suggested that the comparatively weak interaction of A36-U/U36-A in the first position of the anticodon–codon interaction must be stabilized by such hypermodifications at position 37 [13]. However, tRNAs from archaebacteria reading codons starting with U all have m^1G37. Also, some eukaryotic tRNAs as well as one eubacterial tRNA ($tRNA^{Phe}$ from *Mycoplasma* sp.) reading codons starting with U have been shown to have m^1G in position 37 [9]. Furthermore, $tRNA^{Ser(I)}_{GGA}$ and $tRNA^{Ser(V)}_{GGA}$ from *E. coli* have an unmodified A in position 37 [14]. Thus, although the isopentenylation has been shown to be very important for the efficiency of the tRNA from both eubacteria and yeast (see below), a bulky modification in tRNAs reading codons starting with U is apparently not required. Other modifications, like m^1G, found in the tRNAs from archaebacteria, may also work.

Of the transfer RNAs from all three kingdoms reading codons starting with A, only t^6A and its derivatives are present (Table III). In fact, all elongator tRNAs reading codons starting with A have a t^6A or its derivatives in position 37 and no such tRNA has so far been sequenced having an unmodified A37. The only exception to this rule is $tRNA^{Met}$ from *Bacillus subtilis*, which has m^6A37. In tRNAs reading codons starting with C or G, simple methylated derivatives of A or G are present, and this is true for tRNAs from all three kingdoms.

In summary, comparing both codon-reading capacities of tRNA and their pattern of modification at position 37 (3′-side of the anticodon) reveals evolutionary constraints (m^1G, t^6A) as well as great varieties between tRNAs from the three kingdoms. The large variety of modified nucleosides of tRNAs reading codons starting with U might be a reflection of specific functions in the different organisms other than a general improvement in the ability of the tRNA to translate a message.

Function of modified nucleosides in tRNA

Recently, important experimental data have been obtained concerning functional aspects of tRNA modification [1]. Here I will only discuss a few, which have some relevance when comparing tRNA modification in different organisms. The fact that a viable mutant of *E. coli* could be isolated lacking m^5U54 was surprising since this nucleoside is present in all tRNAs in *E. coli* and in most eukaryotic elongator tRNAs [15]. However, careful physiological experiments have shown that the growth rate of the mutant cell is somewhat lower (Table IV) ([16], unpublished results). Experiments *in vitro* have revealed that melting of the TΨC-loop is affected by the modification at U54 [17]. Furthermore, the presence of m^5U reduces the error level and influences the A-site binding *in vitro* [18, 19]. The archaebacterial tRNAs do not contain m^5U54. Instead they have $m^1\Psi54$. Figure 1 shows that the orientation of the methylgroup in relation to the ribose moiety is similar in m^5U and $m^1\Psi$. The data suggest a requirement for a methylgroup at position 54 for proper activity of the tRNA, resulting in m^5U54 and $m^1\Psi54$ in tRNAs from eukaryotes/eubacteria and archaebacteria, respectively.

Archaebacterial tRNAs have frequently m^1G in position 37. Mutants of *E. coli* defective in the synthesis of m^1G has been characterized [20]. When lacking 60% of m^1G in its tRNA, the reduction in growth rate of the mutant is as much as 20% [21]. Furthermore, deficiency in m^1G also influence the regulation of a biosynthetic operon (J. M. Calvo, Ithaca, USA, pers. comm.); therefore, it is likely that the m^1G is important in the anticodon–codon interaction. However, a tRNA constructed *in vitro* with an unmodified G37 is able to carry out translation *in vitro* [22]. Furthermore, two tRNAs, $tRNA^{His}_{GUG}$ from Human HeLa cells and $tRNA^{Leu}_{AAG}$ from *Caenorhabdi elegans*, normally contain G37 [9]. Although these results suggest that the m^1G37 is not essential for the activity of the two tRNAs mentioned, the results obtained

Table III. *Modification at position 37 (3′-side of the anticodon) in different tRNA species and their encoding capacity*

Codon[a] (nucleotide)				Modifications in position 37 in tRNA from:		
1st	2nd	3rd	Amino acid inserted	Eukaryotes	Eubacteria	Archaebacteria
U	N	U, C	Phe, Ser Tyr, Cys	yW; o^2yW; i^6A, m^1G i^6A; m^1G	ms^2i^6A, ms^2io^6 A, m^1G ms^2i^6A; ms^2io^6A, i^6A	m^1G m^1G
		A, G	Leu, Ser, Trp	m^1G	ms^2i^6A, m^1G, ms^2io^6A	m^1G
C	N	U, C	Leu, Pro, His, Arg	m^1G	m^1G, m^2A	m^1G
		A, G	Leu, Pro, Gln, Arg	m^1G	m^1G, m^2A	m^1G
A	N	U, C	Ile, Thr, Asn, Ser	t^6A	t^6A, mt^6A	t^6A
		A, G	Ile, Met Thr, Lys, Arg	t^6A, mt^6A ms^2t^6A	t^6A, ms^2t^6A m^6A	t^6A
G	N	U, C	Val, Ala, Asp, Gly	m^1G, m^1I	m^2A, m^6A	—
		A, G	Val, Ala, Glu, Gly	m^1I	m^2A, m^6A	m^1G

[a] N, any of the four nucleotides.

Table IV. *Function of some modified nucleosides in eubacterial tRNA*[a]

Modified nucleoside	Results obtained *in vitro*	Mutant analyzed	Phenotypic effects of mutants analyzed
m^5U54	Thermal stability; fidelity; A-site binding	*trmA*	Small (4%) reduction in growth rate
m^1G37	—	*trmD*	Reduced (20%) growth rate; derepression
t^6A37	Stabilizes AC–AC interaction	—	—
ms^2i^6A37 ms^2io^6A37	Ribosomal binding; stabilizes AC–AC interaction, decreased UGA suppression	*miaA*	Reduced (up to 50%) growth rate, derepression, antisuppression, increased codon context sensitivity, lower error level, reduced polypeptide chain growth rate

[a] For references, see text.

with the *E. coli* (*trmD*) mutant imply that m^1G37 is of functional significance.

Codons starting with A are read by t^6A37 containing tRNAs from all organisms (cf. Table III). At present no mutant is available why the significance *in vivo* of t^6A cannot be evaluated. However, several results obtained *in vitro* suggest that this modification may play an important role in the anticodon–codon interaction by stabilizing the interaction, most probably through stacking [23, 24].

Transfer RNAs lacking ms^2i^6A have a reduced ability to bind to ribosomes *in vitro* [25]. From measurements of anticodon–anticodon interaction it was shown that ms^2i^6A stabilizes such an interaction, possible by improving stacking [26]. The *miaA1* mutant of *S. typhimurium* has a large (up to 50%) reduction in growth rate, a reduced polypeptide chain elongation rate *in vivo* and an impaired regulation of several biosynthetic operons [27]. Furthermore, the efficiency of several nonsense suppressor tRNAs is reduced when they lack ms^2io^6A37 [28]. However, perhaps more important is that the presence of ms^2io^6A37 make the tRNAs less sensitive to the codon context, i.e. the nucleosides surrounding the codon [28]. Yeast mutants lacking i^6A in the tRNA grow as well as the wild-type cells, but the efficiency of suppression is reduced, indicating that, also in eukaryotic tRNAs, i^6A37 is important in the anticodon–codon interaction [29, 30]. However, in eukaryotic tRNAs that read codons starting with U, other modified nucleosides are also present besides i^6A. Therefore the need as such for a modification in tRNAs reading codons starting with U is more complex and the requirement has been fulfilled in different ways in the different kingdoms, unlike the case for tRNAs, reading codons starting with an A (cf. Tables III and IV).

In conclusion, tRNA modification is present in all kingdoms and in some cases it has been shown to be of functional significance. It seems that the modified nucleoside improves in a subtle but evolutionarily important way the performance of the tRNAs.

Acknowledgement

This paper is dedicated to the memory of Dr Ingvar Svensson, Uppsala, who died in May 1985.

This work was supported by the Swedish Cancer Society (proj. no. 680), Swedish National Science Foundation and the Swedish Board for Technical Development. I am grateful for the critical reading of the manuscript by Drs T. Hagervall Umeå, S. Normark Umeå and K. Stråby Umeå.

References

1. Björk, G. R., in *Procession of RNA* (ed. D. Apirion), p. 291. CRC Press, Boca Raton, Florida (1984).
2. Smolar, N., Hellman, U. and Svensson, I., *Nucleic Acid Res.* **2**, 993 (1975).
3. Singer, C. E., Smith, G. R., Cortese, R. and Ames, B. N., *Nature New Biol.* **238**, 76 (1982).
4. Tsang, T. H., Buck, M. and Ames, B. N., *Biochim. Biophys. Acta* **741**, 180 (1983).
5. Ny, T., Thomale, J., Hjalmarsson, K. J., Nass, G. and Björk, G. R., *Biochim. Biophys. Acta* **607**, 227 (1980).
6. Byström, A. S., Hjalmarsson, K. J., Wikström, P. M. and Björk, G. R., *EMBO J.* **2**, 899 (1983).
7. Arps, P. J., Marvel, C. C., Rubin, B. C., Tolan, D. A., Penhoet, E. E. and Winkler, M. E., *Nucleic Acid Res.* **13**, 5297 (1985).
8. Lindström, R. R., Stüber, D. and Björk, G. R., *J. Bacteriol.* **164**, 1117 (1985).
9. Sprinzl, M., Moll, J., Meissner, F. and Hartman, T., *Nucleic Acid Res.* **13**, r1 (1985).
10. Gupta, R., *J. Biol. Chem.* **259**, 9461 (1984).
11. McCloskey, J. A., *System. Appl. Microbiol.* (in the press).
12. Gupta, R. and Woese, C. R., *Current Microbiol.* **4**, 245 (1980).
13. Nishimura, S., in *Transfer RNA: Structure, Properties and Recognition* (ed. P. R. Schimmel, D. Söll and J. N. Abelson), p. 547. Cold Spring Harbor Laboratory, Cold Spring Harbor, New York (1979).
14. Grosjean, H., Nicoghosian, K., Haumont, E., Söll, D. and Cedergren, R., *Nucleic Acid Res.* **13**, 5697 (1985).
15. Björk, G. and Isaksson, L. A., *J. Mol. Biol.* **51**, 83 (1970).
16. Björk, G. R. and Neidhardt, F. C., *J. Bacteriol.* **124**, 99 (1975).
17. Davenloo, P., Sprinzt, M., Watanabe, K., Albani, M. and Kersten, H., *Nucleic Acid Res.* **6**, 1571 (1979).
18. Marcu, K. B. and Dudock, B. S., *Nature (London)* **162**, 159 (1976).
19. Kersten, H., Albani, M., Männlein, E., Praister, R., Wurmbach, P. and Nierhaus, K. H., *Eur. J. Biochem.* **114**, 451 (1981).
20. Björk, G. R. and Kjellin-Stråby, K. S., *J. Bacteriol.* **133**, 508 (1978).
21. Hjalmarsson, K. J., thesis, Umeå University (1983).
22. Bruce, A. G., Atkins, J. F., Wills, N., Uhlenbeck, O. and Gesteland, R. F., *Proc. Natl. Acad. Sci. USA* **79**, 7127 (1982).
23. Weissenbach, J. and Grosjean, H., *Eur. J. Biochem.* **116**, 207 (1981).
24. Miller, J. P., Hussain, Z. and Schweizer, M. P., *Nucl. Acid. Res.* **3**, 1185 (1976).
25. Gefter, M. L. and Russel, R. L., *J. Mol. Biol.* **39**, 145 (1969).
26. Vacher, J., Grosjean, H., Houssier, C. and Buckingham, R. H., *J. Mol. Biol.* **177**, 329 (1984).
27. Ericson, J. E. and Björk, G. R., *J. Bacteriol*, in press.
28. Bouadloun, F., Srichaiyo, T., Isaksson, L. A. and Björk, G. R., *J. Bacteriol*, in press.
29. Laten, H., Gorman, J. and Bock, R. M., *Nucl. Acid. Res.* **5**, 4329–4342 (1978).
30. Janner, F., Vögeli, G. and Fluri, R., *J. Mol. Biol.* **139**, 207–219 (1980).
31. Lake, J. A., Clark, M. W., Henderson, E., Shawn, P. F., Oakes, M., Schneinman, A., Thornber, J. P. and Mah, R. A., *Proc. Natl. Acad. Sci. USA* **82**, 3716 (1985).
32. Kuchino, Y., Ihara, M., Yabusaki, Y. and Nishimura, S., *Nature (London)* **298**, 684 (1982).

Chemica Scripta 1986, **26B**, 97–101

Structure and Function of RNA

Olke C. Uhlenbeck

Department of Chemistry and Biochemistry, Campus Box 215, University of Colorado, Boulder, Colorado 80309, USA

Paper presented at the Conference on 'Molecular Evolution of Life', Lidingö, Sweden, 8–12 September 1985

Abstract

RNA molecules, like proteins, have a precisely defined tertiary structure in solution that is essential for its biological function. Research in our laboratory involves making changes in the primary structure of an RNA and measuring the resulting change in its function with an appropriate biochemical assay. In this paper several examples of such 'structure–function' experiments will be given. Although their motivation was primarily directed at understanding the biochemical mechanism of how RNAs are 'recognized' by proteins, some insight into how such interactions might evolve can be obtained.

RNA bacteriophage translational operators

Early in the course of phage infection the 14000 Da bacteriophage R17 coat protein regulates the expression of the R17 replicase gene by binding to a small hairpin loop that contains the gene replicase initiator (boxed in Fig. 1) and thereby prevents ribosomes from binding [1]. This site of binding is called the translational operator. Since the interaction has a relatively high association constant ($K_a = 10^9$) and is quite specific, we chose it for study as an example of a sequence-specific RNA–protein interaction. A 21-nucleotide synthetic RNA fragment (capitals in Fig. 1) was found to bind coat protein with a K_a identical to intact R17 RNA [2]. This indicates that, unlike several of the ribosomal proteins, all the structural elements defining the R17 coat protein binding site lie entirely within a relatively short, contiguous sequence of nucleotides.

By measuring the association constant of coat protein to a large number of variants of the wild-type fragment we have developed a general understanding of which structural features of the fragment are important for coat protein binding [3]. To begin with, it is clear that maintaining the hairpin loop secondary structure is essential for the interaction. Variants which cannot form one or more of the upper five base pairs bind very poorly to coat protein. The lower two pairs are much less important. On the other hand, it does not seem to be important what kind of base pairs make up the helix. Several variants which have one or more of the base pairs changed bind to the coat protein with an affinity constant close to that of the natural sequence.

While all seven of the single-stranded residues in the 21mer are important for the RNA–protein interaction, the consequences of changing nucleotides at each position can be quite different. Three general categories are observed. First, while the three uncircled single-stranded residues in Fig. 1 cannot be deleted without greatly reducing the affinity of coat protein binding, any nucleotide can be tolerated at these positions with little or no effect on K_a. In contrast, the three circled A

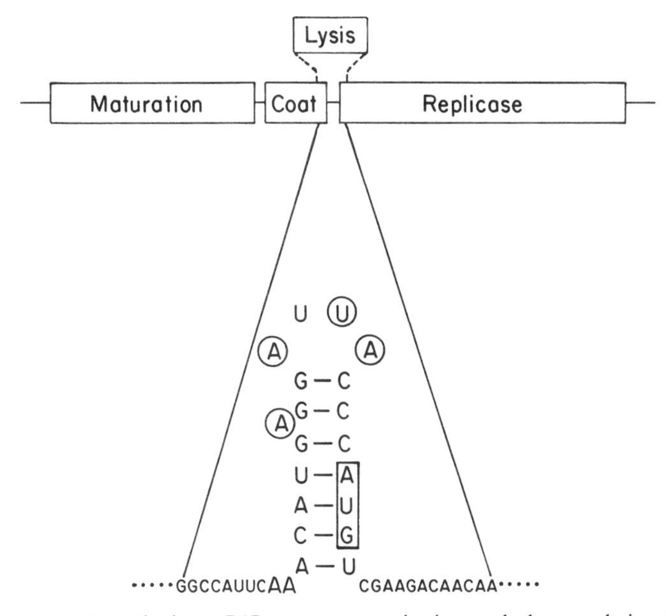

Fig. 1. Bacteriophage R17 genome organization and the translational operator of the replicase gene. The synthetic 21 nucleotide fragment is in capital letters. The replicase initiation codon is boxed. The four circled nucleotides cannot be replaced without abolishing coat protein binding.

residues all appear essential for protein binding. If any one is deleted or substituted with any other residue, binding virtually disappears. Finally, the result of substituting the circled U residue with different nucleotides is unique. When a purine is substituted in this position, the binding is weaker, but when C is substituted the binding is nearly 100-fold tighter [4]. The fact that this tighter binding sequence is not used *in vivo* makes it clear that the equilibrium constant for a protein–nucleic acid interaction evolves to a value that optimizes the regulation and not to one that maximizes the strength of the interaction. The information obtained from all the variant binding sites can be summarized by proposing a model for the R17 coat protein binding site (Fig. 2). The hairpin-loop secondary structure is held together by 5 base pairs of arbitrary sequence. Four of the single-stranded positions must have the nucleotides indicated, while the other three are essential but arbitrary. The model suggests a large number of sequences quite different from the replicase initiator which should be able to bind R17 coat protein. One of these (Fig. 2) that differs from the wild-type sequence in 9 positions was synthesized and found to bind coat protein with the same affinity as the wild-type fragment. This test therefore supports the binding site model.

Several features of the R17 coat protein–RNA interaction

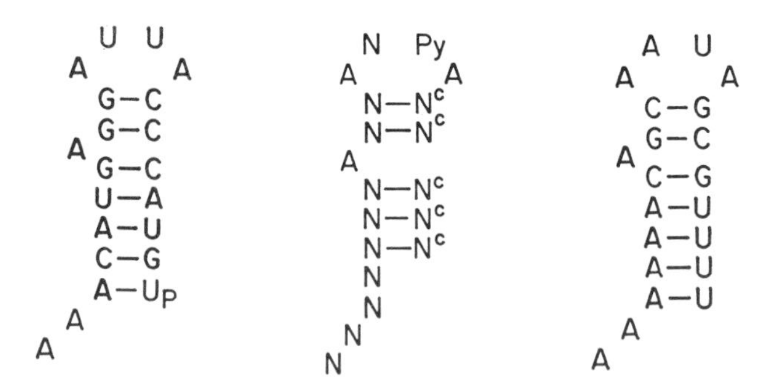

Fig. 2. R17 coat protein binding sites. Left: the natural sequence. Center: a general model for the site. N is an arbitrary nucleotide. Right: a variant which differs at 9 positions from wild-type but binds coat protein.

are worth mentioning. First, the necessity for an intact RNA secondary structure and the substantial number of essential nucleotides make it clear that the two macromolecules contact one another over a large area, similar to two protein subunits. Presumably, 'recognition' is achieved by the proper alignment of weak bonds spread over the surface of interaction. Secondly, the fact that coat protein binding is not sensitive to base-pair changes may be a general feature of RNA binding proteins not shared by DNA binding proteins. Indeed, the observation that the sequence of ribosomal RNAs from closely related species often differ in such a way that helix structure is maintained supports this point [5]. The inability of amino acid side-chains to penetrate the deep groove of the RNA helix may provide a rationale for this. Finally, two structural features which are quite common in the 16S and 23S rRNA secondary structures are bulged A residues and the presence of two purines at the end of helices [5]. It is interesting to note that not only do both features appear in the R17 fragment, but both are essential for protein binding. The possible role of the bulged A in protein binding has been discussed previously [6].

It is tempting to use the binding data of variant RNAs to deduce the location and type of protein–RNA contacts. For example, it would seem reasonable to conclude that the four essential single-stranded residues contact the protein through the nucleotide functional groups whereas the remainder of the molecule (which is not base-specific) contacts the protein through the ribose-phosphate backbone. However, such deductions make the crucial assumption that the structure of the variant fragments only differ from the wild-type at the site of change. Other explanations can exist. For example, when the extrahelical 'bulged' A residue is changed to a C, the position of the nucleotide may change from being intercalated into the helix between the adjacent base pairs to being 'looped out' and the adjacent base pairs stacked upon each other [7, 8]. This difference in structure may be large enough such that, when C is present, the protein–RNA contacts on either side of the residue cannot form, resulting in a reduced K_a. If this view is correct, it is possible that the functional groups of the bulged A are not involved in protein contact at all. Thus, in the absence of detailed structural information on the variants, conclusions on the number and type of contacts should be made with caution.

In one case, however, the precise nature of a coat protein–RNA contact has been proposed [9]. The essential uridine in the hairpin is believed to form a transient covalent complex with one of the two cysteines on the coat protein. The mechanism of the interaction involves attack by the cysteine

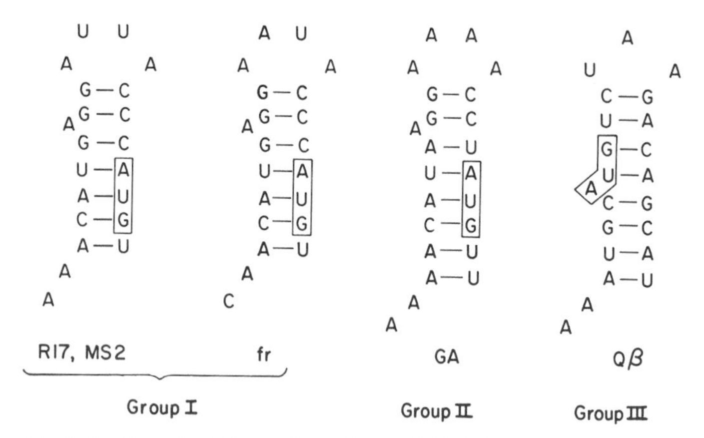

Fig. 3. Replicase initiation regions of several RNA bacteriophage.

anion on the 6-carbon of uridine, which is similar to what has been proposed for the tRNA synthetases [10] and shown for thymidylate synthetase [11]. The tighter binding of the C variant at this position is presumably a consequence of the increased susceptibility of C to nucleophilic attack of the 6 position.

Figure 3 compares possible RNA secondary structure around the replicase initiation codon of four different *E. coli* RNA bacteriophage. In the case of fr [12] and Qβ [13] there is evidence that these structures correspond to the site of translational repression, but the GA [14] site is entirely hypothetical. It is striking that among these three (of four) major seriological groups of phage, the general structure of the initiators are quite similar. All are short hairpin loops with 3 or 4 single-stranded residues and a single bulged A residue on the 5′ side. In three of the four cases a uridine is present in the hairpin as a potential candidate for a transient covalent contact with a cysteine. It is interesting to note that the only virus (GA) that does not have a single-stranded pyrimidine in its hairpin also does not have any cysteines in its coat protein.

We have recently begun a study of the fr coat protein with its translational operator. This system was of interest because the amino acid sequence of fr coat protein differs from that of R17 in 20 of 129 amino acids [15], but the two phage have translational operators which differ at two positions that are not essential for R17 coat protein binding [12]. Thus, the hairpin helix was required in a similar manner and the same

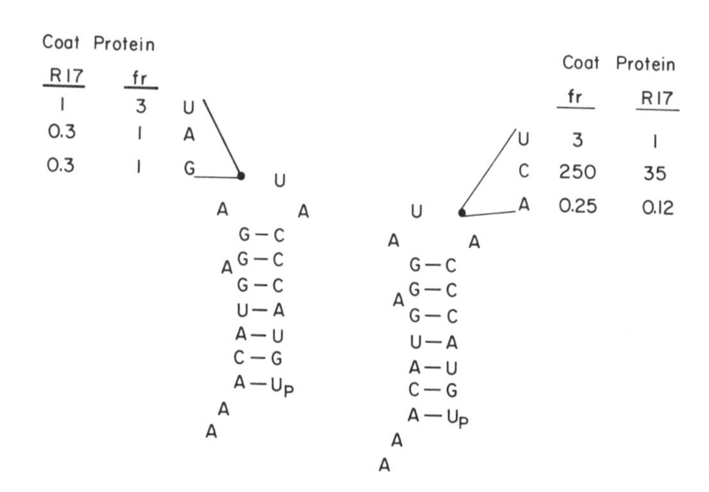

Fig. 4. Binding of fr and R17 coat proteins to several variants. The numbers are K_a values × 10^{-9}. The binding constants were determined in 0.1 M Tris pH 8, 10 mM-MgCl and 0.1 M-KCl at 0 °C as described previously [3]. Unpublished data of K. Kastelic.

single-stranded nucleotides were found to be essential for protein binding. However, one major difference was that the fr coat protein bound to each RNA with a K_a 2–8 times *higher* than R17 coat protein. Some of these data are shown in Fig. 4. The origin of this tighter binding is unclear since all oligomers tested showed the higher K_a.

The higher K_a of fr coat protein for RNA fragments provides an explanation of why A is observed in the place of a U in the fr translational operator. As seen in the data in Fig. 4, the K_a for the A-containing fragment is about threefold lower than the U-containing fragment with both R17 and fr coat proteins. As a result, the K_a of each coat protein for its homologous sequence is the same. In other words, fr compensates for its tighter binding to RNA by using the weaker binding A in the hairpin. This reemphasizes the fact that the equilibrium constant evolved to be the value necessary for the regulation. The fr system evolved slightly differently from the R17 despite the fact that the binding sites are very similar.

Role of the anticodon in aminoacylation

The specific interaction of a tRNA with its cognate aminoacyl tRNA synthetase appears to be a more complex protein–RNA interaction than the R17 system. Both the RNA and the protein are considerably larger and are structurally more complex. In addition, the multistep mechanism of the aminoacylation reaction opens the possibility that different specific contacts between RNA and protein may occur at different stages in the reaction. Significant changes in the conformation of either macromolecule during the course of the reaction may occur as well. Nevertheless, it seems likely that, like the R17 case, contacts between the tRNA and the synthetase will be spread out over a wide portion of the molecule instead of being focused at a few 'recognition' nucleotides. Available data supports this view, with the area of contact being along the inside of the L-shaped tRNA tertiary structure [16].

Our laboratory has been interested in the role of the anticodon nucleotides as part of the contact surface with the synthetase. Despite the fact that the anticodon would appear to be a convenient site to distinguish tRNAs from one another, it has generally been believed that the anticodon plays no role in the synthetase 'recognition' process. Three types of experimental evidence have usually been cited to support this conclusion. First, examples are available where the anticodon was removed without eliminating its ability to be charged [17, 18]. Secondly, anticodon modifications to form suppressor tRNAs do not appear to disrupt the ability of the tRNA to be aminoacylated by the original synthetase. Thirdly, several tRNA isoacceptor groups (such as serine) have quite different anticodons and still are aminoacylated by a single synthetase. In all three of these cases the conclusion that the anticodon is not important is based on the expectation that altering the anticodon should have such a large effect on the overall reaction rate that no aminoacylation will be observed. This expectation is clearly not always the case. Kisselev [19] has collected data from a variety of experiments that suggests that anticodon modification can have a considerable range of effects on the aminoacylation reaction. He lists 9 *E. coli* and 5 yeast synthetases that seem to show some changes in the aminoacylation reaction upon anticodon modification.

Over the past several years our laboratory has developed

Table I. *Aminoacylation of position-34-substituted tRNAs*[a]

tRNA	Nucleotide at position 34	K_m (nM)	V_{max}
Yeast tRNA^{Phe}	G	43	(100)
	U	217	67
	C	250	51
	A	306	56
Yeast tRNA^{Tyr}	G	52	(100)
	U	270	71
	C	230	48
	A	550	80

[a] Data from ref. [22] and [32].

enzymatic procedures which remove anticodon loop nucleotides from yeast tRNA^{Phe} and tRNA^{Tyr} and replace them with arbitrary nucleotides resulting in what we call anticodon-loop-substituted tRNAs. tRNA^{Tyr} and tRNA^{Phe} were chosen primarily for their ease of purification and modification, but also because both synthetases were cited as examples where the anticodon was *not* important in the interaction. tRNA^{Phe} was one of the first where it was shown that the removal of anticodon nucleotides did not eliminate aminoacylation [18] and two different yeast suppressor tRNAs are derived from tRNA^{Tyr} [20, 21].

A large variety of anticodon-substituted tRNA^{Phe} and tRNA^{Tyr} were prepared and tested for aminoacylation with their cognate synthetase. Whereas all substituted tRNAs were active, some of them had a substantially reduced forward reaction rate. Since the rate of spontaneous deacylation remains the same, the level of aminoacylation at equilibrium was reduced unless more enzyme is added. This situation is identical to what was previously observed for the tRNAs missing the anticodons.

The kinetics of the aminoacylation reaction were determined for each substituted tRNA in sufficient detail to obtain a K_m and V_{max} for the reaction. The data for the position 34 (wobble base) substitutions are summarized in Table I. It is clear that both the phenylalanine and tyrosine enzymes interact with their tRNAs in a similar manner at this position. It is primarily the K_m of the reactions which are altered. Depending on the type of substitution, this parameter increases 4- to 10-fold. The data for substitutions at other anticodon positions are also quite similar for the two systems. The effects at position 35 are smaller, but also primarily in K_m, while changing position 33 to any nucleotide has no effect on the aminoacylation reaction in either case. The fact that only the anticodon nucleotides appear important is consistent with the view that it is the inside of the 'L' of the tRNA tertiary structure that contacts the enzyme. U-33 is on the other side of the tRNA and involved in the anticodon loop tertiary structure.

While the above experiments clearly indicate that functional groups on the anticodon nucleotides make contact with their cognate synthetases, it remains unclear the extent to which the anticodons are used by the synthetases to distinguish between the tRNAs. A view on this question can be obtained by examining the aminoacylation properties of tRNAs that have the correct anticodon but differ elsewhere in the molecule. Two examples of such experiments are shown in Fig. 4. In the first, the middle base of anticodon of tRNA^{Tyr} is changed from a ψ to an A, resulting in a tRNA^{Tyr} with a phenylalanine

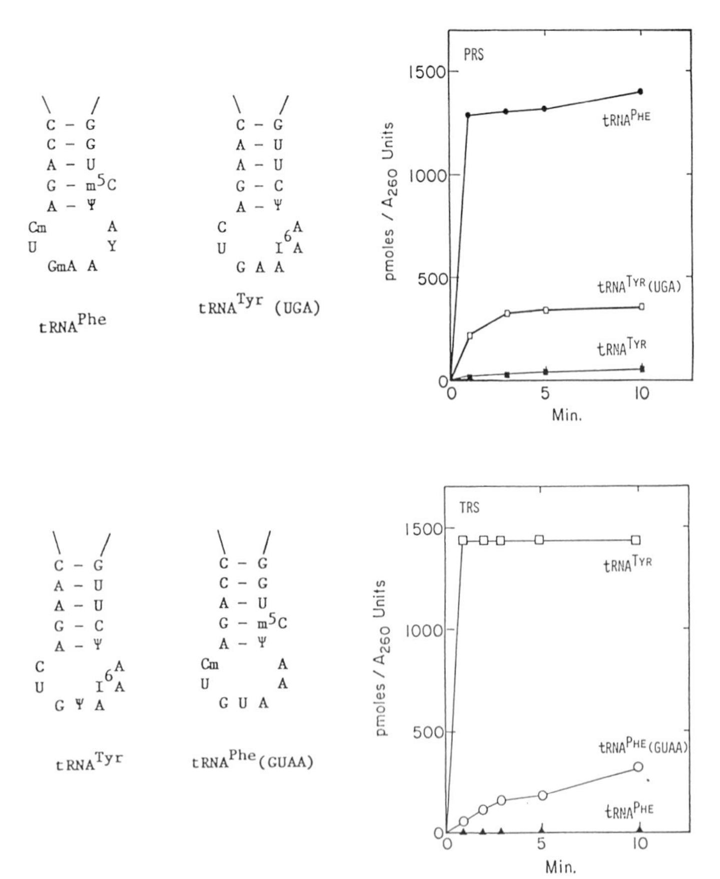

Fig. 5. Misacylation of anticodon-loop-substituted tRNAs. Anticodon structures given at left. Top panel used yeast phenylalanine synthetase. Lower panel used yeast tyrosine synthetase. Data from L. Bare and O. C. Uhlenbeck.

anticodon. Whereas $tRNA^{Tyr}$ does not detectably misacylate with the phenylalanine enzyme, the variant tRNA shows clear misacylation. Kinetic studies [22] show that the K_m is about 13-fold higher and the V_{max} 70-fold lower. The opposite situation was also examined. When $tRNA^{Phe}$ contains the tyrosine codon, misacylation by tyrosine synthetase is stimulated considerably. Since these two tRNAs differ by only the central anticodon nucleotide (position 35), this position must be one of the distinguishing nucleotides used by both synthetases. However, since the misacylation rates are still much lower than the cognate reactions, it is clear that other nucleotides outside the anticodons are not matched correctly. It is unclear how many nucleotides in $tRNA^{Phe}$ would have to be changed in order to get it to aminoacylate with tyrosine enzyme with normal kinetics. Experiments attempting to locate these are in progress.

It seems reasonable to propose that the anticodon originally played a more essential role in the synthetase–tRNA interaction. However, as time progressed, the contact site between the tRNA and the enzyme became more extensive and 'recognition' was achieved using all the contacts. The diminished importance of the anticodon in synthetase recognition would permit isoacceptors with very different anticodon and the possibility of suppressor tRNAs. It is interesting to note that, for at least one synthetase (the *E. coli* methionine enzyme), the anticodon still remains an essential component of the recognition site [23].

Role of uridine-33 in translation

The nucleotide on the 5′ side of the anticodon is one of the most highly conserved nucleotides in tRNA. 253 of 260 tRNAs of prokaryotic, eukaryotic, mitochondrial and archaebacterial sources have a uridine at this position [24]. A structural rationale for this consistency may be that the polynucleotide chain changes direction at this position and uridine is the most favorable nucleotide to carry this out. The conserved nucleotide Ψ-55 lies at the corresponding position in the TψCG loop of tRNA where a similar 'U-turn' structure forms. In most cases the uridine ring is found turned towards the inside of the loop. The crystal structure of yeast $tRNA^{Phe}$ suggests that two intraloop hydrogen bonds stabilize the turn [25]. One extends from the N-3 hydrogen of uridine to an oxygen of phosphate 36. The other, which cannot be present in all tRNAs, is from the O-2′ hydrogen of ribose 33 to the N-7 of adenine 35. However, in the crystal structure of *E. coli* $tRNA^{fmet}$, uridine-33 appears to be pointed in the opposite direction, suggesting that alternative configuration may be possible.

The remarkable conservation of uridine-33 has led to several suggestions that this nucleotide plays an essential role in the translation mechanism. However, experiments involving suppressor tRNAs indicate that this is not the case. Bare *et al.* [26] prepared anticodon-loop-substituted $tRNA^{Tyr}$s with amber anticodons and different nucleotides at position 33 and found that the efficiency of *in vitro* suppression was reduced by no more than twofold when purines were present at position 33. Similarly, Thompson *et al.* [27] showed that the *E. coli* amber suppressor Su_7^+ tRNA was only slightly less effective *in vivo* when a cytidine was inserted at position 33 of the tRNA gene. Thus, while uridine-33 has no obligatory role in the translation process, its presence appears to improve the efficiency of translation. Although the effect appears quite small, it is undoubtedly sufficient for the U containing tRNAs to maintain a selective advantage over mutants in this position. An important conclusion is therefore that just because a nucleotide is highly conserved does not necessarily mean that it has a large selective advantage.

In order to determine in greater detail how uridine-33 functions in the translation system, a variety of U-33 modified yeast $tRNA^{Phe}$ [28] were assayed using kinetic procedures developed by R. Thompson and co-workers to obtain kinetic parameters for several of the early steps of the translation mechanism [29–31]. The system involves simultaneously measuring the rate of dipeptide formation and GTP hydrolysis when ternary complex is added to poly U programmed *E. coli* ribosomes that have acetyl-phe $tRNA^{Phe}$ in the P site. By carrying out the reaction under single turnover conditions in a rapid mixing device, the kinetics of the decoding reaction can be measured. Earlier experiments with *E. coli* phe-$tRNA^{Phe}$ and leu-$tRNA^{Leu}$ indicate that a two-stage recognition process is responsible for the high accuracy of the decoding process [30]. In the first stage, the ternary complex binds ribosomes in one or more reversible binding equilibria and then GTP is hydrolysed to form an activated intermediate. Although elementary rate constants can be obtained for binding and subsequent hydrolysis steps, this stage will be characterized here by K_{GTP}, the measured second-order rate constant of GTP hydrolysis. In the second stage of the reaction, called the proofreading step, the intermediate either

Table II. *Single-turnover kinetics of modified $tRNA^{Phe}s^{a}$*

tRNA	K_{GTP} (10^6 M^{-1} s^{-1})	K_3 (S^{-1})	K_4 (S^{-1})
E. coli $tRNA^{Phe}$	5	0.27	$\leqslant 0.03$
yeast $tRNA^{Phe}$	4.8	0.8	0.2
U	5	0.8	0.2
Ψ	3.6	0.5	0.5
M^3U	2.3	0.5	0.8
Um	2.5	0.5	1.0
C	2.8	0.7	0.3
dC	1.5	>0.7	>2.3
Pu	0.8	>0.8	>3.2

[a] Data from Dix, Wittenberg, Uhlenbeck and Thompson, submitted for publication.

forms dipeptide with rate constant k_3 or dissociates unproductively with rate constant k_4. The values of k_3 and k_4 are calculated from the measured rate of dipeptide formation and the 'proofreading ratio', which corresponds to the number of moles of GTP hydrolyzed per dipeptide bond formed. When a 'near cognate' tRNA is present ($tRNA^{Leu}$) it shows a change in the kinetic parameters at both stages of the reaction [30]. A lower rate of GTP hydrolysis is seen, primarily as a result of weaker binding due to a faster dissociation rate. In addition, a greater proofreading ratio is seen, primarily as a result of an increase in K_m for the noncognate tRNA. By such a two-stage mechanism the wrong amino acid is incorporated at a very low frequency.

A series of yeast $tRNA^{Phe}$ with different modified nucleotides at position 33 were prepared primarily to test whether the two internal hydrogen bonds were critical to the decoding process. Substituting m^3U or C into position 33 would disrupt the N-3 hydrogen bond, while Um should block the 0-2′ hydrogen bond and dC should disrupt both bonds. Purine nucleotides at position 33 would be too bulky to fit into the loop, while inserting U or Ψ should be positive controls. The data for these modified tRNAs are summarized in Table II. It is clear that the tRNAs that cannot form the internal hydrogen bonds are less effective in translation. Even though they have the correct anticodon they show a reduced rate of GTP hydrolysis and an increased proofreading ratio similar to, but not as extreme as, a non-cognate tRNA. The tRNAs that are the least effective contain dC and purine at position 33 and thus are least able to form the correct U-turn structure. Thus, ribosomes are able to detect extremely subtle changes in the anticodon loop structure and reduce the use of those tRNAs.

It is interesting to note that the type of assay used gives a very different view of the selective advantage of U-33. While the overall efficiency of incorporation of a position-33-substituted tRNA is only 2- to 3-fold lower, substantially more GTP must be hydrolysed to give the incorporation.

References

1. Bernardi, A. and Spahr, P. F., *Proc. Natl. Acad. Sci. USA* **69**, 3033 (1972).
2. Krug, M., DeHaseth, P. L. and Uhlenbeck, O. C., *Biochemistry* **21**, 4713 (1982).
3. Uhlenbeck, O. C., Carey, J., Romaniuk, P. J., Lowary, P. T. and Beckett, D., *J. Biomol. Struc. Dynam.* **1**, 539 (1983).
4. Lowary, P. T. and Uhlenbeck, O. C., submitted for publication (1986).
5. Noller, H. F., *Annu. Rev. Biochem.* **53**, 119 (1984).
6. Peattie, D. A., Douthwaite, S., Garrett, R. A. and Noller, H. F., *Proc. Natl. Acad. Sci. USA* **78**, 7331 (1981).
7. Patel, D. J., Kozlowski, S. A., Marky, L. A., Rice, J. A., Broka, C., Itakura, K. and Breslauer, K. J., *Biochemistry* **21**, 445 (1982).
8. Morden, K. M., Chu, Y. G., Martin, F. H. and Tinoco, I., Jr., *Biochemistry* **22**, 5557 (1983).
9. Romaniuk, P. J. and Uhlenbeck, O. C., *Biochemistry* **24**, 4239 (1985).
10. Starzyk, R. M., Koontz, S. W. and Schimmel, P., *Nature (London)* **298**, 136 (1982).
11. Pogolotti, A. L. and Santi, D. V., *Bioorganic Chemistry*, vol. 1, p. 277. Academic Press, New York (1977).
12. Cielers, I. E., Jansone, I. V., Gribanov, V. A., Vishnovski, Y. I., Berzin, V. M. and Green, E. J., *Molekulyarnaya Biologiya* **16**, 1109 (1982).
13. Weber, H., *Biochem. Biophys. Acta* **418**, 175 (1976).
14. Inokuchi, Y., Takahashi, R., Hirose, T., Inayama, S., Jacobson, A. B. and Hirashima, A., *EMBO J.* (in the press).
15. Weber, K. and Konigsberg, W., *RNA Phages* (ed. N. Zinder), p. 51. Cold Spring Harbor Laboratory, Cold Spring Harbor, NY (1975).
16. Rich, A. and Schimmel, P. R., *Nuc. Acids Res.* **4**, 1649 (1977).
17. Haskimoto, S., Kawata, M. and Takemura, S., *J. Biochem.* **72**, 1339 (1972).
18. Thiebe, R. and Zachau, H. G., *Biochem. Biophys. Res. Common.* **33**, 260 (1968).
19. Kisselev, L. L., *Molekulyarnaya Biologiya* **17**, 928 (1983).
20. Piper, P. W., Wasserstein, M., Engback, F., Kaltuft, K., Celis, J. E., Zeuthen, J., Liebman, S. and Sherman, F., *Nature (London)* **262**, 757 (1976).
21. Johnson, P. F. and Abelson, J., *Nature (London)* **302**, 681 (1983).
22. Bare, L. and Uhlenbeck, O. C., *Biochemistry* **24**, 2354 (1985).
23. Schulman, L. H. and Pelka, H., *Nuc. Acid. Res.* **11**, 1439 (1983).
24. Grosjean, H., Cedergren, R. J. and McKay, W., *Biochimie* **64**, 387 (1982).
25. Quigley, G. J. and Rich, A., *Science* **194**, 796 (1970).
26. Bare, L., Bruce, A. G., Gesteland, R. and Uhlenbeck, O. C., *Nature (London)* **305**, 554 (1983).
27. Thompson, R. C., Cline, S. W. and Yarus, M., *Interaction of Translational and Transcriptional Controls in the Regulation of Gene Expression* (ed. M. Grunberg-Manago and B. Safer), p. 189. Elsevier Science, New York (1982).
28. Wittenberg, W. L. and Uhlenbeck, O. C., *Biochemistry* **24**, 2705 (1985).
29. Thompson, R. C., Dix, D. B., Gerson, R. B. and Karim, A. M., *J. Biol. Chem.* **256**, 81 (1981).
30. Thompson, R. C. and Dix, D. B., *J. Biol. Chem.* **257**, 6677 (1982).
31. Thompson, R. C. and Karim, A. M., *Proc. Natl. Acad. Sci. USA* **79**, 4922 (1982).
32. Bruce, A. G. and Uhlenbeck, O. C., *Biochemistry* **21**, 3921 (1982).

Chemica Scripta 1986, 26B, 103–107

Conformational Dynamics and Evolution of tRNA Structure

Rudolf Rigler, Flora Claesens and Lennart Nilsson

Department of Medical Biophysics, Karolinska Institutet, Box 60400, S-104 01 Stockholm, Sweden

Paper presented by Rudolf Rigler at the Conference on 'Molecular Evolution of Life', Lidingö, Sweden, 8–12 September 1985

Introduction

tRNA, a key molecule within the translation machinery, has always been a centre of interest. Although its functional properties are fairly well understood at the biochemical level, the relation between atomic structure and the way it fulfills its task in protein synthesis are still subject to hypothesis and speculation.

Few molecules stimulate the imagination as much as tRNA. With this molecule one hopes to learn the structural basis of its function from the evolution of its structure. The formation of the clover leaf can be understood from the probability of forming hydrogen bonds in polynucleotide chains as suggested by Orgel [1] and later elaborated into an ingenious game by Eigen and Winkler-Oswatitsch [2].

The work of these authors has led to a detailed account of how tRNA sequences may be selected on the basis of base-pair stability and how message and tRNA might have evolved as a negative and positive copy of a nucleic acid sequence [3, 4].

Since decoding is one of the central functions of tRNA, several proposals have been made as to how reading of the code and translation may have evolved in primitive systems. Essential ideas have been put forward notably by Woese [5,6] and Crick *et al.* [7], who have reasoned that a key element in the translation process must be a conformational transition of the anticodon loop. It was suggested that the motion of the message relative to the tRNA was connected with a switch of the stack of the anticodon triplet from the 3′ side of the anticodon loop to its 5′ side [5]. Formation of base pairs on either side of the anticodon loop in addition to the codon–anticodon triplet was proposed in order to stabilize the interaction between tRNA and message during transpeptidation in a primitive translation system [7]. It has also been argued on the basis of experimental evidence that these conformational transitions might trigger changes in the overall conformation of tRNA which might be of importance for its function in peptide-chain elongation.

Investigations on solution structures of biopolymers and their dynamic properties have given evidence of the existence of a variety of conformational states and their functional importance [8]. By the recent development of pulsed laser sources and high-resolution NMR, insight in molecular motion of even complex structures has become possible. For tRNA a picture is emerging of a dynamic structure governed by opening and closing of hydrogen bonds and stacking and unstacking of nucleotide bases which may involve even substantial changes of the overall structure [9].

Like many others we have also focused our attention in particular on the anticodon loop and stem region of tRNA, where the hypermodified nucleotides, particularly at position 37, such as the Y-base wybutine, act as molecular probe-sensing conformational states and their transitions.

The outcome of these investigations provides strong evidence that the anticodon loop of tRNA, when interacting with a codon, must exist in a conformation different from the canonical 3′–5′ stack of the anticodon triplet, and the experimental data regarding the rotational mobility of wybutine are easiest to explain by the existence of a 5′–3′ stack of the anticodon triplet. Both forms exist in a dynamic equilibrium which probably also involves overall conformational changes of the tRNA molecule [10, 11].

We also have started molecular dynamics simulation of the atomic motions within the essential structural element of tRNA-Phe, i.e. a pentadecameric piece containing the anticodon stem and loop. The long-range aspect of this work is to find alternative anticodon conformations for which a potential minimum can be reached. Another aspect is to analyse how motions in the anticodon loop may be related to those in the anticodon stem, a portion of the tRNA molecule which is of relevance for the structure of the anticodon loop region.

Conformational variability of the anticodon loop in tRNA-Phe

Excitation of wybutine at position 37 of yeast tRNA-Phe with picosecond laser pulses [12, 13] leads to a variety of excited states from which the existence of conformational states as well as the mobility of wybutine can be deduced, which depends on the interaction with its immediate environment, i.e. stacking with the neighbouring anticodon base A-36 and A-38 (Fig. 1).

In essence, we have been able to describe two clearly different conformations of wybutine, including one intermediary state which exhibits as their signature different lifetimes of the excited states [14]. In conditions with high neutral salt and Mg^{2+}, wybutine is found predominantly in one state with a lifetime of 7 ns and exhibits no own mobility as deduced from the analysis of the rotational anisotropy.

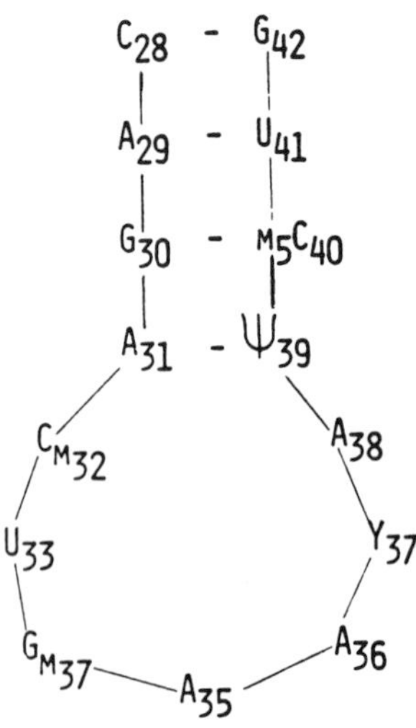

Fig. 1. Sequence of ribonucleotide pentadecamer representing anticodon loop and stem of yeast tRNA^Phe^.

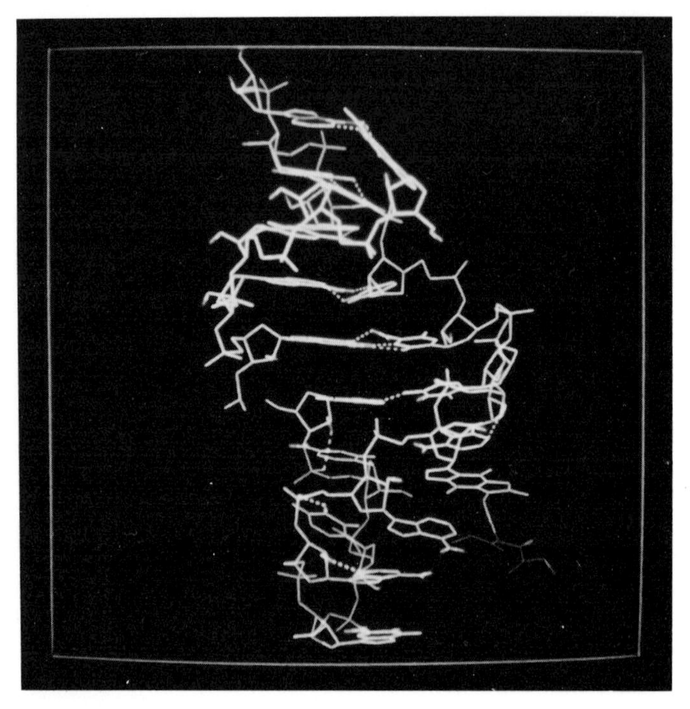

Fig. 2. 3-D structure of the anticodon loop stem region of tRNA^Phe^. Temperature factors (B) colour coded (G. Quigley, personal communication): red – high *B* values; blue – low *B values*. $B = 8\pi^2/3\langle x^2\rangle$ ($\langle x^2\rangle$ mean square fluctuation).

This picture appears to be well in agreement with the crystal structure, where wybutine appears stacked upon the anticodon triplet (Fig. 2). When, however, a codon UUC in the form of the anticodon of *E. coli* tRNA-Glu2 is offered for complementary interaction with the anticodon AAGm of tRNA-Phe, the stacking of wybutine is changed. This large hypermodified base, which in addition has a substantial side-chain spanning almost the entire anticodon triplet, now exhibits a large freedom of rotation, as can be seen from the rotational anisotropy of wybutine fluorescence (Fig. 3). The fast rotation of wybutine of about 370 ps apparently is also the reason for shortening the lifetime of the wybutine fluorescence to about 100 ps. From evaluation of the order parameters of the anisotropy curve in Fig. 3 we deduce that the Y-base must move within a cone of more than 40° half-angle.

Adding the codon UUC to the anticodon AAGm to form a base-paired triplet stack in the manner originally proposed by Fuller and Hodgson [15], the motion of wybutin, considering its long side-chain interfering with the codon–anticodon stack, should be even more restricted rather than enhanced. A solution for this incompatibility would be in fact the existence of a 5′–3′ stacked codon–anticodon triplet as has been proposed before. In this situation (Fig. 4) wybutine positioned at the 'apex' of the anticodon loop would have minimal contacts with its neighbours and could rotate much more freely.

The pentadecameric anticodon-loop-stem fragment – a relevant model

The semisynthetic RNA pentadecamer

5-r(C-A-G-A-Cm-U-Gm-A-A-Y-Ψ-m5C-U-G)

prepared by McLaughlin and Graeser [16], which represents the anticodon stem and loop of yeast tRNA-Phe, has proven to be a relevant and very interesting model for codon–anticodon interactions. Its solution structure has recently been solved from high-resolution NMR studies [17] and it has been shown that the anticodon loop in solution adopts a hairpin loop structure in the 3′–5′ stacked conformation in agreement with the crystallographic picture [18, 19]. It was also found that addition of the codon UpUpC to the anticodon essentially does not alter the structure of the pentadecamer.

We have therefore investigated [20, 21] whether the mobility of wybutine would be in agreement with this result and whether the (anti)codon triplet in the restrained conformation of the anticodon loop of tRNA-Glu2 would be able to induce the substantial change in the anti-codon loop conformation observed with the whole tRNA-Phe molecule. From the rotational anisotropy curves (Fig. 5) it can clearly be seen that wybutine in the presence of saturating amounts of UpUpC is immobile while addition of tRNA-Glu2 leads to an increase in mobility of wybutine which is similar to that found with the whole tRNA-Phe molecule. From the decay of the rotational anisotropy which leads to even larger depolarization of the wybutine fluorescence than in the case of tRNA-Phe and faster rotational motion (rotational relaxation time 100 ps) one must conclude that the codon–anticodon interaction in this case poses even less hindrance for the mobility of wybutine than does the whole tRNA. A detailed evaluation of the wybutine fluorescence and comparison between whole tRNA and pentadecamer [21] shows that the pentadecamer behaves essentially like the tRNA. The difference between the two main conformations of the anticodon loop (stacked and immobile – unstacked and highly mobile wybutine), however, are more pronounced and probably a result of uncoupling the pentadecamer from the rest of the tRNA structure.

The time domain for the motion of individual nucleotide bases which we were able to study experimentally overlaps with that which can be covered by molecular dynamics simulations of motion and dynamics of biological macromolecules. From these simulations information concerning average conformations and the fluctuations of individual atoms and groups of atoms can be obtained [22]. It was therefore of interest to investigate the motion of individual nucleotide bases of the pentadecamer, including wybutine, as well as the local fluctuation around a time-averaged position. A 40 ps molecular dynamics simulation of the pentadecamer

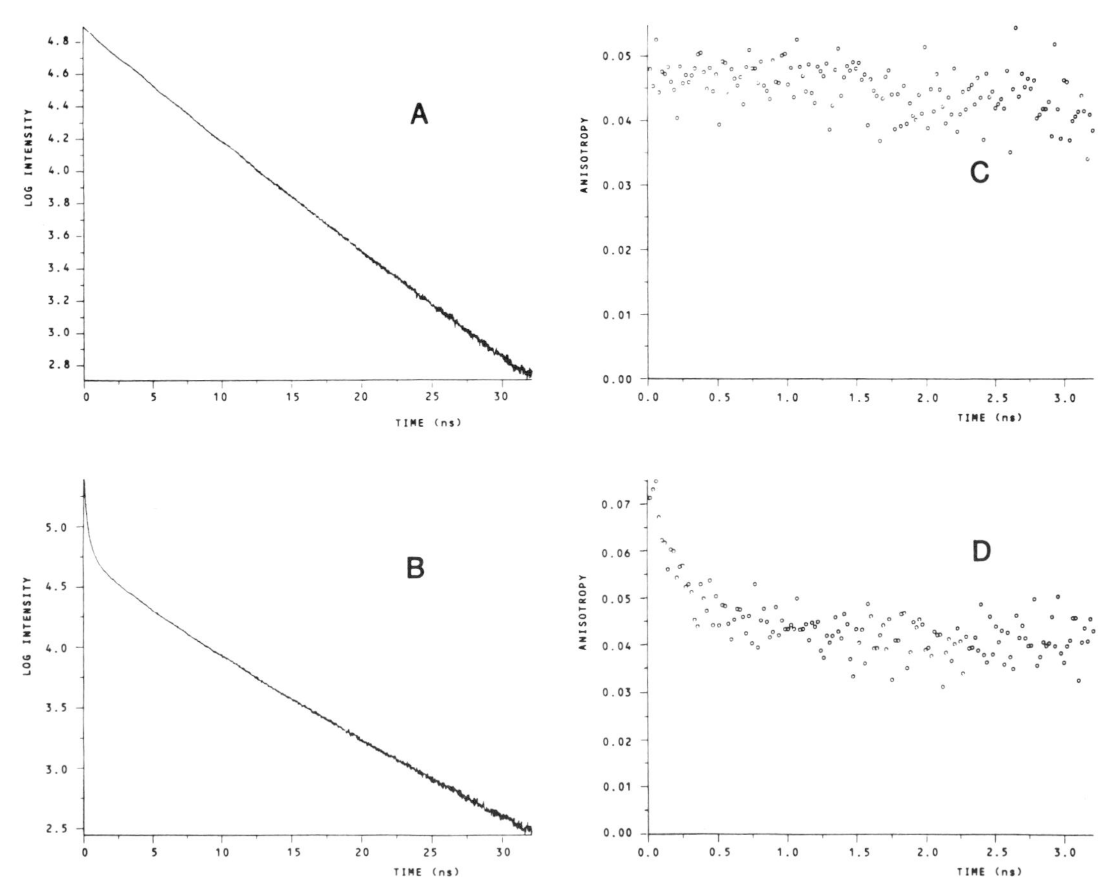

Fig. 3. Decay of intensity and anisotropy of the fluorescence of wybutine in $tRNA^{Phe}$ after a 6 ps laserpulse at 300 nm. In absence of a codon the decay is due to a 7 ns lifetime component (A) while the anisotropy remains unchanged (C). In the presence of the codon S^2UUC of *E. coli* $tRNA^{Glu_2}$ the intensity decay is dominated by a 100 ps component while the anisotropy decays with 370 ps.

TWISTING OF WYBUTINE IN 3' - 5' STACK

ROTATIONAL TUMBLING OF WYBUTINE IN 5'-3' STACK

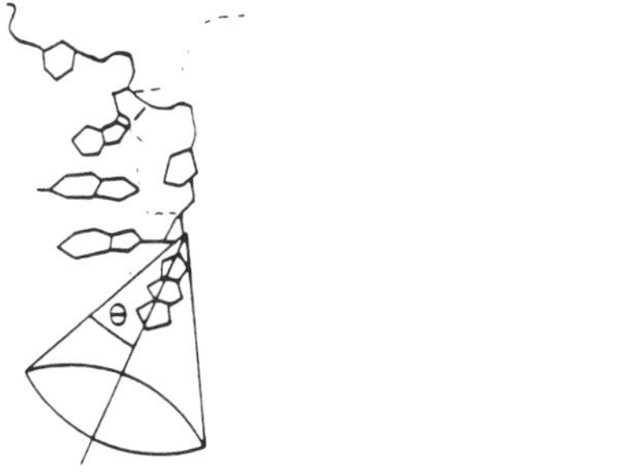

Fig. 4. Models for the anticodon in the 3′–5′ stacked form (above) and in the 5′–3′ stacked form (below).

[23] using a set of potential functions elaborated for nucleotide interactions [24] yielded in particular large fluctuations for G-34 which increased with the time the simulation was carried out (Fig. 6). Although fluctuations were present with wybutine, actual motion was absent in the time scale of the experimental observation, save very fast and restricted motions around 1 ps. Given the limitations of the present molecular dynamics simulation, the data corroborate the picture that wybutine is immobile when stacked on the anticodon triplet in the absence of a codon. Work is continued in order to analyse the situation where the anticodon is interacting with a codon in the 3′–5′ stack as well as in a 5′–3′ stack.

Anticodon loop conformations – a fundamental property?

Our experimental data clearly show that the anticodon loop is able to assume different conformations, which moreover depend on the tertiary structure of the codon offered. Combining all present evidence for the case of tRNA-Phe interaction of its anticodon with the triplet UpUpC does not change the structure in a significant way, i.e. this triplet has to accept the anticodon AAGm in its 3′–5′ stack for hydrogen bonding. However, when offered in a loop-constraint structure as provided by the anticodon loop of tRNA-Glu2 a different anticodon loop conformation of tRNA-Phe is stabilized, leading to an increase of the strength of the codon–anticodon interaction by 3 orders of magnitude [14, 25, 26]. Our data suggest that this conformational variant is significantly different from the canonical anticodon structure and demonstrates that an anticodon conformation with drastically increased codon affinity gives room for a largely unhindered rotational mobility of wybutine. Although no direct stereochemical evidence is available as yet, a 5′–3′ stacked codon–anticodon structure would best accommodate our observations. Indirect evidence for this type of transition has

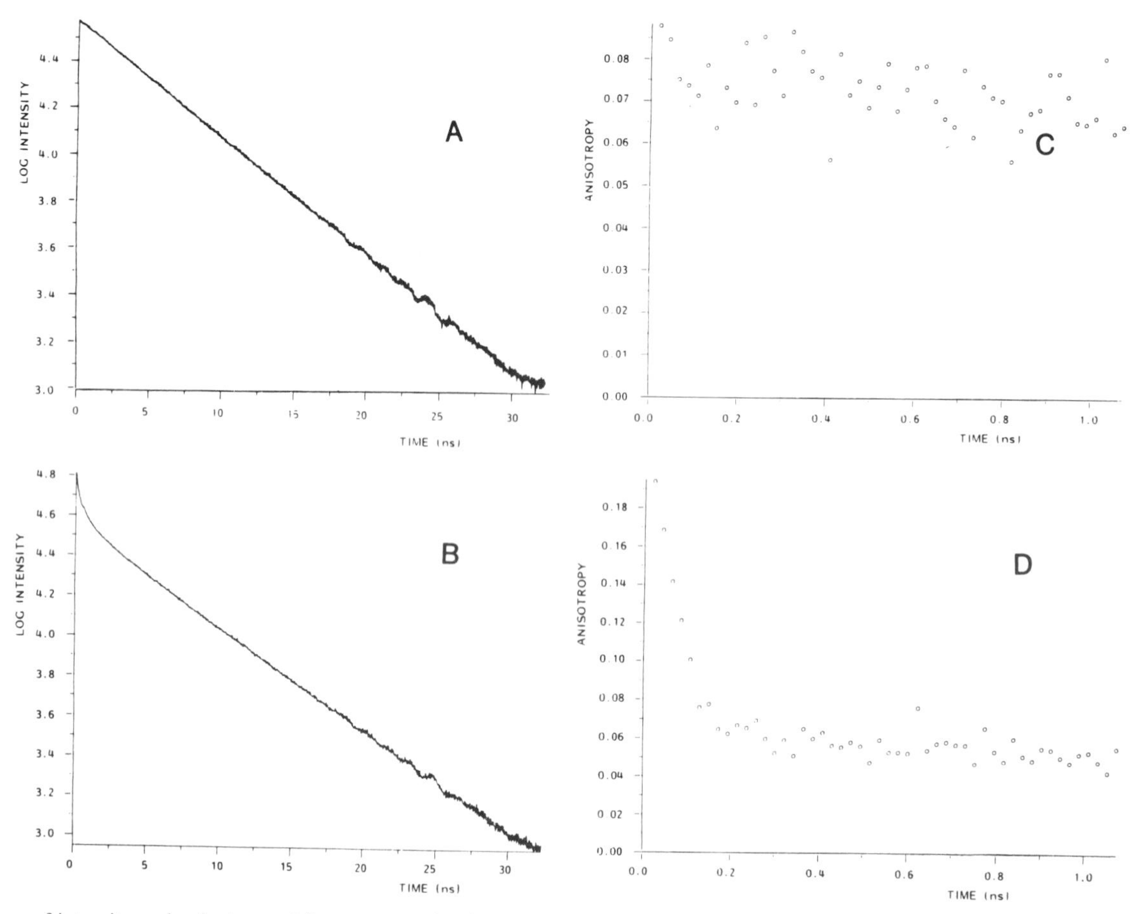

Fig. 5. Decay of intensity and anisotropy of fluorescence of wybutine in the pentadecamer RNA sequence, including the tRNA^{Phe} anticodon loop–stem sequence after a 6 ps laser pulse at 300 nm. Intensity and anisotropy in the presence of the codon rUpUpC (A, C) and of s^2-UpUpC of tRNA^{Glu_2} (C, D).

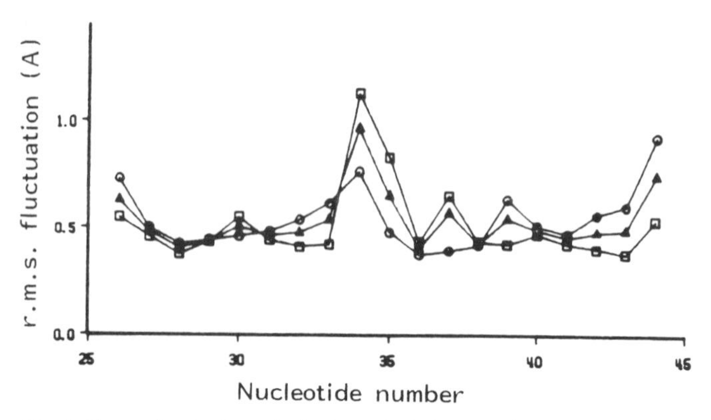

Fig. 6. R.m.s. fluctuation of all atoms (△), bases (□), and backbone (○) averaged for each nucleotide during a 40 ps molecular dynamics simulation.

been given before from temperature relaxation [27] and NMR data [28]; its existence, however, has been questioned on the basis of experimental data which show that hypermodified purines like wybutine can improve the stability of codon–anticodon interaction [29]. Their stabilizing effect has been attributed to the stacking of the large purines with the anticodon an interpretation which has to be modified at least for the case of tRNA-Phe. Likely mRNA occurs in folded loop structures and the anticodon–anticodon loop interactions constitute a relevant model.

In summary, detailed molecular analysis reveals, even in present-day tRNAs, the existence of different anticodon loop conformations, which have been described as necessary requirements for a primitive translation system. If preserved in present tRNA they must constitute a property fundamental to function. Knowledge in particular of the influence of non-cognate codons on the anticodon loop conformation should probide new insight into structural mechanisms of error discrimination.

The results from the complementary interaction of anticodon loop structures suggest on the basis of symmetry arguments that not only one but both loops must have changed their conformation by finding new and even deeper minima in the potential energy surface. Small-angle X-ray studies of the tRNA-Phe-tRNA-Glu2 complex have provided evidence of an overall structure for both tRNAs which is distinctly different from the crystallographic L-shape of the uncomplexed tRNA [11]. The existence of structural and dynamic correlations between different domains of the tRNA structure are most probably essential for its function.

Aside from a direct experimental proof of correlated motions, various domains of tRNA molecular dynamic simulations provide powerful means attacking this problem.

Structural correlations in tRNA

Time-averaged motions have been calculated for anticodon loop and stem structure. Distinct r.m.s. fluctuations around mean positions appear for nucleotides Gm34 and Y37 as well as Ψ39, which forms the first base pair of the anticodon stem (Fig. 6). The existence of intrinsic relationships between anticodon stem as expressed by its base sequence and stability and the anticodon triplet has been discussed by Yarus [30]. Experimental evidence for correlations between anticodon loop and stem structures have been provided for the case of initiator and elongator tRNAs [31,32].

It is likely that these structural correlations are the outcome of molecular evolution selecting for a most effective coding process involving the anticodon itself (two out of three reading [33, 34]) as well as the anticodon stem structure and probably even other parts of the tRNA structure. It is therefore of considerable interest to calculate the correlated motions of individual nucleotides within anticodon loop and stem and ultimately for the whole tRNA. In this manner structural correlations within tRNA probably could be elucidated as well as evolutionary pathways leading to stabilization of certain structures.

Selection of amino acids

One of the enigmas of tRNA is its role as adaptor. The attitude toward this problem varied from describing it as a Rube–Goldberg machine without deeper meaning of its molecular structure on the one hand, and as a molecular structure which has been selected given the rules of molecular forces and interactions on the other hand [6]. A direct connection between anticodon triplet and selection of amino acids has been rejected for a long time. However, more recently signs are accumulating that suggest that the anticodon triplet is probed by the enzyme catalysing the cognate amino-acylation of tRNA [35, 36].

The existence of direct interactions between amino acids and nucleotides has been discussed by Woese [6], based on the observation of chromatographic co-migration. An interesting model for selecting amino acids by anticodon triplets in the form of diketopiperazines has been put forward [37] and recently evidence for preferential interaction of aromatic amino acids with the tRNA-Phe anticodon [38, 39] has been given. In this case wybutine is used as a molecular probe. Thus information could be obtained about the anticodon loop conformation stabilized by interaction with cognate amino acids. Preferential interactions between other amino acids and corresponding anticodons appear to exist (Ponnamperuma, personal communication). The notion of specific nucleotide–amino acid interactions by hydrogen bonding has been corroborated recently by the crystal structure of the Eco RI-DNA complex [40].

An interesting model of amino acid selection by the anticodon triplet involving in addition the fourth nucleotide from CCA end (discriminator base) was presented [41]. Though largely unproven it suggests how the present position of the amino acid at the CCA end could have evolved from the neighbourhood of anticodon loop and CCA end postulated for primitive tRNA structures [4]. Whether such topological relationships exist still today is uncertain as well as is the way aminoacyl-tRNA synthetase fulfils its role as 'correlator'.

Acknowledgement

We acknowledge the support by grants from the Swedish Science Research Council, the Karolinska Institute and the K. and A. Wallenberg Foundation.

References

1. Orgel, L. E., *J. Mol. Biol.* **38**, 381–393 (1968).
2. Eigen, M. & Winkler-Oswatitsch, R., *Das Spiel.* Piper, Munich–Zurich (1975).
3. Eigen, M. and Schuster, P., *Naturwissenschaften*, **64**, 541–565 (1977).
4. Eigen, M. and Winkler-Oswatitsch, R., *Naturwissenschaften* **68**, 282–292 (1981).
5. Woese, C., *Nature (London)*, **226**, 817–820 (1970).
6. Woese, C. R., *Naturwissenschaften* **60**, 447–459 (1973).
7. Crick, F. H. C., Brenner, S., Klug, A. & Pieczenik, G., *Origins of Life*, **7**, 389–397 (1976).
8. Frauenfelder, H., Petsko, G. A. & Tsernoglou, D., *Nature (London)* **280**, 558–563 (1979).
9. Rigler, R. and Wintermeyer, W., *Ann. Rev. Biophys. Bioeng.* **21**, 475–505 (1983).
10. Harvey, S. C. and McCammon, J. A., *Comput. Chem.* **6**, 173 (1982).
11. Nilsson, L., Rigler, R. and Laggner, P., *Proc. Natl. Acad. Sci. USA* **79**, 5891–5895 (1982).
12. Rigler, R., Claesens, F. & Lomakka, G., in *Ultrafast Phenomena*, vol. IV (ed. D. Auston and K. B. Eisenthal), Springer-Verlag, **38**, 472–476 (1984).
13. Rigler, R., Claesens, F. and Kristensen, O., *Analytical Instrumentation* **14**, 525–546 (1985).
14. Claesens, F. and Rigler, R., Conformational dynamics of the anticodon loop in yeast tRNA-Phe as sensed by the fluorescence of wybutine. *Eur. Biophys. J.* (in the press).
15. Fuller, W. and Hodgson, A., *Nature (London)* **215**, 817–821 (1967).
16. McLaughlin, L. W. and Graeser, E., *J. Liqu. Chromatogr.* **5**, 2061–2077 (1982).
17. Clore, G. M., Piper, E. A., Graeser, E. and van Boom, J. H., *Biochem. J.* **221**, 737–751 (1984).
18. Jack, A., Ladner, J. E. and Klug, A., *J. Mol. Biol.* **108**, 619–649 (1976).
19. Holbrook, F. R., Sussman, J. L., Warrant, R. V. and Kim, S. H., *J. Mol. Biol.* **123**, 631–660 (1978).
20. Claesens, F., Rigler, R. and McLaughlin, L. W., in *Biomolecular Stereodynamics*, abstract vol. (ed. R. H. Sarma), p. 263. State University of New York at Albany (1985).
21. Claesens, F., Rigler, R. and McLaughlin, L. W. Submitted for Publication (1986).
22. Karplus, M. and McCammon, A., *Ann. Rev. Biochem.* **52**, 263–300 (1983).
23. Nilsson, L. and Karplus, M., in *3D Structure and Dynamics of RNA* (ed. C. Hilbers and P. van Knippenberg). New York, Plenum Press (1985).
24. Nilsson, L. and Karplus, M., Empirical energy functions for energy minimizations and dynamics of nucleic acids, *J. Comp. Chem.* (in the press).
25. Grosjean, H., Söll, D. G. and Crothers, M., *J. Mol. Biol.* **103**, 499–519 (1976).
26. Labuda, D. and Pörschke, D., *Biochemistry* **19**, 3799–3805 (1980).
27. Urbanke, C. & Maass, G., *Nucleic Acids Res.* **5**, 1551–1560 (1978).
28. Geerdes, H. A. M., Van Boom, J. H. and Hilbers, C. W., *J. Mol. Biol.* **142**, 219–230 (1980).
29. Grosjean, H., Houssier, C. and Cedergren, R., in *3D Structure and Dynamics of Ribonucleic Acids* (ed. C. W. Hilbers and P. H. van Knippenberg). New York, Plenum Press (1985).
30. Yarus, M., *Science* **218**, 646–652 (1982).
31. Woo, N. H., Roe, B. A. and Rich, A., *Nature (London)* **286**, 346–351 (1980).
32. Wrede, P., Woo, N. H. and Rich, A., *Proc. Natl. Acad. Sci. USA* **76**, 3289–3293 (1979).
33. Lagerkvist, U., *Proc. Natl. Acad. Sci. USA* **75**, 1759–1762 (1978).
34. Lagerkvist, U., in *Molecular Evolution of Life*, *Chemica Scripta* **26B**, 85, (1986).
35. Bare, L. and Uhlenbeck, O. C., *Biochemistry* **24**, 2354–2360 (1985).
36. Uhlenbeck, O. C., in *Molecular Evolution of Life*, *Chemica Scripta* **26B**, 97, (1986).
37. Grafstein, D., *J. Theor. Biol.* **105**, 157–174 (1983).
38. Bujalowski, W. and Pörschke, D., *Nucleic Acids Res.* **12**, 7549–7563 (1984).
39. Claesens, F. and Rigler, R. (unpublished observations).
40. Rosenberg, J. M., McClarin, J. A., Frederick, C. A., Wang, Bi-Cheng, Boyer, H. W. and Greene, P., in *Molecular Evolution of Life*, *Chemica Scripta* **26B**, 147, (1986).
41. Shimizu, M., *J. Mol. Evol.* **18**, 297 (1982).

Chemica Scripta 1986, **26B**, 109–119

Evolutionary Aspects of Ribosome–factor Interactions

A. Liljas, S. Thirup

Institute of Molecular Biology, Biomedical Center, University of Uppsala, Box 590, S-751 24 Uppsala, Sweden

and A. T. Matheson

Department of Biochemistry and Microbiology, University of Victoria, Victoria, British Columbia, Canada V8W 2Y2

Paper presented by A. Liljas at the Conference on 'Molecular Evolution of Life', Lidingö, Sweden, 8–12 September 1985

Abstract

The protein synthesis machinery provides a rich source for studies of evolutionary relationships. This paper reviews the relationship between a number of protein factors that bind to the ribosomes at various stages of protein synthesis. The factors involved in GTP binding and hydrolysis clearly have one domain in common which could have originated from a common ancestor. The homology outside this GTP binding domain is less significant. Two release factors that interact with the same region of the ribosome manifest only a low level of homology with the other factors.

One ribosomal protein that interacts with all these factors has been characterized from a variety of organisms. The amino acid sequences from the eucaryotic forms of this protein can easily be aligned with the ones from archaebacteria but not with eubacterial proteins. Furthermore, the localization of a hinge region and a number of characteristic structural features suggests that the C-terminal domain in eubacteria is transposed to the N-terminus in eucaryotes and archaebacteria. This identification may indicate that there was an early divergence of these proteins. Despite this dramatic alteration in the amino acid sequence it is possible to arrange the two forms of the protein in a similar manner within the ribosome.

Introduction

Ribosomes provide a rich source for studies of evolutionary relationships since they are made up of a large number of protein and nucleic acid molecules and because they are found in all living organisms. So far the main effort has been devoted to the analysis of the RNA moieties of the ribosomes. These studies have led to the conclusion that methanogens, extreme halophiles and thermoacidophiles or sulfur-dependent bacteria have characteristics so different from other bacteria that they must be classified as a separate kingdom, the archaebacteria [1]. From topological studies of ribosomal subunits a fourth kingdom, the eocytes has been suggested [2]. This would mainly consist of sulfur-dependent bacteria.

The comparison of amino acid sequence from homologous proteins has been hampered by the difficulty to find ribosomal proteins that correspond to each other over a wide range of species. However, there is one highly characteristic protein that has been identified in most ribosomes analyzed. In *E. coli* this protein is called L7/L12 [3]. Since L7 is a modified form of L12 [4] so far not found in other species we will simply call this protein L12.

We recall that translation of a messenger RNA into protein occurs in three stages – initiation, elongation and termination (see [5] for a review) – and each of these stages is associated with a distinct set of soluble protein factors that perform various functions in the corresponding functional state. Several of these factors bind to the ribosome in complex with a GTP molecule. When the GTP molecule is hydrolyzed to GDP, the protein factor dissociates from the ribosome. Three such factors (IF-2, EF-Tu and EF-G) all depend on the L12 protein for their proper function (see [6] for a review). Furthermore two of the release factors, RF-1 and RF-2, also need L12, even though they do not appear to utilize GTP [7]. For EF-G it is the GTP hydrolysis that is lost when the L12 protein is removed [8].

These factors may bind at the same or overlapping sites [6], indeed, some evidence indicates that they may directly interact with L12. Crosslinking experiments have revealed the proximity between L12 and the factors when they are bound to the ribosome [9]. If ribosomes are treated with trypsin some proteins with exposed surfaces are found to be accessible to digestion. However, the L12 protein in the *E. coli* ribosome is normally highly resistant to digestion by trypsin. Nevertheless, when EF-G is bound to the ribosome with a GTP analogue, L12 becomes highly accessible for digestion [10]. Apparently some conformational rearrangement has occurred in L12 upon binding of EF-G. One possible explanation for this is that an interaction between the factor and protein L12 has exposed a previously masked site for trypsin digestion.

The existence of a number of proteins, the factors, with partly similar functions raises the question whether they have any structural similarity. On the other hand the ribosomal protein, L12, that is needed for the proper function of all these factors, must have structural features that are conserved when comparing widely different species.

This paper will first discuss the structural and evolutionary relationship between the factors interacting with L12. Then we will examine the sequence data available for the L12 protein from a wide range of organisms in relation to the three-dimensional structure. Finally we will briefly relate the observations in this system to the phylogenic relationships between the different kingdoms of organisms.

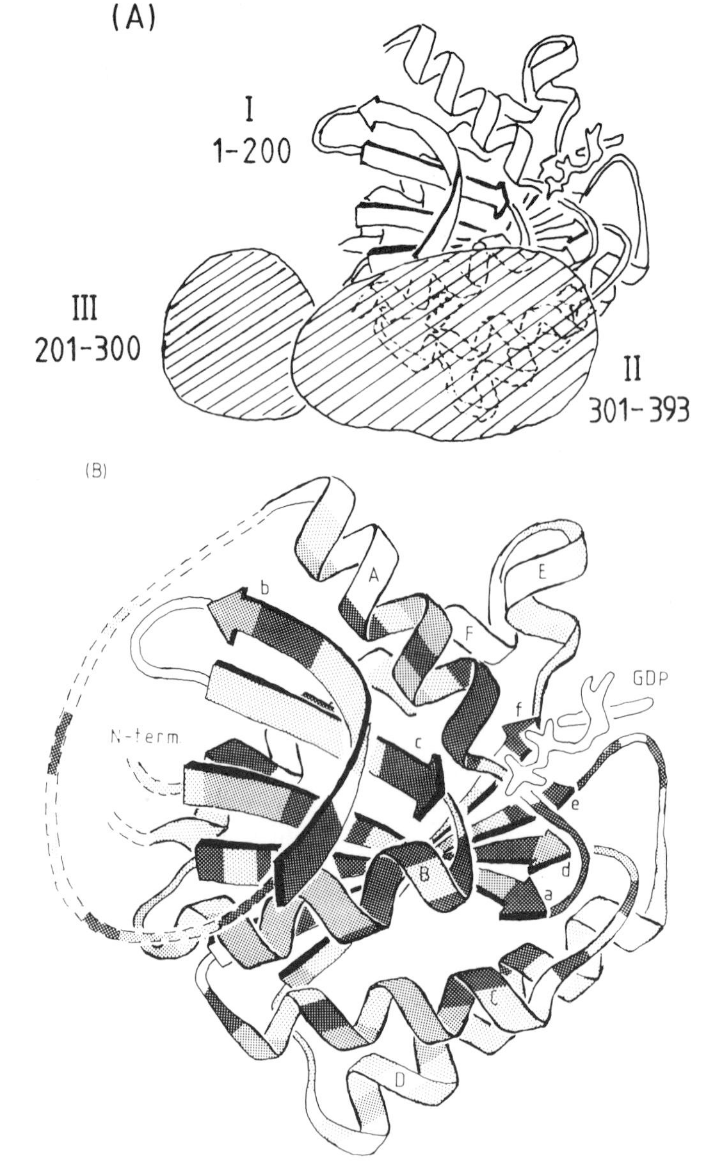

Fig. 1. (A) The relative location of the three domains of EF-Tu [18]. (B) Structural cartoon of domain I of EF-Tu [19]. β-strands are named a, b, c, d, e and f. α-helices are named A, B, C, D, E and F. Broken lines indicate unknown parts of the structure. The connection from helix A to strand b (residues 40–60) is absent owing to proteolytic degradation of EF-Tu prior to crystallization. Light shading indicates homology within the EF-Tu sequences. Medium shading indicates EF-Tu and EF-1α homology. Dark shading indicates the areas where EF-G and IF-2 is homologous with EF-Tu and EF-1α.

Sequence comparisons of protein synthesis factors

At present the amino acid sequences are known for IF-2 [11] and for EF-Tu from *E. coli*; [12] yeast mitochondria [13] and *Euglelia gracilis* chloroplasts [14]. Sequences of EF-1α, the eucaryotic analogue of EF-Tu, are known for yeast [15] and *Artemia salina* [16]. The amino acid sequences of RF-1 and RF-2 from *E. coli* have been published recently [17]. A crystallographic model of EF-Tu shows three distinct domains [18]. Domain I is made up of residues 1–200, domain III of 201–300 and domain II of 301–394 [19]. Figure 1 shows the detailed structure of domain I, which binds GDP [19]. The structural motif and mode of nucleotide binding of this domain is common among nucleotide binding proteins, i.e. a central parallel β-sheet surrounded by α-helices with the nucleotide binding site located at the carboxyl end of the sheet. Other functional aspects of EF-Tu are more difficult to locate in the model, but a number of residues can be related to certain functions from biochemical evidence.

The GTP-binding factors have the property that they interact with tRNA. IF-2 binds f-met-tRNA and positions the f-met-tRNA in the P-site [5]. Similarly EF-Tu binds aa-tRNA and positions this in the free A-site [5]. It has recently been proposed that EF-Tu possesses two tRNA binding sites. Binding site I appears to be specific for the aa-tRNA bound in the ribosomal A-site. Binding site II recognizes peptidyl-tRNA bound at the P-site [20]. EF-G, which is involved in the translocation step of the elongation cycle, also interacts with the tRNAs in both ribosomal sites even though direct binding to the tRNA may not be involved [5]. Even the release factors RF-1 and RF-2, must bind close to the peptidyl-tRNA in the P-site in order to recognize the termination codon at the A-site [7]. Again interactions with the tRNA are considered likely but they have not been demonstrated yet.

Cross-linking experiments have been performed between different tRNAs and EF-Tu. *N*-bromoacetyl-lysyl-tRNA has been photo-crosslinked to His 66 [22]. 3′-oxidized aa-tRNA bound at the ribosomal A-site has been cross-linked to Lys 237 and 3′-oxidized peptidyl-tRNA bound in the ribosomal P-site cross-links to Lys-208 [23]. Though the effects seen may not necessarily reflect an interaction between the residue and the tRNA, they do suggest a close proximity between these components. Studies on mutant species of EF-Tu have shown that Gly 222 → Asp destroys binding site II [20]. Furthermore it is known that Cys 81 is protected against alkylation upon binding of aa-tRNA [24]. Whether this is due to a direct protection of Cys 81 or a conformational change imposed by the tRNA is not known. The residues involved in the GDP binding are strongly conserved not only among the factors [19] but also among other GDP/GTP binding proteins such as YP2, Ras oncogenes [25], transducin [26], and tubulin [27].

Homology among the EF-Tu's

EF-Tu from *E. coli*, yeast mitochondria and *E. gracilis* chloroplasts show a very strong homology over the entire sequence (Fig. 2). Since the structure of EF-Tu is known in detail only for domain I, we will focus on this domain.

The β-strands are strongly conserved in domain I. This is because most of their side-chains are buried in the structure with the consequence that changes in such positions would be expected to have serious steric effects. Loop regions tend to be conserved as well. Thus, at the carboxyl side of the sheet where the GDP binding site is found the conserved residues are needed in order to preserve the binding site. At the amino side of the β-sheet the loops Bd and Ce face the cavity between the domains, whereas loop Df is at the surface. From the steric point of view, mutations in this area would not be harmful. The two helices forming the interface with domain II (B and C) are well conserved. The N-terminal part of helix A which is close to the GDP binding site is conserved whereas the C-terminal part is not. Helix D is conserved though it is not known at present to be involved in any interactions. Helices E and F are not very well conserved. To obtain a reasonable alignment, insertions of 10 residues in helix E of EFTUCEG and 2 residues between helix E and F in EFTUMSC were necessary. The homology is improved towards the end of helix F. The loop that is missing in the crystallographic model (40–60) is very well conserved from residue 50 to the beginning of β-strand b.

EFTUEC vs. EF-1α

When EFTUEC is compared with its eucaryotic analogue EF-1α much of the same general pattern of homology shows up, but the detailed structural homology is in general less than for eubacterial EF-Tu species. The major differences are that strand c, loop dC, helix D and loop Df show much less homology. Loop dC is the only loop at the carboxyl-side of the sheet that is not involved in GDP binding. EF-1α has a large insertion (7 residues) in this loop. In loop Df there is an even larger insertion (22 residues). The homology for domain II and III is in general lower than for domain I but still there are regions within these domains that are strongly conserved.

EFTUEC vs. IF-2

Initiation factor 2 is much larger than EF-Tu, 890 residues, as compared to the 393 residues of EF-Tu. These two sequences are easily aligned since the GDP binding domain of EF-Tu is conserved [11]. Residue 1 of EF-Tu corresponds to residue 379 of IF-2. Strand a, loop aA and the N-terminal part of helix A are well conserved since they are part of the GDP binding site. In IF-2 there is a large deletion of 17 residues after this conserved part, beginning at the C-terminal part of helix A at residue 36 and ending at residue 52. Strand b is conserved in parts though two residues have been deleted at positions 70 and 71. The alignment proceeds up to residue 139 with only a single deletion in IF-2 at residue 130, at the beginning of strand e. The rest of the sequence of domain I can be aligned with a few short deletions and insertions.

In order to align domains II and III of EF-Tu with IF-2 a number of insertions in the IF-2 sequence are needed. The homology is too poor to indicate any convincing relationship between the two sequences. However, a segment around residues 220–230 in EF-Tu is homologous to a segment in IF-2 with a gap of 4 residues in the IF-2 sequence.

EFTUEC vs. EF-G

The GDP binding domain of EF-Tu can easily be located in the amino acid sequence of EF-G [28, 29]. In EF-G residue 1 corresponds to residue 4 of EF-Tu. The sequence is well conserved. At the middle of helix A there is an insertion in EF-G of 3 residues. The sequence from the end of helix A to residue 60 is one residue shorter in EF-G. A few insertions and deletions in both sequences improves the homology slightly. The very strongly conserved residues around position 60 of the EF-Tu's and EF-1α's are conserved in EF-G as well. Just before strand b there is an insertion of 10 residues in EF-G. From strand b to residue 139 there are only single deletions at position 76 and 130. After residue 139 large insertions in the EF-G sequence are required in order to produce some homology with the rest of the sequence of Domain I. Thus in loop eD 29 residues are inserted and 57 residues are inserted in loop Df. An insertion of 9 residues in the EF-G sequence after domain I gives some homology around 220–230. This is, however, not as pronounced as that for the IF-2 sequence in the same area. For the rest of the EF-G sequence the alignment with EF-Tu is not very convincing.

EFTUEC vs. release factors

The two release factors are highly conserved with respect to each other. The overall homology is 51%, including conservative changes as defined in Fig. 2. However, the variation in homology along the sequence is significant: 130 residues at the N-terminal and 40 residues at the C-terminal are not as well conserved as the middle part, where the homology is 89%, including conservative changes [17].

The release factors contain no sequences that are homologous with the sequence of EF-Tu associated with GDP-binding. Using the UWGCG program system [30] for determining the best alignment between the two sequences, the homology was found to be about 30% for both RF-1 and RF-2.

Secondary structure prediction methods [31–33] applied for the release factor sequences show that residues 1–60 are characteristic of α/β structures with alternating α-helixes and β-strands as in domain I of EF-Tu. Comparing this prediction with the secondary structure of EF-Tu, the best sequence alignment also shows a good agreement for the secondary structural elements D, f, E and F, whereas the secondary structure elements B, d, C and e of EF-Tu do not seem to have corresponding secondary structures in the release factors with the present alignment. Furthermore, residues 1 of RF-1 and RF-2 are aligned with residues 60 and 43 respectively of EF-Tu. This means that there is no β-strand corresponding to a and no helix A in RF-1 and RF-2. Since strand a is in the middle of the sheet, strands b and c cannot easily be part of the same type of sheet.

The most homologous regions between EF-Tu and the release factors include helix D and strand f, loop E and a region around residues 220 in EF-Tu, the same region where IF-2 and EF-G were found to be homologous.

Comparison of functional sites in the factors

From the amino acid sequences of the ribosomal factors it can be concluded that the GTP binding factors are members of the same family. Whether or not the release factors are distantly related members is unclear. The aa-tRNA binding site seems to be located at the 'top' of the molecule (Fig. 1). Although cross-links between His 66 and aa-tRNA have been formed this residue may not be directly involved in the aa-tRNA binding, since His 66 is not conserved in the EF-1α's. Residues in the vicinity of this position, which are conserved in all EF-Tu's and EF-1α's, are more likely to be involved in the aa-tRNA binding. These residues include 55–66 and 70–72. The cross-link between aa-tRNA and Lys 237 indicates that the tRNA binding spans an area between domain I and III. The exact location of Lys 237 is presently not known. These residues are also strongly conserved in EF-G, indicating that EF-G could bind to the A-site bound tRNA on the ribosome. In IF-2 this area is not as well conserved as for the other factors, indicating that the mode of tRNA-binding to IF-2 may be different from the mode of aa-tRNA binding in EF-Tu.

The second tRNA binding site (II) of EF-Tu is located in domain III as indicated by the cross-link between P-site bound peptidyl-tRNA and Lys 208 and by the fact that a mutation here (Gly 222 → Asp) destroys this binding site. In the sequence from 208 to 233 the homology between the EF-TU's and EF-1α's is very high. This is suggestive of a tRNA interaction

```
                          GGGGGGGG                                                            AAAAAA..AAAA           AAA
                   .--------.    ////////...//////// ............                                  ..    -.........------
                     .10       .20       .30          .40                    .50        .60                    .70
EFTUEC        1 SKEKFERTK.PHVNVGTI.GHVDHGKTTLTAAI...TTVLAKTYG............GAARAFDQIDNAPEEKARGIT..INTS..........HVEYDT 71
                    11111 11111111 11111111111111   1 111   1             1    1  11 111111111  1 11            111111
EFTUMSC     113 YAAAFDRSK.PHVNIGTI.GHVDHGKTTLTAAI...TKTLAAKGG............ANFLDYAAIDKAPEERARGIT..ISTA..........HVEYET 183
                111111111 11111111 11111111111111   1  11 1                11 11 11 111111111  1111            111111
EFTUCEG       1 ARQKFERTK.PHINIGTI.GHVDHGKTTLTAAI...TMALAATGN............SKAKRYEDIDSAPEEKARGIT..INTA..........HVEYET 71
                 111        1111 11 1111 1111 11 1        1 1             11 1     11     11 1111  1   1              111
EF1AAS        1 GKEK......IHINIVVI.GHVDSGKSTTTGHLIYKCGSIDKRTIEKFEK EAQEMGKGSFKYAWVLDKLKAERERGIT..IDIA..........LWKFET 81
                 111        1111 11 1111 1111 11 1        1 1             11 1     11     11 1111  1   1              111
EF1ASC        1 GKEK......SHINVVVI.GHVDSGKSTTTGHLIYKCGGIDKRTIEKFEK EAAELGKGSFKYAWVLDKLKAERERGIT..IDIA..........LWKFET 81
                             111 1 1111 1111 1  1   1 1  1               111     11    11  1111  1 11             111
EFGEC         1 ...ARTTPIARYRNIG.ISAHIDAGKTTTTERILFYTGVNHKIGEVHD..........GAA....TMDWMEQEQERGIT..I.TSAATTAFWSGMAKQYE. 79
                1         1   1 11 1111111111    1   1 1                           1  1 1 111  1             111  1
IF2EC       379 SDRDTGAAAEPRAPVVTIMGHVDHGKTSLLDYIRS.TKV..........................ASGE.AGGITQHI......GAY....HVE..T 435
                                                                                            1                 1 1
RF1EC         1 ......................................................................M..KPSI..........VAKLEA 11
                                                                            11   11      1  1              111
RF2EC         1 .........................................................MFEINPVNNRIQDLTERSD..VLRG..........YLDYDA 29


                          GGG 222222222222                        22 222222222222    GGGG
                    TT------- /////////////    -------TTTT  ....../// /////////////  -------    .........   //////.///////
                          .80       .90       .100      .110             .120      .130              .140       .150
EFTUEC       72 PTRHYAHVDCPGHADYVKNMITGAAQMDGAILVVAATDGP.......MPQ TREHILLGRQVGVPYIIVFLNKCD.........MVDDEELLE.LVEMEVR 154
                   1111111111111111111111111111111111111           111 11111111111111  1111111 1          111 1111 11111 1
EFTUMSC     184 AKRHYSHVDCPGHADYIKNMITGAAQMDGAIIVVAATDGQ.......MPQ TREHLLLARQVGVQHIVVFVNKVD.........TIDDPEMLE.LVEMEMR 266
                   111111111111111111111111111111111 111           111 11111111111111 1111111 1          111 1111 1111111
EFTUCEG      72 KNRHYAHVDCPGHADYVKNMITGAAQMDGAILVVSAADGP.......MPQ TKEHILLAKQVGVPNIVVFLNKED.........QVDDSELLE.LVELEIR 154
                         11 111 111111111 11 1 1111111  1             11111 111   111   111 111 1            1    1    111
EF1AAS       82 AKYYVTIIDAPGHRDFIKNMITGTSQADCAVLIVAAGVGEFEAGISKNGQ TREHALLAYTLGVKQLIVGVNKMDST..........EPPFSE.ARFEEIK 170
                1        11 111 111111111 11 1 1111111  1             11111 111   111   111 111 1                1     11
EF1ASC       82 PKYQVTVIDAPGHRDFIKNMITGTSQADCAILIIAGGVGEFEAGISKDGQ TREHALLAFTLGVRQLIVAVNKMDS...........VKWDE.SRFQEIV 168
                1 1      11 111 111              1111111  11 1        111       1    11   1  111 1              111 11 1
EFGEC        80 PHR.INIIDTPGHVDFTIEVERSMRVLDGAVMVYCAVGGV.......QPQ SETVWRQANKYKVP.RIAPVNKMDR.145.175.EHFTGVVD.LVKMKAI 190
                        11 1111 11         11   1  111111 11       1111 1 1 1  11    11 111 111 1                1       11
IF2EC       436 ENGMITFLDTPGHAAFTSMRARGAQATDIVVLVVAADDGV.......MPQ TIEAIQHAKAAQVP.VVVAVNKID..............KPE.ADPDRVK 511
                        1        1        1        11 1            11111  1         1   1    1        1 1  1  1 1 1
RF1EC        12 LHERHEEVQALLGDAQTIADQERFRALSREYAQLSDVSRC....FTDWQQ VQEDIETAQMMLDDPEMREMAQDE........LREAKEKSEQLEQQLQV 98
                        11         1 1     1  1       1 11          1 1  1 1   1111                  1 1 1 1 1 111 1
RF2EC        30 KKERLEEVNAELEQPDVWNEPERAQALGKERSSLEAVVDT....LDQMKQ GLEDVSGLLELAVEADDEETFNEA........VAELDALEEKLAQLEFR 116


                                                 G                                                          P          PPP
                  ////.//./    ..........   -----///..........///// ..///////////////........////
                        .160                    .170                 .180        .190               .200     .210     .220
EFTUEC      155 ELLS.QY.DFPG..........DDTPIVRGSAL..........KALEGD. .AEWEAKILELAGFLD.........SYIPEPERAIDKPFLLPIEDVFSIS 221
                   111  11  1 1           11 111 1111           1111       1  11 1    11           111 111 111111111111111
EFTUMSC     267 ELLN.EY.GFDG..........DNAPIIMGSAL..........CALEGRQ PEIGEQAIMKLLDAVD.........EYIPTPERDLNKPFLMPVEDIFSIS 335
                 1 11 11 1111           11 111 1111          1    11      1  111 1    11          1111 1 1  11 111 1111 111
EFTUCEG     155 ETLS.NY.EFPG..........DDIPVIPGSALLSVEALTKNPKITKGE. .NKWVDKILNLMDQVD.........SYIPTPTRDTEKDFLMAIEDVLSIT 231
                   11  1                1  1  11  1                     1  1      1    11         1 11 1 1  111  1111111 1
EF1AAS      171 KEVS.AYIKKID.182..206.DRLPWYKGWNI................E RKEGKADGKTLLDALD.........AILP.PSRPTEKPLRLPLQDVYKIG 256
                   1 11    1           1 1  11                    1      1    1    11         1     1 1  111  1111111 1
EF1ASC      169 KETS.NFIKKVG.180..204.TNAPWYKGWEK................E TKAGVVKGKTLLEAID.........AIEQ.PSRPTDKPLRLPLQDVYKIG 254
                                        1  11 111               1       1  11 1  1    11             1   1 1 11       1 1
EFGEC       191 NWND.ADQGVT.200...259.EIILVTCGSAF..........KNKGVQ. .AMLDAVIDYLPSPVDVPAIDCILKDTPAERHASDDEPFSALAFKIATDP 324
                   11 11   1               11  1  11          11  1     1     11  1  11                                11  1
IF2EC       512 NELS.QYGILPE.......EWGGESQFVHVSA...........KAGTG.. IDELLDAILLQAEVLE.........LKAVR....KGMASGAVIE.SFLDK 576
                  11    1                     1  111           11 1        11  1                1   1         1      11
RF1EC        99 LLLPKDP.DDER.........NAFLEVRAGTGG........DEAALFAGD LFRMYSRYAEARRWRV.........EIMSASEGEHGGYKEIIA....KIS 167
                  1 1 11 1                    1  111                 1          1  1  11           1 1 1  1        1 1      11
RF2EC       117 RMFSGEY.DSAD..........CYLDIQAGSGG........TEAQDWASM LERMYLRWAESRGFKT.........EIIEESEGEVAGIKSVTI....KIS 184


                  PPPPP         A
                        .230      .240        .250        .260                      .270      .280              .290
EFTUEC      222 GRGTVVTGRVERGIIKVGEEV..EIVGIKET.QKSTCTGVEMFRKLLDEG..........RAGENVGVLLRGIKREEI.......ERGQVLAKPG..... 296
                111111111111 11 1111  1111    1  111 11111111 11 1            1111 11111111111            11 111111
EFTUMSC     336 GRGTVVTGRVERGNLKKGEEL..EIVGHNSTPLKTTVTGIEMFRKELDSA..........MAGDNAGVLLRGIRRDQL.......KRGMVLAKPG..... 411
                11111 1111111 11111 1  11111111      11 111111 1 1111            11111111111 1 11          111 11111
EFTUCEG     232 GRGTVATGRVERGTIKVGETV..ELVGLKDT.RSTTITGLEMFQKSLDEA..........LAGDNVGVLLRGIQKNDV.......ERGMVLAKPR..... 306
                1 111 11111 11 1 1       11          1      111      1111               11111  11 1    11          11 1 1
EF1AAS      257 GIGTVPVGRVETGIIKPG.....MIVTFAPANITTEVKSVEMHHESLEQA..........SPGDNVGFNVKNVSVKEL.......RRGYVASDSK..... 329
                1 111 11111 1111 1       11          1      111      1111               11111  11 1    11          1111 1
EF1ASC      255 GIGTVPVGRVETGVIKPG.....MVVTFAPAGVTTEVKSVEMHHEQLEQG..........VPGDNVGFNVKNVSVKEI.......RRGNVCGDAK..... 327
                  1 11  11  111  11 1  1 1 11                 1  1 1            1111     1  11   11           1  1  1
EFGEC       325 FVGNLTFFRVYSGVVNSGDTVLNSVKAARER.FGRIVQMHANKREEIKEV..........RAGD....IAAAIGLKDV.......TTGDCLCDPD.397.
                111 1 1  1  1 1  11 1    1   1   1          11    111 1            111 1 11     1  11          1   11
IF2EC       577 GRGPVATVLVREGTLHKGDIV....LCGFEY.GRVRAMRNELGQEVLEAG.621..636.AAGDEVTVVRDEKKAREVALYRQGKFREVKLARQQ..... 670
                1 11     1 1 1  11                    1 11 1      1 1              1 1 1     1                   11 1   1
RF1EC       168 GDGVYGRLKFESGGHRVQRVP..ATESQGRI.HTSACT.VAVMPELPDAE..........LPDINPADLRIDTFRSSG.......AGGQHVNTTG..... 241
                1       111 11 11                      1  1      1  1 11             1 1 1    1 1               11 1 1
RF2EC       185 GDYAYGWLRTETGVHRVVRKS..PFDSGGRR.HTSFSS.AFVYPEVDDDI..........DIEINPADLRIDVYRTSG.......AGGQHVNRTE..... 268
```

Fig. 2. For legend see facing page.

```
                       .300          .310      .320      .330      .340                     .350     .360
EFTUEC     297 .....TIKPHTK....FESEVYILSKDEGGRHTPFFKGYRPQFYFRTTDV TGTIELPE.........GVE....MVMPGDNIKMVVT........LIHPI 366
                    111 111      1 111111111111 1    1111 1 11 111 11  1              11     1111111 1         1111
EFTUMSC    412 .....TVKAHTK....ILASLYILSKEEGGRHSGFGENYRPQMFIRTADV TVVMRFPK.........EVEDHSMQVMPGDNVEMECD........LIHPT 485
                    11 1111      111111111111111111 1111111 111111 11                 1      111111 111 1        11 11
EFTUCEG    307 .....TINPHTK....FDSQVYILTKEEGGRHTPFFEGYRPQFYVRTTDV TGKIESFR........SDNDNPAQMVMPGDRIKMKVE........LIQPI 381
                                 1 111 11     1        11 1     1 1                              1                1
EF1AAS     330 .....NNPARGS..QDFFAQVIVLN..HPGQ...ISNGYTPVLDCHTAHI ACKFAEIK.........EKC....DRRTGKTTEAEPK........FIKSG 396
                       1       1 1 1 11     1        11 1     1  1    1                       1 1                1
EF1ASC     328 .....NDPPKGC..ASFNATVIVLN..HPGQ...ISAGYSPVLGCHTAHI ACRFDELL.........EKN....DRRSGKKLEDHPK........FLKSG 394
                     1 1 11      1 1   11    11 1 11    1    1                 11             11   1            1
EFGEC          .415.AVEPKTK....ADQEKMGLALGRLAKEDPSFRVWTDEESNQTIIA GMGELHLD........IIVD....RMKREFNVEANVGKPQVAYRETIRQK 493
                       1         1 11 1     1       1        111 1     1          1      11 1 111    1        11
IF2EC      671 .....KSKLGNMFANMTEGEVHEVNIVLKADVQGSVEAISDSLLKLSTDE VKVKIIGS.723.738.ASN....AILVGFNVRADAS.....ARKVIEAE 761
                    1    1       1 1 1   11    1     11             1 1                      11                1
RF1EC      242 .....SAIRITH....LPTGIVVECQDERSQHKNKAKALS.........V LGARIHAA.........EMA....KRQQAEASTRRNL........LGSGD 302
                    1    1       1 1     11    1     1 1                                      11 1    11
RF2EC      269 .....SAVRITH....IPTGIVTQCQNDRSQHKNKDQAMK.........Q MKAKLYNW.........RCR....RKMPRNRRWKITN........PTSAG 319


                    .370                      .380      .390
EFTUEC     367 AMDD..........GLRFA.IREGGRTVGAGVVAKVLS 393
                11            1 11  111111111 111 111
EFTUMSC    486 PLEV..........GQRFN.IREGGRTVGTGLITRIIE 512
               111            11111 111111111111  11
EFTUCEG    382 AIEK..........GMRFA.IREGGRTVGAGVVLSIIQ 408
                                111 111   111 111  1
EF1AAS     397 DAAM.401..421.LGRFA.VRDMRQTVAVGVIKSVNFKDPTAGKVTKAA 454
                                111 111   111 111  1
EF1ASC     395 DAAL.399..419.LGRFA.VRDMRQTVAVGVIKSVD.KTEKAAKVTKAA 452
                 1              1 1    111   1 11  1
EFGEC      494 VTDV..........EGKHA.KQSGGRGQYGHVVIDMYPLEPG 524
               111                  1       1 11  1 1
IF2EC      762 SLDL..........RYYSV.IYNLIDEVKAAMSGMLSPELKQQIIGLAEV 800
                 1               1    111  1 1 1
RF1EC      303 RSDR..........NRTYN.FPQGALPITAST 323
               1               1    1  1 1  1 1
RF2EC      320 AARF..........VLMSL.MTPALKICAPG 339
```

Fig. 2. Alignment of factor sequences. Abbreviations used here and in the text are EFTUEC, EF-Tu from *E. coli*; EFTUMSC, EF-Tu from *Saccharomyces Cerevisiae* mitochondria; EFTUCEG, EF-Tu from *Euglena gracilis* chloroplast; EF1AAS, EF-1 from *Artimia salina*; EF1ASC, EF-1 from *S. cerevisiae*; EFGEC, EF-G from *E. coli*; IF2EC, IF-2 from *E. coli*; RF1EC and RF2EC, RF-1 and RF-2 from *E. coli*. The alignments used are taken in parts or completely from [13] (EFTUMSC); [14] (EFTUCEG); [16] (EF1AAS); [11] (IF2EC and EFGEC). Homology with EFTUEC is indicated by a figure 1 above the sequence in question. Homology includes conservative changes, so that D–E, D–N, E–Q, Q–N, R–K, L–I–M, V–T, T–S, S–A, A–G and F–Y are considered to be homologous pairs. The numbering above the sequences refers to EFTUEC numbering. Secondary structure of EF-Tu is indicated by — for β-strands, / for α-helices and T for turns. The letters A, P, G, 2 indicates the proposed location of binding site I and II, GDP binding site [19] and interactions with domain II, [21], respectively.

site located in this region. In IF-2 this region is conserved to some extent, which may suggest that f-met-tRNA^{met} binds in this area, though it is also known that the N-terminus of IF-2 is important for binding of f-met-tRNA^{met} [11]. The release factors show some homology in this area, but it is not very extensive.

If the factors bind at the same site of the ribosome this could be reflected in conserved areas within the sequences of the factors. Since the homology of EF-Tu, EF-G and IF-2 does not extend very much beyond domain I and since the conserved areas in domain I have been accounted for in terms of GDP and tRNA-binding, it seems unlikely that the ribosome interacting area can be identified in a single conserved domain. Among the EF-Tu's and EF-1α's there are conserved areas in domain II and III, but these could represent sites for ribosome or EF-Ts interactions.

In order to explain the homologies described so far we suggest the following relationship between the factors. We assume that this set of proteins may have evolved from a common ancestral protein with the following common functions: GDP/GTP binding, ribosome binding, interaction with P-site tRNA, and aa-tRNA binding. In EF-Tu all these features exist, while in EF-G the P-site tRNA binding site is modified to some extent. In IF-2 the A site aa-tRNA binding has been lost while the P-site is preserved for binding of f-met-tRNA^{met}. Most of these functions have been lost from the release factors except for a rudimentary P-site tRNA binding site. The ribosome binding site on the factors may be divergent in order to meet different specific needs, such as triggering of the peptidyl transferase activity for the release factors, proofreading for EF-Tu and translocation for EF-G.

Structure of the L12 protein

A number of amino acid sequences of eubacterial L12 have been determined (Fig. 3). They are easily aligned with each other and with spinach chloroplast L12 [41, 44]. Among the characteristic properties of eubacterial L12 are a highly acidic nature, a size corresponding to *c.* 120 amino acids, a high content of alanine and the unusual property of being present in several copies per ribosome. The corresponding protein from eucaryotes and archaebacteria, eL12 share all these properties. The functional role of this protein in eucaryotes also seems to parallel that found for bacterial ribosomes [45]. Even though the amino acid sequences of the eucaryotic and archebacterial eL12 proteins can be aligned with each other

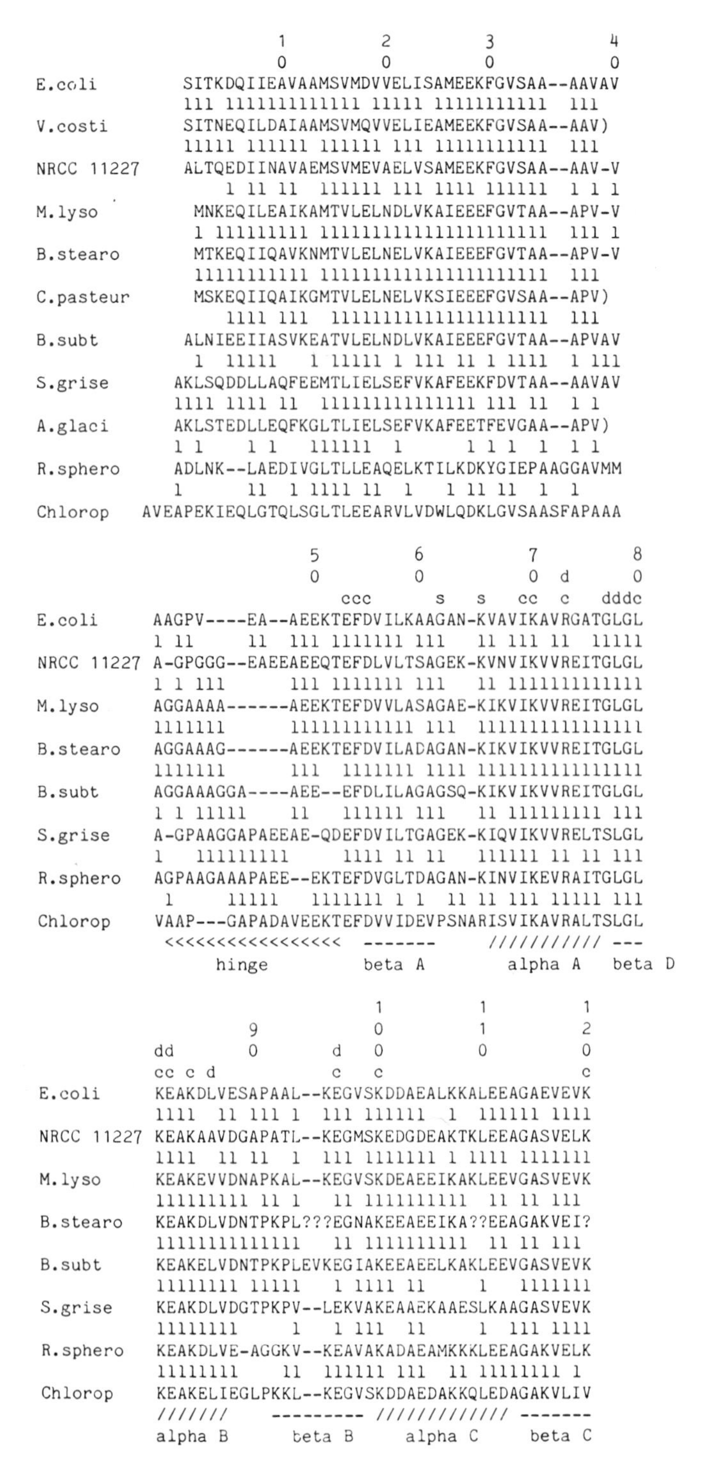

Fig. 3. Amino acid sequences of eubacterial and chloroplast L12. The numbering refers to *E. coli* [4]. The other sequences are from NRCC 41227 [34], *Clostridium pasteurianum* [35], *Bacillus subtilis* [36], *Vibrio costicola* [37], *Bacillus stearothermophilus* (Matheson *et al.*, unpublished data), *Micrococcus lysodeiktikus* [38], *Streptomyces griseus* [39], *Rhodopseudomonas spheroides* [40], *Arthrobacter glacialis* [35] and spinach chloroplasts [41]. The crystallographic determination of secondary structure of the C-terminal region is found below the sequences [42]. Some surfaces of conserved nature are indicated on top of the sequences. A dimer interaction between L12 C-termninal domains is denoted by d, a conserved area on the surface of the molecule is denoted by c and conserved residues that bind a sulfate in the crystal structure is denoted by s [43]. The lines between the sequences denote homologies between neighbouring sequences. The homologies used are: D–E, D–N, E–Q, Q–N, R–K, L–I–V, L–I–M, V–T, T–S, S–A, A–G, F–Y and identical residues. The secondary structure elements are indicated by --- (β-strand), /// (α-helix) and ‹‹‹ (hinge).

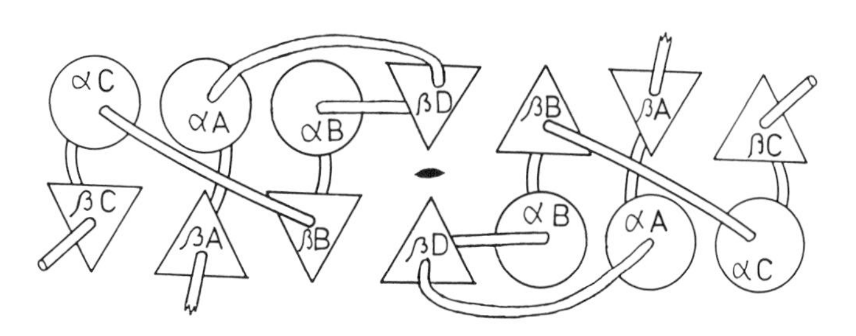

Fig. 4. A simplified representation of the crystallographic structure of the C-terminal domain of L12 from *E. coli* ribosomes. The dimer interaction around the twofold axis in the centre is found in the crystals and may reflect the true molecular structure of L12 dimers. The secondary structure elements are represented with different symbols: the circles represent helices and the triangles β-strands.

it has been difficult to find a simple alignment with eubacterial L12 [46, 47]; however, rearrangement of the eubacterial protein yields the maximum alignment between the eucaryotic and archae-bacterial proteins [48].

The electron micrographs of large ribosomal subunits consistently show a number of protuberances from the main body [2]. The longest one, which is also rather thin, is called the L12 stalk and consists of some or all the four L12 molecules in the ribosome [8, 49, 50]. Since this stalk is a distinctive and conserved protuberance in all three kingdoms it is likely that it represents the site for the L12 proteins in all ribosomes. Since this long protuberance is made up of the highly elongated L12 molecule in eubacteria, one would expect an elongated shape also for the L12-like protein of eucaryotes and archaebacteria.

The eubacterial L12 molecule has both an N- and an C-terminal domain connected by a hinge region [49]. The C-terminal domain consists of slightly less than 70 residues, the crystal structure of which (Fig. 4) has been determined [42]. The N-terminal domain anchors L12-dimers to the large subunit [51–53]. The hinge region can be observed in several ways. First the crystals of the C-terminal domain contain six residues at the N-terminus that could not be observed due to their flexibility. In the sequence alignment of eubacterial L12 (Fig. 3) this region (residues 40–52) is highly variable and contains a stretch rich in alanines and glycines followed by an acidic region. This hinge region is most certainly the cause of the high mobility of the C-terminal domain as observed by ^{1}H-NMR [54, 55] and other methods [56, 57]. This highly mobile domain in large subunits could be a universal feature since it has been observed in a number of very different species [55]. The domain and hinge characteristics provide an opportunity for alignment and comparison of L12 molecules from widely different species.

Structural features of eucaryotic and archaebacterial L12

The amino acid sequences of eucaryotic and archaebacterial eL12 presently available are found in Fig. 5. The homology is good for the N-terminal region (Table I) with only a few deletions. About ten residues at the C-terminus can also be aligned, even though some ambiguity remains between archaebacteria and eucaryotes. The region between residues 90 and 103 (yeast numbering) is highly charged and mainly acidic. Residues 69–89 are primarily alanines and glycines interspersed with occasional prolines, serines and valines. Despite the limited variation in amino acid composition for

```
                    1         2         3         4         5          6
                    0         0         0         0         0          0
M.thermo   MEYIYAAMLL-HTTGKEINEENVKSVLEAAGAEV-DDARVKALIAALEDV
           1111111 11      11 1111  111111 11 1111 1111111111
H.cuti     MEYVYAALIL-NEADEELTEDNITGVLEAAGVDV-EESRAKALVAALEDVD-IEEAV----EEAAAAPA
           1111111111   1  1111111  11 1111 1 11 1 111 1111111 111 1      1 1 1
Sulpho     MEYIYASLLL-HAAKKEISEENIKNVLSAAGITV-DEVRLKAVAAALEEVN-IDEIL----KTATAMPV
           1 11 1 111  11    1  111 11 1 1  1 11 11  1 1111    11111    1     11
S.cervi    MKYLAAYLLLVQGGNAAPSAADIKAVVESVGAEV-DEARINELLSSLEGKGSLEEIIAEGQKKFATVPT
              1 1 111    1   1   1  11 1 1  1 1         1111   1  11         1 1
A.sa' XASKDELAVYAALIL-LDDDVVITTEKVNTILRAAGVSV-EPYSPGLFTKALEGLD-LKSMI-------TNVGS
              1 1 11  1  1   11  1  11 1 11    1         1 1 1 1  11           1
A.salina   MRYVAAYLLAALSGNADPSTADIEKILSSVGIEC-NPSQLQKVMNELKGKD-LEALIAEGQTKLASMPT
           11111 111111 11 111  11 111 1111   1   11111 11 111 11 11111  1111 1
R.liver    MRYVASYLLAALGGNSNPSAKDIAKILDSVGIGADDERKLNKVISELNGKN-IEDVIAQGVGKLASVPA
           1111111111 11111 111 11  111111  1 11 11  11111 11
W.germ     MKFIAAYLLAYLGGNSSPSAADVKDILNAVGAEA-NEEKLEFLLAELKGK


                                                   1          1
            7         8         9                  0          1
            0         0         0                  0          0
H.cuti     AAPAASGSDDEAAADDGDDDEEADADEAAEAEDAGDDDDEEPSGE-GLGDLFG
           11         1        111        11      11  1  11  111
Sulpho     AAV----AAPAG-----QQTQQAAEKKEEKKEEEKKGPSEEEIGG-GLSSLFG
           11     1111        11 1    111111  1  1       1   11
S.cervi    GGA--SSAGPASAGAAAGGGDAA----EEKEEE-AKEES---DDDMGF-GLFD
           1 1    11111111111  11     111 11  1111   111111 1111
A.sa'      GVG----AAPAAGGAAAA-TEAPAA-KEEKKEE-KKEESEEEDEDMGF-GLFD
           111      11111111111 11 1 11 1111 11111111111111 1111
A.salina   GGA------PAAAAGGAAAAPAAEA-KEAKKEE-KKEESEEEDEDMGF-GLFD
           111      11111 11111111        11  111111  11111 1111
R.liver    GGAVAVSAAPGAAAPAAGSAPAAA-------EE--KEESEEKKDEMGF-GLFD
```

Fig. 5. Amino acid sequences of L12-like proteins from eucaryotic and archaebacterial species. The numbering refers to rat liver [58]. The other species are *Methanobacterium thermoautotropicum* [46], *Halobacterium cutribrum* (Oda *et al.*, in preparation), *Sulfalobus acidocaldarius* (Matheson *et al.*, in preparation), *Sacaromyces cerevisiae* [59], *Artemia salina* [60, 61]. The homologies are indicated as in Fig. 3.

Table I. *N-terminal homology of archaebacterial and eucaryotic L12 proteins*[a]

	HC	SA	AS′	SC	AS	RL	WG
MT	56	65	31	38	17	17	29
HC	—	52	32	31	24	23	19
SA	—	—	24	37	29	31	26
AS′	—	—	—	19	24	17	22
SC	—	—	—	—	48	47	51
AS	—	—	—	—	—	58	49
RL	—	—	—	—	—	—	53

[a] Residues 1–68 (rat liver numbering) are compared as aligned in Fig. 5. The number presented is the percentage of identical residues of total number of residues compared. Deletions were ignored.

Table II. *C-terminal homology of archaebacterial and eucaryotic L12 proteins*[a]

	SA	AS′	SC	AS	RL	WG
HC	28	21	16	27	26	22
SA	—	56	41	47	42	47
AS′	—	—	69	74	63	67
SC	—	—	—	68	68	59
AS	—	—	—	—	73	66
RL	—	—	—	—	—	58

[a] Residues 69–111 (rat liver numbering) compared as in Table I.

this region, the sequences are difficult to align (Fig. 5, Table II). With gaps of variable size it is possible to align some of the prolines with their nearest residues. We might add that even though prolines have restricted conformational possibilities, it does not seem fully justified to force all of the prolines into alignment.

The alanine–glycine region and the acidic region correspond to the hinge region in eubacteria. Indeed, the high content of alanine and glycine, and the difficulty to align otherwise similar structures, is strongly suggestive of a region with an undefined and flexible structure, i.e. a hinge. Whether or not the long charged region has any stable structure is unclear. If the hinge includes both the alanine–glycine region and the acidic region it would be considerably longer than in eubacteria. We are then left with no more than ten residues at the C-terminus, which is too few to form a globular domain with the task of interacting with the factors. On the other hand the N-terminal side of the alanine–glycine region is 62–68 residues or about equal to the eubacterial C-terminal domain. It also has an amino acid composition suggestive of a folded structure. We would like to suggest that this is the domain structure that interacts with factors in the eucaryotic and archaebacterial ribosomes. This would imply that the C-terminal residues are left with the task to bind the eL12 protein to the large subunit in order to allow the protein to span the 100 Å of the stalk protuberance. We then assume that the factor interacting domains are at the tip of the stalk, as in eubacteria [50]. The design of the L12 and eL12 proteins would then be similar despite the fact that the amino and carboxyl termini have changed places (Fig. 6). The effect of

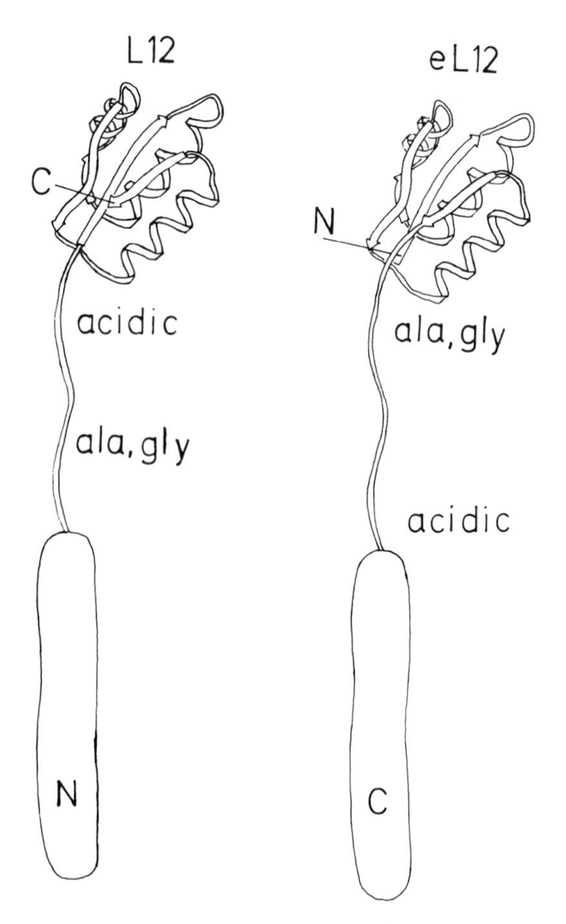

Fig. 6. The possible construction of L12 and eL12 monomers. The hinges are the thin regions connecting the two domains at the N- and C-termini. In L12 the functional domain, for which the structure is known, is located at the C-terminus. Its connection to the hinge is through the N-terminal β-strand. If the N-terminal 65–70 residues of eL12 has a similar structure its continuation into the hinge will be through the C-terminal β-strand. Since the N-and C-termini of these β-strands are close in space the differences in connectivity has very little effect on the over all structure.

this on the position and orientation of the globular domain is marginal since the entry to and exit from this domain are very close on neighbouring strands of the β-sheet.

In order to further test this model three different methods for secondary structure prediction were applied to all available sequences. Figure 7 shows the results of the prediction methods. The method of Robson [32] predicts the secondary structure of bacterial L12 extremely poorly. The methods of Chou and Fasman [31] and Lim [33] predicts some of the known secondary structure elements correctly. For bacterial L12, βA, αA, αC and to some extent αB are reasonably well predicted. Furthermore, the turns between βA and αA as well as between αB and βB are well predicted. In addition, steep changes in predicted α-helix or β-sheet whether correct or incorrect seem to occur at boundaries of observed secondary structure elements. These observations were applied to the predictions of secondary structure of the archaebacterial and eucaryotic eL12 protein with the question whether residues 1–68 can have a fold like the C-terminal domain of *E. coli* L12. Possible turns are found between residues 10–20, 48–52 and 67–70. The last region occurs at the interface between the 'domain' and the 'hinge' regions. The other two should indicate turns between secondary structure elements. If the fold is like the known L12 fold, their position would best fit the regions between βA and αA (residues 10–20) and between

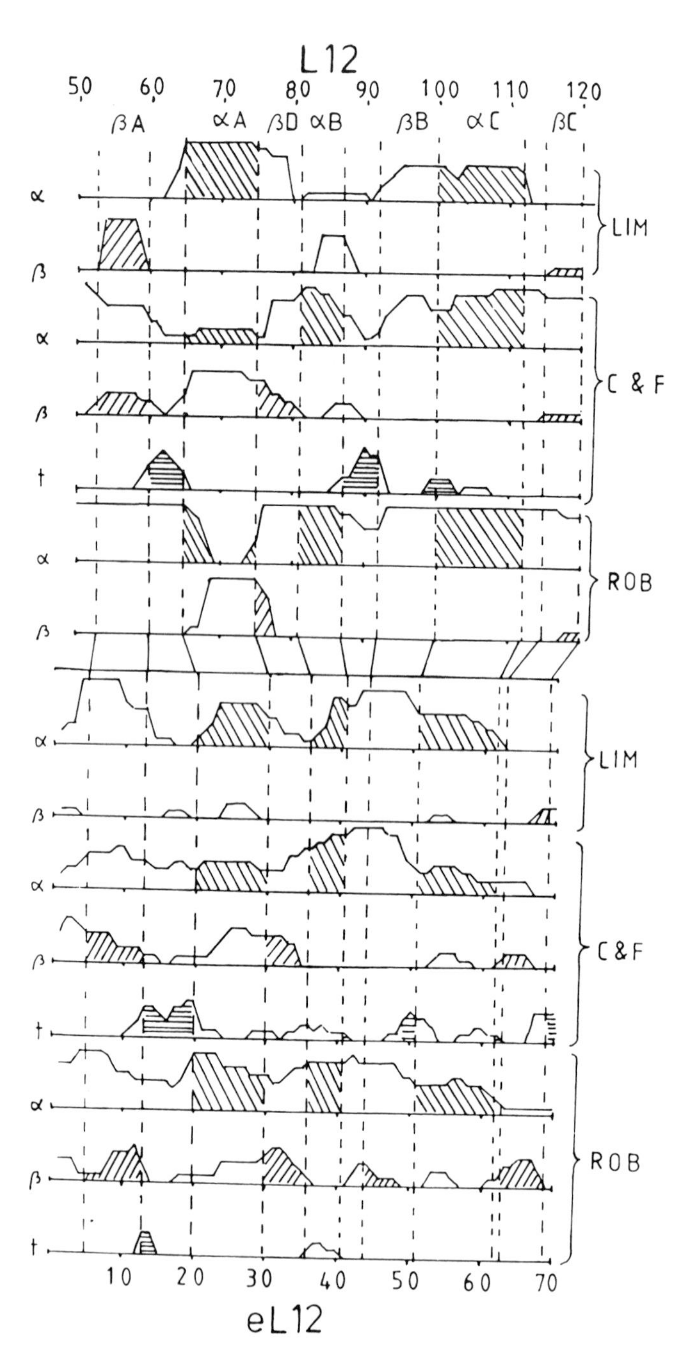

Fig. 7. Secondary structure prediction by three methods [31–33]. Residues 50–120 are displayed for eubacteria and 1–70 for eucaryotes and archaebacteria. The quantity plotted is the number of species for which α-helix, β-sheet and turns respectively have been predicted by one of the methods. The agreement with the crystal structure is indicated as hatched areas of the predicted secondary structure. The vertical lines indicate the possible alignment between L12 from eubacteria and eucaryotes.

βB and αC (residues 48–52). In such a case αA and αB would be placed between residues 20–40 which have reasonable predictions for α-helix with the turn probably around residue 32. Helix C would occur at residues 52–64 but this is less well fit to the predicted structure. The β-structures that were poorly predicted from the bacterial sequences would occur at 4–10, 42–48 and 64–68 in the eucaryotic and archaebacterial sequences. The predictions for these regions as β-strands is poor except for βA.

If we make use of this possible alignment for the L12 and eL12 sequences and allow for deletions or insertions primarily between secondary structure elements, we can obtain the alignment shown in Fig. 8. More than one-third of the

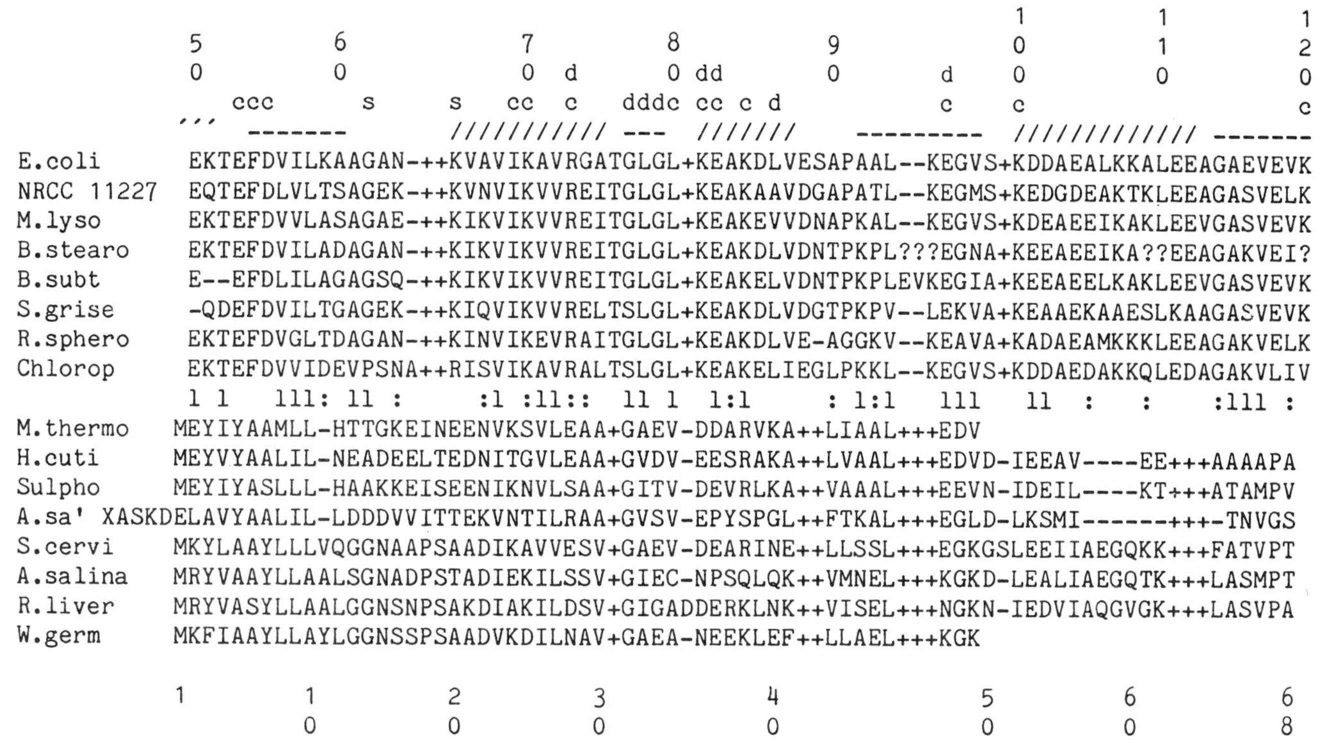

E.coli EKTEFDVILKAAGAN-++KVAVIKAVRGATGLGL+KEAKDLVESAPAAL--KEGVS+KDDAEALKKALEEAGAEVEVK
NRCC 11227 EQTEFDLVLTSAGEK-++KVNVIKVVREITGLGL+KEAKAAVDGAPATL--KEGMS+KEDGDEAKTKLEEAGASVELK
M.lyso EKTEFDVVLASAGAE-++KIKVIKVVREITGLGL+KEAKEVVDNAPKAL--KEGVS+KDEAEEIKAKLEEVGASVEVK
B.stearo EKTEFDVILADAGAN-++KIKVIKVVREITGLGL+KEAKDLVDNTPKPL???EGNA+KEEAEEIKA??EEAGAKVEI?
B.subt E--EFDLILAGAGSQ-++KIKVIKVVREITGLGL+KEAKELVDNTPKPLEVKEGIA+KEEAEELKAKLEEVGASVEVK
S.grise -QDEFDVILTGAGEK-++KIQVIKVVRELTSLGL+KEAKDLVDGTPKPV--LEKVA+KEAAEKAAESLKAAGASVEVK
R.sphero EKTEFDVGLTDAGAN-++KINVIKEVRAITGLGL+KEAKDLVE-AGGKV--KEAVA+KADAEAMKKKLEEAGAKVELK
Chlorop EKTEFDVVIDEVPSNA++RISVIKAVRALTSLGL+KEAKELIEGLPKKL--KEGVS+KDDAEDAKKQLEDAGAKVLIV
1 1 111: 11 : :1 :11:: 11 1 1:1 : 1:1 111 11 : : :111 :
M.thermo MEYIYAAMLL-HTTGKEINEENVKSVLEAA+GAEV-DDARVKA++LIAAL+++EDV
H.cuti MEYVYAALIL-NEADEELTEDNITGVLEAA+GVDV-EESRAKA++LVAAL+++EDVD-IEEAV----EE+++AAAAPA
Sulpho MEYIYASLLL-HAAKKEISEENIKNVLSAA+GITV-DEVRLKA++VAAAL+++EEVN-IDEIL----KT+++ATAMPV
A.sa' XASKDELAVYAALIL-LDDDVVITTEKVNTILRAA+GVSV-EPYSPGL++FTKAL+++EGLD-LKSMI------++-TNVGS
S.cervi MKYLAAYLLLVQGGNAAPSAADIKAVVESV+GAEV-DEARINE++LLSSL+++EGKGSLEEIIAEGQKK+++FATVPT
A.salina MRYVAAYLLAALSGNADPSTADIEKILSSV+GIEC-NPSQLQK++VMNEL+++KGKD-LEALIAEGGQTK+++LASMPT
R.liver MRYVASYLLAALGGNSNPSAKDIAKILDSV+GIGADDERKLNK++VISEL+++NGKN-IEDVIAQGVGK+++LASVPA
W.germ MKFIAAYLLAYLGGNSSPSAADVKDILNAV+GAEA-NEEKLEF++LLAEL+++KGK

Fig. 8. The L12 and eL12 aligned as suggested by the secondary structure prediction. Deletions needed for the alignment of the L12 or eL12 family of sequences is indicated by −. Deletions to align eL12 with L12 are indicated by +. Deletions occur in regions where there are already deletions in either family of sequences or between secondary structure elements. The homologies between the L12 and eL12 sequences are indicated by : (colon) when a few residues agree and by 1 when a majority of the residues in the L12 family agree with a majority of the residues in the eL12 family of sequence. See caption for Fig. 3 for meaning of c, d, s, −−−, /// and ⟨⟨⟨.

residues are highly homologous for a majority of the species from both forms of the protein. For most of the sequences between one-third and one-half of the residues are homologous to any species of the other form of the protein. Some of the details of the alignments deserve special comment. Thus, the conserved Arg 73 in eubacteria is not aligned with any conserved basic residue. The semiconserved Arg 38 in archaebacteria and eucaryotes is aligned with the invariant Lys 84 in eubacteria. The highly conserved glycines in positions 62, 77 and 97 in L12 are aligned with highly conserved glycines in positions 14, 31 and 48 in eL12. The alignment of glycines is partly an effect of the method used to superimpose turns between secondary structure elements, since these parts of protein structures are enriched in glycines.

The analysis of the eubacterial L12 sequences in light of the three-dimensional structure has enabled us to define three conserved surface of different functional potential [43, 44]. Thus we have discussed an anion binding site, a dimer surface between L12 globular domains and a large conserved surface area possibly involved in factor binding (see Fig. 8). When the corresponding positions are examined in eucaryotes and archaebacteria only the dimer surface seems to be conserved to any reasonable extent. Three to five out of seven residues are conserved in most species. This limited conservation of the possible functional surfaces on the globular domain of the L12-like protein could be due to a less than perfect alignment of the sequences or to a true lack of sequence conservation.

Discussion

One of the elongation factors, EF-Tu or, in eucaryotes, EF-1α, is well conserved between eubacteria and organelles on the one hand and on the other its eucaryotic homologues. Four additional factors involved in protein synthesis interact with the same or overlapping sites on the ribosome. Two of them have at least one homologous domain with EF-Tu. This domain is involved in the binding of GTP as well as of tRNA, and it could have originated from a common ancestral protein for all the factors.

The interaction of the factors with one and the same region on the ribosome partly involving L12 would seem to require a conserved surface on the factors. With the present amount of information this surface cannot be located. Conversely the domain on the ribosome interacting with five different protein factors should be highly conserved. The limited changes in EF-Tu over very different species would also suggest this. Nevertheless it is impossible to align the sequences of L12 from all types of ribosomes in a straightforward way. However, one feature is very clearly conserved in all types of L12 molecules. This is a hinge region allowing a great flexibility in the protein.

The C-terminal domain which is involved in factor-related function in eubacteria can be related to the N-terminal region in archaebacteria and eucaryotes. This transposition in the structure of a functionally important ribosomal protein seems at least equally drastic as the observed species differences in rRNA structure [62] or ribosome morphology [63]. Its pattern of occurrence is paralleled by the pronounced 'beak' of the small ribosomal subunit [63–66]. These two features group the archaebacteria and eucaryotes together and separately from the eubacteria. However, from other molecular structures it has been suggested that eubacteria and archaebacteria are relatively close in evolution [67].

Evidently the structural arrangement of L12 discriminates species that belong to the eubacterial kingdom very clearly from other species. Thus L12 from spinach chloroplasts has clearly a eubacterial structure [41, 44]. There are also clear differences between eucaryotes and archaebacteria in the L12 primary structure around residues 56–68. The present data does not provide a clear distinction between archaebacteria and so called 'eocytes' in L12 structure [2].

The homologous factors in protein synthesis (EF-Tu, EF-G, IF-2 and possibly RF-1 and RF-2) seem to have evolved by some divergent process. Too little information is presently available for any detailed analysis. It is not clear whether L12 has evolved in a divergent manner with a dramatic change of structure between eubacteria and other species [60] or whether L12 and eL12 have evolved in a convergent manner from two different proteins [59]. The amount of general homology could argue for a divergent evolution.

Acknowledgement

We would like to thank Prof. C. G. Kurland for support and valuable criticism and C. Sander for useful suggestions. S.T. was the grateful recipient of a long-term EMBO fellowship.

References

1. Woese, C. R. and Fox, G. E., *Proc. Natl. Acad. Sci. USA* **74**, 5088–5090 (1977).
2. Lake, J. A., Henderson, E., Oakes, M. and Clark, M. W., *Proc. Natl. Acad. Sci. USA* **81**, 3786–3790 (1984).
3. Kaltschmidt, E. and Wittmann, H. G., *Proc. Natl. Acad. Sci. USA* **67**, 1276 (1970).
4. Terhorst, C., Möller, W., Laursen, R. and Wittmann-Liebold, B., *Eur. J. Biochem.* **25**, 138 (1973).
5. Weissbach, H., in *Ribosomes, Structure and Genetics* (ed. G. Chambliss, G. R. Craven, J. Davies, K. Davis, L. Kahan and M. Nomura), pp. 255–265. University Park Press, Baltimore (1980).
6. Liljas, A., *Prog. Biophys. Mol. Biol.* **40**, 161–228 (1982).
7. Caskey, C. T., *Biochem. Sci.* **5**, 234–237 (1980).
8. Möller, W., Schrier, P. I., Maassen, J. A., Zantema, A., Shop, E., Reinalda, H., Cremers, A. F. M. and Mellema, J. E., *J. Mol. Biol.* **163**, 553–573 (1983).
9. Traut, R. R., Ferris, R. J., Gavino, G. R., Lambert, J. M., Butler, P. D. and Kenny, J. W., in *Cellular Regulation and Malignant Growth* (ed. Ebashi). Tokyo, Japan Sci. Soc. Press/Springer-Verlag, Berlin (in the press).
10. Gudkov, A. T. and Gongadze, G. M., *FEBS Lett.* **176**, 32–36 (1984).
11. Sacerdot, C., Dessen, P., Hershey, J. W. B., Plumbridge, J. A. and Grunberg-Manago, M., *Proc. Natl. Acad. Sci. USA* **81**, 7787–7791 (1984).
12. Jones, M. D., Petersen, T. E., Nielsen, K. M., Magnusson, S., Sottrup-Jensen, L., Gausing, K. and Clark, B. F. C., *Eur. J. Biochem.* **108**, 507–526 (1980).
13. Nagata, S., Tsunetsugu-Yokota, Y., Naito, A. and Kaziro, Y., *Proc. Natl. Acad. Sci. USA* **80**, 6192–6196 (1983).
14. Montandon, P.-E. and Stutz, E., *Nucl. Acid. Res.* **11**, 5877–5892 (1983).
15. Nagata, S., Nagashima, K., Tsunetsugu-Yokota, Y., Fujimura, K., Miyazaki, M. and Kaziro, Y., *EMBO J.* **3**, 1825–1830 (1984).
16. van Hemert, F. J., Amons, R., Pluijms, W. J. M., van Ormondt, H. and Möller, W., *EMBO J.* **3**, 1109–1113 (1984).
17. Craigen, W. J., Cook, R. G., Tate, W. P. and Caskey, C. T., *Proc. Natl. Acad. Sci. USA* **82**, 3616–3620 (1985).
18. Clark, B. F. C., la Cour, T. F. M., Fontecilla-Camps, J., Morikawa, K., Nielsen, K. M., Nyborg, J. and Rubin, J. R., *FEBS Meeting on Cell Function and Differentiation*, vol. 3, Symp. 12 (1982).
19. la Cour, T. F. M., Nyborg, J., Thirup, S. and Clark, B. F. C., *EMBO J.* **4**, 2385–2388 (1985).
20. Bosch, L., Kraal, B., van Noort, J. M., van Delft, J., Talens, A. and Vijgenboom, E., *TIBS* **10**, 313–316 (1985).
21. Duistervinkel, F. J., Kraal, B., de Graaf, J. M., Talens, A., Bosch, L., Swart, G. W. M., Parmeggiani, A., la Cour, T. F. M., Nyborg, J. and Clark, B. F. C., *EMBO J.* **3**, 113–120 (1984).
22. Johnson, A. E., Miller, D. L. and Cantor, C. R., *Proc. Natl. Acad. Sci. USA* **75**, 3075–3079 (1978).
23. van Noort, J. M., Kraal, B. and Bosch, L., *Proc. Natl. Acad. Sci. USA* **82**, 3212–3216 (1985).
24. Jonak, J., Smrt, J., Holy, A. and Rychlick, I., *Eur. J. Biochem.* **105**, 315–320 (1980).
25. Halliday, K. R. J., *Cycl., Nucl. Prot. Phos. Res.* **9**, 435–448 (1984).
26. Lochrie, M. A., Hurby, J. B. and Simon, M. I., *Science* **228**, 96–99 (1985).
27. Leberman, R. and Egner, U., *EMBO J.* **3**, 339–341 (1984).
28. Laursen, R. A., l'Italien, I. J., Nugarkatti, S. and Miller, D. L., *J. Biol. Chem.* **256**, 8102–8109 (1981).
29. Ovchinikov, Y. A., Alakhov, Y. B., Bundulis, Y. P., Dovgas, N. V., Koslov, V. P., Motuz, L. P. and Vinokurov, L. M., *FEBS Lett.* **139**, 130–135 (1982).
30. Devereux, J., Haeberli, P. and Smithies, O., *Nucl. Acid. Res.* **12**, 387–395 (1984).
31. Chou, P. Y. and Fasman, G., *Adv. Enzymol.* **47**, 45–148 (1978).
32. Garnier, J., Osguthorpe, D. J. and Robson, B., *J. Mol. Biol.* **120**, 97–120 (1978).
33. Lim, V. I., *J. Mol. Biol.* **88**, 873–894 (1974).
34. Falkenberg, P., Yaguchi, M., Roy, C., Zuber, M. and Matheson, A. T., *Can. J. Biochem. Cell. Biol.*, submitted (1985).
35. Visentin, L. O., Yaguchi, M. and Matheson, A. T., *Can. J. Biochem.* **57**, 719–726 (1979).
36. Itoh, T. and Wittmann-Liebold, B., *FEBS Lett.* **96**, 392–394 (1978).
37. Falkenberg, P., Yaguchi, M., Rollin, C. R., Matheson, A. T. and Wydro, R., *Biochim. Biophys. Acta* **578**, 207–215 (1979).
38. Itoh, T., *FEBS Lett.* **127**, 67–70 (1981).
39. Itoh, T., Sugiyama, M. and Higo, K. I., *Biochim. Biophys. Acta* **701**, 164 (1982).
40. Itoh, T. and Higo, K. I., *Biochim. Biophys. Acta* **744**, 105 (1983).
41. Bartsch, M., Kimura, M. and Subramanian, A. R., *Proc. Natl. Acad. Sci. USA* **79**, 6871–6875 (1982).
42. Leijonmarck, M., Eriksson, S. and Liljas, A., *Nature (London)* **286**, 824 (1980).
43. Liljas, A., Kirsebom, L. and Leijonmarck, M., in *Ribosomes* (ed. B. Hardesty). Springer-Verlag, New York (in the press).
44. Leijonmarck, M., Liljas, A. and Subramanian, A. R., *Biochem. Int.* **8**, 69–76 (1984).
45. Sanchez-Madrid, F., Vidales, F. J. and Ballesta, J. P. G., *Biochemistry* **20**, 3263–3266 (1981).
46. Matheson, A. T., Möller, W., Amons, R. and Yaguchi, M., in *Ribosomes, Structure, Function and Genetics* (ed. G. Chambliss, G. R. Craven, J. Davies, K. Davis, L. Kahan and M. Nomura), pp. 297–332. University Park Press, Baltimore (1980).
47. Yaguchi, M., Matheson, A. T., Visentin, L. P. and Zuker, M., in *Genetics and Evolution of RNA Polymerase, tRNA and Ribosomes* (ed. S. Osawa, in H. Ozeki, H. Uchida and Y. Yura), pp. 585–599. University of Tokyo Press (1978).
48. Matheson, A. T., in *The Bacteria*, vol. 8, *The Archaebacteria* (ed. C. R. Woese and R. S. Wolfe), pp. 345–377. Academic Press (1985).
49. Leijonmarck, M., Pettersson, I. and Liljas, A., in *Structural Aspects of Recognition and Assembly in Biological Macromolecules* (ed. M. Balaban, J. L. Sussmann, W. Traub and A. Yonath), pp. 761–777. ISS, Rehovot and Philadelphia (1981).
50. Traut, R. R., Tewari, D. S., Sommer, A., Gavino, G. R., Olsen, H. M. and Glitz, D. G., in *Ribosomes* (ed. B. Hardesty). Springer-Verlag, New York (in the press).
51. van Agthoven, A. J., Maassen, J. A., Schrier, P. I. and Möller, W., *Biochem. Biophys. Res. Commun.* **64**, 1184 (1975).
52. Koteliansky, V. E., Domogatsky, S. P. and Gudkov, A. T., *Europ. J. Biochem.* **90**, 319 (1978).
53. Marquis, D. M. and Fahnestock, S. R., *J. Mol. Biol.* **142**, 161–179 (1980).
54. Gudkov, A. T., Gongadze, G. M., Bushuev, V. N. and Okon, M. S., *FEBS Lett.* **138**, 229–232 (1982).
55. Cowgill, C. A., Nichols, B. G., Kenny, J. W., Butler, P., Bradbury, E. M. and Traut, R. R., *J. Biol. Chem.* **259**, 15257–15263 (1984).
56. Tritton, T. R., *Biochemistry* **17**, 3959–3964 (1978).
57. Lee, C. C., Wells, B. D., Fairclough, R. H. and Cantor, C. R., *J. Biol. Chem.* **256**, 49–53 (1981).
58. Lin, A., Wittmann-Liebold, B., McNally, J. and Wool, I. G., *J. Biol. chem.* **257**, 9189–9197 (1982).

59. Itoh, T., *Biochim. Biophys. Acta* **671**, 16–24 (1981).
60. Amons, R., Pluijms, W. and Möller, W., *FEBS Lett.* **104**, 85–89 (1979).
61. Amons, R., Pluijms, W., Kriek, J. and Möller, W., *FEBS Lett.* **146**, 143–147 (1982).
62. Noller, H. F., *Annu. Rev. Biochem.* **53**, 119 (1984).
63. Lake, J. A., Henderson, E., Clark, M. W. and Matheson, A. T., *Proc. Natl. Acad. Sci. USA* **79**, 5948 (1982).
64. Boublik, M. and Hellmann, W., *Proc. Natl. Acad. Sci. USA* **75**, 2829–2833 (1978).
65. Lutsch, G., Noll, F., Thiese, H., Enzmann, G. and Bielka, H., *Molec. Gen. Genet.* **176**, 281–291 (1979).
66. Frank, J., Verschoor, A. and Boublik, M., *Science* **214**, 1353 (1981).
67. Lake, J. A., *Annu. Rev. Biochem.* **54**, 507–530 (1985).

Chemica Scripta 1986, **26B**, 121–126

Evolution Mapped with Three-dimensional Ribosome Structure

J. A. Lake, E. Henderson, M. W. Clark, A. Scheinman and M. I. Oakes

Molecular Biology Institute and Department of Biology, University of California, Los Angeles, CA 90024, USA

Paper presented by J. A. Lake at the Conference on 'Molecular Evolution of Life', Lidingö, Sweden, 8–12 September 1985

Abstract

Three-dimensional ribosomal structure is highly conserved, even when organisms from different urkingdoms are compared. Hence it is extremely useful as a probe of distant evolutionary events. In general, ribosomal large and small subunits are organized in four structural patterns. By using a parsimony analysis of ribosome structure, we find two results that stand in contrast with the standard evolutionary tree. First we find that the sulfur-dependent bacteria, or eocytes, are topologically nearest neighbors to the eukaryotes rather than to the methanogens. We suggest that the depth of this division is appropriate for a separation at the urkingdom level. Secondly, our data indicate that the halobacteria have diverged from the eubacteria more recently than from any other known group of organisms. We interpret these results to indicate that the halobacteria are incorrectly placed in the archaebacteria, and should probably be included with the eubacteria into a larger group, the photocytes. In brief, the phylogenetic pattern that emerges is intimately linked with the metabolisms of these organisms.

Introduction

Three-dimensional ribosomal structure is coming into its own as a phylogenetic marker. It is particularly useful as a marker of distant evolutionary changes since ribosomal three-dimensional structure is so highly conserved. By the same token, it cannot easily distinguish the more recent phylogenetic divisions that are easily revealed by rRNA sequence and secondary structure analysis. Hence as a technique it nicely complements other methods of phylogenetic determination.

In order to illustrate the usefulness of this technique, we would like to review two findings that are based on structure-determined phylogenies and point out how they differ in quite fundamental ways from the sequence determined phylogenies [14]. The first example will emphasize the relatively close relationship between the eukaryotic cytoplasm and the sulfur-dependent bacteria, or eocytes [6] and the second will form the major focus of this paper and argue that the eubacteria and halobacteria are evolutionarily closest neighbors [8].

Results

Electron micrographs of ribosomal subunits from eubacteria, halobacteria, methanogens, sulfur-dependent bacteria (eocytes) and eukaryotes suggest there are four general classes of ribosome structure. All four are, of course, related and details of their structures are assumed to reflect their separate evolutionary pathways. Electron micrographs of ribosomal subunits from photosynthetic and non-photosynthetic eubacteria and from halobacteria are shown in Fig. 1. Small subunits are shown in the first two columns and large subunits are shown in the last two columns.

The representative photosynthetic eubacteria include a green non-sulfur bacterium (row A), a purple sulfur bacterium (B), a purple non-sulfur bacterium (C) and a cyanobacterium (D). For comparison, micrographs of a non-photo-synthetic Gram-positive bacterium are also included (row E). Ribosomes from *Halobacterium cutirubrum* and from *Halococcus morrhuae* are shown in rows G and E, respectively. For the purposes of this paper, one important point is that the structures of the halobacterial ribosomal subunits are nearly indistinguishable from those of the eubacterial subunits [5] except that the small subunit of halobacterial ribosomes contains a significant bill, whereas the eubacteria contain only a vestigial bill. The structure of the eubacterial–halobacterial ribosome is shown in Fig. 4 (the bill is diagonally shaded).

In contrast, ribosomes of methanogens, eocytes and eukaryotes (Fig. 2A, B, D and F, respectively) are distinctly different from the eubacterial halobacterial pattern. Diagrams of their separate types are illustrated beneath the micrographs in Fig. 2. Additional electron micrographs of eucytes are shown in Fig. 3. Small subunits are shown in 3A and large subunits are in 3B. The representative organisms include *Thermoproteus tenax*, *Sulfolobus acidocaldarius*, *Desulfurococcus mucosus*, *Thermococcus celer* and *Thermofilum pendens*, respectively. Features present in eukaryotic and eocytic small subunits but lacking, or nearly lacking, in eubacterial and halobacterial subunits are important evolutionary markers. These include lobes at the base of the subunit (nearly absent in methanogens, of intermediate size in eocytes, and large in eukaryotes), a prominent bifurcation of the small subunit platform (present in methanogens, eocytes and eukaryotes) and an alternative head shape (in eocytic and eukaryotic ribosomes). In large ribosomal subunits from eocytes and eukaryotes, a lobe is present at the base of the subunits and a prominent bulge is above the lobe. The bulge is separated from the lobe by a gap in eocytes. The gap is filled in eukaryotes. The profiles of the eocytic subunits, where they differ from the eubacterial–halobacterial pattern, are indicated by thin dashed lines in Fig. 4.

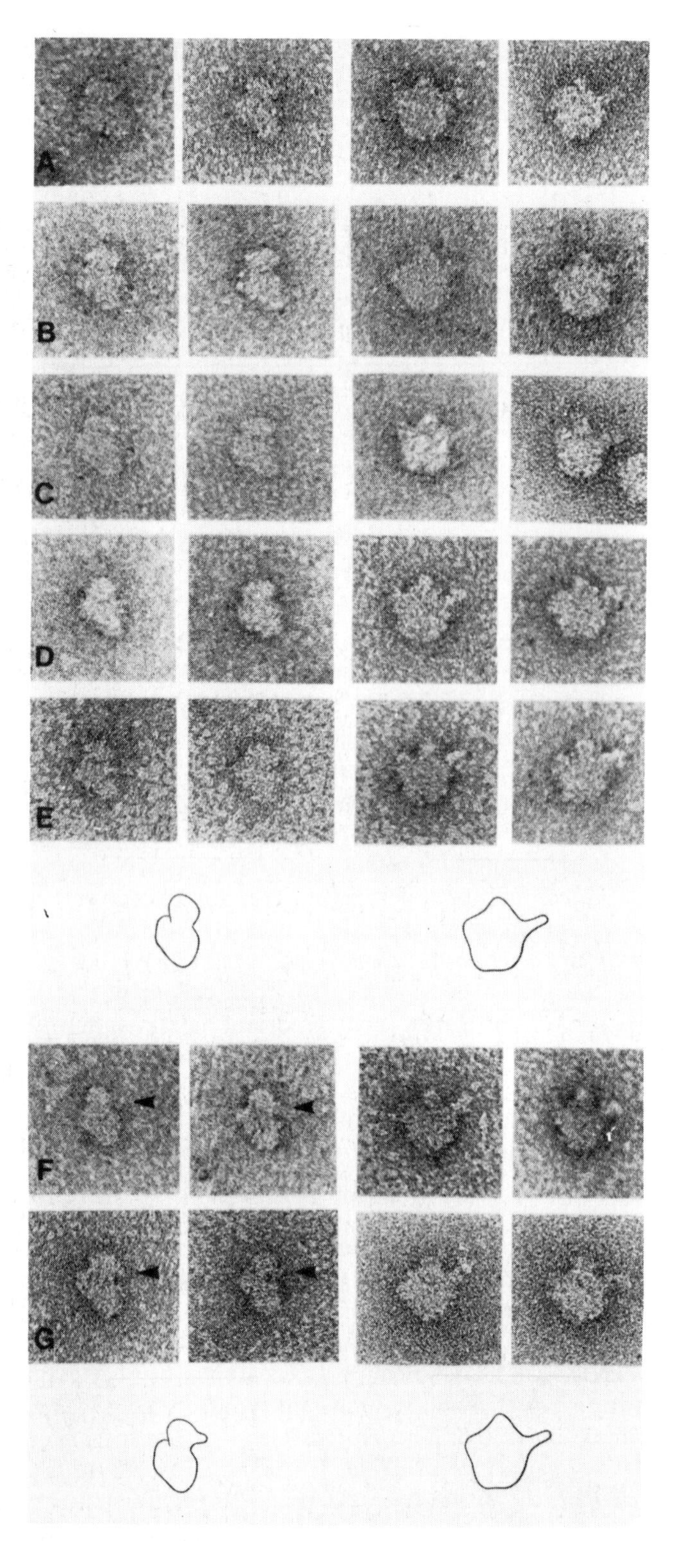

Fig. 1. Electron micrographs of eubacterial (rows A–E) and halobacterial (rows F and G) ribosomal subunits. Small subunits in the asymmetric projection are shown in the first two columns and large subunits in the quasi-symmetric projection (the L7/L12 stalk is to the right of the subunit) are shown in the third and fourth columns. The archaebacterial bills are indicated by arrows in rows F and G. The organisms represented are *Chloroflexus aurantiacus* (a green non-sulfur bacterium, row A), *Thiocapsa pfennigii* (a purple non-sulfur bacterium, row C), *Synechocystis 6701* (a Cyanobacterium, row D), and *Bacillus stearothermophilus* (a non-photosynthetic Gram-positive thermophilic bacterium, row E). The halobacteria are *Halobacterium cutirubrun* (a halobacterium, row F) and *Halococcus morrhuae* (a halococcus, row G). Eubacteria are shown in rows A–E and of those A–D are photosynthetic. Halobacteria are in rows F and G, that in F is photosynthetic and that in G is presumed to be photosynthetic. Magnification: ×250000.

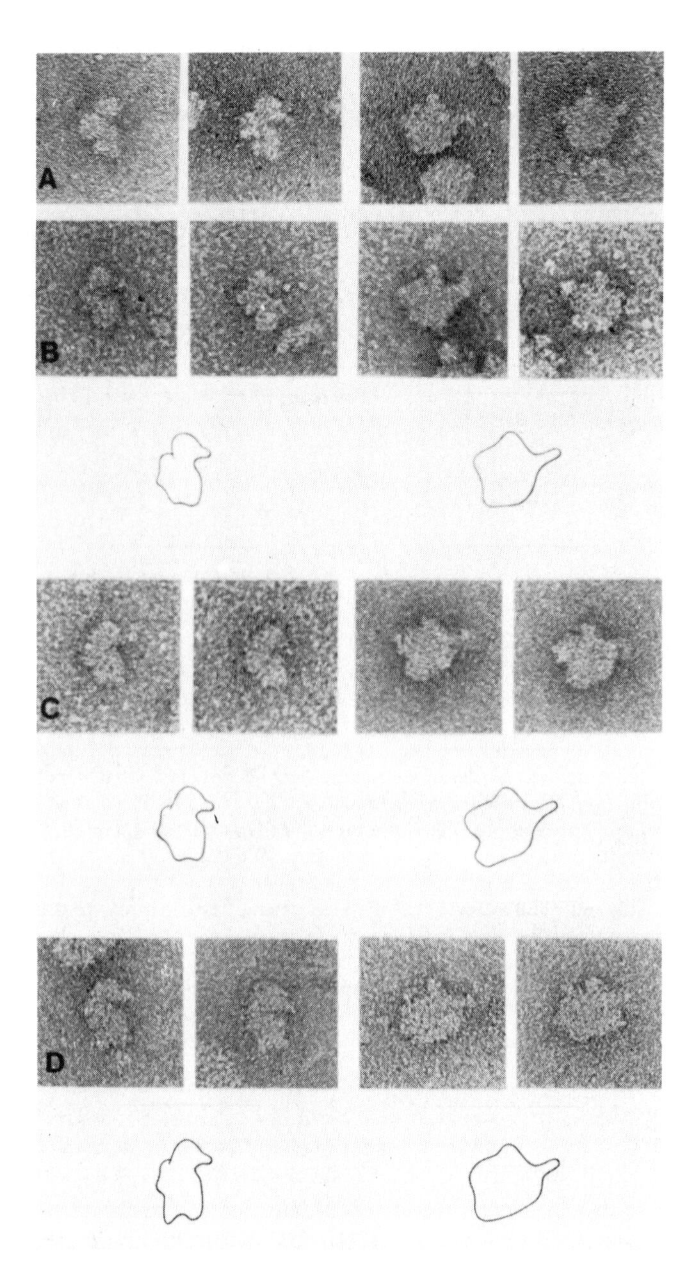

Fig. 2. Ribosomal subunits from the archaebacterial, eocytic and eukaryotic kingdoms (in rows A and B, row C and row D, respectively). As in Fig. 1, small subunits are in the left two panels and large subunits are in the right two panels. The organisms are *Methansarcina barkeri*, a heterotrophic methanogen, in row A; large subunits are from the related strain *M 227*), *Methanobacterium thermoautotrophicum* (a thermophilic, autotrophic methanogen, row B), *Thermoproteus tenax* (a thermophilic, autotrophic, sulfur reducing bacterlium, row C) and *Saccharomyces cerivisiae* (a yeast, row D). The characteristic subunit profiles are diagrammed beneath the groups. The magnification is ×250000.

Discussion

Eukaryotes and eocytes are topologically nearest evolutionary neighbors

Within each of these ribosomal types the three-dimensional structure is relatively constant. Hence variations in structure can provide a phylogenetic basis for relating their evolution. Assuming that ribosome structure evolves relatively slowly, and indeed ribosome structure seems to be evolving more slowly than almost any other molecular property, then a parsimony analysis is appropriate [3]. When the ribosomal features of the methanogens, eubacteria, eocytes and eukary-

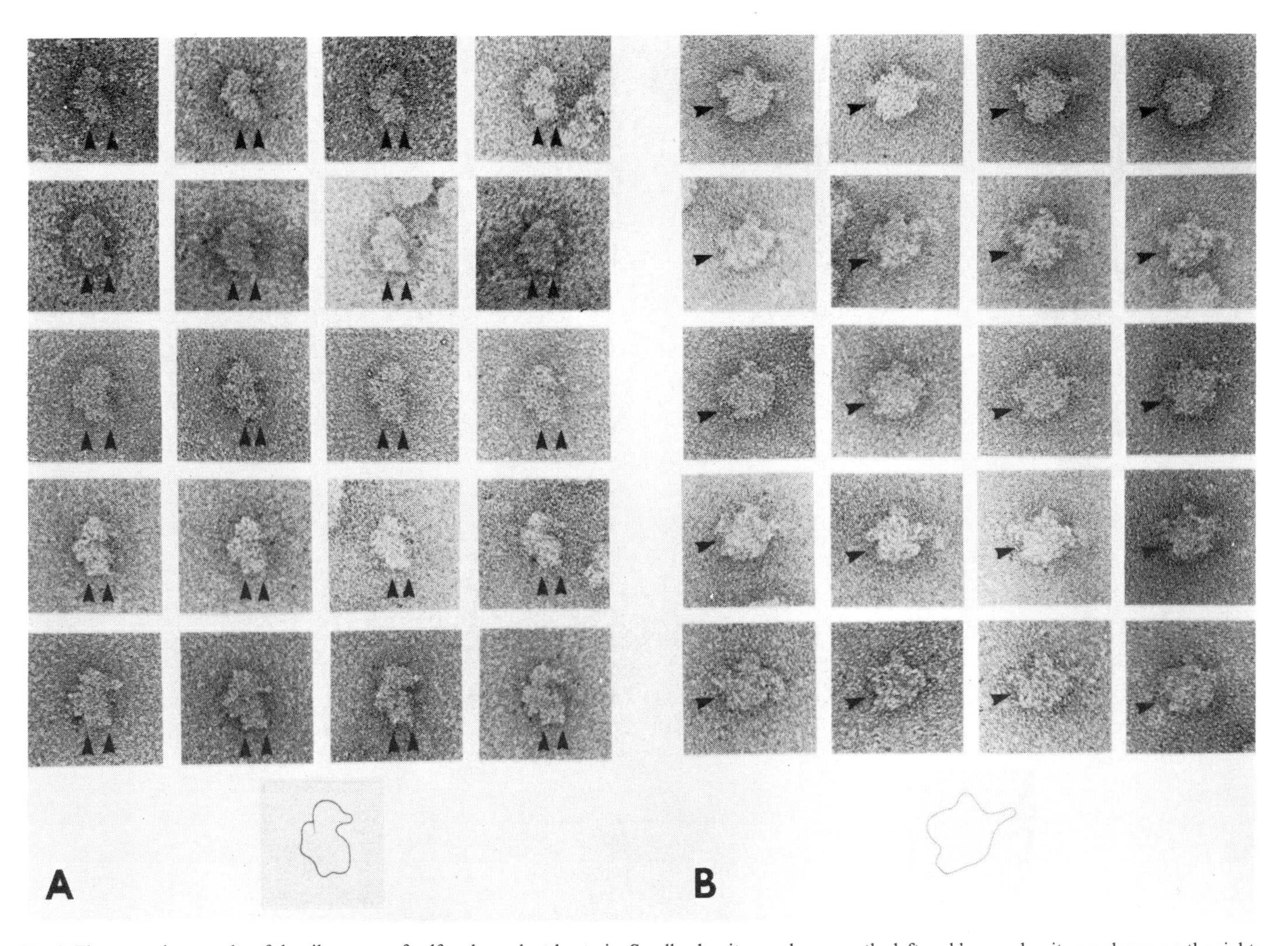

Fig. 3. Electron micrographs of the ribosomes of sulfur-dependent bacteria. Small subunits are shown on the left and large subunits are shown on the right. Magnification × 250 000.

otes were compared [7] then the unrooted molecular tree shown in Fig. 5A was obtained. This tree, the eocyte tree, was quite unexpected since it indicates a fundamentally different topology of early evolutionary events than was accepted at the time. The branching order that had previously been proposed [12], which we call the archaebacterial tree, is shown in Fig. 5B. *The differences between these two trees are significant.* If the archaebacterial tree is correct, then eocytes are a subgroup of the archaebacteria, whereas if the eocyte tree is correct, then eocytes should be given independent status as an urkingdom. Because of the early time at which this division took place, and because both the eubacteria and the eukaryotes appear to have faster molecular clocks than the archaebacteria and the eocytes, the correct tree is extremely difficult to deduce from rRNA sequences. (Indeed it has been shown that under these circumstances most methods of analysis will be positively misleading [2].) Because of the problems associated with sequence analysis in cases where multiple mutations have occurred at a given nucleotide position, we think that the eocytic tree is currently the most strongly favored.

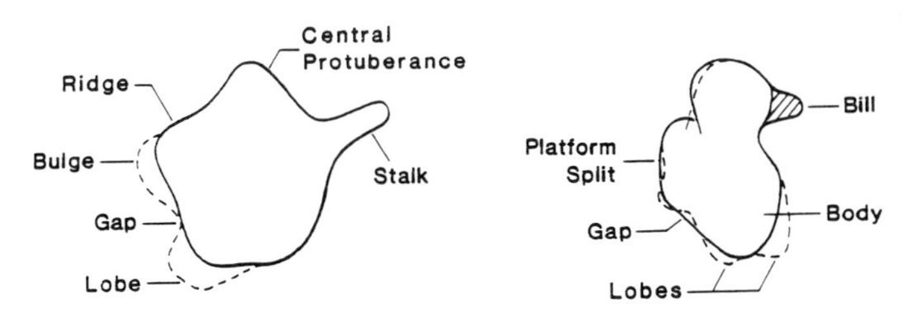

Fig. 4. A summary of the ribosomal features of the eubacteria and halobacteria. The structures common to both are shown by solid lines. The bill, found on the halobacterial and not the eubacterial small subunit, is diagonally shaded. Additional structures found on the eocytic ribosomal subunits, but not on photocytic subunits, are shown as lightly dashed lines. A large subunit is at the top and a small subunit is at the bottom.

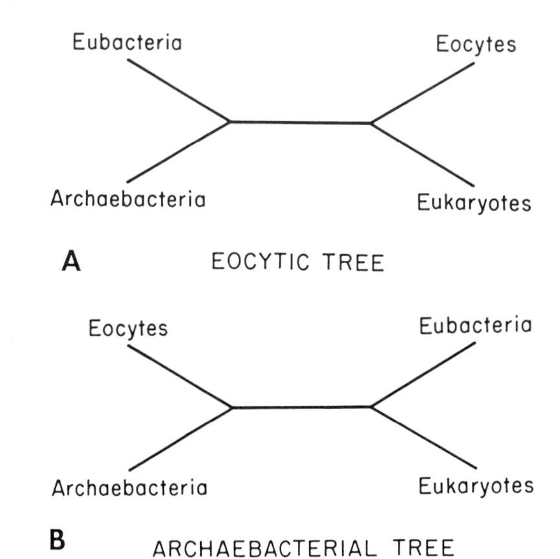

Fig. 5. The unrooted dendogram relating the steps in the evolution of taxa from the eocytes, eubacteria, archaebacteria and eukaryotes. The Eocyte tree corresponds to the most parsimonious interpretation of our data.

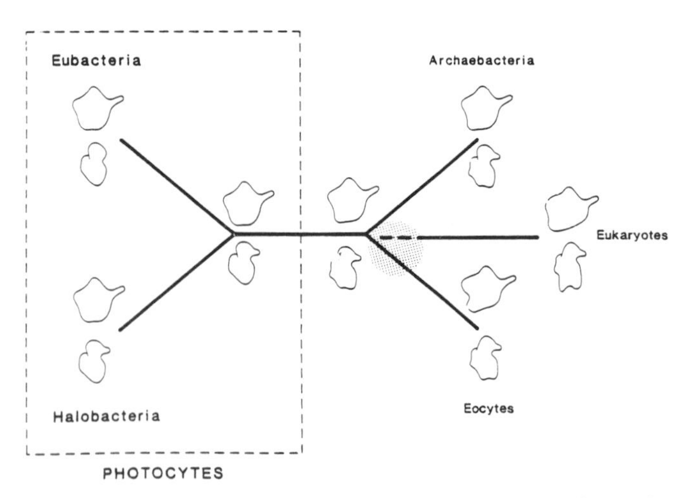

Fig. 6. The unrooted tree relating the steps in evolution indicates that eubacteria and halobacteria are (topologically) closest neighbors. The group corresponding to the photocytes is enclosed by a dashed line. The most likely rootings of this tree are shown in the shaded area.

Determination of the most parsimonious unrooted evolutionary tree places eubacteria and halobacteria as closest neighbors

Once we realized that the eubacteria shared a special relationship with the archaebacteria, we decided to search for subgroups that could reveal the relationship. The answer soon became quite obvious and we realized that halobacteria are more closely related to the eubacteria than they are to the methanogens (archaebacteria). As before, we used the variations in structure between the eubacterial–halobacterial group and the others to provide a phylogenetic basis for relating their evolution. In considering a parsimony analysis of the eubacterial and halobacterial structures, two new 30S structural features were considered. These were a modified head-shape (found in eocytic and eukaryotic subunits) and a split, or bifurcated, platform (found in methanogenic, eocytic, and eukaryotic subunits). The unrooted tree that most parsimoniously fits these data is shown in Fig. 6. It is not the only interpretation of our data, but it is the simplest. Also shown on the tree are the sites of the most probable rootings of the tree suggested by the oligonucleotide catalogues [12] and by DNA–rRNA cross-hybridization data [11]. If the topology of this dendrogram is the correct interpretation of our data, and a body of data on the common biochemical properties of eubacteria and halobacteria (discussed in ref. [5]) support this tree, then one is forced to conclude that the halobacteria are more closely related to the eubacteria than they are to any other currently known organisms. In particular, they should be removed from the archaebacterial urkingdom and placed elsewhere.

Eubacteria and halobacteria form a natural group, the photocytes

If the halobacteria are not archaebacteria, then their altered status raises the question of where they properly belong. There are three alternative classification schemes that would generate monophyuletic groups *sensu* Hennig [15]. These monophyuletic groupings could be created by (A) lumping the archaebacteria, eubacteria and halobacteria into one large urkingdom, (B) splitting the archaebacteria into two urkingdoms, thereby creating a separate urkingdom for the halobacteria, or (C) combining the eubacteria and the halobacteria into an urkingdom. We advocated alternative (C), shown in Fig. 7, since grouping the eubacteria and halobacteria did not 'lump' or 'split', i.e. it did not change the number of highest-level categories. In addition, it would generate a group that was at the same branching level as the archaebacteria (minus the halophiles) and it emphasized the natural biochemical association of the two photosynthetic groups. For this urkingdom we proposed the name photocytes (light + cell) to emphasize the photosynthetic abilities of both subgroups, the eubacteria and the halobacteria. In addition, choosing a classification that naturally accommodated the beginnings of photosynthesis (whether it occurred singly or in numbers) seemed to make good biological sense [9] since photosynthesis clearly represents a landmark in biochemical evolution. The rooted tree corresponding to this interpretation is shown in Fig. 8. A highly schematic tree to illustrate our points in a general, but possibly inaccurate, way is shown in Fig. 8.

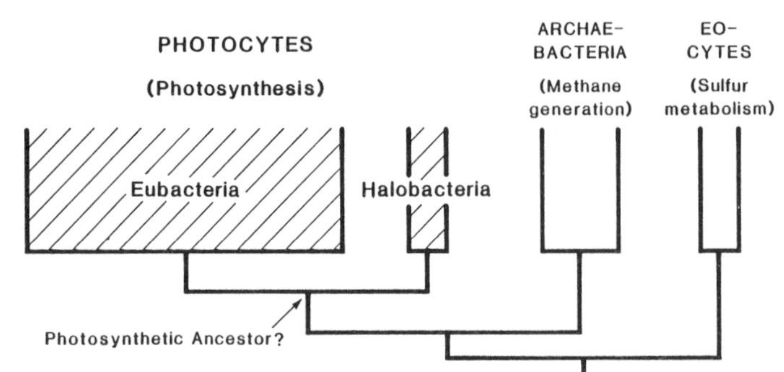

Fig. 7. The rooted evolutionary tree, illustrating eubacteria and halobacteria as sublines of the monophyletic group, the photocytes. This tree shows the phylogenetic relationships among the photocytes, archaebacteria and eocytes. The eubacterial and halobacterial branches of the photocytes are marked by diagonal lines. The possible photosynthetic common ancestor is indicated.

The photocytes might be descended from a common photosynthetic ancestor

If eubacteria and halobacteria are evolutionarily nearest neighbors, then other data on the molecular properties of

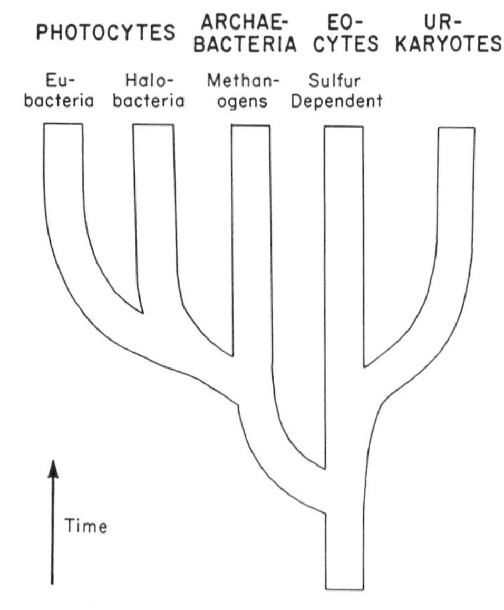

Fig. 8. A highly schematic evolutionary tree illustrating the authors' current best guess about the nature of the evolutionary tree.

these two groups should exist that support this tree. Indeed a relatively large body of data support this relationship. Rather surprisingly, however, many of the properties unique to eubacteria and halobacteria, i.e. not found in the methanogens and eocytes, are related to the molecular details of photosynthesis. This suggests that it might have been possible that the common ancestor of both the halobacteria and eubacteria was photosynthetic. (Very recently Seta *et al.* [10] have reported a simple carotenoporphyrin-quinone that functions as a light-driven transmembrane electron transfer pump. This system has many of the necessary properties predicted for the primitive common ancestor of both the halobacterial and eubacterial photosynthesizing systems. The nature of the primitive system and the data that support our proposal are described elsewhere [8].) If the halobacteria and the eubacteria had a common photosynthetic ancestor, this would then, quite drastically, change the way that one thinks of the eubacteria. For example, it would require that all eubacteria would be descendants of photosynthetic bacteria. To imagine an anaerobic, non-photosynthetic eubacterium such as a clostridium as being descended from a defective photosynthetic bacterium is difficult, but we know of no data that rule this out. Dickerson [1] has, for example, emphasized the large number of non-photosynthetic eubacteria that have descended from photosynthetic lines. (Within the last few weeks the case for all eubacteria being derived from a photosynthetic ancestor has been further strengthened by the finding [15] of a photosynthetic Gram-positive bacterium, thereby demonstrating photosynthesis in a eubacterial group that was formerly thought to be non-photosynthetic.)

All known molecular properties of halobacteria and eubacteria are consistent with their being evolutionarily closest neighbors

A number of molecular properties were previously thought to support the placement of the halobacteria with the methanogens [13, 14]. These include a large number of 'non-eubacterial' characters. Properties that are shared by halobacteria, methanogens and eocytes, but not by the eubacteria, include: ether-linked lipids, methionine carried by initiator tRNA's, elongation factors that react with diptheria toxin, a ribosomal 'bill', significant genealogical closeness as measured by oligonucleotide-catalog-derived S_{AB}'s, similarly designed ribosomal A proteins, and a lack of peptidylglycan cell walls. Unfortunately it has not been appreciated that these characters are phylogenetically uninformative since they are shared by three groups (halobacteria, methanogens and eocytes) and differ in only the eubacteria. In cladistic terms (Wiley, 1981) they represent plesiomorphic characters. The data supporting our proposed photocyte evolutionary tree, in contrast, are shared, derived (or synapomorphic) characters that occur in *both* halobacteria and eubacteria. They include details of ribosome structure and the many common details of photosynthesis in the two groups such as those described by [8]. Thus we know of no data that conflict with our evolutionary proposal.

Eocytes and photocytes as urkingdoms make good biological sense

One of the original reasons for putting the metanogens into a separate urkingdom [13] was that the old divisions of prokaryotic and eukaryotic kingdoms did not make any allowance for phylogenic differentiation among the bacteria. Thus the prokaryotic kingdom consisted of everything that was not eukaryotic (i.e. everything that did not have a nucleus). The proposals in this paper follow the same reasoning pioneered by Woese and Fox. It does not make phylogenetic sense to us to have an archaebacterial urkingdom that contains all bacteria that are not eubacteria (i.e. everything that did not have ester lipids and peptidylglycans). It is obvious that halobacteria, methanogens, and eocytes are extremely different organisms and that the niches they occupy are as different from each other as they are from those occupied by the eubacteria. If the classification 'archaebacteria' is to be useful it should be more than a grab-bag of 'odd creatures'. Clearly our classifications must be based on phylogeny and on major biological properties.

Steven Gould [4] has suggested that one measure of the value of new classification is whether it can redirect our thinking: 'Taxonomy is often regarded as the dullest of subjects, fit only for mindless ordering and sometimes denigrated within science as mere "stamp collecting"... If systems of classification were neutral hat racks for hanging the facts of the world, this disdain might be justified. But classifications both reflect and direct our thinking. The way we order represents the way we think.' In this paper we have tried to develop a classification that can redirect thinking and is faithful to phylogeny. Phylogenetically, it is clear that the halobacteria and the eocytes do not belong in the archaebacteria. If the halobacteria are grouped together with the eubacteria, however, they make good sense cladistically and biologically, for they focus attention on photosynthesis, surely one of the most important biochemical developments in evolution. Likewise the eocytes deserve separate status for it is becoming increasingly clear that the eocytes are one of the deepest and most primitive of the bacterial groups.

In our current view of bacterial evolution (shown very schematically in Fig. 8), each of the three urkingdoms reflect significant biochemical innovation, i.e. photosynthesis (photocytes), methanogenesis (archaebacteria) and sulfur metabolism (eocytes). We believe that a strong case can be made at the highest classification level (i.e. urkingdoms) for using biochemical innovation as an important benchmark. Significantly, these three groups all represent biochemical innovations that are generally thought to have occurred before the last common ancestor of the eukaryotes appeared.

Acknowledgements

We thank J. Washizaki for expert electron microscopy and photography and A. Matheson, B. Pierson, R. Mah, P. Thornber and E. Smith for helpful discussions. We also thank A. Matheson, W. Zillig, G. Fox and B. Pierson for providing cells. This research was supported by research grants from the National Science Foundation (PCM 83-16926 to J.A.L.) and the National Institute of General Science (GM 24034 to J. A. L.). Parts of this paper have been taken rather directly from ref. [8].

References

1. Dickerson, R. E., in *The Evolution of Protein Structure and Function* (ed. Sigman and Brazier). Academic Press, New York (1980).
2. Felsenstein, J., *Syst. Zool.* **27**, 401–410 (1978).
3. Felsenstein, J., *Syst. Zool.* **28**, 49–62 (1979).

4. Gould, S. J., *Hen's Teeth and Horse's Toes*, p. 72. W. W. Norton, New York and London (1984).
5. Henderson, E., Oakes, M., Clark, M. W., Lake, J. A., Matheson, A. T. and Zillig, W., *Science* **225**, 510–512 (1984).
6. Lake, J. A., *Sci. Am.* **245**, August, pp. 84–97 (1981).
7. Lake, J. A., Henderson, E., Oakes, M. and Clark, M. W., *Proc. Natl. Acad. Sci. USA* **81**, 3786–3790 (1984).
8. Lake, J. A., Clark, M. W., Henderson, E., Fay, S., Oakes, M., Scheinman, S., Thornber, J. P. and Mah, R., *Proc. Natl. Acad. Sci. USA* **82**, 3716–3720 (1985).
9. Mayr, E., *Principles of Systematic Zoology*. McGraw-Hill, New York (1969).
10. Seta, P., Bienvenue, E., Moore, A. L., Mathis, P., Bensasson, R. V., Liddell, P., Pessiki, R. J., Joy, A., Moore, T. A. and Gust, D., *Nature (London)* **316**, 653–655 (1985).
11. Tu, J., Prangishvilli, D., Huber, H., Wildgruber, G., Zillig, W. and Stetter, K. O., *J. Mol. Evol.* **18**, 109–114 (1982).
12. Wiley, E. O., *Phylogenetics*. John Wiley, New York (1981).
13. Woese, C. R. and Fox, G., *Proc. Natl. Acad. Sci. USA* **74**, 5088–5090 (1977).
14. Woese, C. R., *Sci. Am.* **244**, 98–122 (1981).
15. Woese, C. R., Debrunner-Vossbrinck, B. A., Oyaizu, H., Stackebrandt, E. and Ludwig, W., *Science* **229**, 762–765 (1985).

Chemica Scripta 1986, **26B**, 127–137

RNA Splicing in Yeast

John N. Abelson, Edward N. Brody*, Soo-Chen Cheng, Michael W. Clark, Phillip R. Green, Gloria Dalbadie-McFarland, Ren-Jang Lin, Andrew J. Newman†, Eric M. Phizicky and Usha Vijayraghavan

Division of Biology, California Institute of Technology, Pasadena, California 91125, USA

Paper presented by John N. Abelson at the Conference on 'Molecular Evolution of Life', Lidingö, Sweden, 8–12 September 1985

Abstract

In eukaryotes, the continuity of information in the gene is frequently interrupted by intervening sequences. This information is removed from an RNA transcript of the gene by a biochemical process called RNA splicing. In this paper, we describe our progress in the study of tRNA and mRNA splicing in yeast.

In tRNA splicing, the splice junctions are brought into close proximity via the secondary and tertiary structure of the mature domain of the precursor. Two enzymes, an endonuclease and a ligase, clip out the intervening sequences and rejoin the ends. Two molecules of ATP are required for each splicing event. The tRNA ligase gene has been cloned. There is a single tRNA ligase gene in yeast and its disruption is lethal to the cell. The ligase protein has a primary localization at or near the nuclear membrane.

In mRNA splicing, the precursor is folded to bring the splice junctions into close proximity by a large complex apparatus called the spliceosome. Included in this complex are proteins and small nuclear RNAs. The effects of mutant precursors are described and a preliminary fractionation scheme is presented which separates components of an extract capable of mRNA splicing *in vitro*.

RNA splicing in yeast

Many eukaryotic genes contain introns, non-coding sequences, which interrupt the continuity of the gene. These sequences are removed from an RNA transcript of the gene by a biochemical process called splicing. In this process, the intervening sequences are clipped out and the ends rejoined to give the mature sequence. All three classes of eukaryotic genes, those coding for tRNA, rRNA, and mRNA, contain introns. There are two general questions that arise concerning the splicing mechanism. First, how are the splice junctions brought into correct alignment in order to insure correct cleavage at the splice junctions? And secondly, what are the biochemical mechanisms of the cleavage and ligation reactions? Research on RNA splicing during the past eight years has revealed that the mechanisms by which the introns are removed from each class of RNA transcript are distinctly different.

Figure 1 symbolically represents each of the three classes of precursor. In the case of mRNA splicing, the problem of correct alignment of the splice junctions is particularly acute. Nuclear genes coding for mRNA frequently contain many introns (50 in the case of the collagen gene) [1] and in some cases they are very large (perhaps as large as 100000 bases in the case of the *Drosophila* homeotic genes) [2, 3]. Most likely, the recognition elements for correct mRNA splicing are found at or near the splice junctions [4]. They are recognized by ribonucleoprotein complexes called SnRNPs [5, 6]. These complexes contain a small nuclear RNA and six or seven proteins [7, 8]. The 5′ splice junction is recognized by the U1 SnRNA [9, 10] and recent evidence suggests that the 3′ splice junction is recognized by the U2 SnRNP [11, 12]. ATP is required for the *in vitro* mRNA splicing reactions, but its role in the reaction is not known.

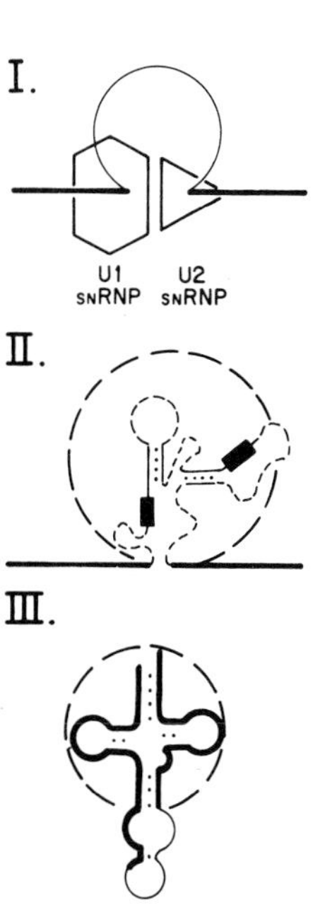

Fig. 1. Recognition of splice junctions in three classes of RNA splicing reactions. (I) Splicing of mRNA precursors probably occurs on a particle called the spliceosome. U1 and U2 snRNP particles are known to be required for splicing. (II) rRNA precursors have extensive secondary structure within their intervening sequences. The resultant folding of the molecule serves to align the splice junctions and to direct the binding of effectors such as GMP which initiate the self-splicing reaction. (III) tRNA precursors have secondary structure within the exon regions which serves to align the splice sites and to direct binding of the tRNA splicing enzymes. In each precursor, exons are indicated by the heavy lines and intervening sequences by light lines. Dashed lines in the IVS of the rRNA precursor indicate large regions of secondary structure which have been omitted for clarity. The large circles indicate the recognition domain. (This figure is taken from Greer and Abelson [1984], ref. 60.)

By contrast, rRNA splicing in *Tetrahymena* as studied by Cech and his colleagues, requires neither ATP nor a protein

Present addresses:
* Institut de Biologie Physico Chemique, 13 rue Pierre et Marie Curie, Paris 75005, France.
† Laboratory of Molecular Biology, Medical Research Council Centre, University Medical School, Hills Road, Cambridge CB2 2QH, England.

catalyst [13]. The splice junctions are brought into conjunction by secondary and tertiary elements of the intron. The splicing reaction is initiated by guanosine and consists of a series of phosphoryl transfer reactions promoted by the RNA itself. Introns with this mode of splicing are found not only in *Tetrahymena* rRNA genes, but in yeast and fungal mitochondrial genes [14] coding for proteins and in the thymidylate synthetase gene of bacteriophage T4 [15].

In tRNA splicing in yeast, the intron is always found one base to the 3′ side of the anticodon [16]. Thus, the splice junctions have a constant position relative to the secondary and tertiary structure of the tRNA domain of the precursor. The splicing reaction is catalyzed by two enzymes and requires two molecules of ATP [17, 18].

We have been studying tRNA and mRNA splicing in yeast. In this article, we summarize our studies on the mechanism of each of these splicing reactions and compare them with the self-splicing reaction.

1. *tRNA splicing in yeast*

(*a*) *Mechanism of the reaction.* There are nine classes of yeast tRNA genes which contain introns [16]. In each gene the introns, which range in size from 13 to 60 base pairs, are found one base to the 3′ side of the anticodon. These genes are transcribed by RNA polymerase III to give a precursor which is extended at both its 5′ and 3′ ends (Fig. 2). These ends are matured by RNase P, by 3′ specific exonucleases and finally by tRNA nucleotidyl transferase which adds the 3′ C, C and A residues. The resulting end-matured precursor is the substrate for the tRNA splicing enzymes. It is these precursors which accumulate at the non-permissive temperature in rna 1 strains of yeast [19]. This made it possible to isolate precursors and to detect the tRNA splicing reaction *in vitro* [20].

An endonuclease cleaves the precursor at both splice junctions. The products of this cleavage have 2′-3′ cyclic phosphates and 5′ hydroxyl termini (Fig. 3). The enzyme which catalyzes this reaction has been purified more than 10000-fold, but it is not yet pure. Its most interesting feature is that it behaves as an integral membrane protein [21]. Since RNA splicing very likely takes place in the nucleus, it is possible that this enzyme is specifically localized in the nuclear membrane. The product of this reaction is a tRNA molecule with a gap in its anticodon loop; this is the substrate for the ensuing ligation reaction. This reaction is catalyzed by a single 90000 Da protein which has been purified to homogeneity [18, 22]. Three distinct biochemical reactions are required of the tRNA ligase and these have all been demonstrated using the pure enzyme. First the 2′-3′ cyclic phosphate is opened by a phosphodiesterase yielding a 3′-OH and a 2′-phosphate. Secondly, a polynucleotide kinase activity transfers the γ-PO_4 of ATP to the 5′-OH at the 3′ splice site. This phosphate is then activated by transfer of AMP from an adenylylated ligase intermediate to form an activated RNA with a 5′-5′ phosphoanhydride bond. The 3′-OH then attacks, AMP leaves and the ligated tRNA is formed. The initial product, however, retains the 2′-phosphate at the spliced junction and this must be removed by an (as yet uncharacterized) phosphatase. Interestingly, the tRNA splicing mechanism is not universal. RNA ligases similar to the yeast tRNA ligase have been found in wheat and in chlamydomonas [23–26], but in mammalian cells a different mechanism seems to pertain. Here the 3′ phosphate is retained in the spliced product [27] (note that it is lost in the yeast mechanism). An enzyme requiring ATP has been found which forms cyclic phosphates at 3′ termini. This enzyme could be involved in tRNA splicing in mammalian cells [28].

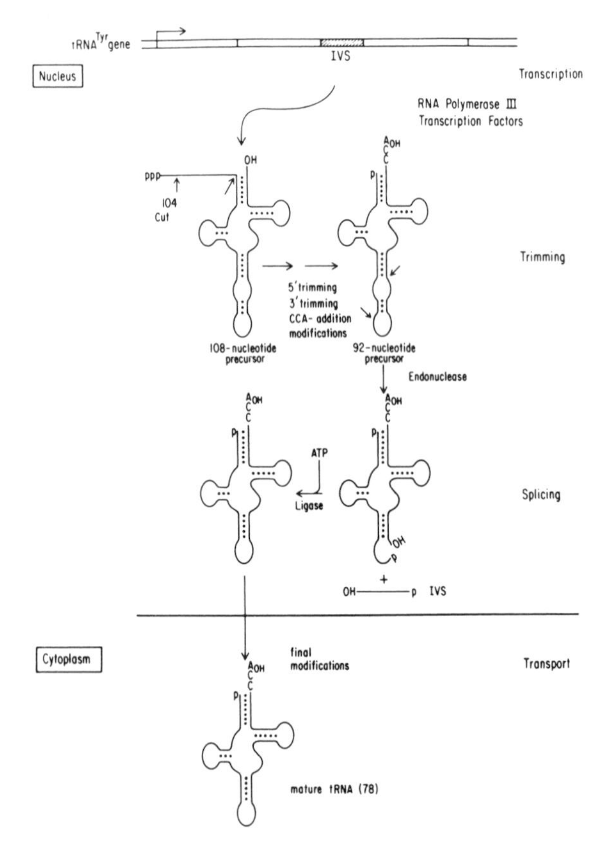

Fig. 2. Major trimming steps in the processing of a tRNA molecule. DNA encoding a tRNAtyr gene from yeast was injected into *X. laevis* oocytes along with α-^{32}P-labeled NTPs, and transcription and processing were followed with time. The initial precursor contains a 5′ leader and a 3′ trailer as well as the intervening sequence. Removal of the 5′ leader and the 3′ trailer sequences occurs first, followed by addition of CCA to the 3′ end and finally the intervening sequence is removed by splicing. All these steps occur in the nucleus. The figure is an adaptation of the data of Melton *et al.* (1980), as described by Guthrie and Abelson (1982).

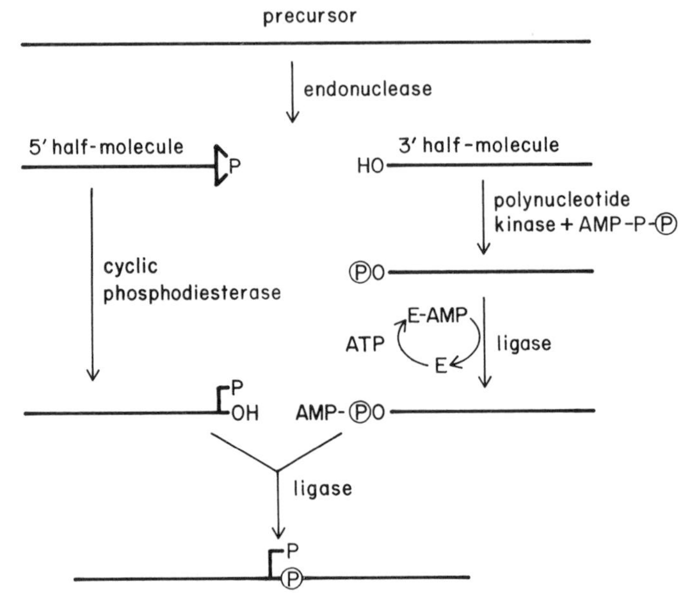

Fig. 3. Mechanism of splicing of tRNA precursors in yeast. A schematic of the steps involved in tRNA splicing in yeast is presented, as first described in Greer *et al.* (1983).

(*b*) *Genetic studies of the tRNA ligase gene.* Using purified tRNA ligase protein, the amino acid sequence of the *N*-terminal 15 residues was determined in the Caltech Microchemical Facility. Knowledge of this sequence allowed us to design a degenerate set of oligonucleotide probes specific for this sequence. Using these probes, we cloned the tRNA ligase gene entirely contained within a 4 kb *Eco* RI fragment [22].

The 4 kb *Eco* RI fragment containing the ligase gene is unique as judged by Southern hybridization of labeled 4 kb fragments to various restriction endonuclease digests of yeast DNA. We therefore conclude that there is a single ligase gene. To provide that this is an essential gene, we inactivated the gene by insertion. This experiment (diagrammed in Fig. 4) demonstrates the power of modern yeast genetics.

A *Bam* HI fragment containing the entire *His 3* gene is cloned into the unique *Bgl* II site in the ligase gene. This altered 4 kb *Eco* RI fragment is transformed into a HIS 3^-/HIS 3^- diploid with selection for growth in the absence of histidine. Rothstein has shown [29] that the ends of linear DNA fragments are recombinogenic in yeast, inserting into the genomes to replace the homologous resident sequence. The HIS^+ diploids are sporulated and tetrads are dissected. If disruption of the ligase gene is lethal to a haploid cell, the segregation will yield two live spores and two dead spores, with the live spores being HIS^-. As can be seen in Fig. 4, the results were as expected. The segregation is 2:2 and the live haploid segregants are HIS^-. Since disruption of the tRNA ligase gene is lethal in yeast, it is likely that the protein is essential. It is now possible to isolate temperature-sensitive mutants in the gene allowing us to assess the complete role of tRNA ligase in the cell.

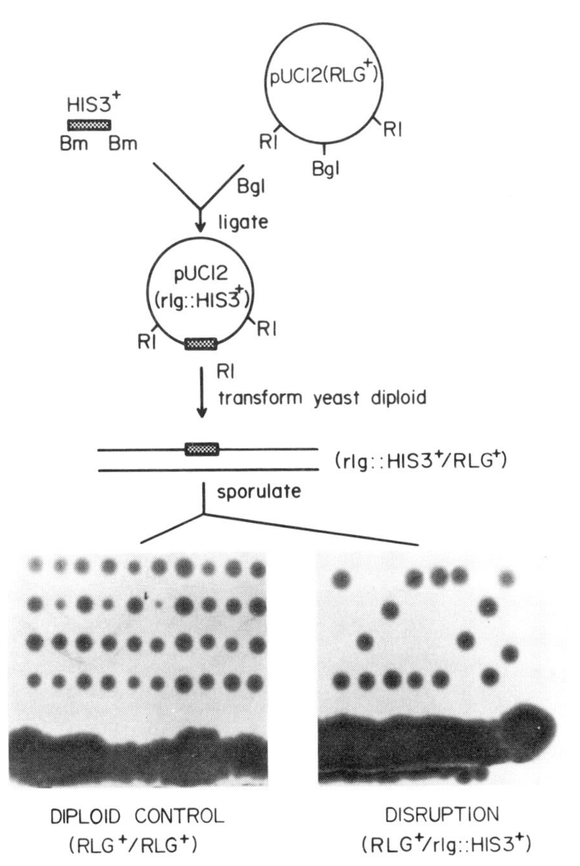

Fig. 4. A diploid carrying one wildtype copy of the ligase gene and one disrupted copy of the ligase gene (*RLG*$^+$/*rlg*::*HIS*3^+) was constructed as shown. This strain and its parent were sporulated and tetrads were dissected.

(*c*) *The cellular location of tRNA ligase.* Since tRNA precursors are synthesized in the nucleus, it is reasonable to assume that the tRNA splicing enzymes are there as well. This has been shown to be the case in *Xenopus* oocytes [30]. The fact that the endonuclease is an integral membrane protein [21], however, suggests that the location of the endonuclease could be at the nuclear membrane. Consequently, we were anxious to determine the location of the tRNA ligase. Using the rabbit antibody to tRNA ligase as probe, indirect immunofluorescence staining of ligase was carried out on permeabilized yeast spheroplasts. Figure 5 presents photographs of selected fields of stained cells. In panel A, the cells have been stained with fluoresceine isothiocyanate conjugated goat anti-rat IgG antibody coupled to a rat monoclonal to yeast tubulin. This is a control which repeats the work of Kilmartin [31] and of Pringle [32] who worked out the indirect immunofluorescence techniques for yeast. Clearly seen is the elongated spindle apparatus in the dividing yeast cells and the characteristic intense dot of tubulin in non-dividing cells. The rest of the panels present results in which rabbit anti-ligase IgG stained cells are visualized by FITC conjugated goat anti-rabbit IgG. By comparing the differential interference contrast micrographs with the fluorescence micrographs on the right, it can be seen that the location is nuclear. Our impression is of patchy staining of the nucleus. This is seen, e.g. in a micrograph of an isolated nucleus (panel G) in which the stain appears to be localized on the periphery of the nucleus.

Immunofluorescence confirms the expected nuclear location of tRNA ligase but it does not have sufficient resolution to answer the question of whether ligase has a nuclear membrane location. Consequently, we have turned to immunoelectron microscopy [33] to obtain higher resolution localizations. In this technique, yeast cells are fixed in glutaraldehyde and the cell wall is removed to form spheroplasts. The spheroplasts are embedded in epon and thin sections are prepared. The thin sections on electron microscope grids are treated first with rabbit anti-ligase and then with goat anti-rabbit conjugated with 150–200 Å colloidal gold spheres. Finally, the preparation is treated with heavy-metal stains and examined in the electron microscope. In this experiment (Fig. 6), 330 cells were examined. Figure 6 shows one such cell. The nuclear membrane is clearly visible, as are other features, notably the vacuoles. Colloidal gold spheres are seen at and near the nuclear membrane as well as at other locations in the nucleus and the cytoplasm. The results for all of the cells (435 gold particles in 89 cells) is plotted below.

In this plot, the closest distance to the nuclear membrane is given for each particle. If a particle is in the nucleus, the distance is plotted to the left of the origin within the cytoplasm to the right. In this experiment, notable clustering is seen at the nuclear membrane and there is also some clustering within the nucleus, close to the nuclear membrane (50–100 nm). This experiment is fraught with experimental difficulties, and it poses a great challenge to achieve objectivity. We have repeated this experiment several times, with variations. In one variation, the anti-ligase is pre-treated with ligase before staining. This should block binding to ligase-specific sites in the sections. This treatment blocks significant staining of the nuclear membrane and of the nuclear region just inside the membrane. We conclude that there is a preferential location for ligase on the nuclear membrane.

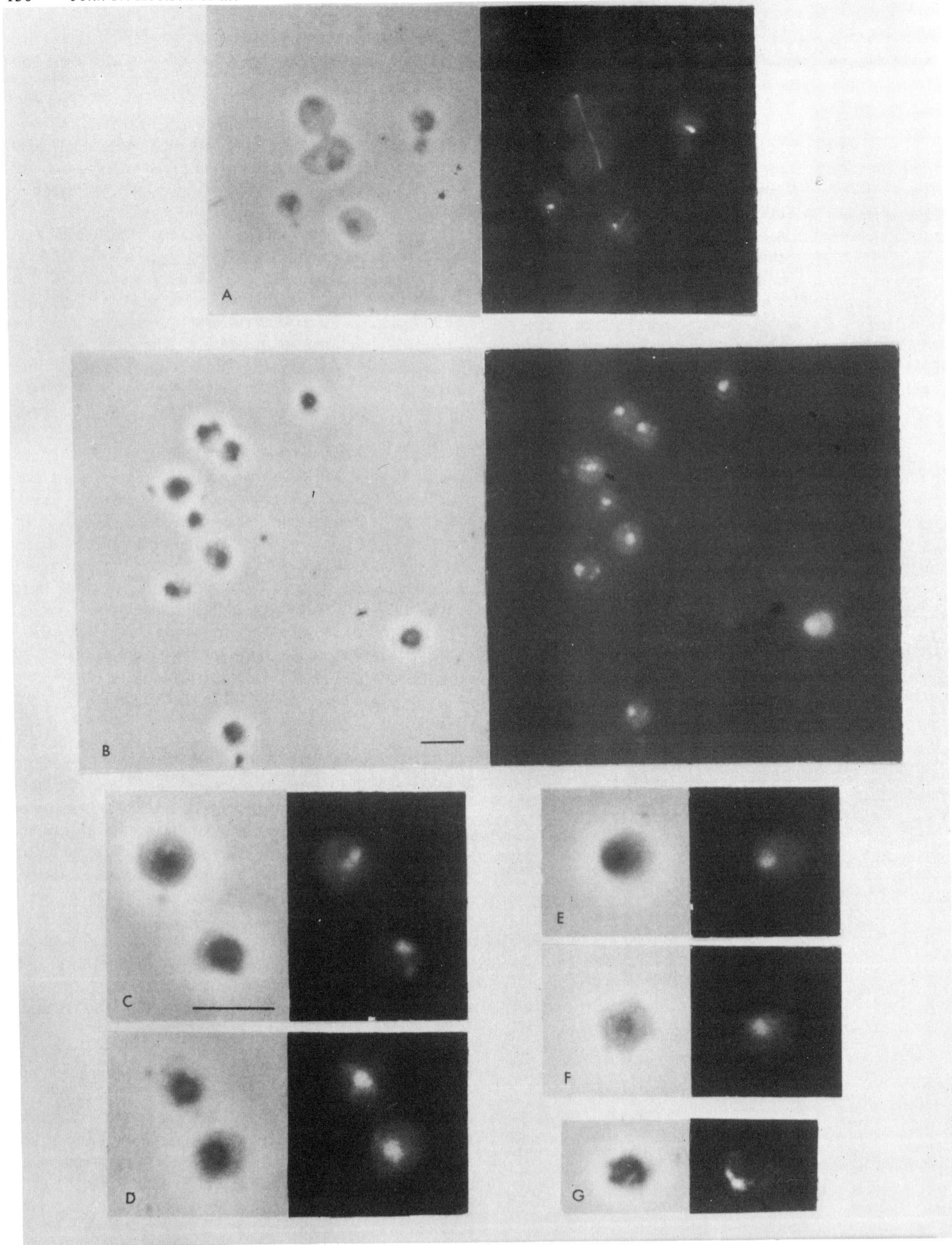

Fig. 5. Indirect immune fluorescence of yeast EJ101. (A) Left, DIC micrograph of yeast cell, first fixed with 3.7% formaldehyde, then cell walls digested with β-glucoronidase, mounted on glass slides, and incubated with antitubulin monoclonal antibodies as the primary antibody, followed by a second antibody against rat IgG that was labeled with FITC. The darkened spot in the cells is the nucleus. Right, the same cells viewed under FITC excitation wavelengths. (B) Left, field of cells under DIC microscopy that first had the cell walls removed, mounted on glass slides and fixed with 5% formaldehyde, followed by incubation with a primary antibody against yeast tRNA ligase (raised in rabbits) and goat anti-rabbit IgG antibody labeled with FITC. Right, the same cells viewed under FITC excitation wavelengths. (C–F) enlargements of individual cells from the fields under DIC (left) and FITC (right) microscopy. (G) Left, DIC micrograph of isolated yeast nucleus that was mounted on a glass slide and fixed with 5% formaldehyde, incubated with anti-tRNA ligase antibody and anti-rabbit IgG antibody labeled with FITC. Right, the same nucleus under FITC excitation wavelength. Bar is 5 μm.

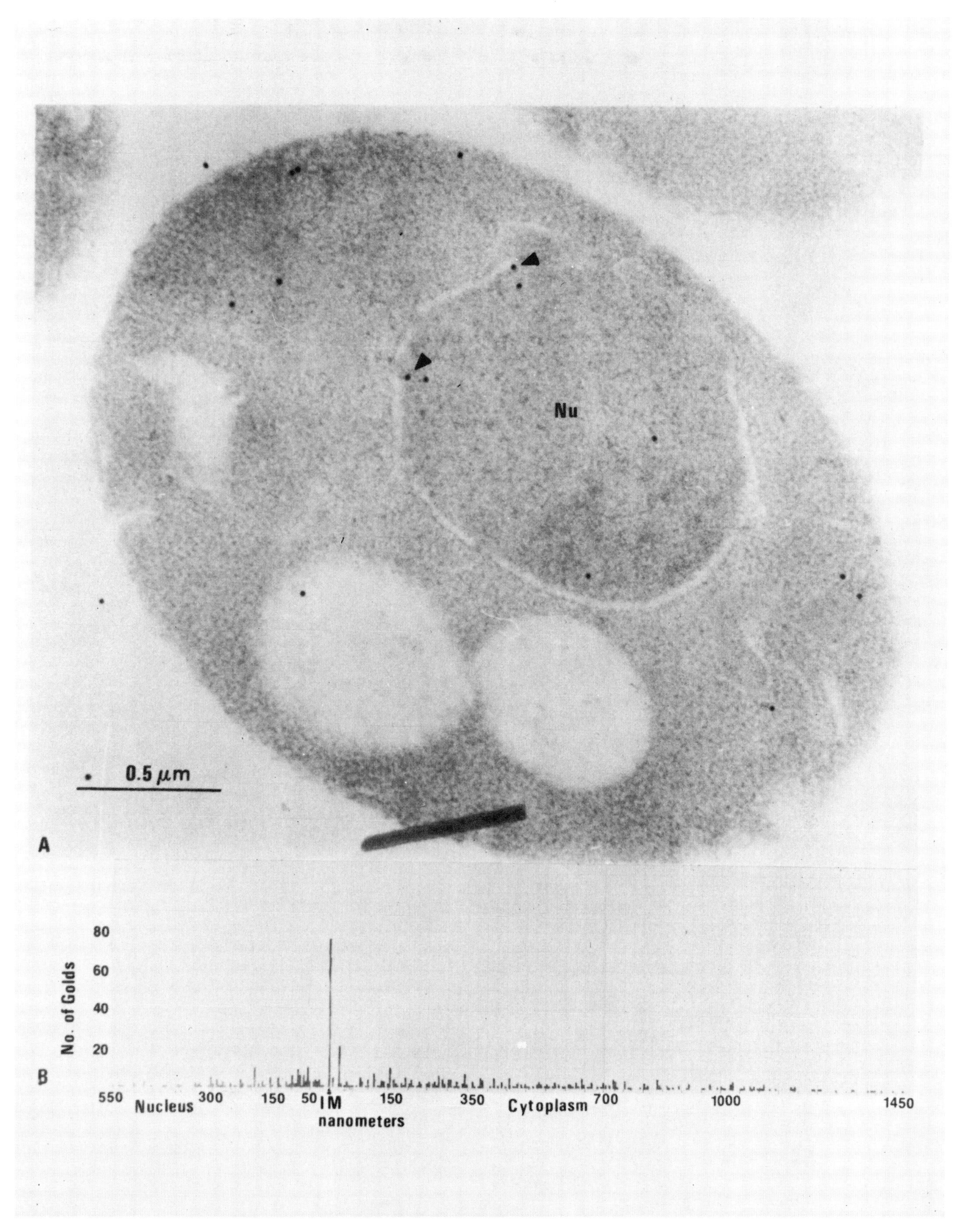

Fig. 6. Indirect immune electron microscopy on EJ101 yeast cell. (A) Section of EJ101 cell that was first fixed with 1% glutaraldehyde, dehydrated through an ethanol series, embedded in Epon 812 resin. The thin sections have been blocked with 8% BSA in Tris–sodium chloride buffer, then incubated overnight at 6 °C in the presence of affinity purified anti-tRNA ligase IgG. The sections were rinsed and further incubated at room temperature in the presence of anti-rabbit IgG antibody conjugated with colloidal gold particles of 15–20 nm in diameter; the sections were rinsed again, fixed with 1% glutaraldehyde, stained with heavy metals and viewed in a Philips 201B microscope. Arrows show two gold particles on nuclear envelope. Nu, nucleus. (B) Histogram of shortest distance from inner membrane to gold particle. IM, inner membrane.

INTRON CONSENSUS SEQUENCES

(MOUNT, 1982; KELLER AND NOON, 1984)

EXON INTRON EXON

$^{C}_{A}$AGGT$^{A}_{G}$AGT.... CT$^{A}_{G}$A$^{C}_{T}$....($^{C}_{T}$)$_{11}$ N$^{C}_{T}$AGG

↑

YEAST INTRON SEQUENCES

(TEEM ET AL., 1984)

GTA$^{T}_{C}$GT.......TACTAAC.......$^{C}_{T}$AG

↑

Fig. 7. Consensus sequences for the metazoan and yeast intron splicing signals. The arrows indicate the branch site nucleotide.

SP64-cyh Eco RI

pre-mRNA — Exon 1 79 nt; IVS 509 nt; Exon 2 301 nt — 889 nt

mRNA — 79 nt 301 nt — 380 nt

SP6-actin Eco RI

pre-mRNA — Exon 1 88 nt; IVS 309 nt; Exon 169 nt — 566 nt

mRNA — 88 nt 169 nt — 257 nt

Fig. 8. Structure of the SP6 polymerase transcripts used as substrates for the *in vitro* splicing reaction.

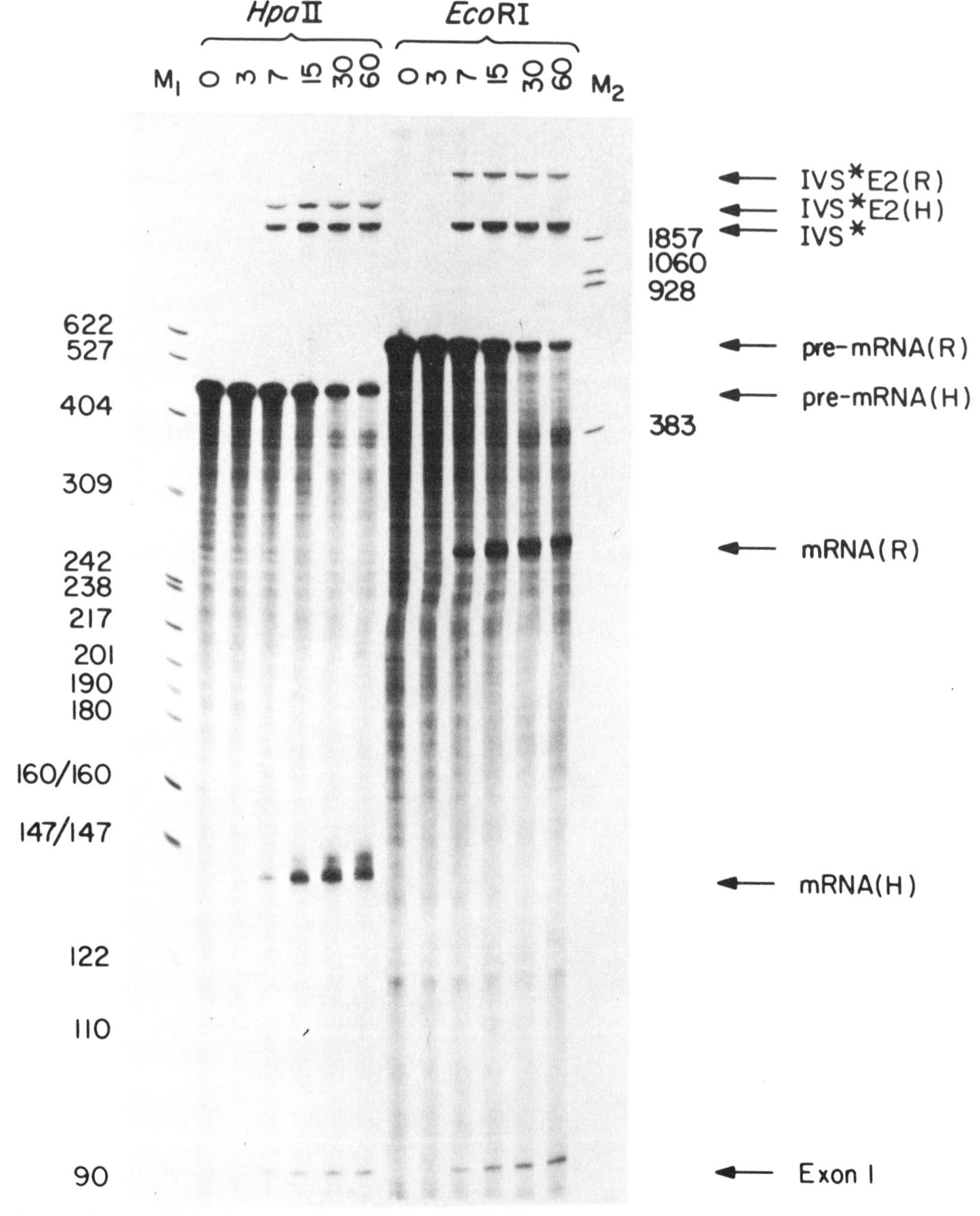

Fig. 9. Splicing of synthetic actin pre-mRNA *in vitro*. Radioactive actin pre-mRNAs synthesized as run-off transcripts from templates linearized with either *Hpa II* or *Eco RI* (see Fig. 4) were incubated in the yeast whole cell extract (see below) under appropriate conditions. Samples were withdrawn at intervals (time of incubation is numbered in minutes), deproteinized, and analyzed by electrophoresis on an 8% polyacrylamide 8 M-urea sequencing gel, followed by autoradiography. End-labeled DNA fragments (pBR322 cut with *Hpa II* (M1) or *Eco RI* (M2) were run alongside as size markers. E1 and E2 refer to the exons, and lariat forms are denoted by an *. R and H refer to RNA molecules produced from *Eco RI* and *Hpa II* transcripts, respectively. The yeast whole cell extract was obtained by lysing yeast spheroplasts in a hypotonic buffer using a Dounce homogenizer. Nuclei were extracted with 0.2 M-KCl and cell debris was removed by centrifugation. The supernatant was centrifuged at 37000 rpm in a 60 Ti rotor, dialyzed against an appropriate buffer and stored in aliquots at −70 °C.

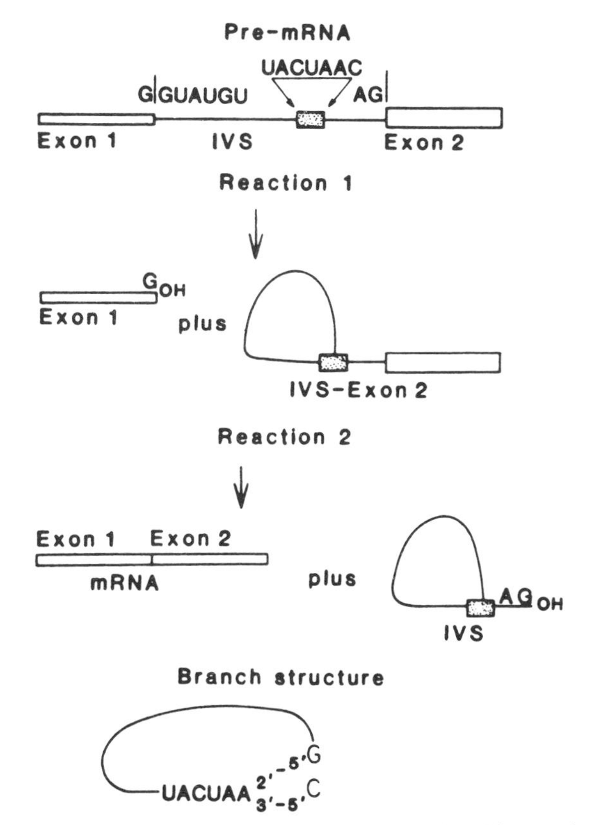

Fig. 10. The splicing pathway of pre-mRNA splicing. The reactions are described in the text. The intron exon junctions are represented by vertical bars. The yeast branch structure is shown.

In the future, we hope to purify the endonuclease to homogeneity, allowing us to prepare antibodies and to determine its location as well. Ultimately, we would hope to be able to test the possibility that these enzymes are connected with the transport of mature tRNA into the cytoplasm.

2. *mRNA splicing in yeast*

(*a*) *Some yeast nuclear genes contain introns*. They are found less frequently than in the genes of higher organisms. We [34], as well as Gallwitz [35], found the first one in the yeast actin gene and we now know that most ribosomal protein genes in yeast also contain an intron [36]. The sequences of these introns revealed that the consensus rules for splice junction sequences in introns are obeyed in yeast. They are, in fact, much more highly specified (Fig. 7). The 5′ splice junction is always GTAPyGT and the 3′ splice junction is pyAG. In addition, a third conserved region, the sequence TACTAAC, is found approximately 30 bp from the 3′ splice junction and is essential for intron function *in vivo*.

(*b*) *mRNA splicing* in vitro. Substrates for mRNA splicing *in vitro* are prepared by transcribing with SP6 polymerase [37], portions of either the yeast actin gene, or the *CYH-2* gene, the gene for ribosomal protein L-29 [38]. The transcripts contain a portion of exon 1, the entire intron and a portion of exon 2 (Fig. 8). We have prepared capped and polyadenylated transcripts, but as it turns out neither of those features are important in the *in vitro* reaction. Originally, splicing *in vitro* was detected by a sensitive S1 nuclease assay [39]. Once the conditions for the reaction had been optimized, however, the products and intermediates of splicing could be separated and detected directly by acrylamide gel electrophoresis (Fig. 9). In this experiment, two actin precursors, differing in the size of exon 2 were used. The *Hpa* II precursor is smaller than the *Eco* RI precursor. Each precursor is spliced to give two products, the mRNAs and the intron. The electrophoretic mobility of the intron is less than that of the precursor because it is topologically circular (see below). Two intermediates of splicing are seen: free exon 1 which is the same size in both cases and a topologically circular intron-exon 2 intermediate which differs in size due to the difference in size of exon 2 in the two precursors. The products and intermediates have been thoroughly characterized. The mRNA is faithfully spliced. The intron has a 3′ hydroxyl end and is circular because the 5′ G is covalently linked to the last A in the TACTAAC sequence via a 2′-5′ linkage.

These results imply a mechanism of yeast mRNA splicing similar to that which has been proposed for splicing in the HeLa cell system [40–44] (Fig. 10). In this two-stage mechanism, the 5′ splice junction is cleaved to give exon 1 terminated with a 3′-hydroxyl end. Coupled to this cleavage, the first G in the intron is covalently linked to an A near the 3′ splice junction in a 2′-5′ linkage [45]. This forms the so-called lariat intermediate. In the second stage, the 3′ splice junction is cleaved and coordinately exon 1 is joined to exon 2 to give the mRNA product and the intron lariat. We have verified that the phosphate in the 2′-5′ linkage is derived from the 5′ splice site and the phosphate in the mRNA spliced junction from the 3′ splice site.

Earlier, we had detected the *in vivo* intron product of actin mRNA splicing [46] and had shown that the lariat is formed by covalent linkage of the 5′ G to the last A in the TACTAAC sequence. Thus, the products and intermediates seen *in vitro* are equivalent to what is detected *in vivo*.

(*c*) *Substrate requirements for splicing*. Using the *in vitro* system, we have been able to examine the consequences of alternatives of the intron consensus sequences [47].

Two mutations were constructed in the *CYH-2* gene. The G at the 5′ splice site was altered by oligonucleotide directed mutagenesis to an A. A second mutation was created in the target for branch formation, an A to C change in the last A of the TACTA<u>A</u>C sequence.

SP6 transcripts of the wild-type gene and the two mutants were prepared and used as substrates for *in vitro* mRNA splicing (Fig. 11). The TACTA<u>C</u>C mutant transcript is not a substrate for splicing. It is remarkable that the 5′ splice site is not cleaved in this mutant despite the fact that it is approximately 500 bases from the site of the mutation. The results are different for the 5′ splice site mutant, <u>A</u>TATGT. Here, the 5′ splice site is cleaved and both intermediates are formed. Both were characterized. Exon 1 is indistinguishable from the wild-type intermediate. In the intron-exon 2 intermediate, the branch is a 2′5′ A-A linkage between the A at the 5′ splice site and the correct A in the TACTAAC sequence. Despite the fact that both Exon 1 and the 3′ splite site are not altered in this mutant, stage 2 of the splicing reaction does not proceed and neither mRNA nor intron products are formed.

Both of these mutant genes were introduced into yeast by transformation and the splicing reaction examined [47]. The results were the same as those seen *in vitro*. Thus, both consensus sequences are absolutely required for splicing. Cryptic 5′ splice sites are not activated as in the case of the equivalent mutation in the human β-globin gene [48].

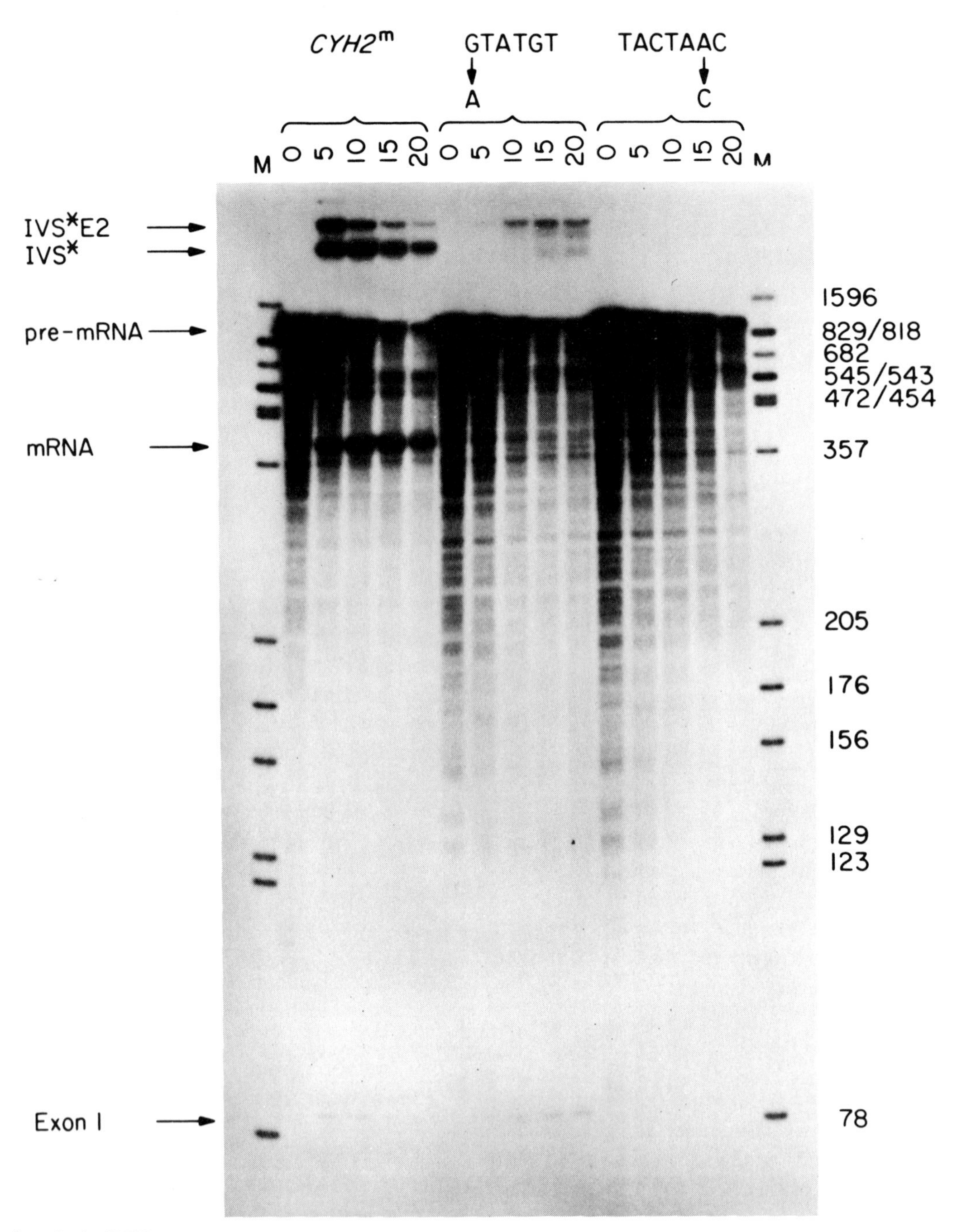

Fig. 11. Splicing of synthetic CYH2^m pre-mRNAs *in vitro*. Pre-mRNAs labeled with ^{32}P containing wild-type intron (CHY2^m) or introns with single-base substitutions at the 5′ splice site (ATATGT) or branch point sequence (TACTACC) were incubated in the yeast whole cell extract under standard conditions (Newman *et al.*, 1985). Samples were withdrawn at intervals, deproteinized, and analyzed by electrophoresis on a 6% polyacrylamide-8 M-urea gel, followed by autoradiography. End-labeled DNA fragments (M13mp8 cut with *Hpa II*) were run alongside as markers).

The results imply that the precursor must be folded to bring the 5′,3′ and TACTAAC consensus sequences into close conjunction in order for splicing to proceed. When an incorrect branch sequence is formed, the second stage of the reaction cannot take place.

(*d*) *Fractionation of the active splicing extract.* In order to understand the mechanism of mRNA splicing, it will be necessary to fractionate the system and characterize the components. We have made preliminary efforts in this direction. Figure 12 diagrams the fractionation scheme.

A 70% $(NH_4)_2SO_4$ (A.S.) precipitate of this extract contains all of the components required for splicing. This precipitate is back-washed with 40% A.S. to give a pellet (40P1) and a supernatant fraction. 40P1 is active in splicing but accumulates the stage 1 intermediates. The supernatant fraction (40 Sup) is inactive. Continued washing of 40P1 with 40% A.S. yields 40P3 (component 1) which is inactive in splicing. 40P3, however, can be complemented to give complete splicing by addition of 40 Sup. The 40 Sup fraction is further fractionated by Heparin agarose to give a flow-through fraction (component II) and a salt-eluted fraction (component III). Component I and component II can carry out the first stage of the reaction. Addition of component III is required to obtain complete splicing.

In the HeLa system, it is known that at least two small nuclear RNAs, U1 [9, 10] and U2 [11, 12] are components of the mRNA splicing system. One way to assess the involvement of RNA in a reaction is to test the sensitivity of the various fractions to micrococcal nuclease. In such an experiment,

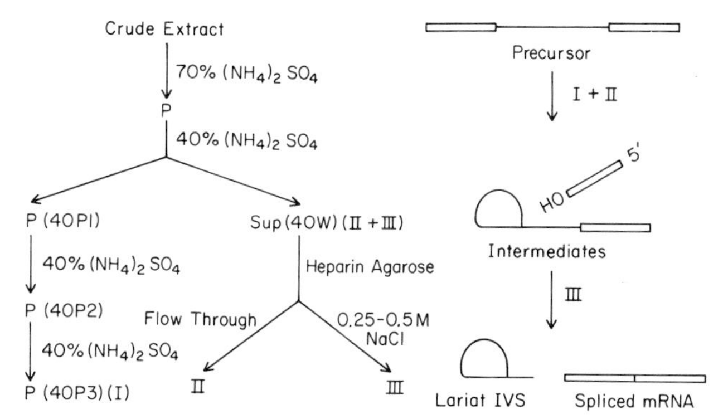

Fig. 12. Ammonium sulfate fractionation of yeast whole cell extract defines a minimum of three components required for splicing. Each component is inactive by itself, components I+II are capable of producing reaction intermediates, but no products and components I+II+III combined reconstitute the system, producing excised intron and mature mRNA. Component I is inactivated by digestion with micrococcal nuclease (Cheng, unpublished). To obtain these components whole cell extract is precipitated with 70% ammonium sulfate. The supernatant is discarded. The pellet is washed with 40% ammonium sulfate. The wash supernatant contains components II and III. The pellet, washed once with 40% ammonium sulfate (PI), contains components I+II; after two more washes, only component I activity remains in the pellet (P3). Components II and III can be separated from each other by heparin agarose chromatography.

components I and II are sensitive to micrococcal nuclease. Component III is not.

Yeast is known to contain small nuclear RNAs [49], but they are not sufficiently homologous to their mammalian counterparts to allow an easy assignment of the U1 and U2 analogues. Nonetheless, it seems likely that the RNA components in the yeast system have an equivalent function and it remains for us to purify and characterize them.

(*e*) *The spliceosome*. The mammalian small nuclear RNAs are found in ribonucleoprotein complexes called SnRNPs. The U1 SnRNP, for example, contains six proteins and one molecule of U1 RNA, a total molecular weight of approximately 250000 Da. If the SnRNPs are components of the splicing machinery, then it is possible that mRNA splicing takes place on a large particle. To test that possibility, we prepared reaction mixtures containing radioactive precursors and following preincubation the reaction components were separated by glycerol gradient sedimentation [50]. Figure 13 shows the sedimentation profile for wild-type actin precursor and the 5′ splice site mutant (G1→C) precursor. A part of the wild-type precursor sediments in a 40S peak that is not seen when mutant precursor is used.

RNA was isolated from each of the gradient fractions and characterized by acrylamide gel electrophoresis (Fig. 14). It can be seen that the precursor, the reaction intermediates, and part of the intron lariat product sediment in the 40S peak. The spliced mRNA is spread through the gradient and must be released from the 40S complex.

We have called this complex the spliceosome. Formation of the spliceosome requires a precursor and conditions which allow the first stage of the splicing reaction to take place. Thus, the CYH-2 G1→A mutant precursor forms the complex, and the TACTA<u>C</u>C mutant precursor does not. Components I and II are required for the stage I reaction and for spliceosome formation.

A 40S complex is large. Since we do not know the partial specific volume of the complex, we cannot determine its

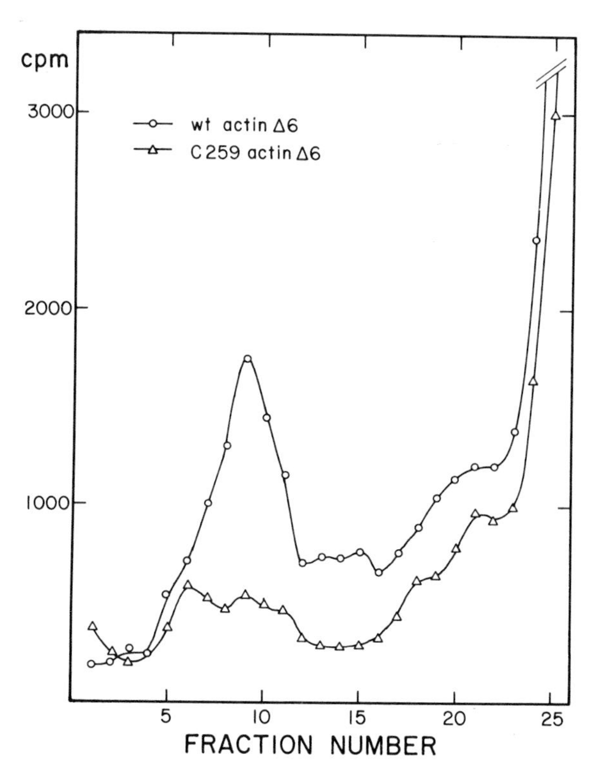

Fig. 13. Glycerol gradient analysis of *in vitro* splicing reactions. Radiolabeled actin Δ6 mRNA precursors having either a wild-type (○—○) or mutant (A259→C; △—△). TACTAAC box were incubated in standard splicing conditions (Newman *et al.*, 1985) and analyzed by centrifugation in glycerol gradients as described (Brody and Abelson, 1985). The sedimentation of 40S markers (not shown) coincides with that of the peak of wild-type precursor (fraction 10).

molecular weight, but the 40S yeast ribosome contains approximately 30 proteins and an RNA molecule of 1650 nucleotides. It has a molecular weight of approximately 1.5×10^6 Da.

Similar complexes have been detected in HeLa mRNA splicing extracts [51, 52] although they appear to be somewhat larger, 50–60S. Clearly, the mRNA splicing machinery is complex.

Summary

The mRNA splicing and tRNA splicing reactions are very different. If there is a common evolutionary origin for these reactions, it is now obscure and it seems more likely that they arose independently. There are similarities, however, between the rRNA self-splicing reaction and mRNA splicing, despite the fact that the former requires no protein and the latter takes place in a ribosome-size particle.

Figure 15 summarizes and compares the two mechanisms. The self-splicing reaction is initiated by the 3′-OH of guanosine. The phosphodiester bond at the 5′ splice junction is transferred to the 3′ hydroxyl of G. In the second step, the 3′-OH of exon I attacks the 3′ splice junction. rRNA is formed and the intron is released as a linear molecule with a G at its 5′ terminus. The linear intron is capable of a number of further reactions including a cyclization reaction in which the 3′-OH terminus of the intron attacks the phosphodiester bond at residue 15 of the intron yielding a circle and an oligomer [13].

A similar scheme can be proposed for mRNA splicing. Here, the 2′-OH of the A in the TACTAAC sequence initiates the reaction by attack at the 5′ splice junction. The phos-

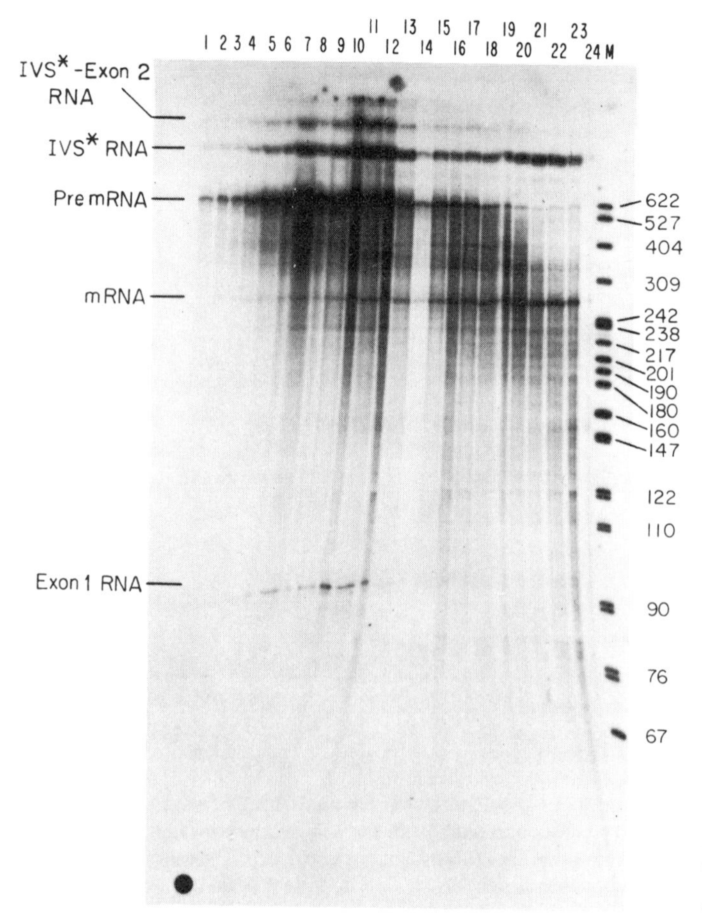

Fig. 14. Polyacrylamide gel analysis of actin Δ6 pre-mRNA splicing products. Fractions from the glycerol gradient shown in Fig. 13 of actin Δ6 pre-mRNA with a wild-type TACTAAC box were deproteinized and subjected to electrophoresis on 8% polyacrylamide, 8 M-urea gels, followed by autoradiography end-labeled DNA fragments (pBR322 cut with *Hpa II* (M) were run alongside as size markers. Lariat forms are denoted by an *).

phodiester bond at the 5′ splice junction is transferred to the 2′-OH of A forming the lariat. The second step is exactly equivalent to the self-splicing reaction. The 3′ hydroxyl of exon I attacks the 3′ splice junction, mRNA is formed and the intron is released as a lariat.

Both schemes accomplish splicing via two concerted phosphoryl transfer reactions. The number of phosphodiester bands in these products is the same as in the precursor so there is not an energetic requirement for ATP. In the self-splicing reaction the intron folds to bring the splice sites into alignment. The 'ribozyme' contains a binding site for G and the activation energy must be obtained by distortion of the splice junction geometry.

The intron cannot perform these functions in mRNA splicing. The components of the spliceosome are required to fold the precursor and catalyze the reactions. It may be that the RNA components of the spliceosome supply in *trans* the functions which the self-splicing intron supplies in *cis*.

mRNA splicing which requires ATP and a complex set of reaction components is clearly not self-splicing, but there is now evidence that it has evolved from a self-splicing reaction. Yeast mitochondrial genomes contain two types of intron [53]. Type I introns are related to the *Tetrahymena* rRNA intron by sequence homology and splicing mechanism. Type II introns, as has recently been discovered, can also undergo a self-splicing reaction, but neither guanosine nor ATP is required and the intron product is excised as a lariat [54, 55]. This exciting result indicates that mRNA splicing and type II intron splicing are derived from a common mechanism. Why then did such a complicated mechanism evolve? It may be that the splicing mechanism allowed shuffling of exons allowing the evolution of even more complex protein domain assemblies [56]. The present splicing mechanism also allows differential use of splice junctions. This mechanism is used in various systems to create multiple products from a single gene,

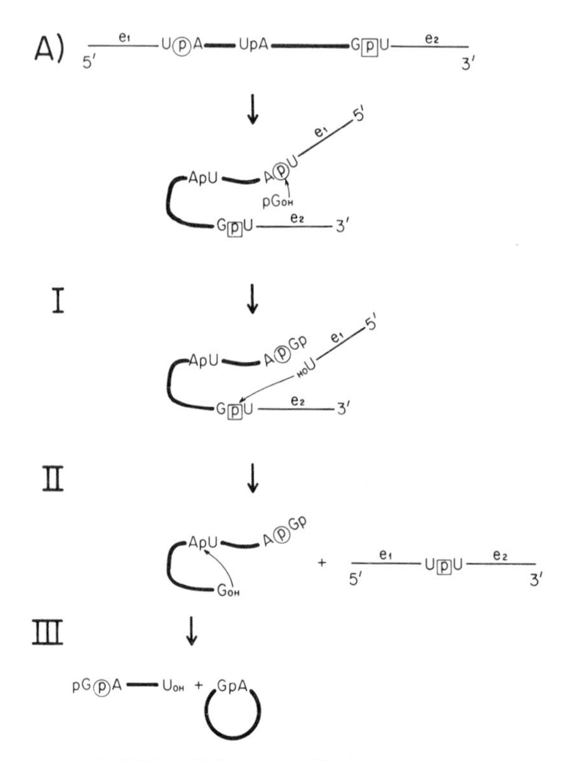

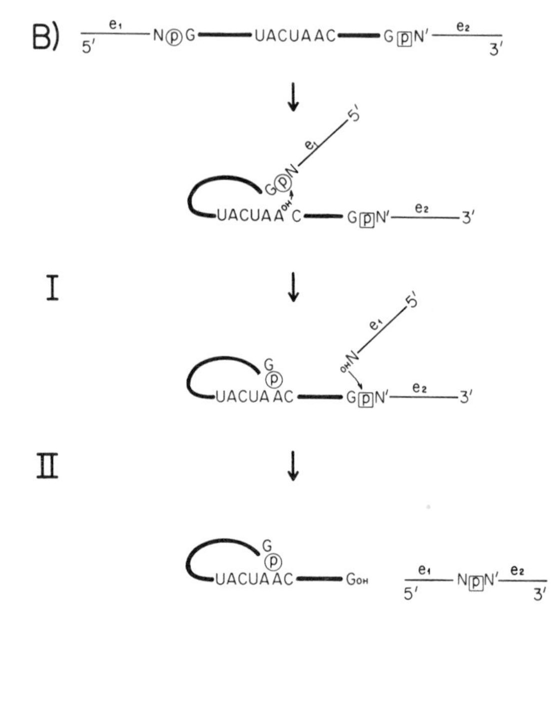

Fig. 15. Comparison of mRNA splicing and rRNA splicing reactions.

for example in the heavy chain immunoglobulins [57], in troponin, and in calcitonin [59].

With regard to the mechanisms of mRNA splicing, we still do not know the role of ATP and most of the components have not been identified. As the black box opens, we will surely be in for more surprises.

References

1. Yamada, Y., Avvedimento, V., Mudryi, M., Ohkubo, Vogeli, G., Meher, I., Pastan I. and de Crombrugghe, *Cell* **22**, 887 (1980).
2. Scott, M. P., Weiner, A. J., Hazelrigg, T. I., Polisky, B. A., Pirrotta, V., Scalenghe, F. and Kaufman, T. C., *Cell* **35**, 763 (1983).
3. Garber, R. L., Kuroiwa, A. and Gehring, W. J., *EMBO J.* **2**, 2027 (1983).
4. Mount, S. M., *Nucl. Acids Res.* **10**, 459 (1982).
5. Lerner, M. R., Boyle, J. A., Mount, S. M., Wolin, S. L. and Steitz, J. A., *Nature* **283**, 220 (1980).
6. Rogers, J. and Wall, R., *Proc. Nat. Acad. Sci. USA* **77**, 1877 (1980).
7. Hinterberger, M., Pettersson, I. and Steitz, J. A., *J. Biol. Chem.* **258**, 2504 (1983).
8. Kinlaw, C. S., Robberson, B. L. and Berget, S. M., *J. Biol. Chem.* **258**, 7181 (1983).
9. Mount, S. M., Pettersson, I., Hinterberger, M., Karmas, A. and Steitz, J. A., *Cell* **33**, 509 (1983).
10. Kraemer, A., Keller, W., Appel, B. and Luhrmann, R., *Cell* **38**, 299 (1984).
11. Black, D. L., Chabot, B. and Steitz, J. A., *Cell* **42**, 737 (1985).
12. Krainer, A. R. and Maniatis, T., *Cell* **42**, 725 (1985).
13. Cech, T. R., *Cell* **34**, 713 (1983).
14. Garriga, G. and Lambowitz, A. M., *Cell* **39**, 631 (1984).
15. Belfort, M., Pedersen-Lane, J., West, D., Ehrenman, K., Malez, G., Chu, F. and Maley, F., *Cell* **41**, 375 (1985).
16. Ogden, R. C., Lee, M. C. and Knapp, G., *Nucl. Acids Res.* **12**, 9367 (1984).
17. Peebles, C. L., Ogden, R. C., Knapp, G. and Abelson, J., *Cell* **18**, 27 (1979).
18. Greer, C. L., Peebles, C. L., Gegenheimer, P. and Abelson, J., *Cell* **32**, 537 (1983).
19. Hopper, A. K., Banks, F. and Evangelidis, V., *Cell* **14**, 211 (1978).
20. Knapp, G., Beckmann, J. J., Johnson, P. F., Fuhrman, S. A. and Abelson, J., *Cell* **14**, 221 (1978).
21. Peebles, C. L., Gegenheimer, P. and Abelson, J., *Cell* **32**, 525 (1983).
22. Phizicky, E., Schwartz, R. C. and Abelson, J., *J. Biol. Chem.* **261**, 2978 (1986).
23. Konarska, M., Filipowicz, W., Domdey, H. and Gross, H. J., *Nature* **293**, 112 (1981).
24. Schwartz, R. C., Greer, C. L., Gegenheimer, P. and Abelson, J., *J. Biol. Chem.* **258**, 8374 (1983).
25. Kikuchi, Y., Tyc, K., Filipowicz, W., Sanger, H. L. and Gross, H. J., *Nucl. Acids Res.* **10**, 7521 (1982).
26. Furneaux, H., Pick, L. and Hurwitz, J., *Proc. Natl. Acad. Sci. USA* **80**, 3933 (1983).
27. Filipowicz, W. and Shatkin, A. J., *Cell* **32**, 547 (1983).
28. Filipowicz, W., Konarska, M., Gross, H. J. and Shatkin, A. J., *Nucl. Acids Res.* **11**, 1405 (1983).
29. Rothstein, R. J., *Meth. Enzymol.* **101**, 202 (1983).
30. De Roberts, E. M., Black, P. and Nishikura, K., *Cell* **23**, 89 (1981).
31. Kilmartin, J. V. and Adams, A. E. M., *J. Cell Biol.* **98**, 922 (1984).
32. Adams, A. E. M. and Pringle, J. R. J., *J. Cell Biol.* **98**, 934 (1984).
33. De Mey, J., *J. Neurosci. Meth.* **7**, 1 (1983).
34. Ng, R. and Abelson, J., *Proc. Natl. Acad. Sci. USA* **77**, 3912 (1980).
35. Gallwitz, D. and Sures, I., *Proc. Natl. Acad. Sci. USA* **77**, 2546 (1980).
36. Teem, J. L., Abovich, N., Kaufer, N. F., Schwindinger, W. F., Warner, J. R., Levy, A., Woolford, J., Leer, R. J., van Raamsdonk-Duin, M. M. C., Mager, W. H., Planta, R. J., Schultz, L., Friesen, J. D., Fried, H. and Rosbash, M., *Nucl. Acids Res.* **12**, 8295 (1984).
37. Melton, D. A., Krieg, P. A., Rebagliati, M. R., Maniatis, T., Zinn, K. and Green, M. R., *Nucl. Acids Res.* **12**, 7035 (1984).
38. Käufer, N. F., Fried, H. M., Schwindinger, W. F., Jasin, M. and Warner, J. R., *Nucl. Acids Res.* **11**, 3123 (1983).
39. Lin, R.-J., Newman, A. J., Cheng, S.-C. and Abelson, J., *J. Biol. Chem.* **260**, 14780 (1985).
40. Krainer, A. R., Maniatis, T., Ruskin, B. and Green, M. R., *Cell* **36**, 493.
41. Ruskin, B., Krainer, A. R., Maniatis, T. and Green, M. R., *Cell* **38**, 317 (1984).
42. Grabowski, P. J., Padgett, R. A. and Sharp, P. A., *Cell* **37**, 415 (1984).
43. Padgett, R. A., Konarska, M. M., Grabowski, P. J., Hardy, S. F. and Sharp, P. A., *Science* **225**, 898 (1984).
44. Hernandez, N. and Keller, W., *Cell* **35**, 89 (1983).
45. Wallace, J. C. and Edmonds, M., *Proc. Nat. Acad. Sci. USA* **80**, 950 (1983).
46. Domdey, H., Apostol, B., Lin, R.-J., Newman, A., Brody, E. and Abelson, J., **39**, 611 (1984).
47. Newman, A., Lin, R.-J., Cheng, S.-C. and Abelson, J., *Cell* **42**, 335 (1985).
48. Treisman, R., Orkin, S. H. and Maniatis, T., in *Globin Gene Expression and Hematopoietic Differentiation* (ed. G. Stamatoyannopoulos and A. W. Nienhuis). Alan R. Liss, Inc., New York (1983).
49. Wise, J. A., Tollervey, D., Maloney, D., Swerdlow, H., Dunn, E. J. and Guthrie, C., *Cell* **35**, 743 (1983).
50. Brody, E. and Abelson, J., *Science* **228**, 963 (1985).
51. Frendeway, D. and Keller, W., *Cell* **42**, 355 (1985).
52. Grabowski, P. J., Seiler, S. R. and Sharp, P., *Cell* **42**, 345 (1985).
53. Michel, F. and Dujon, B., *EMBO J.* **2**, 33 (1983).
54. van der Horst, G. and Tabak, A. F., *Cell* **40**, 759 (1985).
55. Peebles, C. L., Perlman, P. S., Mecklenburg, K. L., Petrillo, M. L., Tabor, J. H., Jarrell, K. A. and Cheng, H.-L., *Cell* **44**, 213 (1986).
56. Gilbert, W., *Nature (London)* **271**, 501 (1978).
57. Early, P., Rogers, J., Davis, M., Calame, K., Bond, M., Wall, R. and Hood, L., *Cell* **20**, 313 (1980); Maki, R., Roeder, W., Traunecular, A., Sidman, C., Wabl, M., Raschke, W. and Tonegawa, S., *Cell* **24**, 353 (1981).
58. Breitbart, R. E., Nguygen, H. T., Medford, R. M., Destree, A. T., Mahdavi, V. and Nadal-Ginard, B., *Cell* **41**, 67 (1985).
59. Rosenfeld, M. G., Amara, S. G. and Evans, R. M., *Science* **225**, 1315 (1984).
60. Greer, C. L. and Abelson, J., *Trends Biochem. Sci.* **9**, 139 (1984).

Chemica Scripta 1986, **26B**, 139–145

Pathways of Information Readout in DNA

Richard E. Dickerson, Mary L. Kopka and Philip Pjura

Molecular Biology Institute, and the Institute of Geophysics and Planetary Physics, University of California at Los Angeles, Los Angeles, CA 90024, USA

Paper presented by Richard E. Dickerson at the Conference 'Molecular Evolution of Life', Lidingö, Sweden, 8–12 September 1985

Abstract

Two modes of information readout exist from double-helical DNA: (1) An *extrinsic* mode, the familiar genetic translation machinery based on triplet codons, in which the amino acid added to the growing chain need bear no intrinsic resemblance to the trio of bases that encoded it, and (2) An *intrinsic* mode, in which control proteins or drugs bind directly to the double helix and make use of intrinsic similarities or complementarity in chemical or physical properties between DNA and control agent. Extrinsic information readout is used for *coding*; intrinsic readout is used for *control*. The complex extrinsic genetic translation machinery that we observe today may have evolved from a simpler system based on intrinsic DNA-protein recognition, but we have scant information as to how this might have occurred. Intrinsic information readout uses two different channels: repressors, restriction enzymes and other control proteins appear to read base sequence primarily via the major groove, whereas most sequence-specific DNA-binding drugs bind within the minor groove.

Introduction

One of the most fundamental concepts in modern biology is that of nucleic acids as informational macromolecules. Molecular biology might reasonably be considered to have begun when Watson and Crick in 1953 proposed a model for the structure of double-helical DNA that immediately suggested a mechanism for conservative replication of the information contained in the base sequences, and when in the ensuing decade the nature of the triplet genetic code was worked out. Yet the familiar genetic coding system, by which three successive bases along the DNA chain are transcribed into messenger RNA, and ultimately are translated into one amino acid along the growing polypeptide chain, is not the only route by which information is read out from DNA. Indeed, it probably is not the most ancient information retrieval mechanism. Much effort has been expended during the past twenty years in trying to see how the present triplet genetic coding system might have evolved from a more primitive precursor. To date no serious proposal has resulted. The genetic coding system as we know it seems to have been so efficient that it wiped all traces of competing systems from the living record.

One puzzling aspect is that the three bases that code for a particular amino acid seem to bear no intrinsic chemical, physical, or structural similarity to the amino acid that they specify. Various proposals have been made for complementarity in shape, or charge, or hydrogen-bonding ability, between codon and amino acid, of a type that could have led to translation of the genetic message in an earlier era without the present-day battery of messengers, ribosomes and enzymes. None of these models for primitive translation has been convincing. The quandary remains.

Intrinsic *v*. Extrinsic Reading of DNA

Setting aside for the moment the dilemma of the evolution of the triplet coding system, there *is* yet another channel of information readout from DNA, that *is* intrinsic, in the sense that the message contained in the DNA is read by agents that bear an intrinsic and complementary resemblance to the section of DNA that they read. This intrinsic information readout channel is used for the control of expression of genetic information. The *lac* repressor of *E. coli* is a good example of this type of reading and control. A protein molecule, the repressor, binds tightly and specifically to one particular 21 bp sequence of DNA known as the *lac* operator, thereby shutting down the following genes that encode three enzymes involved in lactose metabolism. However, if appreciable quantities of lactose are present in the surrounding medium, then a lactose isomer, allolactose, binds to the repressor protein molecule and causes it to fall away from the operator, freeing the genes for transcription and ultimately for translation. The *lac* repressor reads a message in the operator DNA, that says in effect, 'Bind here'. It does so by attaching itself directly to the operator, and the inference is that the repressor molecule in some sense is complementary to the operator sequence, interacting with it by a combination of hydrogen bonds, close van der Waals packing contacts, and

Table I. *Modes of readout of base sequence information from DNA*

1. Extrinsic information readout
   - Multi-step process: DNA, RNA, ribosomes, enzymes
   - Amino acids coded do not have any intrinsic resemblance to their triplet codons: shape, size, charge, reactivity, etc.
   - Critical translation step occurs at matching of tRNA to amino acid on surface of aminoacyl-tRNA synthetase enzyme
   - Elaborate, highly evolved mechanism
   - Base sequence information used for *coding*
2. Intrinsic information readout
   - Single-step process: binding to DNA
   - Protein or other molecule must have intrinsic complementarity to some features of DNA: hydrogen-bonding pattern, shape, orientation, charge, etc.
   - Critical translation step occurs at binding of reading molecule directly to DNA
   - Simple, direct mechanism, possibly primitive
   - Base sequence information used for *control*

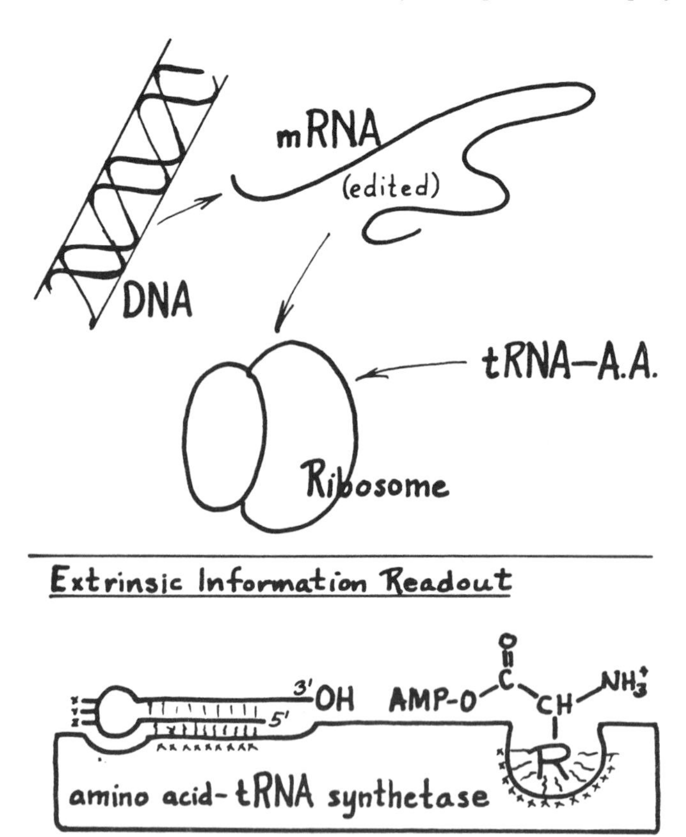

Fig. 1. Extrinsic information readout from DNA. The information first is transcribed from DNA to messenger RNA. After editing (in some cases), this message then is translated at the ribosome into a sequence of linked amino acids, which folds to build the enzyme or other protein molecule. The true 'reading', or matching of base sequence to amino acid sequence, is performed by the amino acid-tRNA synthetase, which has binding sites that hold both a particular transfer or tRNA, and the amino acid for which its anticodon loop codes. Many steps and many components are required, and the ultimate amino acid need bear no intrinsic chemical or physical resemblance to the triplet of bases that directed its addition to the growing chain.

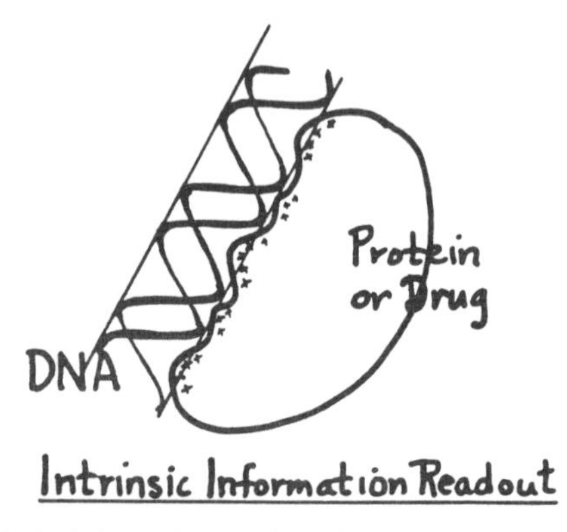

Fig. 2. Intrinsic information readout from DNA. Information as to a particular base sequence in DNA is read directly via contacts with the control protein or drug molecule. These contacts rely on intrinsic similarities or complementarities between DNA and protein or drug, involving hydrogen bonds, van der Waals contacts, and electrostatic charge attraction.

electrostatic charge attractions. The exact nature of this sequence-specific protein–DNA interaction is of considerable interest, but has only been deciphered yet for one example: the complex of the *Eco* RI restriction enzyme with its GAATTC cleavage site as contained within a double-helical oligonucleotide of sequence: TCGCGAATTCGCG [1, 2]. This same type of intrinsic recognition of base sequence is

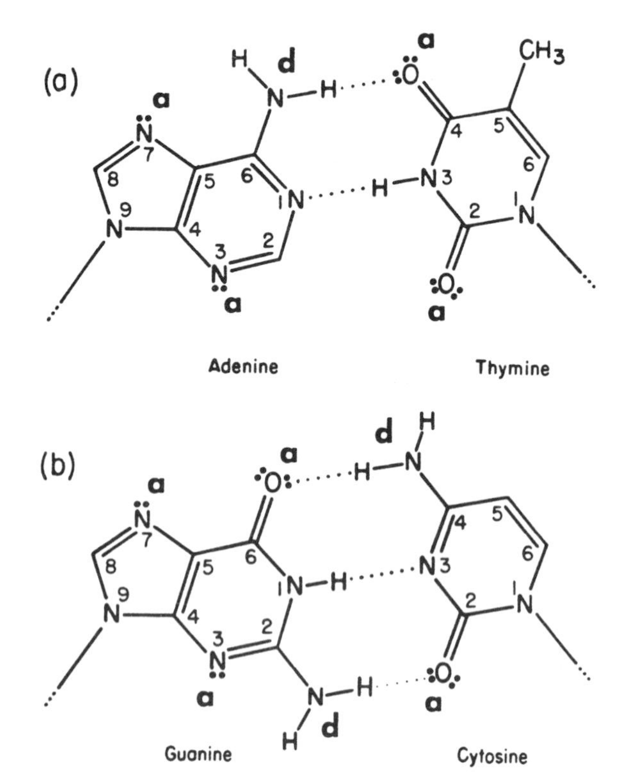

Fig. 3. AT and GC base pairs, showing the pattern of hydrogen bond donors (d) and acceptors (a) on the major groove edge (upper edge) and minor groove edge (lower edge) of each base pair. Repressors or other control proteins and drugs potentially can read the base sequence by sensing the pattern of donors and acceptors along the floors of the major and minor grooves.

believed to be involved with many other control proteins such as repressors and restriction enzymes, and also with certain sequence-specific antibiotics and antitumor drugs.

The distinction between this intrinsic mode of information readout and the extrinsic mode as found in the genetic translation machinery is summarized in Table I and Figs. 1 and 2. The intrinsic mode is simpler, involving many fewer components, and probably is older in an evolutionary sense. An as yet unfulfilled challenge is to see whether the complicated extrinsic readout mode can be derived, in any plausible way, from an earlier and simpler intrinsic mode.

How does intrinsic readout of base sequence information occur? What is it about a repressor protein or a drug molecule that interacts with the DNA in a manner that can differentiate between the four possibilities at each base pair: AT, TA, GC or CG? The accessible sequence information in base pairs is carried on their two edges – that is, on the floors of the major and minor grooves. Seeman *et al.* [3] proposed in 1976 that information readout was accomplished by means of hydrogen bonds between amino acids of the protein and H-bond donors (d) and acceptors (a) on the major and minor edges of the base pairs. As Fig. 3 shows, an AT base pair differs from a GC pair on the major groove edge by having the H-bonding sequence: (a–d–a) for AT, *v.* (a–a–d) for GC. Reversing GC leads to a quite distinct new major groove pattern: (d–a–a) for CG, whereas reversing AT to TA yields (a–d–a) once again, although with a slightly different spatial geometry. Seeman *et al.* proposed that these patterns of hydrogen bond donors and acceptors could be read by H-bonding protein side chains such as lysine or arginine, asparagine or glutamine, and aspartic or glutamic acids. (See Figs. 2 and 3 of ref. 3.)

The minor groove, in contrast, has a lower information

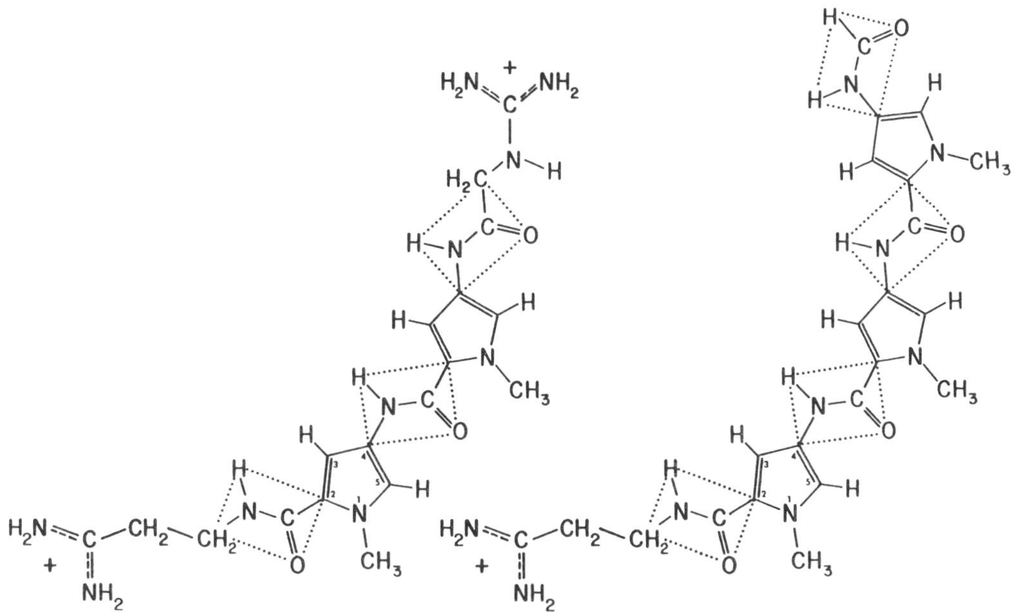

Fig. 4. Netropsin (left) as isolated from *Streptomyces netropsis*, and distamycin (right) as obtained from *Streptomyces distallicus*. Both are built up from repeating (pyrrole–amide) units, and longer and shorter analogues can be made synthetically. Both ends are positively charged in netropsins, but only one end in distamycins. In synthetic molecules of this class, a number sometimes is added after the name to indicate the number of pyrrole rings, as: netropsin-2 and distamycin-3 shown here.

content. GC base pairs can be differentiated from AT pairs by the patterns of donors and acceptors: (a–d–a) for GC *v.* (a–x–a) for AT, but it is unlikely that base pair reversals such as CG for GC could be detected in the minor groove.

Specific binding by control proteins

Only one crystal structure analysis of a complex of a sequence-specific DNA-binding protein with its proper DNA sequence has yet been carried out, that of the *Eco* RI restriction enzyme with TCGCGAATTCGCG, and this is reported elsewhere in this volume [2]. The restriction enzyme is observed to wrap itself around the B-DNA double helix, inserting side chains into the major groove but leaving the minor groove largely untouched. The recognition sequence for the *Eco* RI enzyme is:

```
    5′  3′
    G---C
   A---T
  A---T
 T---A
T---A
C---G
3′  5′
```

Protein side chains appear to interact only with purines in this example. One arginine serves as a hydrogen bond donor to the N7 and O6 of guanine, another is a donor to the N7 of two adjacent adenines, and a glutamic acid is a hydrogen bond acceptor for the C6–NH_2 of the adjacent adenines. A twofold symmetry axis relates both the two ends of the DNA double helix and the two identical subunits of the restriction enzyme. Any combination of base pairs other than the sequence given above would produce a different pattern of donors and acceptors, and would require different amino acid side chains for recognition. This is a clear and beautiful example of a protein using intrinsic chemical properties to 'read' the floor of the major groove of B-DNA.

Specific binding by drug molecules

Two X-ray crystal structure analyses have been carried out to date of sequence-specific drug molecules that read DNA sequence by binding within the minor groove. These are of the antiviral, antitumor antibiotic *netropsin* (Fig. 4) and the carcinogenic DNA stain Hoechst 33258 (Fig. 5). Netropsin and its cousin distamycin are short polymers whose repeating backbone is constructed by alternating methylpyrrole rings and amide groups. One end of the distamycins is cationic, but both ends carry a positive charge in netropsins. Netropsins and distamycins sometimes are differentiated by numbering them according to the number of pyrrole rings. Hence the two naturally occurring compounds depicted in Fig. 4 would be termed netropsin-2 or Nt2 and distamycin-3 or Dst3. These compounds are highly sensitive to DNA geometry and base sequence. They bind only to double-helical DNA in the B form, and not to A or Z DNA, to RNA, or to single stranded polynucleotides. They require a binding site consisting of a series of AT base pairs, and the size of the binding site varies with the length of the drug polymer. Dervan and co-workers [4] have shown that a distamycin or netropsin with *n* pyrroles

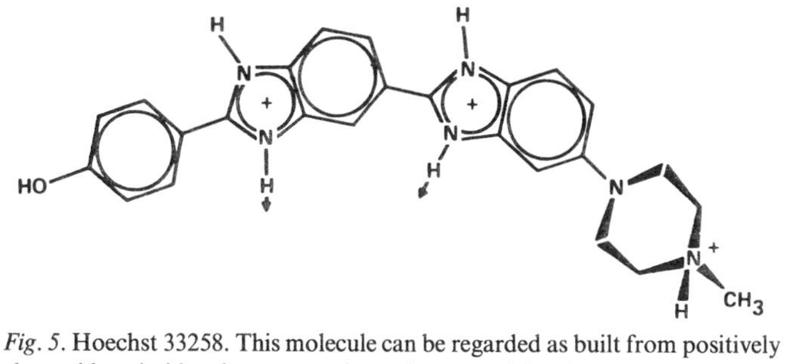

Fig. 5. Hoechst 33258. This molecule can be regarded as built from positively charged benzimidazoles as repeating units, capped by phenol at one end and piperazine at the other. Like netropsin and distamycin, it has a natural curvature, with H-bond donating N–H groups pointing from the concave edge.

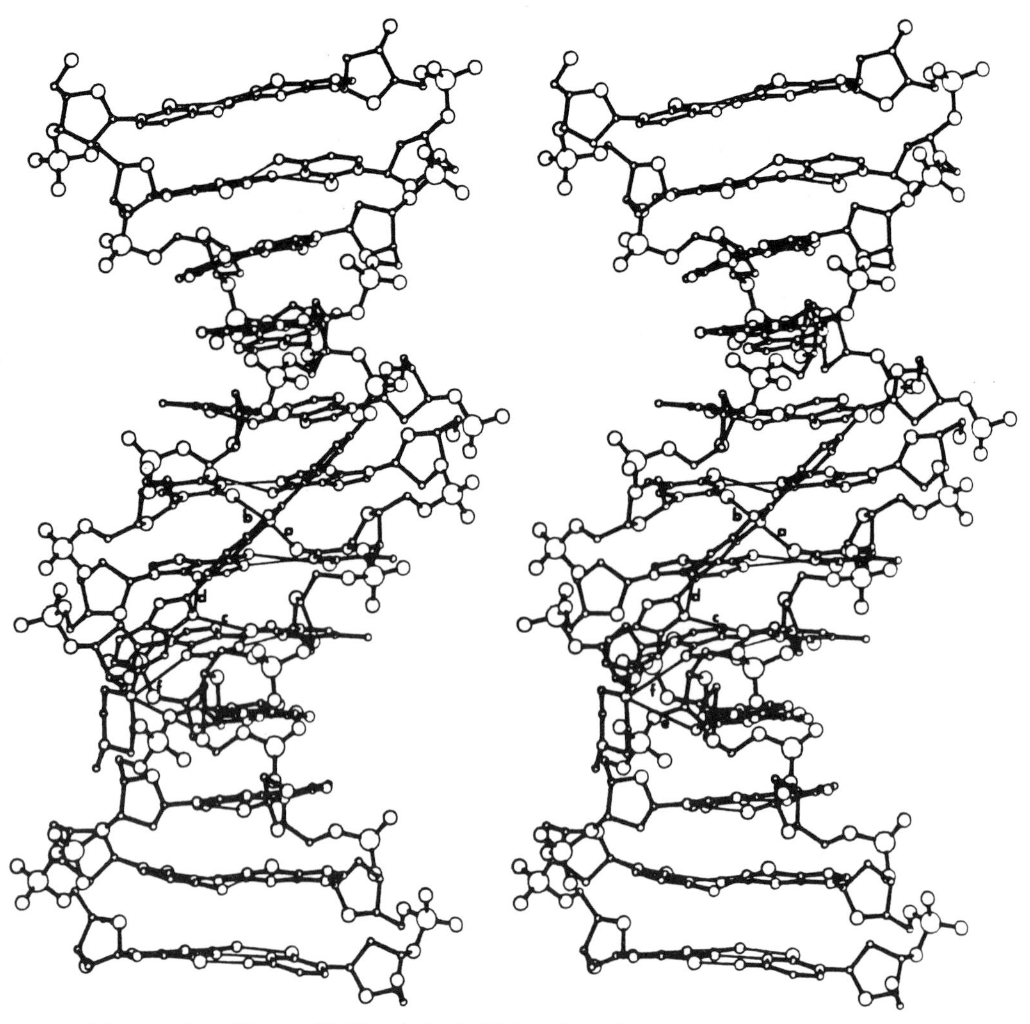

Fig. 6. Stereo pair drawing of the complex of Hoechst 33258 with the B-DNA double helix CGCGAATTCGCG. The drug molecule sits within the minor groove in the central AATT region, making hydrogen bonds to base pair edges, as marked *a* through *f*. In this and subsequent diagrams, bases in the 'down' strand are numbered C1 through G12, and those in the 'up' strand are C13 through G24. Hence hydrogen bond *a* leads from the drug molecule to base T7 of base pair T7·A18, and bond *b* leads to base T19 of base pair A6·T19. The piperazine ring at lower left of the drug is disordered in the crystal. In its alternate conformation it lifts up and out of the groove. (From ref. 8.)

and $n+1$ amides requires a binding site containing $n+2$ successive AT base pairs. They also have shown that distamycin is slightly more tolerant of GC 'errors' near one end of the binding site than is netropsin.

Netropsins and distamycins are antitumor antibiotics, although so toxic that they are of little use in routine cancer chemotherapy. They are believed to act on DNA by stabilizing the B double helix to the extent that they interfere with functions that require opening up the two strands: replication and transcription. Hoechst 33258, in contrast, is a carcinogen. It binds to DNA so tightly that it finds use as a DNA stain in microscopy. It exhibits a preference for AT containing binding sites similar in size to those of Nt2 and Dst3, but is not as base-pair-specific as the others [5].

X-ray crystal structure analyses have been carried out on the complexes of netropsin-2 with CGCGAATT^{Br}CGCG [6, 7], where ^{Br}C is 5-bromocytosine, and of Hoechst 33258 with CGCGAATTCGCG [8]. They bind in a very similar manner within the minor groove, in the AATT center of the double helix. A stereo pair drawing of the complex with Hoechst 33258 is shown in Fig. 6; equivalent drawings for netropsin can be found in ref. 7. Each drug molecule fits snugly down the center of the narrow minor groove, which is barely wide enough in the AATT region to admit it. Indeed, the refined netropsin complex structure shows that the groove is wedged open slightly and the helix axis is bend back 8° at the point of binding.

The drug molecules, when they bind to the minor groove of B-DNA, displace a highly ordered string of water molecules

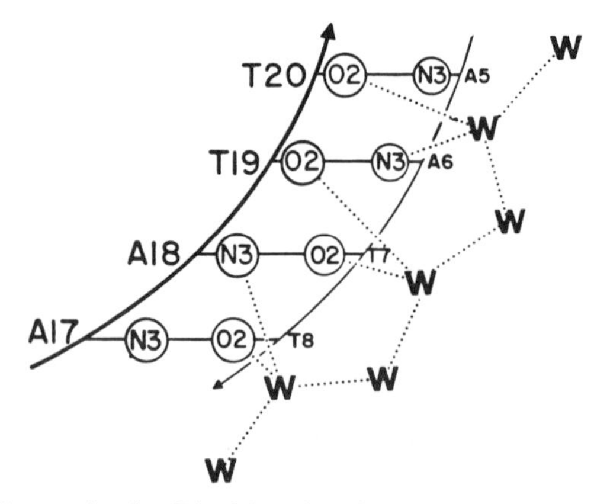

Fig. 7. Water molecules (W) of the spine of hydration that snakes down the minor groove in AT-rich regions of B-DNA. The two sugar-phosphate backbones flanking the minor groove are rendered symbolically. N3 and O2 in circles represent adenine N3 and thymine O2 atoms on the floor of the minor groove. Base pairs are numbered as described in the previous caption. For more information, see refs. 9 and 10.

Fig. 8. The Fire-breathing dragon, trademark of AGIP, the Italian state petroleum corporation. This logo has been adopted within the authors' laboratory as a symbol of the 'dragon's spine' or the spine of hydration in B-DNA.

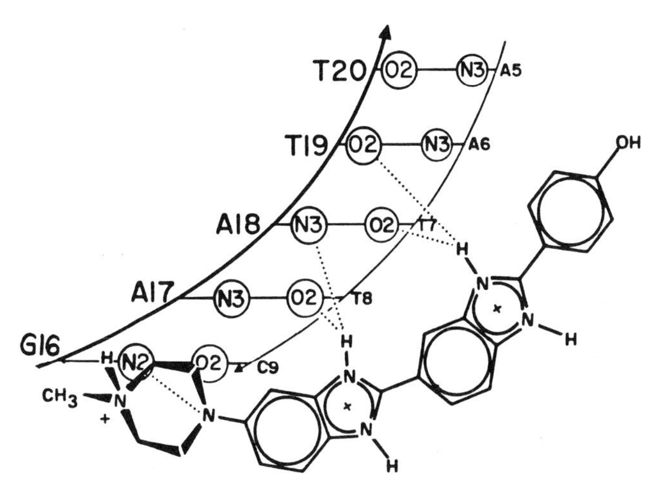

Fig. 10. Binding of Hoechst 33258 within the minor groove, rendered in the style of Figs. 7 and 9. Two of the three bifurcated hydrogen bonds of netropsin recur in Hoechst 33258, along with a possible interaction of the piperazine ring with the C2–NH_2 group of a guanine. (From ref. 8.)

that we believe to be a major contributor to the stability of the B-helix. This so-called 'spine of hydration' is schematized in Fig. 7. First hydration shell water molecules form hydrogen-bonded bridges connecting adenine N3 and thymine O2 atoms that lie on opposite strands of the helix and adjacent base pairs. The bases thus bridged are those that are brought into closer proximity by the rotation of the helix axis [9, 10]. These first-shell water molecules in turn are bridged by a second shell that gives them a local quasi-tetrahedral co-ordination, and that connects them into a long spine down the center of the minor groove. This spine of hydration is characteristic of B-DNA in AT base pair regions, but is disrupted in GC regions because of the intrusive presence of guanine C2–NH_2 groups. These amine groups both introduce hydrogen bond donors where only acceptors had been, and physically displace and disrupt the first-shell water molecules in the spine. Ease of conversion of synthetic polymer fibers from the B to the A form by drying has been observed to correlate very well with the presence of GC base pairs, whereas fibers of polymers containing only AT or IC base pairs stubbornly resist conversion to the A form by drying [9]. The importance of this spine of hydration to the stability of B-DNA has caused it to be nicknamed the 'dragon's spine' after the fire-breathing AGIP trademark (Fig. 8).

Displacement of this spine of hydration, and binding of netropsin or Hoechst 33258, are schematized in Figs. 9 and 10. In both cases, N–H groups from the drug molecule displace waters of the first hydration shell, and make bifurcated hydrogen bonds to the same pairs of adenine N3 and thymine O2. The second shell of the spine of hydration is replaced by the covalent backbone of the drug molecule itself, thus locking the double helix even more firmly into place. Netropsin-2, with three amide N–H, bridges three pairs of O and/or N atoms distributed over four contiguous base pairs, explaining quite simply Dervan's observed $n+2$ rule. Hoechst 33258, in contrast appears to make only two such bifurcated bridges, and these are abnormally long for hydrogen bonds. But its piperazine end also appears to make a possible long hydrogen bond to the C2–NH_2 of the neighboring GC base pair.

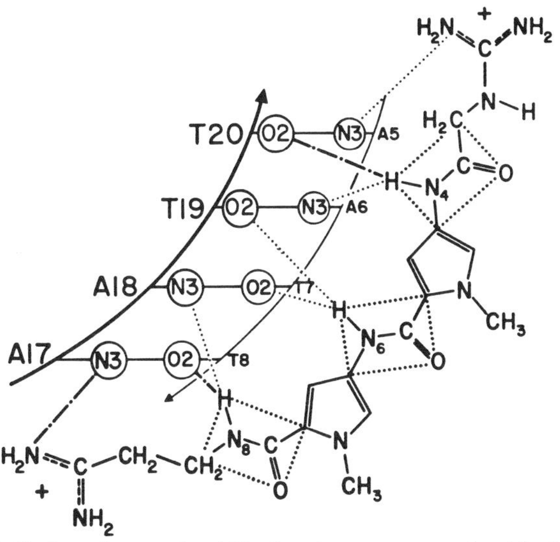

Fig. 9. Binding of netropsin within the minor groove, rendered in the style of Fig. 7. The same bridging H bonds are made, but this time involving N–H groups of amides, rather than water molecules. Dot-dash chain-link bonds are shorter, canonical hydrogen bonds, whereas dashed bonds are somewhat longer. (From ref. 7.)

Reading base sequence in the minor groove: lexitropsins

Unlike the *Eco* RI restriction enzyme, these drug molecules do not read the DNA base sequence and recognize AT base pairs by means of specific hydrogen bonds. The H bonds schematized in Figs. 9 and 10 would be possible, whether the base pairs were AT or GC. Instead, the specificity arises as it did with the spine of hydration – via steric hindrance involving the C2–NH_2 group of guanine.

It is possible to synthesize longer analogues of the natural netropsin and distamycin depicted in Fig. 4, and these require longer regions of AT base pairs as binding sites. We have considered whether it might be possible to create a series of longer netropsin analogues that not only would read AT base pairs, but that would select and bind preferentially to any desired sequence of AT *v.* GC base pairs, without regard to end-for-end turning of a given pair. Such sequence-specific base-reading analogues of netropsin, or *lexitropsins*, might find an important use as site-specific antitumor drugs [6, 7]. A lexitropsin designed to target one particuar sequence of ten successive AT *v.* GC base pairs would bind to random, non-target DNA only one time in $2^{10} = 1024$. This could

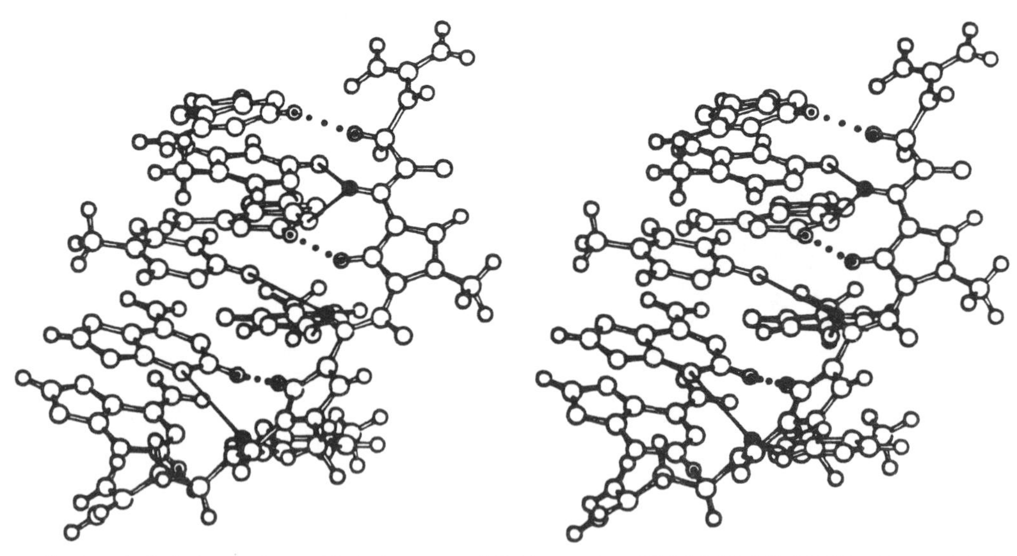

Fig. 11. Stereo closeup of netropsin in its interactions with the four central AT base pairs of CGCGAATT^{Br}CGCG. Bifurcated hydrogen bonds from the three amide N–H (with hydrogens in black) are drawn as thin lines to the DNA. Dotted lines are close non-bonded van der Waals contacts between adenine C2–H of the DNA, and methylene $-CH_2-$ or pyrrole –CH– of netropsin. These contacts are too tight to permit the intrusion of a $C2-NH_2$ group, such as would be present in guanine. Hence these tight non-bonded contacts are the way in which netropsin reads the DNA sequence. Non-hydrogen atoms are as obtained from the refined X-ray crystal-structure analysis. Hydrogen atoms have been added with standard geometries using a program developed by Gilbert Privé.

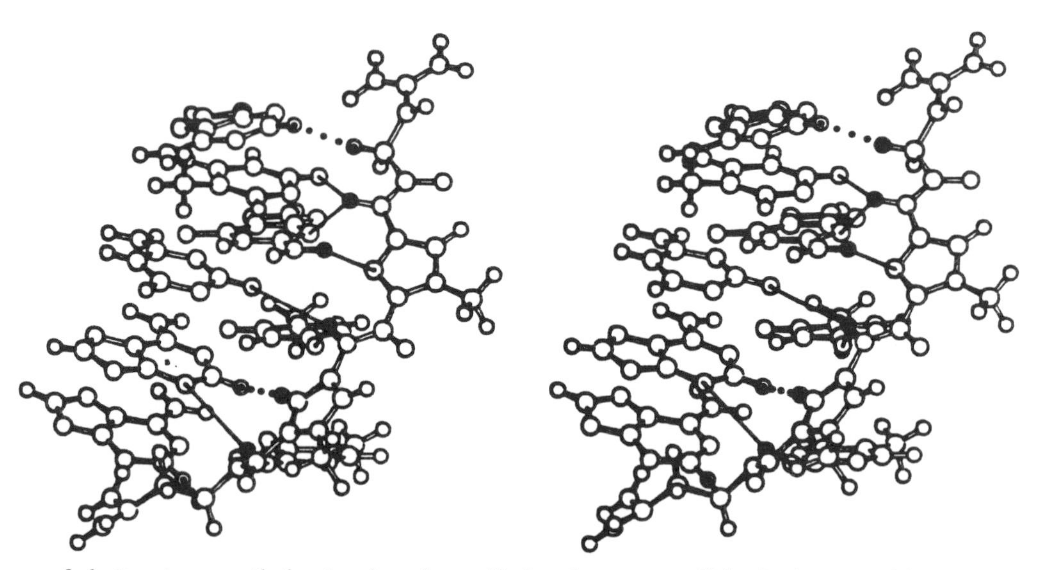

Fig. 12. Stereo close-up of a *lexitropsin*, or a synthetic netropsin analogue with altered sequence specificity. Replacement of the upper pyrrole ring by imidazole, which replaces the –CH– facing the floor of the minor groove by –N–, provides room to accommodate the $C2-NH_2$ of guanine at that point. Moreover, GC is not only tolerated, but is actively favored over AT, because a new hydrogen bond (H atom in black) is formed to strengthen the complex.

mean a three order of magnitude reduction in unwanted side effects of chemotherapy, a goal well worth working towards.

Both our group, and Professor J. William Lown of the University of Alberta arrived simultaneously at the same answer: if one were to replace a pyrrole ring by an imidazole or furan, so that the –CH– nearest the floor of the minor groove became the H-bond acceptor –N– or –O–, then one would both make room for the $-NH_2$ of guanine and would permit an additional stabilizing hydrogen bond between DNA and drug. This replacement of the upper pyrrole of Fig. 11 is shown in Fig. 12. The comparison of netropsin with a lexitropsin is also depicted less accurately but more simply in Figs. 13 and 14. In principle one now can prepare a long netropsin analogue, with AT-recognizing $-CH_2-$ methylene groups at the ends, but with pyrroles *v.* imidazoles at intermediate positions as needed to bind selectively to any desired DNA sequence. Various lexitropsins are being synthesized in Professor Lown's laboratory at Alberta, while the corresponding DNA oligomers are being prepared at UCLA. The complexes will be studied both by nuclear magnetic resonance (at Alberta) and X-ray crystallography (at UCLA), and the individual lexitropsins also will be screened for clinical usefulness. Out of this research may come a better understanding of the molecular basis of chemotherapy, and improved chemotherapeutic agents as well.

Modes and channels of information readout in DNA

In summary there appear to be two distinct modes of information readout from double-helical DNA; an extrinsic mode epitomized by the genetic translation machinery, and an intrinsic mode reflected in binding of control proteins and

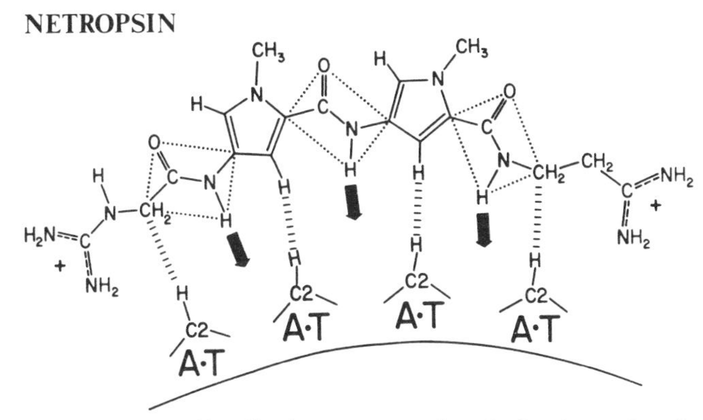

Fig. 13. Schematic of bonding between netropsin and a four base pair region of AT. Dark arrows are bifurcated hydrogen bonds to the floor of the minor groove. The four dashed strips represent close non-bonded van der Waals contacts between adenine C2–H and the drug molecule.

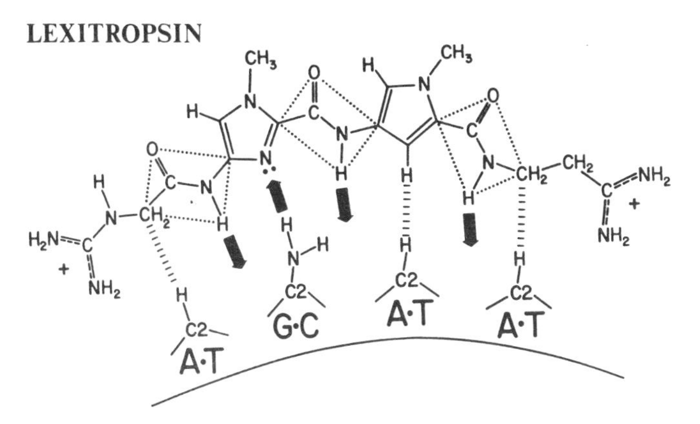

Fig. 14. Replacement of one pyrrole by imidazole makes room for a guanine amine group, and also provides an acceptor for a new hydrogen bond involving the $-NH_2$. In longer lexitropsins, or sequence-reading netropsin analogues, imidazoles and pyrroles could be varied at will in order to match any desired DNA sequence.

some drug molecules. The extrinsic mode is used for *coding*; the intrinsic mode for *control*. One assumes that today's quite complex and multicomponent extrinsic mode must surely have evolved from a simpler precursor, and that this precursor mechanism would have used intrinsically similar properties of base pair triplet codons and the coded amino acids. But what this precursor machinery might be like we cannot say. The presumably less-efficient precursors have long ago lost out in competition with organisms carrying the developed genetic mechanism as we know it. It is instructive, however, to consider the alternative intrinsic readout mode, and try to imagine how something like it might have differentiated into the more complex system.

The intrinsic mode of information readout from double-helical B-DNA appears to involve two quite distinct channels: the major and minor grooves. The major groove is more information-rich, in that the pattern of hydrogen bond donors and acceptors can differentiate between the four possible cases: AT, TA, GC and CG. In contrast, the minor groove can discriminate between AT and GC base pairs, but not between inversions of the same base pair, TA for AT as an example.

X-ray crystal structure analyses of repressors and control proteins alone, without being bound to their respective DNA sequences, all suggested that these control proteins interact mainly with the more information-rich major groove: the *cro* protein [11], catabolite activator protein [12], and DNA-binding domain of the *lambda* repressor [13]. This picture of main use of the major groove by proteins has been confirmed by the only extant high-resolution analysis of a protein–DNA complex, that of the *Eco* RI restriction enzyme [1, 2].

By contrast, drug molecules that bind to DNA and show appreciable base specificity seem to interact with the less information-rich channel, the minor groove [6–8]. Feigon *et al.* [14] have shown by nuclear magnetic resonance studies of more than seventy drug molecules, that those compounds that intercalate between base pairs tend to exhibit preference for GC base pairs if they show specificity at all, whereas those that prefer AT base pairs bind within the minor groove. No examples were found of nonintercalative drugs that bind within the major groove. The AT preference of these minor groove-binding drugs is easily explained: steric hindrance by the C2–NH_2 groups or guanines.

It should be possible to design a drug molecule that can recognize a predetermined succession of AT *v.* GC base pairs, but not one that can take the further step of differentiating between end-for-end inversion of one type of base pair. Within these limits, synthetic organic chemistry should be capable of producing a set of base-specific *lexitropsins*, potentially useful agents in cancer chemotherapy. What nature has evolved to read certain sequences, the synthetist should be able to adapt to read all sequences.

Acknowledgements

We would like to thank Gilbert Privé for developing the hydrogen atom positioning program and preparing stereo pairs of Figs. 11 and 12. This work was supported by NIH Grant GM-31299 and American Cancer Society Grant No. NP-504. This is publication No. 2699 from the Institute of Geophysics and Planetary Physics.

References

1. Frederick, C. A., Grable, J., Melia, M., Samudzi, C., Jen-Jacobson, L., Wang, B. C., Greene, P., Boyer, H. W. and Rosenberg, J. M., *Nature* (*London*) **309**, 327 (1984).
2. Rosenberg, J. M. *et al.*, *Chemica Scripta* **26B**, 147 (1986).
3. Seeman, N. C., Rosenberg, J. M. and Rich, A., *Proc. Natl. Acad. Sci. USA* **73**, 804 (1976).
4. Youngquist, R. S. and Dervan, P. B., *Proc. Natl. Acad. Sci. USA* **82**, 2565 (1985).
5. Harshman, K. D. and Dervan, P. B., *Nucleic Acids Res.* **13**, 4825 (1985).
6. Kopka, M. L., Yoon, C., Goodsell, D., Pjura, P. and Dickerson, R. E., *Proc. Natl. Acad. Sci. USA* **82**, 1376 (1985).
7. Kopka, M. L., Yoon, C., Goodsell, D., Pjura, P. and Dickerson, R. E., *J. Mol. Biol.* **183**, 553 (1985).
8. Dickerson, R. E., Pjura, P., Kopka, M. L., Goodsell, D. and Yoon, C., in *Crystallography in Molecular Biology*, Proceedings of the NATO Advanced Study Institute and EMBO Lecture Course, Bischenberg, Alsace, France, 12–21 September 1985. NATO ASI Series (ed. D. Moras), Plenum, New York (1985).
9. Drew, H. R. and Dickerson, R. E., *J. Mol. Biol.* **151**, 535 (1981).
10. Kopka, M. L., Fratini, A. V., Drew, H. R. and Dickerson, R. E., *J. Mol. Biol.* **163**, 129 (1983).
11. Ohlendorf, D. H., Anderson, W. F., Fisher, R. G., Takeda, Y. and Matthews, B. W., *Nature* (*London*) **298**, 718 (1982).
12. McKay, D. M., Weber, I. T. and Steitz, T. A., *J. Biol. Chem.* **257**, 9518 (1982).
13. Pabo, C. O. and Lewis, M., *Nature* (*London*) **298**, 443 (1982).
14. Feigon, J., Denny, W. A., Leupin, W. and Kearns, D. R., *J. Medicinal Chem.* **27**, 450 (1984).

Chemica Scripta 1986, **26B**, 147–157

The 3 Å Structure of a DNA–*Eco* RI Endonuclease Recognition Complex

John M. Rosenberg, Judith A. McClarin, Christin A. Frederick*, Bi-Cheng Wang†, Herbert W. Boyer‡ and Patricia Greene‡

Department of Biological Sciences, University of Pittsburgh, Pittsburgh PA 15260, USA

Paper presented by John M. Rosenberg at the Conference on 'Molecular Evolution of Life', Lidingö, Sweden, 8–12 September 1985

Abstract

The 3 Å structure of a co-crystalline recognition complex between *Eco* RI endonuclease and the cognate oligonucleotide TCGCGAATTCGCG has been solved by the ISIR method using a platinum isomorphous derivative. Each subunit of the endonuclease is organized into an α/β domain based on a five stranded β-sheet and an extension, called the 'arm', which wraps around the DNA. The primary β-sheet consists of anti-parallel and parallel sub-domains which contain the sites of DNA strand scission and sequence specific recognition, respectively. The hydrolytic active site is located in a cleft which binds the DNA backbone in the vicinity of the scissile bond.

The DNA conformation departs significantly from those which have been observed in crystals containing pure DNA; suggesting that the altered conformation has been stabilized by the enzyme. The conformations seen only upon protein binding are termed neo-conformations to distinguish them from those which are intrinsically stable in the absence of protein.

Sequence specificity is determined by 'modular' interactions based on the crossover α-helices, i.e. those which connect the β-strands of the parallel segment of the principal β-sheet. They are pointing into the major groove of the DNA and amino acid side chains at the amino ends of these helices form bidentate interactions with the bases. The inner recognition module consists of two symmetry-related α-helices which recognize the inner tetranucleotide (AATT), while the two symmetry-equivalent outer recognition modules are single alpha-helices which recognize the GC base pairs.

Introduction

The recognition by a protein of a specific sequence of bases along a strand of double-helical DNA lies at the heart of many fundamental biological processes including the regulation of gene expression by repressors and activators, site specific genetic recombination and host dependent restriction and modification of DNA. The detailed molecular mechanisms of some of these interactions are beginning to be understood due to efforts in many laboratories using genetic, biochemical and structural methods. One of the most intriguing questions in molecular biology today is whether the details of several of these recognition mechanisms will form a small number of simple patterns which would lead to a general understanding of DNA recognition processes. In order to answer this question, structural data is required from representative DNA–protein co-crystalline complexes.

* Permanent address: Dept. of Biology; MIT; 77 Massachusetts Ave, Cambridge, Mass. 02139, USA.

† Biocrystallography Laboratory, Box 12055, VA Medical Center, Pittsburgh, Pennsylvania 15260, USA.

‡ Dept. of Biochemistry and Biophysics, University of California, SF., San Francisco, CA 94143, USA.

The structures of three sequence specific DNA-binding proteins, the *cro* and CI repressors from bacteriophage-λ and the *E. coli* catabolic gene activator protein (CAP) have been solved in the absence of DNA [1–5]. These three proteins share a common 'two-helical motif' at the putative DNA binding site which has led to model building of the recognition complexes [6, 3, 7]. The 7 Å structure of a co-crystalline complex between a tetradecanucleotide and bacteriophage 434 repressor supports the general features of these models [8–10]. The common features in these structures suggests that they are examples of one class of DNA recognition proteins based on the two-helical motif.

Similarly, structural investigations are in progress on other systems, including the Klenow fragment of *E. coli* DNA polymerase I [11], nucleosome core particles [12, 13], the histone octamer [14] and the 'histone like' DNA-binding protein II of *B. stearothermophilus* [15]. These analyses will facilitate the elucidation of the specific structural features of both protein and DNA in these specific protein–DNA interactions as well as general structural principals of DNA–protein interactions.

The highly specific recognition of the double stranded sequence d(GAATTC) by *Eco* RI endonuclease offers compelling advantages as a model system for investigating DNA recognition. It is a small protein of molecular weight 32000 (276 amino acids) of known sequence [16, 17] which hydrolyzes the phosphodiester bond between the guanylic and adenylic acid residues resulting in a 5′-phosphate. This reaction proceeds with inversion of configuration at the reactive phosphorous [18]. The protein forms highly stable catalytically active dimers in solution and will form tetramers at higher protein concentrations [19, 20]. Although *Eco* RI endonuclease requires Mg^{2+} for phosphodiester bond hydrolysis, DNA binding and recognition specificity is retained in the absence of this ion [21–24].

We recently reported [25] our observations based on a preliminary electron density map (see below) of a co-crystalline complex between *Eco* RI endonuclease and a cognate oligonucleotide. Here we report further analysis of this structure.

Methods

Crystallization conditions and methods of data collection were reported previously [25, 26]. It should be noted that hydrolysis of the DNA was prevented by removing the required co-factor, Mg^{2+}, from the crystallization medium.

A platinum derivative was obtained by soaking crystals in 10 mM *cis*-$(NH_3)_2PtCl_2$, dissolved in 40 mM bis-tris-propane (BTP), pH 7.0 plus a mixture of 11% (w/v) polyethylene glycol (PEG) 6000 and 5% (w/v) PEG 400 for 7 d at 25 °C. A mercury derivative was likewise obtained by soaking crystals in 5 mM $Hg(SCH_2CH_2NH_3^+)_2$, dissolved in 40 mM BTP, pH 7.0 plus 16% (w/v) PEG 6000 for 3.5 d at 25 °C. The heavy atom locations were independently determined by difference Patterson methods for both derivatives.* The platinum Difference Patterson Function is shown in Fig. 1. The Cullis R factors for the Pt and Hg derivatives to 3.5 Å are 49.2% and 52.2%, respectively.

An MIR† electron density map was calculated from these two derivatives to 5 Å resolution. The MIR phases were used to calculate a Pt-native Difference Fourier map which revealed the presence of a single minor heavy atom site. Wang's ISIR method (see below) was independently applied to the platinum and mercury data. The general features, such as the solvent regions and the molecular outline, were similar in all three electron density maps; however the MIR and Hg-ISIR maps contained significant amounts of noise while the Pt-ISIR map was clear. We suspect that the noise in both cases is caused by a problem with the mercury derivative.

Statistics for the platinum derivative exhibited evidence of a slight non-isomorphism at high resolution, as can be seen in Fig. 2, which is a plot of intensity differences as a function of resolution. The plot shows a minimum at 3.5 Å, with a slight increase at higher resolution.‡ We felt that the platinum phase information was dubious beyond 3.5 Å and therefore did not utilize it further.

It should be noted in this context that we felt that the accuracy of the phase information used as the starting point for the ISIR procedure was crucial to its successful completion. This can be seen by considering that the ISIR procedure resolves the phase ambiguity initially present in the SIR phases. If however both probable SIR phases are seriously in error for a significant fraction of the data, then the ISIR procedure most likely will converge to a false minimum. This is in contradistinction with the MIR case, where a preponderance of valid phase information can tend to overpower inaccuracies. For ISIR, the initial phase information should be carefully selected to insure that it is accurate.

The positions and occupancies of the major and minor platinum sites were refined as follows: We calculated for each reflection (both general and centric) the absolute value of the difference between the native and derivative structure factors. We retained the largest differences, i.e. those which constituted 30% of the total data set. These data were then used as input

* Even though the platinum compound is a chemotherapeutic agent which appears to act at the DNA level, the location of the major and minor sites turn out to be on the surface of the protein, well away from the DNA.

† We use the following abbreviations: MIR, multiple isomorphous replacement; SIR, single isomorphous replacement; ISIR, iterative single isomorphous replacement.

‡ The source of this slight non-isomorphism was obvious once the structure was solved: The Pt was bonded to the sulfur atoms of two methionyl residues, which fortuitously were in close proximity on the surface of the protein. However, the (mean) positions of the two sulfur atoms in the native structure were not precisely those required by the geometry of the bridging reaction which therefore appears to have required a small structural adjustment in a short segment of the polypeptide chain.

A

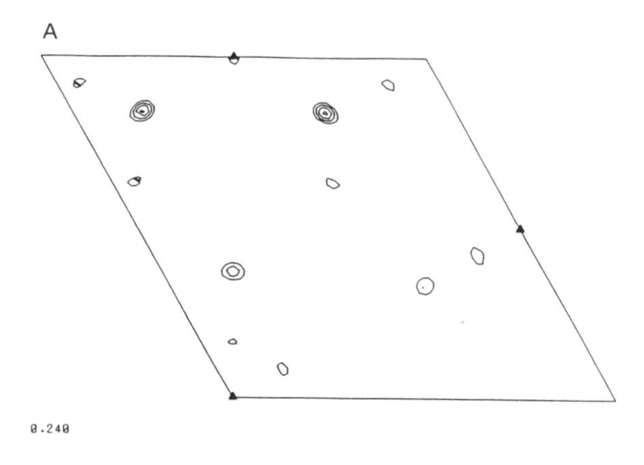

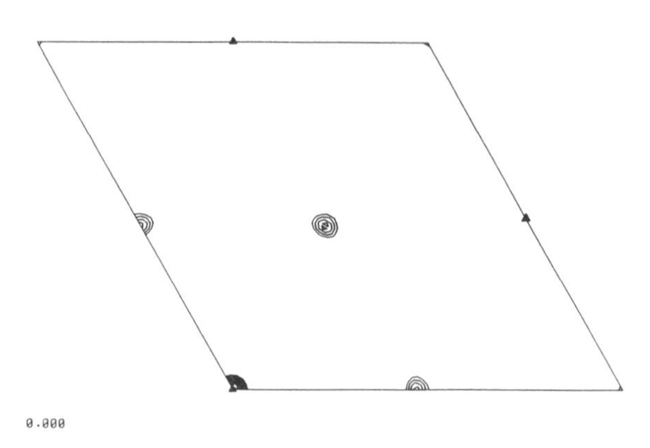

B

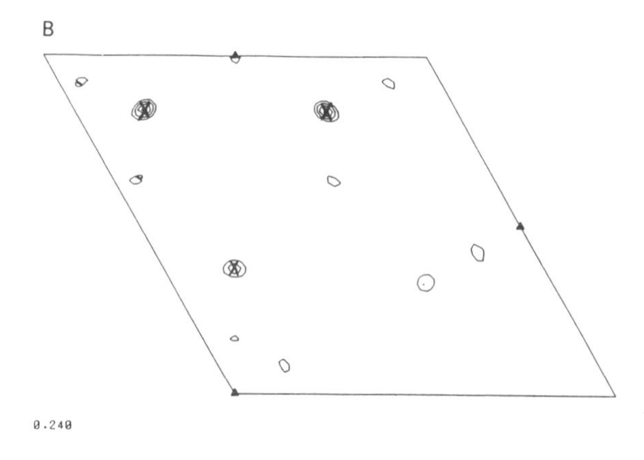

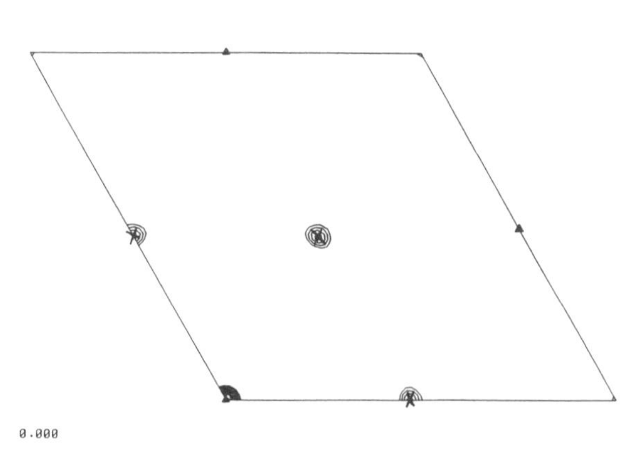

Fig. 1. (A) The $w = 0$ and $w = 0.24$ sections of the platinum difference Patterson function, which contains the Pt-Pt vectors. (B) The indicated positions mark the expected positions for the Pt-Pt vectors based on the coordinates of the major platinum-binding site.

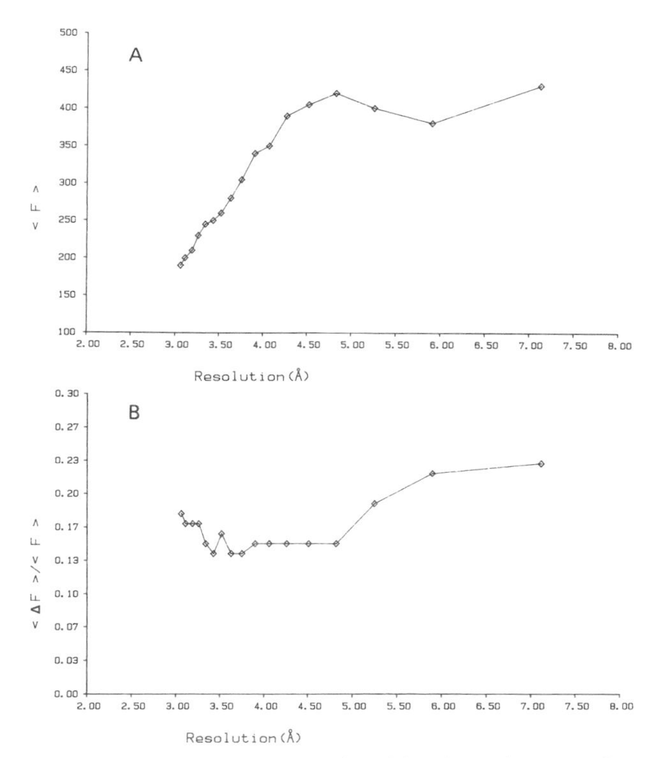

Fig. 2. (A) A plot showing the mean value of the observed structure factors as a function of resolution. (B) A plot showing, as a function of resolution, the ratio of the mean of the absolute values of the differences between native and Pt-derivative structure factors to the mean of the structure factors.

to a conventional full matrix-least squares refinement calculation with the platinum positions and occupancies treated as variables (program QKREF, W. Furey, unpublished).

The ISIR procedure was then used to resolve the phase ambiguity in the platinum SIR data to 3.5 Å; it was used again to extend the data to 3.2 Å and then 3.0 Å resolution, as reported earlier [25]. The average figure-of-merit at the beginning of the process was 0.33 for those 4038 reflections which had both the native and the derivative information and at the end of the process it was 0.84 for all 5880 observed reflections including those 1842 reflections for which the derivative information had been rejected. At this stage, the R-factor (discrepancy index) based on the observed and the map-inversion structure factor amplitudes was 18%. An electron density map, based on these phases was the basis of our earlier report [25].

Although this electron density map was very clear in most places and allowed us to trace the entire DNA double helix and most of the polypeptide backbone, there were a few regions where the electron density was not easily interpretable. We were not able to obtain a continuous, unique tracing of the entire polypeptide chain with convincing assignments of amino-acid side chains of the known sequence to those we could see in the electron density map. We did not invoke disordered regions within the polypeptide chain because Jen-Jacobson had shown that the DNA–*Eco* RI endonuclease complex was extraordinarily resistant to several proteases, including trypsin, chymotrypsin, V8 protease and endoprotease lys C (Jen-Jacobson *et al.*, manuscript in preparation). This striking protease resistance, which included all but 29 amino-acid residues at the amino terminus of the sequence, strongly suggested that the resistant 248 residues were well ordered in the complex. We therefore suspected that the ambiguities represented errors in the electron density map.

In hindsight, the difficulties centered primarily in two regions of the map: The β-hairpin which forms two of the three strands in the subsidiary anti-parallel sheet in the 'arm' (see discussion) was not clearly resolved in the initial map. (By definition, a β-hairpin consists of two antiparallel strands of β-sheet connected by a short turn.) A region surrounding a crystallographic threefold symmetry axis is very densely packed with protein, where three dimers form a tight complex. This region also, probably, contained some noise in the first map with the result that the molecular boundary was unclear in this localized area. In addition, several large ripples were present along some of the threefold axes located in the solvent region, indicating some error. In summary, the first map was difficult to interpret in a region of the molecule which projected out into the solvent and in a second region around the threefold axes where the protein molecules were densely packed.

A fraction of the data were missing from the original data sets and we suspected that the absence of this information was interfering with the ISIR procedure. Three factors led to the absence of data: First, a few reflections at very low resolution were obscured by the beam stop of our Arndt–Wonacott camera. Second, a few reflections were saturated even on the third film of the film packs and were deleted from the data sets by the computer programs we used to process the film data [27, 28]. Third, these programs also deleted a significant proportion of the weakly observed data because they were deemed statistically unreliable.

Efforts were then made to estimate the missing amplitudes and phases and incorporate these estimates in the electron density calculations. This was initiated for reflections within a 5 Å resolution limit and iterated for four cycles (all the observed data to 3.0 Å were used during this process). At each iteration, the structure factor amplitudes and phases were estimated for the missing reflections by Fourier inversion of the modified electron density map. These estimates were used in electron density calculations during the next cycle. At the end of the fourth cycle, a new solvent mask was calculated based on the 5880 originally observed reflections and the 298 estimates generated by this process. The process was repeated in a similar manner to estimate the missing reflections to 4.0 Å, then to 3.5 Å and finally to 3.0 Å. At the end of the process, the average figure-of-merit for the 5880 observed reflections increased from 0.84 to 0.87 and their R-factor dropped from 18 to 15%. This process produced 2394 estimated structure factor amplitudes and phases.

At this stage, an electron density map was calculated using all the observed and estimated reflections (8274 in total). This map showed considerable improvement over the original ISIR map based on the 5880 observed reflections only. This map however, still showed small ripples around some of the three-fold axes, although their magnitudes had been diminished considerably from the first map. These ripples around the threefold were finally levelled by recalculating a solvent mask using a 10 Å radius in the masking function instead of the 5.1 Å radius used earlier. After twelve cycles of iterations and two recalculations of the solvent mask, the final figure-of-merit and map inversion R-factor remained at 0.87 and 15%, respectively. The electron density based on the third set of phases exhibited a slight improvement in clarity and was used for the fitting of the chemical sequence of the enzyme as described below.

We compared the electron density maps which preceded and followed both of the 'extension' steps (phase extension from 3.5 to 3.0 Å using observed amplitudes as well as estimation of missing intensities) and concluded in both cases that the extensions reduced noise and improved the clarity of the maps while maintaining the basic features which were present in the initial 3.5 Å map. These features included the DNA (the phosphate positions were striking and immediately obvious features of all the maps), as well as several prominent α-helices and strands of β-sheet. The improvements in detail were most noticeable in the problematic regions described above, in some of the loops connecting secondary structure elements and in some of the side chains. They enabled us to distinguish possibilities which had been ambiguous before the extension.

The final electron density map was displayed on plexiglass sheets. The DNA and protein secondary structure elements were very clear, as noted above. Over two-thirds of the amino-acid side chains were distinctly visible along with main chain density for all but four amino-acid residues. The missing residues were in the immediate vicinity of the major heavy atom site, and it appears likely that their movement is associated with the small non-isomorphism noted previously. Almost all of the large hydrophobic side chains, tryptophans, phenylalanines and tyrosines were clearly recognizable. Many basic residues, especially arginines, which were located at the DNA–protein interface were also easily identifiable. Most of the poorly visualized side chains were located at the protein–solvent interface. Both the DNA–protein and protein subunit–subunit interfaces were well ordered and provided useful constraints when we assigned the known amino-acid sequence to the electron density map. These assignments were made via inspection of the electron density map, aided by model building, distance measurements and the known stereochemistry of proteins. This process led to a unique tracing of the polypeptide chain through the protein–DNA complex.

Coordinates for an α-carbon atom and for either β-carbon atom, or a terminal side chain atom for larger amino acids were taken from the ISIR map on plexiglass sheets and used to generate atomic coordinates for the entire molecule with the program FRODO [29–31]. Electron density fitting continued with FRODO on an Evans and Sutherland PS340 computer graphics system. The coordinates were regularized to approximately ideal geometry alternately with improving the fit to the electron density. At this point, the model has been fit to all of the electron density features noted above.

It should be noted that the structure reported here represents an intermediate stage of a complete crystallographic structure determination that ultimately will include extensive refinement, i.e. adjustment of the molecular model to optimize the fit between the experimentally measured diffraction pattern and that calculated from the model. It is highly improbable that these adjustments will alter the basic conclusions reported here, however the fine details of the current model should be considered preliminary results.

Description and discussion of the structure

General features of the complex

The co-crystalline asymmetric unit contains one protein subunit of 276 amino acid residues of known sequence [16, 17] and one strand of the oligonucleotide TCGCGAATTC GCG (*Eco* RI site underlined). The complex is a twofold symmetric dimer in which the protein–protein inter-subunit diad, the principal diad of the symmetric DNA double helix and the crystallographic twofold axis all coincide [26]. The molecular boundary, as seen in the Pt-ISIR electron density map, clearly encloses this complex in a well defined globular structure, 50 Å across, see Fig. 3.

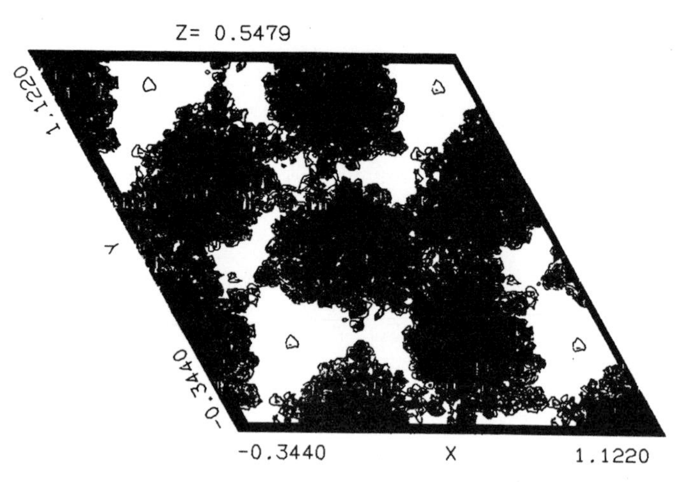

Fig. 3. A projection down the *c*-axis of the Pt-ISIR electron density map of the DNA–*Eco* RI endonuclease complex.

The DNA–protein complexes are packed within the crystalline lattice so that the DNA forms a continuous rod parallel to the *c*-axis. The unpaired 5′ thymine residues at each end of the double helix appear to be stacked upon each other leading to a continuous series of stacked bases across a crystallographic two-fold axis. (The space group is P321.) Therefore, although the oligonucleotide is thirteen nucleotides long there are fourteen stacked nucleotides per unit cell. While this is consistent with a prediction by Harrison that oligonucleotides 14 bp long should be particularly useful in the growth of DNA–protein co-crystals [8], it should be noted that Harrison's prediction was based on assumptions regarding the structure of the DNA which are violated in this case by the neo-kinking reported below (i.e. Harrison assumed B-DNA with 10.5 bp/turn). The oligonucleotide is actually somewhat larger than the DNA-binding face of the protein. However, its length almost exactly matches the net width of the protein, which tapers slightly at the DNA interface leaving a solvent gap separating the ends of the oligonucleotide from the neighboring protein.

There are three major areas of protein–protein interaction which, together with the DNA–DNA interaction, form the crystalline lattice. First is the subunit–subunit interface within the dimeric complex which contains the determinants of dimer formation. (*Eco* RI endonuclease forms highly stable dimers in solution both in the presence and absence of DNA [19, 20].) Second is the region around the threefold symmetry axis, where three dimers are tightly packed (see Fig. 3). Third is a smaller region of limited protein–protein interactions along the direction of the *c*-axis. These involve contacts between loops at the molecular surface of the protein. The DNA–DNA interaction comprises a significant fraction of the net intermolecular interactions along the *c*-axis. This confirms the concept that stability in DNA–protein co-crystals requires compatibility between the DNA–DNA, protein–protein and protein–DNA contacts in the direction of the average DNA

T C G C G A A T T C G C G
G C G C T T A A G C G C T

Fig. 4. The sequence of the synthetic oligonucleotide with the recognition sequence underlined. * Denotes the location of phosphodiester bond hydrolysis resulting in a 5′ phosphate. The hydrolysis reaction requires Mg^{2+} as a cofactor. ......, denotes the location of the type I neo-kink which is coincident with the crystallographic and molecular twofold symmetry axis. The type I neo-kink unwinds the DNA by 25° and introduces a bend between the two central blocks, GAA and TTC, of 12° toward the minor groove. Λ Denotes the location of the asymmetric type II neo-kink, which separates the terminal blocks from the central blocks of nucleotide pairs. This kink bends the helical axis by 23°.

helix axis and suggests that the length and terminal sequence of the cognate oligonucleotide should be treated as a critical variable in future attempts to form sequence-specific DNA–protein co-crystals.

The conformation of the DNA in this complex departs from classical B-DNA. There are three abrupt dislocations in the helix, termed neo-kinks (see below), which divide the DNA into four blocks of 3 bp each, as in Fig. 4. In comparison to the neo-kinks, the helical parameters within each block are relatively more regular. The recognition sequence is contained in the central two blocks of the structure and the flanking sequence, CGC/GCG, is contained in the terminal blocks.

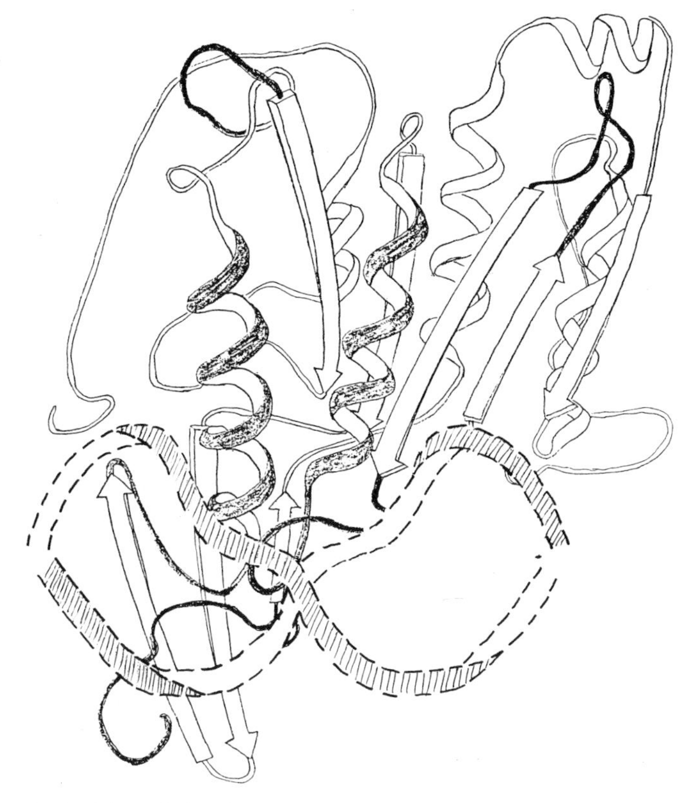

Fig. 5. Schematic backbone drawing of one subunit of (dimeric) *Eco* RI endonuclease and both strands of the DNA in the complex. The arrows represent β-strands, the coils represent α-helices and the ribbons represent the DNA backbone. The helices in the foreground of the diagram connect the third β-strand to the fourth and the fourth to the fifth. They also interface with the other subunit. The amino-terminus of the polypeptide chain is in the arm near the DNA.

Structural organization of the protein subunit

Eco RI endonuclease is an α/β protein consisting of a five-stranded β-sheet surrounded on both sides by α-helices (see Fig. 5). Four of the five strands are parallel, however the location of the single anti-parallel strand divides the sheet into parallel and anti-parallel three-stranded segments (see Fig. 6). Each of these segments forms a sizeable structural unit constructed on a simple three-dimensional pattern in which the physically adjacent elements of secondary structure are essentially contiguous within the primary sequence; i.e. a sub-domain. It is interesting that the parallel sub-domain is the locale for the direct contacts between the protein and DNA bases as well as subunit–subunit interaction while the antiparallel sub-domain contains the site of DNA strand scission. We also note that the parallel sub-domain is topologically very similar to one-half of the well-known nucleotide binding domain [32]. (The nucleotide binding domain is a six-stranded parallel β-sheet, constructed out of two three-stranded sub-domains, which are very similar to each other.)

Following the course of the polypeptide chain, we find the amino terminus of the polypeptide chain located in an extension of the principal α/β domain of the protein, referred to as the 'arm', which wraps around the DNA. The polypeptide chain then forms a long α-helix on the surface of the molecule which is followed by a loop into the first strand of the β-sheet. This β-sheet is formed sequentially starting from the outside of the anti-parallel sub-domain (see Fig. 6). The next loop, which connects the first and second β-strands, contains another α-helix situated on the surface of the molecule. The loop between the second and third (anti-parallel) β-strands projects somewhat into the solvent. The third β-strand is a common element of both the anti-parallel and parallel sub-domains. The parallel sub-domain is formed next, sequentially from the middle of the β-sheet to the fifth strand at the edge of the sheet. The α-helices found at the subunit interface are the crossover helices [33] of the parallel sub-domain, i.e. those connecting the third β-strand to the fourth and the fourth to the fifth. After exiting the fifth β-strand, the polypeptide chain loops around the surface of the complex, placing the carboxy terminus in the proximity of the DNA backbone.

As can be seen in Fig. 5, all the α-helices in the protein are aligned so that their amino terminal ends are pointing toward the DNA. This orients the α-helix dipoles [34] so that they

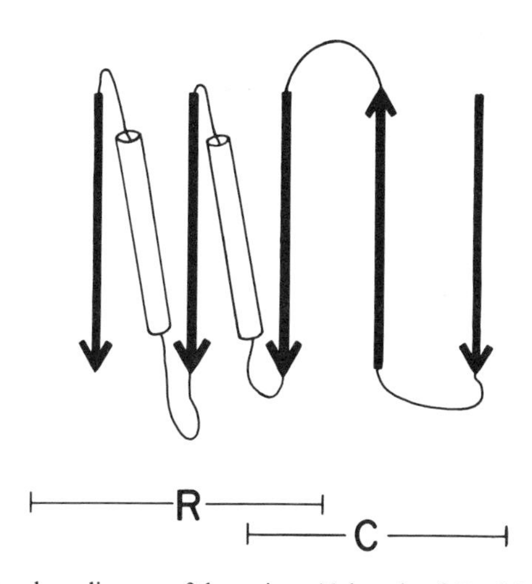

Fig. 6. Topology diagram of the major α/β domain of *Eco* RI endonuclease. The β-sheet is divided into two overlapping topological segments, the parallel and anti-parallel sub-domains which correspond to the functional division of the β-sheet into a sub-domain primarily responsible for recognition of the specific DNA sequence, R, and a sub-domain primarily involved in catalytic activity, C.

interact favorably with the electrostatic field generated by the negatively charged phosphates on the DNA backbone. The two crossover helices of the parallel sub-domains are actually oriented so that their amino terminal ends project into the major groove of the DNA. The amino-acid side chains which interact with the DNA bases are located at the ends of these helices.

The β-sheet exhibits the conventional twist [33, 35–39] with the individual β-strands approximately perpendicular to the DNA helical axis.

Eco RI *endonuclease possesses an arm which wraps around the DNA*

This 'arm' is an extension of the α/β domain (see Fig. 7) which wraps around the DNA partially encircling it, thereby clamping it into place on the surface of the enzyme. Due to the twofold symmetry of the complex there are two arms, each of which interacts with the DNA directly across the double-stranded helix from the scissile bond. Jen-Jacobson has demonstrated that these nonspecific contacts between the arm and the DNA are required for DNA cleavage by selective proteolysis in which portions of the arm are selectively removed (Jen-Jacobson *et al.*, manuscript in preparation). Many of the resulting 'deletion derivatives' retain sequence-specific DNA binding but lack strand scission capability.

The arm has a structural 'identity' of its own. It is composed of the amino terminus of the protein and a β-hairpin sequentially located between the fourth and fifth strands of the primary β-sheet. (A β-hairpin is a structure consisting of two anti-parallel β-strands connected by a short turn.) Part of the amino terminal portion of the arm adds a third β-strand so that the structural foundation of the arm is a three stranded anti-parallel β-sheet. Thus, there are two β-sheets in each *Eco* RI endonuclease subunit; the primary five-stranded sheet described above and the subsidiary three-stranded sheet described here. The first fourteen amino acid residues of the polypeptide chain form an irregular structure which is sandwiched in between the subsidiary β-sheet and the DNA; many of the non-specific DNA–protein contacts mediated by the arm are located here. Additional DNA backbone contacts are located in the short segment of polypeptide chain which connects the third subsidiary β-strand with the α-helix which follows it in the primary sequence (this α-helix is the 'outer recognition module' described below).

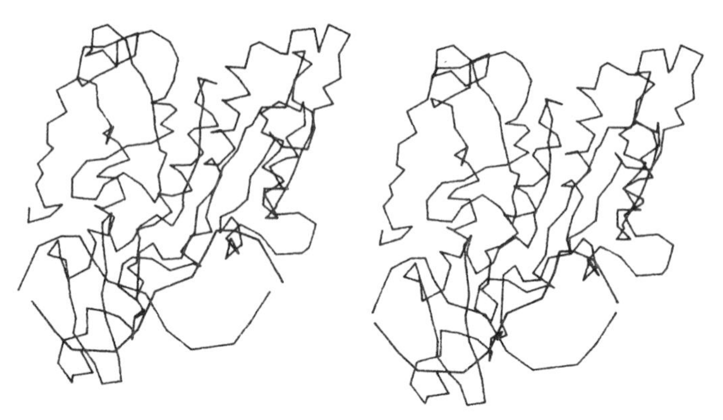

Fig. 7. Stereo drawing of the C_a coordinates of one subunit of *Eco* RI endonuclease and the phosphorous cordinates of the double-stranded DNA. This drawing was generated with the program FRODO.

Even though the arm has the structural features described above, it is not a domain, as defined by Richardson [33]. It does not appear to have a fully developed hydrophobic core and it is composed of two passes of the polypeptide chain, rather than a single chain segment. Indeed, it is doubtful that it could assemble or maintain its tertiary structure in the absence of the principal domain. It is therefore an extension on the principal domain, but one which has an important functional role.

Structural features of the DNA

We reported [25] that the DNA conformation in the recognition complex departs from the B-motif in a way which suggests that the DNA is now adopting conformations which would be unstable in the absence of protein. The most striking of these departures are kinks which occur every three base pairs, as summarized in Fig. 4.

The term 'kink' has not been used consistently in the literature: We have been following the usage originally conceived by Crick and Klug [40], who defined a kink as an abrupt change in helical properties. They specifically included not only sharp bends (abrupt changes in the direction of the helix axis) but also lateral displacements (slip dislocations) of the helix axis, highly localized under- or over-windings (torsional dislocations) and combinations of these elements. They proposed a specific model for the wrapping of DNA around nucleosomes which invoked severe kinks in which the base pairs became unstacked at the kink. Subsequently, the term kink has also been used to describe localized unstacking (see the review by Saenger [41] for several examples). This structure strongly suggests that highly localized changes in helical parameters are likely to be a significant feature of nucleic acid–protein interactions even though they may not involve actual unstacking of adjacent base pairs. A word is therefore required to describe any abrupt change in helical parameters and we feel that 'kink' is the most appropriate term. It should be noted that while there is no absolute discrimination between a 'sharp bend' and a 'kink' we would prefer to use the term 'bend' to refer to a structure in which helical parameters are changed smoothly over several base pairs and 'kink' to refer to one in which the parameters change abruptly and/or irregularly over one or two base pairs. In both cases, it is expected that the DNA (or RNA) be approximately helical on either side of the kink or bend.

It is now known that some DNA sequences are intrinsically bent, even in the absence of proteins [42] whereas the kinks we have noted in the DNA–*Eco* RI endonuclease complex only occur in the presence of protein (the oligonucleotide used in these co-crystals is virtually identical to that studied by Dickerson and colleagues [43–45], which was not kinked in their structures). The intrinsic bends and kinks are probably structurally different from those which require the binding of a specific protein and these two situations are certainly thermodynamically distinct. We feel that it is important to discriminate between these situations and refer to the intrinsically stable kinks and bends as such (or simply as kinks and bends). Those which require a specific binding protein are termed neo-kinks and neo-bends. Thus, Richmond *et al.* observed that the DNA in their nucleosome structure contained 'sharp bends' and/or possible kinks [46] which would therefore be neo-bends.

We feel that one of the more intriguing observations to emerge from the *Eco* RI endonuclease structure is that the repertoire of conformational states intrinsically accessible to DNA has been expanded to include these additional 'neo-conformations' which are stabilized by the binding of a protein. Specifically, we define a neo-conformation as a structural distortion which is imposed on the double helix by a binding protein and which is not seen in the absence of protein.* (This should not be taken to exclude the possibility that thermally transient fluctuations in DNA structure would include neo-conformations in the absence of protein. Indeed, fluctuations of this sort may well be important intermediates in the formation of DNA–protein complexes. However the bulk of DNA molecules at any instant would not be in a neo-conformation according to this definition unless they were bound to a protein.) We suspect that neo-conformations will be a general feature of many nucleic acid–protein interactions. We also suspect that there will be a finite number of well defined, structurally feasible neo-conformations, analogous to the currently known intrinsic elements of nucleic acid and protein secondary structure. Neo-conformations may also have a role in sequence specificity because some sequences may accept the distortion imposed by the protein more readily than other sequences.

The neo-kink which coincides with the crystallographic and molecular two-fold axis will be called a 'type I neo-kink' (referred to in the previous work [25] as a neo-1 kink). It is located between the two (symmetry equivalent) central blocks of DNA, i.e. the GAA and TTC blocks. (The blocks are referred to by the sequence along one strand of the DNA.) The type I neo-kink effectively rotates the entire GAA double helical block with respect to the TTC block. It is possible to imagine the kink as a relative twisting of the GAA and TTC blocks about the average helix axis so as to unwind the DNA by about 25°. This corresponds very well with measurements of the unwinding of supercoiled plasmid DNA in solution by Kim and co-workers [48] who obtained a value of 25° per endonuclease dimer bound in the absence of Mg^{2+}. The principal effect of this twisting motion is that the major groove becomes wider even though the bases are not unstacked across the neo-kink (however, there are significant displacements in the base-pair planes). The phosphate–phosphate distance across the major groove is increased by approximately 3.5 Å. The type I neo-kink also introduces a small bend of approximately 12° between the two central blocks toward the minor groove, away from the protein. The effect of a type I neo-kink on B-DNA can be seen in Fig. 8, where a single type I neo-kink was placed in between segments of DNA which have the helical parameters associated with 'standard' B-DNA, i.e. 10.3 residues/turn and 3.2 Å/residue.

The type II neo-kinks are different from the type I in several respects. A type II neo-kink is asymmetric: The helix axis bends by 23°, however, the net bend consists of rotations about two axes. The first would coincide with the helix diad between the CGC and GAA blocks in unkinked DNA (rotation about this axis is responsible for the asymmetry). The second axis is perpendicular to both the first rotation axis and the average helix axis (which generates a small bend toward the minor groove). The asymmetry of the first rotation

* This is based on the definitions of neo as 'in a new or different form or manner' and 'new chemical compound isomeric with or otherwise related to (such) a compound' [47].

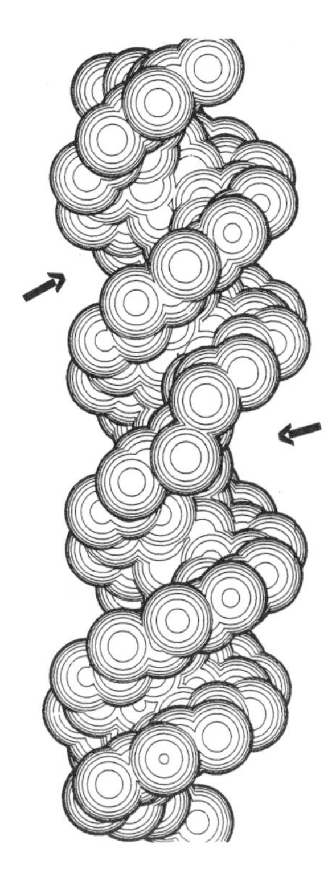

Fig. 8. A drawing of double-stranded B-DNA with a type I neo-kink inserted. The helical parameters used to generate the DNA double helix were the parameters determined for the central block, GAA, of the oligonucleotide bound to *Eco* RI endonuclease in the co-crystals. The arrows indicate the location of the major groove widened by the presence of the type I neo-kink and a standard width B-DNA major groove.

is indicated by the fact that the phosphate–phosphate distance on one side (CGC–GAA) of the neo-kink is significantly shorter than the comparable distance on the other side (TTC–GCG, see Fig. 4). The type II neo-kink does not appear to induce substantial unwinding of the DNA.

Helical symmetry was determined for each block of three base pairs formed by the kinks, by analysis of the geometrical relationships of the centroids of the phosphate and internal deoxyribose moieties [49]. The central blocks exhibited B-type parameters with 10.3 residues per turn and a helical displacement of 3.2 Å. The terminal blocks exhibit A-type parameters of 11.3 residues per turn and a helical displacement of 2.7 Å per residue. The 10.3 bp per turn helical repeat of the central blocks is within the values observed for DNA in solution, bound to crystalline surfaces and in nucleosomes [50–54]. The A-type parameters are reminiscent of the A-like features of the CGC segment in the dodecamer structure of the same sequence [43]. The tendency to wind the double helix with A-like parameters in the terminal blocks should introduce an additional unwinding on the order of 10° into a long DNA molecule (depending on how quickly the DNA reverted to B-type winding).

DNA–protein interactions

The major groove of the recognition hexanucleotide (GAATTC) appears to be filled with protein, forming a large, complementary interface. All the base–amino acid interactions appear to be located in the major groove of the DNA while the edges of the bases which are exposed in the minor groove are open to the solvent. Thus, the direct interaction of complementary surfaces in the major groove is a major determinant of the recognition specificity. However, there is the additional possibility that the energy required to drive DNA into the neo-conformations noted above depends on the base sequence. This would provide an additional, indirect mechanism of sequence recognition which may also contribute to *Eco* RI endonuclease specificity.

A crucial component of the DNA–protein interface is formed by the amino terminal ends of the crossover α-helices in the parallel sub-domain. Since the DNA–protein complex possesses two-fold symmetry, α-helices from both subunits participate in the formation of a four-helix bundle which inserts into the major groove of the DNA. One of the roles of the neo-kinks is to make room for this bundle, which would not fit into the major groove of conventional B-DNA. Both neo-kinks have this steric role of facilitating the DNA–protein interaction. The type I neo-kink is dramatic, significantly enlarging the major groove. The type II neo-kink also appears to expand the groove somewhat. However, it should be noted that the type II neo-kink is very close to the scissile bond and it may also have a role in the hydrolytic mechanism.

The α-helix which connects the third and fourth β-strands makes an angle of approximately 60° with the average DNA helix axis as shown in Fig. 9. The polypeptide chain turns sharply at the end of the α-helix so that the amino acid residues at the amino-terminal end of the helix, those in the bend and in the adjacent stretch of chain are in close proximity to the DNA. This α-helix is also adjacent to the molecular twofold axis and the amino terminus of the helix is physically adjacent to the amino terminus of the symmetry related helix from the other subunit. These two α-helices form a symmetric module which is responsible for the direct interactions between the endonuclease and the bases of the inner tetranucleotide (AATT). This structural unit will be referred to as the inner recognition module. Two symmetry related pairs of amino acid side chains make contact with two symmetry related pairs of sequential adenine residues. Interestingly, each pair of adjacent adenines interact with one amino acid from each subunit, see Fig. 10. In our current interpretation of the electron density map, these residues are glutamic acid 144 and arginine 145. The position of the arginine side chain in the current model is consistent with a hydrogen-bonding interaction with the N7 moieties of adjacent adenines while the glutamic acid is probably hydrogen bonding to the exocyclic N6 groups of both adenines.

Separate α-helical modules are responsible for the direct contacts between the protein and the two outermost GC pairs of the canonical hexanucleotide, GAATTC. These modules are called the outer recognition modules. They are identical by virtue of the two-fold symmetry and each independent module consists of the crossover α-helix which connects the fourth and fifth strands of the principal β-sheet. This helix has many interactions with both α-helices of the inner recognition module and is thereby positioned so that it projects its amino terminus into the major groove of the DNA. At this stage of our analysis, it appears likely that arginine 200 interacts with the guanine in the manner predicted by Seeman, Rosenberg and Rich [55].

In addition to the recognition of and precise requirement for the canonical site, GAATTC, *Eco* RI endonuclease also exhibits a dependence on the sequence of the nucleotides flanking this hexanucleotide [22, 56, 57]. These workers noted that the flanking sequence environment can effect the overall rate of hydrolysis by at least an order of magnitude. Furthermore, Modrich and co-workers showed that *Eco* RI endonuclease can dissociate from DNA after making only a single strand nick and that the frequency of such nicking depended on the flanking sequences [58]. It is unlikely that these effects are due to direct contacts between the enzyme and bases outside of the canonical hexanucleotide because we have not noted any such interactions in the electron-density maps. However, there are extensive contacts between the enzyme and the DNA backbone which extend well beyond the hexanucleotide. This suggests that the conformational free energy of the type II neo-kink and/or the A-like terminal block depends on the sequence of oligonucleotides immediately flanking the *Eco* RI site.

There appear to be interactions between the protein and

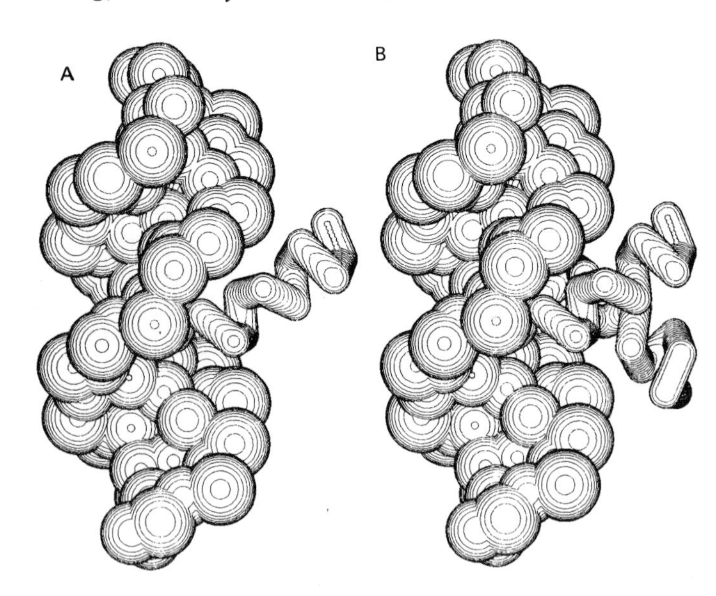

Fig. 9. (A) The double-stranded DNA in the *Eco* RI endonuclease–DNA complex and the α-helix which connects the third and fourth β-strands. (B) The same view as in (A) with the symmetry related α-helix from the other subunit included. The two α-helices form the inner module, which recognizes the inner tetranucleotide, AATT.

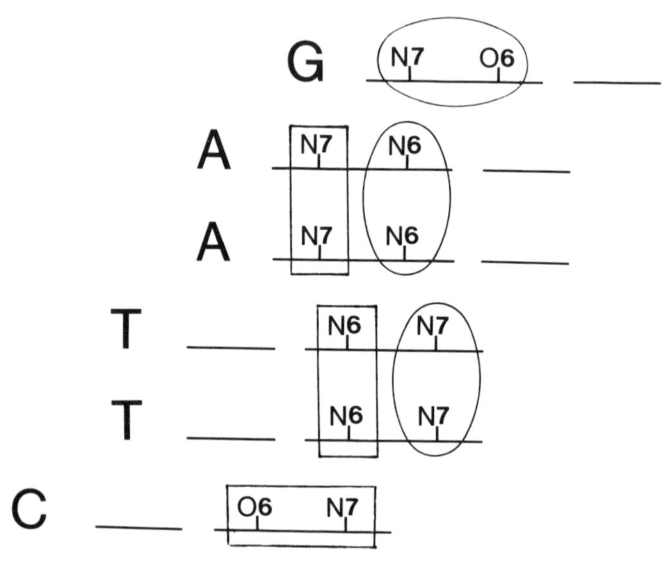

Fig. 10. A schematic drawing depicting the interactions between base pairs in the *Eco* RI recognition site and amino-acid side chains of *Eco* RI endonuclease. Rectangles denote interactions from one subunit and ovals denote interactions from the symmetry related subunit. The proposed interactions are arginine 200 hydrogen bonding to the N7 and O6 of the GC base pairs, arginine 145 hydrogen bonding to the two N7 moieties of adjacent adenines and glutamic acid 144 hydrogen bonding to the N6 groups of adjacent adenines.

the backbone of the DNA from the second through the ninth phosphates on each strand. Counting from the 5′ end of the oligonucleotide, TpCpGpCpGpApApTpTpCpGpCpGp, the third, fourth and seventh phosphates are buried in the protein, i.e. they appear to be inaccessible to solvent. The remaining phosphates in the indicated region interact with the enzyme even though they are partially exposed to solvent. The third, fourth and fifth phosphates are bound in a large cleft in the protein which forms the active site for DNA hydrolysis. This cleft is partially open to solvent in the vicinity of the fifth phosphate, where the scissile bond is located. It is through this channel that magnesium probably enters the active site.*

The cleft surface contains many basic amino acid residues which interact electrostatically with the phosphates. These interactions contribute to the overall stability of the complex. There are, of course, two identical clefts in the surface of the twofold symmetric enzyme. They are too far apart to fit regular B-DNA and we suspect that this separation promotes the formation of the type I neo-kink via long range electrostatic attraction between the basic clefts and phosphates on the incoming DNA molecule.

The contacts noted above in the X-ray structure of the tridecamer–protein complex probably contain all of the major DNA–protein contacts between the endonuclease and larger natural DNA substrates. The association constant measured for the dodecamer, CGCGAATTCGCG, is within experimental error of that measured for plasmid DNA [17, 20, 24, 59]. The lower association constant for an octanucleotide substrate as compared with dodecameric or larger substrates [20, 60] suggests strongly that interactions between the enzyme and the flanking regions of the DNA backbone make significant contributions to the net stability of the complex.

Inferred and observed contact points

There is good general agreement between the results of ethylation interference experiments [61] and the phosphates contacts described previously. The largest effects observed by Lu *et al.* correspond to the third, fourth and seventh phosphates, which as noted above are buried in the protein and protected from solvent. The next largest effect is observed for the reactive phosphate at the fifth position. Small effects are noted for the sixth phosphate, which is probably forming interactions to the protein even though it is partially exposed to the solvent. (We suspect that a stronger ethylation interference would have been observed at lower protein concentrations where the equilibrium is more sensitive to smaller reductions in the protein–DNA association constant.)

Methylation protection and interference experiments implicated both the major groove and the minor groove as points of DNA–protein contact. It is clear that the predicted contacts in the major groove of the recognition sequence is in good agreement with the X-ray structural data. The implications of the minor groove data must be reevaluated in light of the structural results. The N3 positions on the central adenine, in the minor groove, were protected from dimethyl-sulphate by endonuclease and prior methylation at the N3 blocked subsequent binding of the enzyme. Since there is no density observed in the minor groove of the DNA, and there are no sections of polypeptide chain left unaccounted for in the chain tracing of the protein, the observed effects at the N3 are probably related to the conformational changes induced in the DNA by protein binding.

Another approach used to identify contacts on the DNA is based on the hydrogen-bonding degeneracy rules of Seeman *et al.* [55, 62, 63]. In *Eco* RI endonuclease such degeneracy can be induced by conditions of elevated pH (8–9.5), Mn^{2+}, low ionic strength and by the addition of organic compounds such as glycerol or ethylene glycol [62, 64–68]. This has been called the *Eco* RI* reaction [64] and the non-canonical sites so recognized are referred to as *Eco* RI* sites. We previously showed that hierarchies of base preferences could be discerned by qualitative estimations of the rate of cleavage at *Eco* RI* sites in plasmid DNA molecules. These observed hierarchies are consistent with a recognition model which correctly predicted the recognition contacts on the DNA we reported here. The original model assumed two major groove hydrogen bonds per base pair under standard conditions [63], which turned out to be those identified above. An additional assumption was that one or more of these hydrogen bonds are randomly replaced by a water–DNA hydrogen bond under *Eco* RI* conditions. The position of any *Eco* RI* sequence within the appropriate hierarchy is determined by the number of remaining protein–DNA hydrogen bonds [63]. Woodbury *et al.* [62] reached similar conclusions from a somewhat different perspective. This method of 'degeneracy analysis' therefore appears to be a promising method for the elucidation of DNA–protein contact points.

Structural suggestions for conformational mobility

The formation of the *Eco* RI endonuclease–DNA complex requires conformational changes in both substrate DNA and the enzyme in order to achieve specific binding. The DNA adopts neo-kinks in the specific DNA–protein complex. The protein must also alter its conformation during the binding event because the arms encircle the DNA to such an extent that it appears unlikely that substrate DNA could enter the active site in the absence of some movement.

There are four general possibilities:

(1) The arms may be 'rigidly' mobile, retaining their structure while they move with respect to the rest of the molecule.

(2) The arms may have two stable structures, one in the presence and one in the absence of DNA. One detailed example of this relates to the fact that the amino terminal fourteen residues of the arm (which are sandwiched between the rest of the arm and the DNA) appear to be rather loosely associated with the β-hairpin; suggesting that these residues fold against the DNA when it is present and refold in a tighter association with the β-hairpin when DNA is absent.

(3) Part of the arms (probably that consisting of the amino terminal fourteen residues) may undergo an order–disorder transition in which they are disordered in the absence of DNA and condense on it during complex formation.

(4) The dimeric endonuclease could undergo a quaternary conformational change in which the subunits move with respect to each other.

These possibilities are not all mutually exclusive and the

* We have recently demonstrated that Mg^{2+} can be perfused into the crystals and the hydrolytic reaction carried out *in situ* (J. Picone, manuscript in preparation). The enzyme-product co-crystals still diffract X-rays and their structure analysis is in progress.

actual changes probably involves a combination of several of these factors. We have also grown crystals of the protein in the absence of DNA and that structure is in progress.

There is probably an intermediate conformational state, namely that of the non-specific DNA–protein complex, which presumably forms when *Eco* RI endonuclease binds non-specifically to DNA of random sequence. It has been suggested that this binding accelerates formation of the specific complex via facilitated diffusion of the protein along the DNA [69].

Another aspect of the conformational mobility of the complex may be related to the fact that non-specific DNA is not hydrolyzed under physiological conditions. Indeed, there is very strong selective pressure to prevent cleavage at sites other than the canonical *Eco* RI site. They would be lethal because such sites would not be protected by the *Eco* RI modification methylase. One possible mechanism to achieve this could involve linked conformational changes involving both the recognition and cleavage sites such that a functional active site does not form until the correct three dimensional arrangement of residues has already occurred in the recognition region. (An alternative statement of this possibility is that the presence of an incorrect sequence in the recognition site would result in 'distortions' which upon propagation into the cleavage site would then be render the latter non-active.) The overlap between the parallel and anti-parallel sub-domains represents one possible route for this structural communication.

Conclusions

Eco RI endonuclease specifically binds the canonical sequence, GAATTC, through DNA–protein interactions in the major groove of the DNA. The minor groove of the canonical sequence is not directly involved in sequence specificity. Upon binding to the endonuclease, the DNA adopts new (neo) conformational states not previously seen in protein-free DNA. We have defined a neo-conformation as a structural distortion which is imposed on DNA by a binding protein and which is not seen in the absence of protein. Two neo-kinks have been observed in this structure, the type I neo-kink and the type II neo-kink. The type I neo-kink unwinds the DNA by 25° and renders the major groove accessible to the protein. The type II neo-kink's major effect is to bend the DNA which also is required for protein accessibility and may have a role in the hydrolytic activity of the protein.

Eco RI endonuclease is a dimer with identical, symmetry related subunits. Each subunit is an α/β domain with an extension called the 'arm'. The α/β domain is organized into topological sub-domains which have identifiable functional roles. The 3-stranded parallel sub-domain of the β-sheet is associated with sequence recognition and the subunit interface. The 3-stranded anti-parallel sub-domain of the β-sheet is associated with phosphodiester bond cleavage. The two segments overlap to form a 5-stranded β-sheet. The 'arms' which embrace the DNA and clamp it into place are based on a subsidiary three-stranded anti-parallel sheet.

DNA sequence recognition is broken down into modular elements based on the crossover α-helices in the α/β domains. The inner recognition module which recognizes the sequence AATT consists of two symmetry related α-helices, one from each subunit. This inner recognition module appears to hydrogen bond with the adenines in the four adjacent AT base pairs. The outer base pairs at either end of the canonical sequence are recognized by the two outer recognition modules. Each outer module consists of one α-helix which appears to hydrogen bond to the guanines.

References

1. Anderson, W. F., Ohlendorf, D. H., Takeda, Y. and Matthews, B. W., *Nature (London)* **290**, 754 (1981).
2. Anderson, W. F., Takeda, Y., Ohlendorf, D. H. and Matthews, B. W., *J. Mol. Biol.* **159**, 745 (1982).
3. Pabo, C. O. and Lewis, M., *Nature (London)* **298**, 443 (1982).
4. McKay, D. B. and Steitz, T. A., *Nature (London)* **290**, 744 (1981).
5. Steitz, T. A., Ohlendorf, D. H., McKay, D. B., Anderson, W. F. and Matthews, B. W., *Proc. Natl. Acad. Sci. USA* **79**, 3097 (1982).
6. Ohlendorf, D. H., Anderson, W. F., Fisher, R. G., Takeda, Y. and Matthews, B. W., *Nature (London)* **298**, 718 (1982).
7. Sauer, R. T., Yocum, R. R., Doolittle, R. F., Lewis, M. and Pabo, C. O., *Nature (London)* **298**, 447 (1982).
8. Anderson, J., Ptashne, M. and Harrison, S. C., *Proc. Natl. Acad. Sci. USA* **81**, 1307 (1984).
9. Anderson, J. E., Ptashne, M. and Harrison, S. C., *Nature (London)* **316**, 596 (1985).
10. Bushman, F. D., Anderson, J. E., Harrison, S. C. and Ptashne, M., *Nature (London)* **316**, 651 (1985).
11. Ollis, D. L., Brick, P., Hamlin, R., Xuong, N. G. and Steitz, T. A., *Nature (London)* **313**, 762 (1985).
12. Richmond, T. J., Finch, J. T., Rushton, B., Rhodes, D. and Klug, A., *Nature (London)* **311**, 532 (1984).
13. Bentley, G. A. and Lewit-Bentley, A., *J. Mol. Biol.* **176**, 55 (1984).
14. Burlingame, R. W., Love, W. E., Wang, B.-C., Hamlin, R., Xuong, N. H. and Moudrianankis, E. N., *Science* **228**, 546 (1985).
15. Tanaka, I., Appelt, K., Dij, K. L., White, S. W. and Wilson, K. S., *Nature (London)* **310**, 376 (1984).
16. Greene, P. J., Gupta, M., Boyer, H. W., Brown, W. E. and Rosenberg, J. M., *J. Biol. Chem.* **256**, 2143 (1981).
17. Newman, A. K., Rubin, R. A., Kim, S.-H. and Modrich, P., *J. Biol. Chem.* **256**, 2131 (1981).
18. Connolly, B. A., Eckstein, F. and Pingoud, A., *J. Biol. Chem.* **259**, 10760 (1984).
19. Modrich, P. and Zabel, D., *J. Biol. Chem.* **251**, 5866 (1976).
20. Jen-Jacobson, L., Kurpiewski, M., Lesser, D., Grable, J., Boyer, H. W., Rosenberg, J. M. and Greene, P. J., *J. Biol. Chem.* **258**, 14638 (1983).
21. Modrich, P., *Quart. Rev. Biophys.* **12**, 315 (1979).
22. Halford, S. E. and Johnson, N. P., *Biochem. J.* **191**, 593 (1980).
23. Rosenberg, J. M., Boyer, H. W. and Greene, P. J., in *Gene Amplification and Analysis*, vol. 1: *Restriction Endonucleases*, pp. 131–164 (ed. J. G. Chirikjian). Elsevier/North-Holland (1981).
24. Jack, W. E., Rubin, R. A., Newman, A. and Modrich, P., in *Gene Amplification and Analysis*, vol. 1: *Restriction Endonucleases*, pp. 165–179 (ed. J. G. Chirikjian). Elsevier/North-Holland (1981).
25. Frederick, C. A., Grable, J., Melia, M., Samudzi, C., Jen-Jacobson, L., Wang, B.-C., Greene, P. J., Boyer, H. W. and Rosenberg, J. M., *Nature* **309**, 327 (1984).
26. Grable, J., Frederick, C. A., Samudzi, C., Jen-Jacobson, L., Lesser, D., Greene, P. J., Boyer, H. W., Itakura, K. and Rosenberg, J. M., *Journal of Biomolecular Structure and Dynamics* **1**, 1149 (1984).
27. Rossmann, M. G., *J. Appl. Crystallogr.* **12**, 225 (1979).
28. Rossmann, M. G., Leslie, A. G. W., Abdel-Meguid, S. S. and Tsukihara, T., *J. Appl. Crystallogr.* **12**, 570 (1979).
29. Jones, T. A., *J. Appl. Crystallogr.* **11**, 268 (1978).
30. Jones, T. A., in *Computational Crystallography* (ed. D. Sayre), pp. 303–317. Clarendon Press (1982).
31. Pflugrath, J. W., Saper, M. A. and Quiocho, F. A., in *Papers presented at the International Summer School on Crystallographic Computing, Kyoto Japan* (1983).
32. Rossmann, M. G., Liljas, A., Branden, C.-I. and Banaszak, L. J., in *The Enzymes* (ed. P. Boyer), Vol. 11, pp. 61–102 (1975).
33. Richardson, J. S., *Adv. in Protein Chem.* **34**, 167 (1981).
34. Hol, W. G. S., *Prog. Biophys. Molec. Biol.* **45**, 149 (1985).
35. Chothia, C., *J. Mol. Biol.* **75**, 295 (1973).
36. Quiocho, F. A., Gilliland, G. L. and Philligs, G. N., *J. Biol. Chem.* **252**, 5142 (1977).

37. Shaw, P. S. and Muirhead, H., *J. Mol. Biol.* **109**, 475 (1977).
38. Weatherford, D. W. and Salemme, F. R., *Proc. Natl. Acad. Sci. USA* **76**, 19 (1979).
39. Schulz, G. E., Elzinga, M., Marx, F. and Schirmer, R. H., *Nature* **250**, 120 (1974).
40. Crick, F. H. C. and Klug, A., *Nature* (*London*) **255**, 530 (1975).
41. Saenger, W., *Principles of Nucleic Acid Structure*. Springer-Verlag (1984).
42. Wu, H. and Crothers, D. M., *Nature* (*London*) **308**, 509 (1984).
43. Dickerson, R. E. and Drew, H. R., *J. Mol. Biol.* **149**, 761 (1981).
44. Dickerson, R. E., *J. Mol. Biol.* **166**, 419 (1983).
45. Dickerson, R. E. and Drew, H. R., *Proc. Natl. Acad. Sci. USA* **78**, 7318 (1981).
46. Richmond, T. J., Finch, J. T., Rushton, B., Rhodes, D. and Klug, A., *Nature* **311**, 532 (1985).
47. Gove, P. B., *Webster's Seventh New Collegiate Dictionary*. G. & C. Merrian Co. (1963).
48. Kim, R., Modrich, P. and Kim, S.-H., *Nucl. Acids. Res.* **12**, 7285 (1984).
49. Rosenberg, J. M., Seeman, N. C., Day, R. O. and Rich, A., *Biochem. Biophys. Res. Commun.* **69**, 979 (1976).
50. Wang, J. C., *Proc. Natl. Acad. Sci. USA* **76**, 200 (1979).
51. Peck, L. J. and Wang, J. C., *Nature* (*London*) **292**, 375 (1981).
52. Rhodes, D. and Klug, A., *Nature* (*London*) **292**, 378 (1981).
53. Rhodes, D. and Klug, A., *Nature* (*London*) **286**, 573 (1980).
54. Lutter, L. C., *Nucl. Acids. Res.* **6**, 41 (1979).
55. Seeman, N. C., Rosenberg, J. M. and Rich, A., *Proc. Natl. Acad. Sci. USA* **73**, 804 (1976).
56. Thomas, M. and Davis, R. W., *J. Mol. Biol.* **91**, 315 (1975).
57. Alves, J., Pingoud, A., Haupt, W., Langowski, J., Peters, F., Maass, G. and Wolff, C., *Eur. J. Biochem.* **140**, 83 (1984).
58. Jack, W. E., Terry, B. J. and Modrich, P., *Proc. Natl. Acad. Sci. USA* **79**, 4010 (1982).
59. Lillehaug, J. R., Kleppe, R. K. and Kleppe, K., *Biochemistry* **15**, 1858 (1976).
60. Greene, P. J., Poonian, M. S., Nussbaum, A. L., Tobias, L., Garfin, D. E., Boyer, H. W. and Goodman, H. M., *J. Mol. Biol.* **99**, 237 (1975).
61. Lu, A.-L., Jack, W. E. and Modirch, P., *J. Biol. Chem.* **256**, 13200 (1981).
62. Woodbury, C. P., Jr., Hagenbuchle, O. and von Hippel, P. H., *J. Biol. Chem.* **255**, 11534 (1980).
63. Rosenberg, J. M. and Greene, P. J., *DNA* **1**, 117 (1982).
64. Polisky, B., Greene, P., Garfin, D. E., McCarthy, B. J., Goodman, H. M. and Boyer, H. W., *Proc. Natl. Acad. Sci. USA* **72**, 3310 (1975).
65. Hsu, M. and Berg, P., *Biochemistry* **17**, 131 (1978).
66. Malyguine, E., Vannier, P. and Yot, P., *Gene* **8**, 163 (1980).
67. Woodhead, J. L., Bhave, N. and Malcolm, A. D. B., *Eur. J. Biochem.* **115**, 293 (1981).
68. Gardner, R. C., Howarth, A. J., Messing, J. and Shepherd, R. J., *DNA* **1**, 109 (1982).
69. Terry, B. J., Jack, W. E. and Modrich, P., *J. Biol. Chem.* **260**, 13130 (1985).

Chemica Scripta 1986, **26B**, 159–163

The Balbiani Ring Gene Family – An Example of Satellite-like Evolution of Coding Sequences

L. Wieslander, C. Höög, U. Lendahl and B. Daneholt

Department of Medical Cell Genetics, Karolinska Institutet, Box 60400, S-104 01 Stockholm, Sweden

Paper presented by Lars Wieslander at the Conference on 'Molecular Evolution of Life', Lidingö, Sweden, 8–12 September 1985

Abstract

The large Balbiani ring (BR) genes in *Chironomus tentans*, BR 1, BR 2 and BR 6 code for secretory proteins which together form a supramolecular structure, the larval tube. The genes have internally highly repetitive structures similar to many satellite DNAs. The sequence similarities between the various BR genes show that they are related and the hierarchic repetitive structures allow analysis of events in the evolution of the BR gene family starting from a common short ancestor sequence.

Recent evolutionary changes in the BR genes, deduced from comparisons between repeat units within a BR gene and between the corresponding genes in closely related *Chironomus* species, are of the same type as those believed to have been involved in the evolution of the BR gene family as a whole. The observed changes suggest that the BR genes acquire mutations, mainly of a segmental nature at an exceptionally high rate due to their repetitive structure.

Introduction

Sequence duplications have been an important factor in the evolution of genomes. The most conspicuous examples are the highly repetitive, non-coding, satellite sequences in eukaryotes [1]. Duplications are evident also in coding DNA and manifest themselves in principally two ways. Firstly, entire genes can be duplicated which forms the basis for the evolution of gene families. Secondly, individual genes can often be shown to be built entirely or partly from smaller building blocks, which have duplicated to form the present day genes. Such internally repetitious genes are found in a variety of organisms and code for proteins with quite different functions, e.g. immunoglobulin [2], alfa-fetoprotein [3], LDL-receptor [4] and transcription factor IIIA [5]. The degree of repetition is often most pronounced in genes coding for proteins with structural functions, such as the collagen [6], silk fibroin [7], glue protein [8] and Balbiani ring (BR) genes [9]. To study the evolution of the eukaryotic genome, we have chosen the BR gene family as experimental material. These genes are internally highly repetitive in an hierarchic fashion, much in the same way as some non-coding satellite DNAs. By studying the hierarchic internal repetition of the BR genes we hope to better understand the amplification processes which have led to the evolution of the coding BR genes. Furthermore, by comparing the genes in the BR gene family, whose protein products all contribute to the same supramolecular structure (see below), we can evaluate the functional demands put on the repetitive sequences. Finally, comparisons of the repeated sequences within one gene and between corresponding genes in closely related species, provide information on the changes that can occur and are accepted in the coding repeated sequences during evolution.

BR genes and their function

The BR genes, which all code for giant secretory proteins, are found in the dipteran genus *Chironomus* and are only transcribed in the salivary gland cells. The number and chromosomal location of the BR genes may vary from species to species, but here we will describe the situation in *C. tentans*.

The BR genes are located on chromosome IV in three large puffs, BR 1, BR 2 and BR 3. One additional BR locus, BR 6, is present on chromosome III. The BR 6 gene is transcribed at a very low level during standard culturing conditions but may be induced to attain similar high transcription rates as seen in the BR 1 and BR 2 genes. The BR 1, BR 2 and BR 6 genes all produce similarly large RNA transcripts, about 37 kb in size [10]. The BR 2 locus most likely harbour two BR 2 genes so that altogether at least four different 37 kb sized RNA transcripts exist. The product of the BR 3 gene has not yet been characterized.

The primary, 37 kb transcripts are packaged in RNP complexes of defined structure at the gene level and are transported through the nuclear pores into the cytoplasm [11]. Here they end up in huge polysomes and are translated into secretory proteins, Sp I a, b, c and d, which all have molecular weights in the range of $8–10 \times 10^5$ Da [12–15]. The proteins are modified by glycosylation and extensive phosphorylation of serine and threonine residues [12, 16] and are secreted into the salivary gland lumen. Here they remain water soluble but upon excretion through the gland duct the proteins polymerize into water insoluble fibres which are spun into a tube-like structure. This tube serves as protection and as feeding device for the aquatic larvae [17].

Structure of the BR genes

The BR genes share a common architecture as deduced from studies of cloned cDNA molecules. These were obtained by random priming on purified 75 S RNA transcripts using calf DNA oligonucleotides. The cDNA molecules were used for sequence determination and as probes in various hybridization experiments. From these results we have determined the structure of a considerable part of the BR 1, BR 2 and BR 6

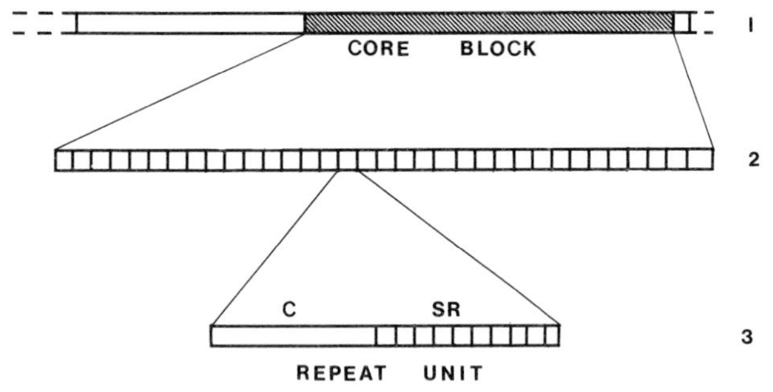

Fig. 1. Schematic representation of the structure of a BR gene. On top (1) a complete BR gene, about 37 kb in length is depicted. A large part, 15–22 kb (shaded area) consists of tandemly arranged repeat units and is called the core block. This block probably extends to the 3′-end of the gene while the 5′-end of the gene is not yet characterized (white area). In the middle (2) is shown the tandem arrangement of the 150–300 bp long repeat units in the core block. Each repeat unit (3) has a constant (C) region in the 5′-half and a 3′-half consisting of tandemly arranged subrepeats, the SR region.

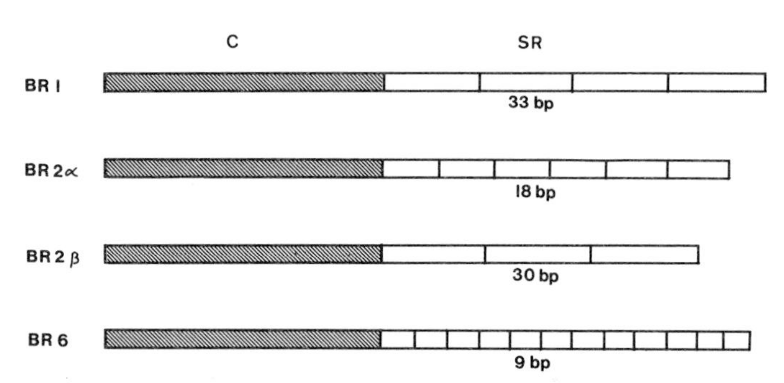

Fig. 2. Schematic representation of the repeat units specific for each BR gene. The repeat units all have homologous C regions, differing by bp substitutions. The SR regions differ in number, length and sequence of the subrepeats.

genes. Figure 1 shows the principle structure of a BR gene. The gene contains a 15–22 kb-long uninterrupted core block. This block is built from 150–300 bp building blocks, the repeat units, which are arranged in tandem. In one core block the repeat units are almost identical, i.e. the core block is highly homogeneous. Each repeat unit displays a characteristic substructure and can be divided in two roughly equally long regions. The 5′-part, the C region (C for constant) is characterized by the precise positioning of four cysteine codons and is not obviously internally repetitious. The 3′-part, the SR region (SR for subrepeat), consists of a number of tandemly arranged subrepeats.

In Figure 1 we have placed the core block towards the 3′ end of the gene based on the known structure of the 3′ end of the BR 1 gene. From the estimated size of a whole BR gene, about 37 kb, and the size of a core block it is evident that a large part of the BR genes have not yet been determined.

BR genes constitute a gene family

The BR genes studied in *C. tentans* are related to each other. They show the same overall structure in that they all contain a large core block. The genes differ because each gene has a characteristic repeat unit. We have thus determined the structure of the BR 1 [18, 19], BR 2α [9], BR 2β [20, 21] and BR 6 [22] repeat units (Fig. 2). Sequence comparison between the repeat units of the various BR genes reveal that the C regions are very similar between the genes. They differ only through bp substitutions and at the nucleotide level the percent homology is between 49 and 81% with the BR 6 C region being the most divergent sequence. At the nucleotide level the SR regions seem less homologous, differing both in sequence, number and length of the subrepeats. Upon close examination similar 9 bp elements can however be detected.

At the amino-acid level the homologies between the repeat units are even more striking. In the C-regions all the four cysteine residues as well as the single methionine residue and an arginine–phenylalanine pair are conserved throughout the genes (Fig. 3). Many of the amino-acid substitutions are also conservative to their nature.

In the SR regions conserved properties can also be recognized. All SR regions are rich in charged residues and proline residues, and a common reoccurring combination is a proline residue with a negatively charged residue on one side and a positively charged residue on the other side (Fig. 4). In this respect it should be noted that in the BR-encoded proteins the serine and threonine residues are phosphorylated to a large extent [12, 16]. Assuming that the serine residues are phosphorylated the similarity between the various SR regions is obvious.

The conserved properties in the repeat units provide clues to the functional constraints on these sequences. The three dimensional protein structures of the BR proteins are unknown. However, it is conceivable that the conserved properties of the C and SR regions contribute to the polymerization process of the BR proteins [23, 24]. The cysteine residues could then form disulfide bridges between different proteins. In the SR regions the proline residues could stabilize the folded structure in such a way that the negatively and positively charged residues are distributed on opposite sides of the folded SR region. Such a distribution could then contribute to the interaction between different proteins through electrostatic interactions.

Evolutionary model

The hierarchic repetitive structure of the BR genes implies that the genes have evolved principally through cycles of duplications and divergence [9]. Furthermore, the similarities between the different BR genes suggest that they share a common origin. This is particularly evident for the C regions which only differ by bp substitutions. A closer approximation

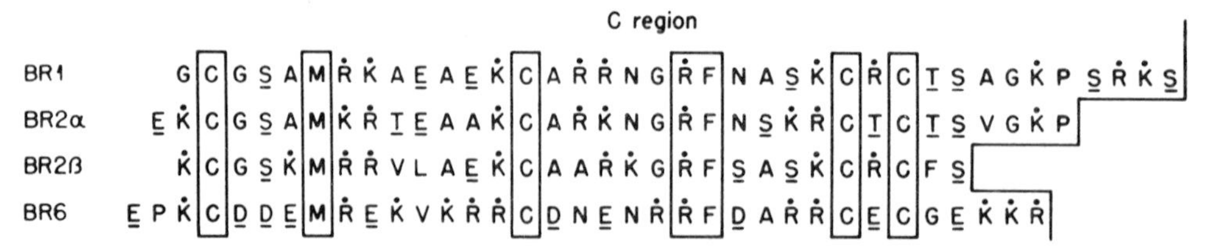

Fig. 3. Detailed comparison between the different BR gene C regions. The amino-acid sequences is given in one-letter code. Completely conserved amino-acid residues are boxed. Positively charged residues are indicated by a dot and negatively charged residues are underlined. The serine and threonine residues are assumed to be phosphorylated.

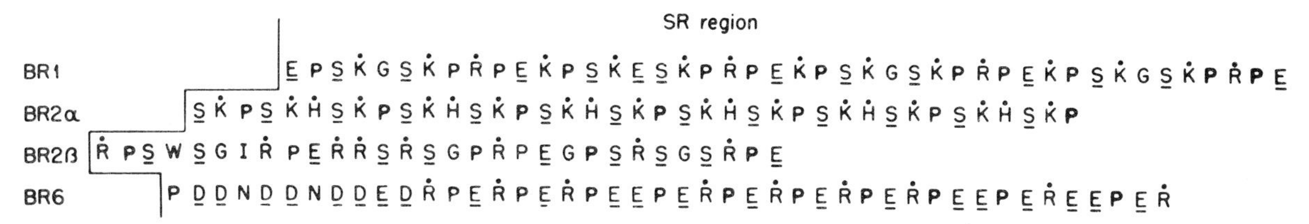

Fig. 4. Detailed comparison between the different BR gene SR regions. The amino-acid sequences is given in one-letter code. Positively charged residues are indicated by a dot and negatively charged residues are underlined. The serine and threonine residues are assumed to be phosphorylated.

to a postulated original C region can thus be derived as the concensus sequence. Furthermore, this concensus sequence shows distantly related internal elements and can therefore be divided into three similar regions, which aligns three of the four cysteine residues [25]. It can therefore be assumed that one such element, 39 bp long, was the ancestor sequence for the C region.

The SR regions contain shorter sequence elements probably because they have partly evolved by reduplications. However, their structure is often more complex suggesting that deletions and substitutions have occurred as well. Thus a more complex evolution than for the C regions is likely. The similarities between the various types of SR regions indicate that these regions also share a common origin, and a concensus sequence from which all known SR regions could have evolved can be constructed [25]. This ancestor sequence is 33 bp long and a pleasing fact is that it actually shows distant homology to the proposed ancestor sequence for the C regions.

Based on the considerations presented above, the evolutionary model shown in Fig. 5 can be proposed. The model starts with a 39 bp ancestor sequence encoding a hydrophilic, cysteine-containing peptide, that could be crosslinked to form a primitive tube, insoluble in water. Tandem reduplications gave rise to a four unit gene producing a longer peptide with the same functions. Various mutations and selection produced the putative concensus gene with a C region and a primitive SR region. Reduplications then led to a multiunit gene before non-tandem duplications gave rise to separate BR chromosomal loci. The evolution of the different types of repeat units then took place within such multiunit structures. We favor this order of events because we have, at the very end of a BR 1 gene core block, found a few C regions which are different from the ones inside that core block. The various SR regions were formed by internal reduplications, deletions and substitutions: the kinds of events that can be envisaged are shown in Fig. 6. The initial events may have been internal reduplications leading to additional segmental mutations. Such mutations are known to occur at higher frequencies in regions with short tandem repeats through, e.g. mistakes in DNA replication [26, 27]. The C regions on the other hand stayed intact because functional constraints did not allow changes.

The BR gene blocks are essentially homogenous throughout their lengths. The evolutionary model therefore require that the retained mutations have been spread by homogenization mechanisms. Such an effect would be the result of unequal-crossing over and/or gene conversion events.

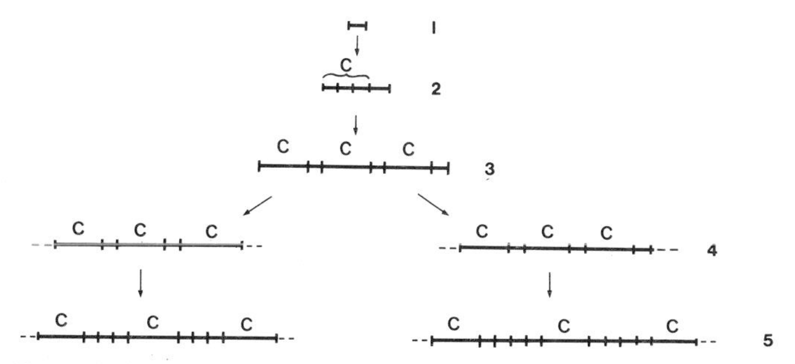

Fig. 5. Model for the evolution of the BR gene family. The model is explained in the text. Briefly, the model starts with a 39 bp ancestor sequence (1). Through tandem reduplications a four-unit gene was formed (2). Various mutations, selection and fixation produced a primitive repeat unit with a C region and a short SR region. Tandem reduplications led to a multiunit structure (3), before duplications and translocations resulted in several separate BR gene loci (4). Segmental mutations, bp substitutions and homogenization mechanisms then formed the different SR regions and produced the homogenous core blocks (5).

Recent evolutionary events

We find support for several of the mechanisms suggested in the evolution model when the repeat units *within* a given BR gene are compared. These mechanisms thus most likely still operate and the BR genes therefore seem to be in a dynamic state in which rapid changes may occur within the limits of preserving the protein function.

The SR regions seem to change at a higher rate than the C regions, which can be demonstrated very clearly in the BR 2β core block [28]. In this block the length of the repeat units may differ between different repeat units. We have detected six such length variants and sequenced four of them. It then turns out that the length variation is due to different numbers of complete subrepeats, varying from two to six which we interpret as the result of *duplication events*. The most common repeat unit variant amounting to about 50% of all the BR 2 repeat units, differ from this pattern in the sense that it has a length corresponding to 3.5 subrepeats. From the determined sequence we conclude that this is the result of a 15 bp *deletion* from within one subrepeat. In the SR regions we also detect a substantial number of *bp substitutions*. In the BR 2β SR regions the majority of these are silent or

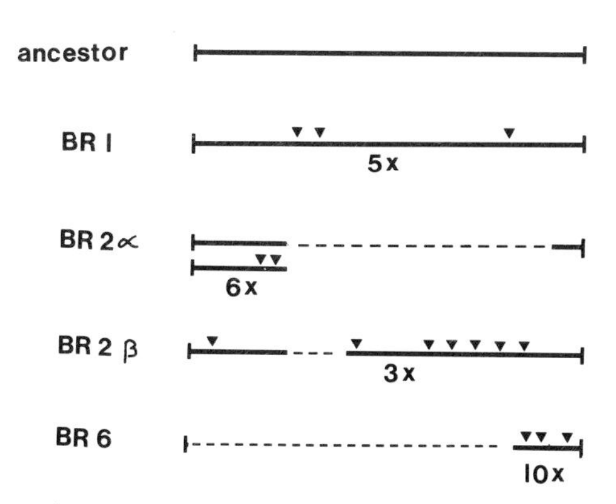

Fig. 6. Schematic representation of the evolution of different SR regions from a common ancestor. The different types of SR regions could have been derived from a common ancestor sequence through bp substitutions (▼), deletions (---) and varying numbers of tandem reduplications ($n \times$).

conservative substitutions but a fair amount leads to amino acid replacements. The described variations in the SR regions are so far best studied in the BR 2β block but are also found in SR regions in the other BR gene core blocks to various extent.

We can also find evidence for *homogenization* mechanisms in the present day BR genes. By comparing whole repeat units and individual subrepeats within one SR region it seems as if the homogenization mechanisms favor the correction of whole repeat units. The same bp substitutions, several of which occur in silent positions, are seen in many repeat units within a core block. Also individual subrepeats within an SR region show the same features but to less extent. The homogenization mechanisms furthermore show a distance effect, adjacent repeat units are often identical or at least more similar than are repeat units further apart.

In order to further study recent evolutionary events, we have specifically compared the corresponding genes in two sibling species, *C. tentans* and *C. pallidivittatus*. These two sibling species are morphologically almost indistinguishable and differ at the cytological level only by a number of chromosomal inversions [29]. The two species live in the same areas but no hybrids are found in nature although they can be crossed in the laboratory and give rise to viable offspring. We have compared the known BR repeat unit sequences in the two sibling species by Southern blot hybridizations and when available by direct comparison of determined sequences. Our results show that the BR 2α, BR 2β and BR 6 repeat units differ between the two species merely by a few bp substitutions as would be expected for two closely related species. In the BR 1 locus however, we record substantial differences which must have occurred after separation of the two species, i.e. during an evolutionarily short time span.

The first difference is quite drastic: a 16 kb core block in *C. pallidivittatus* is completely missing in *C. tentans* as judged from hybridization experiments [30]. It is possible that an original core block present at the separation of the species has diverged rapidly to form two non-homologous core blocks in the two species, different enough to exclude cross-hybridization. If this is the case however, we would have to assume that such mutation and homogenization processes have taken place in this particular BR gene core block at much higher rates than in the BR 2α, BR 2β and BR 6 core blocks investigated. Another equally plausible explanation is that the repetitive structure might provide structural properties which increase the chance of a large deletion, excluding the core block from one of the species.

The second difference between the two sibling species demonstrate that the core blocks can evolve rapidly through segmental mutations in combination with a spread of the mutations by homogenization. A BR 1 core block in *C. tentans* was characterized and compared with the corresponding core block in *C. pallidivittatus*. Each block was found to be homogenous but the type of repeat units in the two blocks differed. The changes could be explained as segmental mutations in the SR regions, which have affected the sequence, length and number of subrepeats in the SR region. At the same time the C regions have changed much less, differing only by a few bp substitutions.

Selection on the BR genes at the protein level

In the DNA sequence of the BR genes we find evidence for a selection at the protein level. This is most obvious in the C regions. Certain codons, e.g. the cysteine codons are strictly conserved between the genes. Furthermore, the repetitive nature of the genes makes it possible to detect conservation also within the core blocks. When we compare the repetitive sequences within a core block we can score bp substitutions in terms of silent or replacement substitutions. We then find that in the C regions there are virtually only silent substitutions. In contrast, the SR regions in the same core block exhibit many replacement substitutions. The SR regions therefore seem to accept larger changes than the C regions in agreement with our findings of frequent segmental mutations in the SR regions.

Interestingly enough, the different BR genes seem to behave differently. The SR regions in the BR 2β core block contain many replacement substitutions but in contrast the SR regions in the BR 2α core block contain only silent bp substitutions. In this context one should recall that in the BR 2α and BR 6 genes, the same structural demand seems to have been solved in two different ways: the BR 2 gene has phosphorylated serine residues, where the BR 6 gene has glutamic acid residues. These differences, regarding selection pressure and protein structure, raise the question why several BR genes have evolved. One possible answer is that the different secretory proteins have slightly different functional properties and that more than one type of protein is necessary to build the proper larval tube. A second possibility is that the different BR gene products are functionally equivalent but for various reasons it is advantageous to construct the larval tube from different BR proteins during variable physiological and/or environmental conditions.

Concluding remarks

The BR genes are examples of protein coding sequences with a highly repetitive, satellite-like structure. This extraordinary sequence organization probably drastically affects the evolutionary rate. The extreme repetitive nature should result in a high number of segmental mutations, e.g. through mistakes during DNA replication or DNA repair. Furthermore, homogenization processes, such as unequal crossing-over and gene conversion, should operate efficiently and spread (or eliminate) the mutations introduced locally, including point mutations, throughout the entire repetitive block. Due to their repetitive structure the BR genes therefore probably acquire mutations at a high rate compared to genes with a unique internal structure.

Acknowledgements

We thank Evy Vesterbäck for typing the manuscript. The work was supported by the Swedish Natural Science Research Council, the Wallenberg Foundation, Magnus Bergvalls Stiftelse and Karolinska Institutet.

References

1. Brutlag, D. L., *Annu. Rev. Genet.* **14**, 121 (1980).
2. Ohno, S., Matsunaga, T. and Wallace, R. B., *Proc. Natl. Acad. Sci. USA* **79**, 1999 (1982).
3. Eiferman, A. F., Young, P. R., Scott, R. W. and Tilghman, S. M., *Nature (London)* **294**, 713 (1981).
4. Russell, D. W., Schneider, W. J., Yamamoto, T., Luskey, K. L., Brown, M. S. and Goldstein, J. L., *Cell* **37**, 577 (1984).

5. Miller, J., McLachlan, A. D. and Klug, A., *EMBO J.* **4**, 1609 (1985).
6. Yamada, Y., Avvedimento, E., Mudryj, M., Ohkubo, H., Vogeli, G., Ivani, M., Pastan, I. and Crombrugghe, B., *Cell* **22**, 887 (1980).
7. Tsujimoto, Y. and Suzuki, Y., *Cell* **18**, 591 (1979).
8. Muskavitch, M. A. T. and Hogness, D. S., *Cell* **29**, 1041 (1982).
9. Sümegi, J., Wieslander, L. and Daneholt, B., *Cell* **30**, 579 (1982).
10. Case, S. T. and Daneholt, B., *J. Mol. Biol.* **123**, 223 (1978).
11. Skoglund, U., Andersson, K., Björkroth, B., Lamb, M. M. and Daneholt, B., *Cell* **34**, 847 (1983).
12. Kao, W.-Y. and Case, S. T., *J. Cell. Biol.* **101**, 1044 (1985).
13. Rydlander, L., Pigon, A. and Edström, J.-E., *Chromosoma* **81**, 101 (1980).
14. Edström, J.-E., Rydlander, L. and Francke, C., *Chromosoma* **81**, 115 (1980).
15. Hertner, T., Eppenberger, H. M. and Mähr, R., *Wilhelm Roux's Arch.* **189**, 69 (1980).
16. Galler, R., Rydlander, L., Riedel, N., Kluding, H. and Edström, J.-E., *Proc. Natl. Acad. Sci. USA* **81**, 1448 (1984).
17. Grossbach, U., in *Results and Problems in Cell Differentiation* (ed. W. Beermann), vol. 8, p. 147. Springer Verlag, Berlin (1977).
18. Wieslander, L., Sümegi, J. and Daneholt, B., *Proc. Natl. Acad. Sci. USA* **79**, 6956 (1982).
19. Case, S. T. and Byers, M. R., *J. Biol. Chem.* **258**, 7793 (1983).
20. Wieslander, L. and Lendahl, U., *EMBO J.* **2**, 1169 (1983).
21. Case, S. T., Summers, R. L. and Jones, A. G., *Cell* **33**, 555 (1983).
22. Lendahl, U. and Wieslander, L., *Cell* **36**, 1027 (1984).
23. Wieslander, L., Höög, C., Höög, J.-O., Jörnvall, H., Lendahl, U. and Daneholt, B., *J. Mol. Evol.* **20**, 304 (1984).
24. Hamodrakas, S. J. and Kafatos, F. C., *J. Mol. Evol.* **20**, 296 (1984).
25. Pustell, J., Kafatos, F. C., Wobus, U. and Bäumlein, H., *J. Mol. Evol.* **20**, 281 (1984).
26. Jones, C. W. and Kafatos, F. C., *J. Mol. Evol.* **19**, 87 (1982).
27. Farabaugh, P., Schmeissner, U., Hofer, M. and Miller, J. H., *J. Mol. Biol.* **126**, 847 (1978).
28. Höög, C. and Wieslander, L., *Proc. Natl. Acad. Sci. USA* **81**, 5165 (1984).
29. Beermann, W., *Chromosoma* **7**, 198 (1955).
30. Lendahl, U. and Wieslander, L., *J. Mol. Evol.* **22**, 63 (1985).

Chemica Scripta 1986, **26B**, 165–170

Evolution of Human Loci for Small Nuclear RNAs

K. Hammarström, C. Bark, G. Westin, J. Zabielski and U. Pettersson

Uppsala University, Department of Medical Genetics, Biomedical Center, Box 589, S-751 23 Uppsala, Sweden

Paper presented by Ulf Pettersson at the Conference on 'Molecular Evolution of Life', Lidingö, Sweden, 8–12 September 1985

Abstract

Eucaryotic cell nuclei contain a class of abundant, metabolically stable small RNAs (snRNAs), which are designated U1 to U7. Based on structural studies of recombinent clones containing sequences complementary to U1, U2, and U4 RNA we have attempted to trace the evolution of human loci for snRNA. One recombinant clone, U2.6, contained three U2 RNA genes. They are arranged in tandemly repeated units of 6.2 kb. The haploid human genome contains approximately twenty U2 gene copies, the majority of which are found at a single chromosomal location, 17q21.

Many loci were found to be pseudogenes, i.e. degenerate gene copies. The structural features of the pseudogenes appear to reflect two distinct evolutionary pathways: gene duplication and the reverse flow of genetic information from RNA to DNA. The former mechanism generates pseudogenes which exhibit flanking sequence homology to the functional genes.

The pseudogenes which seem to be generated by a RNA-mediated mechanism lack homology to the *bona fide* genes outside the coding regions. They are likely to have arisen by reverse transcription of the RNA followed by integration of the resulting cDNA copy into the chromosomal DNA. For one pseudogene, U2.4, it appears that a precursor for U2 RNA rather than the mature U2 RNA itself served as the template in the postulated reverse transcription event.

Introduction

Small nuclear RNAs (snRNAs) comprise a group of abundant and metabolically stable RNA molecules present in all eucaryotic cells so far studied (for a review see [1]). A subset of the snRNAs, designated U1 to U7, are characterized by an unusual cap structure and a high content of uridylic acid. The individual U RNAs are associated with 6–10 polypeptides in distinct ribonucleoprotein particles (snRNPs) with the exception of the U4 and U6 RNAs which are found in the same snRNPs [2, 3]. Patients suffering from certain autoimmune diseases produce antibodies directed against the different snRNPs [4] and these antisera have been extensively used for investigation of the structure and function of snRNPs. The U1 RNP has been shown to participate in the splicing reaction [5–7] as had been suggested previously based on the complementarity between the consensus sequence of splice sites and the 5′ end of U1 RNA [8, 9]. Similarly the U7 RNP has been demonstrated to be required for 3′ end maturation of histone H3 pre mRNA [10]. U2 RNPs may also play a role in the splicing reaction [11, 12] and it has been proposed that the U4/U6 RNPs are involved in the polyadenylation of mRNAs [13] but experimental evidence to support these hypotheses are still lacking.

In recent years many human loci containing snRNA-related sequences have been isolated and characterized [14–33]. An unexpected finding was that most of these loci have the properties of pseudogenes, i.e. degenerated gene copies. In the present communication we attempt to trace the evolutionary history of the snRNA pseudogenes from their structural features.

Results

Ten human loci for U1, U2 and U4 RNA have been studied [25, 26, 30, 31]. These loci were isolated from the human gene library of Lawn *et al.* [34]. The analysis of the clones included restriction enzyme analysis, heteroduplex analysis (U1 and U2), DNA-sequencing and, for U2, determination of the chromosomal localization. A surprising result of the investigation was that only one of the examined loci, U2.6, contained a bona fide gene. The nine other clones contained U RNA-related sequences which did not perfectly match the snRNA sequences, previously reported by others. These loci were classified as pseudogenes on a structural basis. This conclusion has later been questioned for the U1.6 locus since a functional analysis has shown that it is transcriptionally active (Hammarström *et al.*, in preparation). The pseudogenes have various structural features (Fig. 1) which have enabled us to trace their evolutionary histories.

The human U2 genes

The U2.6 locus was studied by heteroduplex/heterotriplex analysis [30]. It was found to contain three U2-related regions with the same orientation and with an equal distance between them [30]. The tandem arrangement was confirmed by restriction enzyme mapping and Southern blot analysis. The repeat unit length is 6.2 kb and all the restriction sites which have been studied are conserved. An approximately 700 bp-long sequence of one gene in the U2.6 locus was determined by the Maxam and Gilbert method [35]. Although the human U2 RNA sequence has not been determined it is likely that the U2.6 genes are true genes because of several reasons: first, there are only four base differences between the U2.6 genes and rat U2 RNA and at those four positions, all human U2 pseudogenes have the same nucleotides as U2.6. Secondly, injection of one of the U2.6 genes into *X. laevis* oocyte nuclei resulted in the synthesis of a RNA species whose RNase T1-fingerprint agreed with the previously reported human U2 RNA fingerprint [36, 37].

No TATA box was found in the appropriate position, but boxes of homology with the U1 genes were found in the 5′ region. Another interesting finding was an approximately 150 bp-long homocopolymeric sequence, $(CT)_n$, in the 3′ flanking

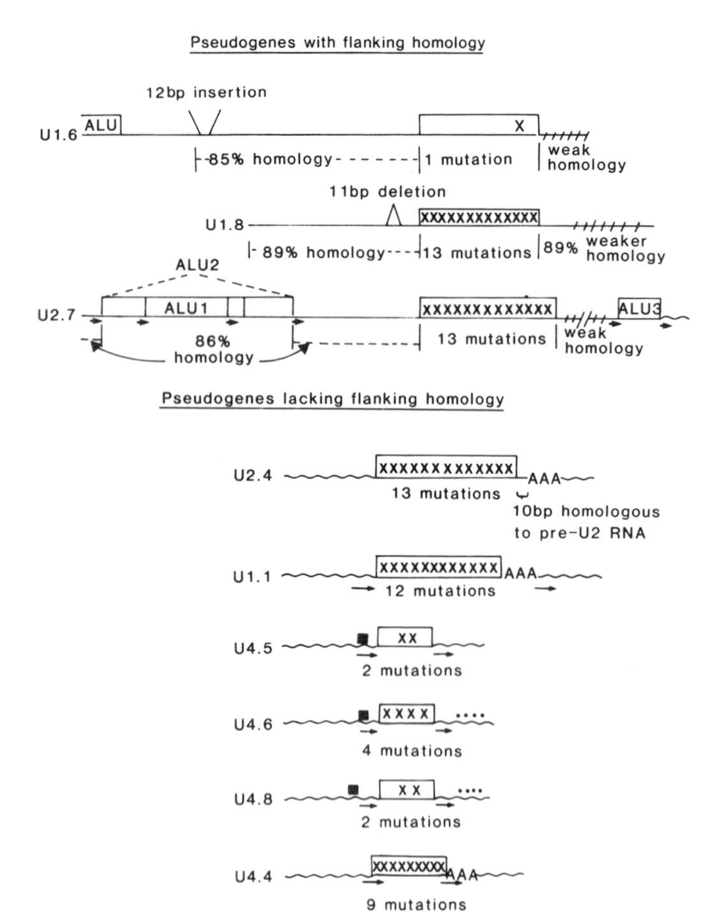

Fig. 1. The structures of the different pseudogenes. The boxes indicate the regions complementary to the snRNAs and the mutations are indicated by ×. Flanking sequence homology to the bona fide genes is shown by a thick line and other sequences are represented by a wavy line. A-rich sequences are shown (AAA) and homology boxes between the U4 pseudogenes are illustrated by a filled box and dots. Arrows indicate flanking direct repeats.

region. Htun *et al.* [38] have shown that this region is particularly sensitive to S1 nuclease.

The copy number of U2 genes was estimated to be approximately 20 per haploid genome.

Two techniques were used in order to map the U2 genes on the human chromosomes. Initially *in situ* hybridization was performed onto metaphase chromosomes using a ^{3}H-labelled probe 500 bp long, derived from the U2 gene [39]. The chromosomes were G-banded and photographed prior to hybridization to facilitate their identification. The analysis of 44 metaphase cells revealed a concentration of grains on the long arm of chromosome 17 (17q21). No other chromosome exhibited a clustering of grains on a single band. Since mouse/human somatic cell hybrids, containing only human chromosome number 17, were available, the *in situ* hybridization result was confirmed by Southern blot analysis of the mouse/human cell hybrids using parental mouse DNA and human DNA as controls. The resolution of the *in situ* hybridization does, however, not allow us to exclude the existence of isolated U2 genes at other locations. There exist several truncated U2 pseudogenes dispersed throughout the genome which contribute to an increase in the background [29].

Similar studies of the human U2 genes at the DNA level and at the chromosome level have been carried out by Van Arsdell *et al.* [28] and Lindgren *et al.* [40]. The results are in complete agreement.

Human pseudogenes for U1, U2, and U4 RNA

The distinction between pseudogenes which have flanking homology to the bona fide genes and those lacking homology, results in two groups consisting of three and six members, respectively. The two groups are proposed to have gone through different evolutionary pathways as will be discussed below. A schematic drawing of the different loci is shown in Fig. 1.

Pseudogenes with flanking homology

The loci belonging to this group are U1.6, U1.8, and U2.7. Both U1.8 and U2.7 have 12 point mutations and one insertion in their coding regions which makes them 93 and 94% homologous to the bona fide genes, respectively. The homology in the upstream region is nearly as good, i.e. 89 and 86%, and extends at least as far as position −225 and −210, respectively. In U2.7 this region has been split by the insertion of a 697 bp-long block of Alu repeats. This invading structure is composed of one Alu sequence which has integrated into another partially duplicated Alu repeat. The U1.8 locus has an 11 bp deletion around position −30. The 3′ flanking region of U1.8 is 89% homologous to the functional U1 gene HU1-I [27] as far as position +200 (the first nucleotide of the coding region being +1) and the homology, although somewhat weaker, continues to around position +350. In U2.7 the downstream homology ends at the border of an Alu repeat at the position where the U2.6 homocopolymeric sequence starts. This $(CT)_n$ stretch is not found within a 130 bp-long region directly downstream of the Alu repeat in U2.7.

The U1.6 locus belongs by definition to this category of pseudogenes since it has one point mutation in its coding region. Its status as a pseudogene is, however, questioned in the light of the finding that it directs the synthesis of U1 RNA in *X. laevis* oocytes. Mutations have accumulated at a much higher rate outside the coding region where the homology is 85% in the 5′ region and 63% in the 3′ region. The homology extends beyond position −300 on the 5′ side. At position −500, U1.6 has an Alu sequence which is not found in HU1-1 [27]. An interesting feature of U1.6 is the presence of a TATA box at position −33 to −37. The bona fide U1 and U2 genes lack the ordinary TATA box found at −29 to −33 in most RNA polymerase II transcribed genes. The promoter elements, the transcriptional activator at −215 and the region necessary for correct initiation at −50, are well conserved containing only one and two mismatches, respectively. In contrast, the consensus sequence for 3′ end formation is more degenerate with three bases which do not agree with the overall consensus sequence established by Hernandez [41].

SnRNA pseudogenes with flanking homology to the functional genes have a reasonable chance of retaining functional promoters for some time as they drift away from their gene families. We have investigated whether the U1.6 locus serves as a template for U1 RNA synthesis in *X. laevis* oocytes (Hammarström *et al.*, in preparation). The injection of a subclone from the U1.6 locus, containing 5 kb of 5′ flanking sequences and 50 bp of 3′ flanking sequences, resulted in synthesis of the mutant U1 RNA encoded by U1.6. The transcriptional rate was 5–10 times lower than that obtained with the cloned functional U1 gene, pHU1-1D, previously studied by Murphy *et al.* [23] and Skuzeski *et al.* [42]. The expression of U1.6 was further enhanced when sequences upstream of position −106 were replaced by the corresponding region from the HU1-1 locus (Hammarström *et al.*, in preparation).

Pseudogenes lacking flanking homology

This category can be subdivided in two groups depending on whether the pseudogenes are full length or truncated at their 3′ ends.

(1) *Full length pseudogenes.* U1.1 and U2.4 are full length pseudogenes lacking flanking homology to their functional counterparts. U2.4 has a 10 bp-long stretch of homology to the U2 genes immediately downstream of the 3′ end of the coding region. This short sequence is present in a human U2 RNA precursor which has been observed *in vivo* [43] as well as in *X. laevis* oocytes injected with the U2.6 gene [37]. The two pseudogenes both have an A-rich sequence immediately following the region that is homologous to U RNA. This feature is shared by many retroposons [44], irrespective of whether the RNA template is polyadenylated or not. U1.1 has another feature which is common among retroposons, in that it has almost perfect direct repeats flanking the pseudogene and the A-rich tail. The repeats are considered to be duplications of the target sequence which result from the integration of retroposons at staggered chromosomal breaks. U1.1 belongs to class III according to the classification system established by Denison and Weiner [19].

(2) *Truncated pseudogenes.* This group is composed of the four U4 pseudogenes which share the following characteristics: they are severely truncated at their 3′ end containing 67–79 nucleotides of the 142–146 nucleotides long U4 RNA sequence [45]; they are flanked by direct repeats and they have scattered point mutations in the U4 coding region. If the U4.4 pseudogene is excluded, the three others form a surprisingly homogeneous group: 67–68 bp in length; 2–4 mutations in the coding region; almost perfect direct repeats of 17–25 bp which in all cases overlap a few nucleotides with the coding region at the 3′ end; and a conserved box of 12–13 nucleotides in the 5′ flanking region. This purine-rich box could be a preferred site for integration, in two of the clones this sequence is part of the direct repeat. U4.6 and U4.8 have four short boxes of homology in the 3′ region just downstream of the direct repeats. These elements may also play a role in conjunction with the pseudogene insertion. Interestingly the 5′ box is shared by a number of retroposons although it is sometimes less well conserved [31].

Discussion

The human loci for snRNA seem to reflect several mechanisms which operate in the evolution of the human genome. The U2 gene cluster provides an example of how the need for a high rate of synthesis of a product has been met by gene amplification to produce a multicopy gene family. The structures of the different pseudogenes reflect two major pathways in evolution: gene duplication and retroposition. These will be discussed in relation to the postulated evolutionary history of the pseudogenes.

The U2 multicopy gene family

It is intriguing that the U2 genes are organized in tandem repeats, whose size exceeds the size of the gene more than 30-fold and yet the entire unit of 6.2 kb is extremely well conserved. An interesting question concerns the role of the remaining part of the repeat unit. Regulatory sequences and promoter elements are generally short and, in the case of U2, it is known that 254 bp of upstream sequence and 94 bp of downstream sequence are sufficient to obtain expression of U2 RNA in *Xenopus laevis* oocytes [37]. It has been shown by Yuo *et al.* [46] that as few as 37 nucleotides at the 3′ end are needed for 3′ end formation of pre-U2 RNA. Northern blots of HeLa cell RNA probed with the U2 repeat unit show hybridization to 7S RNA (unpublished) which is consistent with the mapping of repetitive sequence elements related to the Alu family approximately 3 kb upstream of the U2.6 gene. The role played by the major part of the unit may be purely structural in the sense that it is required for the maintenance of the homogeneity of the gene family. Gene correction by homologous unequal exchange and gene conversion is probably favoured by long homologous sequences. If it was just as efficient on shorter dispersed sequences, middle repetitive elements like Alu repeats would be less divergent. It is worth noting that, although the human U2 gene family displays a remarkable sequence homogeneity, there are great differences in the flanking sequences and the genomic organization between U2 genes in different species. This seems to be a general feature of multicopy gene families [47]. In the absence of homogenization mechanisms operating on the gene families, two randomly chosen members within a family would not be expected to be more homologous than two members of different species [47]. Thus there must be strong gene correction mechanisms operating on the members of the U2 gene family. The S1 sensitive homocopolymeric $(CT)_n$ stretch that is adjacent to the human U2 genes may play a role in conjunction with gene conversion. It is noteworthy that the U2.7 locus, which has drifted away from its family, lacks this sequence.

The majority of the human U2 RNA genes are located in a cluster on chromosome 17q21. Other genes which have been assigned to this cytological region are the genes for thymidine kinase [48, 49], galactokinase [50], alpha type-1 collagen [51, 52] and the gene encoding the cellular tumour antigen p53 [53]. The breakpoints in the 15:17 translocations in patients with promyelocytic leukemia are also found in this region [54], as is one of the major adenovirus 12 modification sites [55]. Lindgren *et al.* [40] have also mapped the human U2 RNA genes to chromosome 17q21 and made the observation that the location of the human U1 genes (1p36) [56, 57] coincides with another major adenovirus 12 modification site. They propose that highly transcribed genes are the primary targets of the viral modification. It is noteworthy that the long arm of chromosome 17 and the tip of the short arm of chromosome 1, containing the human U1 genes, are often involved in cytogenetic abnormalities in leukemias and in certain tumors [58].

Pseudogenes of DNA-mediated origin

The loci U1.8, U2.7 and U1.6, which exhibit flanking homology to the bona fide U genes, are likely to result from the duplication of DNA segments containing U1 or U2 genes [25, 26]. Unequal exchange between chromatids results in a duplication of sequences on one chromatid and deletion of the corresponding sequences from the other chromatid. This mechanism can expand and reduce tandem clusters like the

U2 genes. Unequal exchange between non homologous chromosomes is rare but may occasionally move sequences into new genomic environments. These events generate new genes which may either stay active, if gene correction mechanisms operate on them, or accumulate mutations at a constant rate throughout their length in the absence of correction. The latter seems to be the case for the U1.8 and U2.7 loci, although the 3′ flanking regions are less well conserved. These loci must thus at some stage have been freed from the homogenization mechanisms that operate on multicopy genes. This could probably be achieved in several ways: by moving away from the U1 and U2 clusters; by the integration of foreign sequences in the vicinity of the gene; or by deleting sequences. All these events would reduce the power of gene correction by homologous unequal crossing over and gene conversion. The Alu block which has integrated upstream of U2.7, interrupts the alignment of U2.7 with the U2.6 repeat and U1.8 has an 11 bp-long deletion in the 5′ flanking region. U1.6 differs from U1.8 and U2.7 in the respect that the coding region is remarkably well conserved as if there have been functional constraints for keeping it intact. This feature may reflect that U1.6 in fact is a functional gene encoding a variant U1 RNA species.

Are U1 class I pseudogenes variant U1 genes?

There are two differences between the human U1 and U2 genes which are noteworthy: the U1 genes are less compactly arranged and have a large number of class I pseudogenes whereas the U2 genes are orderly arranged in a tandem array. An interesting feature is that there exist very few class I pseudogenes for U2 RNA as opposed to U1 RNA [28]. The arrangement of the U2 genes is probably more convenient for the homogenization through unequal crossing over and gene conversion than the arrangement of the U1 genes. The U2 genes might therefore be less prone to generate pseudogenes. The present U1 gene family, on the other hand has been suggested to be derived from class I pseudogenes by gene amplification and transposition [32, 59]. The observation that most U1 class I pseudogenes map to a single chromosomal location, 1q12-q22, supports this theory [60]. If the high abundance of U1 class I pseudogenes is not simply a result of the genomic turnover of U1 genes, the interesting possibility emerges that at least some class I pseudogenes are functional. In the light of the finding that the U1.6 locus is transcriptionally active one may have to reconsider the term class I pseudogene (Hammarström *et al.*, in preparation).

Pseudogenes of RNA-mediated origin

The formation of a pseudogene from a snRNA requires two major steps: reverse transcription of the RNA and then insertion of the resulting cDNA copy into the genome.

Reverse transcriptase must be primed by basepairing in order to start to copy the RNA template into a cDNA. The newly synthesized strand is complementary to the region contained between the priming site and the 5′-end of the RNA and has opposite polarity. In the case of the truncated U4 pseudogenes, it has been proposed, in analogy with the U3 truncated pseudogenes studied by Bernstein *et al.* [24], that base-pairing of the 3′ end of the RNA to an internal position provides the priming. Reverse transcription yields a cDNA complementary to the 5′ end of the RNA.

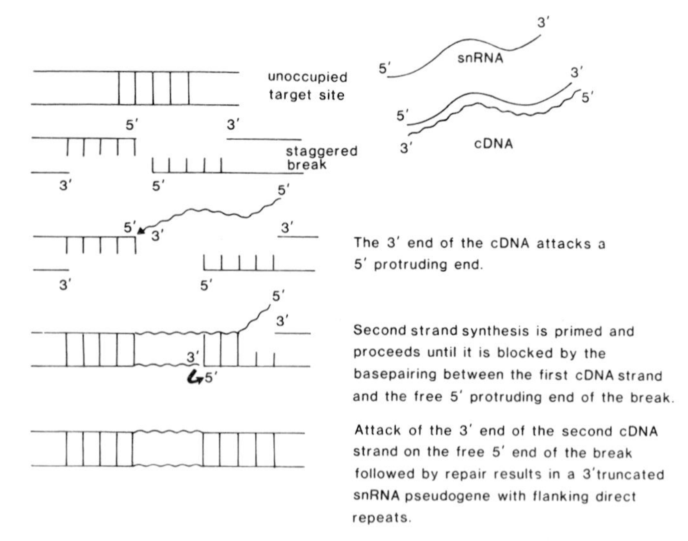

unoccupied target site
staggered break
snRNA
cDNA
5′
3′
The 3′ end of the cDNA attacks a 5′ protruding end.
Second strand synthesis is primed and proceeds until it is blocked by the basepairing between the first cDNA strand and the free 5′ protruding end of the break.
Attack of the 3′ end of the second cDNA strand on the free 5′ end of the break followed by repair results in a 3′truncated snRNA pseudogene with flanking direct repeats.

Fig. 2. Model for the generation of truncated snRNA pseudogenes adapted from Van Arsdell and Weiner (1984).

After integration at staggered breaks and DNA repair, the net result is a 3′ truncated pseudogene with flanking direct repeats. Three of the truncated pseudogenes differ only by one nucleotide in length, which may indicate that priming at a special internal sequence complementary to the 3′ end indeed has occurred [20, 31]. The model originally proposed by Van Arsdell and Weiner [29] which is not based on self-priming (Fig. 2) seems more appropriate for the generation of the truncated U4 pseudogenes. The reason for this is that it takes into account three quite remarkable features shared by U4.5, U4.6 and U4.8: the common site of truncation (at nucleotides 67 or 68); the 3–5 nucleotides overlap between the 3′ end of the snRNA sequence and the downstream repeat (GAAAA, AGAAA, AAA); and the partial conservation of the sequence at the integration sites. The overlap, being a consequence of the basepairing between the first cDNA strand and the 5′ protruding end of the chromosomal break (Fig. 2), would result in the generation of pseudogenes of identical length if the sequences at the integration sites are conserved. A search for the overlap sequence in the U4 RNA sequence [45] reveals that AAA is only found at position 65–68 which gives additional support to the argument.

Self-priming can not account for the full length pseudogenes U2.4 and U1.1 which share the feature of being flanked on the 3′ side by an A-rich sequence. In U1.1 the pseudogene with the A-tail is flanked by direct repeats whereas U2.4 on the other hand lacks direct repeats. It seems likely that the A-tail has its origin in aberrant polyadenylation of snRNAs. In this respect it is interesting that there are small RNAs, like BC1 RNA in brain cells, which are polyadenylated [61]. A model, based on priming of a protruding 3′ end directly on a poly A-tailed U2 RNA precursor, has been proposed for the generation of the U2.4 pseudogene (Fig. 2).

In the case of U1.1 a similar event may have occurred with the exception of the priming of the first cDNA strand. In order to obtain flanking direct repeats one has to postulate insertion at a staggered break. The priming could be performed by a 3′ protruding end on the A-tail of the RNA. It is, in fact, possible to find a 4 bp overlap between the end of the A-tail and the downstream repeat, and interestingly also a 5 bp overlap between the upstream direct repeat and the 5′ end of the pseudogene. A possible model is outlined in Fig. 3.

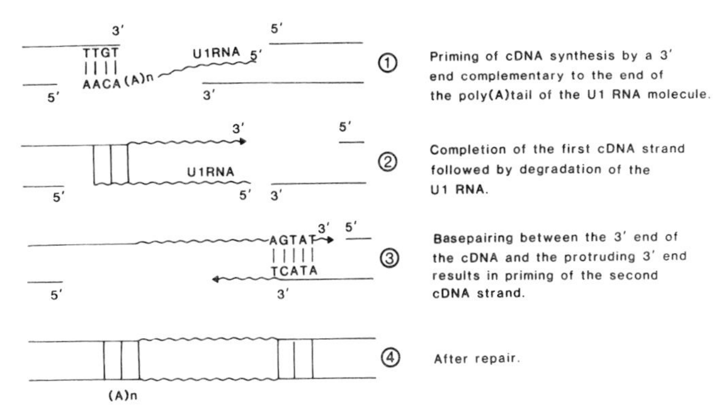

Fig. 3. Model for the generation of the U1.1 pseudogene.

Retroposition – an important factor in evolution

The generation of novel sequences in the chromosomal DNA as the result of reverse flow of genetic information, from RNA to DNA, has been termed retroposition [62]. This process has been estimated to be the origin of as much as 5% of the mammalian genome [63]. Characteristic features for the retroposons are signs of RNA processing; pseudogenes derived from protein-coding genes lack introns and have poly(A) tails (for a review see [44]), a mouse tRNA pseudogene has the terminal CCA sequence which is post-transcriptionally added [64]. For all RNA polymerase II transcribed genes the retroposition event is expected to result in the loss of promoter elements since these are located upstream of the cap site. However, one of the functional preproinsulin I genes of rat and mouse, has been shown to be a retroposon [65]. The reason for it being functional is that the presumed RNA intermediate was aberrantly initiated upstream of the normal cap site. Moreover, a processed pseudogene for calmodulin has been shown to be expressed in a tissue specific manner in chicken [66]. It does not seem unlikely that retroposition may have given rise to other functional genes. In some gene families RNA mediated pseudogenes have been shown to be particularly common, e.g. there are five processed pseudogenes for human β-tubulin [67–69], four for human dihydrofolate reductase [70–72] and six for rat cytochrome c [73, 74]. Human truncated pseudogenes for U2, U3 and U4 seem also to be numerous, which may reflect the abundance of snRNAs in the cell nucleus. Genes transcribed by RNA polymerase III have internal promoters which enable them to undergo several rounds of retroposition. This is believed to be the origin of the high number of Alu sequences in the human genome, since these have been suggested to be pseudogenes, derived from the RNA polymerase III product 7 SL RNA [75].

Retroposition must occur in the germ line in order to be transmitted. The reverse transcriptase could be encoded by an endogenous retrovirus [76, 77] or a retrovirus-like element such as an A-particle [78, 79]. Reverse flow of genetic information has been demonstrated to mediate the transposition of the yeast mobile element Ty [80]. Sequence homology to the *pol* gene, encoding reverse transcriptase, has been found in the *Drosophila* transposable element *copia* [81] and recently it has been reported for protein-coding sequences in introns of fungal mitochondria [82]. The contribution of reverse flow of genetic information in shaping the eukaryotic genome thus seems to be far more important than previously believed.

The mechanism of insertion is less well understood than the reverse transcription event. Chromosomal breaks have to be induced in some way. We have observed that some retroposons have at least distantly related A-rich integration sites [31], but there is no strict sequence dependence. The breaks appear sometimes to be blunt and sometimes staggered. It has been shown that the 3′ end of the *pol* gene encodes an endonuclease which is necessary for the integration of the provirus into the host genome [83, 84]. Assuming that the reverse transcriptase is encoded by the *pol* gene of an endogenous retrovirus or a retroviruslike particle is that the associated endonuclease mediates the integration process. There is, however, a major difference in the length of the direct repeats generated on the insertion by such endonucleases (4–6 bp) and those flanking RNA-mediated pseudogenes (8–25 bp) [44]. The cellular enzyme topoisomerase I induces single stranded breaks in duplex DNA (for a review see [85]). Two such breaks on opposite strands, separated by 8–25 bp would generate the stagger needed in order to account for the length of the direct repeats. Topoisomerase II introduces double stranded breaks in chromosomal DNA (for reviews see [85, 86]). It has been demonstrated that the predominant cleavage sites for topoisomerase II coincide with regions of micrococcal nuclease hypersensitivity, such as in spacers between genes and 5′ flanking regions, but not the coding regions of genes [87, 88]. This last observation probably holds true for most retroposons since only few retroposons, inserted into the coding region of a gene have been reported [72, 89] although there are numerous examples of retroposons found in the flanking regions of genes or in introns (for a review see [44]). Retroposons are particularly prone to insert into pre-existing retroposons, retroposon-tails and other A-rich regions [44]. U2.7 provides a typical example in this respect with one Alu repeat inserted into the AT-rich spacer of a pre-existing Alu sequence.

It should, however, be emphasized that the insertion of a cDNA copy into the genome is an extremely rare event and not until the process can be artificially induced may the different steps be elucidated.

Acknowledgements

This work was financially supported by grants from the Swedish Medical Research Council, The Swedish National Board for Technical Development, and by funds at the University of Uppsala. We thank Linda Baltell and Jeanette Backman for secretarial aid.

References

1. Busch, H., Reddy, R., Rothblom, L. and Choi, Y. C., *Ann. Rev. Biochem.* **51**, 617 (1982).
2. Hashimoto, C. and Steitz, J. A., *Nucl. Acids Res.* **12**, 3283 (1984).
3. Bringmann, P., Appel, B., Rinke, J., Reuter, R., Theissen, H. and Lührmann, R., *EMBO J.* **3**, 1357 (1984).
4. Lerner, M. R. and Steitz, J. A., *Proc. Natl. Acad. Sci. USA* **76**, 5495 (1979).
5. Yang, V. W., Lerner, M. R., Steitz, J. A. and Flint, S. J., *Proc. Natl. Acad. Sci. USA* **78**, 1371 (1981).
6. Padgett, R. A., Mount, S. M., Steitz, J. A. and Sharp, P. A., *Cell* **35**, 101 (1983).
7. Krämer, A., Keller, W., Appel, B. and Lührmann, R., *Cell* **38**, 299 (1984).
8. Lerner, M. R., Boyle, J. A., Mount, S. M., Wolin, S. L. and Steitz, J. A., *Nature* **283**, 220 (1980).
9. Rogers, J. and Wall, R., *Proc. Natl. Acad. Sci. USA* **77**, 1877 (1980).
10. Galli, G., Hofstetter, H., Stunnenberg, H. G. and Birnstiel, M. L., *Cell* **34**, 823 (1983).
11. Ohshima, Y., Itoh, M., Okada, N. and Miyata, T., *Proc. Natl. Acad. Sci. USA* **78**, 4471 (1981).
12. Keller, E. B. and Noon, W. A., *Proc. Natl. Acad. Sci. USA* **81**, 7417 (1984).

13. Berget, S. M., *Nature* (*London*) **309**, 179, (1984).
14. Denison, R. A., Van Arsdell, S. W., Bernstein, L. B. and Weiner, A. M., *Proc. Natl. Acad. Sci. USA* **78**, 810 (1981).
15. Hayashi, K., *Nucl. Acids Res.* **9**, 3379 (1981).
16. Manser, T. and Gesteland, R. F., *J. Mol. Appl. Genet.* **1**, 117 (1981).
17. Van Arsdell, S. W., Denison, R. A., Bernstein, L. B., Weiner, A. M., Manser, T. and Gesteland, R., *Cell* **26**, 11 (1981).
18. Westin, G., Monstein, H. J., Zabielski, J., Philipson, L. and Pettersson, U., *Nucl. Acids Res.* **9**, 6323 (1981).
19. Denison, R. A. and Weiner, A. M., *Mol. Cell. Biol.* **2**, 815 (1982).
20. Hammarström, K., Westin, G. and Pettersson, U., *EMBO J.* **1**, 737 (1982).
21. Manser, T. and Gesteland, R. F., *Cell* **29**, 257 (1982).
22. Monstein, H.-J., Westin, G., Philipson, L. and Pettersson, U., *EMBO J.* **1**, 133 (1982).
23. Murphy, J. T., Burgess, R. R., Dahlberg, J. E. and Lund, E., *Cell* **29**, 265 (1982).
24. Bernstein, L. B., Mount, S. M. and Weiner, A. M., *Cell* **32**, 461 (1983).
25. Monstein, H.-J., Hammarström, K., Westin, G., Zabielski, J., Philipson, L. and Pettersson, U., *J. Mol. Biol.* **167**, 245 (1983).
26. Hammarström, K., Westin, G., Bark, C., Zabielski, J. and Pettersson, U., *J. Mol. Biol.* **179**, 157 (1984).
27. Lund, E. and Dahlberg, J. E., *J. Biol. Chem.* **259**, 2013 (1984).
28. Van Arsdell, S. W. and Weiner, A. M., *Mol. Cell. Biol.* **4**, 492 (1984).
29. Van Arsdell, S. W. and Weiner, A. M., *Nucl. Acids Res.* **12**, 1463 (1984).
30. Westin, G., Zabielski, J., Hammarström, K., Monstein, H.-J., Bark, C. and Pettersson, U., *Proc. Natl. Acad. Sci. USA* **81**, 3811 (1984).
31. Bark, C., Hammarström, K., Westin, G. and Pettersson, U., *Mol. Cell. Biol.* **5**, 943 (1985).
32. Bernstein, L. B., Manser, T. and Weiner, A. M., *Mol. Cell. Biol.* **5**, 2159 (1985).
33. Theissen, H., Rinke, J., Traver, C. N., Lührmann, R. and Appel, B., *Gene* **36**, 195 (1985).
34. Lawn, R. M., Fritsch, E. F., Parker, R. C., Blake, G. and Maniatis, T., *Cell* **15**, 1157 (1978).
35. Maxam, A. M. and Gilbert, W., in *Methods in Enzymology* (ed. L. Grossman and K. Moldave), vol. 65, pp. 499–560. Academic Press, New York (1980).
36. Nohga, K., Reddy, R. and Busch, H., *Cancer Res.* **41**, 2215 (1981).
37. Westin, G., Lund, E., Murphy, J. T., Pettersson, U. and Dahlberg, J. E., *EMBO J.* **3**, 3295 (1984).
38. Htun, H., Lund, E., Westin, G., Pettersson, U. and Dahlberg, J. E., *EMBO J.* **4**, 1839 (1985).
39. Hammarström, K., Santesson, B., Westin, G. and Pettersson, U., *Exp. Cell. Res.* **159**, 473 (1985).
40. Lindgren, V., Ares, M. Jr., Weiner, A. M. and Francke, U., *Nature* (*London*) **314**, 115 (1985).
41. Hernandez, N., *EMBO J.* **4**, 1827 (1985).
42. Skuzeski, J. M., Lund, E., Murphy, J. T., Steinberg, T. H., Burgess, R. R. and Dahlberg, J. E., *J. Biol. Chem.* **259**, 8345 (1984).
43. Eliceiri, G. L. and Sayavedra, M. S., *Biochem. Biophys. Res. Commun.* **72**, 507 (1976).
44. Rogers, J. H., *Int. Rev. Cytol.* **93**, 187 (1985).
45. Krol, A. and Branlant, C., *Nucl. Acids Res.* **9**, 2699 (1981).
46. Yuo, C.-Y., Ares, M. and Weiner, A. M., *Cell* **42**, 193 (1985).
47. Dover, G., *Nature* (*London*) **299**, 111 (1982).
48. Miller, O., Allerdice, P. W., Miller, D. A., Berg, W. R. and Migeon, B. R., *Science* **173**, 244 (1971).
49. Kucherlapti, R. S., McDougall, J. K. and Ruddle, F. H., *Cytogenet. Cell Genet.* **13**, 108 (1979).
50. Elsevier, S. M., Kucherlapti, R. S., Nichols, E. A., Creagan, R. P., Giles, R. E., Ruddle, F. H., Willecke, K. and McDougall, J. K., *Nature* (*London*) **251**, 633 (1974).
51. Church, R. L., Sundar Raj, N. and McDougall, J. K., *Cytogen. Cell Genet.* **27**, 24 (1980).
52. Huerre, C., Junien, C., Weil, D., Chu, M.-L., Morabito, M., Nguyen, Van Cong, Myers, J. C., Foubert, C., Gross, M.-S., Prockop, D. J., Buoe, A., Kaplan, J.-C., de la Chapelle, A. and Ramirez, F., *Proc. Natl. Acad. Sci. USA* **79**, 6627 (1982).
53. Le Beau, M. M., Westbrook, C. A., Diaz, M. O., Rowley, J. D. and Oren, M., *Nature* (*London*) **316**, 826 (1985).
54. Rowley, J. D., *Proc. Natl. Acad. Sci. USA* **74**, 5729 (1977).
55. McDougall, J. K., *J. Gen. Virol.* **12**, 43 (1971).
56. Lund, E., Bostock, C., Robertson, M., Christie, S., Mitchen, J. L. and Dahlberg, J. E., *Mol. Cell. Biol.* **3**, 2211 (1983).
57. Naylor, S. L., Zabel, B. U., Manser, T., Gesteland, R. F. and Sakaguchi, A. Y., *Somat. Cell Molec. Genet.* **10**, 307 (1984).
58. Mitelman, F., *Cytogen. Cell Genet.* **36**, 1 (1983).
59. Weiner, A. M. and Denison, R. A., *Cold Spring Harbor Symp. Quant. Biol.* **47**, 1141 (1983).
60. Lindgren, V., Bernstein, L. B., Weiner, A. M. and Francke, U., *Mol. Cell. Biol.* **5**, 2172 (1985).
61. Sutcliffe, J. G., Milner, R. J., Gottesfeld, J. M. and Lerner, R. A., *Nature* (*London*) **308**, 237 (1984)
62. Rogers, J. H., *Nature* (*London*) **301**, 460 (1983).
63. Baltimore, D., *Cell* **30**, 481 (1985).
64. Reilly, J. G., Ogden, R. and Rossi, J. J., *Nature* (*London*) **300**, 287 (1982).
65. Soares, M. B., Schon, E., Henderson, A., Karathanasis, S. K., Cate, R., Zeitlin, S., Chirgwin, J. and Efstratiadis, A., *Mol. Cell. Biol.* **5**, 2090 (1985).
66. Stein, J. P., Munjaal, R. P., Lagace, L., Lai, E. C., O'Malley, B. and Means, A. R., *Proc. Natl. Acad. Sci. USA* **80**, 6485 (1983).
67. Wilde, C. D., Crowther, C. E., Cripe, T. P., Lee, M. G.-S and Cowan, N. J., *Nature* (*London*) **297**, 83 (1982).
68. Wilde, C. D., Crowther, C. E. and Cowan, N. J., *Science* **217**, 549 (1982).
69. Lee, M. G.-S., Lewis, S. A., Wilde, C. D. and Cowan, N. J., *Cell* **33**, 477 (1983).
70. Chen, M.-J., Shimada, T., Moulton, A. D., Harrison, M. and Nienhuis, A. W., *Proc. Natl. Acad. Sci. USA* **79**, 7435 (1982).
71. Masters, J. N., Yang, J. K., Cellini, A. and Attardi, G., *J. Mol. Biol.* **167**, 23 (1983).
72. Snimada, T., Chen, M. J. and Nienhuis, A. W., *Gene* **31**, 1 (1984).
73. Scarpulla, R. C. and Wu, R., *Cell* **32**, 473 (1983).
74. Scarpulla, R. C., *Mol. Cell Biol.* **4**, 2279 (1984).
75. Ullu, E. and Tschudi, C., *Nature* (*London*) **312**, 171 (1984)
76. Weinberg, R. A., *Cell* **22**, 643 (1980).
77. Jaenisch, R., *Cell* **32**, 5 (1983).
78. Lueders, K. K. and Kuff, E. L., *Cell* **12**, 963 (1977).
79. Ono, M., Cole, M. D., White, A. T. and Huang, R. C. C., *Cell* **21**, 466 (1980).
80. Boeke, J. D., Garfinkel, D. J., Styles, C. A. and Fink, G. R., *Cell* **40**, 491 (1985).
81. Saigo, K., Kugimiya, W., Matsuo, Y., Inouye, S., Yoshioka, U. and Yoki, S., *Nature* (*London*) **312**, 659 (1984).
82. Michel, F. and Lang, B. F., *Nature* (*London*) **316**, 641 (1985).
83. Donehower, L. A. and Varmus, H. E., *Proc. Natl. Acad. Sci. USA* **81**, 6461 (1984).
84. Schwarzberg, P., Colicelli, J. and Goff, S. P., *Cell* **37**, 1043 (1984).
85. Wang, J. C., *Ann. Rev. Biochem.* **54**, 665 (1985).
86. North, G., *Nature* (*London*) **316**, 394 (1985).
87. Udvardy, A., Schede, P., Sander, M. and Hsieh, T.-S., *Cell*, **40**, 933 (1985).
88. Yang, L., Rowe, T. C., Nelson, E. M. and Liu, L. F., *Cell* **41**, 127 (1985).
89. Lemischka, I. and Sharp, P. A., *Nature* (*London*) **300**, 330 (1982).

Proteins and Enzymatic Functions

Chemica Scripta 1986, **26B**, 173–178

Evolution of Ionic Channels

Shosaku Numa

Departments of Medical Chemistry and Molecular Genetics, Kyoto University Faculty of Medicine, Yoshida, Sakyo-ku, Kyoto 606, Japan

Paper presented at the Conference on 'Molecular Evolution of Life', Lidingö, Sweden, 8–12 September 1985

Abstract

The nicotinic acetylcholine receptor and the sodium channel represent ligand-gated and voltage-gated ionic channels, respectively, which are essential for neural signalling. The primary structures of all subunits of the acetylcholine receptor and that of the sodium channel have been elucidated by recombinant DNA techniques. The genes encoding some of the acetylcholine receptor subunits have been characterized. On the basis of the protein and gene structures, the evolution of these channel molecules is discussed.

Introduction

A new approach to understanding molecules essential for neural signalling has been provided by recombinant DNA technology. Thus, the structure and function of proteins constituting neural elements are being elucidated. These proteins include the nicotinic acetylcholine receptor (AChR) and the sodium channel. The AChR and the sodium channel represent ligand-gated and voltage-gated ionic channels, respectively, which mediate neural signalling by modulating the ion permeability of electrically excitable membranes. The primary structures of all subunits of the AChR and that of the sodium channel have been deduced from the nucleotide sequences of the cloned cDNAs. Furthermore, the genomic DNAs encoding some of the AChR subunits have been isolated and characterized. Comparison of the protein and gene structures suggests that the AChR subunits evolved from a common ancestor by gene duplications. The sodium channel contains four homologous repetitive units, which apparently have been generated by internal duplications. The evolutionary aspects of these ionic channels are discussed in the present article.

Acetylcholine receptor

The nicotinic AChR is a transmembrane protein complex that contains both the acetylcholine binding site and the cationic channel. The AChR from fish electric organ is the best characterized neurotransmitter receptor and is known to consist of four kinds of subunits assembled in a molar stoichiometry of $\alpha_2\beta\gamma\delta$ (reviewed in [1–3]). The primary structures of the α- [4], β- [5], γ- [6,7] and δ-subunits [5] of the *Torpedo californica* AChR were deduced by cloning and sequencing the cDNAs. A large portion of the amino acid sequence of the *Torpedo marmorata* AChR α-subunit was deduced by a similar approach [8], and this sequence was completed subsequently [9]. The primary structures of the α- [10], β- [11], γ- [12] and δ-subunits [13] of the calf muscle AChR and of the δ-subunit of the mouse AChR [14] were also deduced from the cDNA sequences. In addition, a novel subunit (named ε-subunit) of the calf muscle AChR was discovered by cloning the cDNA encoding it, and its primary structure was elucidated by sequencing the cDNA [15]. Furthermore, the primary structures of the α-[10] and γ-subunits [16] of the human AChR and of the γ-and δ-subunits of the chicken AChR [17] were predicted from the genomic DNA sequences.

The amino acid sequences of the mammalian, avian and fish AChR subunits are aligned in Fig. 1. The AChR subunits exhibit marked sequence homology and share common structural features. The hydropathy profiles [19] and the predicted secondary structures [20] of all subunit polypeptides are similar [4–7, 10–13, 15, 16, 21]. This suggests that all AChR subunits have the same transmembrane topology, being oriented in a pseudosymmetric fashion across the membrane. The carboxy-terminal half of each subunit molecule contains five putative transmembrane segments [4–7, 9–17, 22–24], that is, four hydrophobic segments (M1–M4) and an amphipathic segment (MA). Expression studies using site-directed mutagenesis of the α-subunit cDNA [24, 25] suggest that these five presumably α-helical segments are involved directly or indirectly in the formation of the ionic channel. The region lying between segments M3 and MA is assigned to the cytoplasmic side of the membrane, whereas the amino-terminal half of each subunit molecule, which contains one or more potential *N*-glycosylation sites [26], is assigned to the extracellular side of the membrane.

Three cysteine residues (positions 130, 144 and 243 in the aligned sequences) and one potential *N*-glycosylation site (position 143) are conserved in all the subunit polypeptides. There is evidence that the α-subunit carries the acetylcholine binding site and that a disulphide bridge is present in close proximity to this site [1, 27]. The amino-terminal half of the α-subunit preceding segment M1 has four cysteine residues (residues 128, 142, 192 and 193 or positions 130, 144, 209 and 210). The cDNA expression studies indicate that all the four cysteine residues of the α-subunit are required for AChR function [24]. Furthermore, the results of these studies are consistent with the presence of a disulphide bridge between the cysteine residues 128 and 142 as well as between the cysteine residues 192 and 193 of the α-subunit; the disulphide bridge between two adjacent cysteine residues seems unlikely [28], but there are precedents for it [29]. The cysteine residues

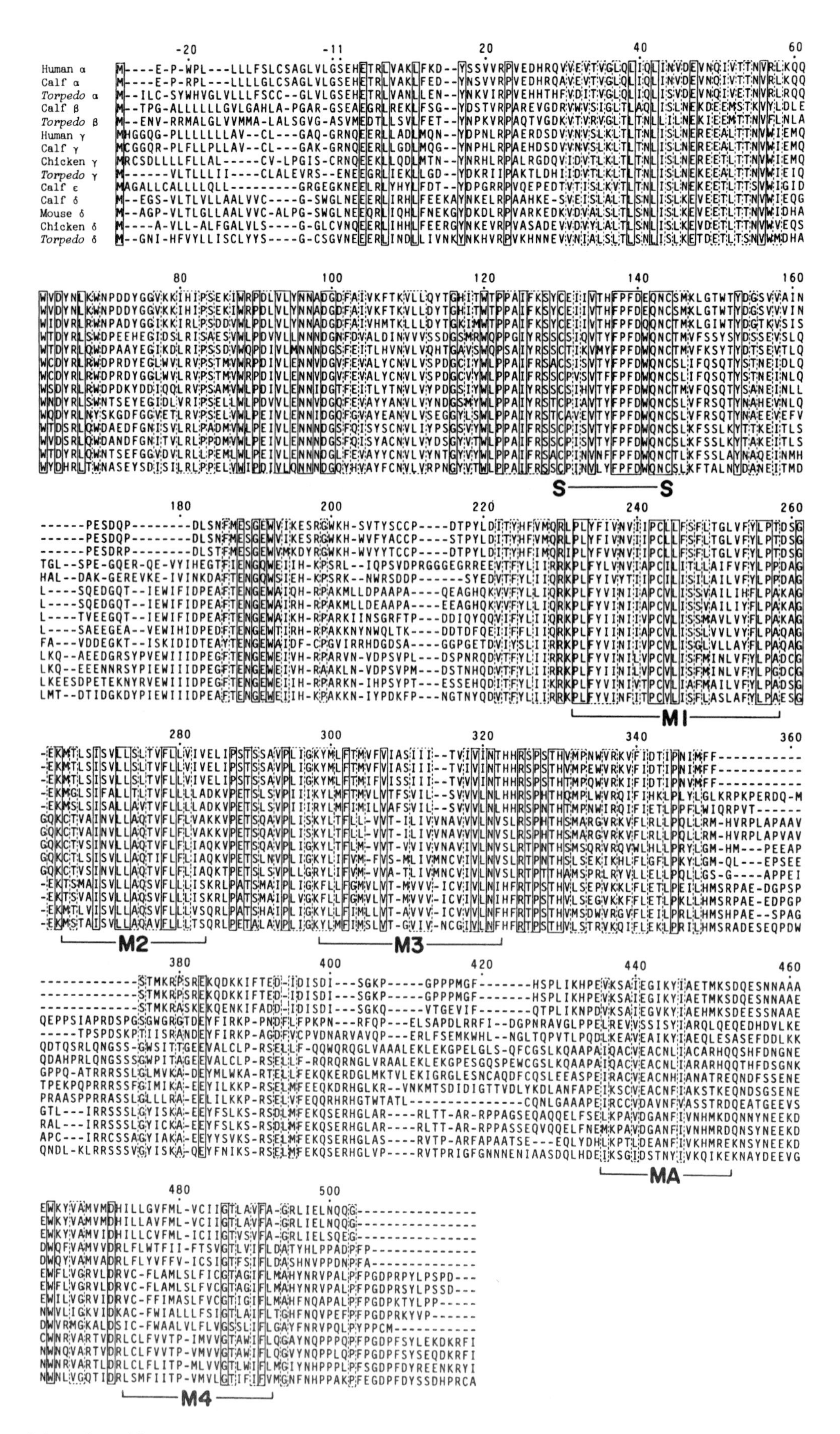

Fig. 1. Alignment of the amino acid sequences of mammalian, avian and fish AChR subunit precursors. The one-letter amino acid notation is used. The 14 sequences from top to bottom have been taken from refs. [10, 10, 4, 11, 5, 16, 12, 17, 6, 15, 13, 14, 17 and 5], respectively. Sets of 14 identical residues at one aligned position are enclosed with solid lines, and sets of 14 identical or conservative residues at one aligned position with dotted lines. Conservative amino acid substitutions are defined as pairs of residues belonging to one of the following groups: S, T, P, A and G; N, D, E and Q; H, R and K; M, I, L and V; F, Y and W [18]. Gaps (–) have been inserted to achieve maximum homology. The positions in the aligned sequences including gaps are numbered beginning with the amino-terminal residue of the mature subunits, and the preceding positions (corresponding to the prepeptide) are indicated by negative numbers. The putative disulphide bridge shared by all subunits (S–S) and the putative transmembrane segments M1–M4 and MA are indicated.

	Human α	Calf α	*Torpedo* α	Calf β	*Torpedo* β	Human γ	Calf γ	Chicken γ	*Torpedo* γ	Calf ε	Calf δ	Mouse δ	Chicken δ	*Torpedo* δ
Human α		97	79	37	40	33	32	33	35	30	34	35	35	34
Calf α	97		79	37	39	33	33	33	35	30	35	35	36	34
Torpedo α	80	81		36	40	32	31	34	34	30	36	35	35	34
Calf β	38	38	37		58	41	40	42	41	37	40	39	40	39
Torpedo β	41	40	42	59		40	40	42	41	38	41	40	40	40
Human γ	33	34	33	41	41		92	65	55	53	46	45	46	44
Calf γ	32	33	32	40	41	92		65	55	53	47	46	47	45
Chicken γ	34	34	35	42	43	67	67		61	53	50	49	49	47
Torpedo γ	35	36	35	41	41	55	56	62		56	48	49	49	48
Calf ε	30	30	30	37	39	53	54	54	57		44	45	44	43
Calf δ	35	36	37	41	42	46	48	50	49	45		87	68	58
Mouse δ	36	36	36	39	41	45	46	49	49	45	87		67	58
Chicken δ	36	36	36	41	41	46	48	50	50	44	69	68		61
Torpedo δ	34	34	35	40	41	45	46	49	49	44	60	60	62	

Fig. 2. Homology matrix for mammalian, avian and fish AChR subunits. The calculations are based on the alignment in Fig. 1. The percentage homology of each pair of the subunit precursors is shown on the upper right side of the diagonal, and that of each pair of the mature subunits on the lower left side of the diagonal.

192 and 193 are unique to the α-subunit, being conserved in all the mammalian and fish AChR α-subunits analysed, whereas the cysteine residues 128 and 142 are conserved in all AChR subunits. It is assumed that the adjacent cysteine residues 192 and 193 of the α-subunit have a specific role in agonist binding and possibly in signal transduction, whereas the cysteine residues 128 and 142, probably forming a disulphide bridge in every subunit, are essential for maintaining the proper conformation of the extracellular region of the AChR molecule [24]. This view is supported by the affinity labelling experiments reported recently by Karlin's group [30].

On the basis of the alignment in Fig. 1, the degrees of homology among the fourteen polypeptide sequences have been calculated, gaps being counted as one substitution regardless of their length. Figure 2 shows the percentage homology of each pair of the AChR subunit precursors including the hydrophobic prepeptide (upper right side of the diagonal) and that of each pair of the mature AChR subunits (lower left side of the diagonal). The degrees of sequence homology of the β-, γ-, and δ-subunits between calf and *Torpedo* are comparable (56–60% for the mature subunits), whereas that of the α-subunits between the two species is much higher (81% for the mature subunits). This is also valid for the γ-subunits (55%) and the α-subunits (80%) of the human and *Torpedo* AChRs. These observations suggest that the α-subunit evolved more slowly than did the other subunits.

The rates of evolution of the individual AChR subunits have been evaluated by comparing the corresponding calf and *Torpedo* sequences. By using the amino acid difference K (per residue) derived from the homology values in Fig. 2 and correcting the K values for multiple substitutions during evolution by $K^c = -\ln(1-K-(1/5)K^2)$ [31], the rate of evolution v (per residue per year) is calculated by $v = K^c/2t$, where t represents the time that has elapsed since the divergence of the species compared. Adopting 500 million years for t [32], the v values for the α-, β-, γ- and δ-subunit precursors or the mature subunits (in parentheses) are calculated to be 0.25×10^{-9} (0.22×10^{-9}), 0.61×10^{-9} (0.59×10^{-9}), 0.68×10^{-9} (0.66×10^{-9}) and 0.60×10^{-9} (0.57×10^{-9}), respectively.

By using the percentage amino acid differences derived from the homology matrix in Fig. 2, a dendrogram showing the degrees of sequence relatedness among the AChR subunit precursors has been constructed by the simple clustering method [33] (Fig. 3). It is not certain at present whether this dendrogram represents the actual evolutionary tree for the AChR subunits, but the γ- and ε-subunits are likely to have diverged by the latest duplication, which may have occurred around the time of separation of mammals/birds and fish.

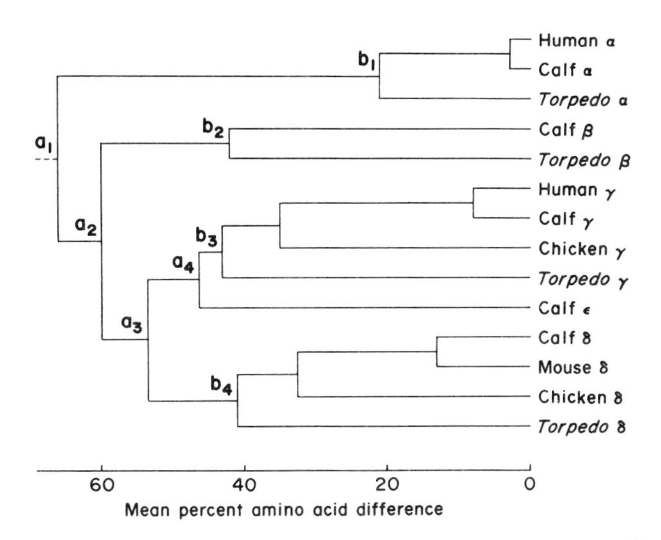

Fig. 3. Dendrogram representing the sequence relatedness among mammalian, avian and fish AChR subunit precursors. Each branch point represents the mean percentage amino acid difference between the sequences connected at that point. a_1–a_4, Branch points corresponding to divergence by gene duplications; b_1–b_4, branch points corresponding to divergence of mammals/birds and fish.

Thus, it is possible that fish also have a second, undetected gene encoding a γ/ε-family subunit, which may or may not now be active.

The finding that the β-, γ- and δ-subunit precursors evolved with comparable rates allows us to estimate reasonably the divergence times for the individual subunits. The divergence time for the γ- and δ-subunits (t_3) is calculated to be about 700 million years by $t_3 = -\ln(1-K_{\gamma\delta}-(1/5)K_{\gamma\delta}^2)/(v_\gamma+v_\delta)$, where $K_{\gamma\delta}$ represents the mean amino acid difference between the γ- and δ-subunit precursors. This suggests that the subunit structure of $\alpha_2\beta\gamma\delta$ was generated about 700 million years ago. Similarly, the divergence time for the β- and γ/δ-subunits is estimated to be about 890 million years.

The α-subunit seems to have evolved with varying rates. The rate of evolution of the α-subunit is estimated to be considerably low when the amino acid sequences are compared between calf and *Torpedo*. A phylogenetic tree constructed by a more elaborate method [34] exhibits a long branch length between the nodes a_1 and b_1 in Fig. 3, implying that the α-subunit had accumulated many amino acid substitutions at a rapid rate during this period of evolution. Because no data are available at present as to the evolutionary rate of the α-subunit in the rapid phase, the date of the earliest divergence (corresponding to the node a_1) cannot be estimated reliably. According to the tree topology in Fig. 3, however, the divergence time corresponding to the node a_1 predates that corresponding to the node a_2. It is therefore safe to state that the earliest divergence occurred more than 890 million years ago.

The protein-coding sequences of the human genes encoding the AChR α- [10] and γ-subunits [16] are divided into nine (P1–P9) and twelve exons (P1–P12), respectively, as shown schematically in Fig. 4. The two genes exhibit a generally similar exon/intron arrangement. The protein regions encoded by distinct exons seem to correspond to different structural and functional domains of the subunit precursor molecules. The chicken AChR γ- and δ-subunit genes [17] show essentially the same exon/intron arrangement as the human γ-subunit gene. Furthermore, the human [16] as well as the chicken AChR δ- and γ-subunit genes [17] are juxtaposed in the same orientation.

Figure 4 also shows the degrees of amino acid sequence homology in the different protein regions encoded by distinct exons, observed among the human, calf and *Torpedo* counterparts. Some protein regions, including the region containing the putative disulphide bridge (positions 130 and 144) and that encompassing the clustered hydrophobic segments M1, M2 and M3, are relatively well conserved among the different species. The sequence homology for the amphipathic segment MA is relatively low, but the amphipathic character of this segment is well conserved (Fig. 1). The relative pattern of regional homology is similar for all subunits [13].

The observations described above suggest that the primordial gene from which all subunit genes diverged had an exon/intron arrangement similar to that of the present-day genes. The marked correlation between exons and protein domains supports the view that the primordial gene may have evolved by exon shuffling [35].

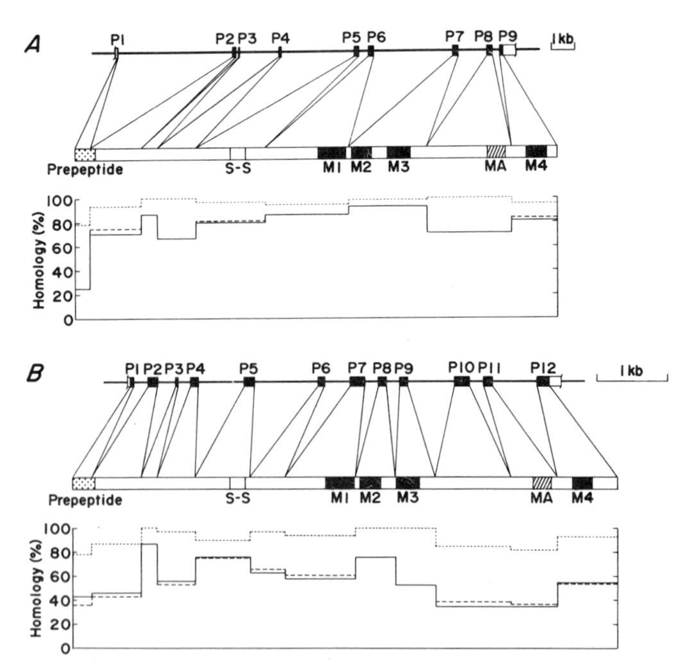

Fig. 4. Arrangement of the protein-coding regions in the human AChR α- (*A*) and γ-subunit genes (*B*) and amino acid sequence homology in the protein regions encoded by distinct exons. The exons comprising the protein-coding sequence (closed boxes) are shown on the upper line; the lengths of the 5′- and 3′-terminal exons, which extend beyond the protein-coding region, have not been determined; a scale of 1000 base pairs (1 kb) is given. The protein regions encoded by the respective exons are shown in the middle diagram. The degree of amino acid sequence homology in each of these protein regions is given below for the human/calf (dotted lines), human/*Torpedo* (solid lines) and calf/*Torpedo* (dashed lines or solid lines where no dashed lines are shown) pairs; the amino acid residues whose codons are split by introns have not been included in the calculation of homology, and gaps have been counted as one substitution regardless of their length. The positions of the prepeptide, the putative disulphide bridge shared by all subunits (S–S) and the putative transmembrane segments M1–M4 and MA are indicated. Reprinted from *Nature*, © 1983. Macmillan Journals Ltd. (ref. [10]) (*A*) and *Eur. J. Biochem.*, © 1985. *FEBS* (ref. [16]) (*B*).

Sodium channel

The sodium channel is a transmembrane protein that mediates the voltage-dependent modulation of the sodium ion permeability [36–38]. The primary structure of the sodium channel from the electric organ of the eel *Electrophorus electricus* was elucidated by cloning and sequencing the cDNA [39]. This protein consists of 1820 amino acid residues, including the initiating methionine. Homology matrix comparison [40] of the amino acid sequence revealed the presence of four internal repeats with homologous sequences, referred to as repeats I, II, III and IV (Fig. 5). This strongly suggests that the four repeated homology units evolved from a single common ancestor by internal duplications. Figure 6 shows the hydropathy profile [41] and the predicted secondary structures [20] of the *Electrophorus* sodium channel. Each repeat contains five hydrophobic segments (S1, S2, S3, S5 and S6) and one positively charged segment (S4) located between segments S3 and S5 (Figs. 5, 6).

Recently, the primary structures of two distinct sodium-channel large polypeptides (designated as sodium channels I and II) from rat brain were deduced by cloning and sequencing the respective cDNAs [42]. The structurally distinct sodium channels may be responsible for the different types of sodium currents observed in excitable membranes (for example, [43, 44]). Rat sodium channels I and II also contain four homo-

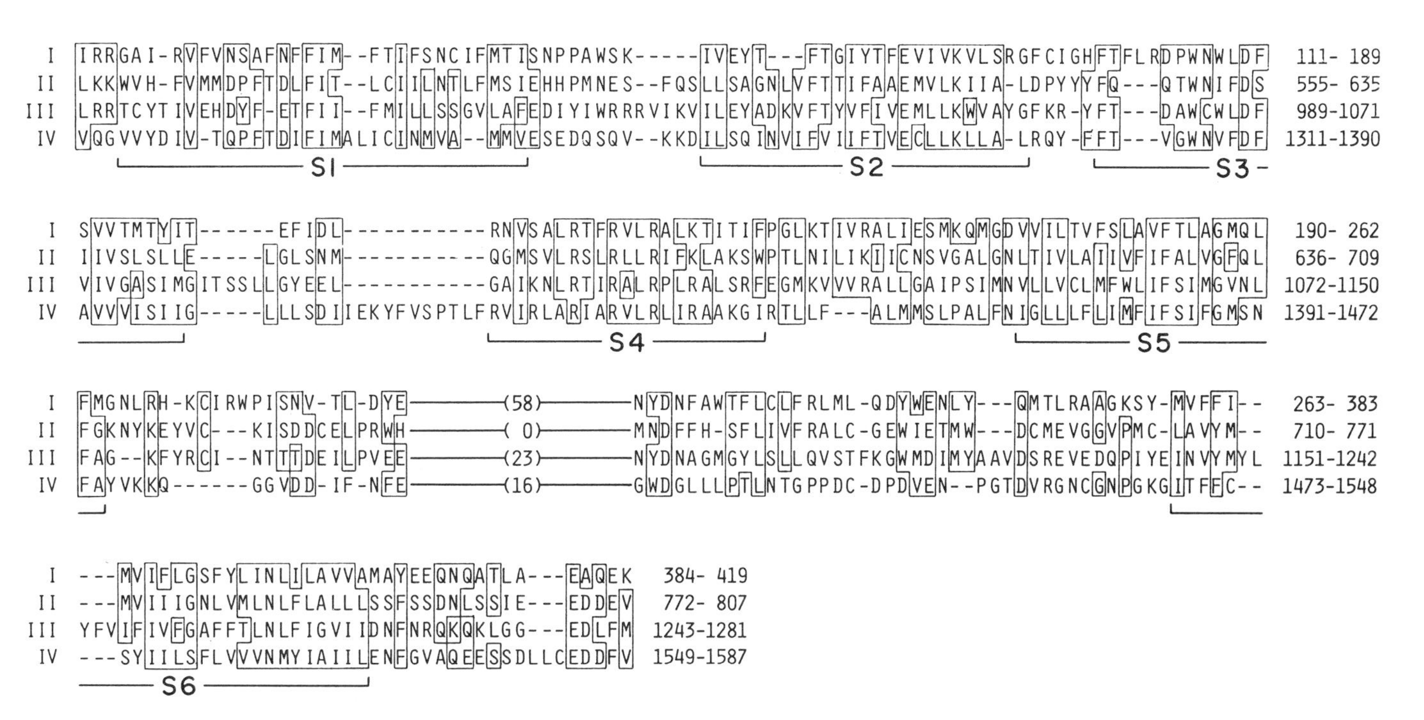

Fig. 5. Alignment of the amino acid sequences of the four internal repeats of the *Electrophorus* sodium channel. The sequences of repeats I (top), II (second row), III (third row) and IV (bottom) are aligned. The residue numbers of the amino acids on each line are given on the right side. Sets of three or four identical or conservative residues at one aligned position are boxed. Gaps (–) have been inserted to achieve maximum homology. The non-homologous segments in repeats I, III and IV are shown by lines, and the sum of residues present in each non-homologous segment is given in parentheses. Segments S1–S6 are indicated. Reprinted from *Nature*, © 1984. Macmillan Journals Ltd. (ref. [39]).

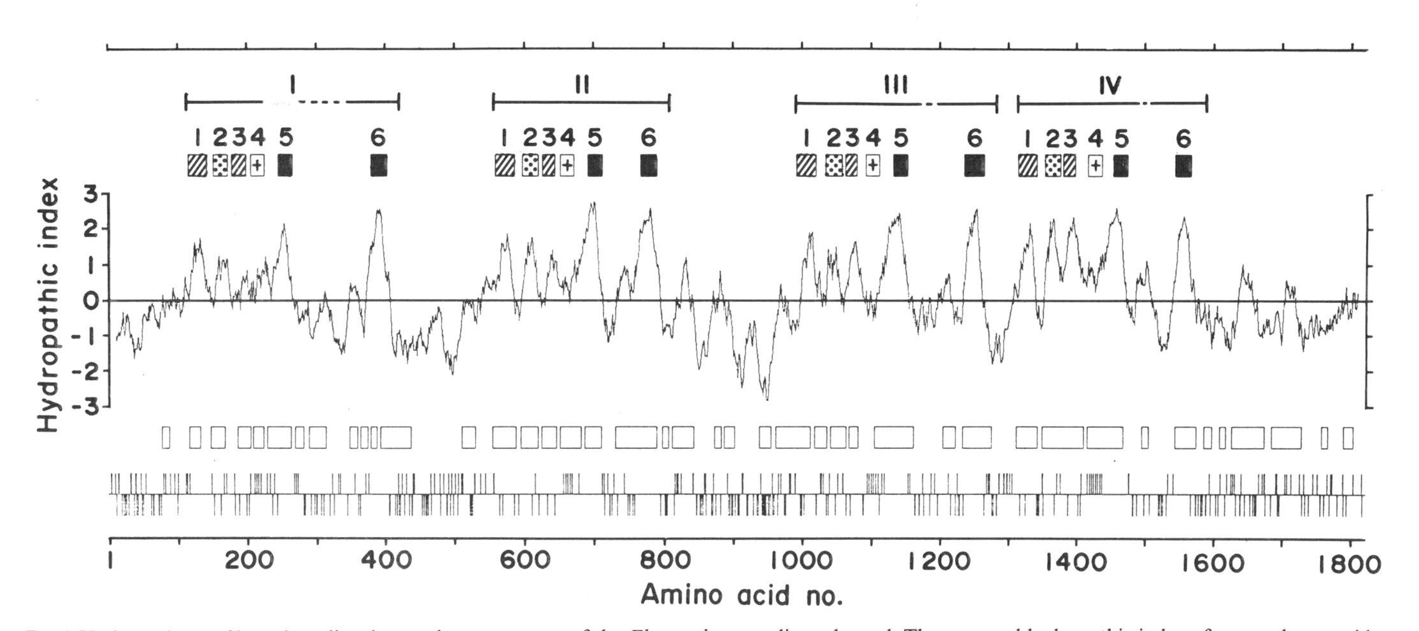

Fig 6. Hydropathy profile and predicted secondary structures of the *Electrophorus* sodium channel. The averaged hydropathic index of a nonadecapeptide composed of amino acid residues $i-9$ to $i+9$ has been plotted against i, where i represents amino acid number. The positions of the predicted structures of α-helix and/or β-sheet that have a length of ten or more residues are shown by open boxes. The positions of positively charged residues (lysine and arginine) and negatively charged residues (aspartic acid and glutamic acid) are indicated by upward and downward vertical lines, respectively. The locations of repeats I–IV are shown by bars, and those of the non-homologous segments within the repeats by dashed lines. The positions of segments S1–S6 in each repeat are indicated. Reprinted from *Nature*, © 1984. Macmillan Journals Ltd. (ref. [39]).

logous internal repeats, each of which has six segments (S1–S6) similar to those of the *Electrophorus* sodium channel. The regions corresponding to the four repeats are highly conserved among the three sodium channels, whereas the remaining regions, all of which are assigned to the cytoplasmic side of the membrane [39, 42], are generally less well conserved.

It is assumed that the four repeated homology units are oriented in a pseudosymmetric fashion across the membrane, segments S1–S6 representing transmembrane segments [39, 42]. Segment S4 is unique in that it contains four to eight arginine or lysine residues located at every third position. The residues intervening between the basic residues are mostly non-polar. This characteristic structure of segment S4 is strikingly well conserved among the three sodium channels.

The voltage-dependent operation of the sodium channel implies that the structure responsible for gating is charged and is able to move in the membrane field, as postulated by

Hodgkin and Huxley [36] and subsequently confirmed by measurement of the gating current [45]. In view of the high degree of conservation of segment S4, it seems most likely that the positive charges present in this segment, many of which presumably form ion pairs with the negative charges in other segments such as segments S3 and S1, act as voltage sensor, thus being involved in activation of the sodium channel [39, 42]. The clusters of acidic residues present between segment S6 of repeat II and segment S1 of repeat III are not so conspicuous in the rat sodium channels as in the *Electrophorus* counterpart. The presence of four homologous repeats in a single sodium channel molecule would be consistent with the sigmoid activation kinetics characteristic of this channel.

It is of interest to compare the molecular structures of the sodium channel and the AChR, both of which contain a transmembrane ionic channel. The AChR is a pentameric protein complex consisting of homologous subunits, while the sodium channel is a single large polypeptide with a molecular size comparable to that of the whole AChR complex and contains four homologous internal repeats. The structural repeats of the sodium channel may correspond functionally to the AChR subunits in that both contain about an equal number of transmembrane segments and are oriented presumably in a pseudosymmetric fashion across the membrane, forming an ionic channel. Thus, the two channel proteins, though their gating is effected in different ways, may be similar in basic structural organization. Moreover, the AChR and the sodium channel share common evolutionary features. The AChR subunits are encoded by separate genes that apparently have been generated from a single common ancestor by gene duplications, while the four repetitive units of the sodium channel are encoded by a single gene that presumably evolved from a primordial gene by internal duplications.

Acknowledgements

The author is indebted to Dr Takashi Miyata for helpful discussions. This investigation was supported in part by research grants from the Ministry of Education, Science and Culture of Japan, the Mitsubishi Foundation and the Japanese Foundation of Metabolism and Diseases.

References

1. Karlin, A., in *Cell Surface Reviews* (ed. C. W. Cotman, G. Poste and G. L. Nicolson), vol. 6, p. 191. North-Holland, Amsterdam (1980).
2. Changeux, J.-P., *Harvey Lect.* **75**, 85 (1981).
3. Conti-Tronconi, B. M. and Raftery, M. A., *A. Rev. Biochem.* **51**, 491 (1982).
4. Noda, M., Takahashi, H., Tanabe, T., Toyosato, M., Furutani, Y., Hirose, T., Asai, M., Inayama, S., Miyata, T. and Numa, S., *Nature (London)* **299**, 793 (1982).
5. Noda, M., Takahashi, H., Tanabe, T., Toyosato, M., Kikyotani, S., Hirose, T., Asai, M., Takashima, H., Inayama, S., Miyata, T. and Numa, S., *Nature (London)* **301**, 251 (1983).
6. Noda, M., Takahashi, H., Tanabe, T., Toyosato, M., Kikyotani, S., Furutani, Y., Hirose, T., Takashima, H., Inayama, S., Miyata, T. and Numa, S., *Nature (London)* **302**, 528 (1983).
7. Claudio, T., Ballivet, M., Patrick, J. and Heinemann, S., *Proc. Natl. Acad. Sci. USA* **80**, 1111 (1983).
8. Sumikawa, K., Houghton, M., Smith, J. C., Bell, L., Richards, B. M. and Barnard, E. A., *Nucleic Acids Res.* **10**, 5809 (1982).
9. Devillers-Thiery, A., Giraudat, J., Bentaboulet, M. and Changeux, J.-P., *Proc. Natl. Acad. Sci. USA* **80**, 2067 (1983).
10. Noda, M., Furutani, Y., Takahashi, H., Toyosato, M., Tanabe, T., Shimizu, S., Kikyotani, S., Kayano, T., Hirose, T., Inayama, S. and Numa, S., *Nature (London)* **305**, 818 (1983).
11. Tanabe, T., Noda, M., Furutani, Y., Takai, T., Takahashi, H., Tanaka, K., Hirose, T., Inayama, S. and Numa, S., *Eur. J. Biochem.* **144**, 11 (1984).
12. Takai, T., Noda, M., Furutani, Y., Takahashi, H., Notake, M., Shimizu, S., Kayano, T., Tanabe, T., Tanaka, K., Hirose, T., Inayama, S. and Numa, S., *Eur. J. Biochem.* **143**, 109 (1984).
13. Kubo, T., Noda, M., Takai, T., Tanabe, T., Kayano, T., Shimizu, S., Tanaka, K., Takahashi, H., Hirose, T., Inayama, S., Kikuno, R., Miyata, T. and Numa, S., *Eur. J. Biochem.* **149**, 5 (1985).
14. LaPolla, R. J., Mayne, K. M. and Davidson, N., *Proc. Natl. Acad. Sci. USA* **81**, 7970 (1984).
15. Takai, T., Noda, M., Mishina, M., Shimizu, S., Furutani, Y., Kayano, T., Ikeda, T., Kubo, T., Takahashi, H., Takahashi, T., Kuno, M. and Numa, S., *Nature (London)* **315**, 761 (1985).
16. Shibahara, S., Kubo, T., Perski, H. J., Takahashi, H., Noda, M. and Numa, S., *Eur. J. Biochem.* **146**, 15 (1985).
17. Nef, P., Mauron, A., Stalder, R., Alliod, C. and Ballivet, M., *Proc. Natl. Acad. Sci. USA* **81**, 7975 (1984).
18. Dayhoff, M. O., Schwartz, R. M. and Orcutt, B. C., in *Atlas of Protein Sequence and Structure* (ed. M. O. Dayhoff), vol. 5, suppl. 3, p. 345. National Biomedical Research Foundation, Silver Spring, Maryland (1978).
19. Hopp, T. P. and Woods, K. R., *Proc. Natl. Acad. Sci. USA* **78**, 3824 (1981).
20. Chou, P. Y. and Fasman, G. D., *A. Rev. Biochem.* **47**, 251 (1978).
21. Numa, S., Noda, M., Takahashi, H., Tanabe, T., Toyosato, M., Furutani, Y. and Kikyotani, S., *Cold Spring Harbor Symp. Quant. Biol.* **48**, 57 (1983).
22. Guy, H. R., *Biophys. J.* **45**, 249 (1984).
23. Finer-Moore, J. and Stroud, R. M., *Proc. Natl. Acad. Sci. USA* **81**, 155 (1984).
24. Mishina, M., Tobimatsu, T., Imoto, K., Tanaka, K., Fujita, Y., Fukuda, K., Kurasaki, M., Takahashi, H., Morimoto, Y., Hirose, T., Inayama, S., Takahashi, T., Kuno, M. and Numa, S., *Nature (London)* **313**, 364 (1985).
25. Mishina, M., Kurosaki, T., Tobimatsu, T., Morimoto, Y., Noda, M., Yamamoto, T., Terao, M., Lindstrom, J., Takahashi, T., Kuno, M. and Numa, S., *Nature (London)* **307**, 604 (1984).
26. Marshall, R. D., *Biochem. Soc. Symp.* **40**, 17 (1974).
27. Karlin, A., *J. Gen. Physiol.* **54**, 245s (1969).
28. Dayhoff, M. O., in *Atlas of Protein Sequence and Structure* (ed. M. O. Dayhoff), vol. 5, suppl. 2, p. 77. National Biomedical Research Foundation, Silver Spring, Maryland (1976).
29. Ovchinnikov, Yu. A., Lipkin, V. M., Shuvaeva, T. M., Bogachuk, A. P. and Shemyakin, V. V., *FEBS Lett.* **179**, 107 (1985).
30. Kao, P. N., Dwork, A. J., Kaldany, R.-R. J., Silver, M. L., Wideman, J., Stein, S. and Karlin, A., *J. Biol. Chem.* **259**, 11662 (1984).
31. Kimura, M., *The Neutral Theory of Molecular Evolution.* Cambridge University Press, Cambridge (1983).
32. Dayhoff, M. O., in *Atlas of Protein Sequence and Structure* (ed. M. O. Dayhoff), vol. 5, suppl. 3, p. 1. National Biomedical Research Foundation, Silver Spring, Maryland (1978).
33. Sokal, R. R. and Sneath, P. H., *Principles of Numerical Taxonomy.* Freeman, San Francisco (1963).
34. Li, W.-H., *Proc. Natl. Acad. Sci. USA* **78**, 1085 (1981).
35. Gilbert, W., *Nature (London)* **271**, 501 (1978).
36. Hodgkin, A. L. and Huxley, A. F., *J. Physiol. (London)* **117**, 500 (1952).
37. Agnew, W. S., *A. Rev. Physiol.* **46**, 517 (1984).
38. Catterall, W. A., *Science* **223**, 653 (1984).
39. Noda, M., Shimizu, S., Tanabe, T., Takai, T., Kayano, T., Ikeda, T., Takahashi, H., Nakayama, H., Kanaoka, Y., Minamino, N., Kangawa, K., Matsuo, H., Raftery, M. A., Hirose, T., Inayama, S., Hayashida, H., Miyata, T. and Numa, S., *Nature (London)* **312**, 121 (1984).
40. Toh, H., Hayashida, H. and Miyata, T., *Nature (London)* **305**, 827 (1983).
41. Kyte, J. and Doolittle, R. F., *J. Molec. Biol.* **157**, 105 (1982).
42. Noda, M., Ikeda, T., Kayano, T., Suzuki, H., Takeshima, H., Kurasaki, M., Takahashi, H. and Numa, S., *Nature (London)* **320**, 188 (1986).
43. Gilly, W. F. and Armstrong, C. M., *Nature (London)* **309**, 448 (1984).
44. Benoit, E., Corbier, A. and Dubois, J.-M., *J. Physiol. (London)* **361**, 339 (1985).
45. Armstrong, C. M. and Bezanilla, F., *Nature (London)* **242**, 459 (1973).

Chemica Scripta 26B, 179–190

Evolution of Hormones and Their Actions

J. R. Tata

Laboratory of Developmental Biochemistry, National Institute for Medical Research, The Ridgeway, Mill Hill, London NW7 1AA, United Kingdom

Paper presented at the Conference on 'Molecular Evolution of Life', Lidingö, Sweden, 8–12 September 1985

Abstract

Hormones are chemical messengers that help coordinate all major metabolic and growth activities of different populations of cells of an organism. Hormones of higher animals have been detected in primitive organisms, such as unicellular eukaryotes, microbes and even plants. Thus hormones arose before many of the functions they control. The acquisition of a hormonal function most likely coincided with the appearance of specific receptors and there is indirect evidence for peptide hormones that the hormone–receptor complex has co-evolved as a unit. Recent recombinant DNA work has shown a high degree of sequence homology between oncogene products and both peptide hormones and their membrane associated receptors. For non-protein hormones such as steroids, iodothyronines and catecholamines, their structure has remained mostly invariant and the receptors highly conserved during evolution. Many of these hormones, as well as small peptide hormones, occur in nervous tissue, possess neurotransmitter activities, and are often chemically related to neurotransmitters and some plant hormones. It has been suggested that early in evolution hormones served a neurocrine or paracrine function and that the present-day nervous and endocrine systems have a common origin. The above aspects of hormones and their actions will be discussed in the context of the evolution of a system of molecular linguistics in intercellular communication.

Introduction

All living organisms, from the most primitive unicellular forms to higher vertebrates possess mechanisms of chemical communication between individual cells as well as for sensing the changes in the immediate environment. This means of communication is essential for adaptation to nutritional changes, defence against noxious agents, coordinating reproductive activity, etc. Examples of primitive systems of communication are amoeboid streaming via cyclic AMP gradients, chemotaxis in bacteria and sex factors in yeast. The dependence on chemical communication must have increased greatly with the advent of multicellular organisms and increasingly complex physiological functions performed by specialized groups of cells. This review deals mainly with how the hormonal system of chemical communication and cellular responses to the signals generated have evolved in complex multicellular organisms.

Hormones can be best defined as chemical messengers that help coordinate the activities of one group of cells with those of another in response to environmental signals. There is virtually no metabolic, growth or developmental process in higher organisms that is not regulated by hormones. Although the term 'hormone' was originally introduced to explain the regulation of physiological functions of higher animals, it is now amply clear that most hormones initially discovered in higher eukaryotes are found in simple, primitive and often unicellular organisms. Hormones are therefore primitive substances, often highly conserved in evolution but not always exerting the same action. It is essential that in considering the evolution of hormones one also considers simultaneously the diversification of their function. The reader is referred to several reviews on comparative or evolutionary endocrinology [1–5].

Major characteristics of hormones

Before considering the questions relating to the evolution of hormones and the processes they control, it is useful to summarize the major features that characterize hormones and distinguish them from other informational molecules such as chemoattractants, pheromones, autocrine growth factors, etc. Table I lists a few well-known hormones and physiological or biochemical processes regulated by them in multicellular plants and animals. The following important characteristics can be generalized:

(1) Hormones are not restricted to any one class of chemicals. They range from simple two-carbon compounds to steroids, amino acids, simple peptides and complex proteins.

(2) Most hormones elicit multiple biochemical responses in their target cells. This raises the important question of whether or not the multiplicity is due to the interaction between the hormone and a single or multiple site(s) in the target cell.

(3) There is a high degree of target-cell specificity for each hormonal response, which highlights the evolutionary acquisition of receptors to a given hormone and functional specialization.

(4) Growth and developmental hormones do not regulate early embryonic development or determine phenotypic specialization. Rather, they modulate a pre-determined, but as yet unexpressed, programme in partially differentiated cells.

(5) The metabolic activity, growth and development of many tissues is often regulated by the coordinated action of multiple hormones, e.g. the control of lactation and other activities of the mammary gland by prolactin, oestrogen and glucocorticoids. This type of multiple control has two implications: (*a*) that each hormone regulates a different cellular function and (*b*) that one hormone regulates the level or activity of receptor for another.

Evolution of hormones

A commonly held view is that present-day hormones first appeared as by-products or waste-products of primitive

Table I. *Some hormones and their actions*

Hormone	Chemical nature	Major physiological and biochemical actions
Adrenaline	Catecholamine	Cardiac activity; thermogenesis; glycogenolysis; regulation of cyclic AMP level
Insulin	Protein	Carbohydrate metabolism; sugar transport; DNA synthesis
Growth hormone	Protein	Regulation of growth in vertebrates; control of overall protein synthesis; lipolysis; amino acid uptake
Vasopressin	Nonapeptide	Water and ion transport in bladder, regulation of cyclic AMP levels; modulation of Na^+ pump
Prostaglandins	Fatty-acid derivatives	Inflammatory responses; uterine tonicity and contraction; regulation of phosphodiesterase and breakdown of cyclic AMP
Ecdysone	Steroid	Insect metamorphosis; salivary-gland regression; sclerotization and pigmentation; gene puffing; regulation of transcription and enzyme induction
L-Thyroxine	Iodothyronine	Amphibian metamorphosis; brain development in mammalian embryos; regulation of basal metabolic rate and growth in neonatal mammals; control of RNA and protein synthesis
Oestradiol	Steroid	Growth and maturation of accessory sexual tissues; induction of egg white and yolk proteins in egg-laying vertebrates; regulation of prolactin synthesis; control of gene expression
Prolactin	Protein	Control of lactation and milk protein synthesis; juvenilization and control of 'water drive' in amphibia

metabolic systems. This may be particularly true for many peptide hormones derived by partial digestion of larger proteins in the gut and other tissues with a specialized digestive function, e.g. salivary glands, pancreas, etc. However, animal hormones are also present in the most primitive forms of plants and even micro-organisms [6]. Thyroid hormones could have arisen as a consequence of protein iodination and cleavage, and indeed thyroid hormone-like substances have been detected in algae and sponges [7]. Similarly, steroid hormones such as oestrogen have been detected in ferns and could have been the consequence of acquisition of the ability to synthesize cholesterol [6, 8], but it has not been possible to assign a hormonal function for the large amounts of oestrogens found in some plants. It is almost certain that the catecholamines, adrenaline and noradrenaline are by-products of melanin or pigment formation in primitive organisms. What is particularly important to realize is that the same substance with a hormonal function in higher organisms may have no identifiable function as a hormone or in chemical communication in primitive organisms. In other words, the acquisition by hormones of a role as chemical messengers must have arisen later through the evolution of receptors in what we now know as target cells. This raises the important question, to be considered later, of the dependence of hormonal function on the evolution or co-evolution of receptors to pre-existing substances.

Evolution of protein hormones

A considerable volume of data on amino acid sequences, and more recently nucleic acid sequences, has provided valuable insight into the evolution of genes encoding protein and peptide hormones [4, 5, 9–12]. These include such hormones as insulin, growth hormone, adrenocorticotropin (ACTH), thyroid-stimulating hormone (TSH), and smaller peptide hormones such as oxytocin, vasopressin, etc. Since other authors are dealing with this question in detail in this volume (see articles by Numa, Mutt, Falkmer, Blundell), I shall restrict myself to a few aspects of evolution of protein hormones particularly in the context of their function.

Perhaps the most intensively studied protein hormone from an evolutionary point of view is insulin (see Falkmer, this volume). Proinsulin may be the primitive substance, perhaps non-hormonally involved in the digestive process and with no insulin-like activity, and that the insulin receptor only recognizes the A and B chains when correctly joined and folded. That those regions of insulin in contact with the receptor constitute the most highly conserved region of the hormone molecule was first convincingly established by X-ray crystallographic analysis of insulin and glucagon [13, and see Blundell, this volume].

Roth and his colleagues have recently published extensively on the very early evolutionary origin of insulin and several other peptide hormones [6, 14–16]. These workers have detected somatostatin, insulin and ACTH in such primitive organisms as unicellular fungi, *Tetrahymena*, and even bacteria. In most instances the hormone-like material in unicellular organisms has been detected immunologically, or in some cases by bioassay. The latter suggests that a corresponding receptor-like molecule is also as primitive as the hormone and that the two must have co-evolved – an important question which will be considered in detail later. More convincing evidence for the presence of peptide hormones in primitive organisms in which they have retained their biological activity is the isolation and chemical characterization of the hormone in question. Where this has been achieved, the evolutionary span is more restricted than that from micro-organisms to mammals, as mentioned above. Ultimately, such studies on phylogenetic distribution will have to be substantiated with the identification and sequencing of genes and messenger RNA encoding protein hormones.

A large number of protein hormones, and all of the smaller peptide hormones, are now known to be derived from a larger precursor, termed 'pre-prohormone'. In some instances two or more peptide hormones or hormone-like agents may be derived from the same precursor, especially true for several neurohormones synthesized in the pituitary and hypothalamus [4, 5, 12]. Perhaps the best example of the latter group of hormones is pro-opiomelanocortin (POMC) which is synthesized in the neurointermediate lobe of the pituitary [12, and see Numa, this volume]. The endogenous opioid peptides, β-endorphin, enkephalins and dynorphins are present within a single polypeptide precursor, POMC, which also contains the hormones adrenocorticotropin ACTH, β-lipotropin (β-LPH) and melanophore-stimulating hormone (MSH). Detailed analysis of the POMC gene and its final peptide products have provided interesting insight into the evolution of the opioid peptide family. For example the enkephalin sequence is found seven times (six met-enkephalins and one leu-enkephalin) in the proenkephalin sequence while met-enkephalin also occurs once in POMC and leu-enkephalin sequence three times in dynorphin fragments. The free enkephalins are, however, not derived from these latter fragments [17]. Melanotropin is repeated three times (α-, β- and γ-MSH) within POMC, which is due to intragenic duplications [18]. These and the repeated enkephalins are found within the single large third exon. Since there are three POMC genes, it can be deduced that the present-day structure of POMC is due to both duplication of a common ancestral gene and duplication within the gene, with a primitive RNA splicing mechanism possibly eliminating the spacer between duplicated units.

Protein and DNA sequencing have further elucidated the phylogenetic variations and overall conservation of the POMC precursor molecule. Permutations of cleavages at different sites generate different products in different species, while the specificity and multiplicity of proteolytic enzymes involved in the processing reaction would determine the diversity of opioid peptides and various hormones that are generated [12, 19]. What is most intriguing is the conserved glycosylation pattern of the individual components that accompanies differential proteolytic cleavage at the N-terminal part of POMC in different species [19]. The evolutionary significance of conserved glycosylation and other post-translational modifications of cleaved hormonal peptides with such different physiological functions remains to be elucidated.

Another molecular mechanism generating functional diversity has also been described in the anterior pituitary, as illustrated in Fig. 1. The glycoprotein hormones, thyroid-stimulating hormone (TSH), luteinizing hormone (LH) and follicle-stimulating hormone (FSH) each contain two 14 kDa subunits termed TSH-α and TSH-β and CI, CII chains for the two gonadotropins. TSH-α and CI are identical in amino acid sequence, implying that the distinct functions, and hence distinct receptor interactions, are mediated by the TSH-β and CII subunits. It has been argued that the α-type chain was a primitive ancestral hormone and that gene duplication gave rise to α and β chains, followed by subsequent duplications and sequence divergence of the β chains [20, 21]. It is

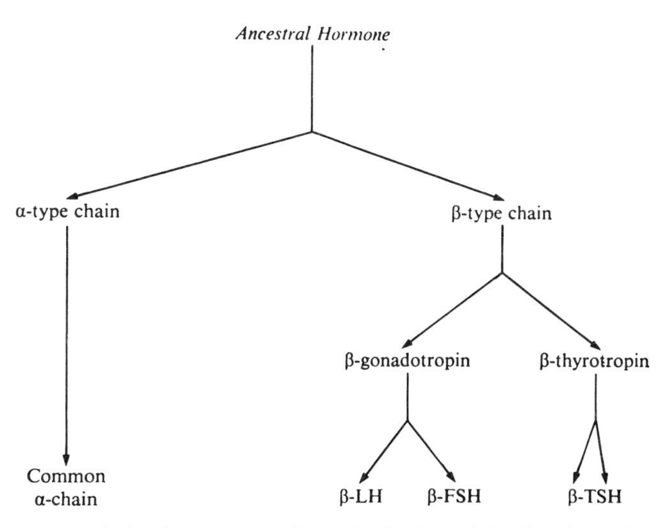

Fig. 1. Evolution by a process of gene duplication of the dimeric pituitary hormones lutropin (LH), thyrotropin (TSH) and follicle stimulating hormone (FSH). It is proposed that the duplication of an ancestral gene led to the α- and β-chains, followed by the successive duplication of the β-chains of the three hormones. From Acher [20].

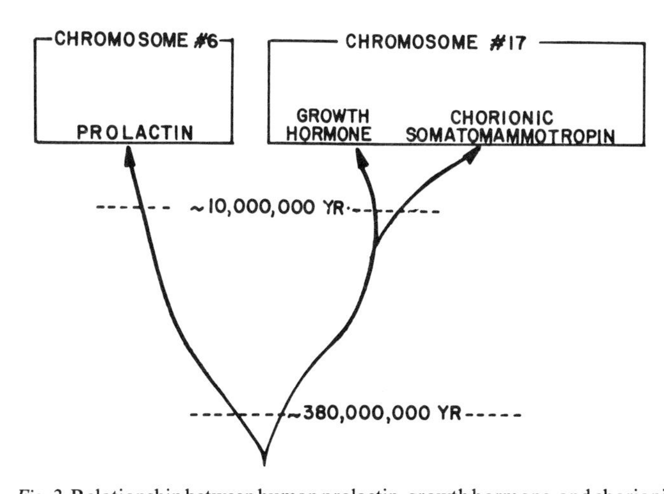

Fig. 2. Relationship between human prolactin, growth hormone, and chorionic somatomammotropin and their chromosomal segregation. Based on Cooke *et al.* [10].

important to note that each β chain is functionally associated with an α chain, suggesting that the chain–chain interactions have been highly conserved. Such separate conservation of structure in one chain with variation in another during functional evolution has also been encountered with other glycoprotein hormones and may offer a clue to the nature of receptors for these hormones in different tissues.

A different approach to evolutionary diversification of protein hormones is exemplified by an analysis using recombinant DNA methods (Fig. 2). Sequencing of cloned cDNA has made it possible to describe with some accuracy the evolutionary relationship between the individual hormone coding sequences in the human growth hormone (hGH), prolactin and chorionic somatomammotropin multigene family [10, 11]. Thus, chromosomal segregation of human prolactin and hGH occurred around 4×10^8 years ago and that hGH and HCS underwent an intrachromosomal recombination more recently within the last 10^7 years. This approach based on recombinant DNA and DNA sequencing not only allows one to establish the evolutionary diversification of peptide hormones but can also reveal the presence of as yet undiscovered hormones. A good example illustrating the

Table II. *Structures of the neurohypophysial hormones*

Oxytocin-like peptides 1 2 3 4 5 6 7 8 9	Vasopressin-like peptides 1 2 3 4 5 6 7 8 9
Cys-Tyr-Ile-Gln-Asn-Cys-Pro-Leu-Gly(NH_2) Oxytoxin (prototherian, metatherian and eutherian mammals)	Cys-Tyr-Phe-Gln-Asn-Cys-Pro-Arg-Gly(NH_2) Arginine vasopressin (prototherian, metatherian and most eutherian mammals)
Cys-Tyr-Ile-Gln-Asn-Cys-Pro-Ile-Gly(NH_2) Mesotocin (Australian marsupials, birds, reptiles, amphibians, lungfish)	Cys-Tyr-Phe-Gln-Asn-Cys-Pro-Lys-Gly(NH_2) Lysine vasopressin (pig; metatherians)
Cys-Tyr-Ile-Ser-Asn-Cys-Pro-Ile-Gly(NH_2) Isotocin (bony fish)	Cys-Phe-Phe-Gln-Asn-Cys-Pro-Arg-Gly(NH_2) Phenypressin (metatherians: macropodids)
Cys-Tyr-Ile-Ser-Asn-Cys-Pro-Gln-Gly(NH_2) Glumitocin (cartilaginous fish; rays)	Cys-Tyr-Ile-Gln-Asn-Cys-Pro-Arg-Gly(NH_2) Vasotocin (nonmammalian vertebrates)
Cys-Tyr-Ile-Gln-Asn-Cys-Pro-Val-Gly(NH_2) Valitocin (cartilaginous fish; sharks)	
Cys-Tyr-Ile-Gln-Asn-Cys-Pro-Val-Gly(NH_2) Aspargtocin (cartilaginous fish; sharks)	

From Acher [9].

latter is the discovery of 'calcitonin gene related peptide' (CGRP) with a novel hormone activity made in the course of analysis of the gene and mRNA encoding the hormone calcitonin [22, 23]. Another interesting example is the recent demonstration by expression of cloned genes in bacteria that gonadotropin releasing hormone (GnRH) and prolactin inhibiting factor (PIF), synthesized in the hypothalamus, are encoded by the same gene [24]. It explains why the enhanced synthesis of gonadotropin is so precisely co-ordinated with inhibition of that of prolactin in the pituitary, an essential requirement for the correct timing of changes in ovarian function during reproductive activity.

Another group of peptide hormones which have been intensively studied, particularly by Acher [5, 9, 20] in the context of structure–function relationship during evolution, are the small neurohypophysial peptide families of oxytocin and vasopressin [5, 9]. These are peptides with nine amino acids, five of which are invariant. These can be divided according to the variable amino acids into six oxytocin-like and four vasopressin-like hormones (Table II). The phylogenetic distribution of these variants is of particular evolutionary interest since usually only one type of peptide is found in a given species. A thorough analysis of these neurohypophysial hormones has made it possible to deduce the evolutionary divergence of these hormones (Fig. 3). In line with other small peptide hormones, recombinant DNA methods have shown vasopressin to be synthesized also within a larger protein precursor and which contains the peptide hormone neurophysin and an unidentified glycopeptide [25]. More recently, the rat vasopressin gene has been isolated and the intron–exon structure of the domains encoding these three components established [26]. The evolution of these precursor genes in the light of their different functions has not yet been elucidated. Since neurohypophysial- and neurophysin-like peptides have been detected in invertebrates [27], it is most likely that an ancestral protogene encoding a multifunctional protein was formed before the invertebrate–vertebrate divergence. As shown in Fig. 4, it is assumed that a gene duplication occurred before the emergence of fish and that parallel evolution led to the appearance of eutherian hormones, oxytocin and vasopressin, along with their respective neurophysins. Since all neurohypophysial hormones have nine amino acids, it is

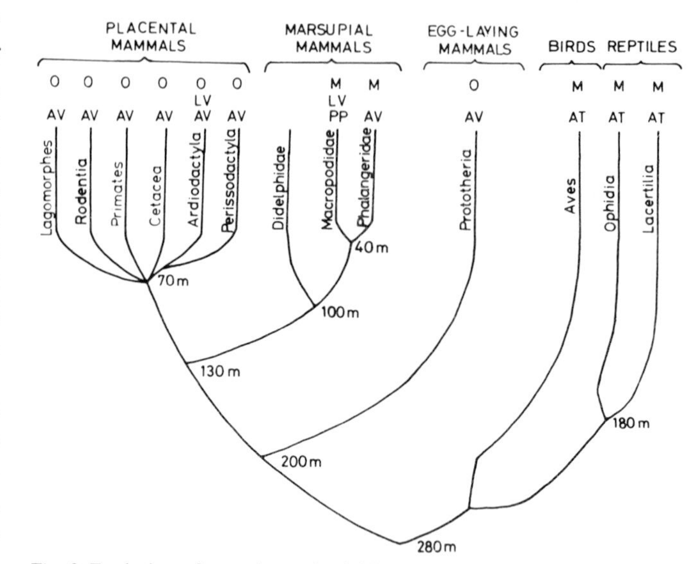

Fig. 3. Evolution of neurohypophysial hormones as expressed in millions of years (m) since the divergence of tetrapods. O, oxytocin; AV, arginine vasopressin; LV, lysine vasopressin; M, mesotocin; PP, phenypressin; AT, arginine vasotocin. From Acher [9].

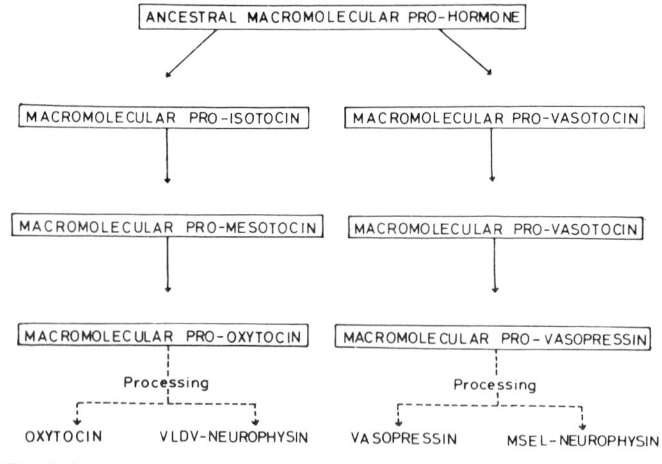

Fig. 4. Scheme suggested by Acher [9] to explain how a duplication of an ancestral gene gave rise to two macromolecular precursors of common neurohypophysial hormone–neurophysin precursors. It is proposed that subsequent mutations led to separate oxytocin-like VLDV- and vasopressin-like MSEL-neurophysin lines and that a highly conserved processing of these precursors finally gave rise to the nonapeptides oxytocin and vasopressin and the corresponding neurophysins.

the processing step, presumably dependent on two-enzyme systems, that has been highly conserved during evolution. The conservation of enzymes involved in the 'tailoring' of the final hormonally active compound is a particularly remarkable feature of the evolution of non-protein hormones.

Evolution of non-protein hormones

The structure of most non-protein hormones, such as steroids, prostaglandins, iodothyronines, etc., has remained virtually unchanged through evolution. All contemporary non-protein hormones can therefore be found in very primitive organisms. Any phylogenetic variation in the nature of such substances is most often due to evolutionary changes in the enzymes involved in their biosynthesis, modification or metabolism.

Progenitors of many non-protein hormones may have served the function of important nutrients or vitamins in primitive organisms. For example, vitamin D is a prohormone in higher animals for the hormone 25-hydroxyergo-calciferol. The conversion takes place in the kidney by an enzyme which is itself regulated by another hormone [28–30]. It is the acquisition of this enzyme that is the key evolutionary event in the conversion of a nutritional factor to a hormone involved in a highly specific manner in regulating calcium metabolism.

In the case of other steroid hormones, the important step in their evolution was not only the capacity to synthesize cholesterol, the parent compound for steroids, but also the capacity to metabolize sterols. It seems that the ability to form polar steroids from cholesterol arose as a signal substance rather early in evolution, as judged from the detection of 20-hydroxyecdysone in pulmonates [31]. The earliest hormone or chemical messenger could have been a hydroxylated cholesterol, such as 3β-, 14α-, 20,22-tetrahydroxycholestane. Thus it is not surprising that oestrogens and ecdysteroids, which play a key role in the reproductive and developmental physiology of all vertebrates and invertebrates, have been detected in many lower eukaryotes, primitive plants, yeast, and even bacteria [32, 33]. In some ferns these steroids are present in very high concentrations but their function is not known. However, the utilization of steroids as regulators of reproduction and development of other primitive organisms is known. Good examples to illustrate this feature are the male and female sex hormones antheridiol and oogoniol (also known as gamones because of their role in gametogenesis) in the water mould *Achlya* in which their function is analogous to that of androgen and oestrogen in vertebrates [8, 34]. The structures of these steroids are shown in Fig. 5. There is limited phylogenetic diversity among other steroids. A well-known example is that of the glucocorticoid hormones, cortisol and corticosterone. In some organisms, both forms are easily detected, but cortisol is the major glucocorticoid produced in the guinea pig and hamster whereas it is corticosterone in rat and man [2]. The reason for this type of diversity remains unknown, especially as we ignore if it reflects an adaptation to changes that may have possibly evolved in receptor structure in different organisms.

There is no phylogenetic variation in many hormones of low molecular weight. For example, the iodothyronines, L-thyroxine and triiodo-L-thyronine, are the only active principles in all organisms in which they regulate developmental and metabolic functions. Only in higher organisms are they synthesized within the thyroglobulin molecule and exclusively in the thyroid gland [34]. In invertebrates and in plants these are produced by non-specialized cells, often in large amounts [35, 36]. Again, as for the sex steroids, they seem not to have a hormone-like function in plants and invertebrates but may serve as a structural or metabolic element. At the other end of the spectrum of phylogenetic diversity of non-protein hormones are the prostaglandins. A wide range of prostaglandins of different structures have now been identified in organisms ranging from horny corals and jelly fish to arthropods, molluscs and mammals [37]. The evolutionary significance of this diversification is not clear, especially as prostaglandins exert a wide range of effects in different tissues and organisms and also because of considerable interconversion of prostaglandins within a given cell type. Be as it may, the fact that these and most other non-protein hormones, such as steroids, iodothyronines and catecholamines, have been retained as important elements of chemical communication and physiological regulatory processes further enhances the importance of considering the evolution of hormonal responses.

Evolution of hormonal responses

A characteristic feature of many hormones is that the same hormone may elicit quite different responses in different organisms or in different tissues of the same organism. Despite this diversity of responses, the major cellular regulatory mechanisms have been retained through evolution and can be classified according to the following three major regulatory processes:

(*a*) The control of permeability or transport of nutrients and other small molecules [3].

(*b*) The ability to generate or transduce secondary intracellular signals or messengers, such as cyclic AMP, Ca^{2+} and inositol triphosphate, which in turn control multiple metabolic processes [38, 39].

(*c*) The regulation of synthesis of enzymes or other proteins to meet developmental, growth or metabolic demands [3, 40].

It is in analyzing the multiplicity of biochemical responses to a given hormone that a consideration of the above classification acquires considerable importance in deciding which one of these would be most intimately related to the hormone–receptor interaction. For example, the physiological actions of rapidly acting hormones, such as adrenaline, vasopressin, etc., are derived from the first two basic processes listed above. On the other hand, the relatively slower actions of growth-promoting and developmental hormones, such as thyroid hormones and sex steroids, are mediated via regulation of nucleic acid and protein synthesis. However, all growth and developmental hormones elicit multiple responses, some of which occur with extreme rapidity and are independent of regulation at the level of nucleic acid and protein synthesis. This multiplicity and the question of whether or not, or to what extent, such rapid effects are in any way related to the relatively slow biosynthetic responses to the hormone, pose a major challenge to the simple biochemical concepts of cellular regulation.

A good example of a protein hormone with varied function during evolution is that of prolactin. As illustrated in Fig. 6, mammalian prolactin, whose major function in mammals is the regulation of lactation and luteotropic activity, also

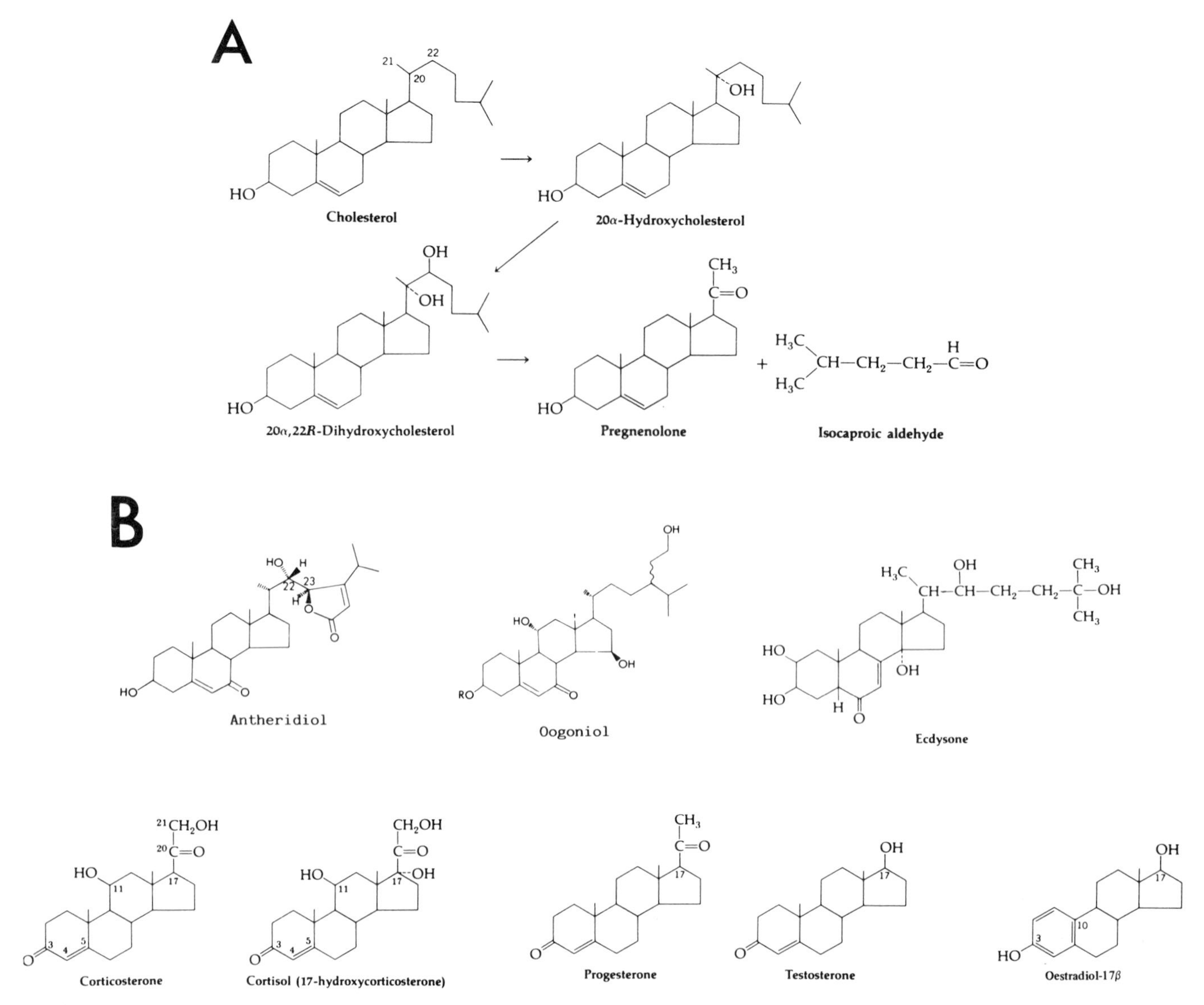

Fig. 5. (A) An abbreviated scheme showing the conversion of cholesterol to pregnenolone from which all steroid hormones are derived [80]. It is the acquisition of enzymes in specialized cells to interconvert steroids that has given rise to present-day hormones. (B) Some steroid hormones that are active in primitive and higher organisms with very specific functions. These include the fungal gametogenic steroids antheridiol and oogoniol, the invertebrate metamorphosis hormone ecdysone, the vertebrate sex hormones oestradiol, testosterone and progesterone, and the glucocorticoids, cortisol and corticosterone. Each of these hormones has very distinct receptors.

controls important physiological functions in non-mammalian vertebrates [41]. In particular, it stimulates crop-sac growth in birds, controls growth and metamorphosis in amphibian larvae, induces 'water drive' in terrestrial forms of urodeles, and regulates salt adaptation and melanogenesis in fish. Prolactins from these non-mammalian species will not, however, induce lactation in all mammals. Since the amino acid sequence of some prolactins is known to vary, the latter fact reflects an evolutionary variation in receptor structure as well. This is further illustrated by the early studies on species specificity of growth promoting activity of growth hormone (Table III) [42]. Furthermore, the growth-promoting and lipolytic activities of growth hormone reside in different parts of the molecule and that a given amino acid substitution can have quite different repercussions on the two biological activities. Indirect evidence for the simultaneous change or conservation of hormone structure and function has also been provided by bioassays of ovine and amphibian TSH in mammals and amphibia [43]. A comparison of the effects of 23 different insulins and analogues of insulin on growth (DNA synthesis) and metabolic activity (glucose oxidation) suggests that the two activities have evolved in non-parallel fashion and may involve separate functional domains of the molecule [44].

The smaller neurohypophysial peptides reveal another interesting aspect of the evolution of response to hormones [5, 9]. Two of the most striking effects of oxytocin and vasopressin are the control of water and salt balance in amphibia and of milk ejection in mammals. The former is directly related to the emergence from water of terrestrial vertebrates and the failure of amphibia to be fully independent of an aquatic habitat. Since this function in amphibia is regulated by arginine vasopressin, a primitive form of the hormone already established in cyclostomes (see Table II), the same polypeptide molecule exerts this function even following a major evolutionary adaptation that ensured efficient utilization of water. Similarly, placental mammals utilize the same primitive oxytocin as in fish and amphibia, so that a newly evolved function like lactation is regulated by a primitive polypeptide regulating a quite different function in pre-mammalian organisms.

As regards responses to the small-molecular-weight non-peptide hormones, it is clear that they have been put to very

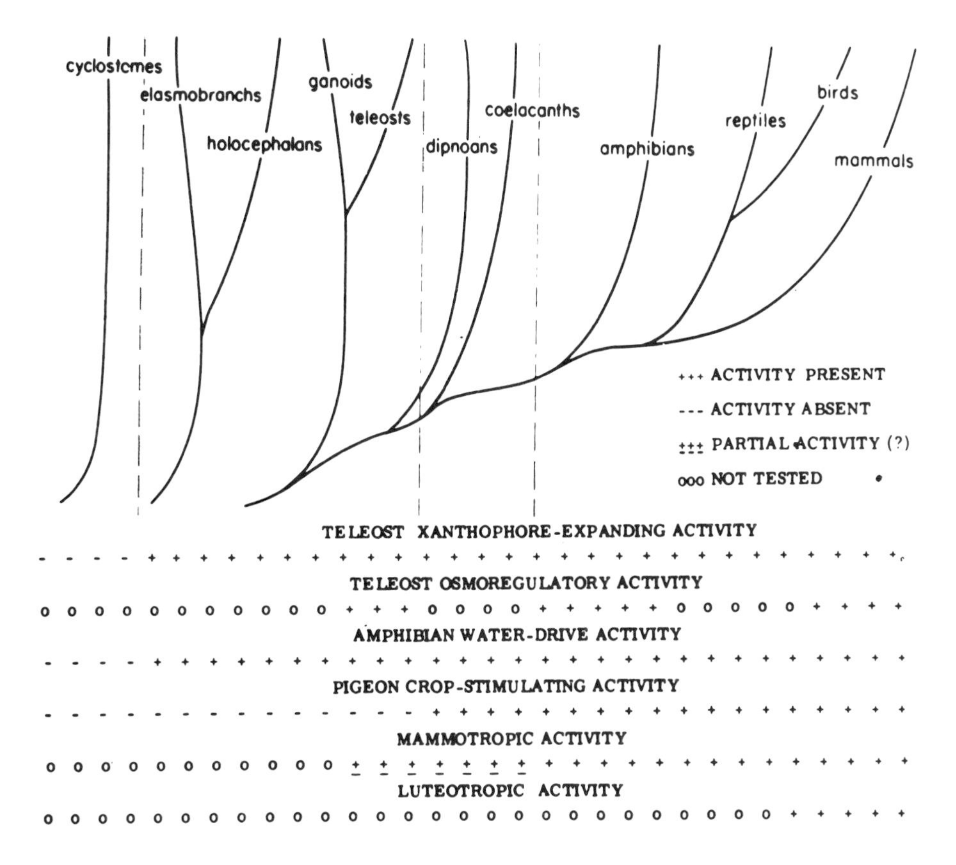

Fig 6. Acquisition by prolactin of different physiological regulatory functions during evolution [41].

different uses during evolution [3, 36]. Thus, for example, thyroid hormones [35, 45, 46] control metamorphosis in amphibia, the maturation of central nervous system in most vertebrates, basal metabolic rate in homeotherms, plumage formation and growth in birds, etc. (Table IV). The significance of the presence of these hormones in primitive organisms in which they seem not to play any physiological role is not always clear. Of special interest is the presence of the two iodothyronines in less primitive organisms where some species are sensitive to a given hormone but related species are not. The neotenic salamanders, *Necturus* and *Proteus*, synthesize substantial amounts of thyroid hormones but do not undergo metamorphosis, whereas the related facultatively neotenic amphibian axolotl will metamorphose when given exogenous hormone [47]. It would be interesting to determine if a loss of response to a hormone is due to loss of receptor under unusual conditions of adaptation.

Table III. *Phylogenetic variations in specificity of response to growth hormones*

	Source of hormones						
Test animal	Fish	Chicken	Pig	Sheep	Cow	Monkey	Man
Fish	+	.	+	+	+	+	+
Chicken	.	+	.	.	–	.	.
Rat	–	+	+	+	+	+	+
Dog	.	.	+	.	+	+	+
Monkey	–	.	–	.	–	+	+
Man	.	.	–	–	–	+	+

+, Growth response; –, no response. From Tata [3].

Steroid hormones also exert very varied actions while retaining an invariable chemical structure, e.g. ecdysone, oestradiol, progesterone, cortisol, corticosterone and aldosterone. All of these steroids have quite distinct action in different tissues. Thus, oestradiol, progesterone and corticosteroids can all induce the synthesis of egg-white proteins in avian oviduct but only oestradiol is effective in inducing the synthesis of egg yolk proteins in the liver [48–50]. It is most likely that such overlap and distinctiveness of response resides in the acquisition or loss of specific hormonal receptors during evolution. Hormones generally arose before the function they control and the latter are indirectly the manifestation of the acquisition of receptors.

Evolution of hormone receptors

There are two functional entities of the receptor: one that recognizes and interacts with the signal or hormone and the other transmitting the consequence of this interaction to bring about the primary biochemical action underlying the physiological effect. Experimentally it has only been possible so far to study the recognition function by following the binding of the radioactive hormone to the receptor or, more recently, with anti-receptor antibodies, but there is no experimental method yet available to establish unequivocally the linkage of hormone–receptor interaction with its primary action.

Two aspects of evolution of receptors have been studied. First, the detection of receptor-like substances in primitive organisms in which the hormone may or may not have any recognizable effect. Second, the possible variation of receptor structure and distribution as related to the increasing com-

Table IV. *Multiple physiological actions of thyroid hormones*

Growth-promoting and developmental actions	Metabolic effects
Growth of most vertebrates	Regulation of basal metabolic rate in mammals
Requirement for all processes of amphibian metamorphosis	Regulation of water and ion transport
Control of plumage and moulting in birds	Calcium and phosphorus metabolism
Maturation of central nervous system and bones in vertebrates	Regulation of cholesterol and fat metabolism in mammals
Regulation of synthesis of mitochondrial enzymes and membrane components in all vertebrates	Nitrogen (urea, creatine) metabolism in poikolotherms
Cell death in amphibian tissue regression	

From Tata [81].

plexity or multiplicity of response to a given hormone during evolution. Thus, receptor-like or hormone-binding substances have been detected in primitive organisms, such as those binding TSH in *Yersinia* [51], human chorionic gonadotropin (hCG) in *Pseudomonas* [52] and opioid peptides in amoeba [53]. Besides these receptors located in plasma membranes, Feldman and his colleagues have described intracellular binding substances in yeast and *Candida* for oestradiol and corticosterone [54, 55]. In most instances, the identification of 'receptor' was based on competition of binding of labelled hormone to the micro-organism by non-labelled hormone and its analogues. It should be stated that there is no evidence so far that these hormones, which have such marked physiological actions in vertebrates, regulate any physiological processes in microbes.

The above findings should be contrasted with specific endogenous, functional peptides and steroids found in unicellular and multicellular fungi. Well-known examples of these are the peptide mating factor in yeast, whose structure resembles that of LH [56] and can activate gonadotroph cells in mammals [57], and the fungal sex steroids antheridiol and oogoniol in the water mold *Achlya* [8, 58] for which specific receptors have been detected in the same organism [59]. This then raises some important conceptual issues concerning the evolution of receptors and the acquisition of novel biological functions for which there is as yet no experimental verification. It can be argued that the first receptors arose from pressure to establish intercellular communication among unicellular organisms, especially in establishing sexual reproduction. Once receptors arose, a co-evolution of hormone and its receptor (with variations introduced by mutational events and gene duplication) can easily account for the diversity. More direct evidence comes from the X-ray crystallographic studies on insulin and glucagon in which stringent spatial constraints reinforce the view of a hormone-receptor co-evolution [13, and see Blundell, this volume]. Whereas this co-evolution of hormone–receptor unit can be understood for protein or peptide hormones [60], the evolution of receptors for steroid and other non-protein hormones cannot similarly be explained.

Some rather unexpected features of possible hormone–receptor co-evolution have recently come to light as a consequence of the rapid advances in recombinant DNA and sequencing techniques. A computer search for DNA and protein sequences related to *c-mos* oncogene revealed a homology for the precursor protein for epidermal growth factor but not for EGF [61]. The oncogene *v-erb-B* which is evolutionarily and functionally related to *c-mos* was found to have substantial homology to EGF receptor. Thus the pre-pro EGF would be related to both EGF and its receptor. Protein hormone receptors can be subdivided into three domains: an extracellular ligand-binding domain, a trans-membrane hydrophobic region, and a cytoplasmic domain which carries the protein kinase activity. Fig. 7 shows the sequence homology between the cytoplasmic domains of insulin and EGF receptors and four members of the *src* oncogene products [62]: an even higher degree of homology between the protein kinase domain of the insulin receptor and avian sarcoma virus UR2 *V-ros* oncogene. The high degree of similarity of the extracellular ligand-binding regions of insulin and EGF receptors, as well as of LDL receptor, the transmembrane domains of these receptors, and the cytoplasmic protein kinase domains of the two peptide hormones and the *src* family of oncogene products is illustrated in Fig. 8.

The relationship between oncogene products and both the hormone and its receptor has led to a novel proposal to explain the evolution of protein hormones and their corresponding membrane-associated receptor. It involves the possibility that a single polypeptide progenitor later evolved to become a two-polypeptide complex of receptor and hormone or any other effector [63]. This suggestion is particularly attractive since it can be generalized. Recently, a high degree of homology has been found in portions of receptors for EGF, insulin and low-density lipoprotein (LDL) [64, 65]. Thirteen of eighteen exons of LDL receptor are homologous to sequences found in not only other peptide hormone receptors but also in EGF, EGF precursor, C9 complement, blood clotting factors, etc. Interestingly, there is also significant sequence duplication within the ligand-binding domain of the different receptors. Sudhof *et al.* [66] have proposed that these hormones and receptors constitute mosaic proteins derived by exon shuffling [67] and belong to a group of supergene families. It has been argued that protein–protein interactions generated within a single peptide chain as it folds would be responsible for the evolution of a hormone–receptor complex and the two separable by proteolytic cleavage as depicted in Fig. 9 [63]. The progenitor gene could have duplicated, followed by a separation of a gene each coding for the hormone and receptor moieties.

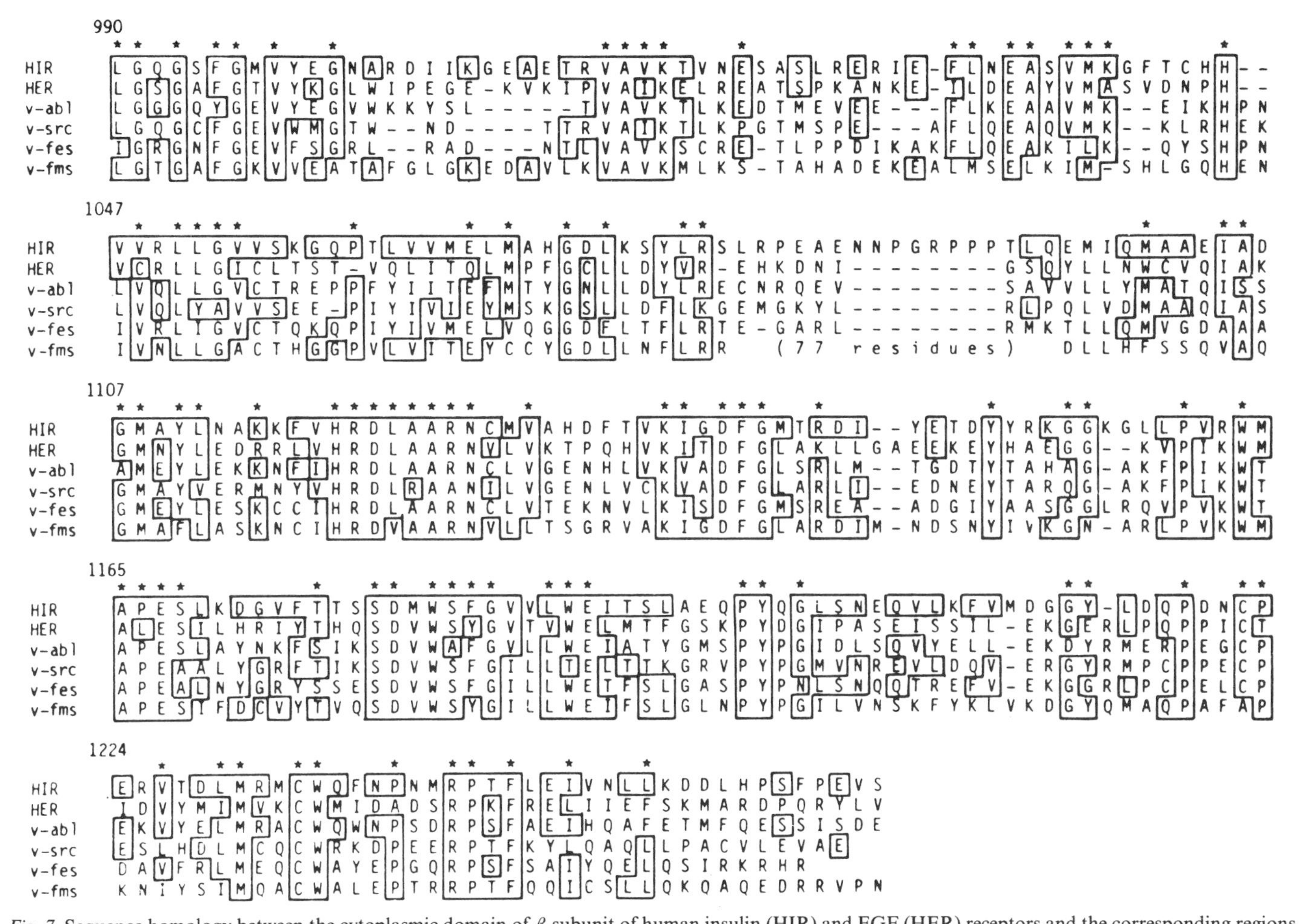

Fig. 7. Sequence homology between the cytoplasmic domain of β-subunit of human insulin (HIR) and EGF (HER) receptors and the corresponding regions of the *src* family (v-abl, v-src, v-fes and v-fem) oncogene proteins. Boxed areas represent matching amino acid residues; * Four or more matches. From Ullrich *et al.* [62].

Whereas gene splitting is the novel feature of the above arguments, gene duplication for diversification of receptor genes, and thus evolution of multiple functions, has been previously suggested by other workers. A duplication of receptor genes and subsequent divergence so that products of the different genes would be associated with different actions of the hormone, or differentially expressed in different tissues, would facilitate evolution of protein hormones themselves [4]. It follows then that receptors in different tissues may be different or somehow associated with different hormonal actions. Although there is no experimental evidence for this prediction, one consequence of this deduction is that evolution of new hormones would follow that of new types of receptors. Another consequence would be the generation of increasing complexity of hormonal effects, especially if there are cross-interactions between related hormones and receptors. Of course, these possibilities do not apply to non-protein hormones, although there are examples of limited cross-interactions between different steroid hormones and their receptors.

N-Terminal Cap
Cross-Linking Domain
LIGAND BINDING & SIGNAL GENERATION
Tyrosine Kinase
SIGNAL TRANSDUCTION
Specific Tail
IR EGFR LDLR v-ROS

Fig. 8. Schematic representation of the similarity in organization of extracellular ligand binding domains of receptors for human insulin (IR), EGF (EGFR), LDL (LDLR) and the cytoplasmic domains with protein kinase activity of IR, EGF and V-ROS oncogene product. The transmembrane regions of all four membrane proteins also show a high degree of structural similarity. From Ebina *et al.* [64].

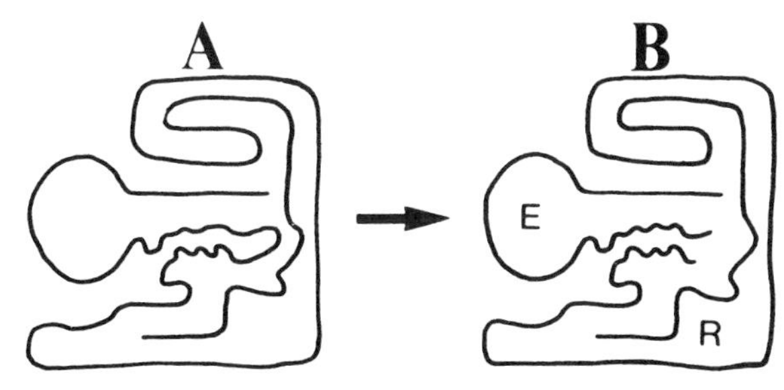

Fig. 9. Scheme proposed by Kaback [63] to explain how a primitive single polypeptide progenitor (A) evolves to become a two-polypeptide unit (B) of hormone or effector (E) and its receptor (R).

Evolution of hormones as a model of evolution of chemical communication

The evolution of hormones and their actions can be visualized as the continuing evolution of an element of an intricate network of chemical communication. This would generate an expanding system of linguistics and as such has been analysed [68, 69]. The necessity for such communication may have arisen as increasingly cells had to adapt to changing nutritional patterns and to derive the advantages of sexual reproduction. As regards cell–cell communication, one has to distinguish between hormones and substances like chemo-attractants, pheromones, humoral agents, etc. – a distinction that is not always clear-cut. It is easier to define the uniqueness of hormones as agents of chemical communication in the context of their role in transmitting information from the environment to the whole organism and in co-ordinating the activities of different populations of cells within the organism. Obviously, the evolution of the nervous and endocrine systems must have been intimately linked to derive the maximum benefits of adaptation to environmental stimuli.

Many reviewers have emphasized the possible common starting-point for the evolution of the nervous and endocrine systems. Perhaps the most striking feature is the close relationship between many plant hormones, neurotransmitters, neurohormones and catecholamines. It is perhaps best illustrated by the pineal gland hormone melatonin [70–72] and the hypothalamic hormone TRH [74, 75]. These neuropeptides serve to link environmental stimuli (e.g. light, temperature, pheromones) via the pituitary or neurotransmitter substances to the other endocrine or hormone-producing tissues. TRH itself has some neurotransmitter activity and could therefore be considered to have arisen with a neurocrine or paracrine function and that only later did it acquire the role of regulating adenohypophyseal hormone secretion. Melatonin also serves as a regulator of biological clocks and, in birds and aquatic species particularly, controls reproductive activity. The pineal gland itself is innervated by noradrenaline-containing nerves which in turn regulate the conversion of tryptophan to the indoleamine melatonin, which is 5-methoxy-*N*-acetyl-tryptamine. The conversion involves the formation of 5-hydroxytryptophan and serotonin which are neurotransmitters and also biologically active in some plants and closely related to the plant hormone, indoleacetic acid [73]. There are also examples of non-peptide hormone-like substances that are shared by both plants and the nervous system. An appropriate example is the formation in the brain of catechol oestrogen (2-hydroxyoestradiol-17β), also found in plants and parts of whose structure resembles, the neutrotransmitters, noradrenaline and dopamine [74]. As regards the first appearance of melatonin, it must be the result of the acquisition of the enzyme serotonin-*N*-acetyl transferase by the pineal gland that is the crucial evolutionary step leading to formation of melatonin and later playing the role of an animal hormone. An intriguing question would be to know whether receptor-like substances capable of interacting with melatonin already existed in pre-pineal organisms or whether its function as a hormone dates from the appearance of receptors in their ultimate target cells.

Another neuroendocrine complex which reinforces the view of a parallel evolution of the primitive nervous and endocrine systems stems from a consideration of the phylogenetic distribution of the hypothalamic releasing hormones [74, 75]. Hypothalamic releasing hormones, which in higher vertebrates regulate the formation and release of pituitary hormones, are found in all primitive nervous systems. Thus, the tripeptide TRH and the decapeptide LH releasing hormone (LHRH) are found in nervous tissue, skin neuroectoderm, gut etc., of lower vertebrates and invertebrates, and similar peptides have been described even in plants [74, 76, 77]. The function of hypothalamic releasing hormones in extra-pituitary nervous tissue, or even in other tissues, is not certain, although there is some evidence suggesting that they have neurotransmitter activity [74]. It is the appearance of receptors for these peptides in the pituitary that must be considered as coincident with their acquisition of a hormonal function as releasing hormones during evolution. Thus, the releasing hormones are part of a family of widely distributed neuropeptides that initially constituted a diffuse, primitive neuroendocrine system and only later in evolution did they acquire the role of regulating pituitary function [78, 79].

Roth and his colleagues have however expressed some reservations about the obligatory co-evolution of primitive nervous and endocrine systems which later gave rise to the present-day hormonal system [14, 15, 78]. They suggest that the present-day hormones existed long before the nervous system made its appearance in multicellular invertebrates. The claims of this and other groups of detecting peptide hormones in plants, unicellular organisms and bacteria [6, 14–16, 74, 76, 77] have yet to be substantiated by chemical characterization and identification of the genes encoding their precursors in these forms of life. On the other hand, non-protein hormones have been convincingly demonstrated in lower organisms [30–32, 36]. Thus proteins containing a large amount of iodothyronines (thyroid hormones) have been detected in algae and sponges, while oestrogens and other steroid hormones are present in high concentrations in some ferns. Their function in plants is not known but the former may serve as structural components and the latter as 'detergents' in membrane organization. But could these substances, and perhaps even biologically active peptides, be called 'hormones' in these primitive forms of life? Without delving into semantics or the merits of different possible definitions, the term 'hormone' should only be applicable when a function for it is established. In other words, the presence of the same substance in primitive organisms is irrelevant to considering it as a hormone later in evolution. What matters is the appearance of a functional receptor (in contrast to a binding element) and, especially in the case of peptide hormones, the co-evolution of hormone–receptor as a unit of biological regulation.

To conclude, however one applies the definition of hormones, in taking an evolutionary approach to hormone action, we have a valuable model for the evolution of a major system of chemical communication. In it we see the evolution of a system of linguistics which has evolved to co-ordinate increasingly complex metabolic and developmental processes [68, 69]. Not all the early forms of chemical signalling elements have necessarily evolved into what are now recognized as hormones. Other bio-regulatory substances, such as neurotransmitters, pheromones, autocrine growth factors, etc., may also share the same origin. Once a receptor system has appeared and then diversified rapidly through evolution, one sees the wide range of hormonal activities, often interrelated, forming an intricate network of co-ordination of even more complex physiological processes. The establishment of hormonal signalling and control has gone hand in hand with the increasing complexity of nervous communication and autocrine growth control, thus giving rise to superimposed layers of positive and negative feedback loops. The application of techniques of gene cloning and DNA sequencing to the study of both hormones and their receptors in different organisms is bound to enhance enormously our present limited knowledge of the wider issues concerning the evolution of the language of chemical communication.

Acknowledgements

I would like to thank Mrs Ena Heather for the preparation of the manuscript.

References

1. Barrington, E. J. W., *Proc. R. Soc. London Ser.* B **199**, 361–375 (1977).
2. Gorbman, A., and Bern, H. A., *A Textbook of Comparative Endocrinology*. John Wiley, New York (1962).
3. Tata, J. R., in *Biological Regulation and Development* (ed. R. F. Goldberger and K. R. Yamamoto), vol. 3B, pp. 1–58. Plenum Publishing, New York (1984).
4. Wallis, M., *Biol. Rev.* **50**, 35–98 (1975).
5. Acher, R., *Proc. R. Soc. London Ser.* B **210**, 21–43 (1980).
6. Roth, J., Le Roith, D., Shiloach, J. and Rubinovitz, C., *Clin. Res.* **31**, 354–363.
7. Barrington, E. J. W., *An Introduction to General and Comparative Endocrinology*. Clarendon Press, Oxford (1963).
8. Timberlake, W. E. and Orr, W. C., in *Biological Regulation and Development* (ed. R. F. Goldberger and K. R. Yamamoto), vol. 3B, pp. 225–283. Plenum Press, New York (1984).
9. Acher, R., in *Evolution and Tumour Pathology of the Neuroendocrine System* (ed. S. Falkmer, R. Hakanson and F. Sundler), pp. 181–201. Elsevier Science Publishers, Amsterdam (1984).
10. Cooke, N. E., Coit, D., Shine, J., Baxter, J. D. and Martial, J. A., *J. Biol. Chem.* **256**, 4007–4016 (1981).
11. Moor, D. D., Conkling, M. A. and Goodman, H. M., *Cell* **29**, 285–286 (1982).
12. Herbert, E., Civelli, O., Douglass, J., Martens, G., and Rosen, H., in *Biochemical Actions of Hormones* (ed. G. Litwack), vol. XII, pp. 1–36. Academic Press, Orlando (1985).
13. Blundell, T. L. and Humbel, R. E. *Nature* (*London*) **287**, 781–787 (1980).
14. LeRoith, D., Shiloach, J., Berelowitz, M., Frohman, L. A., Liotta, A. S., Krieger, D. T. and Roth, J., *Fed. Proc.* **42**, 2602–2607 (1983).
15. Roth, J., LeRoith, D., Shiloach, J., Rosenzweig, J. L., Lesniak, M. A. and Havrankova, J., *New England J. Med.* **306**, 523–527 (1982).
16. LeRoith, D., Pickens, W., Vinik, A. I. and Shiloach, J., *Biochem. Biophys. Res. Commun.* **127**, 713–719 (1985).
17. Mains, R. E., Eipper, B. A., Glembotski, C. C. and Dores, R. M. *Trends Neurosci* **6**, 229–235 (1983).
18. Takahashi, H., Teranishi, Y., Nakanishi, S. and Numa, S. *FEBS Lett.* **135**, 97–102 (1981).
19. Smyth, D. G., in *Endorphins, Chemistry, Physiology, Pharmacology and Clinical Relevance* (ed. J. B. Malick and R. M. S. Bell), pp. 9–22. Marcel Dekker, New York (1982).
20. Acher, R., in *Polypeptide Hormones: Molecular and Cellular Aspects*, Ciba Foundation Symposium 41, pp. 31–55. Elsevier/Excerpta Medica/North-Holland, Amsterdam (1976).
21. Pierce, J. G. and Parsons, T. F., *Ann. Rev. Biochem.* **50**, 465–495 (1981).
22. Rosenfeld, M. G., Amara, S. G. and Evans, R. M., *Science* **225**, 1315–1320 (1984).
23. Edbrooke, M. R., Parker, D., McVey, J. H., Riley, J. H., Sorenson, G. D., Pennengill, O. S. and Craig, R. K., *EMBO J.* **4**, 715–724 (1985).
24. Nikolics, K., Mason, A. J., Szonyi, E., Ramachandran, J. and Seeburg, P. H., *Nature* (*London*) **316**, 511–517 (1985).
25. Land, H., Grez, M., Ruppert, S., Schmale, H., Rehbein, M., Richter, D. and Schutz, G., *Nature* (*London*) **302**, 342–344 (1983).
26. Schmale, H., Heinsohn, S. and Richter, D., *EMBO J.* **2**, 763–767 (1983).
27. Remy, C., in *The Evolution of the Hormonal Systems* (ed. M. Gersh and P. Karlson). Leopoldina Symposium (1985) (in the press).
28. De Luca, H. E., *Fed. Proc.* **33**, 2211–2219 (1974).
29. Norman, A. W. and Henry, H., *Recent Prog. Horm. Res.* **30**, 431–473 (1974).
30. MacIntyre, I., Colston, K. W., Evans, I. M. A., Lopez, E., MacAuley, S. J., Piegnoux-Deville, J., Spanos, E. and Szelke, M., *Clin. Endocrinol.* **5** 85s–95s (1976).
31. Whitehead, D. L. and Sellheyer, K., *Experientia* **38**, 1249–1251 (1982).
32. Karlson, P., *Hoppe-Seyler's Z. Physiol. Chem.* **364**, 1067–1087 (1983).
33. Feldman, D., Do, Y., Burshell, A., Stathis, P. and Loose, D. S. *Science* **218**, 297–298 (1982).
34. Horgen, P. A., in *Cell Communication and Morphogenesis* (ed. D. H. O'Day and P. A. Horgen), pp. 272–294. Marcel Dekker, New York (1977).
35. Pitt-Rivers, R. and Tata, J. R., *The Thyroid Hormones*. Pergamon Press, London (1959).
36. Barrington, E. J. W., *Experientia* **18**, 1–10 (1962).
37. Bergstrom, S., *Recent Prog. Horm. Res.* **22**, 153–169 (1966).
38. Berridge, M. V., in *Insect Biology in the Future* (ed. M. Locke and D. S. Smith), pp. 463–478. Academic Press, New York (1980).
39. Berridge, M. J. and Irvine, R. F., *Nature* (*London*) **312**, 315–321 (1984).
40. O'Brien, R. L., Parker, J. W. and Dixon, J. F. P., *Prog. Mol. Subcell. Biol.* **6**, 201–270 (1978).
41. Nicoll, C. S., in *Handbook of Physiology* (ed. E. Knobil and W. H. Sawyer), section 7, vol. 4, pp. 253–292. American Physiological Society, Washington, D.C. (1974).
42. Geschwind, I. I., *Am. Zoologist* **7**, 89–108 (1967).
43. Fontaine, Y. A., *Nature* (*London*) **202**, 1296–1298 (1964).
44. King, G. L. and Kahn, C. R. *Nature* (*London*) **292**, 644–647 (1981).
45. Tata, J. R., *Adv. Metab. Disorders* **1**, 153–189 (1964).
46. Tata, J. R., in *Handbook of Physiology*, section 7, *Endocrinology* **3**, pp. 469–478. American Physiological Society, Washington, D.C. (1974).
47. Tata, J. R. *Metamorphosis* (ed. J. J. Head). Carolina Biology Readers, no. 46, Carolina Biological Supply Co., Burlington, NC (1983).
48. O'Malley, B. W., McGuire, W. L., Kohler, P. O. and Korenman. S. G., *Recent Prog. Horm. Res.* **25**, 105–153 (1969).
49. Tata. J. R., and Smith, D. F., *Recent Prog. Horm. Res.* **35**, 47–95 (1979).
50. Gruber, M., Boss, E. S. and AB, G., *Mol. Cell Endocrinol.* **5**, 41–50 (1976).
51. Weiss, M., Ingbar, S. H., Winblad, S., Kasper, D. L., *Science* **219**, 1331–1333 (1983).
52. Richert, N. D. and Ryan, R. J., *Proc. Natl. Acad. Sci. USA* **73**, 878–882 (1977).
53. Josefsson, J. O. and Johansson, P., *Nature* (*London*) **282**, 78–80 (1979).
54. Loose, D. S. and Feldman, D., *J. Biol. Chem.* **257**, 4925–4930 (1982).
55. Feldman, D., Stathis, P. A., Hirst, M. A., Stover, E. P. and Do, Y. S., *Science* **224**, 1109–1111 (1984).
56. Hunt, L. T. and Dayhoff, M. D., in *Peptides: Structure and Biological Function* (ed. E. Gross and J. Meienhofer), pp. 757–760. Proceedings of the Sixth American Peptide Symposium, Rockford, IL, Pierce Chemical Co. (1979).
57. Loumaye, E., Thorner, J. and Catt, K. J., *Science* **218**, 1324–1325 (1982).
58. Horgen, P. A., in *Sexual Interactions in Eukaryotic Microbes* (ed. D. H. O'Day and P. A. Horgen), pp. 155–178 (1981).
59. Riehl, R. M., Toft, D. O., Meyer, M. D., Carlson, G. L. and McMorris, T. C., *Exp. Cell Res.* **153**, 544–549 (1984).
60. Hales, C. N. *Nature* (*London*) **314**, 20 (1985).
61. Baldwin, G. S. *Proc. Natl. Acad. Sci. USA* **82**, 1921–1925 (1985).
62. Ullrich, A., Bell, J. R., Chen, E. Y., Herrera, R., Petruzzelli, L. M., Dull, D. J., Gray, A., Coussens, L., Liao, Y.-C., Tsubokawa, M., Mason, A., Seeburg, P. H., Grunfeld, C., Rosen, O. M. and Ramachandran, J. *Nature* (*London*) **313**, 756–761 (1985).
63. Kaback, D. B., *Nature* (*London*) **316**, 490 (1985).
64. Ebina, Y., Ellis, L., Jarnagin, K., Edery, M., Graf, L., Clauser, E., Ou, J.-H., Masiarz, F., Kan, Y. W., Goldfine, I. D., Roth, R. A. and Rutter, W. J., *Cell* **40**, 747–748 (1985).
65. Sudhof, T. C., Russell, D. W., Goldstein, J. L., Brown, M. S., Sanchez Pescador, R. and Bell, G. I., *Science* **228**, 893–895 (1985).
66. Sudhof, T. C., Goldstein, J. L., Brown, M. S. and Russell, D. W., *Science* **228**, 815–822 (1985).
67. Blake, C. C. F. *Internat. Rev. Cytol.* **93**, 149–185 (1985).
68. Hechter, O. and Calek, A. Jr. *Acta Endocrinol.* (*Copenhagen*) **77**, Suppl. 191, 39–66 (1974).

69. Yates, F. E., in *Biological Regulation and Development* (ed. R. F. Goldberger and K. R. Yamamoto), vol. 3A, pp. 25–97. Plenum, New York (1982).
70. Lerner, A. B., Case, J. D., Takahashi, Y., Lee, T. H. and Moore, W., *J. Am. Chem. Soc.* **80**, 2587 (1958).
71. Axelrod, J., *Science* **184**, 1341–1348 (1974).
72. Axelrod, J., Fraschini, F. and Velo, G. P. (eds.), *The Pineal Gland and its Endocrine Role*. Plenum Press, New York (1983).
73. Galston, A. W. and Davies, P. J., *Science* **163**, 1288–1297 (1969).
74. Jackson, I. M. D. and Mueller, G. P., in *Biological Regulation and Development* (ed. R. F. Goldberger and K. R. Yamanoto), vol. 3A, pp. 127–202, Plenum Press, New York (1982).
75. Jackson, I. M. D., *Fed. Proc.* **40**, 2545–2552 (1981).
76. Fukushima, J., Wanatabe, S. and Kaushima, K., *Tohuku J. Exp. Med.* **119**, 115–122 (1976).
77. Morley, J. E., Meyer, N., Pekary, A. E., Melmed, S., Carlson, H. E., Briggs, J. E. and Hershman, J. M., *Biochem. Biophys. Res. Commun.* **96**, 47–53 (1980).
78. Le Roith, D., Shiloach, J. and Roth, J., *Peptides* **3**, 211–215 (1982).
79. Acher, R., in *Brain Peptides* (ed. D. Krieger, M., Brownstein and J. Martin), pp. 153–163. John Wiley, New York (1983).
80. White, A., Handler, P., Smith, E. L., Hill, R. L., and Lehman, I. R., *Principles of Biochemistry*, McGraw-Hill, New York (1978).
81. Tata, J. R., in *Cellular Receptors for Hormones and Neurotransmitters* (ed. D. Schulster and A. Levitzki), pp. 127–146, John Wiley, New York (1980).

Chemica Scripta 1986, **26B**, 191–207

Questions Answered and Raised by Work on the Chemistry of Gastrointestinal and Cerebrogastrointestinal Hormonal Polypeptides

Viktor Mutt

Department of Biochemistry II, Karolinska Institute, Box 60400, S–104 01 Stockholm, Sweden

Paper presented at the Conference on 'Molecular Evolution of Life', Lidingö, Sweden, 8–12 September 1985

Abstract

Work in different laboratories has led to the isolation from mammalian gastrointestinal tissues of, to date, 20 different types of hormonal polypeptides. The complete amino acid sequences have been published for representatives of 17 of them. A few of these polypeptides were isolated first from cerebral and later from gastrointestinal tissue, while for others initial isolation from gastrointestinal tissue was followed by isolation also from cerebral tissue. On the basis of intraspecies structural similarities, presumably reflecting evolutionary relationships, among each other, or/and with polypeptides of other tissues, 12 of the 17 polypeptide types of known amino acid sequences can unequivocally be combined into 6 different groups, each containing a minimum of 2 types of mammalian hormonal polypeptides, of the same species, while the remaining 5 either lack known, structurally related polypeptides in the same species or else have similarities to polypeptides belonging to more than one group, making classification ambiguous.

In several cases, comparisons of peptides within a group shed some light on structure–activity relationships, and also raise questions as to the direction in which evolutionary changes are modifying the activities of the peptides, while comparisons of polypeptides of identical type in different species, and of their precursor proteins, where their amino acid sequences have been deduced from the corresponding nucleotide sequences, disclose some rather puzzling structural relationships.

Amino acid sequences of polypeptides change during evolution and so do languages. Words acquire new meanings, and new words are created. At the beginning of this century the above title would not have made much sense since neither the word 'peptide' nor the word 'hormone' existed. The former was created by Emil Fischer in 1902 in analogy with the earlier known 'saccharide', and with allusion to the old word 'peptone' for fragments of proteins, itself derived from the Greek *pepsis*, for digestion [1], and the latter was a consequence of the discovery of secretin, by Bayliss and Starling, also in 1902 [2]. Bayliss and Starling showed that the stimulatory effect of upper intestinal acidification, studied in Pavlov's laboratory [3], was not due to a nervous reflex from the intestine to the pancreas but to the release from the intestine into the blood-stream of a chemical messenger to the pancreas. This messenger was named 'secretin', and since it appeared obvious that other analogous messengers acting between various organs would be discovered, Bayliss and Starling suggested that a word be created to describe such messengers. This suggestion led to the word 'hormone', derived from the Greek and implying arousal to activity [4]. Secretin is an intestinal hormone: with the discovery of gastrin by Edkins in 1905 [5], the term gastrointestinal hormone came into use.

For a long time it was taken for granted that gastrointestinal hormones originated only in cells of the gastrointestinal tract, and the discovery in the 1930s of substance P, a peptide of hormonal nature common to both brain and intestine [6], did not shake this belief. Substance P was held to be something peculiar, not a gastrointestinal hormone. During the 1970s it was, however, found that the presence of identical peptides in the CNS and the gastrointestinal tract is the rule rather than the exception and therefore substance P may be regarded as the first discovered cerebrogastrointestinal hormonal polypeptide. Indeed it may be asked whether any hormonal polypeptide that occurs in some section of the gastrointestinal tract does not occur also in the CNS.

The discovery of secretin and of gastrin was followed by the discovery of a large number of gastrointestinal hormones, so that by the end of the 1950s more than a dozen had been named, and described, on the basis of their physiological or pharmacological activities [7,8]. Indeed, already between the discoveries of secretin and of gastrin the name 'motilin' had been given to a hypothetical hormone of the intestinal mucosa, with the function of stimulating intestinal peristalsis [9]. None of these various hormones had, however, been isolated in a form sufficiently pure to permit structural analysis, but circumstantial evidence suggested that at least some of them were polypeptides.

In the 1960s gastrin, secretin, cholecystokinin [10] and pancreozymin [11] (the latter two, unexpectedly, in the form of a single hormone cholecystokinin-pancreozymin (CCK-PZ) [12], cholecystokinin (CCK) for short [13]) were isolated from porcine gastric antral [14] and intestinal [15–17] tissue, respectively. These hormones proved to be polypeptides, and other hormonal polypeptides were isolated from gastrointestinal tissues during the 1970s and in the 1980s. Naturally occurring peptides may be formed either from a variety of the more than 500 different types of amino acids known to occur in nature [18] by the involvement of peptide-specific enzymes [19], or by enzymatic cleavages of precursor proteins [20]. As far as is known, all gastrointestinal hormonal polypeptides are formed in the latter fashion.

The 1970s saw the discovery and/or isolation, from porcine intestine, of GIP (the gastric inhibitory or glucose-dependent insulinotropic polypeptide) [21,22], VIP (the vasoactive intestinal polypeptide) [23], motilin [24] and GRP (the, bombesin-related, gastrin-releasing polypeptide) [25]. In addition, substance P was isolated from extracts of bovine brain [26], neurotensin discovered in and isolated from extracts of bovine brain [27] and the tetradecapeptide somatostatin discovered in and isolated from extracts of ovine brain [28], whereupon substance P was isolated also from equine intestine [29], and neurotensin from bovine [30] intestine. Somatostatin-like immunoreactivity was found to be present not only in the CNS but also in the (guinea-pig) pancreas [31]. This led to the isolation from porcine intestine of somatostatin, although of a form in which the previously known tetradecapeptide was extended from its N-terminus by a tetradecapeptide of dissimilar amino acid sequence [32]. Soon thereafter somatostatin-28 was isolated also from ovine [33] and porcine [34] CNS.

The isolation of substance P from both CNS and intestine, and the demonstration of immunoreactive somatostatin in both the CNS and the pancreas was followed by the demonstration of gastrin-like immunoreactivity in the CNS of several species [35] and of VIP-like immunoreactivity in both the CNS and the peripheral nervous system of several mammalian species [36–38]. The C-terminal octapeptide amide of CCK, CCK-8, was isolated from ovine brain by Dockray *et al.* [39].

In the 1980s the opioid peptide dynorphin, originally discovered in and isolated from extracts of porcine pituitary gland [40], was isolated also from porcine intestine [41].

Glicentin [42] and oxyntomodulin [43], admittedly precursor forms to glucagon, but nevertheless with characteristic hormonal properties of their own were isolated from porcine intestine, as were PHI (*p*eptide with N-terminal *h*istidine and C-terminal *i*soleucine amide) [44], PYY (peptide with N-terminal tyrosine and C-terminal tyrosine amide) [45,46] and galanin (peptide with N-terminal glycine and C-terminal alanine amide) [47]. VIP and PHI were subsequently shown to be present in porcine brain in identical form as in intestine [48,49], and a polypeptide named NPY (*n*euro*p*eptide Y, Y for tyrosine) which was originally discovered in and isolated from extracts of porcine brain [50] was shown to be present also in porcine intestine [51]. An endecapeptide with morphogenetic properties in Hydra attenuata, has been isolated from the latter but also from the sea anemone *Anthopleura elegantissima* and from mammalian brain and intestine [52].

Several of the above-mentioned hormonal peptides have been isolated from additional species and a few of them have been found to occur in forms of different chain lengths. Gastrin is exceptional in occurring naturally in both sulphated or non-sulphated forms [53]. Post-translational modifications, other than chain-length variability, have been described for also a few other of the peptides but in these cases it is not always clear whether or not such modifications may represent preparational artifacts. In addition to the 17 types of hormonal gastrointestinal polypeptides mentioned above some structural information has been published for two additional types, for an entero-oxyntin, of 126 amino acid residues, named PIP (*p*orcine *i*leal *p*olypeptide) [54], entero-oxyntin being the name suggested by M. Grossman for non-gastrin intestinal peptide(s) with gastric acid secretion-stimulating activities [55], and for a porcine intestinal hexacosapeptide provisionally designated as VQY on the basis of its N-terminal tripeptide [56]. Also a polypeptide named sorbin, which stimulates water absorption from rat duodenum and guinea-pig gallbladder, has been isolated from porcine intestine [57]. PIP, VQY and sorbin will not be discussed further in the present connection.

An interesting development during recent years has been the elucidation, mainly by the use of recombinant DNA methodology, but to some extent also by direct isolation and sequencing, of the amino acid sequences of the pre- pro-forms of several gastrointestinal hormonal polypeptides.

All in all, the amount of structural information available on gastrointestinal hormonal polypeptides is already quite large, but mostly restricted to amino acid sequences. None of the peptides has yet been crystallized, at least not in forms allowing determination of three-dimensional structures by X-ray diffraction techniques. Nevertheless there are indications that at least some of the polypeptides exhibit some degree of ordered structure, also in aqueous solution. This has been found for secretin from analyses of its chiroptical properties [58, 59] and by dark-field electron microscopy [60], for VIP by NMR studies [61], and suggested for secretin, VIP, and GIP from sequence comparisons with the structurally related glucagon, the three-dimensional structure of which has been determined by X-ray diffraction studies of crystals [62]. Similarly, comparisons of PYY and NPY with the structurally related turkey pancreatic polypeptide (PP), the structure of which has been determined by X-ray diffraction, has led to suggestions concerning the conformations of the two former [63].

Investigations of intramolecular tyrosine-tryptophan distances by fluorescence analysis has suggested a preferred folded structure in the desulphated C-terminal heptapeptide of CCK [64]. A folded conformation for both the unsulphated and the sulphated heptapeptide has been indicated by NMR studies [65]. For the C-terminal octapeptide, NMR analyses indicated a rigid conformation in dimethylsulphoxide, but a more relaxed conformation in aqueous solution [66]. The conformation, in aqueous and in dimethylsulphoxide solution, of the C-terminal tetrapeptide amide of gastrin, and of certain N-acyl derivatives of it, has been investigated by NMR methodology [67, 68], analyses of chiroptical properties [69, 70] and by energy optimization calculations [71]. Using chiroptical analyses, Peggion *et al.* [70] have found that in the longer forms of gastrin the conformation of the tetrapeptide is influenced by amino acid sequences outside of it, with probable implications for biological activity. For somatostatin-14, analyses of chiroptical properties suggested the presence of β-structure but not of α-helix [72]. The conformation of the C-terminal hexapeptide of somatostatin has been studied by NMR methodology and by energy calculations [73]. Also by chiroptical analyses it was found that the conformation of the ring structure of somatostatin-14 was retained in somatostatin-28 [74].

Empirical energy calculations indicated that the C-terminal heptapeptide of neurotensin occurs in numerous conformations but over a rather limited energy range [75]. Energy calculations have suggested several probable conformational states for substance P [76, 77]. For C-terminal sequences of the latter, studies of circular dichroism and of infra-red absorption, in aqueous solution, have provided evidence for intramolecular hydrogen bonds as well as for intermolecular association [78]. For the amphibian tachykinin physalaemin, NMR analyses suggested a stable conformation in aqueous solution [79].

Recently, interest has been focused on investigations of conformational changes that may take place in hormonal peptides when they are transferred from aqueous solutions into aqueous solutions of phospholipids or of synthetic anionic detergents, or to water-lipid interphases, such as may surround plasma membrane receptors for such peptides. For instance, analyses of circular dichroism spectra of dynorphin-13 indicated that its helix content was 5% in water but 17% in dilute aqueous sodium dodecyl sulphate [80], while on the surface of a neutral phospholipid membrane it was, as evidenced by infra-red attenuated total reflection spectroscopy, found to assume a helical structure oriented perpendicularly to the membrane surface [81].

For secretin, glucagon and VIP, analyses of circular dichroism spectra indicated that addition of phospholipids or of anionic detergents to aqueous solutions of them induces helix formation in all three [82].

For glucagon, the three-dimensional structure has been determined in crystals [83] and there has been extensive discussion of how this structure compares to its structure in aqueous solution or in lipid micells (e.g. [84–86]).

Taken as a whole, investigations of three-dimensional structures of gastrointestinal hormonal peptides have not yet contributed anything substantial to an understanding of phylogenetic relationships. This in contrast to the situation in proteins, where crystallographic investigations of haemoglobins were extensively used for studies of such relationships already early in the present century [87], and, at a more sophisticated level, protein structures determined by X-ray crystallography [88, 89] in several cases have been found to be more useful than amino acid sequence comparisons for the tracing of distant evolutionary relationships (e.g. [90, 91]). Such investigations have also been carried out in the pancreatic polypeptide hormone insulin [92].

Twelve of the above-mentioned 17 types of polypeptides, with known amino acid sequences, clearly fit into six groups where each group contains a minimum of two types of mammalian hormonal peptides, gastrointestinal or other. For the remaining five polypeptide types, group belongings are either obscure or not apparent at all. For polypeptides in three of the groups, related non-mammalian polypeptides, species variant or other, are known. Two of the groups have names firmly established by work, mainly, with non-gastrointestinal peptides. These are the opioid peptide group and the tachykinin group. The others are named here by the two mammalian peptides which first showed that there was a group. This gives a secretin–glucagon group, a cholecystokinin–gastrin group, a PYY–pancreatic polypeptide group and a neuromedin B-GRP group. In addition, as discussed below, neurotensin possibly belongs to a seventh group.

The secretin-glucagon group

This, at present, the largest group contains nine types of polypeptides that have actually been isolated from mammalian tissues, and two others the existence of which has been predicted from the amino acid sequence of one of the preprohormones. Three of the nine, are, however, obviously fragments of one of the remaining six. The nine are: glicentin, glicentin-related pancreatic peptide (GRPP), oxyntomodulin, glucagon, secretin, GIP, VIP, PHI, somatocrinin.

Glicentin with its component polypeptides GRPP, glucagon, and oxyntomodulin

Glicentin was first isolated from porcine intestine and found to be composed of 69 amino acid residues, the sequence of which was determined [42]. From amino acid sequences deduced for preproglucagons, from the corresponding experimentally determined nucleotide sequences the glicentin sequences are known for the human [93], bovine [94], hamster [95] and rat [96] forms. The amino acid sequences have been deduced also for two different preproglucagons from the angler-fish *Lophius americanus* [97], but in these the relationship to the mammalian glicentins is somewhat obscure (see below).

The amino acid sequence of human glicentin, as seen from the amino acid sequence of human preproglucagon in which it is immediately preceded by a signal peptide of 20 amino acid residues [93] is:

R S L Q D T E E K S R S F S A S Q A D P L S D P D Q
M N E D K R H S Q G T F T S D Y S K Y L D S R R A Q
D F V Q W L M N T K R N R N N I A

Of this, residues 1–30 represent GRPP [42], residues 33–61 glucagon [98] and residues 33–69 oxyntomodulin [43]. All glucagons for which the amino acid sequences have been determined are composed of 29 amino acid residues. The sequences are identical for the human [93], porcine [98], rat [96], hamster [95], bovine [94] and presumably camel and rabbit [99] glucagons, but guinea-pig glucagon differs from this common mammalian form of glucagon [99] by having Gln-21, Leu-23, Lys-24, Leu-27, and Val-29 [100].

In birds, the turkey and chicken glucagons are identical but differ from the common mammalian glucagon by having serine instead of asparagine in position 28. Duck glucagon differs from turkey/chicken glucagon by having threonine instead of serine in position 16 [99]. The two angler-fish preproglucagons have each their own form of glucagon. These are identical with each other over their N-terminal pentadecapeptides where they differ from the common mammalian glucagon by having Glu-3, Ser-7, Asn-8 and Glu-15 instead of Gln-3, Thr-7, Ser-8 and Asp-15. In their C-terminal tetradecapeptides the angler-fish glucagons are non-identical with each other in 5 positions and with mammalian glucagon in 5 (angler-fish glucagon I) respectively 3 (angler-fish glucagon II) positions. None of the latter differences between the fish glucagons and mammalian glucagon are the same for the two fish glucagons although two of them occur in the same positions: angler-fish glucagon I has Asp-16 and Asn-29, where angler-fish glucagon II has Thr-16 and Ser-29 instead of the Ser-16 and Thr-29 of the mammalian form. It may be recalled that also duck glucagon has Thr-16. The only additional difference between angler-fish glucagon II and mammalian glucagon is in position 27 where the former has a residue of lysine and the latter, as well as angler-fish glucagon I, has a residue of methionine. The three additional differences between angler-fish glucagon I and mammalian glucagon, and angler-fish glucagon II, are at positions 18, 21 and 24, where the former has residues of lysine, glutamic acid and arginine and the latter two have residues of arginine, aspartic acid and glutamine. Whereas the amino acid sequences of the angler-fish glucagons have been deduced from their preprohormones, a glucagon has actually been isolated from the pancreas of the cat-fish *Ictalurus punctata* [101] and shown to be identical to angler-fish glucagon II for 26 of its 29 residues starting from the N-terminus. Residue 27, however, was not one of lysine, as in angler-fish glucagon II, but one of methionine, as in angler-fish glucagon I and in mammalian glucagon. The residues in the 28–29 sequence were the same as in angler-fish glucagon II but their sequence was not determined in the cat-fish. All mammalian preproglucagons for which the amino acid sequences have been determined have a signal sequence of 20 amino acid residues followed by a GRPP sequence of 30 residues which is separated from the glucagon sequence by a lysylarginyl sequence. The sequences of both the signal peptides and of GRPPs are fairly well conserved among these mammalian species although not to the extent of the complete identity of the glucagon sequences. Thus, for instance, the human and hamster signal peptides differ in 7 positions and the GRPPs in 5, and the human and bovine signal peptides in one position and the GRPPs in 7. In the two angler-fish preproglucagons, like in the mammalian ones, the glucagon sequences are separated by lysylarginyl sequences from N-terminal sequences. The latter are of 50 amino acid residues in preproglucagon I and of 49 residues in preproglucagon II but homologies to the mammalian signal peptides and GRPs are scarcely distinguishable.

As to the oxyntomodulins: in the human [93], hamster [95], rat [96] and guinea-pig [100] forms the extension of the glucagon sequence is by the octapeptide Lys-Arg-Asn-Arg-Asn-Asn-Ile-Ala while in the porcine [42] and bovine [94] forms it is by Lys-Arg-Asn-Lys-Asn-Asn-Ile-Ala. In the angler-fish the extension of the glucagon sequences, using the following paired basic amino acid residues as a landmark, are Lys-Arg-Ser-Gly-Val-Ala-Glu for glucagon I and Lys-Arg-Asn-Gly-Leu-Phe for glucagon II.

Consequently, going from fish to mammals, neither the GRPP nor the oxyntomodulin sequences are extensively conserved.

The angler-fish preproglucagons have been found to each contain an additional glucagon-like peptide, GLP, of 34 amino acid residues situated C-terminally to the oxyntomodulin sequences, and separated from these by a lysylarginyl sequence in the case of preproglucagon I and an arginylarginyl in the case of preproglucagon II [97].

The two angler-fish GLPs are highly homologous to each other, the amino acid residues being identical in 27 of the 34 positions, and to other members of the secretin-glucagon group [97]. In the GLP form angler-fish preproglucagon II there is a residue of tyrosine instead of phenylalanine in position 6. This is remarkable since Phe-6 is highly characteristic for the secretin-glucagon group of peptides: recently a GLP has been isolated from cat-fish pancreas and found to be identical with the angler-fish GLP from preproglucagon II in 27 of the 34 residues, including Tyr-6.

In mammalian preproglucagons too there are GLP sequences C-terminal to the oxyntomodulin sequences, but unlike in the fish there are not one but two, non-identical, GLPs arranged in tandem in each preproglucagon. This could mean either that mammals acquired an additional GLP after their divergence from fish or else that fish deleted their second GLP subsequent to that event. This problem has been recently discussed by Lopez *et al.* [102] who conclude that the second alternative must be the case. The mammalian GLP, which in preproglucagon is closest to the glucagon sequence, is referred to as GLP-1, and the GLP C-terminal to it as GLP-2. In all known mammalian preproglucagons the GLP-1 sequences are separated from the oxyntomodulin sequences by lysylarginyl sequences and from C-terminally situated 'intervening peptide' sequences by arginylarginyl sequences. The intervening sequences are separated from also the C-terminally situated GLP-2 sequences by arginylarginyl sequences.

Human GLP-1 [93] has the sequence

HDEFERHAEGTFTSDVSSYLEGQAAKE FIAWLVKGRG

and this is identical to the sequences of the bovine [94], hamster [95] and rat [96] GLP-1s.

It is still not clear to what extent GLP-1 is secreted into the circulation in mammalian species. In rat pancreatic islets it appears to be retained together with GLP-2 in a 'major proglucagon fragment' (MPGF) on release of glicentin, and glucagon, from (pre)proglucagon [103]. It has been pointed out [104] that the similarity to glucagon starts with residue 7 of GLP-1 and that synthetic GLP-1 [1–37] has no hyperglycaemic action in rabbits. Since a characteristic feature of the secretin-glucagon group of peptides is a residue of phenylalanine at position 6 (a feature which is absent from GLP-1 but present in its 7–37 sequence) it might be of interest to investigate whether the latter polypeptide may not be released naturally from some precursor form(s). It may be remembered that the cat-fish GLP that has been isolated [101] would correspond to such a peptide, besides for consisting of 29 amino acid residues, like glucagon. So perhaps GLP-1 7–35 should also be looked for!

The peptide intervening between the pairs of arginyl residues C-terminal to GLP-1 and N-terminal to GLP-2 consists of 13 amino acid residues. It is acidic, and strongly conserved but not identical, for the human, bovine, hamster and rat species. The human form has the sequence [93]

DFPEEVAIVEELG

The GLP-2 sequences too are strongly conserved in the human, bovine, hamster and rat species, but again not identical. The sequence of human GLP-2 is

HADGSFSDEMNTILDNLAARDFINWLI QTKITDR

It may be noted that if GLP-2 would be cleaved so as to release a polypeptide representing its residues 1–29, i.e. a polypeptide of the length of glucagon, this peptide would, like the common mammalian glucagon, have C-terminal threonine.

Secretin

The amino acid sequence of human secretin is [105]

HSDGTFTSELSRLREGARLQRLLQGLV*

(The asterisk here and elsewhere in this article indicates that the C-terminal amino acid residue is in α-amide form.) The mutally identical sequences of the porcine [106] and bovine [107] secretins differ from it by having Asp-15 and Ser-16.

Chicken secretin has the amino acid sequence [108]

HSDGLFTSEYSKMRGNAQVQKFIQNLM*

Compared with the sequences of the known mammalian secretins, most of the 13 substitutions are conservative, but in position 15 the mammalian secretins have acidic residues, and basic ones in position 18, whereas chicken secretin has neutral residues in these positions. The amino acid sequence is not yet known for any preprosecretin but two, bioactive, forms of secretin in which the above given sequence does not have C-terminal -Val-NH_2 but -Val-Gly and -Val-Gly-Lys-Arg respectively have been isolated from porcine intestine [109, 110].

Human *GIP* has the amino acid sequence [111]

Y A E G T F I S D Y S I A M D K I H Q Q D F V N W L L A E K G K K N D W K H N I T Q

The sequence of porcine GIP differs from it by having Arg-18 and Ser-34 [112, 113]. Bovine GIP is identical to porcine GIP except for having Ile-37 instead of Lys-37 [114]. It has been pointed out [115] that the -Gly_{31}-Lys_{32}-Lys_{33}-sequence of GIP is a typical amidation signal in polypeptide hormones [116, 117]. Since, however, in the presently known forms of GIP this sequence is preceded by a residue of lysine, and basic amino acid residues do not seem to be amidated in mammals [116], it could be of interest to find out whether there are mammalian species in which the Lys-30 of the hitherto known forms of GIP is replaced by a neutral residue in amide form. Like for secretin no preproform is yet known for GIP.

VIP and PHI

VIP is identical in the human [118, 119], porcine [120], bovine [121], murine [122] and canine [123] species. The amino acid sequence of this common form of VIP is:

H S D A V F T D N Y T R L R K Q M A V K K Y L N S I L N *

The sequence of guinea-pig VIP [124] differs from it by having Leu-5, Thr-9, Met-19 and Val-26.

The sequence of chicken VIP differs from the common mammalian form by having Ser-11, Phe-13, Val-26 and by having C-terminal Thr-NH_2 [125]. The substitution of isoleucine by valine in position 26 is consequently common for guinea-pig and chicken. VIP has been isolated, in identical form, from porcine intestine [120] and brain [48].

PHI was discovered in extracts of porcine small intestine by way of its C-terminal α-amide structure [126] and this structural feature was used to follow it during the isolation procedure [44].

Subsequently, the amino acid sequence for human PHI, or PHM unless PHI is considered to be a generic name, was deduced from the nucleotide sequence of its precursor protein [119] and a polypeptide with such a sequence was actually isolated from human colonic tissue [127].

The amino acid sequence of human PHI (PHM) is

H A D G V F T S D F S K L L G Q L S A K K Y L E S L M *

The sequence of porcine PHI differs from it by having Arg-12 and, definitionally, Ile-27 [44].

Bovine PHI differs from porcine PHI by having Tyr-10 [128]. It was originally found by Itoh and co-workers [119] that the preparation of human PHI (PHM) contains also the sequence of VIP. In the precursor, the PHI sequence is preceded N-terminally by -Ala-Arg- and is followed C-terminally by, in turn, -Gly-Lys-Arg-, an acidic dodecapeptide, -Lys-Arg-, and the VIP sequence, which is separated by -Gly-Lys-Arg- from the acidic, C-terminal pentadecapeptide of the prepro-VIP/PHI (PHM). Two groups of workers have recently elucidated the gene structure for human prepro-VIP/PHI (PHM) [129, 130].

It is remarkable that in the three species in which both the VIP and the PHI sequences are known, the former are identical but the latter all different, although highly similar. It may be asked whether in an evolutionary perspective VIP may have reached a stable final form whereas PHI is either on its way to disappearing or else to become a hormone, more specific than present-day PHI, which by and large seems to be VIP-like in its actions, although generally weaker [131].

Somatocrinin or the, hypothalamic, growth hormone-releasing factor

Human somatocrinin was originally isolated from pancreatic tumour tissue in the form of two polypeptides, one with 44 amino acid residues in the sequence.

Y A D A I F T N S Y R K V L G Q L S A R K L L Q D I M S R Q Q G E S N Q E R G A R A R L *

[132] and the other with the same sequence except for absence of the four C-terminal residues [133]. The N-terminal nonacosapeptide of human somatocrinin was found to have the full intrinsic activity and potency of the whole molecule, as measured in an *in vitro* system. The amino acid sequences have been elucidated also for the rat [134], porcine [135], bovine [136], caprine [137] and ovine [137] somatocrinins.

Porcine somatocrinin differs from the human form by having Arg-34, Gln-38 and Val-42. Bovine somatocrinin differs from the porcine by having Asn-28 and Lys-41. The caprine and bovine somatocrinins are identical and differ from the ovine by having Val-13. Rat somatocrinin is unexpectedly different from the other mammalian forms: from the human form it differs in position 13 as the ovine form does, in position 28 as the caprine/ovine and bovine forms do, and in positions 34 and 38 as do the caprine/ovine, bovine and porcine forms. In positions 41, however, it is identical (unlike the ovine/caprine and bovine forms) to the human form, and in position 42 it has Phe-42 instead of the Ala-42 of the human and the Val-42 of the caprine/ovine, bovine and porcine forms. Additional differences to the human form are the His-1, Glu-25, Arg-39, Ser-40 and Asn-43 of the rat form. On the basis of the sequences elucidated to date it would seem that changes often occur in positions 34, 38 and 42, and especially so in position 42.

Preprosomatocrinin may be composed of either 107 or 108 amino acid residues, as has been shown by nucleotide sequence determination of its cDNA [138] and of its gene [139]. Contrary to preproglucagon, which in addition to the glucagon sequence contains the sequences of GLP-1 and GLP-2, and to prepro-VIP-PHI/PHM, it contains only a single sequence of the secretin-glucagon group of peptides, i.e. of somatocrinin itself.

There are various other proteins for which some degree of sequence similarity to members of the secretin-glucagon group of polypeptides may be discerned. For instance, a tridecapeptide segment of human somatotropin is in 8 positions identical to a tridecapeptide segment of porcine secretin

[140], in 7 to the corresponding tridecapeptide of human secretin.

Sequence similarities between members of the secretin-glucagon group and β-lipotropin and epidermal growth factor/urogastrone have been noted [141, 142]. The characteristic His... Phe_6- sequence occurs in certain murine histocompatability antigens [143]. Sequence similarities between secretin-glucagon group peptides and plasma prealbumin [144] are of some special interest in view of the finding that prealbumin occurs in the islets of Langerhans [145], and there apparently in the same cells as those which produce glucagon [146]. Another interesting finding is that secretin and the oncogen *v-sis* have 8 positional identities in a dodecapeptide sequence [147].

Recently, three closely related polypeptides, helospectins I and II and helodermin, have been isolated from the venom of the Gila monster, *Heloderma suspectum* [148, 149].

The amino acid sequence of helospectin I is

H S D A T F T A E Y S K L L A K L A L Q K Y L E S I L G S S T S P R P P S S

The sequence of helospectin II differs from it by lacking Ser-38, and the sequence of helodermin by lacking both Ser-38, Ser-37 and Pro-36 and by having Ile-5, Gln-8, Gln-9, Ala-24, Arg-30, Pro-34 and Pro-NH_2 in position 35. It is evident that the structures of the helospectins and of helodermin are related to those of the secretin-glucagon group of peptides. Recent work by Christophe and co-workers [150] suggests that helodermin is neither a lizardial form of one of the known mammalian polypeptides of this group nor a lizard-specific peptide, but rather a special type of hormonal peptide that yet remains to be isolated from mammalian tissues.

The cholecystokinin(-pancreozymin)-gastrin group

It is not clear why this group should contain only two types of mammalian hormonal polypeptides, gastrin and cholecystokinin(-pancreozymin), CCK, when the secretin-glucagon group contains six, or more if the GLPs should prove to be hormonal in their own right, and/or if it will be confirmed that helodermin does occur in mammals. One possibility is, of course, that additional types of peptides exist in this group but have not yet been discovered. The existence of only two types does not, however, mean that this is a simple group. The situation here is complicated by a profuse heterogeneity of the active forms. This has been discussed in several review articles (e.g. [151, 152]) and need be only briefly recapitulated here. Gastrin has been isolated from porcine stomach antral tissue in forms of 6 [153], 14, 17 and 34 [151] amino acid residues, the longer ones being N-terminal elongations of the shorter. The heptadecapeptide has been isolated also from human, dog, cat, cow, sheep [151] and rat [154] antral mucosa, and the tetratriaconta- and tetradecapeptides also from human gastrinoma tissue [151]. The amino acid sequence of rat G-34, has been deduced from the corresponding, experimentally determined, nucleotide sequence of its cDNA [155]. All these chain-length variants of gastrin may exist in forms in which the phenolic group of the tyrosine residue in position 6 from the C-terminus is either free (non-sulphated, -ns) oesterified with sulphuric acid (-s) [53]. The nomenclature here is not quite consistent, instead of -ns and -s [156], the older 'I' and 'II' respectively [157] are still often used.

Quite recently a tetrapeptide and a pentapeptide corresponding to the identical C-terminal tetra- and pentapeptide amide sequences of the known forms of gastrin and of CCK, have been isolated from porcine cerebral cortex and their structures confirmed by sequencing [158]. The pentapeptide has been isolated also from dog intestine and brain [159]. The gastrins from all the species from which they have hitherto been isolated are identical in the C-terminal parts of their amino acid sequences, up to the heptapeptides. Further towards the N-termini there are differences, although mostly rather small, between all the gastrins except for those of the cow and sheep which are mutually identical, over the length to which they have been determined [160].

The amino acid sequence of human G-34 (161, 162) is

$^{+}$E L G P Q G P P H L V A D P S K K Q G P W L E E E E E A Y G W M D F *

where $^{+}$E is pyroglutamyl.

Porcine G-34 [163] differs from it by having Leu-4, Leu-14, Ala-15 and Met-22, and rat G-34 [154, 155] differs from it by having Gln-8, Phe-10, Ile-11, Leu-14, Arg-19, Pro-21 and Met-22. For canine, feline, bovine and ovine gastrins the sequences are known only for the 'little gastrins', i.e. for the 18–34 sequences of the G-34s [160]. Using the G-34 numbering, canine gastrin differs in this sequence from the human by having Met-22 and Ala-25; feline gastrin by having Ala-27; and bovine and ovine gastrins, both, by having Val-22 and Ala-27. From the nucleotide sequence of its cDNA, porcine preprogastrin has been deduced to consist of 104 amino acid residues [164], and human preprogastrin of 101 [165, 166]. The nucleotide sequence has been determined also for the human gastrin gene [167–169].

Cholecystokinin has, in forms of various chain lengths, 58 amino acid residues, and shorter, isolated from porcine [170, 171], canine [172, 159], rat [173], bovine [174] and guinea-pig [175] intestine as well as from ovine [39], human [176, 177], porcine [178, 179] and canine [159] brain. The amino acid sequences of the longer forms of human CCK have been deduced from the nucleotide sequence of the gene for its prepro-CCK (180) and the amino acid sequence of mouse CCK has been deduced from the partially determined nucleotide sequence of the cDNA for mouse prepro-CCK [181]. For rat prepro-CCK the nucleotide sequences have been determined for the cDNA [182, 183] and for the gene [184]. For porcine prepro-CCK the nucleotide sequence of the cDNA has been determined [185].

The amino acid sequence of human CCK-58 is

V S Q R T D G E S R A H L G A L L A R Y I Q Q A R K A P S G R M S I V K N L Q N L D P S H R I S D R D Y^{ø} M G W M D F *

where ø indicates that the tyrosine is sulfated. Porcine CCK-58 differs from the human by having Ala-1, Val-2, Lys-4, Val-5, Val-32, Met-34, Ile-35 and Ser-40. Rat CCK differs from the human by having Ala-1, Val-2, Pro-5, Ser-7, Pro-9, Arg-12, Val-24, Val-34, Leu-35 and Gly-40. For mouse CCK only the 10–58 sequence of CCK-58, i.e. of CCK-49 is known. This is identical to the corresponding rat sequence except for a residue of serine rather than one of glycine in its position 31.

Canine CCK-58 has been reported to differ from the human form by having Ala-1, Val-2, Lys-4, Val-5, Val-34 and

```
Porcine prepro-

CCK       M N G G L C L C V L M A V L A A G T L A Q P V P P A D S A V P G A Q E E E A
gastrin                       M Q R L C A Y V L I H V L A L A A C S E A S W K P G F Q

          H R R Q L R A V Q K V D G E S R A H L G A L L A R Y I Q Q A R K A P S G R V
          L Q D A S S G P G A N R G K E P H E L D R L G P A S H H R R Q L G L Q G P P

          S M I K N L Q S L D P S H R I S D R D YØM G W M D F G R R S A E E Y E Y T S
          H L V A D L A K K Q G P W M E E E E E A Y G W M D F G R R S A E E G D Q R P
```

Fig. 1. Amino acid sequences of porcine prepro-CCK [185] compared to porcine preprogastrin [164].

Ile-35 [159]. According to another report, in addition, by having Pro-9 [186].

For bovine CCK the longest known form is CCK-39. Its sequence differs from that of human CCK by having Val-15 and Ile-16. The CCKs from the six species mentioned consequently have CCKs with identical C-terminal octadecapeptides. This high degree of conservation is in keeping with what is known about the functional importance of this part of the molecule. Less evident is the reason for the complete conservation of the 13–23 sequence of CCK-58 (not established for the bovine form) or indeed of the 13–31 sequence with the sole exception of the conservative alanine-valine substitution in position 24.

For guinea-pig CCK the amino acid sequence is known for the C-terminal docosapeptide [175]. This differs from the corresponding sequence of human CCK by having Ser-1, Gly-4, Asn-8 and Val-17, i.e. it is the only hitherto known mammalian CCK which is not identical to the others in its C-terminal octadecapeptide. It may be recalled that guinea-pig VIP and glucagon too differ from other mammalian mutually identical VIPs and glucagons. Still earlier it had been shown that the amino acid sequences of insulins of the guinea-pig [187] and of several of its hystricomorph relations [188] are far more different from those of other mammalian insulins than is usual for interspecies sequence variability among mammalian insulins.

The substitution of valine for methionine in position 17 of guinea-pig CCK-22, i.e. in position 6 from the C-terminal phenylalanine amide is interesting since it is in this position only that the C-terminal octapeptide of the decapeptide caerulein, originally isolated from the skin of the amphibian *Hyla caerulae* differs [184] by a substitution of threonine for methionine from the mutually identical C-terminal octapeptides of the human, porcine, bovine, canine, rat and mouse CCKs. The nonapeptide phyllocaerulein from *Phyllomedusa sauvagi* has its C-terminal heptapeptide identical to that of caerulein [190].

The remarkable similarity of the C-terminal parts of caerulein, phyllocaerulein, CCK and gastrin would seem to be due to a common evolutionary origin of these polypeptides. Unexpectedly, the amino acid sequences of the preprohormones seem to indicate that while gastrin and CCK indeed have evolved from a common precursor, the relationship of caerulein to gastrin and to CCK is not a simple one. While gastrin and CCK each occur in a single copy in their respective preprohormones the caerulein sequence [191, 192] occurs in four: the second and the third of these, counting from the N-terminus are separated from each other by the pentapeptide Gly-Arg-Arg-Asp-Gly-. The first is separated from the second and the third from the fourth by highly homologous intercaerulein polypeptide sequences of 54 amino acid residues each, with the Gly-Arg-Arg 'amidation signal' sequence at their N-termini. A third polypeptide of 47 amino acid residues, highly homologous over its length with the intercaerulein sequences, starting from their C-termini, separates the first caerulein sequence from the 26 amino acid residues long signal peptide of the 234 amino acid residue preprocaerulein [192].

The evolutionary relationship between (rat) CCK and (porcine) gastrin as reflected by the amino acid sequences of their preproforms, and of the nucleotide sequences coding for the latter, has been discussed by Deschenes *et al.* [193]. Computer-assisted maximation of homologies, involving insertions and deletions, and considerations of secondary structure requirements revealed a striking degree of sequence homology between the two preproforms at both the amino acid and the nucleotide level strongly suggesting evolution from a common ancesterial protein.

Looking at the actual amino acid sequences of porcine preprocholecystokinin [185] and porcine preprogastrin [164] as they stand today (Fig. 1) it is seen that if these sequences are aligned, starting from the identical C-terminal pentapeptides of gastrin and CCK, that the two proteins have identical dodecapeptides near their C-termini but that the sequence homology towards the N-termini is rather weak, but nevertheless distinguished even in the signal peptides. It is still weaker between the human preprohormones, since human prepro-CCK has one more amino acid residue than the porcine, and this residue appears to be inserted in the heterogenous region near residue 30 [180], displacing the N-terminal part of the sequence by one residue position. A remarkable sequence homology between gastrin and the middle antigen of polyoma virus has been pointed out by Baldwin [194]. More recently a somewhat less extensive homology between gastrin and the oncogen *v-myc* has been observed [147].

The PYY-pancreatic polypeptide group

The 'pancreatic polypeptide' (PP) was first discovered in

extracts of chicken [195] and bovine [196] pancreas. Subsequently it has been isolated from several other mammalian and avian species including the human [197], as well as from the alligator [198]. In all cases it is composed of 36 amino acid residues. All the known mammalian and avian forms have C-terminal arginyltyrosine amide while the alligator polypeptide has arginylphenylalanine amide.

Species variability is modest inside the classes but rather extensive between them, the avian and reptile polypeptides showing a much higher degree of sequence homology to each other than to the mammalian. In the mammalian group, human PP has the sequence [197]

A P L E P V Y P G D N A T P E Q M A Q Y A A D L R R Y I N M L T R P R Y *

The mutually identical porcine and canine PPs differ from the human by having Asp-11 and Glu-23. Bovine PP has Glu-6 and Glu-23, and the ovine is like the bovine except for having Ser-2. Rat PP is surprisingly different from all the others, differing from all of them in 8 positions, by having Met-6, Tyr-11, His-14, Arg-17, Glu-21, Thr-22, Gln-23 and Thr-30. In position 2 it, unlike sheep PP, has the more usual proline residue [199].

Human prepro-PP consists of 95 amino acid residues. Its signal peptide of 29 residues is followed by the PP sequence which in turn is followed by -Gly-Lys-Arg-, an icosapeptide with C-terminal arginine and a heptapeptide with N-terminal glutamic acid (200–202). The icosapeptide is, like PP, released from the proform, since it has been isolated from pancreatic tissue in various species. Its sequence has been found to exhibit considerable species variability more so in its N-terminal than in its C-terminal part. All known mammalian icosapeptides have C-terminal prolylarginine. The ovine icosapeptide is unique in having a single residue of cysteine [203]. For the human form intraspecies variation has been noted. Two groups of workers found the sequence

H K E D T L A F S E W G S P H A A V P R

[200, 201], whereas a third found a residue of isoleucine instead of one of valine [202].

The amino acid sequence of PYY is at present known for only the porcine form. It is [46]

Y P A K P E A P G E D A S P E E L S R Y Y A S L R H Y L N L V T R Q R Y *

For NPY the sequences are known for both the human [204, 205] and porcine [206] forms. That of the human form is

Y P S K P D N P G E D A P A E D M A R Y Y S A L R H Y I N L I T R Q R Y *

The porcine sequence differs from the human only in position 17, where it has a residue of leucine. Consequently, porcine NPY and PYY are identical in 25 out of their 36 positions. Yet they have a strikingly different anatomical localization, PYY being present in cells of an endocrine type and NPY in neurones ([207] and references therein). Too, porcine NPY and PYY are each identical to porcine PP in 18 positions. Remarkably, no fewer than 16 of these are such in which the PYY and the NPY are mutually identical. A detail worth noting may be that whereas porcine PYY and NPY both have a residue of leucine in position 17, both porcine and human PP have a residue of methionine there.

Recently, a heptatriacontapeptide clearly related to NPY and PYY, although to a lesser extent to PP, has been isolated from the endocrine pancreas of the angler-fish *Lophius americanus* [208]. Instead of having C-terminal tyrosine amide, however, it has tyrosylglycine, raising interesting questions as to the location of the amidating enzyme(s) in fish. Angler-fish peptide YG is identical to both porcine NPY and to porcine PYY in 23 positions, of which 19 are such where PYY and NPY are mutually identical. In 16 of these 19 the identity extends also to porcine PP. The longest stretch of unbroken identity between porcine NPY and angler-fish peptide YG is the octapeptide N L I T R Q R Y, at the C-terminus of NPY, and preceding the C-terminal glycine of angler-fish peptide YG. The longest stretch of unbroken identity between PYY and angler-fish YG is the C-terminal pentapeptide of this octapeptide. Human prepro-NPY has been shown to consist of 97 amino acid residues [205]. A signal peptide of 28 amino acid residues is followed by the NPY sequence, a -Gly-Lys-Arg- sequence and a C-terminal sequence of 30 residues. On direct alignment of the latter with the C-terminal heptacosapeptide of human prepro-PP [200], starting with the N-terminal, only three identities are observed. However, the C-terminal -Val-Pro-Arg- sequence of human 'pancreatic icosapeptide' [203], also occurs in the prepro-NPY sequence, although at the C-terminal end of an encosapeptide rather than of an icosapeptide, suggesting that during evolution of the NPY and PP genes, a deletion has taken place in the latter sequence, or an insertion in the former. The physiological significance of PP, PYY and NPY is still far from clear. Pharmacologically, PP has been found to have a variety of effects on gastrointestinal functions, such as stimulation of gastric acid secretion in the chicken and inhibition of pancreatic exocrine secretion and gall-bladder contraction in mammals [197]. PYY inhibits pancreatic secretion in cats and pentagastrin-stimulated gastric acid secretion in humans. In cats, it causes intestinal vasoconstriction and inhibition of jejunal and colonic motility [207]. NPY has been found to exhibit vasoconstrictory activities in a variety of mammalian species [209], and to stimulate food intake in cats at least under certain experimental conditions [210, 211].

The neuromedin-B-GRP group

A tetradecapeptide with the sequence

⁺E Q R L G N Q W A V G H L M *

was isolated from the skin of the amphibian *Bombina bombina* by Erspamer and co-workers and named bombesin [212]. A structurally related tetradecapeptide, Alytesin, was isolated from the skin of *Alytes obstetricans* [212]. It differs from bombesin only by having Gly-2 and Thr-6. Two other bombesin-related amphibian peptides are ranatensin from the skin of *Rana pipiens* [213] and litorin, from the skin of *Litoria (Hyla) aurea* [214]. The amino acid sequence of ranatensin is

⁺E V P Q W A V G H F M *

and that of litorin is

⁺E Q W A V G H F M *

Bombesin was found to stimulate the release of gastrin and cholecystokinin in dogs, and bombesin-like immunoreactive material could be demonstrated in extracts of dog and pig antral mucosa [215].

A heptacosapeptide structurally related to bombesin was isolated from porcine non-antral gastric tissue [25] and intestine [216] and named gastrin-releasing polypeptide (GRP).

The amino acid sequence of the human form of GRP was determined in part by sequencing of GRP isolated from metastatic bronchial carcinoid tissue [217] and subsequently deduced in whole from its corresponding cDNA [218]. It is

V P L P A G G G T V L T K M Y P R G N H W A V G H L M *

Porcine GRP differs from it by having Ala-1, Val-3, Ser-4, Val-5 and Ala-12 [25]. Canine GRP differs from it by having Ala-1, Val-3, Gly-5, Gln-7 and Asp-12 [219] and chicken GRP differs from it by having Ala-1, Gln-4, Pro-5, Ser-8, Pro-9, Ala-10, Ile-14 and Ser-19 [220].

Human prepro-GRP consists of a signal peptide of 23 amino acid residues, followed in turn by the GRP sequence, -Gly-Lys-Lys- and a C-terminal sequence of 95 residues [218].

A decapeptide related to ranatensin, bombesin and GRP was isolated from porcine spinal cord and named neuromedin B (B for bombesin) [221]. Subsequently, two longer forms of neuromedin B in which the decapeptide is extended N-terminally have been isolated from porcine spinal cord and brain [222]. In one of them the extension is by an icosapeptide and in the other the triacontapeptide thus formed is further extended by a dipeptide.

The amino acid sequence of the dotriaconta-form of neuromedin B is

A P L S W D L P E P R S R A G K I R V H P R G N L W A T G H F M *

The opioid peptide group

The opioid peptides are known to be synthesized in three different prepro-proteins.

Prepro-opiomelanocortin contains the corticotropin [223, 224] and β-lipotropin [225, 226] sequences [227–229]. The complete amino acid sequence was first deduced from the nucleotide sequence of the corresponding DNA for bovine prepro-opiomelanocortin [230], and later for other pro-opiomelanocortins, inclusive of the human [231]. In all cases the β-lipotropin sequence has been found to constitute the C-terminal part of the prepromelanocortin. The C-terminal entriacontapeptide of β-lipotropin is β-endorphin. Other endorphins are formed by different C-terminal truncations of β-endorphin [232]. The N-terminal pentapeptide of the endorphins has the sequence (Met)enkephalin: Tyr-Gly-Gly-Phe-Met [233]. It is, however, not known whether (Met)-enkephalin is ever released from endorphin(s).

The amino acid sequence of *preproenkephalin A* was first deduced for the bovine form [234, 235] and then for the human [236]. In both species it includes six copies of the (Met)enkephalin sequence, and one of the (Leu) Met-enkephalin. The first (Met)enkephalin sequence, counting from the N-terminus of the protein, is flanked at its N-terminus by a polypeptide of 96 residues with C-terminal Lys-Lys in the bovine, and one with 99 residues and C-terminal Lys-Arg in the human protein. From the second (Met)enkephalin sequence it is separated by Lys-Arg, in both species. The second and the third (Met)enkephalin sequences are separated by tetracosapeptides highly homologous but not identical in the two species, both with N-terminal Lys-Lys and C-terminal Lys-Arg. The third and the fourth (Met)enkephalin sequences are separated by a tetratetracontapeptide in the bovine and a pentatetracontapeptide in the human protein. These intervening polypeptide sequences are again mutually highly homologous and both have N-terminal Lys-Lys and C-terminal Lys-Arg. The fourth and the fifth (Met)enkephalin sequences are separated by nonadecapeptides with N-terminal Arg-Gly-Leu-Lys-Arg- and C-terminal Lys-Arg, suggesting that this fourth (Met)enkephalin sequence is not released as (Met)enkephalin but rather as (Met)enkephalinyl-Arg-Gly-Leu [237, 238]. The fifth (Met)enkephalin sequence is separated from the (Leu)enkephalin sequence by a pentadecapeptide, identical for the bovine and the human protein, with N-terminal Arg-Arg and C-terminal Lys-Arg. The (Leu)enkephalin sequence, finally, is separated from the sixth (Met)enkephalin sequence by hexacosapeptides, highly homologous but non-identical in the two species, both with N-terminal and C-terminal Lys-Arg. The sixth (Met)enkephalin sequence is extended by -Arg-Phe in both species and (Met)enkephalinyl-Arg-Phe may be released as such from preproenkephalin.

The high degree of interspecies homology between the peptide sequences intervening between the enkephalin sequences is rather remarkable. Some recent work suggests that besides (Met)enkephalinyl-Arg-Gly-Leu and (Met)enkephalinyl-Arg-Phe still other C-terminally extended forms of the enkephalins may be released from preproenkephalin A [239].

The amino acid sequences for *preproenkephalin B*, alternatively called *(pre)prodynorphin* [238] are also known for the porcine [240] and human [241] forms, and in part for the rat form [242]. Preproenkephalin B does not contain any (Met)enkephalin sequence but contains three (Leu)enkephalin sequences, separated by intervening peptides of different lengths and sequences. The first (Leu)enkephalin sequence is flanked N-terminally by polypeptides of 174 amino acid residues with C-terminal Lys-Arg in both the human and the porcine forms. The first and second (Leu)enkephalin residues are separated by a heptacosapeptide with N-terminal Arg-Lys and C-terminal Lys-Arg in the human and by a nonacosapeptide with the same terminal dipeptide sequences in the porcine form. The second and third (Leu)enkephalin sequences are separated by tetradecapeptides with N-terminal Arg-Arg and C-terminal Lys-Arg. These tetradecapeptides are identical in the human, porcine and rat species. The third (Leu)enkephalin sequence is, in the preprohormone, extended C-terminally by a tetracosapeptide in the human and porcine proteins and by a tricosapeptide in the rat. These are highly homologous and all three have N-terminal Arg-Arg. Besides possible release as from (Leu)enkephalin all three (Leu)enkephalin sequences are also known to be released in C-terminally extended forms giving opioid peptides with their own names [238]. The only one of these that has actually been isolated from any gastrointestinal tissue is dynorphin-17 representing the second of the (Leu)enkephalin sequences extended by the N-terminal dodecapeptide of the tetradecapeptide connecting it to the third [41]. The sequence of dynorphin-17 is

Y G G F L R R I R P K L K W D N Q [40]

A longer form of dynorphin, dynorphin-32, in which the sequence of dynorphin-17 is extended from its C-terminus by

K R Y G G F L R R Q F K V V T,

i.e. the rest of the tetradecapeptide, the third (Leu)enkephalin sequence and the N-terminal octapeptide of the peptide extending the latter sequence in the preprohormone, has as yet been isolated only from porcine pituitary [243]. There is, however, extensive circumstantial evidence for the presence in gastrointestinal tissues of opioid peptides other than only dynorphin-17 (e.g. [244, 245]).

The tachykinin group

In 1962 Erspamer and co-workers [246] isolated the endecapeptide eledoisin from the posterior salivary glands of the mollusc *Eledone moschata* and found its amino acid sequence to be

+E P S K D A F I G L M *

A few years later this group of workers isolated a structurally related peptide physalaemin from the skin of *Physalaemus fuscumaculatus* [247]. Physalaemin has the sequence

+E A D P N K F Y G L M *

On the basis of their pharmacological actions Erspamer classified physalaemin and eledoisin as tachykinins [248]. Other amphibian tachykinins have subsequently been isolated [248, 249]. A characteristic structural feature of tachykinins is their C-terminal -Gly-Leu-Met-NH_2 sequence, and the occurrence of a residue of phenylalanine in position 5 from the C-terminus. Before the sequence of substance P had been disclosed, Erspamer had predicted from the pattern of pharmacological actions that substance P would be found to have the above mentioned characteristic structural features. This proved indeed to be the case. The amino acid sequence of substance P, as isolated from either bovine brain [26] or equine intestine [24] is [250]

R P K P Q Q F F G L M *

For many years substance P remained the only known mammalian tachykinin. In 1983, however, Kimura *et al.* [251] isolated two other tachykinins from bovine spinal cord and named them neurokinin α, with the sequence

H K T D S F V G L M *

and neurokinin β, with the sequence

D M H D F F V G L M *

These tachykinins were subsequently isolated also by Matsuo and co-workers, who refer to them as neuromedin L [252] and neuromedin K [253] respectively, the K standing for kassinin, an amphibian tachykinin [254] with the sequence

D V P K S D Q F V G L M *

The nomenclature is further confused by neurokinin α, or neuromedin L, and not neurokinin β or 'neuromedin K', having been named 'substance K' by Nawa *et al.* [255]. These workers discovered that bovine brain has two types of preproforms for substance P: one, α preprotachykinin of 112 amino acid residues, containing a single substance P sequence, flanked N-terminally by a polypeptide of 57 amino acid residues with C-terminal arginine, and C-terminally by a tetratetracontapeptide with N-terminal -Gly-Lys-Arg; and the other, β-preprotachykinin, identical to α-preprotachykinin up to its residue 96 but with an insertion of the octadecapeptide Arg-neurokininyl α-Gly-Lys-Arg-Ala-Leu-Asn-Ser- between residues 96 and 97. The latter residue consequently becomes residue 1 of β-preprotachykinin. In addition it is converted from one of methionine to one of valine. It has been shown that α-preprotachykinin and β-preprotachykinin are products of the same gene, the primary RNA transcript being spliced in a tissue-specific manner giving the mRNAs for the two preprotachykinins [256].

A hexatriacontapeptide in which the neurokinin α-sequence of *β*-preprotachykinin is extended N-terminally by the nonacosapeptide sequence intervening between it and the substance P sequence, but lacking the N-terminal Gly-Lys-Arg sequence of the nonacosapeptide, has been isolated from porcine cerebral tissue and named 'neuropeptide K' [257].

Lazarus and co-workers have made observations suggesting the presence of a yet unidentified physalaemin-like tachykinin in extracts of various mammalian tissues, particularly of the stomach [258].

A neuromedin N–neurotensin group?

The amino acid sequence of neurotensin was first determined for the tridecapeptide isolated from bovine hypothalamic tissue [259]. It is

+E L Y E N K P R R P Y I L

The same structure was found for the human intestinal [260] and bovine intestinal [30] neurotensins. In the intestine, neurotensin has, despite its name, been found to occur in cells of endocrine type, named N cells [261, 262].

The sequence of chicken intestinal neurotensin was found to differ from that of the above-given mammalian form by having His-3, Val-4 and Ala-7 [263]. A hexapeptide identical to the C-terminal hexapeptide of neurotensin except for having Lys-1 and Asn-2 (i.e. Lys^8, Asn^9, NT^{8-13}) has been isolated from chicken intestine and named LANT-6 [264].

Recently, Matsuo and co-workers have isolated from porcine spinal cord a hexapeptide identical to the C-terminal hexapeptide of bovine neurotensin except for having Lys-1 and Ile-2, and have named it neuromedin N [265]. This seems to mean that a second member of a mammalian group of peptides has been discovered. For this to be so it will, however, be necessary to isolate the porcine form of neurotensin and exclude the, admittedly unlikely, possibility that neuromedin N is the C-terminal hexapeptide of it.

Neurotensin is structurally distinctly related to xenopsin, an octapeptide from the skin of *Xenopus laevis* [266].

The amino acid sequence of xenopsin is

+E G K R P W I L

Recently, Carraway and Feurle [267] have made observations suggesting that yet unidentified-xenopsin-related peptides are enzymically released from high-molecular-weight precursor(s) in extracts of mammalian and avian stomach.

Non-grouped peptides

The amino acid sequence of porcine intestinal motilin is [268]

F V P I F T Y G E L Q R M Q E K E R N K G Q

Canine intestinal motilin differs from it by having His-7, Ser-8, Lys-12, Ile-13, Arg-14 [269]. In human and dog intestine motilin has been found to occur in cells of an endocrine type [270].

According to a recent report, rat, cow and pig brain extracts contain a peptide, with motilin-like immunoreactivity, of the same size as porcine intestinal motilin, but with distinctly different properties under certain chromatographic conditions [271].

Galanin, so named because of its N-terminal glycine and C-terminal alanine, has as yet been isolated only from porcine intestinal tissue, in the form of a nonacosapeptide with the sequence [47]

G W T L N S A G Y L L G P H A I D N H R S F H D K Y G L A *

Its evolutionary relationships are consequently still quite obscure. Pharmacologically it causes hyperglycaemia in dogs [47, 272] and stimulates contraction of rat smooth muscle *in vitro* [47]. In the gastrointestinal tract of several mammalian species galanin-like immunoreactivity has been shown to be present in neurons [273], and in the CNS-marked regional differences in galanin-like immunoreactivity have been described [274, 275].

The hydra head activator

This endecapeptide, in the hydra with morphogenic activity, was first isolated from *Hydra attenuata* and from *Anthopleura elegantissima*. Later it was found to be present also in mammalian brain and intestine [52]. Its amino acid sequence is

+E P P G G S K V I L F

The conservation of structure over such long evolutionary distances suggests that it has some important function also in mammals, but this is no clearer than its evolutionary relationships.

Somatostatin

Somatostatin was first isolated from ovine hypothalamus as a tetradecapeptide of the sequence [28]

A G C K N F F W K T F T S C

with the two cysteine residues joined by a disulphide bridge. Tetradecapeptides identical to this were subsequently isolated from porcine hypothalamus [276], human pancreatic tumour tissue [277], rat pancreas [278], guinea-pig gastric tissue [279], pigeon pancreas [280] and angler-fish pancreatic islets [281], demonstrating a remarkable conversion of sequence. A polypeptide in which somatostatin-14 was N-terminally extended by another tetradecapeptide of different sequence was isolated from porcine intestine [32] as well as from ovine [33] and porcine [34] hypothalamus. In both species the extending tetradecapeptide too had the identical sequence

S A N S N P A M A P R E R K

Two other forms of somatostatin (somatostatin-25 and somatostatin-20) in which the 3 respectively 8 N-terminal amino acids of somatostatin-28 are missing have been isolated from ovine hypothalamus [33] and porcine intestine [282] respectively.

The picture that the above studies seemed to draw of somatostatin-14 being completely conserved throughout (at least, in the vertebrates) became, however, more complex by further work in several laboratories on fish somatostatins and on their prepro-forms.

On the protein side, a somatostatin of 22 amino acid residues was isolated from the pancreatic islets of the channel cat-fish, *Ictalurus punctata*, and found to differ in its C-terminal tetradecapeptide in no fewer than 7 positions from mammalian somatostatin-14. In its N-terminal extensioned octapeptide there was only a single identity to the corresponding sequence in mammalian somatostatin-28 [283] and even this disappeared when another group of workers subsequently published a revised sequence for cat-fish somatostatin-22 [284], which in 2-positions differs from that in [283]. Later, a somatostatin-14, identical to the mammalian, was also isolated, however, from cat-fish pancreas [285]. Yet another fish somatostatin, a somatostatin-28, was isolated from the Brockmann organs of the angler-fish *Lophius piscatorius*. This differed in its C-terminal tetradecapeptide in only 2 positions from mammalian somatostatin-14, by having Tyr-7 and Gly-10 [286], a change which apparently has brought the Lys-9 of somatostatin-14 into a collagen-like sequence leading to its hydroxylation in the angler-fish. This seems to be the first instance that hydroxylysine, typical of collagen, has been found in any hormonal polypeptide [287, 288]. The N-terminal tetradecapeptide of the somatostatin-28 was found to be identical to the corresponding mammalian peptide over its C-terminal pentapeptide sequence and by having N-terminal serine. Not only the presence of hydroxylysine in this angler-fish somatostatin-28 but also the recent finding of glycosylated forms of cat-fish somatostatin-22 [284] stand as a warning against neglecting peptide isolation and sequencing, and instead determining amino acid sequences of naturally occurring peptides solely by deduction from the experimentally determined corresponding nucleotide sequences. The recombinant DNA methodology has, nevertheless, been of immense importance also in the somatostatin area, especially for an understanding of somatostatin biosynthesis. Using it, Rutter and co-workers [289] found that the Brockmann bodies of the angler-fish *Lophius americanus* have two different preprosomatostatins: one of 121 amino acid residues, preprosomatostatin I, and the other, preprosomatostatin II with 125. In both cases somatostatin-14 sequences were found to be at the C-termini of the preprohormones. In preprosomatostatin I, the sequence of which was also independently reported [290], and later slightly revised [291] by Goodman and co-workers, the tetradecapeptide was identical to the mammalian form, while in preprosomatostatin II it differed from the mammalian by having Tyr-7 and Gly-10. These two preprohormones obviously explain why two somatostatins, differing in their somatostatin-14 sequences, have been isolated from angler-fish pancreatic islet tissue [286, 291]. The isolation of the form with Tyr-7, Gly-10 [286] actually took place after its sequence had been predicted from the nucleotide sequence of its cDNA.

In the N-terminal extensions of the somatostatin-14 moieties of the two preprosomatostatins the C-terminal hexapeptide was identical to the corresponding sequence of mammalian somatostatin-28 in preprosomatostatin I and the C-terminal pentapeptide in preprosomatostatin II. In the latter, however, the amino acid residue corresponding in position to that of the N-terminal of mammalian somatostatin-28 was serine, as in the latter, while it was alanine in preprosomatostatin I. Also the cat-fish has been shown to have two preprosomatostatins, with the somatostatin-14 moieties at the C-termini. In one of them this moiety was found to be identical to the mammalian form [292] and in the other to that of cat-fish somatostatin-22 [293]. The cat-fish preprosomatostatin with the mammalian type of somatostatin-14 has, as in angler-fish preprosomatostatin I, an N-terminal extension of this sequence with its C-terminal hexapeptide identical to that of mammalian somatostatin-28. Indeed it has one additional positional identity to mammalian somatostatin-28 in position 6 of the latter. The other cat-fish preprosomatostatin, however, shows no direct resemblance at all to either mammalian somatostatin-28 or to the other cat-fish preprosomatostatin in the tetradecapeptide sequence preceding the somatostatin-14 moiety.

Recently the amino acid sequences have been deduced from the cDNA sequences for both the human [294] and the rat [291, 295] preprosomatostatin or, as mentioned by the authors, for human 'preprosomatostatin I'. This in view of the angler-fish having two preprosomatostatins and the human preprosomatostatin, the sequence of which has been established corresponding to angler-fish preprosomatostatin I [294]. It may, however, be pointed out that a special search for two different somatostatins-14 in a mammalian tissue was negative [296]. From the rat and human 116 amino-acid-residues-long preprosomatostatin sequences, it is evident that the rat and human somatostatins-28, again found at the C-termini of the preprohormones, are identical to each other and to the earlier known ovine and porcine somatostatins-28. Perhaps more surprising is the high degree of sequence homology between the two preprosomatostatins N-terminal to the somatostatin-28 sequences. Indeed, there are differences in only four positions [295]–positions 14, 21, 43 and 74–and of these the first two lie in the signal peptide sequence. Recently a dotriacontapeptide identical to the one comprising residues 26–56 of human preprosomatostatin has been isolated from extracts of porcine intestine. This suggests that the human and porcine preprosomatostatins may be even more like each other than are the human and rat ones [297]. The nucleotide sequences have been determined not only for the cDNAs but also for the whole genes of the human [298] and rat [299, 300] preprosomatostatins. An interesting observation is that these genes are flanked, both 5′ and 3′, by Z-DNA-forming nucleotide sequences. These sequences are not homologous in the two species, suggesting that here is a case of structural, although not sequence conservation [301].

Somatostatin inhibits the release of growth hormone and of several other, but not all, hormones, and in addition exhibits activities apparently not at all related to inhibition of hormone release [302–304]. In the gastrointestinal tract somatostatin is present both in endocrine cells, called D cells, and in neurons, in the peripheral and central nervous systems in neurons only [302]. The coexistence, in mammalian neurons of a peptide hormone and of a neurotransmitter of the classical type seems first to have been demonstrated for somatostatin [302]. Somatostatin is not structurally clearly related to any other, yet known, mammalian peptide. As is often the case with peptides, some slight degree of structural similarity with other peptide(s) may, however, be found. The identity of the

$-Thr_{10}-Phe_{11}-Thr_{12}-Ser_{13}-$

sequence of mammalian somatostatin to the

$Thr_5-Phe_6Thr_7-Ser_8-$

sequence of mammalian secretin is interesting, and has even been discussed in connection with the biological activities of the two hormones [305]. However, this homology is much too restricted to motivate inclusion of somatostatin into the secretin-glucagon group.

Slightly more impressive is the sequence homology of somatostatin and urotensin II, a peptide hormone with, among other things, osmoregulatory activities, from the caudal neurosecretory system of the teleost *Gillichthys mirabilis*. The amino acid sequence of the dodecapeptide urotensin II [306] is

A G T A D C F W K Y C V

A disulphide bridge joins the two cysteine residues.

Like urotensin II, somatostatin is a peptide hormone with a disulphide bridge, although the size of the ring thereby formed is different in the two hormones. The sequence homology is restricted to residues 1–2 and 7–9.

It is a surprising fact that somatostatin is the only hormonal polypeptide actually isolated from gastrointestinal tissues, and sequenced, that contains residues of cysteine. Chymodenin, a polypeptide affecting pancreatic enzyme secretion, has been found to contain such residues but the complete sequence of chymodenin has not yet been disclosed [307]. Urogastrone [308], identical to the epidermal growth factor [309, 310], exhibits, at least pharmacologically, a hormonal action by inhibiting gastric acid secretion. No fewer than 6 of its 53 amino acid residues are cysteines, forming three disulphide bridges. Epidermal growth factor–urogastrone, EGF-URO, as it has been called [311]–is, seen from several viewpoints, an interesting substance [311, 312]. Currently much interest has been aroused by the finding that pro-EGF-URO shows a high degree of sequence homology to the translation product of the Moloney sarcoma virus oncogene *mos* [313]. Mouse submaxillary gland prepro-EGF-URO was predicted from its cDNA sequence to consist of 1168 amino acid residues by one group of workers [314] and of 1217 by another [315], the difference lying in the part, C-terminal to the EGF-URO, moiety, which by both groups was found to be represented by the 977–1029 sequence of the precursor. N-terminally to the EGF-URO sequence the precursor was found to contain no fewer than 7 sequences having homology in varying degree, but not identity, to each other and to EGF-URO.

Urogastrone has, however, not been isolated from any gastrointestinal tissue although its presence in the glands of Brunner [316] and in the stomach [317] is, at least in some species, highly probable on immunocytochemical evidence. A more detailed description of EGF-URO, such as of its species variability, would therefore carry the contents of this chapter too far afield.

Conclusions

Much of what has been discussed in this chapter reflects the picture of the field of the gastrointestinal and cerebrogastrointestinal hormonal peptides as it appears in 1985. The picture is, however, still swiftly changing! Statements such as that the secretin-glucagon group contains 6 types of hormonal polypeptides but the gastrin-CCK group only 2, and that somatostatin has none to form a group with, may soon turn obsolete through the discovery of additional types of peptides. In addition to the various similarities between peptides discussed here there are many others, that have not been mentioned and it is possible that in the light of future work some of these similarities will take on increased significance.

There are details that can be compiled from the various articles referred to that may be of evolutionary significance, although this remains to be proved. Why is, for instance, GRP in its precursor extended by Gly-Lys-Lys, when members of the tachykinin group, the secretin-glucagon group and the PYY-PP group are extended by Gly-Lys-Arg and members of the gastrin-CCK group, including caerulein, by Gly-Arg-Arg, in all cases where the precursor sequences are known.

Of interest may also be that the two polypeptides VIP and NPY with their very different amino acid sequences are in their precursors (at least in their human forms [119, 205], both extended by the identical pentapeptide Gly-Lys-Arg-Ser-Ser.

The discussion in this chapter has rather strictly focused on amino acid sequence homologies as they stand today. By allowing deletions and/or insertions, other homologies than the ones discussed may be brought to light, but possibly at the expence of probability. This applies also to such sequence homologies that may appear when codon frameshift mutations are introduced [318].

In the future, new insights into evolutionary relationships of peptides may be expected from elucidations of solution conformations of peptides and, not least, of the sequences and conformations of peptide hormone receptors, in different species and in different organs of the same species, i.e. from work that is gaining momentum in several laboratories.

References

1. Zahn, H., *Naturwissenschaften* **54**, 396 (1967).
2. Bayliss, W. M. and Starling, E. H., *Proc. R. Soc. London* **69**, 352 (1902).
3. Becker, N. M. *Arch. Sci. Biol. St Petersburg* **2**, 433 (1893).
4. Bayliss, W. M., in *Principles of General Physiology*, 1st ed. Longmans, Green, London (1915).
5. Edkins, J. S., *Proc. R. Soc. Lond Ser. B* **76**, 376 (1905).
6. v. Euler, U. S. and Gaddum, J. H., *J. Physiol.* (*London*) **72**, 74 (1931).
7. Greengard, H., in *The Hormones, Physiology, Chemistry and Applications* (ed. G. Pincus and K. V. Thimann), p. 201. Academic Press, New York (1948).
8. Mutt, V., *Scand. J. Gastroenterol. Suppl.* **77**, 133 (1982).
9. Hallion, M. by Delezenne, C. and Frouin, A. *C. R. Soc. Biol.* (*Paris*) **56**, 322 (1906).
10. Ivy, A. C. and Oldberg, E., *Am. J. Physiol.* **86**, 599 (1928).
11. Harper, A. A. and Raper, H. S., *J. Physiol.* (*London*) **102**, 115 (1943).
12. Jorpes, J. E. and Mutt, V., *Acta Physiol. Scand.* **66**, 196 (1966).
13. Grossman, M. I., *Gastroenterology* **58**, 128 (1970).
14. Gregory, R. A., in *Surgical Physiology of the Gastro-Intestinal Tract, Symposium* (ed. A. N. Smith), p. 57. Edinburgh Royal Coll., Edinburgh (1962).
15. Jorpes, J. E. and Mutt, V., *Acta Chem. Scand.* **15**, 1790 (1961).
16. Jorpes, J. E., Mutt, V., Magnusson, S. and Steele, B. B., *Biochem. Biophys. Res. Commun.* **9**, 275 (1962).
17. Mutt, V. and Jorpes, J. E., *Eur. J. Biochem.* **6**, 156 (1968).
18. Wagner, I. and Musso, H., *Angew. Chem.* **22**, 816 (1983).
19. Richman, P. G. and Meister, A., *J. Biol. Chem.* **250**, 1422 (1975).
20. *Precursor Processing in the Biosynthesis of Proteins* (ed. M. Zimmerman, R. A. Mumford and D. F. Steiner). *Ann. N.Y. Acad. Sci.* **343** (1980).
21. Brown, J. C., Mutt, V. and Pederson, R. A., *J. Physiol.* (*London*) **209**, 57 (1970).
22. Brown, J. C. and Pederson, R. A., *Endocrinology*, vol. 2 (Proc. Vth Int. Congr. Endocrinology), Hamburg, 1976 (ed. V. H. T. James), p. 568 (1976).
23. Said, S. I. and Mutt, V., *Eur. J. Biochem.* **28**, 199 (1972).
24. Brown, J. C., Mutt, V. and Dryburgh, J. R., *Can. J. Physiol. Pharmacol.* **49**, 399 (1971).
25. McDonald, T. J., Jörnvall, H., Nilsson, G., Vagne, M., Ghatei, M., Bloom, S. R. and Mutt. V., *Biochem. Biophys. Res. Commun.* **90**, 227 (1979).
26. Chang, M. M. and Leeman, S. E., *Fed. Proc. Fed. Am. Soc. Exp. Biol.* **29**, 282/a 202 (1970).
27. Carraway, R. and Leeman, S. E., *J. Biol. Chem.* **248**, 6854 (1973).
28. Brazeau, P., Vale, W., Burgus, R., Ling, N., Butcher, M., Rivier, J. and Guillemin, R., *Science* **179**, 77 (1973).
29. Studer, R. O., Trzeciak, H. and Lergier, W., *Helv. Chim. Acta* **56**, 860 (1973).
30. Carraway, R., Kitabgi, P. and Leeman, S. E., *J. Biol. Chem.* **253**, 7996 (1978).
31. Luft, R., Efendić, S., Hökfelt, T., Johansson, O. and Arimura, A., *Med. Biol.* **52**, 428 (1974).
32. Pradayrol, L., Jörnvall, H., Mutt, V. and Ribet, A., *FEBS Lett.* **109**, 55 (1980).
33. Böhlen, P., Brazeau, P., Benoit, R., Ling, N., Esch, F. and Guillemin, R., *Biochem. Biophys. Res. Commun.* **96**, 725 (1980).
34. Schally, A. V., Huang, W.-Y., Chang, R. C. C., Arimura, A., Redding, T. W., Millar, R. P., Hunkapiller, M. W. and Hood, L. E., *Proc. Natl. Acad. Sci. USA* **77**, 4489 (1980).
35. Vanderhaeghen, J. J., Signeau, J. C. and Gepts, W., *Nature* **257**, 604 (1975).
36. Bryant, M. G., Bloom, S. R., Polak, J. M., Albuquerque, R. H., Modlin, I. and Pearse, A. G. E., *Lancet* i, 991 (1976).
37. Larsson, L.-I., Fahrenkrug, J., Schaffalitzky de Muckadell, O., Sundler, F., Håkansson, R. and Rehfeld, J. F., *Proc. Natl. Acad. Sci. USA* **73**, 3197 (1976).
38. Said, S. I. and Rosenberg, R. N., *Science* **192**, 907 (1976).
39. Dockray, G. J., Gregory, R. A., Hutchison, J. B., Harris, J. I. and Runswick, M. J., *Nature* (*London*) **274**, 711 (1978).
40. Goldstein, A., Fischli, W., Lowney, L. I., Hunkapiller, M. and Hood, L., *Proc. Natl. Acad. Sci. USA* **78**, 7219 (1981).
41. Tachibana, S., Araki, K., Ohya, S. and Yoshida, S. *Nature* (*London*) **295**, 339 (1982).
42. Thim, L. and Moody, A. J., *Regul. Pept.* **2**, 139 (1981).
43. Bataille, D., Tatemoto, K., Gespach, C., Jörnvall, H., Rosselin, G. and Mutt, V., *FEBS Lett.* **146**, 79 (1982).
44. Tatemoto, K. and Mutt, V., *Proc. Natl. Acad. Sci. USA* **78**, 6603 (1981).
45. Tatemoto, K. and Mutt, V., *Nature* (*London*) **285**, 417 (1980).
46. Tatemoto, K. *Proc. Natl. Acad. Sci. USA* **79**, 2514 (1982).
47. Tatemoto, K., Rökaeus, Å., Jörnvall, H., McDonald, T. J. and Mutt, V., *FEBS Lett.* **164**, 124 (1983).
48. Carlquist, M., Jörnvall, H., Tatemoto, K. and Mutt, V., *Gastroenterology* **83**, 245 (1982).
49. Tatemoto, K., Carlquist, M., McDonald, T. J. and Mutt, V., *FEBS Lett.* **153**, 248 (1983).
50. Tatemoto, K., Carlquist, M. and Mutt, V., *Nature* (*London*) **296**, 659 (1982).
51. Tatemoto, K., Siimesmaa, S., Jörnvall, H., Allen, J. M., Polak, J. M., Bloom, S. R. and Mutt, V., *FEBS Lett.* **179**, 181 (1985).
52. Schaller, H. C. and Bodenmüller, H., *Proc. Natl. Acad. Sci. USA* **78**, 7000 (1981).
53. Gregory, R. A. and Tracy, H. J., in *Gastrointestinal Hormones* (ed. J. C. Thompson), p. 13. University of Texas Press, Austin and London (1975).
54. Wider, M. D., Vinik, A. I. and Heldsinger, A., *Endocrinology* **115**, 1484 (1984).
55. Grossman, M. I., and others, *Gastroenterology* **67**, 730 (1974).
56. Schmidt, W. E., Mutt, V., Konturek, S. J. and Creutzfeldt, W., *Dig. Dis. Sci.* **29**, Aug. Suppl. 75S (1984).
57. Vagne, M., Pansu, D., Guignard, H., Carlquist, M., Jörnvall, H. and Mutt, V., *Dig. Dis. Sci.* **29**, Aug. Suppl. 92S (1984).
58. Bodanszky, M., Fink, M. L., Funk, K. W. and Said, S. I., *Clin. Endocrin. Oxford* **5**, Suppl. 195S (1976).
59. Jaeger, E., Filippi, B., Knof, S., Lehnert, P., Moroder, L. and Wünsch, E., in *Hormone Receptors in Digestion and Nutrition* (ed. G. Rosselin, P. Fromageot and S. Bonfils), p. 25. Elsevier, North Holland Biomedical Press, Amsterdam (1979).
60. Korn, A. P. and Ottensmeyer, F. P., *J. Ultrastruc. Res.* **79**, 142 (1982).
61. Fournier, A., Saunders, J. K. and St-Pierre, S., *Peptides* **3**, 345 (1982).

62. Blundell, T. L., Dockerill, S., Sasaki, K., Tickle, I. J. and Wood, S. P., *Metabolism* **25**, Suppl. 1, 1331 (1976).
63. Glover, I. D., Barlow, D. J., Pitts, J. E., Wood, S. P., Tickle, I. J., Blundell, T. L., Tatemoto, K., Kimmel, J. R., Wollmer, A., Strassburger, W., and Zhang, Y.-S., *Eur. J. Biochem.* **142**, 379 (1985).
64. Schiller, P. W., Natarajan, S. and Bodanszky, M., *Int. J. Pep. Protein Res.* **12**, 139 (1978).
65. Durieux, C., Belleney, J., Lallemand, J.-Y., Roques, B. P. and Fournie-Zaluski, M.-C., *Biochem. Biophys. Res. Commun.* **114**, 705 (1983).
66. Koizuka, I., Watari, H., Yanaihara, N., Nishina, Y., Shiga, K. and Nagayama, K., *Biomed. Res.* **5**, Suppl. 161 (1984).
67. Bleich, H. E., Cutnell, J. D. and Glasel, J. A., *Biochemistry* **15**, 2455 (1976).
68. Feeney, J., Roberts, G. C. K., Brown, J. P., Burgen, A. S. V. and Gregory, H. *J. Chem. Soc. Perkin Trans. II*, 601 (1972).
69. Pham Van Chuong, P., Penke, B., de Castiglione, R. and Fromageot, P., in *Hormone Receptors in Digestion and Nutrition* (ed. G. Rosselin, P. Fromageot and S. Bonfils), p. 33. Elsevier North Holland Biomedical Press, Amsterdam (1979).
70. Peggion, E., Foffani, M. T., Wünsch, E., Moroder, L., Borin, G., Goodman, M. and Mammi, S., *Biopolymers* **24**, 647 (1985).
71. Abillon, E., Pham Van Chuong, P. and Fromageot, P. *Int. J. Pept. Protein Res.* **17**, 480 (1981).
72. Holladay, L. A. and Puett, D., *Proc. Natl. Acad. Sci. USA* **73**, 1199 (1976).
73. Knappenberg, M., Brison, J., Dirkx, J., Hallenga, K., Deschrijver, P. and Van Binst, G., *Biochim. Biophys. Acta* **580**, 266 (1979).
74. Wünsch, E. W., Jaeger, E., Moroder, L., Peggion, E. and Palumbo, M. *Biopolymers* **20**, 1741 (1981).
75. Cotrait, M., *Int. J. Pept. Protein Res.* **23**, 355 (1984).
76. Cotrait, M. and Hospital, M., *Biochem. Biophys. Res. Commun.* **109**, 1123 (1982).
77. Manavalan, P. and Momany, F. A., *Int. J. Pept. Protein Res.* **20**, 351 (1982).
78. Mehlis, B., Böhm, S., Becker, M. and Bienert, M., *Biochem. Biophys. Res. Commun.* **66**, 1447 (1975).
79. Bernier, J.-L., Hénichart, J.-P. and Helbecque, N., *Eur. J. Biochem.* **142**, 371 (1984).
80. Maroun, R. and Mattice, W. L., *Biochem. Biophys. Res. Commun.* **103**, 442 (1981).
81. Erne, D., Sargent, D. F. and Schwyzer, R., *Biochemistry* **24**, 4261 (1985).
82. Robinson, R. M., Blakeney, Jr., E. W., and Mattice, W. L., *Biopolymers* **21**, 1217 (1982).
83. Sasaki, K., Dockerill, S., Adamiak, D. A., Tickle, I. J. and Blundell, T., *Nature (London)* **257**, 751 (1975).
84. Moran, E. C., Chou, P. Y. and Fasman, G. D., *Biochem. Biophys. Res. Commun.* **77**, 1300 (1977).
85. Bösch, C., Brown, L. R. and Wüthrich, K., *Biochim. Biophys. Acta* **603**, 298 (1980).
86. Korn, A. P. and Ottensmeyer, F. P., *J. Theor. Biol.* **105**, 403 (1983).
87. Reichert, E. T. and Brown, A. P., *The Differentiation and Specificity of Corresponding Proteins and other Vital Substances in Relation to Biological Classification and Organic Evolution: The Crystallography of Hemoglobins.* Carnegie Institution, Washington (1909).
88. Perutz, M. F. in *Les Prix Nobel en 1962*, p. 82. P. A. Norstedt, Stockholm (1963).
89. Kendrew, J. C., in *Les Prix Nobel en 1962*, p. 103. P. A. Norstedt, Stockholm (1963).
90. Grütter, M. G., Weaver, L. H. and Matthews, B. W., *Nature (London)* **303**, 828 (1983).
91. Jörnvall, H., in *Evolution and Tumour Pathology of the Neuroendocrine System* (ed. S. Falkmer, R. Håkanson and F. Sundler), p. 165. Elsevier, Amsterdam (1984).
92. Bajaj, M., Blundell, T. L., Pitts, J. E., Wood, S. P., Tatnell, M. A., Falkmer, S., Emdin, S. O., Gowan, L. K., Crow, H., Schwabe, C., Wollmer, A. and Strassburger, W., *Eur. J. Biochem.* **135**, 535 (1983).
93. Bell, G. I., Sanchez-Pescador, R., Laybourn, P. J. and Najarian, R. C., *Nature (London)* **304**, 368 (1983).
94. Lopez, L. C., Frazier, M. L., Su, C.-J., Kumar, A. and Saunders, G. F., *Proc. Natl. Acad. Sci. USA* **80**, 5485 (1983).
95. Bell, G. I., Santerre, R. F. and Mullenback, G. T., *Nature (London)* **302**, 716 (1983).
96. Heinrich, G., Gros, P. and Habener, J. F., *J. Biol. Chem.* **259**, 14082 (1984).
97. Lund, P. K., Goodman, R. H., Montminy, M. R., Dee, P. C. and Habener, J. F., *J. Biol. Chem.* **258**, 3280 (1983).
98. Bromer, W. W., Sinn, L. G. and Behrens, O. K., *J. Am. Chem. Soc.* **79**, 2807 (1957).
99. Sundby, F. *Metabolism* **25**, Suppl. 1, 1319 (1976).
100. Conlon, J. M., Hansen, H. F. and Schwartz, T. W., *Regul. Pept*, **11**, 309 (1985).
101. Andrews, P. C. and Ronner, P., *J. Biol. Chem.* **260**, 3910 (1985).
102. Lopez, L. C., Li, W.-H., Frazier, M. L., Luo, C.-C. and Saunders, G. F., *Mol. Biol. Evol.* **1**, 335 (1984).
103. Patzelt, C. and Schiltz, E., *Proc. Natl. Acad. Sci. USA* **81**, 5007 (1984).
104. Ghiglione, M., Uttenthal, L. O., George, S. K. and Bloom, S. R., *Diabetologia* **27**, 599 (1984).
105. Carlquist, M., Jörnvall, H., Forssmann, W.-G., Thulin, L., Johansson, C. and Mutt, V., *IRCS Med. Sci. Biochem.* **13**, 217 (1985).
106. Mutt, V., Jorpes, J. E. and Magnusson, S., *Eur. J. Biochem.* **15**, 513 (1970).
107. Carlquist, M., Jörnvall, H. and Mutt, V., *FEBS Lett.* **127**, 71 (1981).
108. Nilsson, A., Carlquist, M., Jörnvall, H. and Mutt, V., *Eur. J. Biochem.* **112**, 383 (1980).
109. Carlquist, M. and Rökaeus, Å., *J. Chromatogr.* **296**, 143 (1984).
110. Gafvelin, G., Carlquist, M. and Mutt, V., *FEBS Lett.* **184**, 347 (1985).
111. Moody, A. J., Thim, L. and Valverde, I., *FEBS Lett.* **172**, 142 (1984).
112. Brown, J. C. and Dryburgh, J. R., *Can. J. Biochem.* **49**, 867 (1971).
113. Jörnvall, H., Carlquist, M., Kwauk, S., Otte, S. C., McIntosh, C. H. S., Brown, J. C. and Mutt, V., *FEBS Lett.* **123**, 205 (1981).
114. Carlquist, M., Maletti, M., Jörnvall, H. and Mutt, V., *Eur. J. Biochem.* **145**, 573 (1984).
115. Carlquist, M., *Isolation and Chemical Characterization of Gastrointestinal Hormones.* Thesis, Sundt Offset, Stockholm (1984).
116. Bradbury, A. F. and Smyth, D. G., *Biochem. Biophys. Res. Commun.* **112**, 372 (1983).
117 Eipper, B. A., Mains, R. E. and Glembotski, C. C., *Proc. Natl. Acad. Sci. USA* **80**, 5144 (1983).
118 Carlquist, M., McDonald, T. J., Go, V. L. W., Bataille, D., Johansson, C. and Mutt, V., *Horm. Metab. Res.* **14**, 28 (1982).
119 Itoh, N., Obata, K.-I., Yanaihara, N. and Okamoto, H., *Nature (London)* **304**, 547 (1983).
120. Mutt, V. and Said, S. I., *Eur. J. Biochem.* **42**, 581 (1974).
121. Carlquist, M., Mutt, V. and Jörnvall, H., *FEBS Lett.* **108**, 457 (1979).
122. Dimaline, R., Reeve, Jr., J. R., Shively, J. E. and Hawke, D., *Peptides* **5**, 183 (1984).
123. Wang, S. C., Du, B. H., Eng, J., Chang, M., Hulmes, J. D., Pan, Y.-C. E. and Yalow, R. S., *Life Sci.* **37**, 979 (1985).
124. Du, B.-H., Eng. J., Hulmes, J. D., Chang, M., Pan, Y.-C. E. and Yalow R. S., *Biochem. Biophys. Res. Commun.* **128**, 1093 (1985).
125. Nilsson, A., *FEBS Lett.* **60**, 322 (1975).
126. Tatemoto, K. and Mutt, V., *Proc. Natl. Acad. Sci. USA* **75**, 4115 (1978).
127. Tatemoto, K., Jörnvall, H., McDonald, T. J., Carlquist, M., Go, V. L. W., Johansson, C. and Mutt, V., *FEBS Lett.* **174**, 258 (1984).
128. Carlquist, M., Kaiser, R., Tatemoto, K., Jörnvall, H. and Mutt, V., *Eur. J. Biochem.* **144**, 243 (1984).
129. Tsukada, T., Horovitch, S. J., Montminy, M. R., Mandel, G. and Goodman, R. H., *DNA* **4**, 293 (1985).
130. Bodner, M., Fridkin, M. and Gozes. I., *Proc. Natl. Acad. Sci. USA* **82**, 3548 (1985).
131. Edvinsson, L. and McCulloch, J., *Regul. Pept.* **10**, 345 (1985).
132. Guillemin, R., Brazeau, P., Böhlen, P., Esch, F., Ling, N. and Wehrenberg, W. B., *Science* **218**, 585 (1982).
133. Rivier, J., Spiess, J., Thorner, M. and Vale, W., *Nature (London)* **300**, 276 (1982).
134. Böhlen, P., Wehrenberg, W. B., Esch, F., Ling, N., Brazeau, P. and Guillemin, R., *Biochem. Biophys. Res. Commun.* **125**, 1005 (1984).
135. Böhlen, P., Esch, F., Brazeau, P., Ling, N. and Guillemin, R., *Biochem. Biophys. Res. Commun.* **116**, 726 (1983).
136. Esch, F., Böhlen, P., Ling, N., Brazeau, P. and Guillemin, R., *Biochem. Biophys. Res. Commun.* **117**, 772 (1983).
137. Brazeau, P., Böhlen, P., Esch, F., Ling, N., Wehrenberg, W. B. and Guillemin, R., *Biochem. Biophys. Res. Commun.* **125**, 606 (1984).
138. Gubler, U., Monahan, J. J., Lomedico, P. T., Bhatt, R. S., Collier, K. J., Hoffman, B. J., Böhlen, P., Esch, F., Ling, N., Zeytin, F.,

Brazeau, P., Poonian, M. S. and Gage, L. P., *Proc. Natl. Acad. Sci. USA* **80**, 4311 (1983).
139. Mayo, K. E., Cerelli, G. M., Lebo, R. V., Bruce, B. D., Rosenfeld, M. G. and Evans, R. M., *Proc. Natl. Acad. Sci. USA* **82**, 63 (1985).
140. Weinstein, B., *Experientia* **28**, 1517 (1972).
141. Gráf, L. *Acta Biochim. Biophys. Acad. Sci. Hung.* **11**, 267 (1976).
142. Scheving, L. A., *Arch. Biochem. Biophys.* **226**, 411 (1983).
143. Maloy, W. L., Nathenson, S. G. and Coligan, J. E., *J. Biol. Chem.* **256**, 2863 (1981).
144. Jörnvall, H., Carlström, A., Pettersson, T., Jacobsson, B., Persson, M. and Mutt, V., *Nature (London)* **291**, 261 (1981).
145. Jacobsson, B., Pettersson, T., Sandstedt, B. and Carlström, A., *IRCS Med.Sci. Endocrine System* **7**, 590 (1979).
146. Liddle, C. N., Reid, W. A., Kennedy, J. S., Miller, I. D. and Horne, C. H. W., *J. Pathology* **146**, 107 (1985).
147. Korec, E., Hlozánek, I., Korcová, H. and Simúnek, J., *Folia Biol. (Prague)* **31**, 161 (1985).
148. Parker, D. S., Raufman, J.-P., O'Donohue, T. L., Bledsoe, M., Yoshida, H. and Pisano, J. J., *J. Biol. Chem.* **259**, 11751 (1984).
149. Hoshino, M., Yanaihara, C., Hong, Y.-M., Kishida, S., Katsumaru, Y., Vandermeers, A., Vandermeers-Piret, M.-C., Robberecht, P., Christophe, J. and Yanaihara, N., *FEBS Lett.* **178**, 233 (1984).
150. Robberecht, P., De Graef, J., Woussen, M.-C., Vandermeers-Piret, M.-C., Vandermeers, A., De Neef, P., Cauvin, A., Yanaihara, C., Yanaihara, N. and Christophe, J., *Biochem. Biophys. Res. Commun.* **130**, 333 (1985).
151. Gregory, R. A., *Bioorg. Chem.* **497**, 511 (1979).
152. Rehfeld, J. F., *J. Neurochem.* **44**, 1 (1985).
153. Gregory, R. A., Dockray, G. J., Reeve, Jr., J. R., Shively, J. E. and Miller, C., *Peptides* **4**, 319 (1983).
154. Reeve, Jr., J. R., Dimaline, R., Shively, J. E., Hawke, D., Chew, P. and Walsh, J. H., *Peptides* **2**, 453 (1981).
155. Schaffer, M. H., Agarwal, K. L. and Noyes, B. E., *Peptides* **3**, 693 (1982).
156. Rehfeld, J. F., *Gastroenterology* **70**, 146 (1976).
157. Gregory, R. A. and Tracy, H. J., *Gut* **5**, 103 (1964).
158. Rehfeld, J. F. and Hansen, H. F., *J. Biol. Chem.* (in the press).
159. Reeve, Jr., J. R., Eysselein, V. E., Deveney, D., Shively, J. E., Miller C. and Walsh, J. H., *Dig. Dis. Sci.* **29**, Aug., Suppl. 68S (1984).
160. Grossman, M. I., *Nature (London)* **228**, 1147 (1970).
161. Choudhury, A. M., Kenner, G. W., Moore, S., Ramachandran, K. L., Thorpe, W. D., Ramage, R., Dockray, G. J., Gregory, R. A., Hood, L. and Hunkapiller, M., *Hoppe-Seyler's Z. Physiol. Chem.* **361**, 1719 (1980).
162. Wünsch, E., Wendlberger, G., Mladenova-Orlinova, L., Göhring, W., Jaeger, E., Scharf, R., Gregory, R. A. and Dockray, G. J., *Hoppe-Seyler's Z. Physiol. Chem.* **362**, 179 (1981).
163. Kenner, G. W., Moore, S., Ramachandran, K. L., Ramage, R., Dockray, G. J., Gregory, R. A., Hood, L. and Hunkapiller, M., *Bioorg. Chem.* **10**, 152 (1981).
164. Yoo, O. J., Powell, C. T. and Agarwal, K. L. *Proc. Natl. Acad. Sci. USA* **79**, 1049 (1982).
165. Boel, E., Vuust, J., Norris, F., Norris, K., Wind, A., Rehfeld, J. F. and Marcker, K. A., *Proc. Natl. Acad. Sci. USA* **80**, 2866 (1983).
166. Kato, K., Himeno, S., Takahashi, Y., Wakabayashi, T., Tarui, S., and Matsubara, K., *Gene* **26**, 53 (1983).
167. Kato, K., Hayashizaki, Y., Takahashi, Y., Himeno, S. and Matsubara, K., *Nucleic Acids Res.* **11**, 8197 (1983).
168. Wiborg, O., Berglund, L., Boel, E., Norris, F., Norris, K., Rehfeld, J. F., Marcker, K. A. and Vuust, J., *Proc. Natl. Acad. Sci. USA* **81**, 1067 (1984).
169. Ito, R., Sato, K., Helmer, T., Jay, G. and Agarwal, K., *Proc. Natl. Acad. Sci. USA* **81**, 4662 (1984).
170. Mutt, V. and Jorpes, E., *Biochem. J.* **125**, 57P (1971).
171. Mutt, V. *Clin. Endocrin.* **5**, Suppl, 175S (1976).
172. Eysselein, V. E., Reeve, Jr., J. R., Shively, J. E., Hawke, D. and Walsh, J. H., *Peptides* **3**, 687 (1982).
173. Eng, J., Du, B.-H., Pan, Y.-C. E., Chang, M., Hulmes, J. D. and Yalow, R. S., *Peptides* **5**, 1203 (1984).
174. Carlquist, M., Mutt, V. and Jörnvall, H., *Regul. Pept*, **11**, 27 (1985).
175. Zhou, Z.-Z., Eng, J., Pan, Y.-C. E., Chang, M., Hulmes, J. D., Raufman, J.-P. and Yalow, R. S., *Peptides* **6**, 337 (1985).
176. Miller, L. J., Jardine, I., Weissman, E., Go, V. L. W. and Speicher, D., *J. Neurochem.* **43**, 835 (1984).
177. Reeve, Jr., J. R., Eysselein, V. E., Walsh, J. H., Sankaran, H., Deveney, C. W., Tourtellotte, W. W., Miller, C. and Shively, J. E., *Peptides* **5**, 959 (1984).
178. Eng., J., Shiina, Y., Pan, Y.-C. E., Blacher, R., Chang, M., Stein, S. and Yalow, R. S., *Proc. Natl. Acad. Sci. USA* **50**, 6381 (1983).
179. Tatemoto, K., Jörnvall, H., Siimesmaa, S., Halldén, G. and Mutt, V., *FEBS Lett.* **174**, 289 (1984).
180. Takahashi, Y., Kato, K., Hayashizaki, Y., Wakabayashi, T., Ohtsuka, E., Matsuki, S., Ikehara, M. and Matsubara, K. *Proc. Natl. Acad. Sci. USA* **82**, 1931 (1985).
181. Friedman, J., Schneider, B. S. and Powell, D. *Proc. Natl. Acad. Sci. USA* **82**, 5593 (1985).
182. Deschenes, R. J., Lorenz, L. J., Haun, R. S., Roos, B. A., Collier, K. J. and Dixon, J. E. *Proc. Natl. Acad. Sci. USA* **81**, 726 (1984).
183. Kuwano, R., Araki, K., Usui, H., Fukui, T., Ohtsuka, E., Ikehara, M. and Takahashi, Y. *J. Biochem. (Tokyo)* **96**, 923 (1984).
184. Deschenes, R. J., Haun, R. S., Funckes, C. L. and Dixon, J. E., *J. Biol. Chem.* **260**, 1280 (1985).
185. Gubler, U., Chua, A. O., Hoffman, B. J., Collier, K. J. and Eng, J., *Proc. Natl. Acad. Sci. USA* **81**, 4307 (1984).
186. Eysselein, V. E., Reeve, Jr., J. R., Shively, J. E., Miller, C. and Walsh, J. H., *Proc. Natl. Acad. Sci. USA* **81**, 6565 (1984).
187. Smith, L. F., *Am. J. Med.* **40**, 662 (1966).
188. Blundell, T. and Horuk, R., *Hoppe-Seyler's Z. Physiol. Chem.* **362**, 727 (1981).
189. Anastasi, A., Erspamer, V. and Endean, R., *Arch. Biochem. Biophys.* **125**, 57 (1968).
190. Anastasi, A., *Experientia* **25**, 8 (1969).
191. Hoffman, W., Bach, T. C., Seliger, H. and Kreil, G., *EMBO J.* **2**, 111 (1983).
192. Wakabayashi, T., Kato, H. and Tachibana, S., *Nucleic Acids Res.* **13**, 1817 (1985).
193. Deschenes, R. J., Narayana, S. V. L., Argos, P. and Dixon, J. E., *FEBS Lett.* **182**, 135 (1985).
194. Baldwin, G. S., *FEBS Lett.* **137**, 1 (1982).
195. Kimmel, J. R., Pollock, H. G. and Hazelwood, R. L., *Endocrinology* **83**, 1323 (1968).
196. Lin, T.-M., Chance, R. and Evans, D., *Gastroenterology* **64**, 865 (1973).
197. Chance, R. E., Johnson, M. G., Hoffman, J. A. and Lin, T.-M., in *Proinsulin, Insulin, C-Peptide* (ed. S. Baba, T. Kaneko and N. Yanaihara), p. 419. Excerpta Medica, Amsterdam-Oxford (1979).
198. Lance, V., Hamilton, J. W., Rouse, J. B., Kimmel, J. R. and Pollock, H. G., *Gen. Comp. Endocrinol.* **55**, 112 (1984).
199. Kimmel, J. R., Pollock, H. G., Chance, R. E., Johnson, M. G., Reeve, Jr., J. R., Taylor, I. L., Miller, C. and Shiveley, J. E., *Endocrinology* **114**, 1725 (1984).
200. Boel, E., Schwartz, T. W., Norris, K. E. and Fiil, N. P., *EMBO J.* **3**, 909 (1984).
201. Leiter, A. B., Keutmann, H. T. and Goodman, R. H., *J. Biol. Chem.* **259**, 14702 (1984).
202. Takeuchi, T. and Yamada, T., *Proc. Natl. Acad. Sci. USA* **82**, 1536 (1985).
203. Schwartz, T. W., Hansen, H. F., Håkanson, R., Sundler, F. and Tager, H. S., *Proc. Natl. Acad. Sci. USA* **81**, 708 (1984).
204. Corder, R., Emson, P. C. and Lowry, P. J., *Biochem. J.* **219**, 699 (1984).
205. Minth, C. D., Bloom, S. R., Polak, J. M. and Dixon, J. E., *Proc. Natl. Acad. Sci. USA* **81**, 4577 (1984).
206. Tatemoto, K., *Proc. Natl. Acad. Sci. USA* **79**, 5485 (1982).
207. Adrian, T. E., Ferri, G.-L., Bacarese-Hamilton, A. J., Fuessl, H. S., Polak, J. M. and Bloom, S. R., *Gastroenterology* **89**, 1070 (1985).
208. Andrews, P. C., Hawke, D., Shively, J. E. and Dixon, J. E., *Endocrinology* **116**, 2677 (1985).
209. Lundberg, J. M., Saria, A., Franco-Cereceda, A. and Theodorsson-Norheim, E., *Acta Physiol. Scand.* **124**, 603 (1985).
210. Levine, A. S. and Morley, J. E., *Peptides* **5**, 1025 (1984).
211. Stanley, B. G. and Leibowitz, S. F., *Proc. Natl. Acad. Sci. USA* **82**, 3940 (1985).
212. Anastasi, A., Erspamer, V. and Bucci, M., *Experientia* **27**, 166 (1971).
213. Nakajima, T., Tanimura, T. and Pisano, J. J., *Fed. Proc.* **29**, 282/a205 (1970).
214. Anastasi, A., Erspamer, V. and Endean, R., *Experientia* **31**, 510 (1975).
215. Erspamer, V. and Melchiorri, P., in *Gastrointestinal Hormones* (ed. J. C. Thompson), p. 575. University of Texas Press, Austin and London (1975).

216. McDonald, T. J., Jörnvall, H., Tatemoto, K. and Mutt, V., *FEBS Lett.* **156**, 349 (1983).
217. Orloff, M. S., Reeve, Jr., J. R., Miller Ben-Avram, C., Shively, J. E. and Walsh, J. H., *Peptides* **5**, 865 (1984).
218. Spindel, E. R., Chin, W. W., Price, J., Rees, L. H., Besser, G. M. and Habener, J. F., *Proc. Natl. Acad. Sci. USA* **81**, 5699 (1984).
219. Reeve, Jr., J. R., Walsh, J. H., Chew, P., Clark, B., Hawke, D. and Shively, J. E., *J. Biol. Chem.* **258**, 5582 (1983).
220. McDonald, T. J., Jörnvall, H., Ghatei, M., Bloom, S. R. and Mutt, V., *FEBS Lett.* **122**, 45 (1980).
221. Minamino, N., Kangawa, K. and Matsuo, H., *Biochem, Biophys. Res. Commun.* **114**, 541 (1983).
222. Minamino, N., Sudoh, T., Kangawa, K. and Matsuo, H., *Biochem. Biophys. Res. Commun.* **130**, 685 (1985).
223. Shepherd, R. G., Willson, S. D., Howard, K. S., Bell, P. H., Davies, D. S., Davis, S. B., Eigner, E. A. and Shakespeare, N. E., *J. Am. Chem. Soc.* **78**, 5067 (1956).
224. Gráf, L., Bajusz, S., Patthy, A., Barát, E. and Cseh, G., *Acta Biochim. Biophys, Acad. Sci. Hung.* **6**, 415 (1971).
225. Li, C. H., Barnafi, L., Chrétien, M. and Chung, D., *Nature (London)* **208**, 1093 (1965).
226. Hsi, K. L., Seidah, N. G., Lu, C. L. and Chrétien, M., *Biochem. Biophys. Res. Commun.* **103**, 1329 (1981).
227. Mains, R. E., Eipper, B. A. and Ling, N., *Proc. Natl. Acad. Sci. USA* **74**, 3014 (1977).
228. Roberts, J. L. and Herbert, E., *Proc. Natl. Acad. Sci. USA* **74**, 4826 (1977).
229. Nakanishi, S., Inoue, A., Taii, S. and Numa, S. *FEBS Lett.* **84**, 105 (1977).
230. Nakanishi, S., Inoue, A., Kita, T., Nakamura, M., Chang, A. C. Y., Cohen, S. N. and Numa, S., *Nature (London)* **278**, 423 (1979).
231. Takahashi, H., Hakamata, Y., Watanabe, Y., Kikuno, R., Miyata, T. and Numa, S., *Nucleic Acids Res.* **11**, 6847 (1983).
232. Ling, N., Burgus, R. and Guillemin, R. *Proc. Natl. Acad. Sci. USA* **73**, 3942 (1976).
233. Hughes, J., Smith, T. W., Kosterlitz, H. W., Fothergill, L. A., Morgan, B. A. and Morris, H. R., *Nature (London)* **258**, 577 (1975).
234. Noda, M., Furutani, Y., Takahashi, H., Toyosato, M., Hirose, T., Inayama, S., Nakanishi, S. and Numa, S., *Nature (London)* **295**, 202 (1982).
235. Gubler, U., Seeburg, P., Hoffman, B. J., Gage, L. P. and Udenfriend, S., *Nature (London)* **295**, 206 (1982).
236. Comb, M., Seeburg, P. H., Adelman, J., Eiden, L. and Herbert, E., *Nature (London)* **295**, 663 (1982).
237. Morley, J. S., *Br. Med. Bull.* **39**, 5 (1983).
238. Hughes, J., *Br. Med. Bull.* **39**, 17 (1983).
239. Nyberg, F., Wahlström, A., Sjölund, B. and Terenius, L., *Brain Res.* **259**, 267 (1983).
240. Kakidani, H., Furutani, Y., Takahashi, H., Noda, M., Morimoto, Y., Hirose, T., Asai, M., Inayama, S., Nakanishi, S. and Numa, S., *Nature (London)* **298**, 245 (1982).
241. Horikawa, S., Takai, T., Toyosato, M., Takahashi, H., Noda, M., Kakidani, H., Kubo, T., Hirose, T., Inayama, S., Hayashida, H., Miyata, T. and Numa, S., *Nature (London)* **306**, 611 (1983).
242. Civelli, O., Douglass, J., Goldstein, A. and Herbert, E., *Proc. Natl. Acad. Sci. USA* **82**, 4291 (1985).
243. Fischli, W., Goldstein, A., Hunkapiller, M. W., and Hood, L. E., *Proc. Natl. Acad. Sci. USA* **79**, 5435 (1982).
244. Smyth, D. G., *Br. Med. Bull.* **39**, 25 (1983).
245. Jingami, H., Nakanishi, S., Imura, H. and Numa, S., *Eur. J. Biochem.* **142**, 441 (1984).
246. Erspamer, V. and Anastasi, A., *Experientia* **18**, 58 (1962).
247. Anastasi, A., Erspamer, V. and Cei, J. M., *Arch. Biochem. Biophys.* **108**, 341 (1964).
248. Erspamer, V., *Ann. Rev. Pharmacol.* **11**, 327 (1971).
249. Erspamer, V. and Melchiorri, P. *Trends Pharmacol Sci.* **1**, 391 (1980).
250. Chang, M. M., Leeman, S. E. and Niall, H. D., *Nature (London) New Biol.* **232**, 86 (1971).
251. Kimura, S., Okada, M., Sugita, Y., Kanazawa, I. and Munekata, E., *Proc. Japan Acad. Ser. B.* **59**, 101 (1983).
252. Minamino, N., Kangawa, K., Fukuda, A. and Matsuo, H., *Neuropeptides* **4**, 157 (1984).
253. Kangawa, K., Minamino, N., Fukuda, A. and Matsuo, H., *Biochem. Biophys. Res. Commun.* **114**, 533 (1983).
254. Anastasi, A., Montecucchi, P., Erspamer, V. and Visser, J., *Experientia* **33**, 857 (1977).
255. Nawa, H., Hirose, T., Takashima, H., Inayama, S. and Nakanishi, S., *Nature (London)* **306**, 32 (1983).
256. Nawa, H., Kotani, H. and Nakanishi, S., *Nature (London)* **312**, 729 (1984).
257. Tatemoto, K., Lundberg, J. M., Jörnvall, H. and Mutt, V., *Biochem. Biophys. Res. Commun.* **128**, 947 (1985).
258. Lazarus, L. H., Linnoila, R. I., Hernandez, O. and DiAugustine, R. P., *Nature (London)* **287**, 555 (1980).
259. Carraway, R. and Leeman, S. F., *J. Biol. Chem.* **250**, 1907 (1975).
260. Hammer, R. A., Leeman, S. E., Carraway, R. and Williams, R. H., *J. Biol. Chem.* **255**, 2476 (1980).
261. Helmstaedter, V., Taugner, Ch., Feurle, G. E. and Forssmann, W. G., *Histochemistry*, **53**, 35 (1977).
262. Polak, J. M., Sullivan, S. N., Bloom, S. R., Buchan, A. M. J., Facer, P., Brown, M. R. and Pearse, A. G. E., *Nature (London)* **270**, 183 (1977).
263. Carraway, R. and Bhatnagar, Y. M., *Peptides* **1**, 167 (1980).
264. Carraway, R. E. and Ferris, C. F., *J. Biol. Chem.* **258**, 2475 (1983).
265. Minamino, N., Kangawa, K. and Matsuo, H., *Biochem. Biophys, Res. Commun.* **122**, 542 (1984).
266. Araki, K., Tachibana, S., Uchiyama, M., Nakajima, T. and Yasuhara, T., *Chem. Pharm. Bull.* **23**, 3132 (1975).
267. Carraway, R. E. and Feurle, G. E., *J. Biol. Chem.* **260**, 10921 (1985).
268. Schubert, H. and Brown, J. C., *Can. J. Biochem.* **52**, 7 (1974).
269. Reeve, Jr., J. R., Ho, F. J., Walsh, J. H., Ben-Avram, C. M. and Shively, J. E., *J. Chromatogr.* **321**, 421 (1985).
270. Usellini, L., Buchan, A. M. J., Polak, J. M., Capella, C., Cornaggia, M. and Solcia, E., *Histochemistry* **81**, 363 (1984).
271. Beinfeld, M. C. and Korchak, D. M., *J. Neurosci.* **5**, 2502 (1985).
272. McDonald, T. J., Dupre, J., Tatemoto, K., Greenberg, G. R., Radziuk, J. and Mutt, V., *Diabetes* **34**, 192 (1985).
273. Melander, T., Hökfelt, T., Rökaeus, Å., Fahrenkrug, J., Tatemoto, K. and Mutt, V., *Cell Tissue Res.* **239**, 253 (1985).
274. Rökaeus, Å., Melander, T., Hökfelt, T., Lundberg, J. M., Tatemoto, K., Carlquist, M. and Mutt, V., *Neurosci. Lett.* **47**, 161 (1984).
275. Skofitsch, G. and Jacobowitz, D. M. *Peptides* **6**, 509 (1985).
276. Schally, A. V., Dupont, A., Arimura, A., Redding, T. W., Nishi, N., Linthicum, G. L. and Schlesinger, D. H., *Biochemistry* **15**, 509 (1976).
277. Böhlen, P., Brazeau, P., Esch, F., Ling, N., Wehrenberg, W. B. and Guillemin, R., *Regul. Pept.* **6**, 343 (1983).
278. Benoit, R., Böhlen, P., Brazeau, P., Ling, N. and Guillemin, R., *Endocrinology* **107**, 2127 (1980).
279. Conlon, J. M., *Life Sci.* **35**, 213 (1984).
280. Spiess, J., Rivier, J. E., Rodkey, J. A., Bennett, C. D. and Vale, W., *Proc. Natl. Acad. Sci. USA* **76**, 2974 (1979).
281. Noe, B. D., Spiess, J., Rivier, J. E. and Vale, W., *Endocrinology* **105**, 1410 (1979).
282. Arakawa, Y. and Tachibana, S., *Life Sci.* **35**, 2529 (1984).
283. Oyama, H., Bradshaw, R. A., Bates, O. J. and Permutt, A., *J. Biol. Chem.* **255**, 2251 (1980).
284. Andrews, P. C., Pubols, M. H., Hermodson, M. A., Sheares, B. T. and Dixon, J. E., *J. Biol. Chem.* **259**, 13267 (1984).
285. Andrews, P. C. and Dixon, J. E., *J. Biol. Chem.* **256**, 8267 (1981).
286. Morel, A., Chang, J.-Y. and Cohen, P., *FEBS Lett.* **175**, 21 (1984).
287. Andrews, P. C., Hawke, D., Shively, J. E. and Dixon, J. E., *J. Biol. Chem.* **259**, 15021 (1984).
288. Spiess, J. and Noe, B. D., *Proc. Natl. Acad. Sci. USA* **82**, 277 (1985).
289. Hobart, P., Crawford, R., Shen, L. P., Pictet, R. and Rutter, W. J., *Nature (London)* **288**, 137 (1980).
290. Goodman, R. H., Jacobs, J. W., Chin, W. W., Lund, P. K., Dee, P. C. and Habener, J. F., *Proc. Natl. Acad. Sci. USA* **77**, 5869 (1980).
291. Goodman, R. H., Jacobs, J. W., Dee, P. C. and Habener, J. F., *J. Biol. Chem.* **257**, 1156 (1982).
292. Minth, C. D., Taylor, W. L., Magazin, M., Tavianini, M. A., Collier, K., Weith, H. L. and Dixon, J. E., *J. Biol. Chem.* **257**, 10372 (1982).
293. Magazin, M., Minth, C. D., Funckes, C. L., Deschenes, R., Tavianini, M. A. and Dixon, J. E., *Proc. Natl. Acad. Sci. USA* **79**, 5152 (1982).
294. Shen, L.-P., Pictet, R. L. and Rutter, W. J., *Proc. Natl. Acad. Sci. USA* **79**, 4575 (1982).
295. Goodman, R. H., Aron, D. C. and Roos, B. A., *J. Biol. Chem.* **258**, 5570 (1983).
296. Morel, A., Nicolas, P. and Cohen, P., *J. Biol. Chem.* **258**, 8273 (1983).

297. Schmidt, W. E., Mutt, V., Kratzin, H., Carlquist, M., Conlon, J. M. and Creutzfeld, W., *FEBS Lett.* **192**, 141–146, 1985.
298. Shen, L.-P. and Rutter, W. J., *Science* **224**, 168 (1984).
299. Montminy, M. R., Goodman, R. H., Horovitch, S. J. and Habener, J. F., *Proc. Natl. Acad. Sci. USA* **81**, 3337 (1984).
300. Tavianini, M. A., Hayes, T. E., Magazin, M. D., Minth, C. D. and Dixon, J. E., *J. Biol. Chem.* **259**, 11798 (1984).
301. Hayes, T. E. and Dixon, J. E., *J. Biol. Chem.* **260**, 8145 (1985).
302. Luft, R., Efendić, S. and Hökfelt, T., *Diabetologia* **14**, 1 (1978).
303. Reichlin, S., *New Engl. J. Med.* **309**, 1495 (1983).
304. McIntosh, C. H. S., *Life Sci.* **37**, 2043 (1985).
305. Deschodt-Lanckman, M., Robberecht, P., Pector, J. C. and Christophe, J. *Arch. Int. Physiol. Biochim.* **83**, 960 (1975).
306. Pearson, D., Shively, J. E., Clark, B. R., Geschwind, I. I., Barkley, M., Nishioka, R. S. and Bern, H. A., *Proc. Natl. Acad. Sci. USA* **77**, 5021 (1980).
307. Adelson, J. W., Nelbach, M. E., Chang, R., Glaser, C. B. and Yates, G. B., in *Gastrointestinal Hormones* (ed. G. B. J. Glass), p. 387. Raven Press, New York (1980).
308. Gregory, H., *Nature (London)* **257**, 325 (1975).
309. Cohen, S., *J. Biol. Chem.* **237**, 1555 (1962).
310. Cohen, S., and Carpenter, G., *Proc. Natl. Acad. Sci. USA* **72**, 1317 (1975).
311. Hollenberg, M. D. *Vitam. Horm. N.Y.* **37**, 69 (1979).
312. Carpenter, G. and Cohen, S., *Ann. Rev. Biochem.* **48**, 193 (1979).
313. Baldwin, G. S., *Proc. Natl. Acad. Sci. USA* **82**, 1921 (1985).
314. Gray, A., Dull, T. J. and Ullrich, A., *Nature (London)* **303**, 722 (1983).
315. Scott, J., Urdea, M., Quiroga M., Sanchez-Pescador, R., Fong, N., Selby, M., Rutter, W. J. and Bell, G. I., *Science* **221**, 236 (1983).
316. Elder, J. B., Williams, G., Lacey, E. and Gregory, H., *Nature (London)* **271**, 466 (1978).
317. Kasselberg, A. G., Orth, D. N., Gray, M. E. and Stahlman, M. T., *J. Histochem. Cytochem.* **33**, 315 (1985).
318. Track, N. S., *Comp. Biochem. Physiol.* **45B**, 291 (1973).

Chemica Scripta 1986, **26B**, 209–212

Phylogeny of Insulin

'Primitive' Insulins and the Cell Cycle

Sture Falkmer*, Eva Dafgård and Wilhelm Engström

Department of Tumor Pathology, Karolinska Hospital, S-104 01 Stockholm, Sweden

Paper presented by Sture Falkmer at the Conference on 'Molecular Evolution of Life', Lidingö, Sweden, 8–12 September 1985

Abstract

The insulin family consists of hormones (insulin, relaxin, PTTH) and growth factors (IGF I, IGF II, NGF). A common ancestor molecule from which these members of the insulin family diverged during evolution has been suggested. The hypothetical divergence between insulin and the IGFs has been surmised to have occurred about 500 million years ago, i.e. at the time period when the first vertebrates appeared on earth, namely the cyclostomes.

In the extant cyclostomes, namely the lamprey and the hagfish, a separate islet organ appears for the first time in evolution. In the invertebrates, both Protostomian and Deuterostomian, the insulin immunoreactive cells usually occur in the central nervous system or as endocrine cells of open type in the gut mucosa. In the Atlantic hagfish, *Myxine glutinosa*, its separate islet parenchyma consists of 99% insulin cells and 1% somatostatin immunoreactive cells. Myxine insulin has been comprehensively analyzed as regards its amino-acid sequence, its tertiary structure, its receptor binding affinity, its synthesis and release, as well as its biological activity and efficiency as regards carbohydrate, lipid and protein metabolism.

Now, we have also made some pilot studies as to its ability to act as a growth factor in Swiss 3T3-cells. So far, no observations have been made, indicating that Myxine insulin is more effective as a factor regulating cell proliferation and cell growth than most mammalian insulins.

The neuroendocrine system and the insulin family

Insulin is the best known of all the peptide hormones that constitute the messenger substances in the large neuroendocrine system [1]. These neurohormonal peptides are known to form so-called hormone families [2]. The insulin family consists not only of hormones, such as insulin (proinsulin, preproinsulin), relaxin and the prothoracicotropic hormone (PTTH) [3], but also of peptides stimulating cell growth and cell proliferation, such as the Insulin-like Growth Factors (IGF I and IGF II) or Multiplication Stimulating Activity (MSA) and Nerve Growth Factor (NGF) [4, 5]. Although there is some controversy whether or not the amino-acid homologies are close enough to justify such a family formation, the structural similarities between the proposed members of the insulin family are big enough to suggest a common ancestor from which these molecules diverged during evolution; the divergence between insulin (proinsulin) and the IGFs has been surmised to have occurred before the appearance of the first vertebrates, namely, the Ostracoderms, some 500 million years ago [2], whereas the divergence between the two IGFs has been timed at the appearance of the first mammals [4].

* To whom correspondence should be addressed.

It is against this background that we have tried to examine the occurrence of insulin-producing cells in lower vertebrates and in invertebrates and to try to get an idea of whether or not insulins isolated from lower vertebrates have a biological effect more similar to that of IGFs than to those mammalian (ox, pig, man) insulins. As several of the neurohormonal peptides are produced both in the central nervous system (CNS) and in the gastrointestinal tract and some of its associated glands (e.g. the pancreas with its islets of Langerhans), particular attention has been paid to this 'brain-but axis' in our phylogenetic studies of insulin [1].

Criteria for the presence of insulin

Our criteria for detection of cells, producing insulin, are rather strict; they are based on the assumption that the biosynthesis of insulin also in non-mammalian species takes place in the rough endoplasmic reticulum, where the hormone is processed via preproinsulin and proinsulin to hexamers, dimers, or monomers of insulin during the passage through the Golgi apparatus to the secretion granules; the release occurs via emiocytosis (exocytosis) [6]. As a consequence, insulin-producing cells should not only show light-microscopical immunohistochemical evidence for the presence of insulin but also ultrastructural proofs for the presence of secretion granules in their cytoplasm; ideally, these secretion granules by modern immunocytochemical techniques should be shown to contain insulin [1].

The immunocytochemical and ultrastructural observations should, of course, be correlated with the results of radioimmunochemical (RIA) evidence for the presence of an insulin-like substance in acid-ethanol extracts of the same tissues. Before the advent of the RIA techniques, biological assays of the tissue extracts had to be used, where the specificity of the insulin-like activity was confirmed by the facts that the activity was inhibited by an excess of anti-insulin serum and by the addition of low concentrations (0.002 M) of mercapto-ethanol (breaking the cystine –SS– bridges [7]. Of course, the isolation of the peptide and the establishment of its amino-acid sequence is the ultimate proof for the presence of insulin [6]. Additional biochemical support for the production of insulin is that mRNA for insulin can be isolated [8] as well as proinsulin, and that the results of pulse-chase experiments can offer evidence for insulin processing in the cells investigated [9]. Antisera should be raised and a homologous insulin RIA established; *in vitro* and *in vivo* experiments of the release mechanisms as well as receptor binding studies should be performed [10]. Ideally, the three-dimensional structure of the molecule by means of X-ray crystallography should be established [9].

Unfortunately, such a battery of tests has only exceptionally been applied in phylogenetical investigations of the neuroendocrine system. So far, the only neurohormonal peptide in a low-order animal that has been completely investigated according to all these criteria, is the insulin produced by the cyclostome *Myxine glutinosa*, the Atlantic hagfish [7, 9, 10].

Plants, prokaryotes, unicellular eukaryotes, coelenterates, flatworms and annelids

Applying the criteria given above, the recent claims [11] that insulin or an insulin-like peptide occurs in *Escherichia coli*, in some eukaryote unicellular organisms, and some lower Protostomian invertebrates, are difficult to accept. In those immunocytochemical investigations we have performed so far we have been unable to detect any cells that fulfill our criteria for being insulin-producing cells [12; own unpublished observations].

We are, however, aware of the fact that our criteria may be too rigid and that, theoretically, an insulin-like substance can well be produced in these low-order phyla and unicellular organisms in a similar manner to that in which the IGFs in mammalian tissues are synthesized and released [5], i.e. without passing the Golgi apparatus and without being stored in any secretion granules with ultimate exocytosis.

Higher protostomian invertebrates

In insects and mollusks some evidence has been obtained that insulin-producing cells might occur in the CNS and in the gut mucosa, respectively [13–16]. It has been tried to extract and identify the insulin-immunoreactive peptides in the insect brain; one is PTTH, which has been sequenced [3]. Actually, some evidence has been obtained that there is also a slightly modified insulin in insect brain [15]. Although the presence of insulin has recently been reported also in the hepatopancreas of a crustacean [17], our criteria for the actual presence of the insulin-producing cells in the digestive tract mucosa or in its associated glands have not been convincingly fulfilled in many protostomian invertebrates, so far. Therefore, it might be justified to conclude that in these animals insulin, like most other neurohormonal peptides [18], appears preferentially as a 'brain peptide', and that a real brain–gut axis with neuroendocrine cells of the same kind both in CNS and in the alimentary tract usually has not been developed; at least, it has not been shown for insulin.

Deuterostomian invertebrates

In the Deuterostomian evolution line, where the vertebrates predominate, there are some invertebrate groups which are of considerable interest with regard to the phylogeny of the insulin cells [1, 2]. In the tunicates it has been shown immunohistochemically that a brain–gut axis of the neuroendocrine system is fully developed, with dual occurrence in the brain and gut of a large number of neurohormone immunoreactive neurons and endocrine cells, including the insulin cells [19]. Several of them, however, now seem to be ready to leave the brain and to appear as open cells of endocrine type, dispersed in the gastrointestinal mucosa. Still, no actual islet parenchyma has been developed [1, 2].

Vertebrates

In the vertebrates something of a step-wise evolution of the GEP neuroendocrine system can be discerned [1]. Already in the cyclostomes ('Agnatha'), such as the hagfish and the lamprey, a separate islet organ has been developed. It is a two-hormone parenchyma, consisting of insulin (99%) and somatostatin (1%) cells. In the hagfish it has originated as outbuddings from the mucosa of the common bile duct. Somatostatin cells were found to occur both in the bile duct mucosa and in the gut; glucagon and PP cells were present in the gut mucosa only. As to the brain–gut–axis, it has recently been observed that in hagfish and lamprey brains somatostatin and PP cells are present but not glucagon or insulin cells [20].

Of particular interest in this connection is the distribution of the insulin cells. They have not only left the brain but also the gut mucosa and are restricted to the islet organ and the adjacent parts of the bile duct only. This absence of insulin cells from the gut mucosa is a feature that is almost regularly present in all the subsequent evolutionary stages of the vertebrates [1]. One exception is the gut mucosa of a primitive reptile, the turtle *Chrysemys picta* [21].

The next step in the evolution of the GEP neuroendocine system occurs at the level of the appearance of the first gnathostomian vertebrates, namely the Holocephali (the ratfish, the rabbit fish, and the elephant fish) among the cartilaginous fish [1]. Here, also, the first exocrine acinar pancreas appears. Its islets of Langerhans represent a three-hormone endocrine parenchyma, containing, in addition to the insulin and somatostatin cells, also glucagon/glicentin cells. The PP cells are usually lagging behind in the gut mucosa and in the epithelium of the pancreatic duct.

However, already in the plagostomian cartilaginous fish (sharks and rays) the islets of Langerhans in the pancreas constitute a four-hormone islet organ of the same type as in all higher vertebrates, including man [1].

Amino-acid sequences of 'primitive' insulins

In all the invertebrates there is still only one member of the insulin superfamily whose amino-acid sequence has been established, namely PTTH [3]. Thus, all that is known about the evolution of the insulin molecule is obtained from the vertebrates. Here, as regards the 'primitive' insulins (i.e. insulins from low-order species), we have some data about the primary and tertiary structure of the insulin molecule in *Myxine glutinosa* [10] and the amino-acid sequence of that in *Squalus acanthias* [22]. From these data and from what has been published about the molecular structure of insulins from some 40 high-order vertebrate species, it can be concluded that the well-known basic structure of the monomer and the dimer has been carefully conserved throughout the whole vertebrate evolution [23]. As *Myxine glutinosa* is one of the few extant representatives of the phyla that first appeared on earth, the molecular biology of its insulin has been intensively studied [6]. The availability of recombinant plasmids coding for hagfish insulin has made it possible to identify and clone the chromosomal gene for hagfish preproinsulin and to compare its structure with those of insulin and IGF genes of higher organisms. The possibility then also arises to investigate how insulins obtained from different species can act as growth factors and, in particular, how they can affect DNA synthesis, cell division, and cellular enlargement.

The cell cycle

The eukaryote cell cycle can be divided into two fundamental processes [24]. The first of these is named the chromosome cycle, including DNA-replication and mitosis. The second is the growth cycle, i.e. protein accumulation, leading to cellular enlargement. An interrelationship exists between the chromosome cycle and the growth cycle in the sense that cells approximately double in size prior to mitosis under normal physiological growth conditions [24–26]. It has, however, been possible to separate the cycles in a variety of experimental systems [27] in the sense that cells can undergo DNA synthesis and cell division without growing in size.

Insulin effects on the cell cycle

Mouse 3T3-cells (embryonic fibroblasts), starved to quiescence in low serum concentrations, can be preferentially stimulated to initiate DNA-replication by addition of one purified growth factor, Epidermal Growth Factor (EGF) [24]. The effect of DNA synthesis can be greatly enhanced by concomitant addition of pig insulin in a dose-dependent manner [25]. If, however, pig insulin was added alone, even at high concentration, no stimulatory effects on DNA synthesis were observed. This is in line with reports from other laboratories [28] and pig insulin has, therefore, been considered to exert a synergistic effect rather than a primarily mitogenic effect on quiescent 3T3-cells. In pilot studies we have tested the growth-promoting effects of different insulins from various vertebrates on 3T3-cells. The insulins studied were from the hagfish, the dogfish and the turkey. Neither of these three insulins had any stimulatory effect on the DNA synthesis. In these studies we just tested low insulin concentrations. In a subsequent work [29] to a preceding report on dogfish insulin [22] it was found that dogfish insulin had a greater growth effect on 3T3-cells than ox insulin. This effect was supposed to reflect the substitution of tyrosine in dogfish insulin at the B_{25} position, as this is also seen in the IGFs [30]. Further studies have to be done, testing high concentrations of the insulins. We recently reported preliminary results of the marginal effects that hagfish insulin had on the DNA synthesis in EGF-stimulated cells [23]. Preliminary results indicate that both dogfish insulin and turkey insulin have a synergistic effect of EGF's growth promoting effect. Dogfish insulin was found to be less potent than the pig insulin, but turkey insulin was found to be more potent.

In studies on the effects of different insulins on cellular enlargement, continuously proliferating 3T3-cells were depleted of serum during an 8-hour period preceding mitosis [31]. Such cells undergo cell division at a decreased mitotic size. Recent studies have shown that this reduction in cell size could be counteracted by addition of megadoses of pig insulin [26]. So far, analogous experiments with hagfish, dogfish and turkey insulin have not given results of conclusive nature. Since high doses of insulin were required for cellular enlargement in the absence of serum, it is possible that the induction of growth in cell size involves an initial binding of some insulin-like factor to another membrane located receptor. One possible candidate for such a role is the IGF I receptor to which IGF I binds with high affinity and insulin binds but at a lower affinity [32]. It is less likely that the effects on growth in cell size are mediated by the IGF II receptor since it was recently found in another experimental system that insulin cannot bind even at low affinity to this particular receptor. It remains to be determined experimentally, however, whether stimulation of cellular enlargement is mediated via the IGF I receptor, as well as how other well-characterized members of the insulin family, such as MSA, NGF [3], and dogfish and turkey insulins affect the chromosome cycle and the growth cycle, respectively.

Although our studies have been confined to the 3T3-cells, they suggest that growth factors can act in at least three different ways. First, some factors (e.g. EGF) can preferentially stimulate the chromosome cycle, leading to DNA replication and mitosis. Secondly, growth factors (such as insulin) preferentially stimulate growth in cell size. Thirdly, certain growth factors (such as IGF I) stimulate DNA synthesis and cell division as well as cellular enlargement [33]. Thus, the differential effects of purified growth factors may help explain why certain cell lines require a multiple set of growth factors for long-term proliferation under serum-free conditions.

Acknowledgements

This review is based on original works supported by grants from the Swedish Medical Research Council (Project No. 12X–718), the Swedish Diabetes Association, the Cancer Society in Stockholm, the King Gustaf V Jubilee Fund, the Research Funds of Faculty of Medicine at the Karolinska Institute, Stockholm, and the British Council.

References

1. Falkmer, S., El-Salhy, M. and Titlbach, M., in *Evolution and Tumour Pathology of the Neuroendocrine System* (ed. S. Falkmer, R. Håkansson, F. Sundler), pp. 59–87. Elsevier, Amsterdam (1984).
2. Falkmer, S. and Van Noorden, S., *Handbook Exp. Pharmacol.* **66**, 81–119 (1983).
3. Nagasawa, H., Kamito, T., Fugo, H., Suzuki, A. and Ishizaki, H., *Science* **22**, 1344–1345 (1984).
4. Blundell, T. L. and Humbel, R. R., *Nature (London)* **287**, 781–787 (1980).
5. Frosch, E. R. and Zapf, J., *Diabetologica* **28**, 485–493 (1985).
6. Steiner, D. F., Chan, S. Docherty, K., Emdin, S. O., Dodson, G. G. and Falkmer S., in *Evolution and Tumour Pathology of the Neuroendocrine System* (ed. S. Falkmer, R. Håkanson, and F. Sundler), pp. 203–223. Elsevier, Amsterdam (1984).
7. Wilson, S. and Falkmer, S., *Can. J. Biochem.* **43**, 1615–1624 (1965).
8. Giddings, S. J., Chirgwin, J. and Permutt, M. A., *Diabetologia* **28**, 343–347 (1985).
9. Falkmer, S. and Emdin, S. O., in *Structural Studies of Molecules of Biological Interest* (ed. G. Dodson, J. P. Glusker and D. Sayre), pp. 420–440. Oxford University Press, Oxford (1981).
10. Emdin, S. O., Myxine insulin. Amino-acid sequence, three dimensional structure, biosynthesis, release, physiological role receptor binding affinity, and biological activity. Umeå University Medical Dissertation (N.S.), **66**, 1–159 (1981).
11. LeRoith, D. and Roth, J., in *Evolution and Tumour Pathology of the*

Neuroendocrine System (ed. S. Falkmer, R. Håkansson and F. Sundler), pp. 147–164. Elsevier, Amsterdam (1984).
12. Falkmer, S., Gustafsson, M. K. S. and Sundler, F., *Nord. Psykiatr, Tidskr. Suppl.* **11**, 21–30 (1985).
13. Davidson, J. K., Falkmer, S., Mehrotra, B. K. and Wilson, S., *Gen. Comp. Endocr.* **17**, 388–401 (19171).
14. Boquist, L., Falkmer, S. and Mehrotra, B. K., *Gen. Comp. Endocr.* **17**, 388–401 (1971).
15. Duve, H., Thorpe, A. and Lazarus, N. R., *Biochem. J.* **284**, 221–227 (1979).
16. El-Salhy, M., Falkmer, S., Kramer, J. K. and Speirs, R. D., *Cell Tiss. Res.* **232**, 295–317 (1983).
17. Sanders, B., *Gen. Comp. Endcor.* **50**, 366–373 (1983).
18. Van Noorden, S., in *Evolution and Tumour Pathology of the Neuroendocrine System* (ed. S. Falkmer, R. Håkanson and F. Sundler), pp. 7–38. Elsevier, Amsterdam (1984).
19. Thorndyke, M. C. and Falkmer, S. in *The Evolutionary Biology of Primitive Fishes*, pp. 379–400 (ed. R. G. Foreman, A. Gorbman, J. M. Dodd and R. Olsson). Plenum, New York (1985).
20. Falkmer, S., in *Neurosecretion and the Biology of Neuropeptides* (ed. H. Kobayashi, H. A. Bern and A. Urano), pp. 317–325. Japan Sci. Press, Tokyo (1985).
21. Gapp, D. F., Kenny, M. P. and Polak, J. M., *Peptides* **6**, Suppl. **3**, 347–352 (1985).
22. Bajaj, M., Blundell, T. L., Pitts, J. K., Woods, S. P., Tatnell, M. A., Falkmer, S., Emdin, S. O., Gowan, L. K., Crow, H., Schwabe, C., Wollmer, A. and Strassburger, W., *Eur. J. Biochem.* **131**, 535–542 (1983).
23. Falkmer, D., Dafgård, E., El-Salhy, M., Engström, W., Grimelius, L. and Zetterberg, A., *Peptides* **6**, Suppl. **3**, 315–320 (1985).
24. Mitchison, J. M., *The Biology of the Cell Cycle.* Cambridge University Press, Cambridge (1971).
25. Zetterberg, A., Dafgård, E. and Engström, W., *Proc. Int. Confer. Chemother.* **235**, 15–17 (1983).
26. Zetterberg, A., Engström, W. and Dafgård, E., *Cytometry* **5**, 368–375 (1984).
27. Baserga, R., *Exp. Cell Res.* **151**, 1–5 (1984).
28. Jimenez de Asua, L., O' Farreau, M. K., Clingan, D. and Rudland, P. S., *Proc. Natl. Acad. Sci. USA* **74**, 3845–3849 (1977).
29. Bajaj, M., Ph.D. thesis, University of London, UK (1984).
30. Dafgård, E., Bajaj, M., Honegger, A. M., Pitts, J., Wood, S. and Blundell, T., *J. Cell. Sci. Suppl.* **3**, 53–64 (1985).
31. Larsson, O., Dafgård, E., Engström, W. and Zetterberg, A., *J. Cell Physiol.* **127**, 267–273 (1986).
32. Massague, J. and Czech, M. P., *J. Biol. Chem.* **257**, 5048–5045 (1982).

Chemica Scripta 1986, **26B**, 213–219

Diversity and Invariance in the Evolution of Protein Tertiary Structure

Tom Blundell

Laboratory of Molecular Biology, Department of Crystallography, Birkbeck College, University of London, Malet Street, London WC1E 7HX, United Kingdom

Paper presented at the Conference on 'Molecular Evolution of Life', Lidingö, Sweden, 8–12 September 1985

Abstract

The maintenance of tertiary structure in the divergent evolution of a family of proteins with differing functions is usually achieved with the invariance of very few residues. These residues often include buried polar residues such as serines or threonines, bridging cystines and prolines which have special structural roles not easily assumed by other amino acids. The most commonly invariant residue is glycine, which is unique in its lack of a side-chain and its consequent ability to adopt unusual conformations. The characterization of patterns of invariant and conservatively varied residues consistent with a particular tertiary structure is a useful aid in protein structure prediction.

Introduction

Selective pressures in evolution will be mainly on the functions of proteins as manifested in the whole organism. These functions will depend on the arrangement of the polypeptide chain in three-dimensions. Because many differing amino acid sequences are able to attain the same secondary and tertiary structures [1–3], we may conclude that there will be only weak restraints derived from the three-dimensional structure on the replacement of most amino acids during evolutionary time. Nevertheless the tertiary structure must be retained to keep the precise arrangement of the catalytic residues, ligand binding sites and other functionally important groups. Even when gene duplication occurs so that selective pressures on one gene are decreased and completely new functions are evolved, the tertiary structure of any expressed protein must be attained to avoid speedy intracellular degradation. Thus it is at the level of tertiary structure that we should look for evidence for distant evolutionary relationships.

In this article I consider variation of sequences in distantly related proteins in terms of the conservation of their tertiary structures. Mutations are accepted in the majority of residues which constitute the hydrophobic core or maintain a prefernce for certain secondary structures; these are roles that can be assumed by many amino acids with only small resulting differences in three-dimensional structure. Invariant residues are few but tend to be those with a unique structural role. They include buried polar residues such as serines, threonines and aspartates; bridging cystine disulphides, and some prolines – especially *cis*-prolines or those in poly-proline helices. However, in most protein families the uniqueness of glycine [4] – its lack of side-chain and its ability to adopt unusual conformations – make this the most commonly invariant residue.

Sequence variability in protein evolution

Let us consider first the variable residues in families of proteins.

Although in some diverged proteins such as the insulins and insulin-like growth factors (IGF I and II) [5] the residues of the hydrophobic core are completely invariant, the cores of most protein families show considerable latitude for variation of both volume and shape. Consider, for example, the duplicated domains of the vertebrate lens proteins, the crystallins. High-resolution X-ray analysis of γ-II crystallin [6] shows that the protein comprises two domains, each of which may be considered as a sandwich of two similar antiparallel β-pleated sheets with strands badc′. The overall least-squares fitting of one domain to another gives a root-mean-square deviation of 1.42 Å for all Cα atoms. Residues at each edge of the sheet are most variable, with only their Cα and Cβ atoms contributing to the hydrophobic core

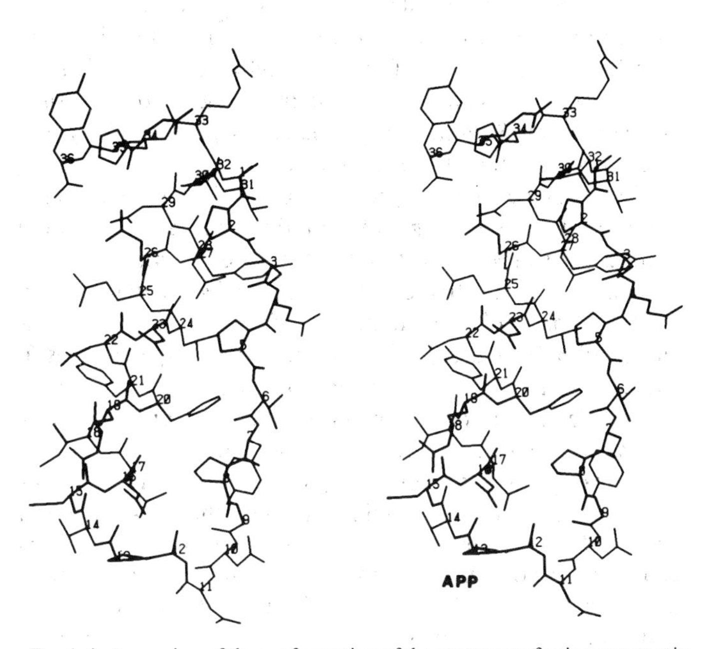

Fig. 1. A stereo view of the conformation of the protomer of avian pancreatic polypeptide (aPP). The sequences of the family are given in Table I. The polyproline helix with prolines at residues Pro 2, Pro 5 and Pro 8 is to the right and is packed against the α-helix (residues 14–32). The COOH-terminal residues (33–36) are free from intramolecular intractions and may exist in several conformations. (Reproduced with permission from ref. [12].

Table I. *The sequences of the divergently evolved family of pancreatic polypeptides from various animals and the sequences of the homologous gut peptide (PYY) and neuropeptide (NPY).*

Residues with the number encircled are involved in intermolecular interactions in the dimer, while those underlined are involved in interactions between the polyproline-helix and α-helix within the protomer (from refs. [15] and [16] and the references therein.)

	1	$\underline{2}$	3	4	$\underline{5}$	6	$\underline{7}$	$\underline{8}$	9	10	11	12	13	(14)	15	$\underline{16}$	(17)	18
Chicken	Gly	Pro	Ser	Gln	Pro	Thr	Tyr	Pro	Gly	Asp	Asp	Ala	Pro	Val	Glu	Asp	Leu	Ile
Alligator	Thr	Pro	Leu	Gln	Pro	Lys	Tyr	Pro	Gly	Asp	Gly	Ala	Pro	Val	Glu	Asp	Leu	Ile
Goose	Gly	Pro	Ser	Gln	Pro	Thr	Tyr	Pro	Gly	Asn	Asp	Ala	Pro	Val	Glu	Asp	Leu	X
Bovine	Ala	Pro	Leu	Glu	Pro	Glu	Tyr	Pro	Gly	Asp	Asn	Ala	Thr	Pro	Glu	Gln	Met	Ala
Ovine	Ala	Ser	Leu	Glu	Pro	Glu	Tyr	Pro	Gly	Asp	Asn	Ala	Thr	Pro	Glu	Gln	Met	Ala
Canine	Ala	Pro	Leu	Glu	Pro	Val	Tyr	Pro	Gly	Asp	Asp	Ala	Thr	Pro	Glu	Gln	Met	Ala
Porcine	Ala	Pro	Leu	Glu	Pro	Val	Tyr	Pro	Gly	Asp	Asp	Ala	Thr	Pro	Glu	Gln	Met	Ala
Human	Ala	Pro	Leu	Glu	Pro	Val	Tyr	Pro	Gly	Asp	Asn	Ala	Thr	Pro	Glu	Glu	Met	Ala
PYY	Tyr	Pro	Ala	Lys	Pro	Glu	Ala	Pro	Gly	Glu	Asp	Ala	Ser	Pro	Glu	Glu	Leu	Ser
NPY	Tyr	Pro	Ser	Lys	Pro	Asp	Asn	Pro	Gly	Glu	Asp	Ala	Pro	Ala	Glu	Asp	Leu	Ala

	$\underline{19}$	$\underline{20}$	(21)	22	$\underline{23}$	(24)	25	26	$\underline{27}$	(28)	29	$\underline{30}$	$\underline{31}$	32	33	34	35	36
Chicken	Arg	Phe	Tyr	Asp	Asn	Leu	Gln	Gln	Tyr	Leu	Asn	Val	Val	Thr	Arg	His	Arg	$Tyr-NH_2$
Alligator	Gln	Phe	Tyr	Asp	Asp	Leu	Gln	Gln	Tyr	Leu	Asn	Val	Val	Thr	Arg	Pro	Arg	$Phe-NH_2$
Goose	Arg	Phe	Tyr	Asp	Asn	Leu	Gln	Gln	Tyr	Arg	Leu	Val Asn	Val	Phe	Arg	His	Arg	$Tyr-NH_2$
Bovine	Gln	Tyr	Ala	Ala	Glu	Leu	Arg	Arg	Tyr	Ile	Asn	Met	Leu	Thr	Arg	Pro	Arg	$Tyr-NH_2$
Ovine	Gln	Tyr	Ala	Ala	Glu	Leu	Arg	Arg	Tyr	Ile	Asn	Met	Leu	Thr	Arg	Pro	Arg	$Tyr-NH_2$
Canine	Gln	Tyr	Ala	Ala	Glu	Leu	Arg	Arg	Tyr	Ile	Asn	Met	Leu	Thr	Arg	Pro	Arg	$Tyr-NH_2$
Porcine	Gln	Tyr	Ala	Aal	Glu	Leu	Arg	Arg	Tyr	Ile	Asn	Met	Leu	Thr	Arg	Pro	Arg	$Tyr-NH_2$
Human	Gln	Tyr	Ala	Ala	Asp	Leu	Arg	Arg	Tyr	Ile	Asn	Met	Leu	Thr	Arg	Pro	Arg	$Tyr-NH_2$
PYY	Arg	Tyr	Tyr	Ala	Ser	Leu	Arg	His	Tyr	Leu	Asn	Leu	Val	Thr	Arg	Gln	Arg	$Tyr-NH_2$
NPY	Arg	Tyr	Tyr	Ser	Ala	Leu	Arg	His	Tyr	Ile	Asn	Leu	Ile	Thr	Arg	Gln	Arg	$Tyr-NH_2$

but with the side-chain often exposed to solvent. The two sheets form a wedge with the bottom filled with conserved isoleucine and tryptophan residues. Few of the other core residues are completely conserved. There is some evidence for conservation of volume by complementary changes, e.g. from Cys to Phe (total volume $= 321$ Å^3) in the NH_2-terminal domains to Leu and Ile (total volume $= 337$ Å^3) in the COOH terminal domains of γ-crystallins. In most β-sheet proteins the distance between the sheets tends to be conserved, but non-complementary core changes can be accommodated by rotations (of up to 15°) of one sheet relative to the other [7]. Similar relative movements occur in α-helical proteins such as the globins [8] even though the core hydrophopicity and overall volume tend to be conserved.

If we assume that the gene is expressed as a functionally useful protein in all stages of evolution, then all steps in the evolutionary process must lead to a thermodynamically stable protein [8]. As the free energy of the folded protein is only 8–15 kcal per mol lower than the unfolded state, then mutations will only be accepted if they cost less. Large changes in the core will not usually lead to a functional protein; small changes leading to readjustments of the sheets or helices may be less energetically unfavourable. Of course where there are families of genes, for example for globins, crystallins, immunoglobulins and histocompatibility antigens, the non-expression of one of the gene family may not be particularly disadvantageous and the organism might survive until a complementary mutation occurs. In this way quite radical changes in the hydrophobic cores may have occurred. Pseudogenes may also provide a mechanism for silent mutations leading to complementary changes.

In order to retain a stable tertiary structure, the protein must retain not only a close-packed hydrophobic core but also a potential to achieve the correct main chain conformation. The low success rates (less than 56% [9]) of even the best secondary predictive schemes indicate that local, predominantly steric factors which lead to preference for particular conformations may often take second place to tertiary interactions. Some of the reasons for the ability of secondary structures to accommodate unusual residues is becoming apparent from detailed analyses of helices. For example, Blundell *et al.* [10] and Barlow [11] have shown that many helices defined by high-resolution X-ray analysis are irregular, with sharp discontinuities, and that most regular helices are curved so that even prolines can be accommodated.

The secondary structure that is least amenable to amino acid variation is the polyproline helix, which characterises the strands of collagen and which is occasionally found in globular proteins. For example, the small pancreatic polypeptide (PP) has a simple tertiary structure comprised of a strand of polyproline-like helix (1–8) packed against an amphipathic α-helix (18–32) (Fig. 1) [12–14]. There are prolines at positions 2, 5 and 8, the side-chains of which form a hydrophobic surface to the polyproline-like helix which has approximate threefold screw rotational symmetry. These prolines are conserved in all mammalian PP molecules (with the exception of position 2 in sheep PP) and also in the homologous gut peptide (PYY) and neuropeptide (NPY) [15, 16] as shown in Table I.

Insertions and deletions

Insertions may introduce a single amino acid (for example, in a β-bulge), extend helices or β-strands by several amino acids, increase the lengths of loops, or introduce new helices, β-strands or even domains.

The aspartic proteinases constitute interesting examples of evolutionary divergence by variation with insertions and

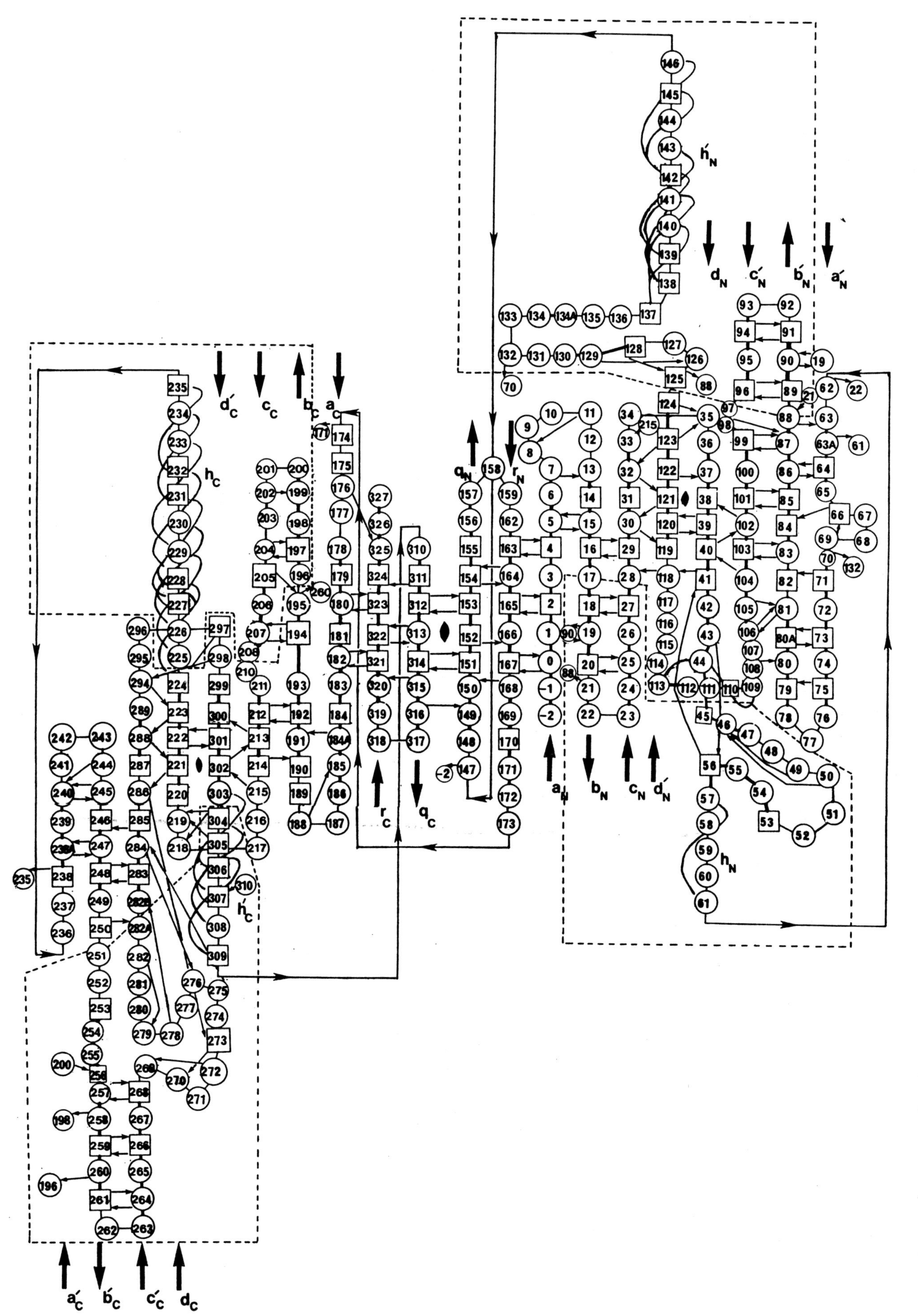

Fig. 2. A schematic representation of the secondary structure of the aspartic proteinase, endothiapepsin, an enzyme which may have evolved by duplications of a smaller ancestral protein. Residues are numbered to emphasize homology with pepsin so that the active site aspartates are Asp 32 and Asp 215 (see Table II). Residues which contribute to the core and are inaccesssible to solvent are boxed. Hydrogen bonds are shown by arrows from the NH to the O of the contributing peptides. Note the general topological equivalence of the NH_2- and COOH-terminal domains. Equivalent residues lie equidistant from the central dyad (⬮) on a straight line. Within each domain there is further internal similarity so that all β-strands *a*, *b c d* and helix *h* are equivalent. (Reproduced with permission from ref. [19].)

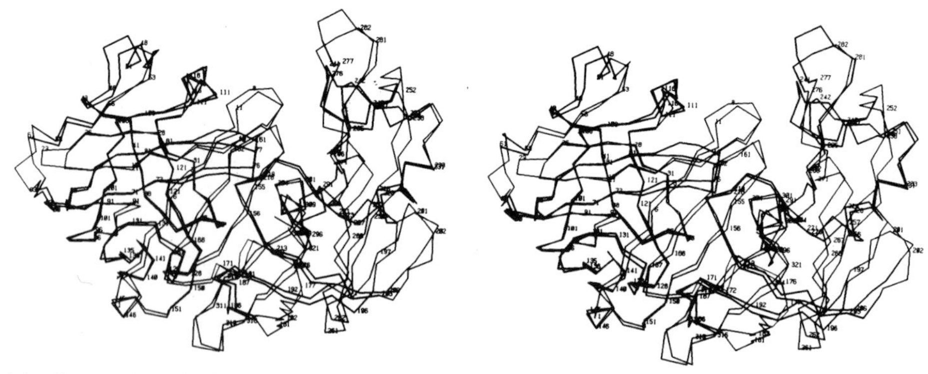

Fig. 3. A stereo view of the divergently evolved tertiary structures of endothiapepsin and penicillopepsin. Note the close similarity, particularly in the vicinity of the active site aspartates at positions 32 and 215.

deletions both in the internally duplicated domains and between homologous proteins. Figure 2 shows the secondary structure of one aspartic proteinase, endothiapepsin, which has been defined by high-resolution X-ray analysis [17], while Fig. 3 compares the tertiary structure with an homologous enzyme, penicillopepsin [18]. Figure 2 shows that the structure contains four repeated $\beta\beta\beta\beta\alpha$ units ($a_N b_N c_N d_N h_N$ and $a_N' b_N' c_N' d_N' h_N'$ in the NH_2-terminal domain, and $a_c b_c c_c d_c h_c$ and $a_c' b_c' c_c' d_c' h_c'$ in the COOH-terminal domain [19]. These structural motifs are each folded (along the dotted lines in Fig. 2) orthogonally on to themselves to form a sandwich. The positions of the folding lines of the β-sheets are often sites of insertions which create a β-bulge [20], for example, at 97 and 98 of strand c'_N (Fig. 2). In a topologically equivalent position in the COOH-terminal domain (strand c_c residues 208–210) there is a bulge corresponding to three extra residues in mammalian aspartic proteinases such as renin, pepsin and chymosin. This is reduced to two extra residues in endothiapepsin and is absent in penicillopepsin, both enzymes being of fungal origin. Interestingly, this insertion is stabilized by a disulphide bridge in mammalian enzymes. An even larger insertion, which is also stabilized by a disulphide bond, occurs at the bend of strand c'_c of endothiapepsin compared to topologically equivalent positions; this is also a region where insertions occur in penicillopepsin, endothiapepsin and renin relative to porcine pepsin.

Figures 2 and 3 show that other insertions and deletions occur between $\beta\beta$ hairpins or between $\alpha\beta$ motifs when internally repeated units and homologous aspartic proteinases are compared. The shorter loop regions appear to have conformations characteristic of their lengths [21]. In many cases these points of insertion and deletion in the aspartic proteinases correspond to exon–intron junctions in the genomic structure [22]. Similar but even larger insertions occur when the serine proteinases are compared (see ref. 1 for review).

Invariant residues in divergently evolved proteins

Although the structural elements may change their relative positions and orientations during divergent evolution of a family of proteins and peripheral loops between strands vary in length and conformation, the arrangement of the side-chains that define the active site or bind a cofactor is remarkably conserved in both identity and three-dimensional structure.

Thus the relative positions of the two catalytic aspartates and the immediately contiguous sequences (see Table II) are perfectly conserved in different aspartic proteinases (see Fig.

Table II. *The sequences around the catalytic residues (Asp* 32 *and Asp* 215 *pepsin numbering) of the aspartic proteinases*

Note the invariance of Asp Thr Gly (DTG) in both sequences which occupy topologically equivalent positions in the NH_2– and COOH-terminal domains.

	27	28	29	30	31	32*	33	34	35	36	37	38	39	40	41	42
Mouse renin (sub. max.)	F	K	V	I	F	D	T	G	S	A	N	L	W	V	P	S
Human renin	F	K	V	V	F	D	T	G	S	S	N	V	W	V	P	S
Endothiapepsin	L	N	L	D	F	D	T	G	S	S	D	L	W	V	F	S
Penicillopepsin	L	N	L	N	F	D	T	G	S	A	D	L	W	V	F	S
Porcine pepsin	F	T	V	I	F	D	T	G	S	S	N	L	W	V	P	S
Human pepsin	F	T	V	V	F	D	T	G	S	S	N	L	W	V	P	S
Chicken pepsin	F	T	V	I	F	D	T	G	S	S	N	L	W	V	P	S
Chymosin	F	T	V	L	F	D	T	G	S	S	D	F	W	V	P	S
	210	211	212	213	214	215*	216	217	218	219	220	221	222	223	224	
Mouse renin (sub. max.)	C	E	V	V	V	D	T	G	S	S	F	I	S	A	P	
Human renin	C	L	A	L	V	D	T	G	A	S	Y	I	S	G	S	
Endothiapepsin	I	D	G	I	A	D	T	G	T	T	L	L	Y	L	P	
Penicillopepsin	—	S	G	I	A	D	T	G	T	T	L	L	L	L	B	
Porcine pepsin	C	Q	A	I	V	D	T	G	T	S	L	L	T	G	P	
Human pepsin	C	Q	A	I	V	D	T	G	T	S	L	L	T	G	P	
Chicken pepsin	C	Q	A	I	V	D	T	G	T	S	L	L	V	M	P	
Chymosin	C	Q	A	I	L	D	T	G	T	S	K	L	V	G	P	

* The catalytic aspartates

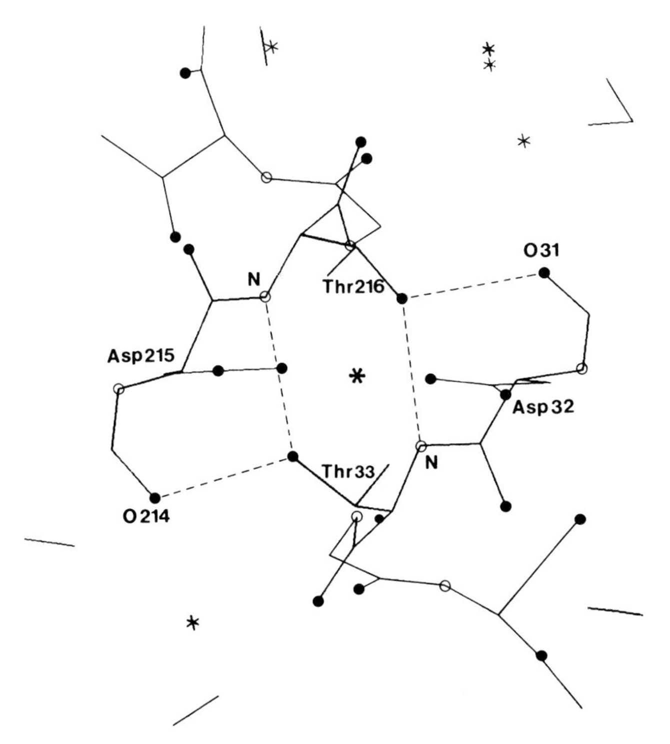

Fig. 4. The conformation of the residues in the vicinity of the active site aspartates (32 and 215) of endothiapepsin. The sequences are those shown in Table 2. They are arranged with a pseudo-dyad symmetry possibly reflecting an ancestral dimer. * is a water molecule on the dyad bound to the aspartates.

3). Likewise, in the serine proteinases the catalytic quartet His 57, Asp 102, Ser 195 and Ser 214 (chymotrypsinogen numbering) is invariant and the arrangement in space is conserved. In the azurins and plastocyanin the geometry of the ligands of the important copper atom is the same within the accuracy of the present coordinates [7] even though the histidines occupy positions on loops of different lengths. In families of divergently evolved proteins of similar function the catalytic residues are invariant in sequence and conserved in their three-dimensional arrangement.

What other residues are invariant? Let us consider first the proteinases. In the microbial and mammalian serine proteinases five glycines and two half-cystines are the invariant residues in addition to the catalytic quartet (see above). In the two topologically equivalent domains of the aspartic proteinases the only invariant residues occur in the two regions shown in Table II. Apart from the active site aspartates (positions 32 and 215) the invariant residues are glycine and threonine. This part of the structure is shown in Fig. 4. The dyad related topologically equivalent strands turn sharply at the glycines (positions 34 and 217) exploiting the ability of glycine to attain a conformation with a positive ϕ torsion angle. The invariant threonines are buried and hydrogen-bonded in a hydrophobic environment to the other main-chain. Thus the two dyad-related strands are held firmly together by a buried polar residue.

A very similar situation occurs in the repeated Greek-key motifs of the lens β and γ-crystallins [6, 23]. Each comprises an extended $\beta\beta$ hairpin (strands a and b) which is folded on to a β-sheet comprising four antiparallel strands (c'dab) (see Fig. 5). The fold is achieved by a glycine (at position 13 of motif 1 in γII) (see Table III). The topologically equivalent glycine is invariant in all (> 60) known sequences. The folded hairpin buries a serine on strand d (position 34 of motif 1 in γII) which is hydrogen-bonded to the peptide NH (position 11) of the folded hairpin. In all the sequences which form such a Greek key, one is an alanine; the rest are conserved as serine. There is no other conserved residue which is common to all the known sequences, although pairs of sequences generally have significant homologies.

This observation has allowed us to search for sequences compatible with a similar tertiary structure. We have recently shown that the surface protein of the sporulating bacterium *Myxococcus xanthus* has such a sequence (see Table III) and have shown that the internally quadruplicated sequence is compatible with a tertiary structure similar to that found in vertebrate lens proteins [24]. Thus for the first time we have identified a β,γ-crystallin family member outside the vertebrates.

In both the aspartic proteinases and the lens crystallins the invariance of a buried and hydrogen-bonded serine or threonine is striking. It appears to indicate that when a polar group is buried and forms a stabilizing hydrogen bond it may

Table III. *Alignment of the four repeated sequences of protein S* (*S*1-*S*4) *with the first two motifs of* γII-crystallin (γ1, γ2).

(The protein S sequence is that defined for gene 2. The residues for gene 1 are given above; there is an inserted alanine between Ser 146 and Gly 147 in gene 1. Each repeated sequence corresponds to a Greek-key motif shown in Fig. 5(*a*) and (*b*). From ref. [24].)

```
                6         11  13                                                    34  36
      1                 10                  20                              30
γ1    G K I T F Y E D R G F Q G H C Y E C S S   D C P N L Q           P Y F S R C N S I R V D S
    48            K     D   T                                                           K   M 86
S2    V K A I L Y Q N D G F A G D Q I E V V A   N A E E L G           P L N N N V S S I R V I S V P V Q
   136  K                             S   T     N     S                                     T P
S4    L A V V L F K N D N F S G D T L P V N S   D A P T L G           A M N N N T S S I R I S

    40                    50                  60                      70                80
γ2    G C W M L Y E R P N Y Q G H Q Y F L R R G D Y P D Y Q     Q W M G F N D S I R S C R L I P Q
    1                       G             K   D E   K   D K   E                                47
S1  M A N I T V F Y N E D F Q G K Q V D L P P G N Y T R A Q L A A L G I E N N T I S S V K V P P G
    91                                                                                    E   135
S3    P R A R F F Y K E Q F D G K E V D L P P G Q Y T Q A E L E R Y G I D N N T I S S V K P Q G
```

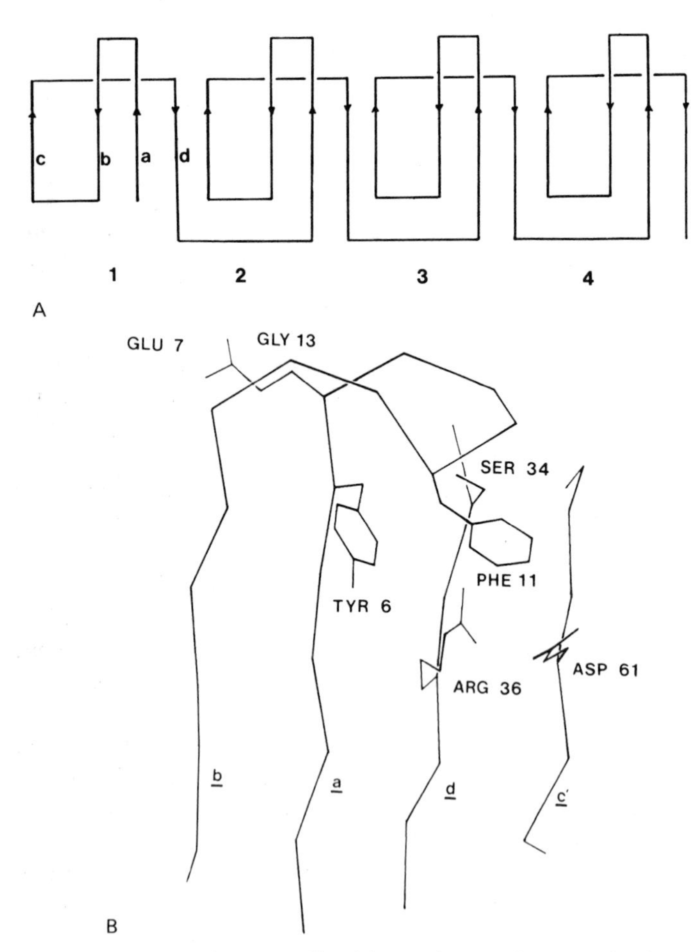

Fig. 5. The structure of γII-crystallin of the vertebrate eye lens comprises four repeated Greek key motifs (A). This may reflect an internal quadruplication of an ancestral protein comprised of one Greek key. The motifs are arranged as four β-sheets, one of which is shown in (B). Only residues equivalent to Gly 13 and Ser 34 are invariant. The surrounding residues for which side-chains are drawn are conservatively varied. (Reproduced from ref. [27].)

be very important, for the tertiary structure and mutations are not accepted at this point without loss of the tertiary structure and therefore function. In the different aspartic proteinases several aspartates and one histidine are conserved because they are internally hydrogen-bonded and structurally important.

The invariance of glycines in divergently evolved proteins is a very common occurrence and their pattern is increasingly recognized as a feature characteristic of certain supersecondary structural motifs. This is particularly marked in many of the oncogene products, where a pattern is characteristic of the $\alpha\beta$ structure of the Rossman fold [25–26]. More recently, Sibanda and Thornton [21] have shown that glycines are conserved at certain positions of $\beta\beta$-hairpins; for example, glycine tends to be at position L4 in a typical five-membered loop. In all places where glycines are conserved in $\beta\beta$, $\beta\alpha$, or $\alpha\alpha$ loops the conformation involves a tight twist with a positive ϕ main-chain torsion angle.

In the insulin family [5] certain glycines are also highly conserved or invariant. For example, B8 glycine at a bend between a β-strand and an α-helix is invariant. The B20 and B23 glycines are also conserved in many insulin-like growth-factor and relaxin sequences. Apart from the glycines the major invariant features are the three cystine disulphides at A7B7, A6A11 and B19A20. Amongst extracellular proteins, especially in hormones, growth factors, plasma proteins and extracellular domains of receptors, cystines are the most conspicuous invariant features. In insulin the disulphides not only stabilize the relative positions of main-chains by forming a covalent bridge, but also contribute to the very small but important hydrophobic core. In the snake neurotoxins this is also the case and the tertiary structure comprises a series of somewhat variable loops pinned together by a core of hydrophobic cystines. Although the cystine-rich sequences of toxins, neurophysins, epidermal growth factors and the LDL, insulin and EGF receptors are not homologous, they all contain between 6 and 8 conserved disulphides in a sequence of 40–50 amino acids which is highly variable elsewhere. It is likely therefore that they contain similar looped structures with a core of cystine sulphurs which carry out a unique structural role which leads to their conservation in evolution.

The pattern of invariant residues and protein structure prediction

The recognition that glycines, cystines and buried polar residues are likely to be more conserved in evolution offers a method for searching for sequences with similar tertiary structures. This becomes important now that there are large numbers of the amino acid sequences derived from cDNA and genomic DNA sequences. Where a family of clearly homologous sequences is available we may identify invariant residues and confirm in the three-dimensional structure that they play an important structural role. If no family is available we can try to identify residues with unique structural roles. We can identify patterns of residues that must contribute to the hydrophobic core and suggest places, usually loops between $\beta\beta$, $\beta\alpha$, $\alpha\beta$ or $\alpha\alpha$ motifs or folds of β-sheets, where insertions or deletions are compatible with a conservation of the tertiary structure. These patterns can then be used to search the data banks for compatible sequences (T. L. Blundell, W. R. Taylor, J. M. Thornton and M. J. E. Sternberg unpublished results). A successful use of this technique was the prediction of the protein S structure mentioned above [24]. Thus an understanding of the conservation and variation of proteins in evolution comprises the most useful information in a rational approach to protein prediction.

Acknowledgements

I am grateful to many colleagues for stimulating and helpful discussions. I thank Miss P. Cleasby for typing the manuscript.

References

1. Bajaj, M. and Blundell, T. L., *Annu. Rev. Biophys. Bioeng.* **13**, 453–492 (1984).
2. Rossman, M. G., Moras, D. and Olsen, K. W., *Nature* (*London*) **250**, 194–199 (1974).
3. Ohlsson, I., Nordstrom, B. and Branden, C. I., *J. Mol. Biol.* **89**, 339–354 (1974).
4. Neurath, H., *J. Am. Chem. Soc.* **65**, 2039–2040 (1943).
5. Blundell, T. L. and Humbel, R. E., *Nature* (*London*) **287**, 781–787 (1980).
6. Blundell, T. L., Lindley, P. F., Miller, L., Moss, D., Slingsby, C., Tickle, I. J., Turnell, W. G. and Wistow, G., *Nature* (*London*) **289**, 771–777 (1981).
7. Chothia, C. and Lesk, A., *J. Mol, Biol.* **260**, 309–323 (1982).
8. Lesk, A. and Chothia, C., *J. Mol. Biol.* **136**, 225–270 (1980).
9. Kabsch, W. and Sander, C., *FEBS Lett.* **155**, 179–182 (1983).

10. Blundell, T. L., Barlow, D., Borkakoti, N. and Thornton, J., *Nature (London)* **306**, 281–283 (1983).
11. Barlow, D., Ph.D. thesis, University of London (1985).
12. Blundell, T. L., Pitts, J. E., Tickle, I. J., Wood, S. P. and Wu, W–C. *Proc. Natl. Acad. Sci USA* **78**, 4175–4180 (1981).
13. Glover, I. D., Haneef, I., Pitts, J. E., Wood, S. P., Moss, D. S., Tickle, I. J. and Blundell, T. L., *Biopolymers* **22**, 293–298 (1983).
14. Glover, I. D., Moss, D. S., Tickle, I. J., Pitts, J. E., Haneef, I., Wood, S. P. and Blundell, T. L., *Adv. Biophys.* **20**, (in the press).
15. Tatemoto, K. and Mutt, V., *Proc. Natl. Acad. Sci. USA* **75**, 4115–4119 (1978).
16. Glover, I. D., Barlow, D. J., Pitts, J. E., Wood, S. P., Tickle, I. J., Blundell, T. L., Tatemoto, K., Kimmel, J. R., Wollmer, A., Strassburger, W. and Zhang, Y–S. *Eur. J. Biochem.* **142**, 379–385 (1984).
17. Pearl, L. H. and Blundell, T. L., *FEBS Lett.* **174**, 96–99 (1984).
18. Sielecki, A. and James, M. N. G., *J. Mol. Biol.* **163**, 299–361 (1983).
19. Blundell, T. L., Jenkins, J. A., Pearl, L. H., Sewell, T. and Pedersen, V., *Aspartic Proteinases and Their Inhibitors* (ed. V. Kostka). pp. 151–161. De Gruyter, Berlin (1985).
20. Richardson, J. S., Getzoff, E. D. and Richardson, D. C., *Proc. Natl. Acad. Sci. USA* **75**, 2574–2578 (1978).
21. Sibanda, B. L. and Thornton, J., *Nature (London)* **316**, 170–174 (1985).
22. Sibanda, B. L., Blundell, T. L., Hobart, P. M., Fogliano, M., Bindra, J. S., Dominy, B. W. and Chirgwin, J. M., *FEBS Lett.* **174**, 102–111 (1984).
23. Summers, L., Wistow, G., Narebor, M., Moss, D. S., Lindley, P. F., Slingsby, C., Blundell, T. L., Bartunik, H. and Bartels, K., *Pept. Prot. Rev.* **3**, 147–167 (1984).
24. Wistow, G., Summers, L. and Blundell, T. L., *Nature (London)* **316**, 771–773 (1985).
25. Wierenga, R. K. and Hol, W. G., *Nature (London)* **302**, 842–844 (1983).
26. Sternberg, M. J. E. and Taylor, W. R., *FEBS Lett.* **175**, 387–392 (1984).
27. Slingsby, C., *Trends, Biochem, Soc.* **10**, 281–284 (1985).

Chemica Scripta 1986, **27B**, 221–229

Limited Proteolysis, Domains, and the Evolution of Protein Structure

Hans Neurath

University of Washington, Department of Biochemistry, Seattle, Washington 98195, USA

Paper presented at the Conference on 'Molecular Evolution of Life', Lidingö, Sweden, 8-12 September 1985

Abstract

Analysis of protein homology by protein sequencing, DNA sequencing, or both, has provided evidence that these linear structures can be partitioned into segments which correspond to protein domains that can be isolated by limited proteolysis or identified by examination of the structure or function of the whole protein. In the present review, the domain structures of representative members of certain protein families and supergene families are examined. They include the digestive and regulatory serine proteases, the protein kinases, receptor proteins, and multi-functional proteins such as certain blood coagulation proteins and fibronectin. The occurrence of homologous domains in seemingly unrelated proteins suggests that they have been recruited in the course of evolution from different sources of the gene pool by a process of exon shuffling and splicing. Attractive as this scenario appears, it raises as-yet unanswered questions of a more fundamental nature, e.g. the relationship of the exon/intron organization to the domain structure of proteins, the size of the 'domain inventory', and, perhaps most important, the mechanism of folding of the nascent polypeptide chains into domains. These and related problems are discussed in this review.

Introduction

The enormous repertoire of biological functions of proteins is paralleled by a multiplicity of chemical structures. While in the past, analyses of structure/function relationships have been directed toward the protein molecule as a whole, a more rational and successful approach arises from the recognition that all but the smallest protein molecules are assemblies of substructures or domains which have been recruited in the course of evolution from various and often unrelated sources of the gene pool. These domains, or 'modules', can be recognized by visual inspection of the three-dimensional models of protein molecules and in many cases can be isolated by limited proteolysis with retention of conformation and function. Analysis for homology by protein or DNA sequencing has provided evidence for the segmentation of these linear structures into component parts that are related to structural domains. The relation of exon/intron distribution of the genomic DNA to the domain structure of the protein is one of the most intriguing problems facing the molecular biologist interested in protein evolution. The related question of the 'code' which dictates the transition of the linear polypeptide chain to a unique three-dimensional conformation is the missing link between the genetic message and its three-dimensional expression in the mature protein. In this discussion I shall emphasize the information that can be gained from an analysis of the domain structure of representative proteins for an understanding of the broader aspects of the molecular evolution of the structure and function of proteins.

Protein folding and domains

The term 'domain' is loosely used to connote a compact substructure of a protein molecule but there is no consensus of its physical meaning. This is not the place to consider the complex and controversial subject of protein folding [1, 2]; it should suffice to differentiate between the major folding units of helices, beta sheets and beta turns, for which the term 'folding element' or 'motif' may be used, and the compact substructures, or 'domains' into which these folding elements coalesce. The driving force is the exclusion of water molecules or, conversely, the formation of hydrophobic cores. Although domains are usually considered as self-assembling segments of polypeptide chains, such a definition is of little practical value since it is generally not experimentally verifiable. It appears more profitable to adhere to the geometric definitions of Wodak & Janin [3], following the work of Lee & Richards [4], according to whom a domain is a group of atoms (or amino acid residues) having a minimum surface/volume ratio, resulting in an autonomous region with the most interactions within and the least without. By virtue of an iterative cleavage algorithm which relates the surface buried between two groups of residues to a cleavage point moving along the polypeptide chain, domain boundaries appear as minima of the interface area. These calculations seem to agree well with the domain regions identified by inspection of protein models [5]. Recently, Rose and coworkers [6, 7] developed an algorithm bases on contour maps similar to those used to geographically represent topographic surface features of irregular objects. The contour map of lysozyme is shown in Fig. 1. While originally designed to identify antigenic determinants in proteins of known structure, this approach shows real promise for the identification of domains in proteins of known structure and for the recognition of surfaces between interacting protein domains.

The problem of domain recognition actually goes back some twenty years, when D. C. Phillips [8] noted that the first 40 residues of lysozyme form a compact structure; several years later, Rossman [9] produced a quantitative comparison of similarities in three-dimensional structures of proteins and proposed that they represent 'domains'.

While the three-dimensional structures obtained by X-ray crystallography are the most accurate models from which to delineate domains [10, 11], a great deal of inferential evidence has been gathered at an ever increasing rate, from the comparison of sequences obtained directly from proteins or derived from genomic or cDNA sequences. With the discovery

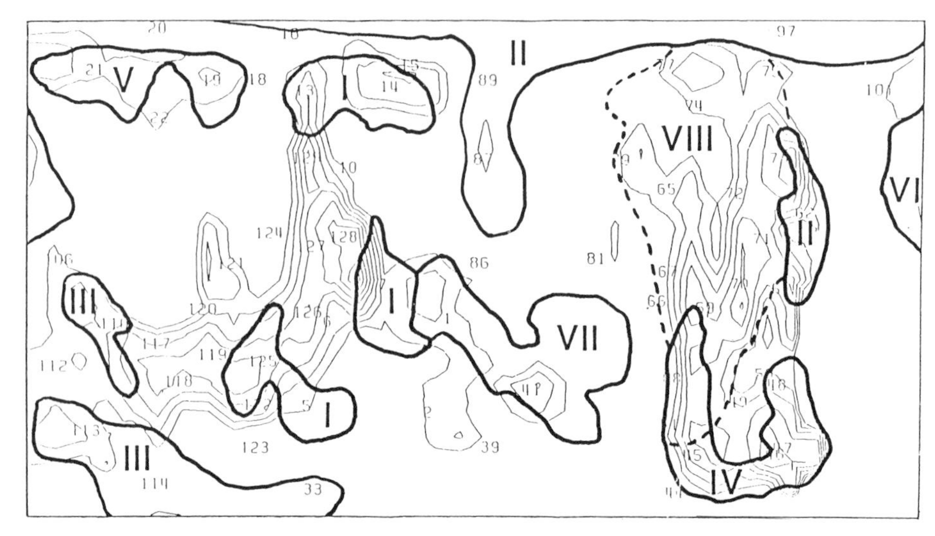

Fig. 1. Contour map of the molecular surface of lysozyme. The contour interval is 1 Å. Only contour lines 8 Å or higher are displayed to show the exposed surface. The heavy contour lines and roman numerals denote antigenic sites. Courtesy of Dr. G. D. Rose. Taken from ref. [7]; for further details, see ref. [6].

of non-coding sequences in genomic DNA, the question of the relationship between protein domains and the organization of the gene has been raised, and has added impetus to the search for domains and their evolutionary origin. These topics will be discussed later in this review in conjunction with specific examples.

The inter-domain regions of a protein molecule, i.e. the 'hinges' and 'fringes' [12], differ from domains proper not only in structural organization but in the constraints to which they have been subjected in the course of evolution [13]. In these inter-domain regions, the variations in amino acid sequence are usually so large as to abolish any recognizable sequence similarity between functionally analogous modules.

Lastly, a word about the upper and lower limits of the sizes of domains. While there is no *a priori* upper limit, it would be surprising if the very large modules were not assemblies of smaller domains. Estimates of the lower limit hover around the range of $M_r = 4000–6000$, corresponding to a chain length of approximately 30–50 amino acid residues (e.g. glucagon, pancreatic trypsin inhibitor). In this connection it is of interest that Fontana and co-workers [14] have recently shown that subfragments of approximately 100 residues, derived from the carboxyl-terminal cyanogen bromide fragment of thermolysin, are able to acquire a stable, native-like conformation, capable of reversible unfolding.

Limited proteolysis

A widely used experimental technique to probe for structural and functional domains in native proteins is that of limited proteolysis, a term which was first introduced by Linderstrøm-Lang and Ottesen [15]. In a general sense, proteolysis is limited by the accessibility of the susceptible peptide bond to the attacking protease and depends primarily on the conformation of the protein substrate. Proteolysis of native proteins usually occurs in inter-domain regions [12]. Physiologically, the most important examples are the post-translational processing reactions of protein precursors as, for instance, the removal of signal peptides from the nascent polypeptide chain in transmembrane processes or the activation of zymogens and peptide hormone precursors [16, 17]. An extreme example is the action of several proteases, differing in substrate specificity, on α_2-macroglobulin, the 'universal' protease inhibitor. All of them cleave the polypeptide chain of 1450 amino acid residues only in a narrow nine-residue segment near the middle of the chain [18]. The published reports of the limited proteolysis of protein substrates are legion and much of the impetus has come from the strategy of protein sequence analysis, designed to simplify the problem by cleaving the starting protein into smaller fragments [19]. Limited proteolysis of proteins of known structure and function by a variety of proteases has enabled the separation of domains with partial retention of function and thus proved an important, though not universally applicable, procedure for the recognition of protein domains. A classical example is the proteolytic cleavage of DNA polymerase (pol I) into a large fragment which catalyses 3′–5′ exonuclease action on both single-stranded and unpaired regions of double-stranded DNA, and a small fragment which retains only 5′–3′ exonuclease activity. These two domains are covalently linked in the native enzyme [20, 21]. Other examples are the separation and isolation of kringles from prothrombin, plasminogen, urokinase and plasminogen activator [22], the isolation of the Gla (γ-carboxyl glutamic acid) domains from coagulation factors, IX, X, and protein C [23], and the proteolytic cleavage of complement C9 producing a cystine-rich amino-terminal domain [24], which occurs also in membrane receptors for LDL (see below). Representative examples of the limited proteolysis of multifunctional proteins are given in Table I.

The domain structure of proteins

Proteases

Among the known supergene families of proteases, the mammalian serine proteases are the most widely distributed and the most versatile physiologically. Analysis for homology suggests that the pancreatic serine proteases are the most direct descendants of the ancestral protease from which the hepatic plasma proteases and the granuloctye serine proteases have subsequently evolved [25, 26]. The pancreatic serine proteases, i.e. trypsin, chymotrypsin, elastase and kallikrein, represent a closely knit family, similar in amino acid sequence, three-dimensional structure, active site configuration and mechanism of action. The recent elucidation of the genomic

Table I. *Representative examples of limited proteolysis of multifunctional proteins*

Protein	Reference
DNA polymerase	1, 2
Muscle glycogen phosporylase	3
Regulatory subunit type I of cAMP-dependent protein kinase	4
Skeletal muscle myosin light chain	5
Pig heart citrate synthase	6
Colicins E2 and E3	7
Fibronectin	8–13
Von Willebrand factor	14, 15

1. Klenow, H. and Hennigen, I., *Proc. Natl. Acad. Sci. USA* **65**, 168 (1970).
2. Ollis, D., Brick, P., Hamlin, R., Xuong, N. G. and Steitz, T. A., *Nature (London)* **313**, 762 (1985).
3. Titani, K., Koide, A., Hermann, J., Ericsson, L. H., Kumar, S., Wade, R. D., Walsh, K. A., Neurath, H. and Fischer, E. H., *Proc. Natl. Acad. Sci. USA* **74**, 4762 (1977).
4. Takio, K., Smith, S. B., Krebs, E. G., Walsh, K. A. and Titani, K., *Proc. Natl. Acad. Sci. USA* **79**, 2544 (1982).
5. Takio, K., Blumenthal, D. K., Edelman, A. M., Walsh, K. A., Krebs, E. G. and Titani, K., *Biochemistry* **24**, 6028 (1985).
6. Bloxham, D. P., Ericsson, L. H., Titani, K., Walsh, K. A. and Neurath, H., *Biochemistry* **19**, 3979 (1980).
7. Ohno-Iwashita, Y. and Imahori, K., *Biochemistry* **19**, 652 (1980); *J. Biol. Chem.* **257**, 6446 (1982).
8. Sekiguchi, K. and Hakamori, S.-I., *Proc. Natl. Acad. Sci. USA* **77**, 2661 (1980).
9. Petersen, T. E., Thogersen, H. C., Skorstengaard, K., Vibe-pedersen, K., Sahl, P., Sottrup-Jensen, L. and Magnusson, D., *Proc. Natl. Acad. Sci. USA* **80**, 137 (1983).
10. Pedersen, T. E. and Skorstengaard, K., in *Fibronectin, Its Role in Coagulation and Fibronlysis* (ed. J. McDonagh). Marcel Dekker, New York (1984).
11. Pande, H., Calaycay, J., Ben-Avram, C. M. and Shively, J. E., *J. Biol. Chem.* **260**, 2301 (1985).
12. Sekiguchi, K., Siri, A., Zardi, L. and Hakamori, S.-I., *J. Biol. Chem.* **260**, 5105 (1985).
13. Paul, J. I. and Hynes, R. O., *J. Biol. Chem.* **260**, 13477 (1985).
14. Girma, J.-P., Chopek, M. W., Titani, K. and Davie, E. W., *Biochemistry* (in the press).
15. Girma, J.-P., Chopek, M. W., Titani, K. and Davie, E. W., *Biochemistry* (in the press).

DNA sequences of these proteases, and of their exon/intron distribution [27–30], renders this group the most completely characterized family of proteases and, thus, a superbly defined reference point to which to relate the more complex structures of the regulatory plasma proteases (see below). Visual inspection and analysis of the hydrogen bonding pattern of X-ray structures had already demonstrated that these proteases can best be described as being composed of two similarly folded domains symmetrically disposed around a twofold axis [31] (Fig. 2). The genes coding for trypsin, chymotrypsin, elastase and kallikrein differ from each other in the number and distribution of introns (Fig. 3) but the exon/intron splice junctions correspond in each case to regions on the protein surface which correlate with sites of sequence variations, thus allowing for insertions and deletions without significantly affecting the global conformation of the molecule [3]. In each case the introns separate the codons of the catalytic triad (aspartic acid, serine and histidine) but the overall pattern of exons is the same for kallikrein and trypsin and related to those of chymotrypsin and elastase, which are similar. In their

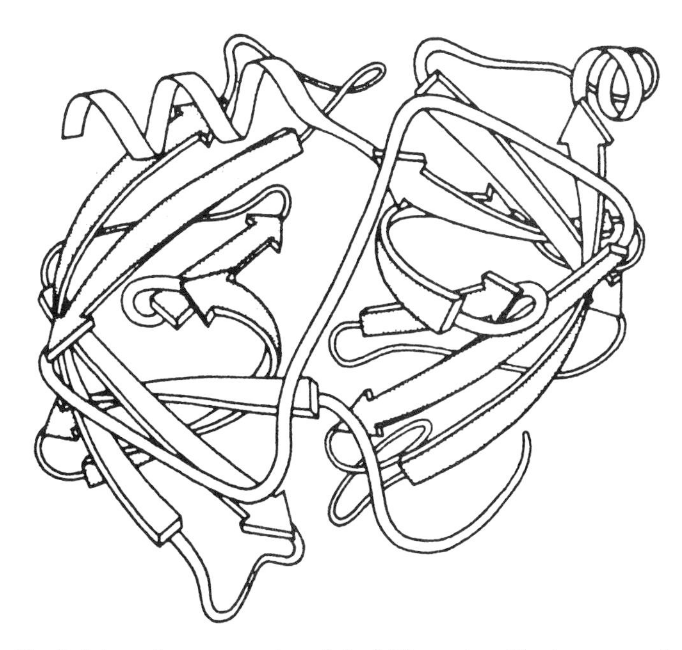

Fig. 2. Schematic representation of the folding units of bovine pancreatic elastase. The molecule consists of two domains, symmetrically disposed about a twofold axis of symmetry. Each domain contains two beta sheets and one helix, interconnected by random segments. Taken from ref. [32].

active form, these proteases may be classified as bi-domain, monofunctional proteins since they serve solely digestive purposes. Comparison of their precursor forms with those of the regulatory plasma serine proteases reveals significant, and in many instances unexpected, differences.

Regulatory proteases

The most thoroughly studied regulatory serine proteases are those of the blood coagulation cascade, all of which have been shown to be related to pancreatic trypsin [35]. Since they will be considered in some detail by Earl Davie, who has made the major contributions to this field, the present discussion will be limited to certain aspects relevant to the more general topic of protein evolution. Fundamentally, blood coagulation is a series of consecutive zymogen activation reactions, wherein the activation product of one reaction catalyses that which follows in the cascade [16, 35]. Regulation is fine-tuned by cofactors, among which proteins V and VIII are of particular interest since they are not serine proteases but require activation for their full effectiveness [36]. The initial phases of the blood coagulation cascade are interrelated to those of the fibrinolytic cascade on the one hand and the kininogen cascade on the other [37]. Analysis of protein sequences for homology clearly demonstrated a high degree of similarity of discrete regions of the polypeptide chains, believed to represent functional domains. The most prominent similarities were found in segments corresponding to the active serine protease, the activation peptide, and a segment containing γ-carboxyl glutamic acid (Gla) residues [38]. A deeper insight into the protein sequence of the precursors of the coagulation of proteases and certain of their cofactors was gleaned from a systematic study of the corresponding cDNAs and, in certain instances, of the genomic DNAs as well. These studies revealed segments homologous to those seen in other coagulation proteases, as well as in seemingly totally unrelated proteins. Major structural domains that have been found in

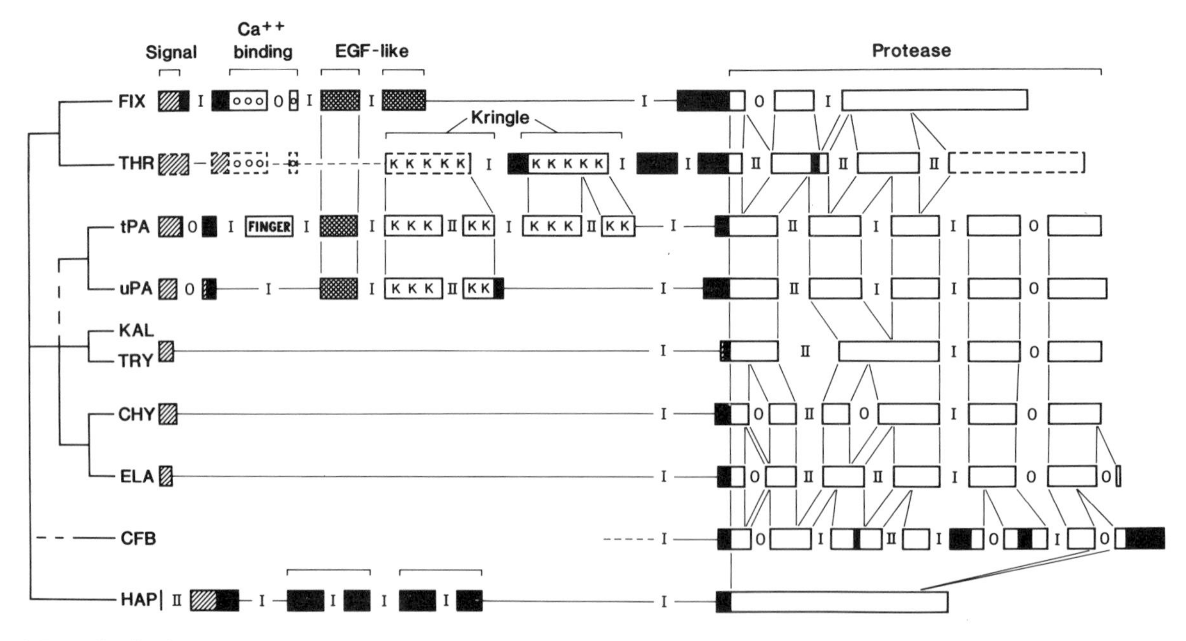

Fig. 3. Exon distribution in mammalian serine proteases. The proteases are denoted, from top to bottom, as follows: FIX, human coagulation factor IX; THR, human thrombin; tPA, human tissue plasminogen activator; uPA, pig urokinase; KAL, mouse kallikrein; TRY, rat pancreatic trypsin; CHY, rat chymotrypsin; ELA, rat elastase; CFB, human complement factor B; HAP, human haptoglobin. Exons are shown as boxes shaded according to their homologies. Sequences with no discernible homology to each other are in black. Introns are not to scale but their phase relative to the reading frame is indicated as to whether they follow the first (I), second (II), or third (III) nucleotide of a codon. Taken from ref. [33].

the sequence analysis of the coagulation and fibrinolytic systems include, besides the protease domain, Gla domains, kringles (i.e. triple-looped, disulfide-bonded 70- to 80-residue peptide segments), EGF (epidermal growth factors) and 'fingers' type I and II, first observed in fibronectin (Fig. 4). A representative list of the type and number of the domains occurring in these proteins is given in Table II. The non-protease factor VIII contains sequences which are related to the plasma copper protein ceruloplasmin to which it has no obvious functional relationship [40–42]. Factors V and VIII both are substrates for yet another protease, activated protein C, which specifically inactivates them by a regulatory mechanism which does not lie on the direct pathway of blood coagulation [43–45]. Two other proteins, Z and S, appear to be related to protein C. Protein Z has no known physiological function, presumably because a double amino acid replacement in the catalytic triad renders it inactive as a protease [46]. A similar explanation has been proposed to explain why streptokinase has no enzymatic function of its own [47]. Protease S is also inactive as a protease but contains the Gla and EGF domains [48].

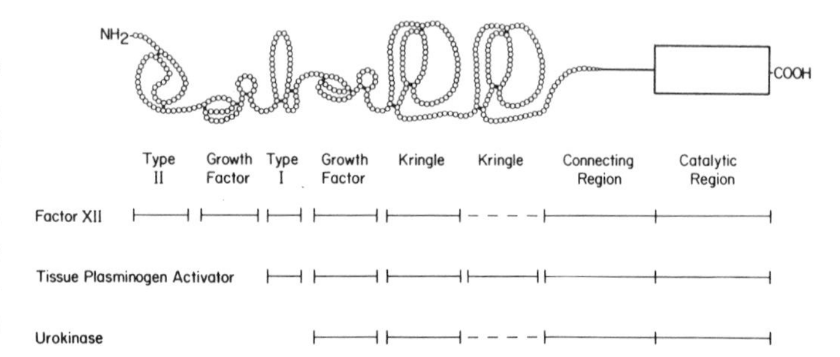

Fig. 4. Comparison of the domain organization of factor XII, tissue plasminogen activator and urokinase. The domains are schematically represented, and identified by name. Their presence is indicated below by solid lines, their absence by dashed lines. Type I and type II refers to the corresponding domains in fibronectin. Taken from ref. [39].

In several instances it has been possible to relate a specific function of a domain to that of the protein as a whole. The Gla domains are the Ca^{2+} ion binding sites, essential for the functions of the vitamin K-dependent coagulation proteases. The kringle domains are the binding sites for the interaction of the coagulation proteins with mediators such as membranes, phospholipids and the cofactor proteins V and VIII. Some of

Table II. *Occurrence and distribution of domains in coagulation proteins*

Domain[a]	PTH	X	IX	C	S	XI[b]	XII	PG	tPA	UK
Protease	1	1	1	1	0	2	1	1	1	1
Gla	1	1	1	1	1	0	0	0	0	0
OH-Asp	0	1	1	1	+	0	0	0	0	0
EGF	0	2	2	2	+	0	2	0	1	1
Kringle	2	0	0	0	0	0	1	5	2	1
Finger I	0	0	0	0	0	0	1	0	1	0
Finger II	0	0	0	0	0	0	1	0	0	0

[a] GLA, The domain containing 9–12 gamma glutamic acid residues. OH-Asp, A residue of beta hydroxy aspartic acid. EGF, Epidermal growth factor. Finger I and II, Homologies occurring in fibronectin. PTH, PG, tPA and UK denote, respectively, prothrombin, plasminogen, tissue plasminogen activator and urokinase.

[b] XI occurs in plasma as a dimer.

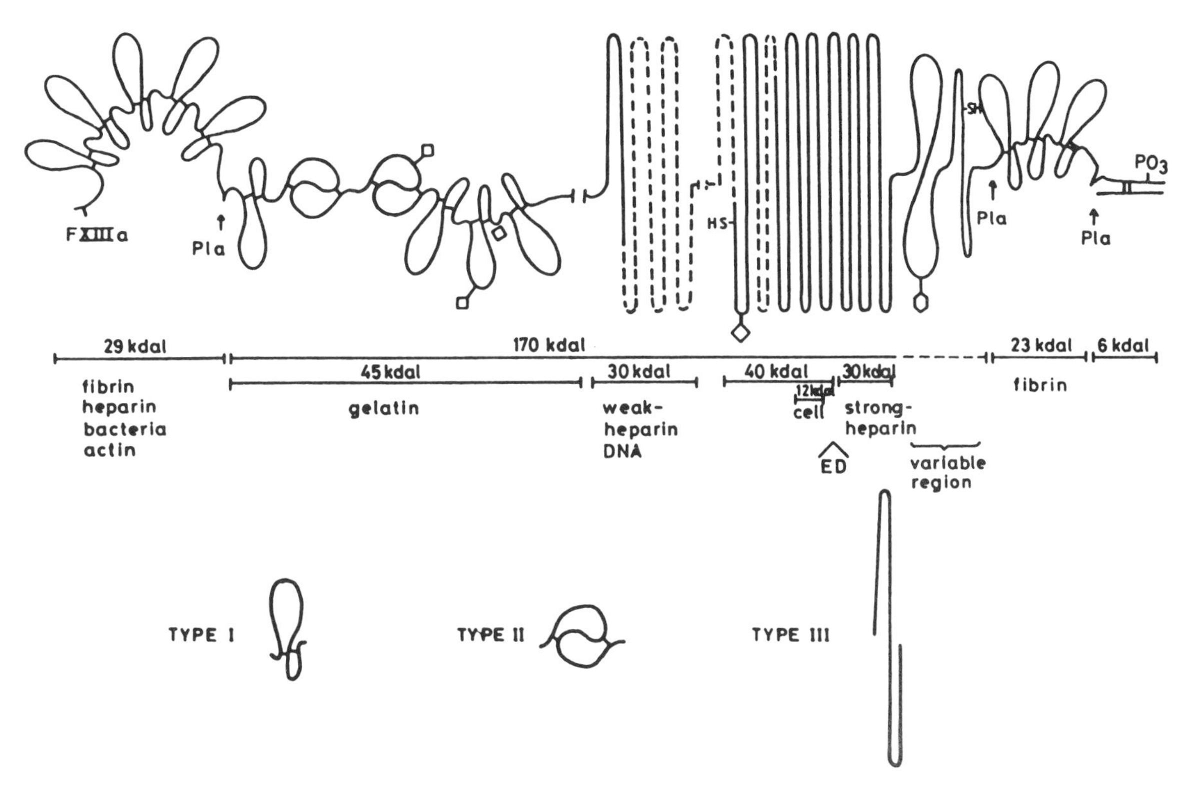

Fig. 5. Domain organization in bovine fibronectin. Only one of the two chains is shown. The 29, 170, 23 and 6 kDa fragments were obtained by limited proteolysis with plasmin, the 45, 30 and 170 kDa fragments by digestion with chymotrypsin and the 12 kDa fragment by digestion with pepsin. Squares denote glucosamine containing carbohydrate residues; hexagons, those that contain galactosamine. SH denotes cysteine, F XIII_a transglutaminase site, and ED the extra domain which has been found in the mRNAs from a human cell line. The types of homologies are schematically represented at the bottom. Taken from [51].

these domains, e.g. the kringles of prothrombin, have been isolated from the parent proteins by limited proteolysis without impairment of function [22]. The kringles of tissue plasminogen activator are the sites of interaction with fibrin. In the case of fibronectin, the various domains have been clearly associated with specific functions characteristic of the whole protein (see below). These findings clearly suggest that these proteins are the products of an evolutionary process of shuffling and splicing of gene segments generated in response to a variety of specific physiological demands. The occurrence and physiological role of the EGF domain in some but not all of these proteins remain enigmatic. It would be of interest to test this domain, isolated either from the parent protein or, by cloning, for growth-factor-like physiological activities of its own.

Other multifunctional proteins

Coagulation factor XII [49], tissue plasminogen activator [50], and fibronectin are complex multifunctional proteins *par excellence*. They share homologous kringle, EGF and fibronectin domains and yet fulfill quite different physiological functions. Fibronectin is perhaps the most versatile of these proteins as it adheres to negatively charged DNA and heparin, to positively charged polyamines, to proteins such as fibrin, collagen and actin, and is covalently incorporated by transaminase into fibrin clots. These multiple functions are carried out by separate structural domains within the intact proteins as well as individually after isolation by limited proteolysis (see Table 1 for references). The genesis of the molecule involves gene duplication of at least three characteristic structural folding units [51], of which types I and II have unique disulfide pairing patterns. The complete cDNA structure of human fibronectin has been recently reported [52] and the arrangement of the folding units in one of the two disulfide-bonded chains (2146–2325 amino acid residues per chain, depending on the splicing pathway of the pre-mRNA) is shown in Fig. 5.

Another large, multifunctional protein is the Von Willebrand factor, a multimeric plasma glycoprotein of 260 kDa subunits that are held together by disulfide bonds. It functions prior to the blood coagulation cascade by forming a bridge between platelets and the damaged vascular subendothelium. It binds to platelet glycoproteins but also to collagen of the subendothelial tissue and to factor VIII. Von Willebrand factor has been recently cloned [53] and its cDNA sequence and protein sequence have been determined. The domain structure has been mapped by limited proteolysis [54].

Lastly, mention should be made of a multi-domain protease which is of interest because regulation is mediated by domains which are part of one and the same covalent structure. Reference is made to the family of calcium-activated proteases, the calpains, which are involved in a variety of cellular processes. The protein is composed of two subunits of 30 and 80 kDa, respectively. The 80 kDa subunit contains four domains [55]. Domain II is the catalytic domain and is homologous to the plant proteases papain and actinidin, as well as to mammalian cathepsins B and H, all of which are sulfhydryl proteases. Domain IV is a regulatory, calcium-binding domain and is homologous to the calcium binding proteins calmodulin, troponin C and myosin light chain; it contains the characteristic E–F hand cage structure which surrounds the calcium ion. It is obvious that the 80 kDa subunit arose by fusion of genes coding for cysteine proteases

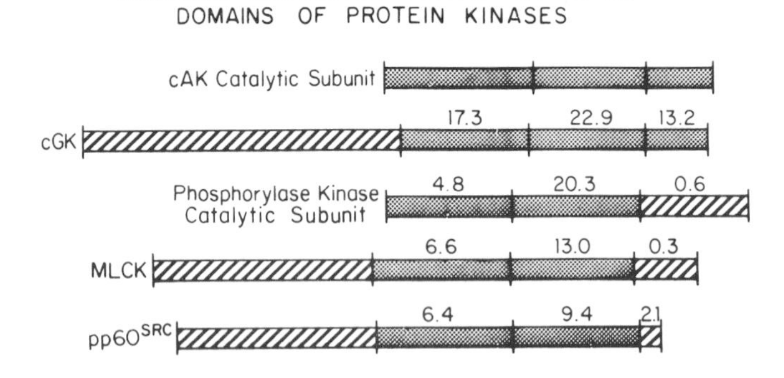

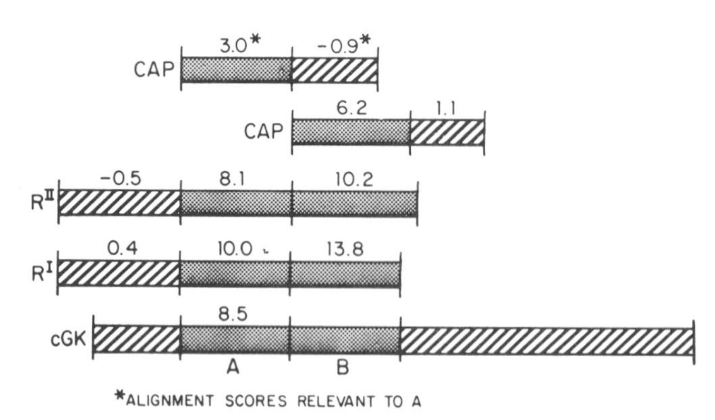

Fig. 6. Homologous domains in protein kinases and related proteins. The numbers denote alignment scores expressed in units of standard deviation from a mean of those from randomly generated sequences of the same composition. The stipled segments denote homologous regions, the striped segments regions which are not homologous. Top: Catalytic domains aligned with respect to cAMP-dependent protein kinases. MLCK denotes myosin light-chain kinase, $pp60^{src}$ transforming phosphoprotein of *Rous sarcoma* virus. Taken from ref. [59]. Bottom: Cyclic nucleotide binding domains aligned with respect to cGMP-dependent protein kinases. CAP denotes catabolite gene activating protein, RI and RII denote types I and II regulatory subunits of cAMP-dependent protein kinase.

on the one hand and for calcium-binding proteins on the other. The exact functions of domains I and III are unknown. Domain II would be expected to be active in isolation but seems to be repressed in combination with domains I and III and derepressed by domain IV. The 30 kDa subunit has been recently cloned [56]. The cDNA sequence reveals two distinct domains: a 98-residue amino-terminal domain which contains unusual polyglycyl sequences and a 267-residue carboxyl-terminal domain which contains four potential calcium-binding site that might have arisen by gene duplication of the calcium-binding domain of the 80 kDa subunit and might serve to enhance its potentiating action.

Protein phosphorylation

Peptide bond hydrolysis and phosphorylation are among the most important post-translational modifications of proteins. Under physiological conditions, peptide bond cleavage is essentially an irreversible reaction which creates a permanent change in the micro-environment of the protein. It plays a major role in the timing of expression, as in zymogen activation, and in protein and tissue remodelling, such as the release of protein hormones from their precursors and in fibrinolysis. Protein phosphorylation–dephosphorylation is the major reversible post-translational mechanism of metabolic regulation, catalyzed by the opposing actions of protein kinases and phosphatases.

Within the context of the present discussion, it should suffice to refer to the two types of protein kinases, i.e. those in which the terminal phosphate of ATP is transferred to a serine or a threonine residue of the protein and those in which a tyrosine residue is the receptor [57, 58]. The activity of the serine/threonine protein kinases is regulated by specific effectors among which cAMP, cGMP, Ca^{2+}, and diacylglycerol are the best understood. The effectors of the tyrosine kinases include the growth factors EGF and PDGF, insulin and other viral oncogene-related factors. The functional components of both types of protein kinases include an effector-binding site and an ATP-binding site and also an autophosphorylation site.

Protein sequences of several serine/threonine protein kinases have been determined by Krebs, Walsh, Titani and coworkers at the University of Washington [e.g. 59]; partial sequences of tyrosine kinases have been reported by other laboratories and have been summarized by Hunter and Cooper [60]. These analyses have provided clear evidence for the existence of homologous functional domains in the various protein kinases. Figure 6 compares these domains in cAMP-, cGMP-, and Ca^{2+} calmodulin-dependent protein kinases [61], as well as in the analogous domains of the gamma chain of phosphorylase kinase [62] and the transforming retroviral tyrosine-specific pp60src [63]. Segmentation into domains is based on alignment scores calculated for the corresponding segments of the protein chains. Two major generalizations may be derived from these analyses: first, domains associated with analogous functions are homologous. Thus the cAMP and cGMP binding sites of the regulatory subunits of the corresponding kinases occur in domains which are homologous to each other as well as to a domain of the cAMP-binding catabolite gene activator protein (CAP) of *E. coli* [64, 65]. Similarly, the catalytic (ATP-binding) sites of all protein kinases, including the calmodulin-dependent myosin light-chain kinase and the tyrosine kinases, occur in homologous domains as do the autophosphorylation sites. The second generalization is a derivative of the first, i.e. these proteins have arisen by duplication and fusion of genetic elements of two ancestral supergene families to generate the allosteric nature of these enzymes [59].

In this connection, it has been suggested that allosteric regulation arose initially from the accidental interaction of an active enzyme with an inhibitor and that the association was weakened by covalent modification of the complex, e.g. by phosphorylation, adenylylation, etc. [66]. It seems of little functional consequence whether the regulatory and catalytic domains are covalently linked, as in the cGMP-dependent protein kinases, or are in separate subunits, as in the cAMP-dependent kinases, since in both cases the corresponding domains are homologous. Studies of the protein kinases have revealed two instances in which a biological function is uniquely associated with an unusually small protein segment. In the case of myosin light chain kinase, a calmodulin docking site resides in a carboxyl-terminal 27-residue fragment which, if removed by limited proteolysis, leaves the kinase calmodulin-independent [65, 67]. The other instance is a 75-residue inhibitor of the cAMP-dependent protein kinase. A 20 amino acid residue peptide fragment, isolated by limited proteolysis

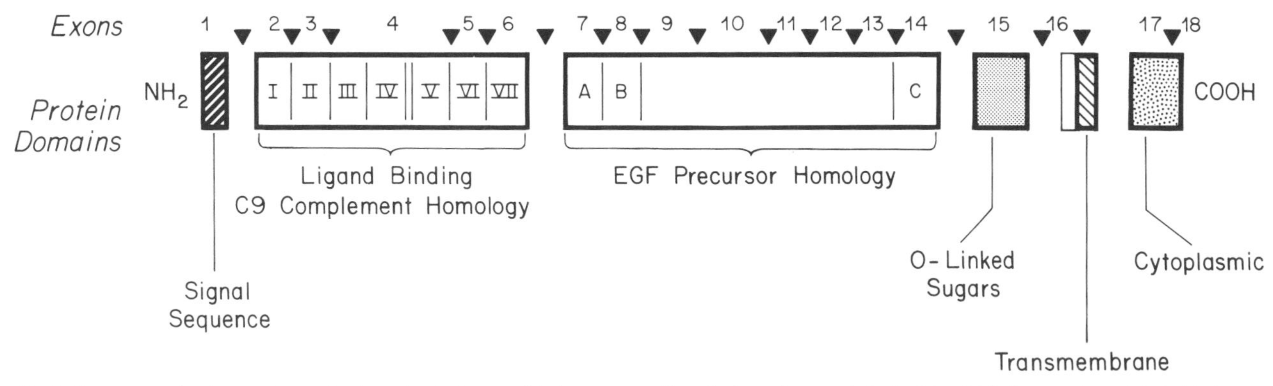

Fig. 7. Exon organization and protein domains in human LDL receptor. The six domains are delimited by thick lines and are labeled below. Domain I is the signal sequence and is absent in the mature protein. The position at which introns interrupt the coding region are indicated by arrowheads. Exon numbers are shown between the arrowheads. For further details, see the original reference. Taken from ref. [74].

with rat mast cell protease II, was as active an inhibitor as the whole protein and showed sequence similarity to the autophosphorylation site of the type II regulatory subunit of the protein kinase [68], whereas the remainder of the sequence shows no similarity to any known protein sequence [69]. These findings should serve as a warning not to conclude that two proteins are homologous just because the sequences around their active sites are similar.

An interesting variation of post-translational modification of proteins involving both phosphorylation–dephosphorylation and peptide bond hydrolysis has been recently found in the processing of the precursor of filaggrin, a highly phosphorylated 500 kDa protein containing multiple copies of filaggrin proper. This reaction occurs during differentiation of epidermal cells and involves both dephosphorylation and proteolysis, yielding filaggrin, a 26 kDa protein which polymerizes with keratins into macrofibrils. The 16 phosphorylated repeat domains of filaggrin are undoubtedly the products of gene multiplication and each of them needs to be phosphorylated prior to fragmentation of the precursor [70].

Receptor proteins

Only brief mention will be made in this discussion of the domain structure of receptor proteins, a subject which has progressed with amazing speed, largely due to the successes of DNA cloning and sequencing. The EFG receptor, for one, has been described as a composite of three principal domains: an extracellular 619 amino acid residue EGF binding domain, separated by a 26-residue transmembrane domain from a 542 intracellular domain which is homologous to the V-erb-B protein kinase, short of the 32 carboxyl-terminal residues which include the autophosphorylation site [71]. A somewhat analogous situation has been described for the insulin receptor [72]. The T lymphocyte antigen receptors and the polypeptide chains of the major histocompatibility antigen complexes contain segments resembling domains of immunoglobulins [37]. Suedhof *et al.* have recently described the structure of the human LDL receptor, a cholesterol transport system that mediates endocytosis through coated pits [74]. This sequence, deduced from the corresponding genomic DNA, can be fragmented into five different domains among which the first and second are of particular interest since they are evidently products of different gene precursors (Fig. 7). Domain 1, comprised of approximately 300 amino acid residues, is rich in disulfide bonds and contains seven repeats of 40 amino acid residues and 3 disulfide bridges each. A similar structure occurs in the C9 component of complement [24, 75] and resembles a four-disulfide-bonded domain of known three-dimensional conformation in wheat-germ agglutinin [76]. Domain 2, approximately 400 amino acid residues, is homologous to the EGF precursor and contains 3 domains homologous to the EGF domains in coagulation factors IX, X and protein C. Each of these is encoded by a single exon. Another repeat sequence in the EGF precursor domain is also found in factors IX, X and Protein C. It is clear, therefore, that the LDL receptor protein is a chimeric protein which arose from two different supergene families.

Conclusions

Evolution is always a timely and fascinating topic of discussion. It is a topic that affords an opportunity to look back into the past through the eyes of today's observer, but therein also lies the risk of illusory reasoning. We rely on the analysis and comparison of relics of the past embodied in current structures but, unlike archaeology, the relics themselves have undergone change during the long periods of biological evolution. Biological archaeology is based on analogies and not on prima facie evidence.

A new topic was added to the discussion of protein evolution when it was recognized that protein molecules are composed of domains and that homologous domains are found in structurally or functionally unrelated proteins. The inventory of such domains is growing at an unprecedented rate, largely due to the 'DNA Express', which provides us with a compendium of translated protein sequences so fast that sometimes we find ourselves with a sequence in search of a protein rather than vice versa.

The analysis of domain structures has established unexpected relationships that were not imagined when the structure of a protein was considered as an entity. One of the earliest observations of the structural similarity of seemingly unrelated proteins was the discovery of a collagen-like 'stem' in the C_{1q} component of the complement system, covalently linked to a globular 'flower' domain [77]. Structural domains and associated functions have been described more recently for the subcomponents Clr and Cls of human complement [78]. The examples cited in this review for representative multi-domain proteins serving a variety of important physiological functions speak for themselves and need not be recapitulated. It has also been shown that tandem gene duplication can account for some of the repetitive sequences and corresponding domains in proteins and that gene duplication and exon shuffling and splicing can explain the recruitment of domains from seemingly unrelated parent genes. However, this scenario, attractive as it appears, raises almost as many questions as it purports to answer. To begin with, how large is the inventory of gene segments coding for domains? The human pool of structural genes is estimated to be of the order of 50×10^3. The number of genes coding for domains must be considerably smaller, perhaps 10^3 or even less, and infinitesimally small when compared to the statistically possible combinations of say 40–60 amino acid residues, assumed to constitute the smallest domain. Whatever the size of the inventory, it is clear that the evolutionary variability of proteins arising from the shuffling of preformed domains is far more restricted than if each protein were the product of a single and unique evolutionary event.

What is the driving force for the selection and combination of domains in the biosynthesis of a protein? The functional necessity is clearly one of the most important elements and dictates the structural requirements for the expression of physiological function. This is particularly evident in multifunctional proteins, such as the coagulation proteins, plasminogen activator, the receptor proteins, the protein kinases and fibronectin, to mention a few. How does one rationalize, by comparison, the presence of apparently non-functional domains, such as the amino-terminal half of myosin light-chain kinase or the EGF domain in the coagulation proteases or in the LDL receptor? Are these evolutionary relics on the way out, so to speak, or harbingers of incoming biological functions?

There are even more fundamental unresolved questions. Consider the premise on which the evidence for homology* is based, i.e. the evidence that two proteins of similar sequences have arisen from a common ancestor by a divergent evolution. The rules of the game defining homology on the basis of sequence comparison [79] (protein or DNA) are generally understood and, for the most part, obeyed. However, translating linear sequences into three-dimensional conformations is a fundamentally different problem. While the dogma that the linear amino acid sequence determines the three-dimensional conformation is unchallenged, the mechanism and rules of protein folding are not understood. In the extreme case, two proteins may have similar domains but mutational events may have erased any statistically significant sequence similarity as, for instance, in the case of the *cro* and *CAP* proteins of *E. coli* [80]. And since molecular conformation is at least as important for the biological function of a protein as its amino acid sequence, if not more so, the acid test for homology requires comparison of the molecular structures of two or more proteins, in relation to their sequence.

This takes us back to the fundamental problem of protein folding and the mechanism of coalescence of the folding units of the nascent polypeptide chain into a domain. Probably the greatest challenge in protein chemistry and molecular biology is to understand the process which leads from the genomic DNA sequence to the three-dimensional conformation of the mature protein. An implicit problem is the organization of the gene itself, i.e. the exon/intron distribution in the genomic DNA and its relation to domains. Arguments have been advanced on both sides of the ledger, i.e. that there is, or is not, a relationship between gene organization and protein domains and it would be beyond the scope of this contribution to evaluate them. Clearly, the problem is unresolved, but I am inclined to agree with the recently expressed views of Gilbert [81], who concluded that the folding principles of proteins may become more apparent if we can understand the structure of the exon products and find the rules by which they are fitted together.

I believe that we are about to enter the 'Second Coming' of protein chemistry. The First Coming brought us the amino acid sequences and the three-dimensional conformations of many proteins. With the advent of DNA cloning and sequencing and the new methodologies of molecular biology, the popularity of protein chemistry waned until it became evident that the two disciplines are mutually dependent. The greatest challenge before us is to discover the principles of protein folding to the point where we can predict the conformation of a protein, or a protein domain, from the organization *and* the sequence of the constituent genomic DNA. If and when that goal is reached, a giant step will have been taken toward our understanding of protein evolution.

* Homology, in the strictest sense, denotes evolution from a common ancestor. It is often used, erroneously, as a synonym for similarity or partial sequence identity, regardless of whether the structures being compared have a common ancestor.

Acknowledgement

Part of the work described in this review was supported by a grant from the National Institutes of Health (GM-15731).

References

1. *Protein Folding* (ed. R. Jaenicke). Elsevier/North Holland Biomedical Press, Amsterdam, New York (1980).
2. Chothia, C., *Ann. Rev. Biochem.* **53**, 537 (1984).
3. Wodak, S. J. and Janin, J., *Biochemistry* **20**, 6544 (1981).
4. Lee, B. K. and Richards, F. M., *J. Mol. Biol.* **55**, 379 (1971).
5. Wetlaufer, D., *Proc. Nat. Acad. Sci., USA* **70**, 697 (1973).
6. Lee, R. H. and Rose, G. D., *Biopolymers* **24**, 1613 (1985).
7. Fanning, D. W. and Rose, G. D., *Proc. 9th American Peptide Symposium*, Pierce Chemical Co., Rockford, Ill. (1985).
8. Phillips, D. C., *Sci. Am.*, November, p. 78 (1966).
9. Rossman, M. and Liljas, A. J., *Mol. Biol.* **85**, 177 (1974).
10. Wetlaufer, D., Rose, G. D. and Taaffe, L., *Biochemistry* **15**, 5154 (1976).
11. Rose, G. D., *J. Mol. Biol.* **134**, 447 (1979).
12. Neurath, H., in *Protein Folding* (ed. R. Jaenicke), p. 501. Elsevier/North Holland Biochemical Press, Amsterdam, New York (1980).
13. Reeck, G. R., in *Chromosomal Proteins and Gene Expression* (ed. G. R. Reeck, G. H. Goodwin and P. Puisdomenick). Plenum Press, New York (1985).
14. Dalzoppo, D., Vita, C. and Fontana, A., *J. Mol. Biol.* **182**, 331 (1985).
15. Linderstrom-Lang, K. U. and Ottsen, M. C. R., *Trav. Lab. Carlsberg* **26**, 403 (1947).
16. Neurath, H. and Walsh, K. A., *Proc. Natl. Acad. Sci. USA* **73**, 3825 (1976).
17. *Precursor Processing in the Biosynthesis of Proteins* (ed. M. Zimmerman, R. A. Mumford and D. F. Steiner). *Ann. N.Y. Acad. Sci.* 343 (1980).
18. Mortensen, S. B., Sottrup-Jensen, L., Hansen, H. F., Petersen, T. E. and Magnusson, S., *FEBS Lett.* **135**, 295 (1981).
19. Walsh, K. A., Ericsson, L. H., Parmelee, D. C. and Titani, K., *Ann. Rev. Biochem* **50**, 317 (1981).
20. Klenow, H. and Hennigen, I., *Proc. Natl. Acad. Sci. USA* **65**, 168 (1970).
21. Ollis, D., Brick, P., Hamlin, R., Xuong, N. G. and Steitz, T. A., *Nature (London)* **313**, 762 (1985).
22. Patthy, L., Trexler, M., Vli, C., Banyai, L. and Varadi, A., *FEBS Lett.* **171**, 131 (1984).
23. Esmon, N. L., DeBault, E. and Esmon, C. T., *J. Biol. Chem.* **258**, 5548 (1983).
24. Stanley, K. K., Kocher, H. P., Luzio, J. P., Jackson, P. and Tschopp, J., *EMBO J.* **4**, 375 (1985).

25. Neurath, H., *Science* **224**, 350 (1984).
26. *Federation Proc.* (in the press).
27. MacDonald, R. J., Swift, G. H., Quinto, C., Swain, W., Pictet, R. L., Nikovits, W. and Rutter, W. J., *Biochemistry* **21**, 1453 (1982).
28. Craik, C. S., Choo, Q. L., Swift, G. H., Quinto, C., MacDonald, R. J. and Rutter, W. J., *J. Biol. Chem.* **259**, 14255 (1984).
29. Swift, G. H., Craik, C. S., Stary, S. J., Quinto, C., Lahaie, R. G., Rutter, W. J. and MacDonald, R. J., *J. Biol. Chem.* **259**, 14271 (1984).
30. Bell, G. I., Quinto, C., Quiroga, M., Valenzuela, P., Craik, C. S. and Rutter, W. J., *J. Biol. Chem.* **259**, 14265 (1984).
31. Hartley, B. S. and Shotton, D. M., in *The Enzymes* 3rd edition (ed. P. Boyer), vol. 3, p. 323 (1971). Academic Press, New York, N.Y.
32. Richardson, J. S., *Adv. Prot. Chem.* **34**, 168 (1981).
33. Roger, J., *Nature (London)* **315**, 458 (1985).
34. Craik, C. S., Rutter, W. J. and Fletterick, R., *Science* **220**, 1125 (1983).
35. Davie, E. W., Fujikawa, K., Kurachi, K. and Kisiel, W., *Adv. Enzymol.* **48**, 277 (1979).
36. Fass, D. N., Hewick, R. M., Knutson, G. J., Nesheim, M. E. and Mann, K. G., *Proc. Natl. Acad. Sci. USA* **82**, 1688 (1985).
37. Kisiel, W. and Fujikawa, K., *Behring Inst. Mitt.* **73**, 29 (1983).
38. Katayama, K., Ericsson, L. H., Enfield, D. L., Walsh, K. A., Neurath, H. and Titani, K., *Proc. Natl. Acad. Sci. USA* **76**, 4990 (1979).
39. McMullen, B. A. and Fujikawa, K., *J. Biol. Chem.* **260**, 5328 (1985).
40. Church, W. R., Jurnigan, R. L., Toole, J., Hewick, R. M., Knopf, J., Knutson, G., Nesheim, M. E., Mann, K. G. and Fass, D. N. *Proc. Natl. Acad. Sci. USA* **81**, 934 (1984).
41. Vehar, G. A., Keyt, B., Eaton, D., Rodriguez, H., O'Brien, D. P., Rotblat, F., Opperman, H., Keck, R., Wood, W. I., Harkins, R. N., Tuddenham, E. G. D., Lawn, R. M. and Capon, D. J., *Nature (London)* **312**, 337 (1984).
42. Gitschier, J., Wood, W. I., Goralka, T. M., Wion, W. L., Chen, E. Y., Eaton, D. H., Vehar, G. A., Capon, D. J. and Lawn, R. M., *Nature (London)* **312**, 326 (1984).
43. Kisiel, W. J., *Clin. Invest.* **64**, 761 (1979).
44. Fernlund, P. and Stenflo, J., *J. Biol. Chem.* **257**, 12170 (1982).
45. Foster, D. and Davie, E. W., *Proc. Natl. Acad. Sci. USA* **81**, 4766 (1984).
46. Højrup, P., Jensen, M. S. and Petersen, T. E., *FEBS Lett.* **184**, 333 (1985).
47. Jackson, K. W. and Tang, J., *Biochemistry* **21**, 6020 (1982).
48. Dahlbäck, B., Lundwall, A. and Stenflo, J., *Abstracts XIIth Int. Congress on Thrombosis and Haemostasis* **331**, 56 (1985).
49. Fujikawa, K. and McMullen, B. A., *J. Biol. Chem.* **258**, 10924 (1983).
50. Pennica, D., Holmes, W. E., Kohr, W. J., Harkins, R. N., Vehar, G. A., Ward, C. A., Bennett, W. F., Yelverton, E., Seeburg, P. H., Heynecker, H. L. and Goeddel, D. V., *Nature (London)* **301**, 214 (1983).
51. Pedersen, T. E. and Skorstengaard, K., in *Fibronectin, Its Role in Coagulation and Fibrinolysis* (ed. J. McDonagh). Marcel Dekker Inc., New York (1984).
52. Kornblihtt, A. R., Umezawa, K., Vibe-Pedersen, K. and Baralle, E. F., *EMBO J.* 1755 (1985).
53. Sadler, J. E., Shelton-Inlocs, B. B., Sorace, J. M., Harlan, J. M., Titani, K. and Davie, E. W., *Proc. Natl. Acad. Sci. USA* (in the press).
54. Girma, J. P., Chopek, M. W., Titani, K. and Davie, E. W., *Biochemistry* (in the press).
55. Ohno, S., Emori, Y., Imajoh, S., Kawasaki, H., Kisaragi, M. and Suzuki, K., *Nature (London)* **312**, 566 (1984).
56. Sakihama, T., Kakidani, H., Zenita, K., Yumoto, N., Kikuchi, T., Sasaki, T., Kannagi, R., Nakanishi, S., Ohmori, M., Takio, K., Titani, K. and Murachi, K., *Proc. Natl. Acad. Sci. USA* (in the press).
57. Pike, L. J. and Krebs, E. G., in *The Receptors* (ed. P. M. Conn), vol. 4. Academic Press, New York (in the press).
58. Sefton, B. M. and Hunter, T., *Adv. Cyclic Nucleotide Protein Phosphorylation Res.* **18**, 195 (1984).
59. Takio, K., Wade, R. D., Smith, S. B., Krebs, E. G., Walsh, K. A. and Titani, K., *Biochemistry* **23**, 4207 (1984).
60. Hunter, T. and Cooper, J. A., *Adv. Cyclic Nucleotide Res.* **17**, 443 (1984).
61. Takio, K., Blumenthal, D. K., Edelman, A. M., Walsh, K. A., Krebs, E. G. and Titani, K., *Biochemistry* (in the press).
62. Reimann, E. M., Titani, K., Ericsson, L. H., Wade, R. D., Fischer, E. H. and Walsh, K. A., *Biochemistry* **23**, 4185 (1984).
63. Barker, W. C. and Dayhoff, M. O., *Proc. Natl. Acad. Sci. USA* **79**, 2836 (1982).
64. Weber, I. T., Takio, K., Titani, K. and Steitz, T. A., *Proc. Natl. Acad. Sci. USA* **79**, 7679 (1982).
65. Blumenthal, D. K., Takio, K., Edelman, A. M., Charbonneau, H., Titani, K., Walsh, K. A. and Krebs, E. G., *Proc. Natl. Acad. Sci. USA* **82**, 3187 (1985).
66. Fischer, E. H., *Bull. Instut. Pasteur* **81**, 7 (1983).
67. Edelman, A. M., Takio, K., Blumenthal, D. K., Hansen, R. S., Walsh, K. A., Titani, K. and Krebs, E. G., *J. Biol. Chem.* (in the press).
68. Scott, J. D., Fischer, E. M., Demaille, J. G. and Krebs, E. G., *Proc. Natl. Acad. Sci. USA* **82**, 4379 (1985).
69. Scott, J. D., Fischer, E. H., Takio, K., Demaille, J. G. and Krebs, E. G., *Proc. Natl. Acad. Sci. USA* **82** (in the press).
70. Resing, K. A., Dale, B. A. and Walsh, K. A., *Biochemistry* **24**, 4167 (1985).
71. Ullrich, A., Coussens, L., Hayflick, J. S., Dull, T. J., Gray, A., Tam, A. W., Lee, J., Yarden, Y., Libermann, T. A., Schlessinger, J., Downward, J., Mayes, E. L. V., Whittle, N., Waterfield, M. D. and Seeburg, P. J., *Nature (London)* **309**, 418 (1984).
72. Ullrich, A., Bell, J. R., Chen, E. Y., Herrera, R., Petruzelli, L. M., Dull, T. J., Gray, A., Coussens, L., Liao, Y. C., Tsubokawa, M., Mason, A., Seeburg, P. H., Grunfeld, C., Rosen, O. M. and Ramachandran, J., *Nature (London)* **313**, 756 (1985).
73. Williams, A. F., *Nature (London)* **308**, 108 (1984).
74. Suedhof, T. C., Goldstein, J. L., Brown, M. S. and Russell, D. W., *Science* **228**, 815 (1985).
75. DeScipio, R. G., Gehring, M. R., Podack, E. R., Kan, C. C., Hugli, T. E. and Fey, G. H., *Proc. Natl. Acad. Sci. USA* **81**, 7298 (1984).
76. Wright, C. S., *J. Mol. Biol.* **111**, 439 (1977).
77. Porter, R. R. and Reid, K. B. M., *Adv. Prot. Chem.* **33**, 1 (1979).
78. Villiers, C. L., Arlaud, G. J. and Colomb, M. G., *Proc. Natl. Acad. Sci. USA* **82**, 4477 (1985).
79. Doolittle, R. F., *Science* **214**, 149 (1981).
80. Takeda, Y., Ohlendorf, D. H., Anderson, W. F. and Matthews, B. W., *Science* **221**, 1020 (1983).
81. Gilbert, W., *Science* **228**, 823 (1985).

Chemica Scripta 1986, **26B**, 231–236

Evolution of Isozymes and Different Enzymes in a Protein Superfamily: Alcohol Dehydrogenases and Related Proteins

Hans Jörnvall

Department of Chemistry I, Karolinska Institute, Box 60400, S–104 01 Stockholm, Sweden

Paper presented at the Conference on 'Molecular Evolution of Life', Lidingö, Sweden, 8–12 September 1985

Abstract

Alcohol dehydrogenases and related enzymes reveal successive changes at several different levels. First, alcohol dehydrogenase exhibits closely related isozymes of one class, with a high extent of residue conservation (> 90% positional identities). The corresponding subunits reflect recent and multiple gene duplications, giving largely species-specific isozyme patterns, and explaining unique properties. At a second level, isozymes of different classes show larger differences, reflecting more distant duplications, and giving forms more common among species. Third, different enzymes – alcohol and polyol dehydrogenases – show clear but more distant similarities (about 25% positional identities), reflecting further duplications. These forms are divided into two evolutionary lines. They also reveal functional convergence in the sense that enzymes with largely different structures and mechanisms work with similar substrates. Finally, they reflect domain and subdomain building units in the architecture.

Introduction

Alcohol dehydrogenases and related enzymes illustrate several principles of protein evolution. On the distant side, they illustrate use of domain-building units in different combinations, early divergence of activities in parallel evolutionary lines [1] and family assignments of enzymes [2]. On the more detailed side, they illustrate exceptionally well isozyme evolution, and allow close structure–function correlations [3, 4]. Several stages of gene duplication can be traced, also during recent divergence [5].

Information on new structures and the genetic organization is rapidly increasing and the field is expanding. However, patterns are already emerging. Presently discernible principles and variations are now summarized, starting with recent details and isozymes, and proceeding to more distant relationships.

Isozyme evolution

Alcohol dehydrogenase, aldehyde dehydrogenase, and sorbitol dehydrogenase are enzymes that illustrate different stages of isozyme evolution [6]. By duplication and accumulation of point mutations, successive divergence is noticed, leading to new enzyme activities. This is summarized in Fig. 1. For comparison, globins and serine proteases are also shown in relation to the dehydrogenases, in order to emphasize the principle. Detailed analysis of human alcohol dehydrogenases (below) reveals several stages that support the general outline

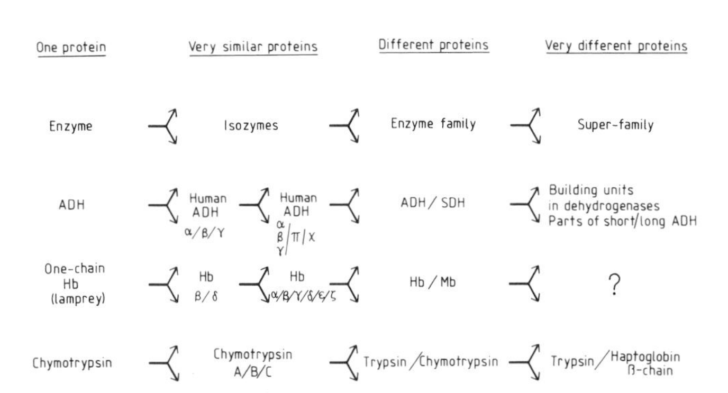

Fig. 1. Representation of isozymes and new enzymes as forms with successive, evolutionary changes. Emphasis on family relationships and divergence. Each arrow represents gene duplication(s) and subsequent accumulation of point mutations. Positions of alcohol dehydrogenases and related enzymes discussed are indicated, as well as well-known globins and serine proteases for comparison. ADH, Alcohol dehydrogenase, SDH, sorbitol dehydrogenase; Hb, hemoglobin; Mb, myoglobin.

suggested in Fig. 1. However, the results also show that the stages are complex, that the distinction between isozymes and new enzymes are arbitrary, and that different conclusions may be obtained, depending on which property is considered.

Thus, the different classes of alcohol dehydrogenase are classically considered as isozymes (because they affect one substrate in common – ethanol), whereas the alcohol/sorbitol dehydrogenases are considered as different enzymes (not active on identical substrates). Anyway, both interclass alcohol dehydrogenase isozymes and the sorbitol/alcohol dehydrogenase pair structurally deviate considerably ([7] and unpublished). Similarly, the two aldehyde dehydrogenases (mitochondrial/cytosolic) structurally also deviate considerably [8] but are enzymatically more or less equivalent, as enzymes towards acetaldehyde. Consequently, there is no clear distinction between isozymes and enzymes. Similarly, there is no direct coupling between the extent of structural divergence and that of functional divergence. The important question is: which residues are exchanged (functional or non-functional) – not how many.

Human alcohol dehydrogenase

Class I. This enzyme system has been divided into three different classes, with subunits that differ considerably in

Table I. *All positions with residue differences between class I major subunits of type* α, β_1 *and* γ_1.
(Boxed residues are unique at each position. Allelic variants β_1/β_2 and γ_1/γ_2 are not listed. Data from [5, 17, 18, 20].

Position	α	β_1	γ_1	Position	α	β_1	γ_1
17	Leu	Val	Leu	134	Arg	Gly	Gly
25	Glu	Asp	Glu	141	Leu	Leu	Val
34	His	Tyr	His	143	Ile	Thr	Val
43	Val	Val	Ala	185	Asn	Asn	Lys
47	Gly	Arg	Arg	207	Ala	Ala	Val
48	Thr	Thr	Ser	208	Ile	Val	Val
50	Asp	Asp	Glu	276			
56	Thr	Asn	Asn	297	Asp	Ala	Asp
57	Met	Leu	Leu	303	Met	Ile	Ile
93	Ala	Phe	Phe	318	Ile	Val	Ile
94	Ile	Thr	Thr	319	Leu	Tyr	Phe
102	Ile	Val	Ile	327	Cys	Gly	Ser
116	Val	Leu	Leu	328	Val	Ile	Val
117	Ser	Gly	Gly	348	His	His	Asn
120	Gln	Arg	Arg	349	Val	Val	Ile
128	Ser	Arg	Arg	363	His	His	Arg
133	Arg	Arg	Ser	371	Ile	Val	Val
				373	Met	Thr	Thr

Sum of positions with unique residues: α, 17; β_1, 10; γ_1, 13.

electrophoretic mobility, immunological properties and substrate specificity [9]. Three types of subunit (α, β, and γ) in dimeric combinations constitute the class I isozymes and are derived from different genetic loci [10]. In addition, further polymorphism by allelic variants have been established for the β and γ subunits [10–12], expanding the number of different dimer combinations, and explaining the complex isozyme patterns observed. Several of these isozymes have been purified and kinetically characterized [13–16].

Structural analyses have been completed on the α_1 [17], β_1 [5, 18], β_2 [12, 19], and γ_1 [5, 20] subunits. These data show that α, β and γ subunits are closely related, with α somewhat more deviating and β somewhat less deviating from the average pattern (Table I), and with γ the form apparently most similar to the originally purified horse enzyme. In total, the α, β_1, β_2, γ_1 and γ_2 subunits differ at 35 of 374 positions.

The pattern is compatible with at least two separate but comparatively recent gene duplications, giving a minimum of three gene loci for α, β and γ, as shown in Fig. 2. The β_1/β_2 and γ_1/γ_2 allelic differences have also been studied. One point mutation (at position 47) defines β_1/β_2 and two point mutations [unpublished together with Höög *et al.*] the γ_1/γ_2 pair. Functionally, amino acid substitutions explain the enzymatic properties and can be utilized to further correlate structure–function relationships for the class I isozymes [12, 17–20, and unpublished together with *Eklund* et al.]. Furthermore, corresponding cDNA sequences have been determined [17, 21–23, Höög *et al.*, unpublished], and analysis of the gene structures have been started. In summary, class I isozymes reveal recent duplications, small differences and divergent functional correlations. All these properties are typical for closely related isozymes in any protein system.

Class II and class III isozymes. These proteins have also been analyzed [unpublished, together with Kaiser and Vallee *et al.*]. Although not yet completely determined, it is evident that these classes have structures deviating much further from each other and from the Class I isozymes than the α, β, and γ chains do from each other. However, class II and class III structures also have recognizable homology towards class I, although at a lower level (25–75% positional identity in various regions as deduced from present peptide data, versus the 93–94% positional identity already known for the divergence within class I). Consequently, it can be concluded that all subunits of all three classes are linked by ancestral gene duplications, and that those separating classes II and III are more distant than those within class I. This situation is summarized in Fig. 3. It therefore follows that human alcohol dehydrogenase exhibits two types of isozyme divergence, both giving multiple forms. One with more recent duplications and less accumulation of subsequent point mutations (the divergence within class I, and possibly also within the other classes although this is yet incompletely known), and one with more distant duplications and a greater accumulation of point mutations (the interclass divergence). These structural observations support and explain the interpretations made from enzymatic and physical properties [9, 24, 25].

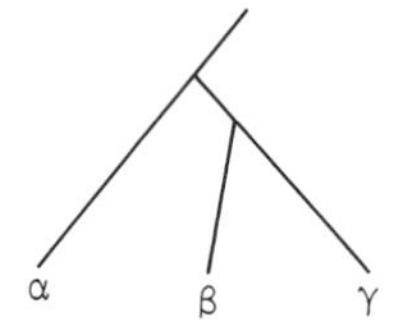

Fig. 2. The duplications corresponding to the three types of subunit (α. β and γ) representing the class I human alcohol dehydrogenase isozymes. Exact branch patterns are unknown, but are arbitrarily drawn to reflect the extent of present residue differences between the three characterized human forms.

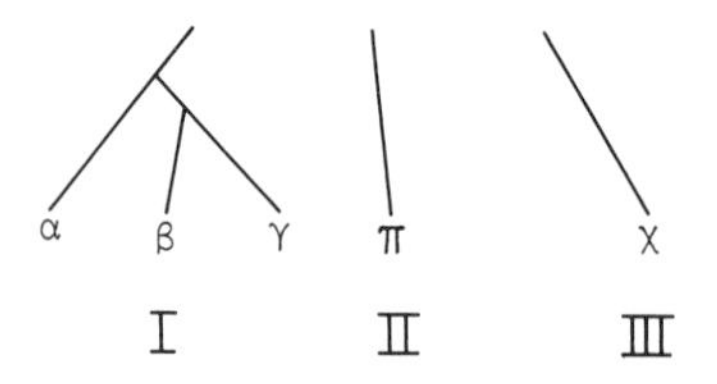

Fig. 3. The duplications corresponding to the three types of subunit class (I, II and III) representing the different classes of human alcohol dehydrogenase isozymes. Exact branch patterns are unknown, but are arbitrarily drawn as in Fig. 2 to reflect the extent of residue difference between the various forms. Additional duplications in each of the lines are not excluded. In fact, data support further duplications, but corresponding structures are yet unsufficiently known to be represented in this figure.

Table II. *Residues in human α, β_1 and γ_1 enzyme subunits and horse E and S enzyme subunits at all positions where the S alternative is known and where any of the subunits differ.*
Unique residues in the horse doublets or human triplets are boxed. Entire structures of α, β_1, γ_1 and E are established, but for S only 51 positions have been analyzed (tentative data for position 115 and 363 are indicated by parentheses and two alternatives, respectively). Consequently, for complete judgement of the final pattern, the S structure is the limit [27]. At present, there appears to be no correlation between $\alpha/\beta/\gamma$ differences on the one hand and E/S differences on the other.)

	Horse		Man		
Position	E	S	α	β_1	γ_1
17	Glu	Gln	Leu	Val	Leu
93	Phe	Phe	Ala	Phe	Phe
94	Thr	Ile	Ile	Thr	Thr
101	Arg	Ser	Arg	Arg	Arg
102	Val	Val	Ile	Val	Ile
110	Phe	Leu	Tyr	Tyr	Tyr
115	Asp	(Ser)	Asp	Asp	Asp
116	Leu	Leu	Val	Leu	Leu
117	Ser	Ser	Ser	Gly	Gly
120	Arg	Arg	Gln	Arg	Arg
363	Arg	Arg/Lys	His	His	Arg
366	Glu	Lys	Lys	Lys	Lys

Sum of these positions with unique residues E/S, 6; $\alpha/\beta/\gamma$, 8.

Isozymes of alcohol dehydrogenase in other species

The structures of horse-liver alcohol dehydrogenase isozymes have also been studied. From enzymatic properties, two isozyme chains have been deduced, called E (for activity versus ethanol) and S (for activity versus steroid alcohols). Structural investigations by peptide mapping have revealed at least six differences between these two subunits [26]. Since both subunits occur in essentially all livers but in different ratios, they have been deduced also to represent a gene duplication [27]. Comparisons of the structures of the human and horse isozymes at the positions where the intra-species isozyme variations occur, reveal that the positional variations this far known are largely different between horse and man, as shown in Table II. Because all isozymes are not completely analyzed in either species, interpretations should not yet be carried into too much detail. However, the present structural data appear to suggest that the origin of the E and S types of subunit in the horse enzyme may be different from both the intra-class-I duplications in man and the interclass duplications in any species. If so, the E/S divergence would be derived from another recent gene duplication in the ancestral line of the horse, followed by conventional accumulation of single point mutations. Consequently, the structures of the horse-liver isozymes may further support the conclusions from those of the human liver isozymes that several and separate gene duplications have occurred. The combined results also show that alcohol dehydrogenase is one of few enzymes where different gene duplications within recent mammalian radiations can now be traced.

Summary of isozyme patterns

Structural data reveal that the characterized E and S types of subunit in the horse enzyme form one family of closely related polypeptide chains, derived from a recent gene duplication. Similarly, the α, β and γ subunits of the human enzyme constitute one family also derived from recent gene duplications, of which at least two have been discerned. These isozyme families show members with closely related structures (positional identities above 90%). Obviously, they correspond to typical, closely related isozymes expected for the first type of divergence shown in Fig. 1.

At the same time, the classes I, II, and III polypeptide structures in man show that families also exist at a more distant level with interrelationships well below the 90% structural identity level. The corresponding duplications are of considerably older origins and are therefore presumably common to the pattern in different mammalian species. Although isozyme patterns are still incompletely investigated in most species, the present conclusions are compatible with the known occurrence of class II and/or class III type isozymes also in other mammalian species [28–33]. The interclass family relationships therefore illustrate isozyme divergence to a further stage, at which new enzyme functions are emerging. This is also compatible with the known large differences in enzymatic properties between these isozymes, approaching the step of new enzymes in Fig. 1, and associated with a large structural divergence.

Alcohol-polyol dehydrogenases

Evolution of new enzymes. The structure of sorbitol dehydrogenase [34] from mammalian liver has been shown to be distantly but clearly related to the alcohol dehydrogenase structures [1, 7]. However, in this case, the overall identity is one stage lower, about 25% [4, 7]. At the same time, a new enzyme activity, polyol dehydrogenase has emerged, with concomitant lack of ethanol dehydrogenase activity. Consequently, the sorbitol dehydrogenase structure reveals evolution of a new enzyme type.

It is concluded that the intra-class variations of alcohol dehydrogenases, the interclass variations of alcohol dehydrogenases, and the alcohol–polyol dehydrogenase variations, all in mammalian liver, illustrate three different levels of evolution of new enzyme activities corresponding to different stages in Fig. 1. The intra-class variations show structural identities above about 90%, the interclass variability shows structural identity well below that level but still above 25%, and the inter-enzyme variability shows structural identities around 25%. Consequently, the successive changes suggested in Fig. 1 are extremely well demonstrated by the different types of alcohol dehydrogenase and polyol dehydrogenase.

At the same time as these correlations are emphasized between enzymatic and structural properties, it should be noticed that at all three levels, the interspecies variations are extensive, and therefore considerably blur the general pattern just outlined. Thus, the intertype relationship between alcohol and polyol dehydrogenases, at about the 25% structural identity level [7], is no more distant than the interspecies variation between different alcohol dehydrogenases when sufficiently divergent members are investigated, such as yeast and mammalian alcohol dehydrogenases [3, 35]. In fact, mammalian polyol dehydrogenases have several features, including the quaternary structure [1] and positions of internal chain deletions [4] that make them seem more similar to yeast alcohol dehydrogenase than yeast and mammalian alcohol dehydrogenases are between themselves. This illustrates that the rate of evolutionary change for alcohol dehydrogenase is considerable, and greater than that for several other dehydrogenases investigated in similar species (glyceraldehyde 3-phosphate dehydrogenase, lactate dehydrogenase). This higher rate of divergence for alcohol dehydrogenase may probably be linked to the divergent substrate specificity [36] and to less strict functional roles as compared to the other dehydrogenases which are locked in specific metabolic chains.

Enzyme nomenclature–distinction of one of several protein properties. The comparisons of enzymatic and structural properties also demonstrate a difficulty in enzyme classification when structures rather than activities are used as the basis of distinction. Thus, whereas sorbitol and alcohol dehydrogenases are distinguished as different enzymes and given different enzyme numbers in conventional nomenclature (EC 1.1.1.1 for alcohol dehydrogenase and EC 1.1.1.14 for sorbitol dehydrogenase), all alcohol dehydrogenases are given the same number (EC 1.1.1.1). However, the structural data reveal that the degree of identity is as high or higher between sorbitol dehydrogenase and alcohol dehydrogenase of different enzyme numbers as between yeast and mammalian alcohol dehydrogenases of the same number [3, 4, 7, 35].

This apparent contradiction is no real problem, but illustrates the somewhat arbitrary nature of enzyme classification. Thus, in enzyme classification, essentially only one property of proteins is utilized, i.e. the enzyme activity. However, proteins have several other properties that also exhibit successive changes in the same way as those in Fig. 1. The distinction between isozymes and new enzymes in Fig. 1 only concerns activities towards substrates. Consequently, proteins considered as closely related isozymes when one property is emphasized (i.e. the enzymatic activity) need not be equally closely related when another property is considered (like quaternary structure, positional homology, or other structural features).

Aldehyde dehydrogenases

The point about non-equal deviations in structure and function is further illustrated by recently characterized aldehyde dehydrogenases. Although this far without known obvious structural relationships to alcohol dehydrogenases, the cytosolic and mitochondrial aldehyde dehydrogenases [8, 37] can now be compared. It is evident then, that they represent quite divergent isozyme structures, showing only 68% structural identity between the two forms within one species [8]. Consequently the corresponding gene duplication is concluded to be comparatively distant. This structural conclusion is compatible with the fact that the isozyme types of aldehyde dehydrogenase are of similar occurrence in several species [38, 39]. These observations are also compatible with the fact that the two types of isozyme of aldehyde dehydrogenase do not cross-hybridize, suggesting considerable structural divergence, as pointed out before [6]. At the same time, the enzymatic activity is largely conserved, and the two types of aldehyde dehydrogenase are clearly distinguished as one common enzyme type with the same classification number.

Therefore, aldehyde dehydrogenases illustrate, exactly like the alcohol and polyol/alcohol dehydrogenases above, that the classification of divergence (Fig. 1) overemphasizes one property of proteins, and that changes are successive. Considering only the enzyme activity, aldehyde dehydrogenases are, like interclass alcohol dehydrogenases, only divergent to a minor extent. However, considering other properties, like structural divergence and several other physico-chemical characteristics, 'typical' isozymes instead behave like new protein. This is in no way contradictory, but supports the concept in Fig. 1 of gradual changes from an isozyme to new enzymes. Depending on which property is considered, stages in the evolution may be differently interpreted.

Functional convergence

Apart from demonstrating successive evolutionary changes, and the divergence of isozymes/new enzymes, alcohol dehydrogenases and polyol dehydrogenases also illustrate functional convergence. Thus, although most characterized alcohol dehydrogenases can be distinguished as clearly related, showing distinct structural homology and similar enzymatic mechanisms, with zinc at the active site, there is also another type of alcohol dehydrogenase in nature. This includes Drosophila alcohol dehydrogenase, which is an alcohol dehydrogenase but has few superficial properties in common with characterized mammalian, yeast, plant or bacterial alcohol dehydrogenases [1, 40, 41].

The Drosophila alcohol dehydrogenase has a considerably smaller subunit (approximately two-thirds in size), is apparently working without zinc at the active site, has other enzymatic mechanisms, and none of the structural features typical of the active site of alcohol dehydrogenase. It may be concluded that evolution of alcohol dehydrogenase activity has occurred twice (Fig. 4), creating one well-spread enzyme family with 'long' alcohol dehydrogenases, working with zinc at the active site, and represented by the enzyme in yeast, plants and mammals. The other type, represented by 'short' enzymes, works without zinc at the active site, and is illustrated by the insect enzyme type [1]. Consequently, the two separate origins demonstrate a case of functional convergence towards a common enzyme activity from largely different solutions of enzyme geometry and mechanism.

Furthermore, this convergence in enzymatic function is known not only in the case of alcohol dehydrogenases, but also in the case of polyol dehydrogenases. Thus, that activity is similarly split into two types of enzyme. Furthermore, each of the two types are also related to one each of the two types of alcohol dehydrogenase activity, illustrating parallel evolutionary changes (Fig. 4). Thus, sorbitol dehydrogenase is clearly related to mammalian alcohol dehydrogenase of the 'long', metalloenzyme type. Bacterial ribitol dehydrogenase is similarly related to the 'short', insect alcohol dehydrogenase type [1]. Subsequent to these initial distinctions, the corresponding enzyme types have been still further extended, and the bacterial glucose dehydrogenase is now known also to be a typical member of the short type of non-zinc dehydrogenase [42], as shown in Fig. 4.

Therefore, alcohol and polyol dehydrogenases illustrate both considerable structural divergence, as represented in Fig. 1, and functional convergence as illustrated in Fig. 5. These wide relationships are, as in the case of recent and repeated duplications for alcohol dehydrogenases, not yet that commonly found in other protein types.

At the same time as the extensive divergence between the

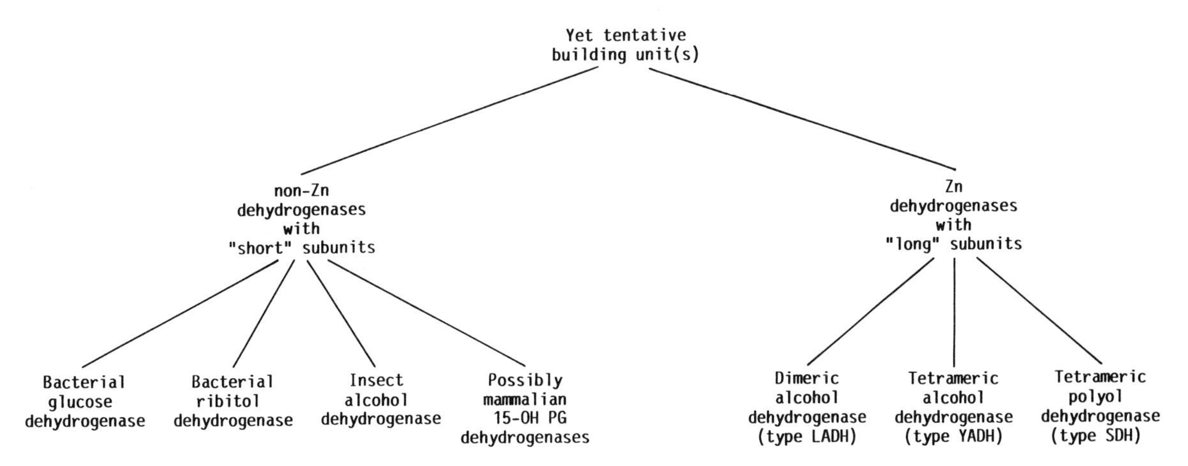

Fig. 4. Divergence of alcohol and polyol dehydrogenases into two evolutionary lines. Similar types of enzymes activity have evolved in both lines. One line works with 'short' subunits without zinc at the active site, the other with longer subunits and a catalytically active zinc atom. Data from [42].

two functionally convergent lines of Fig. 5 are emphasized, it should be remembered that the two lines also show some structural similarities. Thus, large parts of the structures constituting the coenzyme-binding domains have clearly discernible homology [42] but are now found in different parts of the molecule (N-terminal domain of the short type of dehydrogenase, but in the C-terminal half of the long type of alcohol dehydrogenase [2, 3, 41, 42]. This type of structural similarity, noticed for the coenzyme-binding structures in the two lines of polyol/alcohol dehydrogenases discerned in Fig. 5, represents a still more distant evolutionary connection. It reflects the common occurrence of similar building units with related fold in all characterized NAD-linked dehydrogenases and other enzymes [2]. Consequently, this type of relationship constitutes the last stage in Fig. 1, representing a super-family distinction among several widely divergent proteins.

Conclusion

The characterized forms of alcohol dehydrogenases and structurally or metabolically related enzymes illustrate all levels of enzyme relationships outlined in Fig. 1, and emphasize

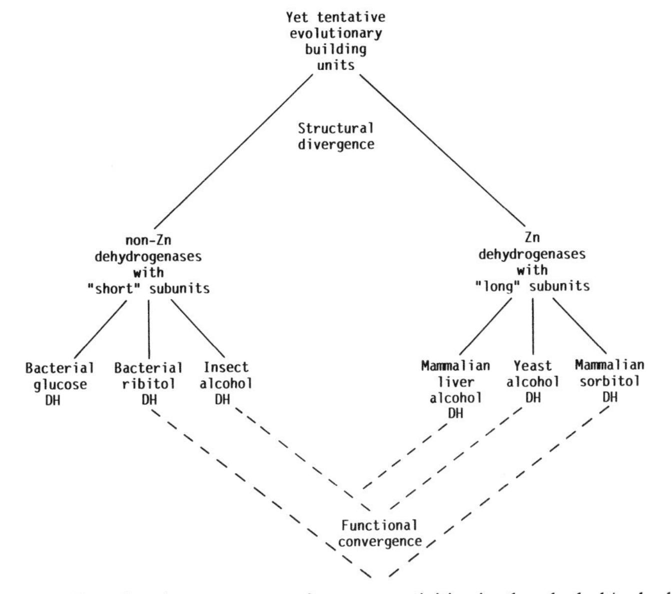

Fig. 5. Functional convergence of enzyme activities in the alcohol/polyol dehydrogenase lines. As shown in Fig. 4, alcohol/polyol dehydrogenases have evolved into two lines, working with different structures and mechanisms. Yet, as shown in the bottom half, several of the activities in both lines functionally converge on the same substrates.

successive changes. The short and long types of enzyme show partial similarities, restricted to regions of the molecules at non-identical positions in different polypeptides. This is a reflection of super-family build-up of proteins from domain or subdomain structures that reflect distant ancestral building units.

The two types of polyol and alcohol dehydrogenases, with short and long subunits, as well as the aldehyde dehydrogenases with still longer chains and non-characterized detailed connections to the alcohol dehydrogenases, illustrate functional divergence. In the case of alcohol dehydrogenase, this divergence also offers an example of evolution of metalloenzymes versus non-metalloenzymes, in the sense that the alcohol and polyol dehydrogenases with long subunits work with zinc at the active site, whereas those with short subunits do not.

At more recent stages, the different classes and enzyme types form families, allowing distinction also of isozymes. Depending on which protein property is emphasized, isozymes may be more or less similar. There is no strict correlation between degree of structural divergence and divergence in enzyme activity. Some enzymes closely related in activity, like the interclass alcohol dehydrogenases or the two types of aldehyde dehydrogenase, are not equally closely related in structure, whereas other enzymes not considered isozymes, like the sorbitol and liver alcohol dehydrogenases, are structurally related.

Less divergent isozyme variabilities are illustrated by recent gene duplications and accumulation of point mutations in intra-class alcohol dehydrogenases. Finally, polymorphism at the allelic level, from recent point mutations, is directly observed in Oriental and Caucasian types of alcohol dehydrogenase [12, 19] in human populations.

In summary, alcohol dehydrogenases illustrate evolutionary changes created by single point mutations, as well as by exchange of most structural features. The correlation with function should be emphasized. Alcohol dehydrogenase and related enzymes reflect a comparatively high rate of evolutionary change. Correspondingly, it may be significant that alcohol dehydrogenase has no known, strictly functional, single role. The metabolic function is presumably broad and variable between species. Interestingly, there is an apparently high rate of evolutionary change, not only in relation to point mutations but also in relation to gene duplications in the alcohol dehydrogenase line. This fact indicates that gene duplications are frequent, but it may also indicate that duplications, like point mutations, could be subject to some functional constraints.

Acknowledgement

Collaborations in structural analyses, as evident from the references quoted, are gratefully acknowledged, as well as help by Carina Palmberg in preparation of the figures. The studies were supported by grants from the Swedish Medical Research Council (project 03X–3532), the Knut and Alice Wallenberg Foundation, the Swedish Cancer Society (project 1806) and the Karolinska Institute.

References

1. Jörnvall, H., Persson, M. and Jeffery, J., *Proc. Natl. Acad. Sci. USA* **78**, 4226 (1981).
2. Rossmann, M. G., Liljas, A., Brändén, C.-I. and Banaszak, L. J., in *The Enzymes* (ed. P. D. Boyer), vol. 11, p. 61. Academic Press, New York (1975).
3. Jörnvall, H., Eklund, H. and Brändén, C.-I., *J. Biol. Chem.* **253**, 8414 (1978).
4. Eklund, H., Horjales, E., Jörnvall, H., Brändén, C.-I. and Jeffery, J., *Biochemistry* **24**, 8005 (1985).
5. Hempel, J., Holmquist, B., Fleetwood, L., Kaiser, R., Barros-Söderling, J., Bühler, R., Vallee, B. L. and Jörnvall, H., *Biochemistry* **24**, 5303 (1985).
6. Jörnvall, H., in *Dehydrogenases Requiring Nicotinamide Coenzymes* (ed. J. Jeffery), p. 126. Birkhäuser Verlag, Basel (1980).
7. Jörnvall, H., von Bahr-Lindström, H. and Jeffery, J., *Eur. J. Biochem.* **140**, 17 (1984).
8. Hempel, J., Kaiser, R. and Jörnvall, H., *Eur. J. Biochem.* **153**, 13 (1985).
9. Vallee, B. L. and Bazzone, T. J., in *Isozymes: Current Topics in Biological and Medical Research* (ed. M. C. Rattazzi, J. G. Scandalios and G. S. Whitt), vol. 8, p. 219. Liss, New York (1983).
10. Smith, M., Hopkinson, D. A. and Harris, H., *Ann. Hum. Genet.* **34**, 251 (1971).
11. Bosron, W. F., Magnes, L. J. and Li, T.-K., *Biochem. Genet.* **21**, 755 (1983).
12. Jörnvall, H., Hempel, J., Vallee, B. L., Bosron, W. F. and Li, T.-K., *Proc. Natl. Acad. Sci. USA* **81**, 3024, (1984).
13. Bosron, W. F., Magnes, L. J. and Li, T.-K., *Biochemistry* **22**, 1852 (1983).
14. Wagner, F. W., Burger, A. R. and Vallee, B. L., *Biochemistry* **22**, 1857 (1983).
15. Dreetz, J. S., Luehr, C. A. and Vallee, B. L., *Biochemistry* **23**, 6822 (1984).
16. Yin, S.-J., Bosron, W. F., Magnes, L. J. och Li, T.-K., *Biochemistry* **23**, 5847 (1984).
17. von Bahr-Lindström, H., Höög, J.-O., Hedén, L.-O., Kaiser, R., Fleetwood, L., Larsson, K., Lake, M., Holmquist, B., Holmgren, A., Hempel, J., Vallee, B. L. and Jörnvall, H., *Biochemistry* (in the press).
18. Hempel, J., Bühler, R., Kaiser, R., Holmquist, B., de Zalenski, C., von Wartburg, J.-P., Vallee, B. L. and Jörnvall, H., *Eur. J. Biochem.* **145**, 437 (1985).
19. Bühler, R., Hempel, J., von Wartburg, J.-P. and Jörnvall, H., *FEBS Lett.* **173**, 360 (1984).
20. Bühler, R., Hempel, J., Kaiser, R., de Zalenski, C., von Wartburg, J.-P. and Jörnvall, H., *Eur. J. Biochem.* **145**, 447 (1984).
21. Duester, G., Hatfield, G. W., Bühler, R., Hempel, J., Jörnvall, H. and Smith, M., *Proc. Natl. Acad. Sci. USA* **81**, 4055 (1984).
22. Ikuta, T., Fujiyoshi, T., Kurachi, K. and Yoshida, A., *Proc. Natl. Acad. Sci. USA* **82**, 2703 (1985).
23. Hedén, L.-O., Höög, J.-O., Larsson, K., Lake, M., Lagerholm, E., Holmgren, A., Vallee, B. I., Jörnvall, H. and von Bahr-Lindström, H., *FEBS Lett.* **194**, 327 (1986).
24. Ditlow, C. C., Holmquist, B., Morelock, M. M. and Vallee, B. L., *Biochemistry*, **23**, 6363 (1984).
25. Wagner, F. W., Parés, X., Holmquist, B. and Vallee, B. L., *Biochemistry* **23**, 2193 (1984).
26. Jörnvall, H., *Nature (London)* **225**, 1133 (1970).
27. Jörnvall, H., *Eur. J. Biochem.* **16**, 41 (1970).
28. Holmes, R. S. Albanese, R., Whitehead, F. D. and Duley, J. A., *J. Exp. Zool.* **215**, 151 (1981).
29. Dafeldecker, W. P., Parés, X., Vallee, B. L., Bosron, W. F. and Li, T.-K., *Biochemistry* **20**, 856 (1981).
30. Dafeldecker, W. P., Meadow, P. E., Parés, X. and Vallee, B. L. *Biochemistry* **20**, 6729 (1981).
31. Dafeldecker, W. P. and Vallee, B. L., *J. Prot. Chem.* **1**, 59 (1982).
32. Algar, E. M., Seeley, T.-L. and Holmes, R. S., *Eur. J. Biochem.* **137**, 139 (1983).
33. Seeley, T.-L., Mather, P. B. and Holmes, R. S., *Comp. Biochem. Physiol.* **78**B, 131 (1984).
34. Jeffery, J., Cederlund, E. and Jörnvall, H., *Eur. J. Biochem.* **140**, 7 (1984).
35. Jörnvall, H., *Eur. J. Biochem.* **72**, 443 (1977).
36. Jörnvall, H., *Biochem. Soc. Transact.* **5**, 636 (1977).
37. Hempel, H., von Bahr-Lindström, H. and Jörnvall, H., *Eur. J. Biochem.* **141**, 21 (1984).
38. Eckfeldt, J. H. and Yonetai, T., *Arch. Biochem. Biophys.* **175**, 717 (1976).
39. Greenfield, N. J. and Pietruszko, R. *Biochim. Biophys. Acta* **483**, 35 (1977).
40. Schwartz, M. F. and Jörnvall, H., *Eur. J. Biochem.* **68**, 159 (1976).
41. Thatcher, D. R. *Biochem. J.* **187**, 875 (1980).
42. Jörnvall, H., von Bahr-Lindström, H., Jany, K.-D., Ulmer, W. and Fröschle, M., *FEBS Lett.* **165**, 190 (1984).

Chemica Scripta 1986, **26B**, 237–240

The Factor VIII Gene and the Molecular Genetics of Hemophilia

Richard M. Lawn

Department of Molecular Biology, Genentech, Inc. 460 Point San Bruno Boulevard, South San Francisco, California 94080, USA

Paper presented at the Conference on 'Molecular Evolution of Life', Lidingö, Sweden, 8–12 September 1985

Abstract

Hemophilia A is caused by a deficiency of clotting factor VIII. The complete 9 kb human factor VIII cDNA and 186 kb gene have been cloned and characterized. DNA sequence analysis confirmed the large size of a single chain precursor protein, and revealed the internal triplication and duplication of protein domains. A large unique protein domain separates the repeated subunits and is entirely encoded by a 3 kb gene exon. Surprisingly, the triplicated protein domains of factor VIII are homologous to the copper-binding protein ceruloplasmin, which is related to a long line of blue copper proteins.

The cloned factor VIII gene has allowed the identification of the molecular defect causing hemophilia in some individuals, and has provided the means of DNA based antenatal diagnosis and carrier detection of the disease.

Introduction

Hemophilia is one of the first human genetic diseases to have been chronicled. Ancient records told of an affliction in certain families whose male children died after circumcision or otherwise failed to control bleeding. The writers of the Talmud glimpsed the inheritance pattern of this disease and waived circumcision of boys whose older brothers or cousins had died after the procedure. The disease they were describing is now called hemophilia. If untreated, severely affected hemophiliacs suffer painful and prolonged bleeding crises after minor bumps and bruises, and usually die at an early age from blood loss or the complications of internal hemorrhage. The most celebrated hemophilic family in history were the descendants of Queen Victoria. A defective gene causing this disease spread from the Queen herself throughout the royal families of Europe. The affliction of the son of Czar Nicholas II enhanced the influence of the infamous monk Rasputin upon the desperate parents and exacerbated the disastrous reign of the last Russian monarch.

Now that the cause of hemophilia has been pinpointed to the lack of specific blood-clotting factors, the disease can be treated with whole plasma, or more recently with impure concentrates of the missing factor. Over the past two decades such treatment has dramatically improved the lifestyle and life expectancy of hemophiliacs. However, current products are mostly prepared from blood pooled from large numbers of donors. As a result, hemophiliacs are exposed to the blood-borne viruses causing hepatitis and AIDS. Hence there has been a strong pharmaceutical as well as scientific motivation to clone the genes for the anti-hemophilic factors and produce these proteins via recombinant DNA techniques. The successful cloning of the factor VIII and IX genes has already provided insights regarding the structure and evolution of these proteins and enabled us to begin identifying the genetic lesions responsible for the various forms of the disease.

The clotting cascade

Blood clots form rapidly and locally as a result of serial amplification of initial signals by a complex network of inactive protease zymogens and cofactors, and multiple positive and negative affectors. The heart of this process is the intrinsic clotting cascade which begins with activation of prekallikrein and factor XII and finally results in the conversion of the soluble protein fibrinogen to fibrin, which forms a filamentous network that stabilizes the platelet plug. Located in the middle of this cascade are the serine protease factor IX and its large protein cofactor, factor VIII. The complex of activated factor IX, factor VIII, calcium ions, and a phospholipid surface activates factor X. The two common forms of hemophilia arise from a deficiency of either factor VIII ('classic' or hemophilia A, comprising ~85% of cases) or of factor IX ('Christmas disease' or hemophilia B). The genes for both of these clotting factors are located on the X chromosome, where a single defective allele will cause hemophilia in males. Female hemophiliacs have only rarely been reported. The X linkage of these factors contributes to the prevalence of these types of hemophilia and to the frequent appearance of newly arising hemophilia-causing mutations.

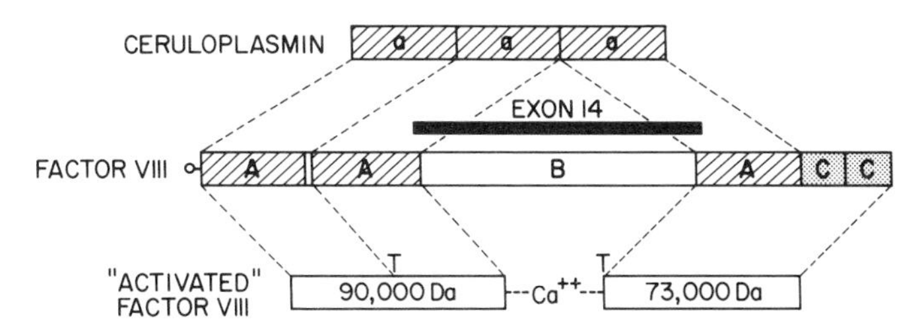

Fig. 1. Single-chain factor VIII, its active form and its relationship to ceruloplasmin. The centre bar depicts the 2351 amino acid single-chain factor VIII protein predicted by the cloned DNA sequence [1–5]. The amino terminus is at the left side, where the small line and circle represent the 19 amino acid hydrophobic leader. The internally homologous A and C domains and the homology of the three factor A domains to the copper binding protein ceruloplasmin (top line) are indicated. The bottom line represents the current model of factor VIII in its most highly activated form. 'T' indicates the sites of thrombin cleavage [6]. The two subunits are probably held together by calcium ions. The bar above the single-chain protein shows the extent of the molecule which is encoded by the unusually large exon 14 of the factor VIII gene.

The cloning of the factor VIII gene

Three properties of the factor VIII protein made the cloning and expression of the gene a difficult task: the extremely low abundance in plasma, its large size, and uncertainty about the site of synthesis. Limited protein sequence data was used to synthesize oligonucleotide probes for the identification of initial genomic clones [1, 2]. These clones derived from fragments of the human genome contained in phage lambda vectors. Eventually we obtained a series of overlapping lambda and cosmid clones which contained the entire gene [1]. Fragments of initial genomic clones were employed to identify sources of factor VIII mRNA from which to obtain cDNA clones. The entire 9000 bases (9 kb) of cDNA was connected to viral promoters and transfected into mammalian tissue culture cells. The cDNA directs the synthesis and secretion of biologically active factor VIII which corrects the clotting time of hemophilic plasma and which interacts with antibodies and proteins that affect plasma-derived factor VIII [2, 3]. The sequence of the cloned factor VIII gene and cDNA revealed some interesting features of the structure, function and evolution of the protein.

The factor VIII gene and protein

The continuous open reading frame of the cDNA sequence codes for a 2332 amino acid mature protein preceded by a 19-residue hydrophobic secretion signal peptide. This is large enough to code for the 330000 molecular weight form of the molecule reported by Tuddenham's group [4]. This settled one uncertainty about the protein. As isolated from plasma, factor VIII usually consists of about a dozen bands on SDS polyacrylamide gels ranging in size from 40000 to 210000. Only rarely has a larger molecule been seen. The initial cloning and subsequent analysis not only confirm the presence of a very large single-chain precursor, but have led to the identification of the smaller species as specific proteolytic activation and degradation products of the 330000 Da precursor [5, 6].

The large size of the protein is reflected in the gene. The factor VIII gene is comprised of 26 exons separated by 25 non-coding introns that span 186000 base pairs, which is about 0.1% of the human X chromosome [1]. It is the largest gene yet cloned and contains the largest exon (3106 bp) and introns (36000 bp) yet reported. The 3106 bp exon number 14 is unusually large; most exons are in the 75–200 bp range. Only about 5% of the gene is comprised of protein encoding exons, which is rather typical for the genes of vertebrates. Although it might take hours to transcribe such a large gene into RNA, there has been no effective selective pressure to reduce the relative intron size of this enormous gene.

The translated cDNA sequence provided several surprises. We found that factor VIII contains two types of internally repeated domains. An 'A' domain of −350 amino acids is repeated three times at about 30% amino acid homology and a different 'C' domain of 150 residues exists twice at about 35% homology [3, 5]. Shared cysteine residues contribute to the significance and structural similarity of these domains. It has been proposed that intragenic duplication of DNA sequences leads to this kind of protein domain repetition. The repeated domains are separated by a unique 'B' domain of about 1000 amino acids. The intermediate B domain is eliminated by a series of proteolytic cleavages during the activation of the molecule. The final 'activated' factor VIII probably consists of a 90000 Da protein (either intact or with one further internal cleavage) which derives from the amino terminal third of the precursor (consisting of two A domains) and a 73000 Da subunit which derives from the carboxy terminal region (consisting of one A and two C domains) [cf. 6].

Unexpectedly, the triplicated A domain bears 30% homology to what was presumed to be an unrelated protein, ceruloplasmin. Ceruloplasmin is a blue copper binding plasma protein of molecular weight 130000. It is comprised of three of the A-like domains of factor VIII, and little else. Ceruloplasmin binds six copper atoms and has an iron oxidizing activity.

Small blue copper proteins have existed for 3000 million years. Limited homology exists in the copper binding regions of the bacterial azurins and the plant plastocyanins. An ancestral oxidase may have been a precursor of the larger multicopper oxidases fungal laccase and ceruloplasmin about 1500 million years ago. A doubling of a 170-residue plastocyanin-like precursor might have prefigured the 350 residue units of the recent blue oxidases [7]. Although neither factor VIII nor its close cousin factor V have oxidase activity, factor V at least is known to bind a single copper [8]. There have been insufficient amounts of purified factor VIII protein to perform similar studies. The role of such metal ions in the stabilization or cofactor activity will have to be the subject of future experimentation.

The location of exon–intron boundaries can sometimes reflect the process of evolution by segmental DNA duplication. However, only some intron locations in the factor VIII gene are shared in its various homologous sections [1]. Intron

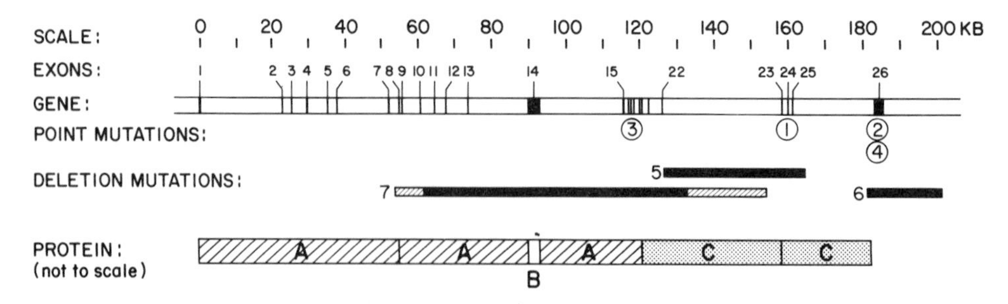

Fig. 2. The factor VIII gene and the location of known mutations. Below a size scale in KB (kilobases) is shown a representation of the human factor VIII gene [1]. Exons are indicated by filled regions of the horizontal bar and are numbered from the 5′ to 3′ end of the gene. Below the gene are indicated the location of known mutations in the factor VIII gene of hemophiliacs [11,12]. Circled numbers represent point mutations in exons 18, 24 and 26. Numbers 1–3 are chain termination mutations and 4 is an amino acid substitution mutation. Filled horizontal bars are partial gene deletions where the cross-hatched bar indicates uncertainty in the boundaries of the deletion. The bottom of the figure shows the corresponding regions of the protein encoded by the exons shown in the top line. Of necessity, the protein depiction is not to scale.

boundaries occur precisely at the borders of the two C repeats, as would be expected for a DNA duplication mechanism where the boundaries fell within the introns. The junction between the final A repeat and the C repeats is also marked by an intron. In contrast, the other repeat boundaries and some of the introns within the aligned repeat units are not in conserved locations. This might reflect the gain or loss of introns since the time of intragenic sequence duplications.

It is interesting to speculate as to the origin of the intermediate B domain. Its nearly 1000 amino acids encompass nearly all of the 3.1 kb exon 14, which also contains short regions of bordering A repeats. One could imagine that a large open reading frame, perhaps part of a processed gene which passed through an RNA intermediate, was inserted into an ancestral gene which predated factors VIII and V. This could account for the unusually large size of this exon. The intermediate region is also interesting because it is entirely removed during the activation of factor VIII by plasma proteases. It is also a region of markedly low homology of amino acid sequence between human and porcine factor VIII and human factor V (Jay Toole and David Fass, personal communication). Factors V and VIII are both 330000 Da proteins which occupy similar roles as cofactors of serine proteases in successive steps of the clotting cascade. They appear to be processed in roughly similar fashion to yield large N-terminal and C-terminal active subfragments [9, 10]. The factor V gene has yet to be cloned, but the similarities in the two proteins that are already apparent would indicate that the factor VIII and V genes will display many features of a relatively recent duplication and divergence.

Mutations in the factor VIII gene of hemophiliacs

Using the cloned factor VIII gene, it is now possible to discern the genetic basis of some hemophilias. Several clinical observations suggest that hemophilia A results from a heterogeneous collection of mutations. First, despite the fact that hemophilia can be fatal without proper treatment, the disease persists in human populations; this implies that new mutations continuously arise. (Fifty years ago Haldane proposed this pattern for all X-linked lethal diseases). Second, hemophilia victims display a wide spectrum of severity, possibly due to different types of factor VIII mutations. Symptoms range from mild to severe, and roughly 10% of hemophiliacs raise inhibiting anti-factor VIII antibodies.

Preliminary results of hemophilic DNA screening confirm this idea. DNA has been prepared from blood samples of about 200 hemophilia A patients and analyzed by Southern blot hybridization with factor VIII probes [11, 12]. To date, seven distinct mutations in the factor VIII gene of hemophiliacs have been characterized. The first apparent point is that each of the seven mutations has only been seen in a particular patient (and immediate family members) and has not yet been seen in the DNA of non-related hemophiliacs examined. Three of the mutations consist of partial gene deletions, the type most easily detected by blot hybridization. A small proportion of point mutations can also be detected by the judicious use of restriction enzymes. The dinucleotide CG is the major site of methylation of human DNA, and methyl cytosine is subject to conversion to thymidine by deamination. Thus CG dinucleotides may be hot spots for mutations. Restriction enzymes like *TaqI* which contain CG in their recognition sequence might be particularly useful in searching for mutations and restriction fragment length polymorphisms [13]. The factor VIII gene contains five *TaqI* sites that would change from arginine to stop codons if this predicted TCGA to TTGA transition in the recognition site occurred. Examples of three of these nonsense mutations have now been found. They are located in exons 18, 24 and 26 of the gene. The resulting truncated factor VIII proteins are presumed to be relatively inactive and/or unstable, and are the likely cause of the severe hemophilia in these three patients. The exon 24 mutation example is of particular interest because it clearly demonstrates a *de novo* hemophilia-causing mutation. Southern blotting shows that the altered *TaqI* site is only contained in the factor VIII gene of the hemophiliac, but not in his unaffected two brothers, sister, and parents. He is also the only reported hemophiliac in his family, and produces large amounts of inhibiting anti-factor VIII antibodies [11]. We have more recently discovered a missense mutation in a mild hemophiliac affecting the same *TaqI* site in exon 26 which led to a nonsense mutation in one of the severe hemophiliacs (Gitschier and Lawn, manuscript in preparation). The methylation mechanism could have led to a C to T transition in the anti-sense strand which would convert the TCGA sequence at the *TaqI* site to TCAA and result in a substitution of glutamine for arginine.

Although only a small number of factor VIII mutations have been characterized, several tentative conclusions can be made. As expected, there seems to be a wide variety of mutations causing factor VIII absence or dysfunction, and the occurrence of 'sporadic' or new hemophilia A cases can be supported by the occurrence of a detectable DNA sequence change in the first affected family member. In addition, partial gene deletions have been found in only two of 21 inhibitor (factor VIII antibody producing) patients analyzed [11, 12]. Thus, in the case of hemophilia A, there is no strong correlation between the inhibitor status and gross gene defects, as there may be for hemophilia B inhibitors and the factor IX gene [14]. Finally, despite the number of different mutations causing hemophilia, it is now possible to detect common restriction fragment length polymorphisms with the cloned factor VIII probe for use in DNA based antenatal diagnosis and carrier detection of hemophilia A [15]. It is hoped that the cloning of the human factor VIII gene will mark the beginning of a fruitful period in the understanding of the structure and evolution of this elusive protein and of the treatment of this classic disease.

References

1. Gitschier, J., Wood, W. I., Goralka, T. M., Wion, K. L., Chen, E. Y., Eaton, D. H., Vehar, G. A., Capon, D. J. and Lawn, R. M., *Nature* (*London*) **312**, 326–330 (1984).
2. Toole, J. J., Knopf, J. L., Wozney, J. M., Sultzman, L. A., Buecker, J. L., Pittman, D. D., Kaufman, R. J., Brown, E., Schoemaker, C., Orr, E. C., Amphlett, G. W., Foster, W. B., Coe, M. L., Knutson, G. J., Gass, D. N. and Hewick, R. M., *Nature* (*London*) **312**, 342–347 (1984).
3. Wood, W. I., Capon, D. J., Simonsen, C. C., Eaton, D. L., Gitschier, J., Keyt, B., Seeburg, P. H., Smith, D. H., Hollingshead, P., Wion, K. L., Delwart, E., Tuddenham, G. G. D., Vehar, G. A. and Lawn, R. M., *Nature* (*London*) **312**, 330–337 (1984).
4. Rotblat, F., O'Brien, D. P., O'Brien, F. J., Goodall, A. H. and Tuddenham, E. G. D., *Biochemistry* **24**, 4294–4300 (1985).
5. Vehar, G. A., Keyt, B., Eaton, D., Rodriguez, H., O'Brien, D. P., Rotblat, F., Opperman, H., Keck, R., Wood, W. I., Harkins, R. N.,

Tuddenham, E. G. D., Lawn, R. M. and Capon, D. J., *Nature* (*London*) **312**, 337–342 (1984).
6. Eaton, D., Rodriguez, H. and Vehar, G. A., *Biochemistry* (in the press).
7. Ryden, L., in *Copper Proteins and Copper Enzymes* (ed. R. Lontie), vol. 1, pp. 157–179. CRC Press, Boca Raton, Florida (1984).
8. Mann, K. G., Lawler, C. M., Vehar, G. A. and Church, W. R., *J. Biol. Chem.* **259**, 12949–12951 (1984).
9. Fass, D. N., Hewick, R. M., Knutson, G. J., Nesheim, M. E. and Mann, K. G., *Proc. Natl. Acad. Sci. USA* **82**, 1688–1691 (1985).
10. Church, W. R., Jernigan, R. L., Toole, J., Hewick, R. M., Knopf, J., Knutson, G. J., Nesheim, M. E., Mann, K. G. and Fass, D. N., *Proc. Natl. Acad. Sci. USA* **81**, 6934–6937 (1984).
11. Gitschier, J., Wood, W. I., Tuddenham, E. G. D., Shuman, M. A., Goralka, T. M., Chen, E. Y. and Lawn, R. M., *Nature* (*London*) **315**, 427–430 (1985).
12. Antonarakis, S. E., Waber, P. G., Kittur, S. D., Patel, A. S., Kazazian, H. H., Mellis, M. A., Counts, R. B., Stamatoyannopoulos, G., Bowie, E. J. W., Fass, D. N., Pittman, D. D., Wozney, J. M. and Toole, J. J., *New England J. Med.* (in the press).
13. Barker, D., Schafer, M. and White, R., *Cell* **36**, 131–138 (1984).
14. Giannelli, F., Choo, K. H., Rees, D. J. G., Boyd, Y., Rizza, C. R. and Brownlee, G. G., *Nature* (*London*) **303**, 181–182 (1983).
15. Gitschier, J., Lawn, R. M., Rotblat, F., Goldman, E. and Tuddenham, E. G. D., *Lancet* i, 1093–1094 (1984).

Chemica Scripta 1986, **26B**, 241–245

The Proteins of Blood Coagulation: their Domains and Evolution

Earl W. Davie

Department of Biochemistry, University of Washington, Seattle, Washington 98195, USA

Paper presented at the Conference on 'Molecular Evolution of Life', Lidingö, Sweden, 8–12 September 1985

Abstract

The proteins participating in blood coagulation and fibrinolysis are multifunctional proteins, many of which are precursors to serine proteases. These proteins share common domains that show considerable amino acid sequence homology. These common domains include Gla structures, kringle structures, growth factor domains, type I and type II finger domains, catalytic domains, and other types of sequences such as the tandem repeats in factor XI and plasma prekallikrein. These multifunctional plasma proteins have evolved by gene duplication and the shuffling, exchange, and insertion of the DNA coding for these domains from one gene to another. In some cases, these recombination events may have involved introns since they often occur between domains in these closely related proteins.

The coagulation cascade that occurs in mammalian blood involves the stepwise participation of a large number of plasma proteins that eventually give rise to the formation of thrombin. This enzyme then converts fibrinogen to an insoluble fibrin clot (Fig. 1). These reactions involve a number of glycoproteins that participate as enzymes as well as cofactors. In this manuscript, we wish to focus on common domains or regions that are present in these proteins that participate in the coagulation process as well as the fibrinolytic process. Also, sequences that are repeated and occur within one or more of these proteins will be reviewed.

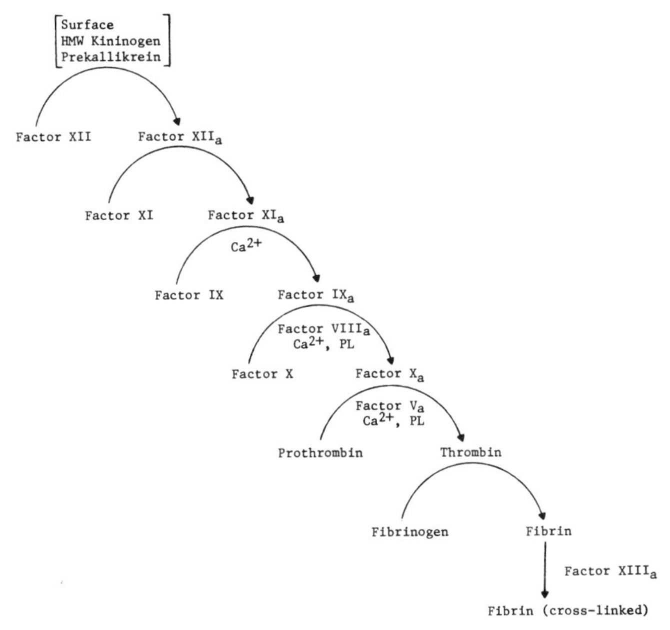

Fig. 1. Tentative mechanism for the initiation of blood coagulation in the intrinsic pathway. PL, Phospholipid.

A protein domain is rather difficult to define since it means different things to different investigators. To the enzymologist, a domain may be a functional region in a protein that has catalytic activity or binds to a substrate, a cofactor, an antibody, a cellular receptor or some other molecule. To those doing protein sequence analyses or crystallography, a domain may be an amino acid sequence that is repeated within a protein, or a sequence that is common to several proteins, or a polypeptide fragment that is generated by limited proteolysis or some particular three-dimensional structure within the protein. In the case of the plasma glycoproteins, domains have been defined that have both functional and/or sequence significance. Very few of these proteins have been crystallized, however, and consequently domains based upon three-dimensional structures have not been identified.

The first group of proteins that will be discussed are a family of vitamin K-dependent glycoproteins, that occur in plasma either as a single chain (prothrombin and factor IX) or as a two-chain molecule (factor X and protein C). Each of these proteins is converted to an active serine protease by the cleavage of one or two internal peptide bonds. This liberates a small activation peptide in the case of factor IX, factor X, and protein C. With prothrombin, a large fragment of 271 amino acids is liberated from the amino terminal portion of the molecule during the generation of thrombin.

The amino acid sequences for these proteins have been established by amino acid sequence analysis, as well as by cDNA and gene cloning. These proteins show considerable amino acid sequence identity (35–40%) and thus it has been suggested that they share a common ancestry [1].

Each of the four vitamin K-dependent proteins is synthesized as a single polypeptide chain with a prepro leader sequence [2–10]. This prepro leader sequence, which is rich in hydrophobic amino acids, is usually about 40 amino acids in length and ends with an arginine residue. The arginine is located just prior and adjacent to the amino-terminal residue present in the mature protein circulating in plasma. The removal of the prepro leader sequence during protein biosynthesis requires a protease called signal peptidase. This enzyme, which is membrane bound, cleaves a 'pre' fragment of about 30 amino acids in length from the amino-terminal region of the prepro leader sequence. This is followed by a second processing protease that cleaves the pro portion of the leader sequence at an arginine residue at the carboxyl end of the leader sequence. This generates the vitamin K-dependent proteins

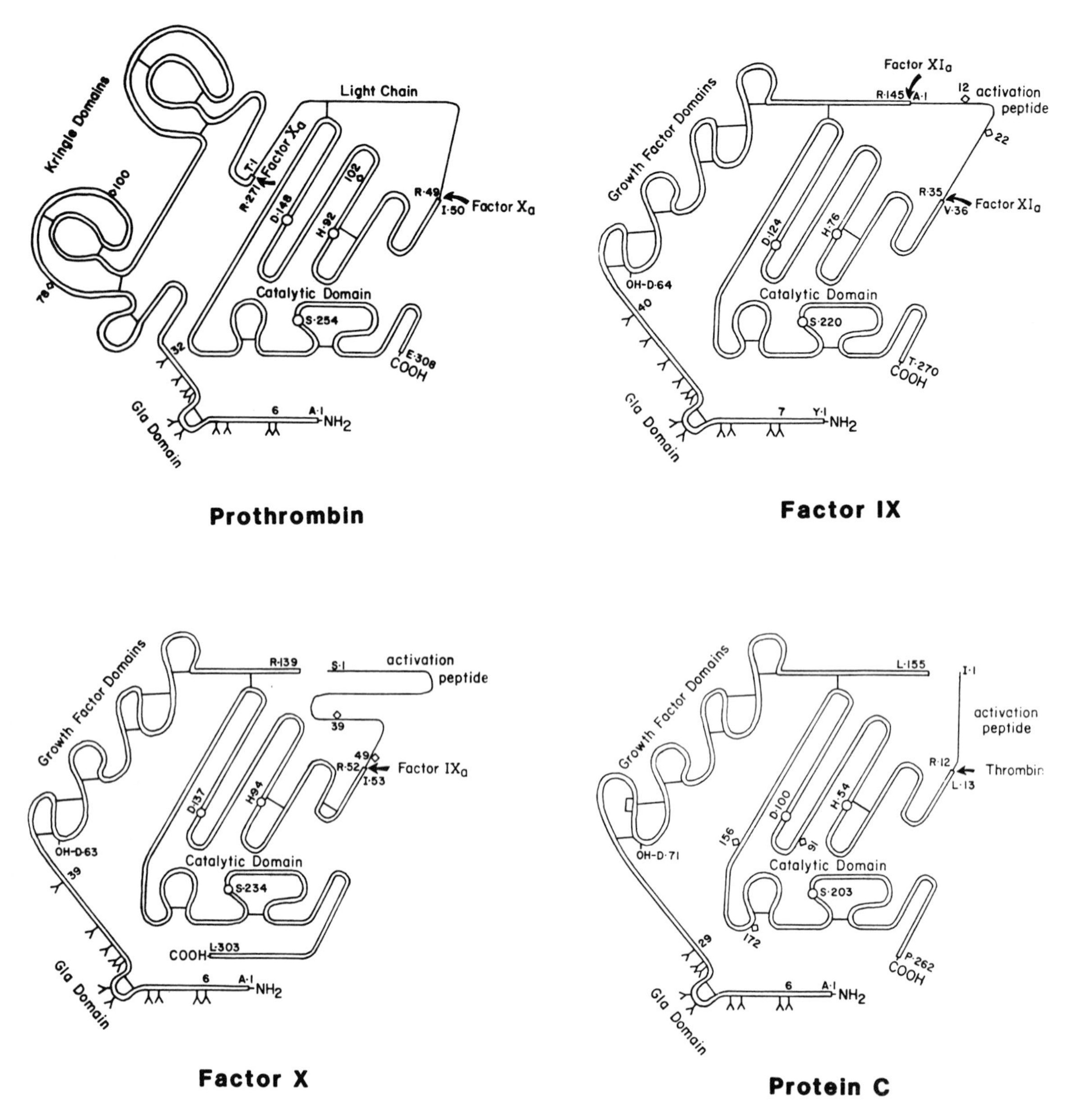

Fig. 2. Tentative structures for human prothrombin, factor IX, factor X and protein C. Y, γ-carboxyglutamic acid residues present in the Gla domains; ◇, potential N-linked carbohydrate chain. Cleavage sites and the enzymes involved during the activation of each protein are identified by a solid arrow. Amino acids involved in catalysis (H, D, S) are also identified. Proposed disulfide bonds have been placed by analogy to those in bovine prothrombin and epidermal growth factor.

that circulate in plasma. These proteins all contain a Gla domain with 9–12 γ-carboxyglutamic acid (Gla) residues (Fig. 2). The Gla residues in the vitamin K-dependent proteins in plasma result from the carboxylation of the first 9–12 glutamic acid residues in the amino terminal end of these proteins by a vitamin K-dependent carboxylase [11]. The Gla residues play an important role in the binding of these proteins to calcium ions and phospholipid during the coagulation process. In prothrombin, the Gla domain is followed by two homologous domains called kringle structures [12]. The kringle domains occur in tandem and are composed of about 80 amino acids (Fig. 2). They are characterized by the presence of highly conserved disulfide bonds. With factor IX, factor X, and protein C, the Gla domain is followed by two potential growth factor domains (Fig. 2). The growth factor domains are homologous to epidermal growth factor [13, 14], and contain about 60 amino acids. They also have highly conserved disulfide bonds. The first potential growth factor domain in factor IX, factor X, and protein C also contains one residue of β-hydroxyaspartic acid [15, 16]. The function of the kringle structures, the growth factor domains and the β-hydroxyaspartic acid is not known.

The tandem pair of kringle domains or growth factor domains in the vitamin K-dependent proteins are followed by a connecting region which links the amino-terminal portion of these molecules to their catalytic domain (Fig. 2). The catalytic domain contains the serine protease, which is very homologous in amino acid sequence to pancreatic trypsin. This domain includes specific serine, histidine, and aspartic acid residues that are involved in catalysis, and these amino acids are located in positions analogous to the catalytic triad present in pancreatic trypsin. The catalytic domain also contains an aspartic acid six amino acid residues prior to the active site serine. This aspartic acid lies in the bottom of the binding pocket in pancreatic trypsin and is responsible for its substrate specificity toward basic amino acids. An aspartic acid in the vitamin K-dependent proteins is located in the same position as that in trypsin and apparently is responsible for their specificity toward arginine-containing proteins.

During the past six or seven years, we have undertaken studies in our laboratory to isolate and characterize the cDNAs and genes for these proteins. These studies were initiated to provide us with additional information regarding the amino acid sequence of these proteins and their mechanism of biosynthesis and regulation. It was also felt that these studies might give us some clues as to the evolutionary relatedness of these proteins and their genes. In the experiments with the human prothrombin gene, Sandra Degen observed the presence of 13 introns and 14 exons in a gene that spans about 25 kb of DNA [2, 17]. The first intron was located in the prepro leader sequence at residue −17 followed by a second intron at residues 37/38 and a third at residue 46. The next introns were located at the base of kringle 1 (residue 63), within kringle 1 (residue 98), between kringles 1 and 2 (residue 144), and just following kringle 2 (residue 249). The remaining 6 introns were located throughout the rest of the prothrombin molecule. From these experiments, it was evident that the location of the 13 introns did not separate the exons of the prothrombin gene into well-defined amino acid domains. For instance, kringle 2 was separated by introns just before and after its occurrence in the protein. Kringle 1, however, contained an intron within its amino acid sequence.

With factor IX, 7 introns and 8 exons have been identified in the coding and 3′ non-coding region of the gene [18, 19]. This gene is located on the X chromosome and spans about 34 kb of DNA. The introns in the gene for factor IX varied considerably in size and DNA sequence. The first intron was located at position −17 in the prepro leader sequence, the second between residues 38 and 39 following the Gla domain and the third at residue 47 just prior to the first potential epidermal growth factor domain. These first three introns were in the same position as those in prothrombin. The fourth intron in factor IX was located between the two growth-factor domains (residue 65), and the fifth one immediately following the second growth factor domain (residue 128). The remaining two introns were present in the catalytic domain, including the sixth between residues 15–16 and the seventh intron at residue 54. The eighth exon contained the remainder of the catalytic domain (amino acid residues 55–235). Accordingly, prothrombin and factor IX shared some common features in their gene structure, i.e. the location of the first 3 introns, but the remainder of their gene structure had diverged considerably during evolution.

The gene for protein C turned out to be extremely similar to that of factor IX in terms of the location of the introns. The gene for this protein also contains seven intervening sequences in about 11 kb of DNA in the coding and 3′ non-coding region [20]. Furthermore, these seven introns were all located in essentially the same position as the seven introns in the coding and 3′ non-coding region of the gene for factor IX. These include the following positions in amino acid sequence of prepro protein C: −19, 37–38, 46, 92, and 137 in the light chain and 15–16 and 55 in the heavy chain of activated protein C. This data suggests that these two proteins have undergone a rather recent gene duplication in terms of their evolutionary origin.

The cloning of human protein C by Don Foster [9] in our laboratory and others [10] also provided information regarding the processing of this protein. As mentioned earlier, protein C is synthesized as a single-chain molecule but occurs in plasma as a two-chain molecule (Fig. 2). The two-chain molecule which is held together by a disulfide bond results from the removal of a connecting dipeptide of Lys-Arg which is located between the carboxyl end of the light chain and the N-terminal end of the heavy chain. The processing enzyme that removes this connecting dipeptide has not been established.

Factor X has a similar domain structure to that of factor IX and protein C (Fig. 2). Furthermore, the five introns that have been identified thus far by Steve Leytus in our laboratory are located at essentially the same position in the gene for factor X as those in the genes for factor IX and protein C. The data provide further evidence to support the concept that these four vitamin K-dependent serine proteases present in plasma in a zymogen form evolved from some common ancestral gene. Furthermore, prothrombin diverged much earlier from this ancestral gene than the other three vitamin K-dependent proteins.

A possible mechanism for the evolution of the vitamin K-dependent proteins is that the genetic information coding for the prepro leader, the Gla domain, the kringle structures, the potential growth-factor domains and catalytic domain in these proteins have undergone a series of translocation events. This could lead to the transfer of the DNA coding for these regions or domains either stepwise or together from one donor chromosome to a second chromosome containing the serine protease or catalytic domain [19, 21]. It also appears possible that, in this family of proteins, the introns could play some role in this shuffling of domains since the introns often appear between domains and in the same relative position in the amino acid sequence of these closely related proteins. Similarly, the prepro leader region and Gla domains in prothrombin could result from a translocation event leading to the transfer of the prepro and Gla domains to a serine protease, such as urokinase, plasminogen, tissue plasminogen activator, or factor XII. These latter proteins, like prothrombin, all have a kringle structure immediately preceding the serine protease portion of the molecule [22]. This type of evolutionary process was suggested several years ago by Gilbert, who pointed out that introns may be involved in the evolution and shuffling of domains in some proteins [23]. This evolutionary process is also similar to that described by Jacob [24], who suggested that the joining of pre-existing DNA fragments can lead to the generation of new genes with fragments or domains that give rise to proteins that are useful to the organism. Thus, the evolution of these multi-functional serine proteases that require vitamin K for their biosynthesis may be the result of the joining of a large number of different fragments during the last 500 million years or so and this has led to the formation of genes coding for multi-functional proteins. In the case of the vitamin K-dependent proteins, these various functions include the evolution of a protein that is secreted (signal sequence), binds to calcium (a Gla domain), contains a tandem repeat of the kringle structures or potential growth factors (precise functions not known), contains a connecting region (permits the serine protease to circulate in blood in a zymogen form), and lastly contains a serine protease capable of the hydrolysis of a peptide bond. The high substrate specificity of these proteins could then result from additional minor changes in and around the catalytic region of the molecule.

Two additional proteins involved in the contact pathway of blood coagulation, including plasma prekallikrein and factor XI (Fig. 1), have been isolated and characterized in our laboratory [25, 26] as well as by many other scientists [27, 28]. Recently, human factor XI and plasma prekallikrein have also been cloned by Dominic Chung and Kazuo Fujikawa in our laboratory. The cDNA for factor XI indicates that this protein is synthesized as a single chain, with a leader sequence of 18 amino acids and a mature protein containing 607 amino acids. In plasma, the protein circulates as a dimer held together by a disulfide bond(s). Factor XI is converted to factor XI_a in the presence of factor XIIa by the cleavage of an internal Arg-Val bond resulting in the formulation of a heavy chain composed of 369 amino acids and a serine protease portion composed of 238 amino acids. The most striking feature of the heavy chain is the presence of four tandem repeats of about 90 amino acids. These tandem repeats show considerable amino acid sequence homology with each other. They do not, however, show any similarity with the kringle structures, growth factor, or domains described in other proteins.

The cDNA for plasma prekallikrein indicates that this protein is synthesized with a leader sequence of 19 amino acids and a mature protein containing 619 amino acids. Plasma prekallikrein is converted to plasma kallikrein by factor XII_a which cleaves an internal Arg-Val bond (Fig. 2). This enzyme has a heavy chain of 471 amino acids and a catalytic domain of 248 residues and the two chains are held together by a disulfide bond. Plasma prekallikrein also contains four tandem repeats of about 90 amino acids and these repeats are highly homologous with each other. Furthermore, they are strongly homologous (about 60%) with the four tandem repeats present in factor XI. Accordingly, this repeat represents another potential domain that is present in the plasma proteins that participate in the intrinsic pathway of blood coagulation. Thus far, we have not characterized the genes for these proteins. It seems highly likely, however, that these two closely related plasma proteins will have a genomic structure very similar to each other.

The next protein that we will consider is factor XII. This protein also participates in the contact phase of blood coagulation. In addition, it participates along with prekallikrein in fibrinolysis. Amino acid sequence analysis by McMullen and Fujikawa [22] has established the complete sequence for the human glycoprotein. It is a single-chain molecule composed of 596 amino acids. Factor XII is converted to factor XII_a by the cleavage of an internal Arg-Val bond giving rise to a heavy chain (353 amino acids) and a light chain (243 amino acids) and these two chains are held together by a disulfide bond. The heavy chain contains four different domains, including a kringle, two growth-factor domains, and a type I and type II domain of fibronectin. The light chain contains the catalytic region or serine protease portion of the molecule. This domain organization is similar to that in several of the proteins involved in fibrinolysis, including urokinase and tissue plasminogen activator. Thus factor XII appears to belong to this subfamily of serine proteases, which has evolved through gene duplication.

Lastly, several of the proteins that participate in the coagulation process as cofactors or fibrin formation should be mentioned. The evolution of fibrinogen and its three chains from a common ancestor is an excellent example of protein evolution and has been reviewed in detail by Doolittle [29] and Henschen and co-workers [30]. Likewise, factor VIII [31–34] and factor V [35] participate as cofactors in blood coagulation (Fig. 1) and share common domains with ceruloplasmin. They are discussed by Richard Lawn in this volume and will not be dealt with here. These proteins and their genes, however, are additional interesting examples of the evolutionary process that has occurred with the plasma proteins that are involved in the coagulation process.

Acknowledgements

The author wishes to thank his colleagues at the University of Washington as well as his former co-workers, who have made so many contributions to the work mentioned in this article. Particular thanks are due to Kazuo Fujikawa, Dominic Chung, Ko Kurachi, Sandra Degen, Shinji Yoshitake, Walt Kisiel, Benito Que, Don Foster, Barbara Schach, Torben Petersen, Mark Rixon, Doug Malinowski, and Evan Sadler. This work was supported in part by a research grant (HL 16919) from the National Institute of Health.

References

1. Katayama, K., Ericsson, L. H., Enfield, D. L., Walsh, K. A., Neurath, H., Davie, E. W., and Titani, K., *Proc. Natl. Acad. Sci. USA* **76**, 4990 (1979).
2. Degen, S. J. Friezner, MacGillivray, R. T. A., and Davie, E. W., *Biochemistry* **22**, 2087 (1983).
3. MacGillivray, R. T. A., and Davie, E. W., *Biochemistry* **23**, 1626 (1984).
4. Kurachi, K., and Davie, E. W., *Proc. Natl. Acad. Sci. USA* **79**, 6461 (1982).
5. Choo, K. H., Gould, K. G., Rees, D. J. G. and Brownlee, G. G., *Nature (London)* **299**, 178 (1982).
6. Jaye, M., De La Salle, H., Schamber, F., Balland, A., Kohli, V., Findeli, A., Tolsloshev, P., and Lecocq, J. P., *Nucleic Acids Res.* **11**, 2325 (1983).
7. Leytus, S. P., Chung, D. W., Kisiel, W., Kurachi, K. and Davie, E. W., *Proc. Natl. Acad. Sci. USA* **81**, 3699 (1984).
8. Fung, M. R., Hay, C. W., MacGillivray, R. T. A., *Proc. Natl. Acad. Sci. USA* **82**, 3591 (1985).
9. Foster, D. and Davie, E. W., *Proc. Natl. Acad. Sci. USA* **81**, 4766 (1984).
10. Long, G. L., Belageje, R. M., and MacGillivray, R. T. A., *Proc. Natl. Acad. Sci. USA* **81**, 5653 (1984).
11. Stenflo, J. and Suttie, J. W., *Annu. Rev. Biochem.* **46**, 157 (1977).
12. Magnusson, S., Petersen, T. E., Sottrup-Jensen, L. and Claeys, H., in *Proteases and Biological Control* (ed. E. Reich, D. B. Rifkin and E. Shaw), p. 123. Cold Spring Harbor Laboratories, Cold Spring Harbor, New York (1975).
13. Young, C. L., Barker, W. C., Tomaselli, C. M. and Dayhoff, M. D., in *Atlas of Protein Sequence and Structure* (ed. M. D. Dayhoff), p. 73. National Biomedical Research Foundation, Silver Spring, Maryland (1978).
14. Doolittle, A. F., Feng, D. F. and Johnson, M. S., *Nature (London)* **307**, 558 (1984).
15. Drakenburg, T., Fernlund, P., Roepstorff, P. and Stenflo, J., *Proc. Natl. Acad. Sci. USA* **80**, 1802.
16. McMullen, B., Fujikawa, K. and Kisiel, W., *Biochem. Biophys. Res. Commun.* **115**, 8 (1983).
17. Degen, S. J. Friezner, Rajput, B., Reich, E. and Davie, E. W., in *Protides of Biological Fluids* (ed. H. Peeters). Pergamon Press, New York (in the press).
18. Anson, D. S., Choo, K. H., Rees, D. J. G., Gianelli, F., Gould, F. A., Huddleston, J. A. and Brownlee, G. G., *EMBO J.* **3**, 1053 (1984).
19. Yoshitake, S., Schach, B. G., Foster, D. C., Davie, E. W. and Kurachi, K., *Biochemistry* **24**, 3736 (1985).
20. Foster, D., Yoshitake, S. and Davie, E. W., *Proc. Natl. Acad. Sci. USA* **82**, 4673 (1985).

21. Patthy, L., *Cell* **41**, 657 (1985).
22. McMullen, B. A. and Fujikawa, K., *J. Biol. Chem.* **280**, 5328 (1985).
23. Gilbert, W., *Nature* (*London*) **271**, 501 (1978).
24. Jacob, F., *The Possible and the Actual.* Pantheon Books, New York (1982).
25. Kurachi, K. and Davie, E. W., *Biochemistry* **16**, 5831 (1977).
26. Heimark, R. L. and Davie, E. W., *Methods Enzymol.* **89**, 157 (1981).
27. Bouma, B. N. and Griffin, J. H., *J. Biol. Chem.* **252**, 6432 (1977).
28. Bouma, B. N., Miles, L. A., Beretta, G. and Griffin, J. H., *Biochemistry* **19**, 1151 (1980).
29. Doolittle, R. F., in *Protides of Biological Fluids* (ed. H. Peeters), p. 41. Pergamon Press, New York (1980).
30. Henschen, A., Lottspeich, F., Topfer-Petersen, E., Kehl, M. and Timpl, R., in *Protides of Biological Fluids* (ed. H. Peeters), p. 47. Pergamon Press, New York (1980).
31. Gitschier, J., Wood, W. I., Goralka, T. M., Wion, K. L., Chen, E. Y., Eaton, D. H., Vehar, G. A., Capon, D. J. and Lawn, R. M., *Nature* (*London*) **312**, 326 (1984).
32. Wood, W. I., Capon, D. J., Simonsen, C. C., Eaton, D. L., Gitschier, J., Keyt, B., Seeburg, P. H., Smith, D. H., Hollingshead, P., Wion, K. L., Delwart, E., Tuddenham, E. G. D., Vehar, G. A. and Lawn, R. M., *Nature* (*London*) **312**, 330 (1984).
33. Vehar, G. A., Keyt, B., Eaton, D., Rodriguez, H., O'Brien, D. P., Rotblat, F., Oppermann, H., Keck, R., Wood, W. I., Harkins, R. N., Tuddenham, E. G. D., Lawn, R. M. and Capon, D. J., *Nature* (*London*) **312**, 337 (1984).
34. Toole, J. J., Knopf, J. L., Wozney, J. M., Sultzman, L. A., Buecker, J. L., Pittman, D. D., Kaufman, R. J., Brown, E., Shoemaker, C., Orr, E. C., Amphlett, G. W., Foster, W. B., Coe, M. L., Knutson, G. J., Fass, D. N. and Hewick, R. M., *Nature* (*London*) **312**, 342 (1984).
35. Church, W. R., Jernigan, R. L., Toole, J., Hewick, R. M., Knopf, J., Knutson, G. J., Nesheim, M. E., Mann, K. G., and Fass, D. N., *Proc. Natl. Acad. Sci. USA* **81**, 6934 (1984).

Chemica Scripta 1986, **26B**, 247–250

The Relation Between Protein Structure in α/β Domains and Intron-Exon Arrangement of the Corresponding Genes

Carl-Ivar Brändén

Swedish University of Agricultural Sciences, Department of Molecular Biology, Uppsala Biomedical Centre, Box 590, S-751 24 Uppsala, Sweden

Paper presented at the Conference on 'Molecular Evolution of Life', Lidingö, Sweden, 8–12 September 1985

Abstract

The relation between protein structure in α/β domains and intron–exon arrangement of the corresponding genes has been examined for the following enzymes: triosephosphate isomerase, pyruvate kinase, alcohol dehydrogenase and glyceraldehyde phosphate dehydrogenase. It was found that, in general, the domains and subdomains are separated by introns. The mononucleotide binding motifs have in most cases extra helical elements at their ends. The individual exons in these α/β domains were analyzed for similarities in their fold. The majority of these exons form single or double $\alpha\beta$-units. It is suggested that such units formed the early building blocks of α/β domains.

Introduction

Gilbert [1] has suggested that the intron/exon structure of eucaryotic genes might provide a record of its evolutionary history. Early genes might have evolved by different combinations of small exonic regions encoding structural or functional domains.

A number of different systems now exist where both the intron/exon arrangement of the gene and the three-dimensional structures of the corresponding or homologous gene products are known. The individual systems have been examined for possible correlation in a number of cases [2–11].

A somewhat confusing picture has emerged, mainly because a specific gene that is examined today might both have lost and acquired introns during recent evolution [12, 13]. Furthermore, introns may have different functional significance. Stone *et al.* [9] have classified introns in two main groups: A, which result from the original assembly of the gene from smaller units, and B, which have never had any function. In addition they divide type A in three different groups: A1 for building up a functional domain from smaller units, A2 for duplication of complete domains and A3 for joining dissimilar domains.

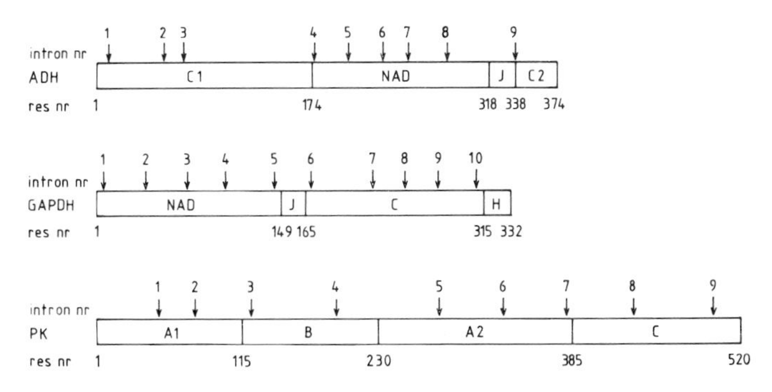

Fig. 1. Schematic diagram of intron positions and domain boundaries along the polypeptide chain in alcohol dehydrogenase (ADH), glyceraldehyde-3-phosphate dehydrogenase (GPDH) and pyruvate kinase (PK).

Proteins of the α/β structural type [14] are particularly good examples of the occurrence of all these types of introns. Recently the gene structures have become available for several such genes where the corresponding 3-dimensional protein structure is also known. Here we will correlate the results obtained for alcohol dehydrogenase [8, 15, 16], glyceraldehyde-3-phosphate dehydrogenase [9, 17, 18], pyruvate kinase [10, 19] and triosephosphate isomerase [11, 20] with particular emphasis on exons of possible type A.

(A) Introns which separate protein chains into dissimilar domains and subdomains

The subunits of NAD^+-dependent dehydrogenases are divided into two structurally and functionally different domains [21]. One of these bind the coenzyme and the other provides the residues necessary for substrate binding and catalysis. The NAD^+-binding domains of alcohol dehydrogenase [22] and glyceraldehyde-3-phosphate dehydrogenase [23] as well as lactate dehydrogenase [24] are all similar in structure and bind the coenzyme in a similar way. The catalytic domains on the other hand, have quite different structures. There is no sequence homology between these enzymes.

Triose phosphate isomerase is a one-domain structure of eight α/β units arranged such that they form an α/β barrel structure with an inner core of eight parallel strands surrounded by eight helices. The subunit of pyruvate kinase is divided into three domains: A is an α/β barrel, B is built up from antiparallel β-strands and C has the mononucleotide binding motif which is one half of the NAD^+ binding domain of the dehydrogenases [19].

Figure 1 gives a schematic representation of the intron positions and domain divisions along the polypeptide chain for alcohol dehydrogenase, glyceraldehyde-3-phosphate dehydrogenase and pyruvate kinase. The catalytic domain of the alcohol dehydrogenase subunit is built up from residues 1–174 and 318–374. Residues 1–174 form a compact domain, C 1, of antiparallel strands. Residues 340–374 form a subdomain of the catalytic domain, C 2, which is linked to the coenzyme binding domain, NAD (residues 175–317) by a 20 residue long helix, J, (residues 320–340). Intron four at position 179 is close to the first junction between the catalytic and the NAD^+-binding domains. Intron 9 at position 338 separates the subdomain C2 from the joining helix. There is no intron close to position 318 which is the last residue of the α/β motif of the NAD^+-binding domain. It may be a matter of definition whether this joining helix should be regarded as belonging to the NAD-binding domain, the catalytic domain or as a separate structural element. A similar joining helix is present in all three dehydrogenases.

The NAD^+-binding domain of glyceraldehyde-3-phosphate dehydrogenase comprises residues 1–148. Residues 150–165 comprise the joining helix, J, 165–315 build up the antiparallel β-structure of the catalytic domain, C, and residues 316–332 form a C-terminal helix, H, which is physically in the NAD^+-binding domain. Intron 6 at residue 145 is close to the junction between the two domains. There is also an intron at position 172 close to the end of the joining helix between the domains. In addition there is an intron at position 310 close to the beginning of the C-terminal helix.

Domain A of pyruvate kinase approximately comprises residues 1–120 (A 1) and 220–385 (A 2). Domain B is built up of residues 120–220 which are inserted between strand β_3 and helix α_3 of the α/β barrel of domain A.

Domain C comprises the C-terminal residues 386–529. Residues 420–475 form the mononucleotide binding motif and 475–495 form a helix corresponding to the joining helix of the NAD^+ binding domains. Residues 386–420 form two helices which form interdomain contact and residues 496–529 form a C-terminal antiparallel loop. The critical domain boundaries are thus around residues 120, 220 and 385. Intron 3 at position 124 is close to the junction between A 1 and B. Intron 7 at position 379 is reasonably close to the junction between A 2 and C. There is, however, no intron in the vicinity of residue 220. Exon 5 (residues 187–277) is, on the other hand, by far the longest exon in pyruvate kinase. It is tempting to suggest that an intron has been lost around residue 220 which would divide this exon into two pieces of similar lengths as the other exons that build up these α/β structures.

It would seem from the comparison of these three enzymes that most of the experimental observation is compatible with the notion that introns of type A 3 have been preserved to a surprisingly large extent in these enzymes. When the three-dimensional structure determinations of the NAD^+-dehydrogenases revealed the domain structure–function relationships it was suggested [25, 26] that they reflected evolutionary relationships and that these and other complex enzymes have evolved by reassorting and joining genes that coded for simple ancestral polypeptide chains. At that time no plausible genetic mechanism was known that could easily account for such rearrangements. Introns of type A3 would have precisely this function.

(B) Introns which separate the mononucleotide binding motifs

The coenzyme binding domain is built up from six parallel strands of pleated sheet surrounded by helices on each side of the sheet. Rao and Rossmann [27] showed that this domain could be divided into two roughly identical units each associated with a mononucleotide binding area. This mononucleotide binding fold (mnf) is a structural module which has been observed in a number of different proteins [28]. This module is defined to comprise three parallel strands of pleated sheet joined by two helices in a right-handed fashion [21]. The part of a polypeptide chain from the beginning of the first strand to the end of the third strand will be called an mnf-unit. The residues that link two such units to an NAD-binding domain will be called the linking region. At the end of the domains these units are usually joined to helices. Figure 2 gives a schematic diagram of the occurrence of introns in relation to these structural elements.

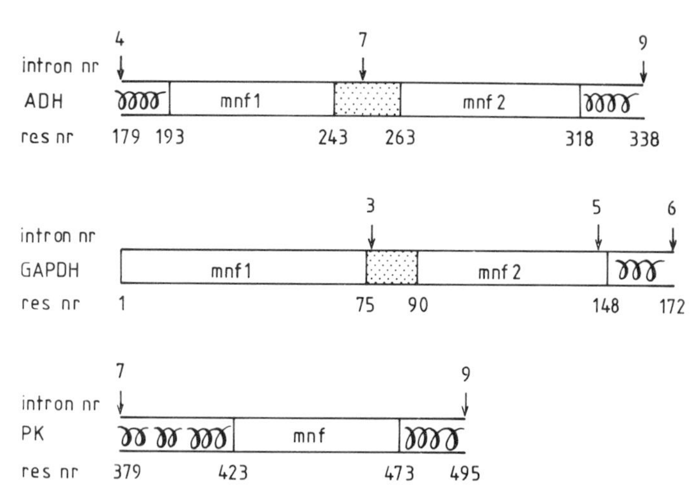

Fig. 2. Schematic diagrams of intron positions in relation to the positions of the mononucleotide binding fold, mnf, along the polypeptide chains in alcohol dehydrogenase (ADH), glyceraldehyde-3-phosphate dehydrogenase (GPDH) and pyruvate kinase (PK). Dotted regions marked the linking regions between two mnfs and the label (helix symbol) mark helices at the beginning or end of the mnfs.

In only one case, mnf 1 of glyceraldehyde-3-phosphate dehydrogenase, is there a strict correlation between exon boundaries and mnf boundaries. However, in all other cases the exons are extended at the boundaries of the mnf region to include short helices or linking regions. In pyruvate kinase there are helices both at the beginning and the end of the unit. In alcohol dehydrogenase there is a helix at the beginning of mnf 1 and at the end of mnf 2. The linking region is part of mnf 2 in both dehydrogenases.

The mnf units are thus excised by introns but in most cases the units contain additional helices or linking regions at the termini. Whether these introns could be of type A2 reflecting ancient gene duplication or type A3 from a later stage in evolution is impossible to deduce from the data at hand owing to these extra additions.

(C) The structure of individual exons in α/β domains

If type A 1 introns are present, individual exons should show structural similarities as building blocks for α/β domains provided that the folding pattern has been preserved. Brändén *et al* [8] noticed a repeating pattern in the first mnf unit of alcohol dehydrogenase. This unit is built up of three exons, each of which is folded into an α-helix followed by a β-strand; an $\alpha\beta$ motif. We will now examine all individual exons within the α/β domains of these enzymes to see how general this repeating pattern is. Figure 3 shows a schematic diagram of the secondary structure elements that comprise those exonic regions that build up the α/β domains in all four enzymes.

There are 24 such exons, exons 5–10 of ADH, all eight exons in TIM, exons 1–5 in GPDH and exons 2, 3, 6, 7 and 9 of PK. These exons can be arranged into two main groups:

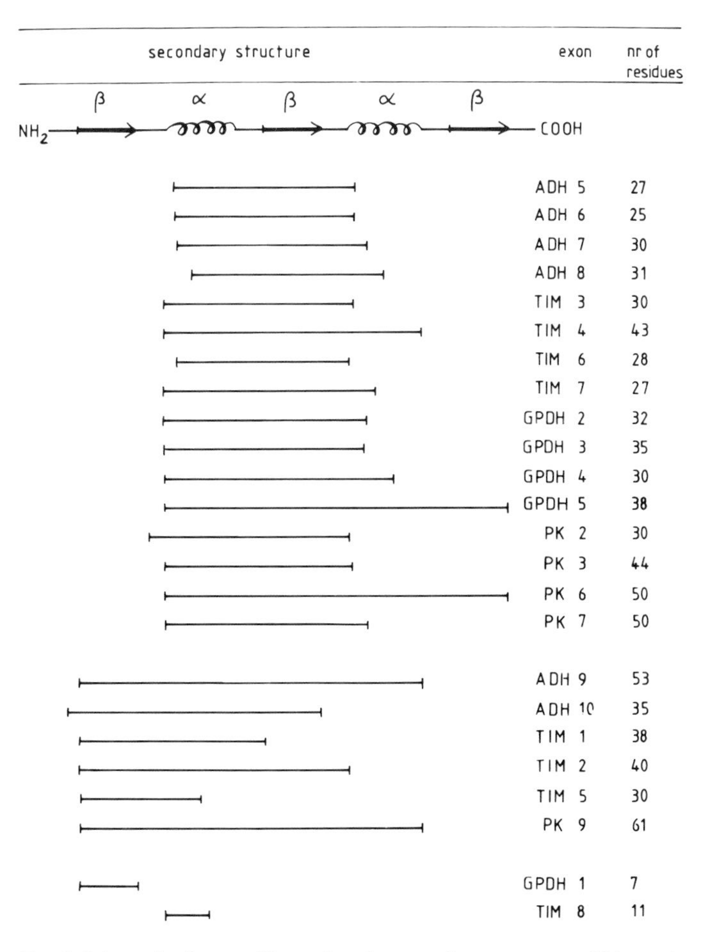

Fig. 3. Schematic diagram illustrating the secondary structure of 24 exonic regions in the α/β domains of alcohol dehydrogenase (ADH), triosephosphate isomerase (TIM), glyceraldehyde-3-phosphate dehydrogenase (GPDH) and pyruvate kinase (PK). The exonic regions are numbered from the amino end of the protein.

group 1, which start with a helix, and group 2, which start with a β-strand. In addition there are two short exons: GPDH1, which is only a β-strand, and TIM 8, which is a short helix of 11 residues.

The majority of the exons belong to group 1. Ten of these can be regarded as an $\alpha\beta$ unit similar to the regular $\alpha\beta$ units that build up mnf 1 of ADH. Most of these have a similar length of about 30 residues. The exceptions are PK3 and PK7, which are somewhat longer. Two additional exons, GPDH5 and PK6, each consist of two such $\alpha\beta$ units. The remaining four exons in group 1 contain part of an additional helical element after the β-strand. Only one of these, TIM4, consists of a complete $\alpha\beta\alpha$ unit.

The exons that belong to group 2 vary considerably in size. Two of them are $\beta\alpha$ units, two are $\beta\alpha\beta$ units and the remaining two are $\beta\alpha\beta\alpha$ units.

It is impossible to draw firm conclusions from these data. However, the fact that most exons correspond to $\alpha\beta$ units indicates that, if introns of type A 1 exist, then an $\alpha\beta$ unit is a strong candidate for a primordial building block. A number of possibilities would then exist for the occurrence of exons of different structural type. Apart from loss of introns of type A 1, and creation of introns of type B with no functional significance, one cannot rule out the possibility that an exon which originally formed an $\alpha\beta$ structure could convert to a $\beta\alpha$ structure by a few point mutations.

Conclusions

By comparing the gene intron/exon arrangement with the protein three-dimensional structures of alcohol dehydrogenase, glyceraldehyde-3-phosphate dehydrogenase, pyruvate kinase and triosphosphate isomerase the following general conclusions can be made:

(1) Introns are present at the boundaries of α/β domains similar to several other systems with domains of different structural type. This supports the hypothesis that introns have been used to form complex enzymes by joining dissimilar simpler domains.

(2) Introns are in general present at the boundaries of the mononucleotide binding domain. This supports the early ideas by Rossmann *et al.* [27] that this domain has been an important building block to create enzymes whose function depend on the ability to bind nucleotides.

(3) The majority of the exonic regions found in these proteins consists of single or double $\alpha\beta$ units. This indicates that an $\alpha\beta$ unit could have been an important early building block for the creation of α/β domains.

References

1. Gilbert, W., *Nature (London)* **271**, 501 (1978).
2. Sakano, H., Roigers, J. H., Hüppi, K., Brack, C., Braunecker, A., Maki, R., Wall, R. and Tonegawa, S., *Nature (London)* **277**, 627–633 (1979).
3. Jung, A., Sippel, A. E., Grez, M. and Schütz, G., *Proc. Natl. Acad. Sci. USA* **77**, 5759–5763 (1980).
4. Go, M., *Nature (London)* **291**, 90–92 (1981).
5. Inaa, G., Piatigarsky, J., Norman, B., Slingsby, C. and Blundell, T., *Nature (London)* **302**, 310–315 (1983).
6. Holm, I., Ollo, R., Panthier, J.-J. and Rougeon, F., *EMBO J.* **3**, 557–562 (1984).
7. Loebermann, H., Tokuoka, R., Deisenhofer, J. and Huber, R., *J. Mol. Biol.* **177**, 531–556 (1984).
8. Brändén, C.-I., Eklund, H., Cambillau, C. and Pryor, A. J. *EMBO J.* **3**, 1307–1310 (1984).
9. Stone, E. M., Rothblum, K. N. and Schwartz, R. J., *Nature (London)* **313**, 498–500 (1985).
10. Loberg, N. and Gilbert, W., *Cell* **40**, 81–90 (1985).
11. Gilbert, W., reported at the Second Bio/Technology-Waksman Institute Symposium (see Bialy, H., *Biotechnology* **3**, 516, 1985).
12. Cornish-Bowden, A., *Nature (London)* **313**, 434–435 (1985).
13. Rogers, J. (1985), *Nature (London)* **315**, 458–459 (1985).
14. Levitt, M. and Chothia, C., *Nature (London)* **261**, 552–558 (1976).
15. Denis, E. S., Gerlach, W. L., Pryor, A. J., Bennetzen, J. L., Inglis, A., Llewellyn, D., Sachs, M. M., Ferl, J. R. and Peacock, W. J., *Nucl. Acids Res.* **12**, 3983–4000 (1984).
16. Eklund, H., Nordström, B., Zeppezauer, E., Söderlund, G., Ohlsson, I., Boiwe, T., Söderberg, B.-O., Tapia, O. and Brändén, C.-I., *J. Mol. Biol.* **102**, 27–49 (1976).
17. Stone, E. M., Rothblum, K. N., Alevy, M. C., Kuo, T. M. and Schwartz, R. J. *Proc. Natl. Acad. Sci. USA* **82**, 1628–1632 (1985).
18. Moras, D., Olsen, K., Sabesan, M., Buehner, M., Ford, G. and Rossmann, M., *J. Biol. Chem.* **250**, 9137–9162 (1975).
19. Stuart, D. L., Levine, M., Muirhead, H. and Stammers, D. K., *J. Mol. Biol.* **134**, 109–142 (1979).
20. Banner, D. W., Bloomer, A. C., Petsko, G. A., Phillips, D. C., Pogson, C. I., Wilson, I. A., Corran, P. H., Furth, A. J., Milman, J. D., Offord, R. E., Priddle, J. D. and Waley, S. G., *Nature (London)* **255**, 609–614.
21. Rossmann, M. G., Liljas, A., Brändén, C.-I. and Banaszak, L. J., in *The Enzymes*, 3rd ed. (ed. P. D. Boyer) New York, Vol XI, Academic Press, pp. 61–102 (1975).
22. Brändén, C.-I., Eklund, H., Nordström, B., Boiwe, T., Söderlund, G., Zeppezauer, E., Ohlsson, I. and Åkeso, Å., *Proc. Natl. Acad. Sci. USA* **70**, 2439–2442 (1973).
23. Buehner, M., Ford, G. C., Moras, D., Olsen, K. W. and Rossmann, M. G., *Proc. Natl. Acad. Sci. USA* **70**, 3052–3054 (1973).

24. Adams, M. J., Ford, G. C., Koekok, R., Letz, P. J. Jr., McPherson, A. Jr., Rossmann, M. G., Smiley, I. E., Schevitz, R. W. and Wonnacott, A. J., *Nature (London)* **227**, 1098–1103 (1970).
25. Ohlsson, I. Nordström, B. and Brändén, C.-I., *J. Mol. Biol.* **89**, 339–354 (1974).
26. Rossmann, M. G., Moras, D. and Olsen, K. W., *Nature (London)* **250**, 194–199 (1974).
27. Rao, S. T. and Rossmann, M. G., *J. Mol. Biol.* **76**, 241–256 (1973).
28. Brändén, C.-I. *Q., Revs. Biophys.*, **13**, 317–338 (1980).

Chemica Scripta 1986, **26B**, 251–255

Use of Protein Sequence and Structure to infer Distant Evolutionary Relationships

R. G. Brennan, L. H. Weaver and B. W. Matthews

Institute of Molecular Biology and Department of Physics, University of Oregon, Eugene, Oregon 97403, USA

Paper presented by Brian W. Matthews at the Conference on 'Molecular Evolution of Life', Lidingö, Sweden, 8–12 September 1985

Abstract

Some problems in the comparison of protein amino acid sequences are discussed and empirical formulae are developed to assess the statistical significance of the agreement between different sequences. Examples are given to illustrate the advantage of evaluating distantly related sequences by direct comparison of amino acids rather than by the use of minimum base change per codon.

The example of DNA-binding proteins is used to show how a family of sequences can be used to establish relationships that are not apparent at the level of pairwise sequence comparisons. Such sequence comparisons, taken together with known three-dimensional structures, have shown that many DNA-binding proteins contain a similar 'helix-turn-helix' DNA-binding unit. The amino acid sequences of many known DNA-binding proteins have been systematically searched in order to locate putative helix-turn-helix units.

During evolution, the three-dimensional structure of a protein changes more slowly than does its amino acid sequence. Therefore structural similarities between proteins may suggest evolutionary relationships that are not apparent at the level of the amino acid sequences. One such example is provided by the lysozymes from hen (chicken) egg-white, goose egg-white and bacteriophage T4.

The use of nucleic acid and amino acid sequence comparison to reconstruct the evolutionary history of contemporary macromolecules is well established. Such methods can be applied directly and reliably to a given family of proteins, such as the globins, which have amino acid sequences that are clearly homologous. However, given sufficient time, the amino acid sequences of two proteins derived from a common evolutionary precursor may have altered to the point that they retain few if any common amino acids. In this paper we briefly review methods of comparing proteins that are of this type, i.e. their amino acid sequences have little if any correspondence yet they may have evolved from a common evolutionary precursor.

Amino acid sequence comparison between two proteins

The comparison of distantly related, or unrelated, amino acid sequences is non-trivial. It may be possible to make two sequences appear to be similar by the introduction of appropriate 'gaps' or 'deletions', but each gap increases the chances that a given correspondence of amino acids is due to a chance event [e.g. refs 1, 2]. As suggested by Doolittle [1], each gap should carry a penalty.

Another difficulty in sequence comparison is the estimation of the probability that a given amino acid correspondence could arise by chance. A common way to establish this likelihood is by repeated comparisons of unrelated or jumbled sequences [e.g. 3, 4]. In this context it is possible to establish some empirical formulae that may be useful in assessing the statistical significance of amino acid sequence correspondence.

Table I summarizes a series of sequence comparisons made by the method of Fitch [3]. In Fitch's method, one chooses a certain 'probe length', L (e.g. $L = 20$ amino acids) and compares all possible 20 amino-acid segments from the first protein with all possible 20 amino-acid segments from the second. The majority of such comparisons will be between amino acid segments that are unrelated but also included among the comparisons are those 20 amino-acid segments in the two proteins that have the best correspondence. By plotting the frequency distribution of all possible correspondences the relationship of the 'best' correspondence to the overall distribution can be evaluated. A specific example, provided by the comparison of the sequences of the lysozymes from bacteriophage T4 and phage P22 is shown in Fig. 1. Fitch [3] evaluates the correspondence between amino acid sequences in terms of the minimum base change per codon (MBC). In addition, Fig. 1 also shows the distribution for comparison of the same sequences based on a direct compa-

Table I. *Dependence of the minimum base change per codon (MBC) and amino acid change per codon (AAC) on sequence length for comparisons between two proteins*

Minimum base change per codon			
Segment length L	Mean $\overline{MBC}$	Standard deviation $\sigma(MBC)$	$\sigma(MBC)\,L^{1/2}$
20	1.489	0.148	0.662
30	1.492	0.122	0.668
66	1.488	0.0827	0.672
94	1.488	0.0667	0.647
Amino acid change per codon			
Segment length L	Mean $\overline{AAC}$	Standard deviation $\sigma(AAC)$	$\sigma(AAC)\,L^{1/2}$
20	0.940	0.0525	0.235
30	0.938	0.0432	0.237
66	0.941	0.0290	0.236
94	0.938	0.0241	0.234

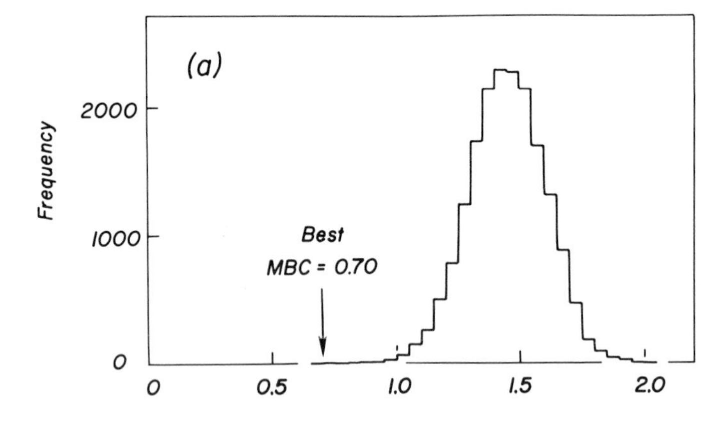

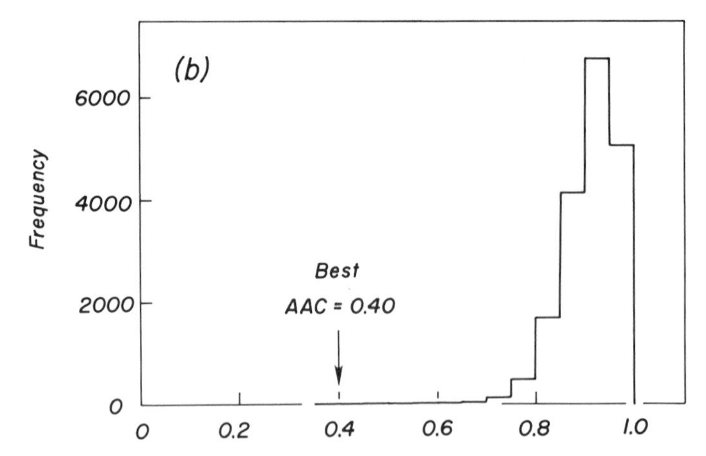

Fig. 1. This figure, together with Table 2, shows the results of applying Fitch's method [3] to search for sequence homology between the amino acid sequence of P22 phage lysozyme and that of T4 phage lysozyme [5]. In the first instance (a), sequence-agreement is scored as the minimum base change per codon. The identical comparison is then repeated but scored as the amino acid change per codon (b). P22 lysozyme has 146 residues and T4 lysozyme 164. The sequences were compared using segment lengths of $L = 20$ residues, resulting in a total of 18415 individual comparisons.

rison of amino acid sequences, i.e. amino acids changed per codon (AAC). (As discussed below, the latter method of scoring is preferable for comparing distantly related sequences [cf. ref. 1].)

Table I gives the mean values and the standard deviations obtained from a large number of sequence comparisons of different proteins. For a given probe length, L, we find that the mean value of the distribution and its standard deviation are constant (within a few per cent) no matter which pair of amino acid sequences is being compared. The values given in Table I are the averages of many comparisons. The mean values are also independent of the probe length, but, as shown in Table I, the standard deviation decreases inversely in proportion to the square root of the probe length.

The results can be summarized as follows:

(1) If two amino acid sequences of length L are compared on the basis of minimum base change per codon, the expected mean value of the minimum base change per codon for sequences chosen at random is

$$\overline{\text{MBC}} = 1.489 \quad (1)$$

and the standard deviation is

$$\sigma(\text{MBC}) = 0.662\ L^{-1/2} \quad (2)$$

(2) If the amino acid sequences are compared directly on the basis of amino acid sequence differences, the expected mean value of the amino acid changes per codon for sequences chosen at random is

$$\overline{\text{AAC}} = 0.939 \quad (3)$$

and the standard deviation is

$$\sigma(\text{AAC}) = 0.236\ L^{-1/2} \quad (4)$$

These empirical formulae provide a quick way to check the significance of an apparent matching between two amino acid sequences. The usual way to evaluate the significance (or the chance expectation) of a given observation is to calculate how many standard deviations (s) it is from the mean value. For comparisons made on the basis of minimum base change per codon, the number of standard deviations that a given score, MBC, is from the mean, is

$$s = \frac{(\overline{\text{MBC}} - \text{MBC})}{\sigma(\text{MBC})} \quad (5)$$

where $\overline{\text{MBC}}$ is the mean value of MBC and σ(MBC) is the standard deviation. Substituting from (1) and (2) we obtain the following empirical expression for the significance:

$$s = \frac{(1.498 - \text{MBC})\ L^{1/2}}{0.662} \quad (6)$$

Similarly, for agreement scores calculated by direct comparison of amino acid sequences, the significance can be estimated as

$$s = \frac{(0.939 - \text{AAC})\ L^{1/2}}{0.236} \quad (7)$$

As an example, suppose that within the two sequences compared there are five consecutive amino acids that are found to correspond exactly in the two sequences. The agreement score (in terms of direct amino acid sequence comparison) is AAC = 0, that the significance (equation (7); $L = 5$) is $s = 8.9$, i.e. 8.9 standard deviations from the 'random' value.

Perhaps a more typical example would be the case where, say, in a stretch of ten consecutive amino acids, five are found to agree in the two sequences being compared. In this case the agreement score (for direct amino acid sequence comparison) is AAC = 0.5, leading to a significance (equation (7), $L = 10$) of $s = 5.9$ standard deviations from the mean.

It is instructive to compare the significance that would have been obtained had the same amino acid correspondence (5 matches out of 10) been evaluated on the basis of the minimum base change per codon. If one takes the most favorable situation, namely each mismatch corresponding to a single base change per codon, the agreement score would be MBC = 0.5, yielding a significance (equation (6); $L = 10$) of $s = 4.7$ standard deviations from the mean. If one considers a more likely case in which the MBC = 0.75 then s would be only 3.5 standard deviations from the mean. The striking result is that the significance based on minimum base change per codon is less than that for direct amino acid comparison.

This difference can be even more dramatic in the actual comparison of two proteins. Figure 1 shows the comparison of the amino acid sequences of phage T4 lysozyme and phage P22 lysozymes with a probe length $L = 20$ residues. As summarized in Table II, the best correspondence between

Table II. *The use of minimum base change per codon and amino acid change per codon to search for sequence homology in two lysozymes. The probe length, L, is 20 amino acids*

	Scoring with minimum base change per codon	Scoring with amino acid change per codon
Mean value	$\overline{MBC} = 1.472$	$\overline{AAC} = 0.937$
Standard deviation	$\sigma(MBC) = 0.156$	$\sigma(AAC) = 0.0575$
Best score	$MBC = 0.70$	$AAC = 0.40$
Significance	$s = 4.9\sigma$	$s = 9.3\sigma$

For explanation of this table, see the caption to Fig. 1.

these two sequences (12 amino acid identities in a consecutive sequence of 20) has a significance $s = 9.3\sigma$ based on AAC, but only $s = 4.9\sigma$ when based on MBC. As can be seen in Fig. 1, the difference is because the width of the distribution based on MBC is much broader than that based on AAC. In other words the 'signal to noise ratio' is less favorable for comparisons based on MBC than those based on AAC. It is sometimes argued that the use of MBC is desirable because pairs of amino acids that are chemically similar tend to have a low value of their MBC and pairs of amino acids that are different have high values. Therefore, it is argued, the use of MBC will allow one to detect sequence segments that are chemically similar even though their amino acid sequences are not identical or close to identical. This may be true, but the same procedure increases the number of ways in which 'good' scores can be obtained and, at the same time, reduces the discrimination between sequences containing identical amino acids and sequences containing similar ones. In comparisons of distantly related amino acid sequences it is clearly preferable to calculate AAC rather than MBC. The latter method may cause one to miss a sequence correspondence that is actually statistically significant.

Multiple sequence comparisons

At first appearance, two amino acid sequences may seem to be unrelated, however in some instances a third sequence that contains elements in common with the two initial sequences will be found, thus establishing a 'linkage' between them. This is one way in which multiple sequence comparisons can be used to establish distant evolutionary relationships.

A good example is provided by the cro repressor and the λ repressor proteins of bacteriophage λ. These two proteins bind to the same sites on the DNA of phage λ yet their amino acid sequences seemed to be unrelated [6]. However, following the determination of the three-dimensional structure of the cro protein [7], as well as the sequences of other DNA-binding proteins, it was realized that the sequences of cro and λ repressor proteins are in fact related (Fig. 2) [8–10]. By making use of multiple sequence comparisons it has, in fact, been possible to show that many DNA-binding proteins have regions of sequence that correspond. This region of correspondence consists of about 22 amino acids and is now known as the 'helix-turn-helix' DNA-binding fold [8–12].

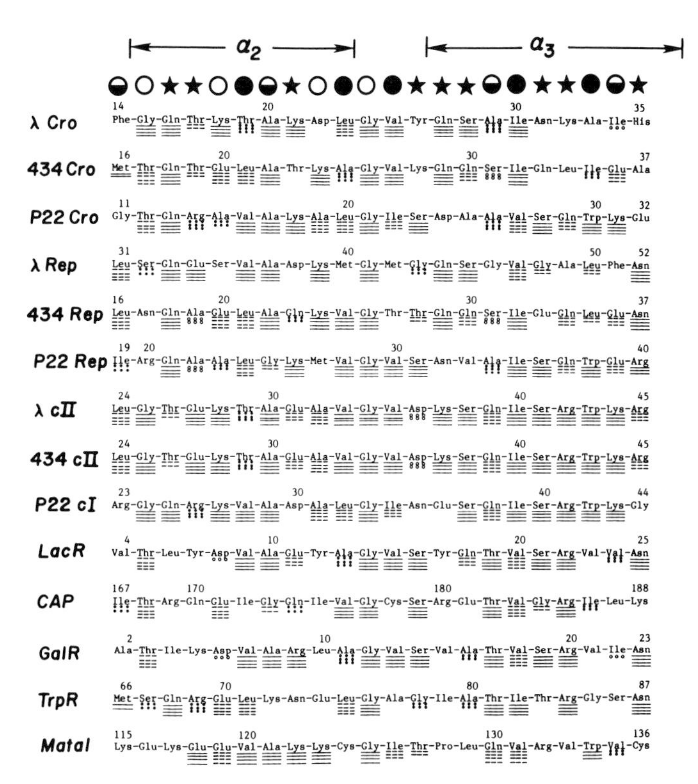

Fig. 2. Segments of the amino acid sequences of a number of gene-regulatory proteins that appear to be homologous with the DNA-binding helix-turn-helix unit of cro, λ-repressor and CAP. Amino acids that are identical in two or more sequences are underlined. The symbols at the top of the figure indicate the locations of the residues in the helix-turn-helix unit in cro; open circles indicate full exposure to solvent, half open indicates part exposure and solid circles indicate buried residues. Stars indicate presumed DNA-contact residues [based on ref. 8–11].

DNA-binding proteins

Because of the interest in locating putative helix-turn-helix units in other DNA-binding proteins, we will describe the procedure that we have adopted.

The basic idea is to compare a new amino acid sequence with a 'master set' of pre-aligned sequences that are taken from proteins known (or assumed) to contain helix-turn-helix units [cf. 11]. Our present 'master set' consists of 10 proteins (Table III). Of these, the X-ray structures of λ cro [7], λ repressor [13], 434 repressor [14], catabolite gene activator protein (CAP) [15] and *trp* repressor [16] are known, and the predicted [9] helix-turn-helix fold in *lac* repressor has been verified by NMR [17, 18]. A new amino acid sequence is searched for potential helix-turn-helix regions by comparing, in turn, every segment of 22 amino acids with the master set. Agreement of each 22 amino-acid segment with the master set is evaluated by summing the amino acid differences between the tested segment and each of the 'master' segments in turn, followed by division by 220 (i.e. 22 amino acids × 10 proteins in the master set) to normalize to a per-codon basis. Initially we calculated the agreement between a new sequence and the master set on the basis of minimum base change per codon, or by comparing DNA sequences [9, 11] but, for the reasons given in the preceding section, we believe that the most sensitive procedure is to count amino acid changes directly. The 22 amino-acid segment with the best score is noted and the scores from all segments can be used to evaluate the mean and standard deviation of the distribution. In practice, the scores from comparison, of many different sequences with the

Table III. *Putative helix-turn-helix units in DNA-binding proteins based on sequence homology with a master set of ten sequences*

Protein	Amino acid segment	Score (AAC)	Significance (AAC)	Expectation of chance occurrence	Score (MBC)	Significance (MBC)	Reference
Master set of 10							
λ cro	14–35*	0.77	5.8	$< 10^{-4}$	1.13	4.0	8–11
434 cro	17–38*	0.73	7.2	$< 10^{-4}$	1.09	4.4	8, 11
P22 cro	11–31*	0.75	6.5	$< 10^{-4}$	1.07	4.7	10
λ Rep(cI)	31–52*	0.78	5.4	$< 10^{-4}$	1.15	3.7	8–11
434 Rep	16–37*	0.72	7.5	$< 10^{-4}$	1.00	5.5	10
P22 Rep(cII)	20–41*	0.74	6.8	$< 10^{-4}$	1.02	5.3	8, 11
CAP	167–188*	0.83	3.7	0.020	1.11	4.2	11, 19
cII	24–45*	0.76	6.1	$< 10^{-4}$	1.15	3.7	8, 11
Lac I	4–25*	0.77	5.8	$< 10^{-4}$	1.05	4.9	9, 11
Trp R	66–87*	0.79	5.1	$< 10^{-4}$	1.15	3.7	11
Other DNA-binding proteins							
GalR	2–23*	0.73	7.2	$< 10^{-4}$	1.11	4.2	11, 19
TetR	25–46*	0.75	6.5	$< 10^{-4}$	1.06	4.8	11
DeoR	22–43*	0.76	6.1	$< 10^{-4}$	1.12	4.1	20
Imm Rep(1)	48–69†	0.77	5.8	$< 10^{-4}$	1.12	4.1	21
FnR	195–216*	0.78	5.4	$< 10^{-4}$	1.06	4.8	22
Imm Rep (2)	19–40*	0.79	5.1	$< 10^{-4}$	1.17	3.5	21
AntP	97–118	0.80	4.7	0.0005	1.26	<2.7	23
MuB	19–40*	0.80	4.7	0.0002	1.14	3.8	24
BirA	20–41*	0.80	4.7	0.0002	1.15	3.7	25
DBP II (RM)	5–26	0.80	4.7	0.0001	1.20	3.1	26
Mata1	114–135*	0.80	4.7	0.0001	1.21	3.0	11
P2R	18–39*	0.80	4.7	0.0004	1.12	4.1	27
MuA	125–146*	0.81	4.4	0.0035	1.15	3.7	28
ad5E1a	289–310	0.81	4.4	0.0014	1.25	<2.7	29
DNA PolI	160–181	0.81	4.4	0.0049	1.20	3.1	30
cmyb(CH)	128–149	0.81	4.4	0.0011	1.31	<2.7	31
vmyb	128–149	0.81	4.4	0.0011	1.33	<2.7	31
TF1(SPO1)	66–87	0.81	4.4	0.0004	1.21	3.0	32
TN21	162–183	0.81	4.4	0.0009	1.17	3.5	33

*† For explanation of indicators see text, p. 000.

master set are pooled. This improves the accuracy of both the mean and the standard deviation.

The results of comparing the amino acid sequences of many DNA-binding proteins with the master set are summarized in Table III. For each protein, the table gives the 22 amino-acid segment that agrees best with the master set. For these segments, the agreement score and the significance (number of standard deviations from the mean) are given both in terms of amino acid changes per codon and minimum base change per codon.

Table III gives the scores for the 10 proteins in the master set (each of these proteins was compared with the remaining 9 members of the set). Then follows the amino acid segments from a number of other DNA-binding proteins ranked according to their agreement with the master set. In consulting Table III it must be remembered that larger proteins contain more 22 amino-acid segments than do small ones. It is, therefore, more likely that a large protein will include an amino acid segment with high score, simply on the basis of chance. To take this into account, one can use standard statistical tables to estimate, from the significance (Table III), the frequency that such an event is expected to occur by chance. Multiplying this frequency by the number of amino acid segments in the protein yields the 'Expectation of chance occurrence' (Table III), i.e. the overall probability that the observed score is expected to occur by chance in a protein of equal size [cf ref. 11].

The survey presented in Table III is based exclusively on amino acid sequence comparison. It is possible to make the search for helix-turn-helix units more discriminatory by rejecting amino acid segments that do not satisfy certain structural criteria [11]. This has not been done here.

Many of the segments presented in Table III have already been identified as possible helix-turn-helix units. In such cases we have marked them with an asterisk and included a reference to the earlier work. In a few cases the segments chosen by others as the most likely helix-turn-helix substructure differ from those suggested by the procedure used here. These are indicated by a dagger. We have arbitrarily restricted Table III to 22 amino-acid segments with AAC scores of 0.81 or better. A more extensive compilation will be presented elsewhere.

Protein structure comparison

Although the emphasis in this article has been on the comparison of amino acid sequences, reference should be made to the use of protein structure comparison to explore distant evolutionary relationships [e.g. see 34, 35]. Since protein structures change more slowly with time than do amino acid

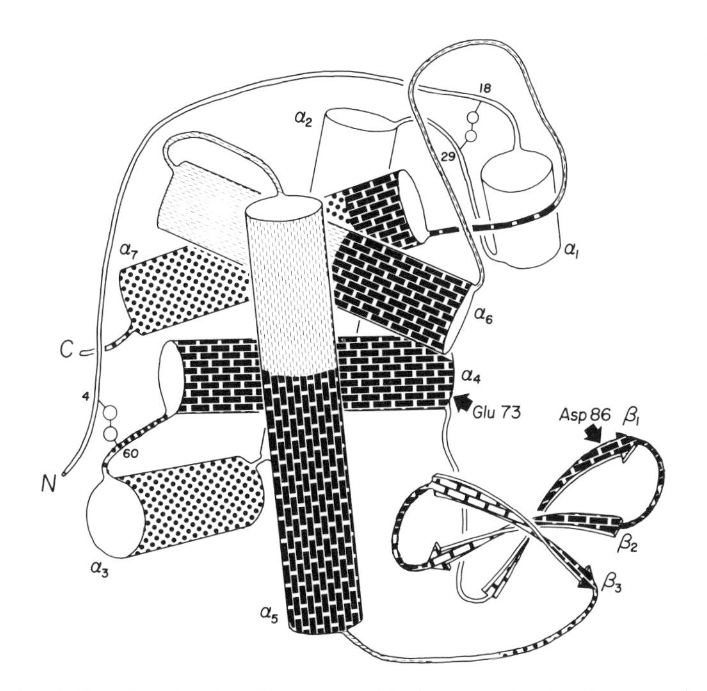

Fig. 3. Simplified drawing of goose egg-white lysozyme showing those parts of the structure that are common to hen and to phage lysozymes. Parts common to all three lysozymes are indicated by a brick-like pattern, parts common to goose egg-white lysozyme and hen (chicken) egg-white lysozyme are dotted, parts common to goose egg-white lysozyme and T4L are dashed, and parts that occur only in goose egg-white lysozyme are shown as open areas [after refs 37, 38].

sequences, the extant protein structures may contain a record of evolutionary relationships that are undetectable at the level of amino acid sequences. On the other hand, two protein structures may be structurally similar because they have independently adopted similar modes of folding – not necessarily because they evolved from the same precursor.

One example of structural similarly in the absence of amino acid sequence correspondence is provided by the structures of hen egg-white lysozyme and the lysozyme from T4 bacteriophage. In this instance, new information has recently been provided by the determination of the structure of goose egg-white lysozyme [37]. The amino acid sequence of this enzyme has no obvious homology with either chicken or phage lysozymes. Nevertheless, the three-dimensional structure of goose lysozyme has some parts that are common with chicken lysozyme, some parts that are in common with phage lysozyme and other parts that are common to both (Fig. 3) [37, 38]. We believe that this pattern of structural correspondence strongly suggests that all three lysozymes have evolved from a common precursor [37]. The lysozyme comparison is just one example of the use of protein structure comparison to evaluate potential distant evolutionary relationships. Other examples are summarized by Weaver *et al.* [38].

Acknowledgements

This work is supported in part by grants to B.W.M. from the National Institute of Health (GM 20066; GM 21967) and the National Science Foundation (PCM 8312151) and by an NIH postdoctoral fellowship (GM 10476) to R.G.B.

References

1. Doolittle, R. F., *Science* **214**, 149 (1981).
2. Haber, J. E. and Koshland, D. E. Jr., *J. Mol. Biol.* **50**, 617 (1970).
3. Fitch, W. M., *J. Mol. Biol.* **16**, 9 (1966).
4. Jue, R. A., Woodbury, N. W. and Doolittle, R. F., *J. Mol. Evol.* **15**, 129 (1980).
5. Weaver, L. H., Rennell, D., Poteete, A. R. and Matthews, B. W., *J. Mol. Biol.* **184**, 739 (1985).
6. Sauer, R. T. and Anderegg, R., *Biochemistry* **17**, 1092 (1978).
7. Anderson, W. F., Ohlendorf, D. H., Takeda, Y. and Matthews, B. W., *Nature (London)* **290**, 754 (1981).
8. Anderson, W. F., Takeda, Y., Ohlendorf, D. H. and Matthews, B. W., *J. Mol Biol.* **159**, 745 (1982).
9. Matthews, B. W., Ohlendorf, D. H., Anderson, W. F. and Takeda, Y., *Proc. Natl. Acad. Sci. USA* **79**, 1428 (1982).
10. Sauer, R. T., Yocum, R. R., Doolittle, R. F., Lewis, M. and Pabo, C. O., *Nature (London)* **298**, 447 (1982).
11. Ohlendorf, D. H., Anderson, W. F. and Matthews, B. W., *J. Mol. Evol.* **19**, 109 (1983).
12. Steitz, T. A., Ohlendorf, D. H., McKay, D. B., Anderson, W. F. and Matthews, B. W., *Proc. Natl. Acad. Sci. USA* **79**, 3097 (1982).
13. Pabo, C. O. and Lewis, M., *Nature (London)* **298**, 443 (1982).
14. Anderson, J. E., Ptashne, M. and Harrison, S. C. *Nature* **316**, 596 (1985).
15. McKay, D. B. and Steitz, T. A., *Nature (London)* **290**, 744 (1981).
16. Schevitz, R. W., Otwinowski, Z., Joachimiak, A., Lawson, C. L. and Sigler, P. B., *Nature (London)* **317**, 782 (1985).
17. Zuiderweg, E. R. P., Kaptein, R. and Wüthrich, K., *Proc. Natl. Acad. Sci. USA* **80**, 5837 (1983).
18. Zuiderweg, E. R. P., Billeter, M., Boeleus, R., Scheek, R. M., Wüthrich, K. and Kaptein, R., *FEBS Lett.* **174**, 243 (1984).
19. Weber, I. T., McKay, D. B. and Steitz, T. A., *Nucl. Acids, Res.* **10**, 5085 (1982).
20. Valentin-Hansen, P., Hojrup, P. and Short, S., *Nucl. Acids, Res.* **13**, 5927 (1985).
21. Dhaese, P., Seurinck, J., DeSmet, B. and Montagu, M. V., *Nucl. Acids Res.* **13**, 5441 (1985).
22. Shaw, D. J. and Guest, J. R., *Nucl. Acids Res.* **10**, 6119 (1982).
23. Levine, M., Rubin, G. M. and Tijan, R., *Cell* **38**, 667 (1984).
24. Miller, J. L., Anderson, S. K., Fujita, D. J., Chaconas, G., Baldwin, D. L. and Harshey, R. M., *Nucl. Acids Res.* **12**, 8627 (1984).
25. Howard, P. K., Shaw, J. and Otsuka, A. J., *Gene* **35**, 223 (1985).
26. Tanaka, I., Appelt, K., Dijk, J., White, S. W. and Wilson, K. S., *Nature (London)* **310**, 376 (1984).
27. Ljungquist, E., Kockum, K. and Bertani, L. E., *Proc. Natl. Acad. Sci. USA* **81**, 3988 (1984).
28. Harshey, R., personal communication.
29. Perricaudet, M., Akusjarvi, G., Virtanen, A. and Petterson, U., *Nature* **281**, 694 (1979).
30. Joyce, C. M., Kelley, W. S. and Grindley, N. D., *J. Biol. Chem.* **257**, 1958 (1982).
31. Klempnauer, K.-H., Gonda, T. J. and Bishop, J. M. *Cell* **31**, 453 (1982).
32. Greene, J. R., Brennan, S. M., Andrew, D. J., Thompson, C. C., Richards, S. H., Heinrikson, R. L. and Geiduschek, E. P., *Proc. Natl. Acad. Sci. USA* **81**, 7031 (1984).
33. Hyde, D. R. and Tu, C.-P. D., *Cell* **42**, 629 (1985).
34. Rossman, M. G., Liljas, A., Brändén, C. I. and Banaszak, L. J., in *The Enzymes* (ed. P. E. Boyer), vol. 11, 3rd edition, pp. 61–102. Academic Press, New York (1975).
35. Matthews, B. W., in *The Proteins* (ed. H. Neurath and R. L. Hill), vol. III, 3rd edition. pp. 403–590. Academic Press, New York (1977).
36. Matthews, B. W., Grütter, M. G., Anderson, W. F. and Remington, S. J., *Nature* **290**, 34 (1981).
37. Grütter, M. G., Weaver, L. H. and Matthews, B. W., *Nature (London)* **303**, 828 (1983).
38. Weaver, L. H., Grütter, M. G. Remington, S. J., Gray, T. M., Isaacs, N. W. and Matthews, B. W., *J. Mol. Evol.* **21**, 97 (1985).

Chemica Scripta 1986, **26B**, 257–258

The Structural Basis of Photosynthetic Light Reactions in Bacteria

Robert Huber

Max-Planck-Institut für Biochemie, 8033 Martinsried, Federal Republic of Germany

Paper presented at the Conference on 'Molecular Evolution of Life', Lidingö, Sweden, 8–12 September 1985

Abstract

In photosynthesis light energy is absorbed and subsequently converted into chemical energy stored in energy-rich compounds. Photosynthesis occurs in certain bacteria and green plants in specialized photosynthetic membranes which are either infoldings of the cell membrane (in prokaryotes) or are localized in chloroplast organelles (in eukaryotes). The primary events of the photosynthetic process consist of light absorption and charge separation. These occur in functional units, the light-harvesting complexes and the reaction centers respectively. The light-harvesting complexes funnel the energy of the absorbed light to the reaction centers with very high efficiency. In the reaction centers the charge separation across the photosynthetic membrane occurs.

A vast amount of functional and structural data of components of the photosynthetic apparatus of bacteria has been accumulated [1] but a detailed understanding requires knowledge of the three-dimensional structures of the macromolecular complexes involved in these reactions at the atomic level. This goal has been achieved recently for both light-harvesting complexes and reaction centers by crystallographic analyses.

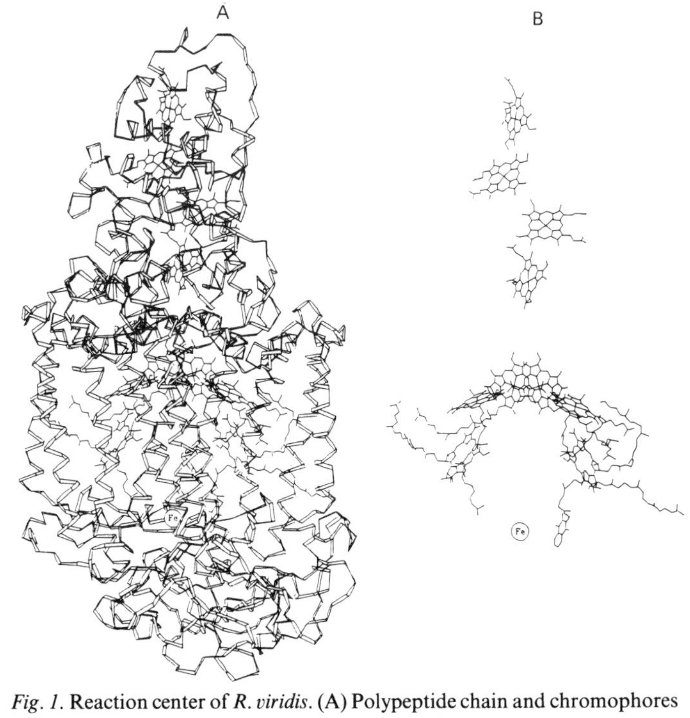

Fig. 1. Reaction center of *R. viridis.* (A) Polypeptide chain and chromophores of cytochrome, L, M, H subunits. (B) Chromophores.

The main component of the light-harvesting organelles of the cyanobacterium *Mastigocladus laminosus* was analysed by Schirmer, Bode, Huber, Sidler and Zuber [2]. The reaction center of the purple bacterium *Rhodopseudomonas viridis* had been crystallized by Michel [3] and its crystal structure was determined by Deisenhofer, Miki, Epp, Huber and Michel [4–6].

The light-harvesting organelles of cyanobacteria (the phycobilisomes) form antenna-like structures that are assembled from stacked hollow double discs. The discs are formed from trimers of ($\alpha\beta$) protein pigment complexes. The pigments are open-chain tetrapyrrole systems to which these protein complexes owe their deep-blue colour. The pigments cover the wall of the antenna and stick into the central channel. They absorb the light and conduct light energy probably by a mechanism of inductive resonance within a few psec along the antenna to the photosynthetic membrane. Energy-transfer with the phycobilisomes occurs almost without loss by fluorescence or other mechanisms of energy dissipation.

The folding of the individual protein subunits is largely α-helical. Very remarkably the arrangement of the helical segments in the globular protein bodies resembles the fold first observed in the globin family of proteins. This may point to

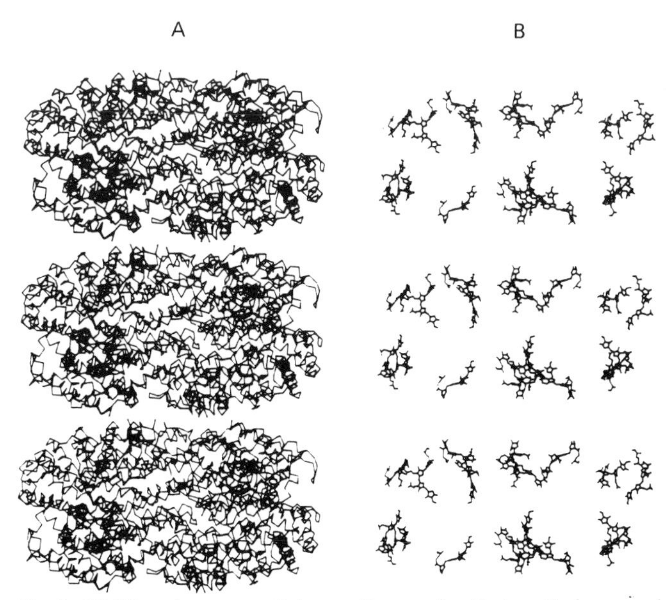

Fig. 2. (A) Three hexamers of *Agmenellum quadruplicatum* C-phycocyanin polypeptide chain. (B) Chromophores.

a very distant evolutionary relationship of biliproteins and globins.

The reaction centre of *R. viridis* is a huge protein pigment complex assembled from four different proteins, the H, L, M, subunits and the cytochrome. *In vivo* it is integrated in the photosynthetic membrane, in the crystals it is associated with detergent. The central part of the complex is made by the L and M subunits. The L and M subunits have similar shapes and foldings and resemble dishes to hold the essential chromophores, four bacteriochlorophylls, two bacteriopheophytins, an inorganic iron and a quinone. These chromophores are arranged in a very special way to form two similar branches of aromatic systems in close contact. The branches meet at two closely associated bacteriochlorophylls, the special pair. This is the primary electron donor which, upon excitation by light releases an electron. The electron is conducted via a bacteriochlorophyll and a bacteriopheophytin to the quinone (Qa), the primary electron acceptor. This process of charge separation occurs in the psec range. The unique arrangement of the chromophoric groups is necessary to prevent back reaction. The special pair is located close to the periplasmic side of the photosynthetic membrane, the quinone is on the cytoplasmic side. The establishment of an electrical charge across the membrane is followed by further reactions to transfer the electron from Qa to a secondary quinone (Qb) and to generate a pH gradient. The oxidized special pair is reduced by the cytochrome component of the reaction center complex and the circuit thus closed. The cytochrome has four heme groups, bound and arranged in a row along which fast electron transfer seems possible. The cytochrome forms the periplasmic and the H-subunit the cytoplasmic cover of the LM pigment complex.

The arrangement of the chromophores in the phycobilisomes and the reaction center indicate a unique pathway for the light energy travelling along the antenna and the photo-induced electrons crossing the membrane, respectively.

References

1. Collection of articles in *The Photosynthetic Bacteria* (ed. R. K. Clayton and W. R. Sistrom). Plenum Press, New York (1978).
2. Schirmer, T., Bode, W., Huber, R., Sidler, W. and Zuber, H., *J. Mol. Biol.* **184**, 257 (1985).
3. Michel, H., *J. Mol. Biol.* **158**, 567 (1982).
4. Deisenhofer, J., Epp, O., Miki, K., Huber, R. and Michel, H., *J. Mol. Biol.* **180**, 385 (1984).
5. Deisenhofer, J., Michel, H., and Huber, R., *Trends Biochem. Sci.* **10**, 243 (1985).
6. Deisenhofer, J., Epp, O., Miki, K., Huber, R. and Michel, H., *Nature (London)* **318**, 618 (1985).

Chemica Scripta 1986, **26B**, 259–262

Inorganic Pyrophosphate and the Molecular Evolution of Biological Energy Coupling

H. Baltscheffsky*, M. Lundin, C. Luxemburg, P. Nyrén and M. Baltscheffsky

Department of Biochemistry, Arrhenius Laboratory, University of Stockholm, S-106 91 Stockholm, Sweden

Paper presented by Herrick Baltscheffsky at the Conference on 'Molecular Evolution of Life', Lidingö, Sweden, 8-12 September 1985

Abstract

In photosynthetic and respiratory systems of living cells the energy coupling between exergonic electron transport reactions and the endergonic phosphorylation reactions, leading to conservation of energy liberated at the oxidation-reduction level in energy-rich ATP molecules, has long been recognized to occur with the membrane-bound and proton-pumping coupling factor ATPase (H^+-ATPase, ATP synthase). The first and only known alternative system for energy coupling in biological membranes was discovered by us in chromatophores from the purple non-sulfur photosynthetic bacterium *Rhodospirillum rubrum*. The alternative energy-rich phosphate compound formed, at the expense of light energy, was inorganic pyrophosphate (PP_i). It has recently been possible to solubilize from chromatophore membranes and purify the corresponding alternative coupling factor, a proton-pumping PPase (H^+-PPase, PP_i synthase).

Functional and structural properties of this alternative energy coupling system indicate that it may have preceded the ATPase in biological evolution. In this presentation the coupling factor PPase from chromatophores will be described. So will some of our new data indicating the existence of a coupling factor PPase in yeast mitochondria. Finally, certain evolutionary and mechanistic implications from the relative simplicity of both enzyme and product in the PP_i synthase system will be considered.

Introduction

The term 'energy coupling' is used in this presentation in a very restricted sense, i.e. to describe the coupling, in biological membrane-bound systems, between exergonic electron transport reactions and endergonic phosphorylation reactions, by means of which the energy liberated at the oxidation-reduction level becomes conserved in energy-rich phosphate compounds.

ATP, which has long been known as the chemical 'energy currency' of living cells, is formed in photosynthetic and respiratory systems in reactions involving the widely investigated membrane bound proton pumping coupling factor ATPase (H^+-ATPase, ATP synthase). Its properties and those of its polypeptide subunits (five or more in the F_1 or 'knob' part and three in the integrally membrane bound F_0 or 'stalk' part) have been described in great detail, also in several review articles [1–5]. Evolutionary aspects of the coupling factor ATPase and subunits have been discussed in recent years [6–8] and are the subject of two presentations at this conference [9, 10].

The first, and so far only known, alternative system for biological, membrane-bound energy coupling gives inorganic pyrophosphate (PP_i) instead of ATP. It was discovered by us in chromatophores from the purple non-sulfur photosynthetic bacterium *Rhodospirillum rubrum* [11, 12] where it was formed at the expense of light energy when only P_i and no ADP had been added to the reaction medium [13]. It has recently been possible to solubilize, purify and characterize a coupling factor PPase (H^+-PPase, PP_i synthase) from the chromatophores [14, 15]. What appears to be a similar coupling factor PPase has now been demonstrated to occur also in, and solubilized from, yeast (*Saccharomyces cerevisiae*) mitochondria [16].

Focusing upon evolutionary aspects, this presentation will review the most pertinent information about the alternative coupling factors, the PPases from photobacterial chromatophores and yeast mitochondria. The molecular evolution of biological energy coupling is assumed to have occurred in two distinct and consecutive steps:

$$AH_2 + B + 2P_i \xrightleftharpoons{\text{coupling factor PPase}} A + BH_2 + PP_i + H_2O \quad (1)$$

$$AH_2 + B + P_i + ADP \xrightleftharpoons{\text{coupling factor ATPase}} A + BH_2 + ATP + H_2O \quad (2)$$

where A/AH_2 and B/BH_2 are neighbour carriers in an electron transport chain yielding in their oxidation-reduction reaction a sufficient amount of free energy to allow, over the respective coupling factor system, the formation of energy-rich phosphate. Support for this assumption will be given and, in addition, some evolutionary and mechanistic implications emerging from the comparative simplicity of the coupling factor PPase system will be considered.

The coupling factor PPase from *Rhodospirillum rubrum*

Three sets of data led to our discovery of the distinct coupling factor PPase in chromatophores from *Rhodospirillum rubrum*. The first indication was the demonstration of an uncoupler stimulated, membrane bound PPase activity in isolated, washed chromatophores [17, 18]. Soon thereafter we found in such chromatophores a photophosphorylation in the absence of added ADP, which yielded PP_i as the energy-rich product [11–12]. And subsequently, it was possible to demonstrate that PP_i when added in the dark to this system could act as an energy donor, as measured with both reversed electron transport at the cytochrome level [19, 20] and energization of endogenous carotenoids [19, 21].

* To whom correspondence should be addressed.

Direct evidence for the existence of this special, alternative factor for energy coupling was obtained by physical separation, after sonication, of the ATPase which became solubilized, from the PPase which remained bound to the membranes [22], and also by selective inhibition with a monospecific antibody of the hydrolysis and the synthesis of ATP [23].

Early attempts to solubilize and purify the integrally membrane-bound coupling factor PPase were made both by us and by others [24], but the enzyme turned out to be very difficult to handle in this respect. However, after long and strenuous years, we now have an almost pure and rather stable solubilized PPase preparation [14, 15], which lends itself to functional studies as well as incorporation into liposomes with retained energy transfer reactions [25, 26]. The molecular weight is still unknown and the subunit composition somewhat ambiguous, due to the very hydrophobic nature and content of added detergent of the isolated enzyme, but preliminary data indicate that the subunit number is in the range 4–6.

Incorporation of the solubilized and purified PPase preparation into soybean phospholipid liposomes has yielded additional information. Stimulation by uncouplers of the membrane bound PPase activity, which had been lost upon solubilization, was regained after incorporation into liposomes [25]. The capacity of PP_i to drive the synthesis of ATP was demonstrated in liposomes to which only the purified coupling factors PPase and F_0F_1 ATPase had been added [26]. Both these experiments clearly indicate that the alternative coupling factor PPase, as the ATPase, can function as a proton pump. Thus, the basic similarities between the PPase- and the ATPase-coupling factors are further underlined.

DCCD (dicyclohexylcarbodiimide), which is known to act at the proton channel level, gives a similar inhibition of the PPase and the ATPase activities in chromatophores [27], which may well be of particular evolutionary significance, as we have pointed out earlier [27]. The 'proteolipid' in the H^+ channel part of the ATPase coupling factor, where DCCD binds [28], has a very high content of glycine and alanine, which appear to have been the two first amino acids coded for by an early, evolving genetic code [29]. We consider a proton channel, linked to a membrane-bound enzyme hydrolyzing PP_i, to be a logical candidate for an original or very early functional unit in prebiological or early biological membranes. The original biological energy coupling may have emerged when a vectorial electron transport system became bound to such a membrane [27, 30].

In Fig. 1 we try to account for both the similarity between the two coupling factors in response to DCCD and their different response to oligomycin, which inhibits only the ATPase. The figure also shows that only the coupling factor ATPase has a pronounced 'knob' part. The slightly stimulating effect of oligomycin on light-induced formation of PP_i in chromatophores [12] may well be due to a closing of 'open' or 'leaky' channels of the ATPase where some of the comparatively easily removable F_1 parts ('knobs') may be uncoupled. Such an apparent closing has long been known to occur in submitochondrial systems [31].

Very recently a method was developed in our laboratory, which allows rapid and sensitive monitoring of PP_i synthesis [32]. With this method we have, for the first time, been able to study in detail the interrelationships between the energy requiring and energy yielding reactions in the energy conversion system of *R. rubrum* chromatophores. The reversibility and characteristics of energy linked reactions in the coupling factor PPase systems, such as PP_i synthesis driven by reversed transhydrogenase [33] or by artificially induced pH-gradients are now investigated.

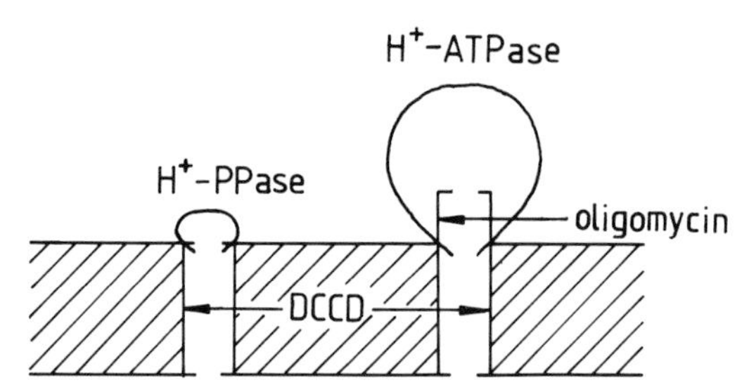

Fig. 1. An illustration of the two membrane-bound coupling factors H^+-PPase and H^+-ATPase, showing proton channel blocking effects of DCCD and oligomycin. The proton channel of ATPase has been extended to differ from the one of PPase in order to account for their different response to oligomycin.

In all the instances studied where one might expect a comparatively low level of energization, both the rate and extent of PP_i synthesis turn out to exceed those of ATP synthesis [34]. This is in contrast to the case under conditions of high energization such as saturating illumination, when the rate of PP_i synthesis only is about 10% of the rate of ATP synthesis. Another recent finding is that the rate of ATP synthesis can be increased by inhibiting specifically PP_i synthesis and/or hydrolysis with NaF [34]. It was discussed above that the specific ATPase inhibitor oligomycin increases the rate of PP_i synthesis. It thus appears that the alternative phosphorylating reactions are competing for the available proton gradient *in vivo*. The data indicate that, under physiological conditions, simultaneous synthesis of both PP_i and ATP usually occurs in *R. rubrum*.

A coupling factor PPase from *Saccharomyces cerevisiae*

Yeast mitochondria have a membrane bound PPase activity which may be more than two-fold stimulated by the ATPase inhibitor Dio-9 [16]. Solubilization of this PPase has been achieved with a method earlier used for heart mitochondria [35]. The solubilized PPase is not stimulated by Dio-9. This agrees with our interpretation that the stimulatory effect obtained with Dio-9 on the membrane bound activity indicates involvement of the PPase, when bound *in situ*, in energy coupling.

The ATPase inhibitor oligomycin was found to stimulate PPase in yeast mitochondria [16]. It may be recalled that oligomycin also in chromatophores inhibited only reactions of the ATPase, whereas formation of PP_i was somewhat stimulated [12].

While attempting to determine formation of PP_i in yeast mitochondria and submitochondrial particles (SMP) with the new and sensitive method for continuous determination of PP_i synthesis [32], it became clear that oligomycin, which was added with the intention to inhibit an existing formation of free ATP in the absence of added ADP, strongly stimulates this ATP formation (Fig. 2). This novel reaction is dependent on added NADH and P_i. Preincubation of SMP with oligomycin up to 40 min did not diminish this stimulation [36]. Maximum stimulation of free ATP formation was obtained with at least 10 μg oligomycin/mg protein, with not more than 50 μg protein/ml, and after energization in the presence

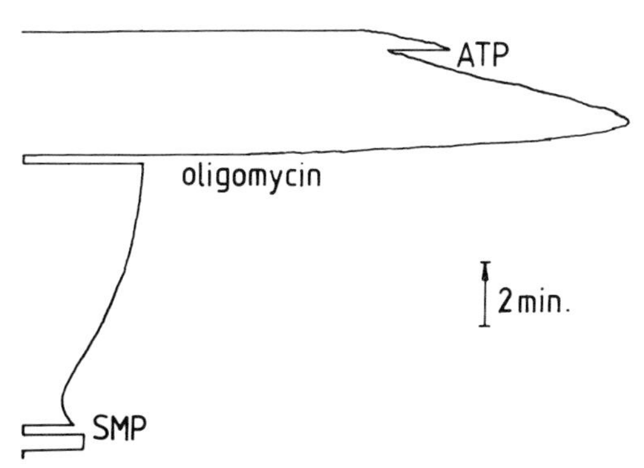

Fig. 2. A typical curve from the luminometric measurements of free ATP formed. The assay mixture contains: 3 mM phosphate, 5 mM NADH, 10 nM DAPP (P^1-P^5-di(adenosine-5′)pentaphosphate), 0.1 M Tris-Ac, 2 mM EDTA pH 7.75 and ATP monitoring reagent containing *i.a.* luciferase and a final concentration of 10 mM Mg^{2+} (commercially available from LKB Wallac, Finland). The total volume is 0.5 ml. The additions are: 71 microgram SMP-protein, 1 microgram oligomycin, 6 nM ATP standard.

of NADH for at least 5–10 min*. With ‘endogenous’ ADP the amount of free ATP formed was usually about 0.3 nmol ATP/mg SMP protein* [36]. This is close to the amount of F_1 in ox heart mitochondria [37].

A new phenomenon is thus that oligomycin can very strongly, in an energy dependent way, stimulate formation of free ATP. A possible explanation for this phenomenon is that one ATP from only one single turnover is released by the effect of oligomycin, that this release is energy dependent and that the ATPase is inhibited after this release.

Discussion

The experimental studies of the two alternative coupling systems in chromatophores from *R. rubrum* have added to our general knowledge about energy coupling in biological membranes and are providing an increasingly solid background for more detailed discussions about its molecular evolution. Early results indicated that not only ATP but also PP_I can drive energy-requiring reactions in both chromatophores and mitochondria from yeast and rat liver [18]. The apparent existence of membrane bound PPase coupling factors in other prokaryotic systems [38] as well as in (eukaryotic yeast) mitochondria clearly indicates that broader knowledge of this so far unique alternative system for energy coupling in biological membranes should be within close reach. The archaebacteria is a group of organisms where it would appear to be of particular value to gain more information about membrane bound energy coupling systems, from both a molecular and an evolutionary point of view.

In Fig. 3 the well known structures of PP_i and ATP are shown, just to remind the reader about how much simpler PP_i is than ATP. PP_i can be obtained by heating of P_i, and may well have been more abundant than ATP on the primitive earth at the time when life began. Due to its greater complexity ATP is a much more versatile compound that PP_i: whereas PP_i can only phosphorylate, ATP is able to phosphorylate, pyrophosphorylate and adenylate, properties which are well in line with an early takeover of a major role in energy metabolism from PP_i.

* These experiments were made in co-operation with Dr Lucia Pereira da Silva, Instituto de Biologia, Departamento de Bioquímica, UNICAMP, Campinas SP, Brasil.

PPi

```
    O   O
    ||  ||
HO-P-O-P-OH
    |   |
   OH  OH
```

ATP

```
                    O   O   O
                    ||  ||  ||
adenyl-ribosyl-O-P-O-P-O-P-OH
                    |   |   |
                   OH  OH  OH
```

Fig. 3. Visualization of PP_i and ATP showing the similarities and differences between these two energy-rich compounds and emphasizing the relative simplicity of PP_i.

PP_i occurs in place of ATP also in other energy and phosphate transfer reactions than those which are linked to biological membranes. Several reactions of intermediary metabolism in proprionic acid bacteria utilize PP_i [39]. Other reactions in these bacteria involve high molecular weight inorganic polyphosphates [40]. A possible sequence of events in the evolution of early reactions with energy-rich phosphates may be:

$$PP_i \longrightarrow (P_i)_n \xrightarrow{ADP} ATP$$

where n is a high number. Two notes of caution are: (1) carbamyl phosphate and other possible candidates for early energy-rich phosphate compounds are not considered here, and (2) no indications have been obtained for a participation of high molecular weight inorganic polyphosphates in membrane bound energy transfer reactions.

As we have discussed recently [41], the apparent relative simplicity of both substrate and enzyme in the coupling factor PPase, as compared with the corresponding ATPase, would not appear to necessitate any assumption of a fundamentally less complex mechanism for energy coupling in the PPase system, even with a postulated mechanism involving the rotational motion of subunits variety of the conformational change hypothesis [42]. Nevertheless, the possibility cannot be excluded that the less complex PPase system uniquely retains basic coupling properties from a long evolutionary past, properties which may be lacking or more difficult to unravel in the more exquisitely elaborated coupling factor ATPase system.

References

1. Pullman, M. E. and Schatz, G., *Annu. Rev. Biochem.* **36**, 539 (1967).
2. Beechey, R. B. and Cattell, K. J., *Curr. Top. Bioenerg.* **5**, 305 (1973).
3. Baltscheffsky, H. and Baltscheffsky, M., *Annu. Rev. Biochem.* **43**, 871 (1974).
4. Amzel, L. M. and Pedersen, P. L., *Annu. Rev. Biochem.* **52**, 801 (1983).
5. Kagawa, Y., in *Bioenergetics* (ed. L. Ernster), in *New Comprehensive Biochemistry* (gen. eds. A. Neuberger and L. L. M. van Deenen), vol. 9, p. 149. Elsevier, Amsterdam, New York and Oxford (1984).
6. Baltscheffsky, H., *BioSystems* **14**, 49 (1981).
7. Harris, D. A., *BioSystems* **14**, 113 (1981).
8. Walker, J. E., Tybulewicz, V. L. J., Falk, G., Gay, N. J. and Hampe, A., in *H^+-ATPase (ATP Synthase): Structure, Function, Biogenesis. The F_0F_1 Complex of Coupling Membranes* (ed. S. Papa, K. Altendorf, L. Ernster and L. Packer), p. 1. ICSU Press, Adriatica Editrice, Bari (1984).

9. Walker, J. E. and Cozens, A. L., *Chemica Scripta* **26B**, 263 (1986).
10. Ernster, L. *et al. Chemica Scripta* **26B**, 273 (1986).
11. Baltscheffsky, H., von Stedingk, L.-V., Heldt, H.-W. and Klingenberg, M., *Science* **153**, 1120 (1966).
12. Baltscheffsky, H. and von Stedingk, L.-V., *Biochem. Biophys. Res. Commun.*, **22**, 722 (1966).
13. Horio, T., von Stedingk, L.-V. and Baltscheffsky, H., *Acta Chem. Scand.* **20**, 1 (1966).
14. Nyrén, P., Hajnal, K. and Baltscheffsky, M., *Biochim. Biophys. Acta* **766**, 630 (1984).
15. Baltscheffsky, M. and Nyrén, P., in *Methods in Enzymology* (ed. S. P. Colowick and N. O. Kaplan), vol. 126. Academic Press, New York and London (1986).
16. Lundin, M. and Baltscheffsky, H. (submitted for publication).
17. Baltscheffsky, M., FEBS 1st Meeting, London, Abstract, 67 (1964).
18. Baltscheffsky, M., in *Regulatory Functions of Biological Membranes* (ed. J. Järnefelt), p. 277. Elsevier, Amsterdam (1968).
19. Baltscheffsky, M., *Nature* (*London*) **216**, 241 (1967).
20. Baltscheffsky, M., *Arch. Biochem. Biophys.* **133**, 46 (1969).
21. Baltscheffsky, M., *Arch. Biochem. Biophys.* **130**, 646 (1969).
22. Johansson, B. C., Baltscheffsky, M. and Baltscheffsky, H., in *Proceedings of the IInd International Congress on Photosynthesis Research* (ed. G. Forti, M. Avron and A. Melandri), p. 1203. Junk Publishers, The Hague (1972).
23. Johansson, B. C., *Dissertation*, Stockholm (1975).
24. Rao, P. V. and Keister, D. L., *Biochem. Biophys. Res. Commun.* **84**, 465 (1978).
25. Shakhov, Y. A., Nyrén, P. and Baltscheffsky, M., *FEBS Lett.* **146**, 177 (1982).
26. Nyrén, P. and Baltscheffsky, M., *FEBS Lett.* **155**, 125 (1983).
27. Baltscheffsky, M., Baltscheffsky, H. and Boork, J., in *Electron Transport and Photophosphorylation* (ed. J. Barber), p. 249. Elsevier, Amsterdam, New York and Oxford (1982).
28. Sebald, W. and Wachter, E., *FEBS Lett.* **122**, 307 (1980).
29. Eigen, M. and Schuster, P., *Naturwissenschaften* **65**, 341 (1978).
30. Mitchell, P., in *Chemiosmotic Coupling and Energy Transduction.* Glynn Research Ltd., Bodmin, England.
31. Lee, C. P. and Ernster, L., *Biochem. Biophys. Res. Commun.* **18**, 523 (1965).
32. Nyrén, P. and Lundin, A., *Anal. Biochem.* **151**, 504 (1985).
33. Nore, B., Husain, I., Nyrén, P. and Baltscheffsky, M. *FEBS Lett.* **200** 733 (1986).
34. Nyrén, P., Nore, B. and Baltscheffsky, M. *Biochim. Biophys. Acta* (in press).
35. Efremovich, N. V., Volk, E. S., Baikov, A. A. and Shakhov, Y. A., *Biokhimiya* **45**, 1033 (1980).
36. Lundin, M., Pereira da Silva, L. and Baltscheffsky, H. (submitted for publication).
37. Vadineanu, A., Berden, J. A. and Slater, E. C., *Biochim. Biophys. Acta* **449**, 468 (1976).
38. Baltscheffsky, M. and Nyrén, P., in *Bioenergetics* (ed. L. Ernster), in *New Comprehensive Biochemistry* (gen. eds. A. Neuberger and L. L. M. van Deenen), vol. 9, p. 187. Elsevier, Amsterdam, New York and Oxford (1984).
39. Wood, H. G. and Goss, N. H., *Proc. Natl. Acad. Sci. USA* **82**, 312 (1985).
40. Liu, C. L., Hart, N. and Peck, H. D. Jr., *Science* **217**, 363 (1982).
41. Baltscheffsky, M., Boork, J., Nyrén, P. and Baltscheffsky, H., *Physiol. Vég.* **23**, 697 (1985).
42. Cox, G. B., Jans, D. A., Fimmel, A. L., Gibson, F. and Hatch, L., *Biochim. Biophys. Acta* **768**, 201 (1984).

Chemica Scripta 1986, **26B**, 263–272

Evolution of ATP Synthase

J. E. Walker and A. L. Cozens

Medical Research Council, Laboratory of Molecular Biology, Hills Road, Cambridge, CB2 2QH, England

Paper presented by John E. Walker at the Conference on 'Molecular Evolution of Life', Lidingö, Sweden, 8–12 September 1985

Abstract

The multi-subunit enzyme ATP synthase is found in eubacteria, mitochondria and chloroplasts. Studies of its subunits and genes are helping to show how it has evolved. The enzymes from these sources are similar in structure and contain a basic core of eight different polypeptides; extra subunits are associated with the mitochondrial enzyme. Genes for bacterial complexes are found in clusters. In *E. coli* the eight genes are arranged in the *unc* operon. The operon divides into two subclusters that correspond to the extrinsic membrane sector, F_0 and the intrinsic membrane sector F_1. The gene order persists in other bacteria although the clusters are not always co-transcribed. Two members of the *Rhodospirillaceae* have a cluster containing F_1 genes only. In a blue green alga gene orders are most closely related to those found in chloroplast DNA where six of the eight polypeptides of ATP synthase are encoded. This provides further evidence of common ancestry between cyanobacteria and chloroplasts. Most of the subunits of the mitochondrial enzyme are nuclear encoded. They are synthesised as precursors that are able to enter the mitochondrion. One subunit, the proteolipid, has two different precursors encoded in two separate genes in bovine DNA. Their functions are not known at present. In addition to the two expressed genes two pseudogenes for this subunit also have been characterized. Further study of these nuclear genes should help to explain their evolution and may provide evidence for or against a symbiotic origin of mitochondria.

Introduction

ATP synthase (proton translocating ATPase, H^+-ATPase, F_1 F_0-ATPase) is found in mitochondria [1], chloroplasts [2] and probably all eubacteria that have been examined for its presence [3–5]. It plays a central role in energy transduction by using the energy generated by respiration in bacteria and mitochondria, or by photosynthesis in chloroplasts and photosynthetic bacteria, to synthesise ATP from ADP and P_i [6].

A number of other H^+-ATPases that are involved in secretory processes have been described [7–11]. These, the H^+-ATPase of the yeast plasma membrane [12] and cation translocating ATPases [3–16] are distinctly different by biochemical and structural criteria from ATP synthase and their physiological role is to hydrolyse ATP and generate ion gradients.

Two alternative views have been advanced to explain the origin of mitochondria and chloroplasts [17]. According to one widely held hypothesis they arose during evolution by endosymbiosis [18, 19]. Thus, it is proposed that mitochondria originated by engulfment by a proto-eukaryotic organism of an early bacterium to which the present-day organisms, *Paracoccus denitrificans* and purple non-sulphur photosynthetic bacteria, may be related [20]. Plant and eukaryotic alga cells are proposed to have arisen by subsequent engulfments of prokaryotic or eukaryotic algae by eukaryotes. Blue green algae (cyanobacteria) are thought to be related to the progenitors of chloroplasts in *Rhodophyta* (red algae); green algae and higher plants are proposed to have ancestors resembling the prokaryote, *Prochloron didemni*; chlorophyte chloroplasts may be derived from eukaryotic algae [20]. The alternative view proposes an autogeneous origin of organelles whereby nuclear and organellar genomes became physically compartmented and functionally specialised within a single cell [21]. Hence, study of ATP synthases in mitochondria, chloroplasts and bacteria may help to distinguish between these two alternatives of the origins of the organelles themselves.

Abbreviations used: DCCD, dicyclohexylcarbodiimide; oscp, oligomycin sensitivity conferral protein.

Mechanistic and structural unity of ATP synthases

According to current ideas, the enzyme functions by coupling the electrochemical potential difference for protons across the inner mitochondrial, thylakoid and cytoplasmic membranes respectively, to the synthesis of ATP from ADP and P_i [6]. The binding change mechanism of Boyer proposes that energy is required for binding of substrates and release of product [22]. Under anaerobic conditions in bacteria the enzyme also may function in the reverse sense by coupling hydrolysis of ATP to proton translocation across the membrane in the outward direction, thereby generating an electrochemical membrane potential essential for driving other cellular functions, such as motility and transmembrane transport [23].

Ultrastructural and biochemical studies have indicated that the enzymes from these diverse sources have closely related structures. They are made of two sectors, commonly called F_1 and F_0. F_1 is an extramembrane assembly attached to the membrane sector F_0. It can be dislodged from the membrane as a soluble assembly, F_1-ATPase, and contains the catalytic sites of the enzyme where substrates ADP and P_i bind. F_0 contains a transmembrane proton channel through which the membrane electrochemical potential gradient for H^+ is coupled to ATP synthesis [1, 2, 4].

Recent studies of proteins and corresponding genes from a variety of sources [24–27] have confirmed the close structural unity of enzymes from eubacteria, mitochondria and chloroplasts. The enzyme from the bacterium *Escherichia coli* contains eight different polypeptides. Five of these α, β, γ, δ and ϵ, assembled in stoichiometries of $\alpha_3\beta_3\gamma_1\delta_1\epsilon_1$, make up the F_1 complex; the remaining three, *a*, *b* and *c* constitute the F_0 assembly in an estimated stoichiometry of $a_1b_2c_{10\text{-}12}$ (see Fig. 1) [4, 28]. Counterparts of all eight polypeptides are found in the chloroplast and probably in the mitochondrial enzymes [5] (see Table I); so these polypeptides may be regarded as a core complex. Supernumerary proteins are associated in particular with the mitochondrial complex. The role of these various proteins in its reguation and assembly will be discussed below.

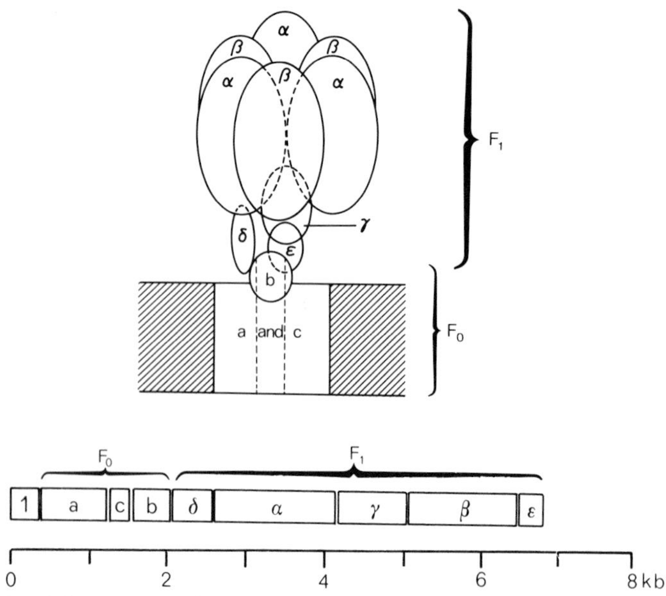

Fig. 1. Arrangement of subunits in *E. coli* ATP synthase and of corresponding genes in the *unc* operon.

The functional properties of the various subunits contain important clues to their evolutionary origins. The α- and β-subunits are both involved in catalysis by the enzyme. However, β appears to have a more intimate role in catalysis than α [22, 29]. They are related to each other in sequence (and therefore structure) [24, 30] and both contain nucleotide-binding sites [22, 31]. This implies that the genes for the α- and β-subunits were derived from a common ancestor and have evolved by gene duplication and divergence. In this context it is of interest that an F_1-ATPase isolated from the obligate anaerobe *Lactobacillus casei* appears to be more primitive, apparently having an α_6 structure rather than the $\alpha_3\beta_3$ structure of other F_1-ATPases [32]. Weak sequence relationships have been found with α and β and other nucleotide-binding proteins [30, 33]. So it appears that the common ancestor itself was derived from a primordial nucleotide binding protein. The functions and origins of the other F_1 components, the γ-, δ- and ε-subunits, are more obscure. They all are probably needed for binding F_1 to F_0 [summarized in 24] and so are likely to be involved in the coupling mechanism. It has been suggested that γ may gate the proton channel [34]. Weak sequence relationships have been noted between these subunits [35]; also the sequences of the *b*-subunit [36] and oscp (equivalent to bacterial δ-subunit [27, 37, 38]) contain internal repeats [38]. Thus, it appears that gene duplication and divergence may have been involved in their evolution. The F_0 subunits *a*, *b* and *c* all seem to be required for functioning of the proton channel [39, 42] although a recent account suggests that the *a*-subunit can conduct protons [40]. A carboxyl group in the *c*-subunit that is buried in the lipid bilayer, seems to be particularly important for proton translocation [41]. Much of the *b*-subunit is extramembranous [36, 42]. Thus, it also appears to provide an important structural link between F_1 and F_0 and to have a role in the coupling mechanism.

Table I. *Equivalent subunits in ATP synthases*

Type	*E. coli*	Beef mitochondria	Chloroplasts	References
F_1	α	α	α	24, 27, 96, 97
	β	β	β	24, 27, 96, 97, 98
	γ	γ	(γ)	24, 27
	δ	oscp	?	24, 35, 37
	ε	δ	ε	24, 27, 96, 98, 99
	—[c]	ε	?	27
F_0	*a*	*a* [or ATPase 6]	X	24, 71[a]
	b	(*b*)	I	24, 56
	c	*c* [or ATPase 9]	III	24, 41, 100
Supernummary	—	inhibitor	?	101[b]
	—	factor 6	?	68, 69
	—	A6L	?	71, 74

a The presence of an *a*-subunit gene has been demonstrated in pea chloroplast DNA [103]. This is probably the 'X' subunit noted in CF_1 F_0 from spinach chloroplasts [58].

b Nelson has suggested that chloroplast ε is the inhibitor protein.

c Subunits on the same horizontal line are believed to be equivalent in bacteria (exemplified by *E. coli*), beef mitochondria, and chloroplasts. Absence of parentheses indicates that the proteins have been sequenced; those in parenthesis have not been sequenced, but equivalence is likely on other grounds. A dash indicates that the subunit is absent from the *E. coli* complex.

This structural and mechanistic unity of ATP synthases leads to an obvious but important conclusion – the ATP synthases of eubacteria, mitochondria and chloroplasts have arisen by divergent evolution from a common progenitor. This progenitor must have contained the eight core polypeptides. Therefore, discussion of the evolution of the enzyme complex can be conveniently divided into the three parts; the evolution of the progenitor and the evolution of the mitochondrial and chloroplast complexes.

Evolution of the progenitor

The evolution of the progenitor core enzyme of eight interacting polypeptides was certainly complex and is obscure. However, it is possible to isolate a number of important separate parts of the process. This rests upon the proposition that components of F_1 evolved independently of components of F_0 and that subsequently, they became associated together in a single complex. This proposition implies that the early evolution of the catalytic mechanism of phosphorylation of ADP (or hydrolysis of ATP) was an event unrelated to the transmembrane transporting mechanism for protons. A plausible account of this early evolution of F_1 and F_0 has been suggested previously [43, 44]. In brief, this account starts with the evolution of the cellular membrane. This is likely to have been an early event as emphasised by Bangham [45] and Woese [46]. Amphiphilic lipid complexes, it has been proposed, could absorb molecules and perform primitive catalysis. In particular they might catalyse fermentations, breaking down compounds anaerobically to produce others with high energy bonds (anhydrides). This in turn could lead to macromolecular synthesis (nucleic acids, proteins). For the development of life a further essential evolutionary step was needed, the standardisation of energy currency, thereby permitting an integrated cell to appear. ATP was selected very early in evolution although it may have been preceded by pyrophosphate [47]. Once a common energy currency had been established, it would be advantageous to limit diffusion and enclose these reactions in a primitive cell. However, fermentation in a closed space would lead to acidification in the cell, slowing synthetic reactions and promoting hydrolysis. This initially could be relieved by adventitious proton leakage, but subsequently proton channels would develop, amongst them an assembly related to the present-day proton channel of ATP synthase.

Evolution of ATP synthase activity

The development in the primitive cell of a cytoplasmic activity to hydrolyse ATP such that hydrolysis was not coupled to other functions would not be advantageous. So it can be argued that the association of nucleotide binding protein with the proton channel took place before an efficient ATPase had developed. It could be that this early enzyme was already multimeric, to prevent loss of enzyme by diffusion through pores in the primitive membrane [48]. The physical coupling of the primitive ATPase with the channel, would convert it to an ATP-driven pump, thereby increasing its efficiency to extrude protons. Under these conditions the ATP-driven pump could improve. Improvement would require a tighter physical association of the channel and ATPase. The development of this tight interaction between the ATPase and proton channel could have led to the acquisition of extra subunits in both extrinsic and intrinsic membrane sectors (γ, δ and ϵ and also *b*).

Conversion of this ATP-driven proton pump into ATP synthase requires the evolution of an electron transport system. This would have to be able to harness more redox energy than is needed to maintain the internal cellular pH. Excess H^+ pumping out of the cell by pumps not coupled to ATP hydrolysis thereby would produce an electrochemical gradient. So the protons could leak back through the ATP driven pump, drive it in reverse and produce ATP. As fermentable substrates became increasingly scarce bacteria with this ability would proliferate [49].

Modular evolution

This account of the separate early evolution of precursors of F_1 and F_0 finds support in the present day arrangements of the genes encoding the present day prokaryotic ATP synthases. The genes for the *E. coli* enzyme are grouped together in the *unc* operon [24]. The operon itself contains two clusters corresponding to F_0 and F_1 genes (Fig. 2). In two of the purple non-sulphur bacteria, *Rhodospirillum rubrum* [26] and *Rhodopseudomonas blastica* [25], F_1 genes also form a cluster and are arranged in the same order as in *E. coli* (the location of F_0 genes in these *Rhodospirillaceae* is at present unknown). Related gene orders are also found in a cyanobacterium (A. L. Cozens and J. E. Walker, unpublished work) and in plant chloroplasts [5] (see Fig. 2).

These strongly related arrangements of genes for ATP synthase that correspond to the two major structural units of the enzyme, are reminiscent of adjacent gene clusters encoding heads and tails in the prokaryotic lambdoid phages [50]. It has been proposed that lambdoid phages have undergone modular evolution whereby genetic modules for DNA encapsidation and for tails (for example) interchange in interbreeding phage populations. Thus, evolution acts primarily not at the level of intact virus but at the level of individual functional units (modules). At present, insufficient information is available to assess whether this mechanism has been important in evolution of ATP synthase, but the existence of genetic modules corresponding to F_1 and F_0 is consistent with the view that it may be so, and the persistence of gene orders may be regarded as a significant evolutionary marker.

Another feature of the morphogenetic loci of bacteriophages that has been noted is that the order of genes within the locus is closely related to their order of action during assembly [51]. This correlation led to the suggestion that the gene orders in the *E. coli unc* operon may be related to assembly [24]. However, current models of the assembly pathway [52] are incompatible with this view.

Chloroplast ATP synthase

Origin of chloroplasts

The evidence for a eubacterial origin for plastids (summarized in [17]) is compelling. However, protein homologies in the 2Fe-2S ferredoxins and in the cytochromes c_6(f) family can be most simply explained by a polyphylectic ancestry for plastids, i.e. the plastids arose at least twice in independent events from different ancestors, themselves not plastids. The cyanobacteria and the group represented by *Prochloron*, which exists only as a symbiont of a protozoon [53], are the only prokaryotes which photosynthesise using similar pathways to eukaryotic algae and plants, i.e. they contain chlorophyll *a* and are able to oxidize water. (For reviews see [17, 20].)

Cyanobacteria, in common with red and brown algal

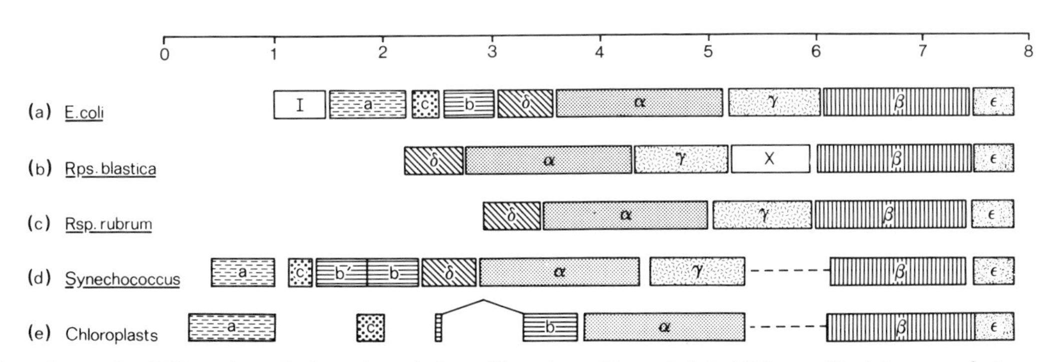

Fig. 2. Organization of genes for ATP synthases in bacteria and plant chloroplasts. The scale is in kilobases. The letters *a*, *c*, *b*, δ, α, γ, β and ϵ indicate the ATP-synthase subunit encoded in the genes. I in the *E. coli unc* operon and X in *Rps. blastica atp* operon encode proteins of unknown function [24, 25]. Chloroplast *c* is subunit III and chloroplast *b* is subunit I; *b* is located between *c* and α and contains a large intron [56]. *b′* in *Synechococcus* is a duplicated and diverged form of *b*. Dashed lines in *Synechococcus* and chloroplasts signify that the two gene clusters are well separated and separately transcribed. For fuller details of the plant species that have been investigated see ref. [5].

chloroplasts have the accessory phycobiliprotein pigments, and lack chlorophyll *b* and so they are proposed to have common ancestry. *Prochloron*, on the other hand, contains chlorophyll *b*, but lacks phycobiliproteins, and so has been suggested as representing the group from which green algae and plant chloroplasts arose. Fox *et al.* [54] have classified prokaryotes and plastids by mapping 16S rRNA with RNAse-T_1. According to this classification, *Prochloron* is more closely related to complex cyanobacteria, such as *Nostoc*, than to the chloroplasts of plants or green algae [55].

Chloroplast ATP synthase genes

Genes for chloroplast ATP synthase are distributed in both the nucleus and chloroplast DNA (for a review see [5]). The number and identity of the chloroplast encoded components of the enzyme has been controversial. Early studies with spinach indicated that three polypeptides were encoded and synthesized in the chloroplast [2]. Subsequently, largely as a result of DNA sequencing studies, five genes for ATP synthase have been shown to be in the chloroplast DNA, derived from a number of plant species (for references see legend to Table I). These are arranged in two clusters. Gene order in these clusters is related to the bacterial gene orders (Fig. 2) [5].

Recent hybridisation studies with DNA derived from the gene for the ATP synthase *a* subunit of the unicellular cyanobacterium *Synechococcus* have indicated that this sixth component is also encoded in pea chloroplasts, a finding confirmed by DNA sequencing and transcriptional mapping [103]. This gene is equivalent to the mitochondrial ATPase-6 subunit. It is located on the 5′ side of the gene for ATP synthase subunit III (or *c* subunit) and is co-transcribed with it.

The genes for the ATP synthase of the unicellular cyanobacterium *Synechococcus* 6301 are arranged in two clusters (see Fig. 2). This arrangement most closely resembles that found in chloroplasts of wheat, pea and tobacco, differing only in absence of the δ and γ genes from the chloroplast clusters. The gene order is also clearly related to those found in both *E. coli* and the purple non-sulphur photosynthetic bacteria. However, the genes are split into two independent transcriptional units in both chloroplast DNA and in *Synechococcus*, indicating that co-transcription of F_1 genes is not essential for coordinate expression. This arrangement is strong evidence that *Synechococcus* and chloroplasts share a more recent common ancestor than do *Synechococcus* and *E. coli* or *Rsp. rubrum*, since α-and β-subunits arose by gene duplication and their separation into two distinct clusters is therefore likely to be a more recent event. This also implies that the arrangement in *E. coli* is closer to the ancestral one.

Examination of the sequences of the proteins encoded by these genes provides further support for the contention that the chloroplast ATP synthase is more closely related to the cyanobacterial ATP synthase than to the related enzymes from other sources (*E. coli*, the purple non-sulphur bacteria or the mitochondria of fungi and mammals). This is illustrated by comparison of subunit-*c* sequences (Table II). Pairwise comparisons show a range of identitites of amino acid sequence from 15–86%; only four amino acid residues are conserved between all fourteen species so far examined. Notably, the most closely related pair are the wheat chloroplast and *Synechococcus* proteins in which 70/81 residues (86%) are identical. A close relationship has also been noted between the *c*-subunits of chloroplasts and in the filamentous cyanobacterium, *Mastigocladus laminosus* [41].

Table II. *Comparison of amino-acid sequences of c-subunits of ATP synthase from bacteria, mitochondria and chloroplasts* [41]. *The values shown are percentage identities of sequences. PS3 is a thermophilic bacterium*

1. *E. coli*	100							
2. PS3	37	100						
3. *Synecho* 6301	34	42	100					
4. Wheat chlorop	29	39	86	100				
5. *Rsp. rubrum*	21	28	28	27	100			
6. Beef mito	21	29	28	31	45	100		
7. Yeast mito	16	26	24	15	37	59	100	
8. *N. crassa* mito	22	31	25	23	44	52	52	100
	1.	2.	3.	4.	5.	6.	7.	8.

Comparison of all six chloroplast coded ATP synthase subunits with their counterparts from *E. coli* and *Synechococcus* (Table III) shows that the *a* and *c* subunits (both components of F_0) are most highly conserved in chloroplasts and blue green algae, and are rather poorly conserved in *E. coli*. The third component of F_0, the *b*-subunit, is weakly conserved in all three species. The high degree of homology observed between β-subunits is a feature of all organisms studied so far; presumably it reflects the constraints placed upon the active site of ATP synthase. These calculations suggest that the evolution of the F_0 portion in cyanobacteria and chloroplasts has been more strictly constrained than has the evolution of F_1.

This conclusion is borne out by comparisons between pairs of chloroplast genes; for example comparison of *c*-subunits in tobacco and wheat, shows that 80/81 residues are conserved whereas α-subunits show only an 87% identity of protein sequence.

Duplication and divergence of the Synechococcus b *subunit*

The DNA sequence of the larger ATP synthase gene cluster contains two open reading frames (putative genes) between the genes for the *c*- and α-subunits (see Fig. 2). One of these, named *b*, is closely related in its amino terminal region to the chloroplast subunit I (see Fig. 3) and in its carboxy terminal region to the *E. coli b*-subunit. The second gene, named *b*′, is weakly related in sequence to *E. coli b* and at best only

Table III. *Comparison of the amino acid sequences of ATP synthase subunits of* Synechococcus *with homologues from* E. Coli *and chloroplasts. The chloroplast subunit-a sequence is from pea* [103], *other sequences of chloroplast subunits are from wheat* [94, 100]. *The values shown are percentage identities of sequences*

Subunit	*E. coli*	Chloroplast
a	23	72
c	34	86
b	21	26
α	54	70
β	65	78
ϵ	26	31

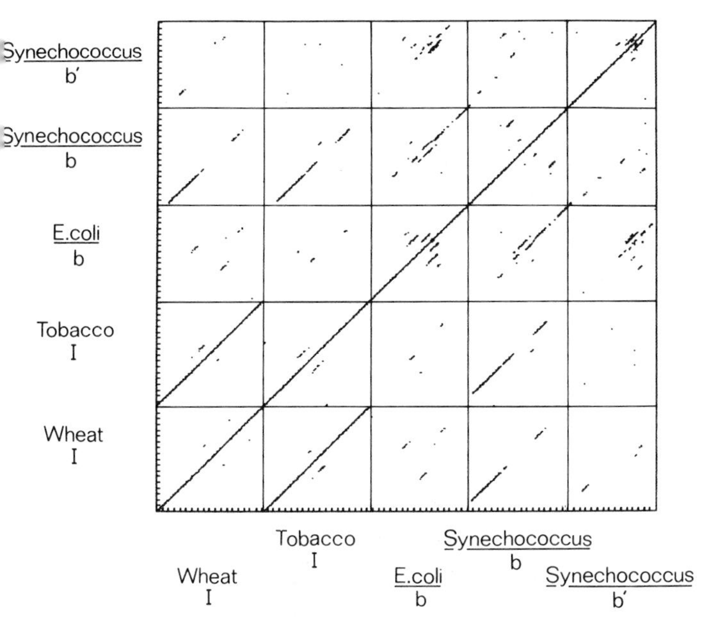

Fig. 3. Pair-wise comparisons of deduced amino acid sequences of *b* subunits of ATP synthase from *E. coli*, *Synechococcus* 6301 and tobacco and wheat chloroplasts. The calculation was made with the aid of the computer program DIAGON [93], with a span length of 25, and a score of 280.

distantly resembles chloroplast subunits I or the *Synechococcus b* protein. However, its distribution of hydrophobic residues closely resembles other members of this family in that it has a hydrophobic amino terminal domain, the rest of the sequence being hydrophilic (see Fig. 4). This implies that all these proteins have similar secondary structures. In the case of the *E. coli b* protein this has been interpreted as suggesting that the protein lies mostly outside of the membrane and is anchored in the membrane by its amino terminal hydrophobic segment.

The sequences of these proteins contain a feature that distinguishes chloroplast subunit I and *Synechococcus b* and *b'* proteins from their *E. coli* counterpart. Alignment of their sequences with the *E. coli b* sequence shows an extra N-terminal extension in the chloroplast subunit I and *Synechococcus b* and *b'* subunits that is absent from the *E. coli b* protein. In the wheat chloroplast counterpart it has been shown that 17 amino acids are removed from the amino terminus by post-translational proteolysis to produce the mature *b* protein which is found assembled in the ATP synthase complex [56] (see Fig. 5). On the basis of these sequence relationships, it seems very possible that a similar cleavage will occur in the cyanobacterial *b*- and *b'*- subunits. However, the biological function of this amino terminal extension is obscure. In both cyanobacteria and chloroplasts, subunit *b* is produced on the F_1 side of the membrane (as in *E. coli*) and so there is no apparent need for a leader sequence to direct this subunit through membrane, as is required by nuclear encoded chloroplast proteins that are synthesized in the cytoplasm. However, it is equally possible that the requirements for correct insertion of *b* in chloroplasts differ from those in *E. coli*.

The *b* subunit of *E. coli* ATP synthase is essential for reconstitution of a proton-permeable F_0 with the *a*- and *c*-subunits, and analysis of *atp* mutants indicates that it plays a central role in coupling of the F_1 to the F_0 [57]. It is probably present in two copies per complex, and assembled with one *a*- and about 10 ± 1 *c*-subunits [4]. Its hydrophilic α-helical domain extends above the membrane and interacts with F_1-subunits [36]. Post-translational processing of chloroplast *b* subunit might therefore be expected to limit the rate at which the complex is assembled. The existence of *b'*, a duplicated and diverged version of the *b*-subunit, in *Synechococcus* raises the possibility that a similar situation exists in the chloroplast enzyme. The most likely candidate for this role is subunit II which is a nuclear coded product [58].

The chloroplast subunit I has an intron

The genes for the ATP synthase subunit I in wheat, tobacco and pea chloroplasts contain an intervening sequence [56]. Introns have been demonstrated in other higher plant chloroplast genes [59], and appear commonly in algal plastid genes. They are smaller (< 1 kb) than the introns generally found in eukaryotic protein-coding genes, and do not follow the GT : AG rule at the intron/exon boundaries [60]. Rather they have a different consensus sequence for the regions around the splice sites [61, 62]. Two different points of view can be advanced to explain the presence of introns in chloroplast genes. One point of view derives from the proposal that introns were present in early prokaryotes and have been deleted from prokaryotes during evolution. Their presence in some chloroplast genes could be regarded as evidence for an intermediate form in this process. An alternative viewpoint is that since introns have not yet been found in protein coding genes in eubacteria, the chloroplast introns have appeared since the establishment of chloroplasts.

Regulation of chloroplast ATP synthase activity

Mitochondria (see below) have an additional inhibitor subunit. Its function is probably to control the activity of the ATP synthase under changing physiological conditions such that ATP hydrolysis is avoided [64–67]. No such subunit has been demonstrated in the chloroplast enzyme, but Nalin and McCarty [63] have demonstrated that control of the enzyme is mediated by a different mechanism. This involves the light dependent reduction of a disulphide bond of the γ-subunit of ATP synthase by a thioredoxin system on illumination of intact chloroplasts or isolated thylakoid membranes. It produces an active ATP synthase, and reactivates photophosphorylation. *Synechococcus* and many other cyanobacteria are unable to grow in the absence of light, unlike the purple non-sulphur photosynthetic bacteria, and lack α-ketoglutarate dehydrogenase from their citric acid cycle, so it is essential that they too have a mechanism to shut down the ATPase in the dark. It is not yet clear where the relevant cysteine residues occur in the spinach ATP synthase γ-subunit, or whether these are conserved in *Synechococcus*, but it is clear that this regulatory system has evolved independently from the mitochondrial or *E. coli* systems.

Mitochondrial ATP synthase

Composition of the mitochondrial ATP synthase complex

In addition to the eight core subunits, a number of other proteins are associated with the mammalian mitochondrial complex (Table I). For example, both fungal and mammalian enzymes have a small basic inhibitor protein that binds to the

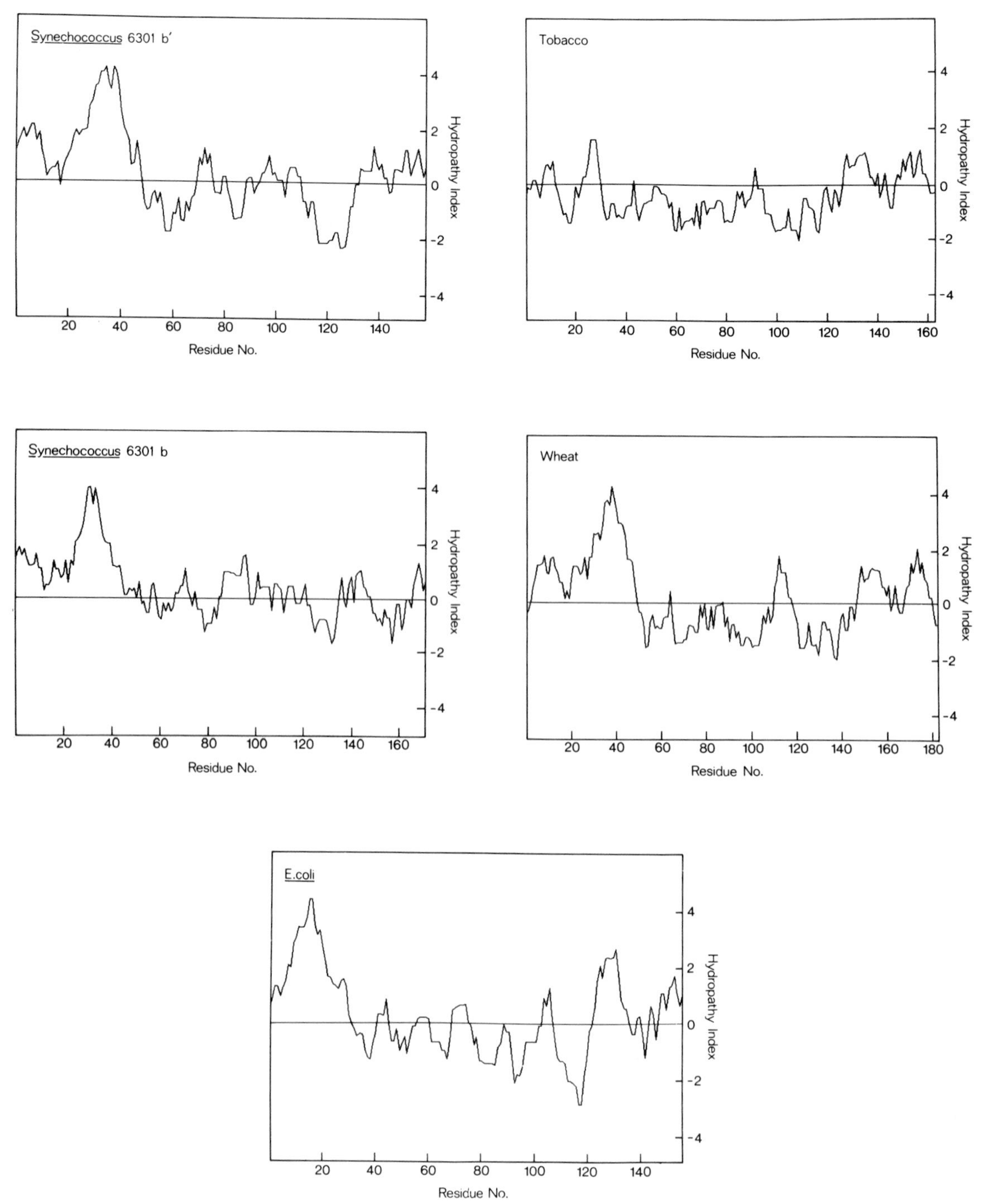

Fig. 4. Comparison of the hydropathy profiles of *b* subunits from bacteria and chloroplasts. These were calculated with the computer program HYDROPLOT [94].

β-subunits. As discussed above, this may well have a physiological function by serving to regulate the enzymic activity under changing conditions of pH in the mitochondrion [64–67]. In addition, two other proteins without apparent chloroplast or bacterial equivalents have been identified in bovine complex; these are F_6 [68, 69] and ϵ [27]. They may be important for correct (tight) binding of F_1 to F_0. A further supernummary protein, *aap*I, a component of the membrane sector of yeast ATP synthase, has been shown by genetics to be involved in assembly of F_0 [70]. A similar role may be played by its putative bovine counterpart A6L, a small hydrophobic protein encoded in mitochondrial DNA [71]. It seems possible that these extra subunits were acquired during evolution to satisfy the particular needs of mitochondrial ATP synthase (e.g. import of some subunits from the cytoplasm and subsequent assembly of these subunits with those produced inside the mitochondrion regulation).

Distribution of genes for ATP synthase

As in the chloroplast, genes for mitochondrial ATP synthase are also divided, some being in the nucleus and others in mitochondrial DNA [5]. With one exception, yeast, a constant distribution of genes is found in mammals, insects, and fungi; the mitochondrial DNA in all species (including yeast) contains the gene for the F_0-subunit *a* (or ATPase 6). In addition, in yeast and *Aspergillus*, this gene is immediately preceded by a gene for a small hydrophobic protein, *aapl*. As discussed above, this protein seems to be important in assembly of the enzyme complex. A gene for a small hydrophobic protein

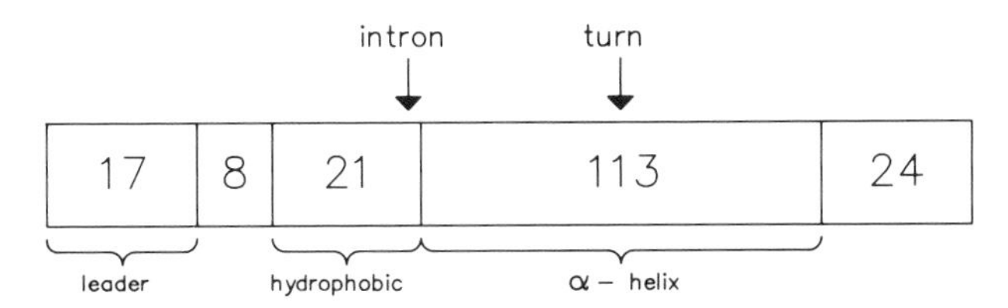

Fig. 5. Structure of the wheat chloroplast subunit I of ATP synthase. Numbers refer to amino acids. *E. coli b* lacks the 17 base-pair leader sequence, but has a similar arrangement of hydrophobic amino acids followed by an extramembranous α-helical domain [36].

(A6L) also precedes the ATPase-6 gene in mammals and *Drosophila* [71–73]; this may be an analog of *aapI* although their protein sequences are at best very weakly related. However, it remains to be shown that the protein is a component of the ATP-synthase. Yeast mitochondrial DNA also contains the ATPase-9 gene encoding the F_0 component known as the proteolipid, or dicyclohexylcarbodiimide-binding protein [75]. This arrangement is exceptional, the equivalent gene being in nuclear DNA in all other species so far examined. Thus, it is of particular interest that the mitochondrial DNAs of *Neurospora crassa* and *Aspergillus nidulans* each contain a gene encoding a proteolipid that is homologous to the proteolipid (nuclear coded) that is assembled into the ATP synthase complex. The biological significance of these genes is not clear, nor has it been demonstrated that they are expressed [76, 77].

Plant mitochondria provide a third pattern of distribution of genes encoding ATP synthase subunits. The wheat α-subunit [78] and the cucumber [79] and maize [80] proteolipid subunits are encoded in mitochondrial DNA. In *Nicotiniana plumbagifolium* a gene for the β-subunit has been characterized in nuclear DNA [102].

Origin of mitochondria

The evidence for and against an endosymbiotic origin of mitochondria has been assessed by Gray and Doolittle [17]. They conclude that 'the case for a xenogenous (endosymbiotic) origin of mitochondria is clearly less compelling than that for a xenogenous origin of plastids and we can in no way consider it proven'. Nonetheless, the symbiotic origin of mitochondria does receive strong support from a number of sources, amongst them the close evolutionary affinity between respiratory biochemistry and physiology of mitochondria and certain extant prokaryotes, notably *Paracoccus denitrificans* and members of the *Rhodospirillaceae* [20]. However, comparison of protein sequences of F_1-ATPase from *Rps. blastica* and bovine mitochondrial counterparts with those from *E. coli* gives ambiguous results. The well conserved α- and β-subunits are closer in *Rps. blastica* and mitochondria than either is to *E. coli*, but in the case of the poorly conserved γ, δ and ε polypeptides, the bacterial proteins are closer to each other than either set is to the mitochondrial subunits [25].

A xenogenous origin of mitochondria has clear implications for the evolution of ATP synthase. It implies that at some stage in the evolution of the enzyme, genes must have left the early mitochondrion and entered the nucleus or, alternatively, that the nuclear genes are derived from the engulfing microorganism. However, *Thermoplasma acidophilus*, which apparently does not have an ATP synthase of the F_1F_0 type, has been proposed as a descendant of the proto-eukaryote [81] and so this would argue against this latter hypothesis. In order for the nuclear encoded protein products to assemble with mitochondrially synthesised proteins to make the ATP synthase complex, they must have evolved a mechanism for entry into the mitochondrion.

The alternative hypothesis, the autogenous origin of mitochondria, has different implications for evolution of ATP synthase. Firstly, all of the genes would be derived from the prokaryotic ancestor of protoeukaryotes and a small number would have to partition into the mitochondria. The partitioning would imply the evolution of an efficient mechanism for import of the many proteins encoded in the nucleus that enter the mitochondrion. The difficulties of nominating an extant descendant of the prokaryote ancestor have been discussed by Gray and Doolittle [17]. Again, were *T. acidophilus* to be this ancestor, the absence of an F_1 F_0-ATPase would argue against this hypothesis.

In present-day mitochondria the imported proteins are synthesized as longer precursors with an amino terminal extension. They are released into the cytosol and then enter the mitochondria. In the process this extension is removed by proteolysis by a matrix protease [82, 83].

Nuclear genes for bovine mitochondrial ATP synthase

In order to gain a further understanding of the process of evolution of the mitochondrial ATP synthase it is necessary to investigate the nuclear genes for the enzyme complex. To this end we have cloned c-DNAs for the α-, β-, γ-, oscp, inhibitor and proteolipid subunits. In turn these c-DNA clones are being used to isolate the corresponding structural genes. These findings are summarized in the following sections.

Bovine mitochondrial import sequences. From the sequences of the precursors of the oscp, inhibitor and proteolipid it is evident that, as in other mitochondria, precursors fall into at least two classes. In one class, embracing the extrinsic membrane proteins as exemplified by the oscp and the ATPase inhibitor, the precursors are approximately 25 amino acids in length. The second class has currently two related members, the P1 and P2 proteolipids. It is characterised by longer import sequences about 65–70 amino acids long.

These import sequences are weakly related to each other

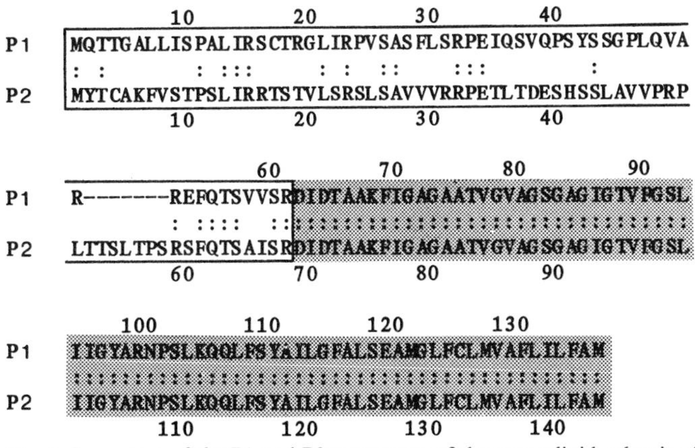

Fig. 6. Sequences of the P1 and P2 precursors of the proteolipid subunit of bovine ATP synthase.

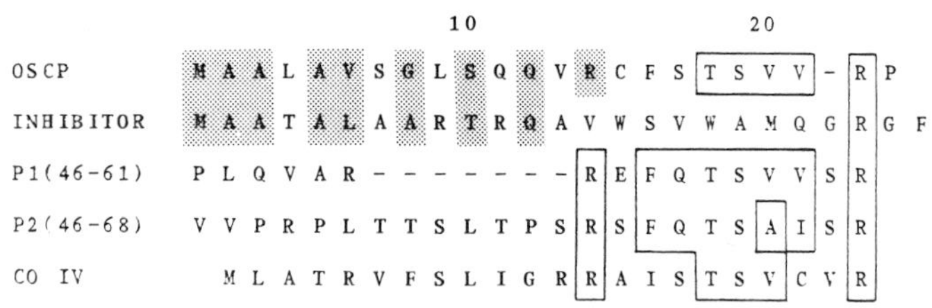

Fig. 7. Relationships between precursor sequences of imported bovine mitochondrial proteins. The mature proteins (not shown) abut these sequences. The cytochrome oxidase subunit IV sequence, COIV, is from [95]. The remainder were deduced in the present work.

particularly in the region adjacent to the mature subunits. This region contains a conserved arginine residue and may be a recognition site for the processing protease.

Two genes for the bovine proteolipid. Two different c-DNAs, named P1 and P2, were isolated from the bovine c-DNA library [84]. DNA sequence analysis showed that they encode proteins of 136 and 143 amino acids respectively. These proteins are identical in their C-terminal 71 amino acids, all but the first two being the sequence of the bovine ATP synthase proteolipid subunit (Fig. 6). To produce the mature subunit, 61 and 68 amino acids respectively are removed from the P1 and P2 precursors. The bovine proteolipid pre-sequences are significantly related to each other (see Fig. 6) and also much more weakly related to the pre-sequence of the *N. crassa* pre-proteolipid. These similarities suggest that the presequences have related secondary structures. For reviews see [82, 83].

In common with *N. crassa*, the bovine precursors contain a relatively large number of basic residues; P1 has 7, P2 9 (plus one histidine) and *N. crassa* 12. Both P1 and P2 also contain 2 glutamic acid residues, one of them, Glu-34, being in equivalent positions in both sequences; in addition, P2 contains an aspartic acid residue. Acidic residues have only rarely been seen in mitochondrial import sequences, e.g. yeast cytochrome c_1 which contains three acidic amino acids at the C-terminal end of the pre-sequence [85].

In the case of the *N. crassa* pre-proteolipid it has been suggested that the highly charged pre-sequence helps to solubilize the lipophilic mature protein in an aqueous environment. In addition, it has been proposed that the large net positive charge of the protein helps to drive it across the inner mitochondrial membrane, aided by the net negative electrochemical potential inside [86]. The observations reported here with the bovine pre-proteolipid sequences are consistent with these ideas.

Processing the precursors. Strictly conserved residues in P1 and P2 pre-sequences group in three local environments, around residues 10–15, 32–34 and 59–68 (P2 numbering). The first two of these clusters contain a conserved proline residue which may be important in the formation of bends in the secondary structure of the pre-sequence. The third cluster is found immediately before the mature protein sequence, this region being related also to pre-sequences of other bovine ATP synthase subunits and of cytochrome oxidase subunit IV; it probably represents a recognition site for a proteolytic processing enzyme. However, the pre-sequences are remarkably diverse (see Fig. 7). A common feature is that cleavage occurs at one or two residues to the C-terminal side of an arginine residue. A Lys-Arg sequence is associated with the recognition site for processing of *N. crassa* pre-proteolipid and the precursor of cytochrome *c* peroxidase contains a Lys-Arg sequence [87]. An arginine residue in the leader peptide of ornithine carbamoyltransferase has been shown to be required both for import and proteolytic cleavage [88].

It has been shown that the *N. crassa* protein is processed in two stages [89]. The first cleavage takes place at an internal site in the pre-sequence, also characterised by a Lys-Arg recognition sequence. A similar site is also found in the middle of the cytochrome *c* peroxidase pre-sequence, which also may be processed in two stages [87]. The bovine P2 pre-sequences contain two Arg-Arg sequences. The sequence of the region preceding one of these (P2, 31–32) is related to the equivalent sequence in P1 and also to the sequences preceding the cleavage sites at the start of the mature protein. So it may be that the bovine pre-proteolipids are also cleaved in an analogous fashion to their *N. crassa* counterpart.

Destination of the P1 and P2 precursors. It could be argued *a priori* that the two pre-sequences might serve to direct the proteolipid subunit to different organelles, for instance one to the mitochondrion, the other to coated vesicles. This would imply that coated vesicles (or some other organelle) contain an H^+-ATPase with a mature proteolipid subunit, and possibly other subunits, identical to that in the mitochondrion. However, the available evidence is contrary to this hypothesis, all known secretory H^+-ATPases being relatively insensitive to the effect of DCCD.

It seems more likely that both precursors enter mitochondria, a supposition supported by the significant relationship of the two precursor sequences. Further experimentation is required to determine if they are able to enter the same mitochondria (for example from a particular tissue).

Ψ-genes for the proteolipid. Recent investigations (M. Dyer and J. E. Walker, unpublished results) have revealed an additional complexity in the evolution of the bovine proteolipid, namely the existence of at least two Ψ-genes.

Conclusion and prospects

1. *Origin of chloroplasts*

Comparisons of both the order and sequences of genes for subunits of ATP synthase from chloroplasts and *Synechococcus* are consistent with the hypothesis that chloroplast genomes are part of the same evolutionary and taxonomic group as the cyanobacteria; they are more closely related to them than the cyanobacteria are to photosynthetic bacteria of the purple non-sulphur type. In general these data support the phylogenies drawn up by Fox *et al.* [54] using comparative analysis of 16S ribosomal RNA sequences. These phylogenies differ in important respects from the classifications made using mainly morphological, ie. phenotypical, not genotypical, characters. Examination of the 16S rRNA classifications could also explain the apparently anomalous results of Ambler *et al.* [90, 91] obtained by comparing cytochrome sequences from the purple non-sulphur bacteria. Cross-species transfer of genetic information may be an important evolutionary mechanism; for example in the case of *nif* gene clusters it has been shown that nitrogen-fixing eubacteria contain DNA sequences that hybridise to *nif* genes from *Klebsiella pneumoniae* but are not present in related bacteria that do not fix nitrogen [92]. However, it seems that the ATP synthase is ubiquitous in the eubacteria, as well as in mitochondria and chloroplasts. In other words, whereas the *nif* gene functions are not essential to the cells which use them, it is likely that the *atp* gene clusters are a more ancient genetic element and encode an enzyme essential in many, if not all, eubacteria.

2. *Gene clusters*

The conservation of gene order for genes of ATP synthase subunits is very striking. Occasionally genes such as the *Rps. blastica* X-gene have been inserted into the basic eight genes exemplified by the *E. coli unc* operon. Others, for example the genes for δ- and γ-subunits of chloroplasts, have been deleted from the plastid DNA and are found in the nucleus. In contrast to the tryptophan biosynthetic operons, hitherto inversions or fusions are notably absent.

In order to ascertain if other arrangements are to be found of genes for ATP synthase, it would be particulary interesting to study their organization in other major groups of eubacteria, particularly in the aerobic and anaerobic Gram-positive bacteria. It is also important to establish whether F_1F_0-ATPases are to be found amongst the *Archaebacteria*; preliminary investigations in *Halobacterium halobium* and *Sulpholobus* indicate that an enzyme closely related to the eubacterial type is not present.

3. *Nuclear genes*

Another area relevant to evolution of ATP synthases that is being actively investigated concerns nuclear genes for the mitochondrial (and the chloroplast) complex. Many questions remain unanswered. Are the genes widely scattered over several chromosomes or are they localised? Are other genes for ATP synthase duplicated as in the case of the proteolipid? What is the role of the pseudogenes in the evolutionary process? Further investigation of these and other questions may help to provide more convincing molecular evidence for a symbiotic origin of mitochondria.

Acknowledgements

A.L.C. is supported by an M.R.C. Research Studentship. Figure 1 is reprinted by permission from the *Biochemical Journal* **12**, 234–235, copyright © 1984, The Biochemical Society, London. Figures 2 and 3 are from the *EMBO J*, **5**, 217–222 (1986) and Figure 6 from *EMBO J*. **4**, 3519–3524 (1985); they are reprinted with permission from IRL Press.

References

1. Senior, A. E., in *Membrane Proteins in Energy Transduction* (ed. R. A. Capaldi) pp. 233–278, Dekker Inc., New York and Basel (1979).
2. Nelson, N., *Current Topics Bioenerg*, **11**, 1 (1981).
3. Downie, J. A., Gibson, F. and Cox, G. B., *Ann. Rev. Biochem.* **48**, 103 (1979).
4. Fillingame, R. H., *Current Topics Bioenerg.* **11**, 35 (1981).
5. Walker, J. E. and Tybulewicz, V. L. J., in *The Molecular Biology of the Photosynthetic Apparatus* (ed. C. Arntzen, L. Bogorad, S. Bonitz and K. Steinbach) pp. 141–153, Cold Spring Harbor (1985).
6. Nicholls, D., *Bioenergetics. An Introduction to the Chemiosmotic Theory*. Academic Press, London (1982).
7. Apps, D. K., Pryde, J. G., Sutton, R. and Phillips, J. H., *Biochem. J.* **190**, 273 (1980).
8. Anderson, D. C., King, S. C. and Parsons, S. M., *Biochemistry* **21**, 3037 (1982).
9. Hutton, J. C. and Peshavaria, M., *Biochem. J.* **204**, 161 (1982).
10. Ohkuma, S., Moriyama, Y. and Takano, T., *Proc. Natl. Acad. Sci. USA*. **79**, 2758 (1982).
11. Forgac, M., Cantley, L., Wiedenmann, B., Altstiel, L. and Branton, D., *Proc. Natl. Acad. Sci. USA*. **80**, 1300 (1983).
12. Goffeau, A. and Slayman, C. W., *Biochim. Biophys. Acta* **639**, 197 (1981).
13. Hesse, J. E., Wieczorek, L., Altendorf, K., Reicin, A. S., Dorus, E. and Epstein, W., *Proc. Natl. Acad. Sci. USA* **81**, 4746 (1985).
14. Kawakami, K., Noguchi, S., Noda, M., Takahashi, H., Ohta, T., Kawamura, M., Nojima, H., Nagano, K., Hirose, T., Inayama, S., Hayashida, H., Miyata, T. and Numa, S., *Nature (London)* **316**, 733 (1985).
15. MacLennan, D. H., Brandl, C. J., Korczak, B. and Green, N. M., *Nature (London)* **316**, 696 (1985).
16. Shull, G. E., Schwartz, A. and Lingrel, J. B., *Nature (London)* **316**, 691 (1985).
17. Gray, M. W. and Doolittle, W. F., *Microbiol. Rev.* **46**, 1 (1982).
18. Margulis, L., *Symbiosis in Cell Evolution*. W. H. Freeman and Co. London, New York (1981).
19. Stanier, R. Y., in *Evolution in the Microbial World*. Society for Gen. Microbiol. Symp. **24**, pp. 219–240. Cambridge University Press (1974).
20. Whatley, J. M., John, P., Whatley, F. R., *Proc. R. Soc. Lond.* B **204**, 165 (1979).
21. Cavalier-Smith, T., *Nature (London)* **256**, 463 (1975).
22. Cross, R. L., *Annu. Rev. Biochem.* **50**, 681 (1981).
23. Thauer, R. K., Jungermann, K. and Decker, K., *Bacteriol. Rev.* **41**, 100 (1977).
24. Walker, J. E., Saraste, M. and Gay, N. J., *Biochem. Biophys. Acta* **768**, 164 (1984).
25. Tybulewicz, V. L. J., Falk, G. and Walker, J. E., *J. Mol. Biol.* **179**, 185 (1984).
26. Falk, G., Hampe, A. and Walker, J. E., *Biochem. J.* **228**, 391 (1985).
27. Walker, J. E., Fearnley, I. M., Gay, N. J., Gibson, B. W., Northrop, F. D., Powell, S. J., Runswick, M. J., Saraste, M. and Tybulewicz, V. L. J., *J. Mol. Biol.* **184**, 677 (1985).
28. Foster, D. L. and Fillingame, R. H., *J. Biol. Chem.* **257**, 2009 (1982).
29. Amzel, L. M. and Pedersen, P. L., *Annu. Rev. Biochem.* **52**, 801 (1983).
30. Walker, J. E., Saraste, M., Runswick, M. J. and Gay, N. J., *EMBO J.* **1**, 945 (1982).
31. Harris, D. A., *Biochim. Biophys. Acta* **463**, 245 (1978).
32. Biketov, S. F., Kasho, V. N., Kozlov, I. A., Mileykovskaya, Y. I., Ostrovsky, D. M., Skulachev, V. P., Tikhonova, G. V. and Tsuprun, V. L., *Eur. J. Biochem.* **129**, 241 (1982).
33. Gay, N. J. and Walker, J. E., *Nature (London)* **301**, 262 (1983).
34. Yoshida, M., Sone, N., Hirata, H. and Kagawa, Y., *Proc. Natl. Acad. Sci. USA* **74**, 939 (1977).
35. Ovchinnikov, Yu. A., Modyanov, N. N., Grinkevich, V. A., Aldanova, N. A., Koestetsky, P. V., Trubetskaya, O. E., Hundal, T. and Ernster, L., *FEBS Lett.* **175**, 109 (1984).
36. Walker, J. E., Saraste, M. and Gay, N. J., *Nature (London)* **298**, 867 (1982).
37. Walker, J. E., Runswick, M. J. and Saraste, M., *FEBS Lett.* **146**, 393 (1982).
38. Ovchinnikov, Yu. A., Modyanov, N. N., Grinkevich, V. A., Aldanova, N. A., Trubetskaya, O. E., Nazimov, I. V., Hundal, T. and Ernster, L., *FEBS Lett.* **166**, 19 (1984).
39. Schneider, E. and Altendorf, K., *EMBO J.* **4**, 515 (1985).
40. von Meyenburg, K., Jorgensen, B. B., Michelson, O., Sorensen, L. and McCarthy, J. E. G., *EMBO J.* **4**, 2357 (1985).
41. Sebald, W. and Hoppe, J., *Current Topics Bioenerg.* **12**, 1 (1981).
42. Hoppe, J. and Sebald, W., *Biochim. Biophys. Acta* **768**, 1 (1984).
43. Harris, D. A., *Biosystems* **14**, 113 (1981).
44. Raven, J. A., *Biophys. Membr. Transp.* **2**, 151 (1977).
45. Bangham, A. D., *Prog. Biophys. Mol. Biol.* **18**, 29 (1968).
46. Woese, C. R., *J. Mol. Evol.* **13**, 95 (1979).
47. Baltscheffsky, H., in *Energy Conservation in Biological Membranes* (ed. G. Shäfer and M. Klingenberg), pp. 3–18. Springer Verlag, Berlin, Heidelberg, New York (1978).
48. Koshland, D. E., *Fed. Proc.* **35**, 2104 (1976).
49. Alberts, B., Bray, D., Lewis, J., Raff, M., Roberts, K. and Watson, J. D., in *The Molecular Biology of the Cell*, pp. 522–528. Garland Publishing Inc., New York and London (1983).
50. Reanny, D. C. and Ackermann, H.-W., *Adv. Virus Research* **27**, 205 (1985).

51. Casjens, S. and Hendrix, R., *J. Mol. Biol.* **90**, 20 (1974).
52. Gibson, F., *Proc. R. Soc. London* B, **215**, 1 (1982).
53. Lewin, R. A., *Ann. N.Y. Acad. Sci.* **361**, 325 (1981).
54. Fox, G. E., Stackebrandt, E., Hespell, R. B., Gibson, J., Maniloff, J., Dyer, T. A., Wolfe, R. S., Balch, W. E., Tanner, R. S., Magrum, L. J., Zablen, L. B., Blakemore, R., Gupta, R., Bonen, L., Lewis, B. J., Stahl, D. A., Luehrsen, K. R., Chen, K. N. and Woese, C. R., *Science* **209**, 457 (1980).
55. Seewaldt, E. and Stackebrandt, E., *Nature* (*London*) **295**, 618 (1982).
56. Bird, C. R., Koller, B., Auffret, A. D., Huttly, A. K., Howe, C. J., Dyer, T. A. and Gray, J. C., *EMBO J.* **4**, 1381 (1985).
57. Jans, D. A., Hatch, L., Fimmel, A. L., Gibson, F. and Cox, G. B., *J. Bacteriol.* **162**, 420 (1985).
58. Westhoff, P., Alt, J., Nelson, N. and Herrmann, R. G., *Mol. Gen. Genet.* **199**, 209 (1985).
59. Zurawski, G., Bottomley, W. and Whitfeld, P. R., *Nucl. Acids Res.* **12**, 6547 (1984).
60. Nevins, J. R., *Annu. Rev. Biochem.* **52**, 441 (1983).
61. Koller, B., Gingrich, J. C., Stiegler, G. L., Farley, M. A., Delius, H. and Hallick, R. B., *Cell* **36**, 545 (1984).
62. Karabin, G. D., Fooley, M. and Hallick, R. B., *Nucl. Acids Res.* **12**, 5801 (1984).
63. Nalin, C. M. and McCarty, R. E., *J. Biol. Chem.* **259**, 7275 (1984).
64. Pullman, E. and Monroy, G. C., *J. Biol. Chem.* **238**, 3762 (1963).
65. Asami, K., Juntii, K. and Ernster, L., *Biochim. Biophys. Acta* **205**, 307 (1970).
66. Ernster, L., Juntii, K. and Asami, K., *J. Bioenerget.* **4**, 149 (1973).
67. van der Stadt, R. J. and van Dam, K., *Biochim. Biophys. Acta* **347**, 240 (1974).
68. Fang, J., Jacobs, J. W., Kanner, B. I., Racker, E. and Bradshaw, R. A., *Proc. Natl. Acad. Sci. USA* **81**, 6603 (1984).
69. Grinkevich, V. A., Aldanova, N. A., Kostetsky, P. V., Modyanov, N. N., Hundal, T., Ovchinnikov, Yu. A. and Ernster, L., *Eur. Bioenerget. Congress Reports* **3**, 307 (1984).
70. Macreadie, I. G., Novitski, C. E., Maxwell, R. J., John, U., Ooi, B. G., McMullen, G. L., Lukins, H. B., Linnane, A. W. and Nagley, P., *Nucl. Acids Res.* **11**, 4435 (1983).
71. Anderson, S., de Bruijn, M. H. L., Coulson, A. R., Eperon, C., Sanger, F. and Young, I. G., *J. Mol. Biol.* **156**, 638 (1982).
72. Anderson, S., Bankier, A. T., Barrell, B. G., de Bruijn, M. H. L., Coulson, A. R., Drouin, J., Eperon, I. C., Nierlich, D. D., Roe, B. A., Sanger, F., Schreier, P. H., Smith, A. J. H., Staden, R. and Young, I. G., *Nature* (*London*) **290**, 457 (1981).
73. de Bruijn, M. H. L., *Nature* **304**, 234 (1983).
74. Mariottini, P., Chomyn, A., Attardi, G., Trovato, D., Strong, D. D. and Doolittle, R. F., *Cell* **32**, 1269 (1983).
75. Macino, G. and Tzagoloff, A., *J. Biol. Chem.* **254**, 4617 (1979).
76. van den Boogaart, P., Samallo, J. and Agsterribbe, E., *Nature* (*London*) **298**, 187 (1982).
77. Brown, T. A., Ray, J. A., Waring, R. B., Scazzocchio, C. and Davies, W. R., *Current Genetics* **8**, 489 (1984).
78. Hack, E. and Leaver, C. J., *EMBO J.* **2**, 1783 (1983).
79. Hack, E. and Leaver, C. J., *Current Genetics* **8**, 537 (1984).
80. Dewey, R. E., Schuster, A. M., Levings III, C. S. and Timothy, D. H., *Proc. Natl. Acad. Sci. USA* **82**, 1015 (1985).
81. Searcy, D. G., Stein, D. B. and Searcy, K. B., *Ann. N.Y. Acad. Sci.* **361**, 312 (1981).
82. Schatz, G. and Butow, R. A., *Cell* **32**, 65 (1984).
83. Hay, R., Böhm, P. and Gasser, S., *Biochim. Biophys. Acta* **779**, 65 (1985).
84. Gay, N. J. and Walker, J. E., *EMBO J.* **4**, 3519 (1985). (1985).
85. Sadler, I., Suda, K., Schatz, G., Kaudewitz, F. and Haid, A., *EMBO J.* **3**, 2137 (1984).
86. Viebrok, A., Perz, A. and Sebald, W., *EMBO J.* **1**, 565 (1982).
87. Kaput, J., Goltz, S. and Blobel, G., *J. Biol. Chem.* **257**, 15054 (1982).
88. Horwich, A. L., Kalonsek, F. and Rosenberg, L. E., *Proc. Natl. Acad. Sci. USA* **82**, 4930 (1985).
89. Schmidt, B., Wachter, E., Sebald, W. and Neupert, W., *Eur. J. Biochem.* **144**, 581 (1984).
90. Ambler, R. P., Daniel, M., Hermoso, J., Meyer, F. E., Bartsch, R. G. and Kamen, M. D., *Nature* (*London*) **287**, 661 (1979).
91. Ambler, R. P., Meyer, T. E. and Kamen, M. D., *Nature* (*London*) **278**, 661 (1979).
92. Ruvkun, C. B. and Ausubel, F. M., *Proc. Natl. Acad. Sci. USA* **77**, 191, (1980).
93. Staden, R., *Nucl. Acids Res.* **10**, 2951 (1982).
94. Kyte, J. and Doolittle, R. F., *J. Mol. Biol.* **157**, 105 (1982).
95. Lomax, M. I., Bachman, N. J., Nasoff, M. S., Caruthers, M. H. and Grossman, L. I., *Proc. Natl. Acad. Sci. USA* **81**, 6295 (1984).
96. Howe, C. J., Fearnley, I. M., Walker, J. E., Dyer, T. A. and Gray, J. C., *Plant Molecular Biol.* **4**, 333 (1985).
97. Deno, H., Shinozaki, K. and Sugiura, M., *Nucl. Acids Res.* **11**, 2185 (1983).
98. Krebbers, E. T., Larrinua, I. M., McIntosh, L. and Bogorad, L., *Nucl. Acids Res.* **10**, 4985 (1982).
99. Zurawski, G., Bottomley, W. and Whitfeld, P. R., *Proc. Natl. Acad. Sci. USA* **79**, 6260 (1982).
100. Howe, C. J., Auffret, A. D., Doherty, A., Bowman, C. M., Dyer, T. A. and Gray, J. C., *Proc. Natl. Acad. Sci. USA* **79**, 6903 (1982).
101. Frangione, B., Rosenwasser, E., Penefsky, H. S. and Pullman, M. E., *Proc. Natl. Acad. Sci. USA* **78**, 7403 (1981).
102. Boutry, M. and Chua, N. H., *EMBO J.* **4**, 2159 (1985).
103. Cozens, A. L., Walker, J. E., Phillips, A. L., Huttly, A. K. and Gray, J. C., *EMBO J.* **5**, 217 (1986).

Chemica Scripta 1986, **26B**, 273–279

Structural, Functional and Evolutionary Aspects of Proton-translocating ATPase

L. Ernster, T. Hundal, B. Norling, G. Sandri* and L. Wojtczak†

Department of Biochemistry, University of Stockholm, Arrhenius Laboratory, S-106 91 Stockholm, Sweden

and V. A. Grinkevich, N. N. Modyanov and Yu. A. Ovchinnikov

Shemyakin Institute of Bioorganic Chemistry, USSR Academy of Sciences, Ul. Vavilova 32, 117 988 GSP-1 Moscow, V-334 USSR

Paper presented by Lars Ernster at the Conference on 'Molecular Evolution of Life', Lidingö, Sweden, 8–12 September 1985

Abstract

This paper is a summary of structural, functional and evolutionary aspects of proton-translocating F_0F_1-ATPases, with special reference to two subunits of the mitochondrial ATPase, the oligomycin sensitivity conferring protein (OSCP) and coupling factor F_6, which are not found in chloroplasts and bacteria. Evidence is reviewed which indicates that OSCP is homologous to both the δ and b subunits of *E. coli* F_0F_1-ATPase as well as to mitochondrial F_6. Both OSCP and F_6 contain repeating sequences. OSCP also shows certain homology to the mitochondrial ATP/ADP carrier, and F_6 to the mitochondrial ATPase-inhibitor protein as well as to the c subunit of *E. coli* F_0. These observations are discussed in terms of a possible common evolutionary origin of those subunits of the mitochondrial F_0F_1-ATPase which are of special functional importance for the mitochondrial energy-transducing system as compared to those of chloroplasts and bacteria. Some recent data which reveal a striking homology between the catalytic centers of F_0F_1- and E_1E_2-ATPases are also briefly reviewed.

Introduction

Mitochondria, chloroplasts and bacteria contain a proton-translocating ATPase, which is the universal enzyme for electron-transport-linked ATP synthesis (for recent reviews; see refs 1–4). It consists of a membrane-bound, proton-translocating (F_0), and a peripheral, catalytic (F_1) component, both of which are multisubunit proteins. F_1 from mitochondria, chloroplasts and bacteria contains five types of subunits (α–ϵ), with a stoichiometry of $\alpha_3\beta_3\gamma_1\delta_1\epsilon_1$. F_0 from bacteria consists of three subunits (a–c), with a stoichiometry of $a_1b_2c_{6-12}$. Although the compositions of chloroplast and mitochondrial F_0 are less well established, both contain polypeptides similar to the bacterial a and c subunits.

Despite these striking similarities, the mitochondrial, chloroplast and bacterial F_0F_1-ATPases differ from each other in several respects. In particular, the mitochondrial enzyme (Fig. 1) differs from the other F_0F_1-ATPases in containing three types of subunits that are not found in chloroplasts or bacteria. One, referred to as the ATPase inhibitor protein, is reversibly associated with F_1, and its function seems to be to regulate ATP synthesis and ATP hydrolysis in relation to other energy-linked functions of the respiratory chain, such as the electrogenic transport of Ca^{2+}.

* Permanent address: Dipartimento di Biochimica, Biofisica e Chimica delle Macromolecole, Università di Trieste, Trieste, Italia.

† Permanent address: Department of Cellular Biochemistry, Nencki Institute of Experimental Biology, Warsaw, Poland.

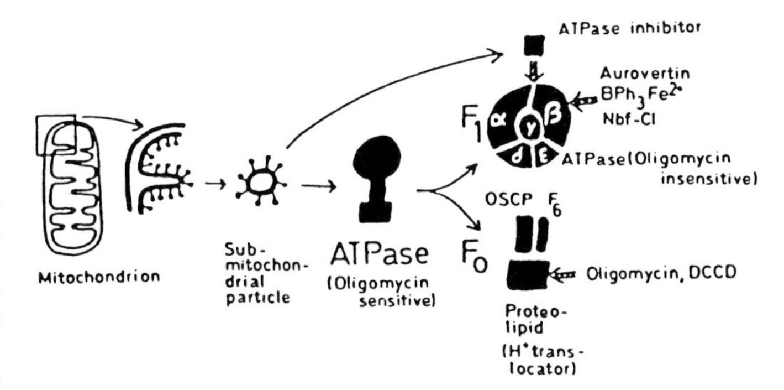

Fig. 1. Topology and components of the mitochondrial F_0F_1-ATPase. Hatched arrows indicate the sites of action of inhibitors. From Ernster [49]. For further details, see also ref. 50.

A second F_0F_1-ATPase subunit peculiar to mitochondria is the 'oligomycin sensitivity conferring protein' (OSCP). This protein is bound to both F_0 and F_1, and constitutes a structural and functional link between the proton-translocating and catalytic components of the mitochondrial F_0F_1-ATPase. A third, specifically mitochondrial ATPase subunit is the coupling factor F_6, which is bound to F_0 and is required for the proper anchorage of OSCP.

This paper is a survey of recent information concerning the OSCP and F_6 subunits of the mitochondrial F_0F_1-ATPase, with special emphasis on their structural, functional and evolutionary relationship to other subunits of the ATPase systems of mitochondria and bacteria. Some recent comparative aspects of F_0F_1-ATPases and various transport ATPases (so-called E_1E_2-ATPases) will also be briefly reviewed.

Topology of the OSCP and F_6 subunits of mitochondrial ATPase based on resolution-reconstitution studies

It is generally believed that F_6, together with OSCP, is required for the binding of F_1 to F_0 and for the conferral of oligomycin sensitivity. Three models for the role of F_6 have been proposed (Fig. 2). According to the first model (Fig. 2A), F_6 serves as the connecting link between the membrane sector of F_0 and OSCP, which, in turn, serves as a link to F_1. This model was based on the finding [5] that, in submitochondrial particles depleted of F_1, OSCP and F_6, by treatment with urea ammonia, and silicotungstic acid, respectively, the addition of both F_6 and OSCP is necessary in order to rebind F_1 in an

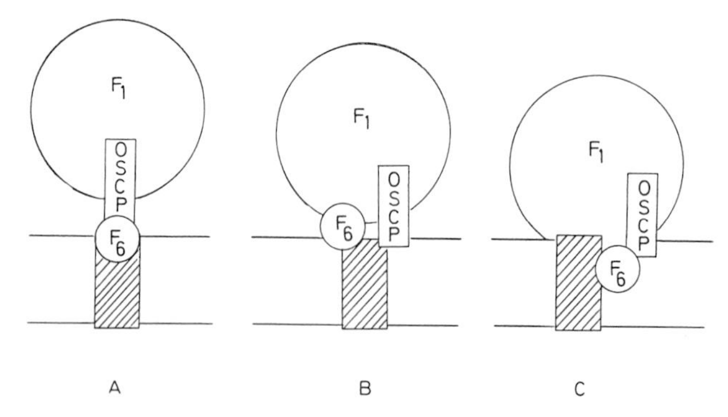

Fig. 2. Proposed topology of F_6 and OSCP in relation to the membrane sector of F_0 (hatched bar) according to (A) Vàdineanu *et al.* [5]; (B) Norling *et al.* [7]; (C) Sandri *et al.* [8].

oligomycin-sensitive manner. A second model (Fig. 2B) suggests that F_6 and OSCP constitute separate links between F_1 and the hydrophobic portion of F_0. This model is based on the findings [6, 7] that F_6 alone can rebind F_1 to F_6-, OSCP- and F_1-depleted particles in an oligomycin-insensitive manner, and that subsequent addition of OSCP renders the bound F_1 oligomycin-sensitive.

The third model (Fig. 2C) has been proposed on the basis of recent findings [8] that, in the presence of suitable cations (NH_4^+, Rb^+ or Cs^+), F_1 can be rebound to F_1-, OSCP- and F_6-depleted particles, in amounts equal to the native content of F_1, and that the F_1 so bound becomes partially oligomycin-sensitive upon the subsequent addition of OSCP and fully oligomycin-sensitive upon the addition of F_6 and OSCP. Based on these findings, this model postulates that F_1 possesses two binding sites on F_0, one directly to the membrane sector, and another, through F_6 and OSCP, where F_6 is needed for the anchorage of OSCP on the membrane. This model bears some similarity to the situation in *E. coli*, where there is evidence that F_1 is bound to F_0 partly through a direct link between its α and/or β subunit(s) and the c subunit of F_0 [9] and partly between its δ and ϵ subunits and the b subunit of F_0 [10]; it is the latter link that seems to be replaced in the case of mitochondria by F_6 and OSCP (see below). It should be pointed out that the proposed model does not intend to specify the location of F_6 and OSCP in relation to the membrane lipids or the surrounding water-phase. In fact, there is evidence [11] that a large portion of the surface of the OSCP molecule in the native F_0F_1 complex is exposed to the water-phase.

The scheme in Fig. 3 summarizes the various steps of resolution and reconstitution of mitochondrial F_0F_1-ATPase as described above. It should be added that, whenever oligomycin sensitivity is mentioned, it also refers to sensitivity to *N,N′*-dicyclohexylcarbodiimide (DCCD).

Structure of OSCP and F_6

Considerable progress has been made in recent years concerning the primary structure of the F_0F_1-ATPase of *E. coli*. This progress is mainly due to the identification of the genes coding for the F_0F_1-ATPase [13] – the *unc* (or *atp* or *pap*) operon – and the determination of the nucleotide sequence of this operon. Parallel to this development the primary structures of subunits of mitochondrial F_1 and of the a and c subunits of mitochondrial F_0 as well as of the mitochondrial

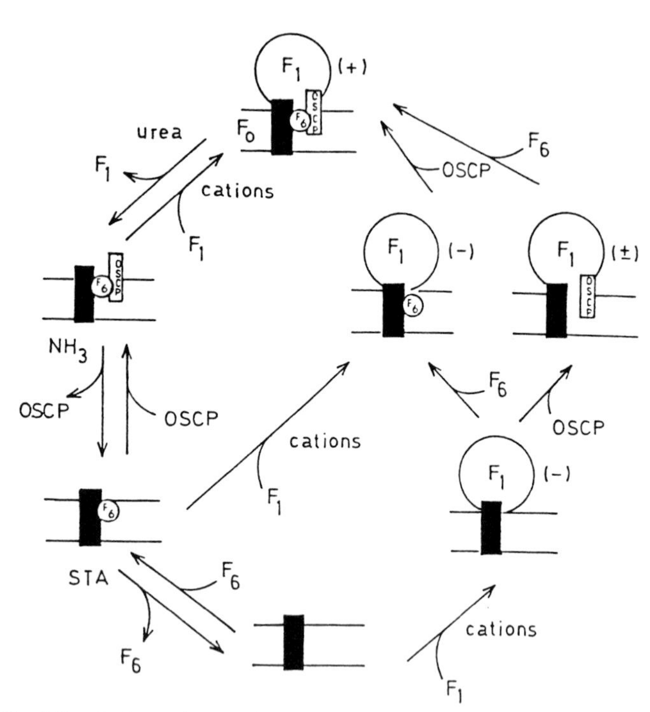

Fig. 3. Resolution and reconstitution of the F_0F_1-ATPase in submitochondrial particles. (From ref. 12): +, oligomycin-sensitive; ±, partially oligomycin-sensitive; -, oligomycin-insensitive; STA, silicotungstic acid.

ATPase-inhibitor protein have been determined (for review, see ref. 14). Likewise, the primary structures of several subunits of the chloroplast F_0F_1-ATPase have been established (cf. ref. 14).

In our laboratories we have been engaged in studies of the structure and function of OSCP and F_6 from beef heart mitochondria. Fig. 4 shows the amino acid sequences of OSCP and F_6. OSCP (Fig. 4A) contains 190 amino acids and has a molecular weight of 20968 Da. F_6 (Fig. 4B) consists of

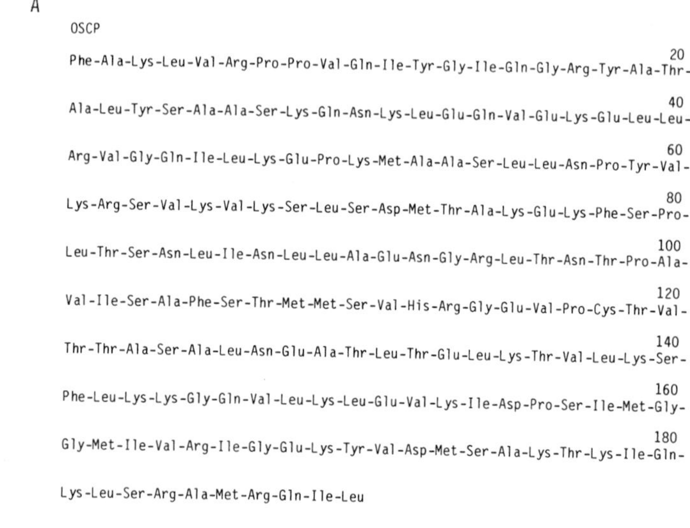

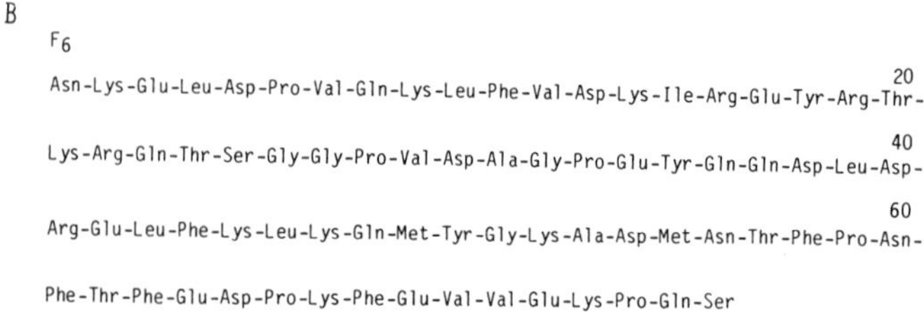

Fig. 4. Primary structures of (A) OSCP and (B) F_6 from beef heart mitochondria. (From refs. 23, 32.)

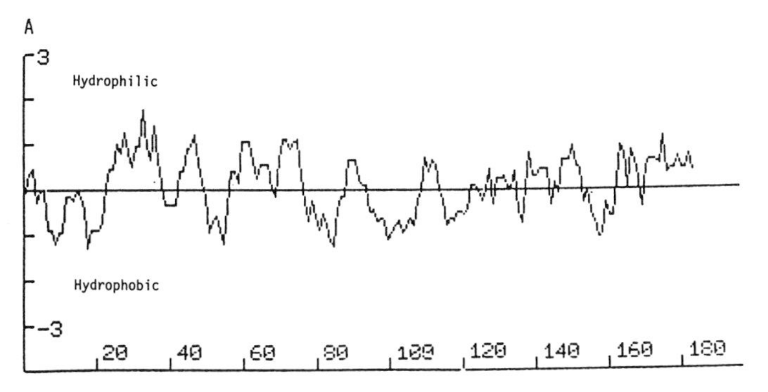

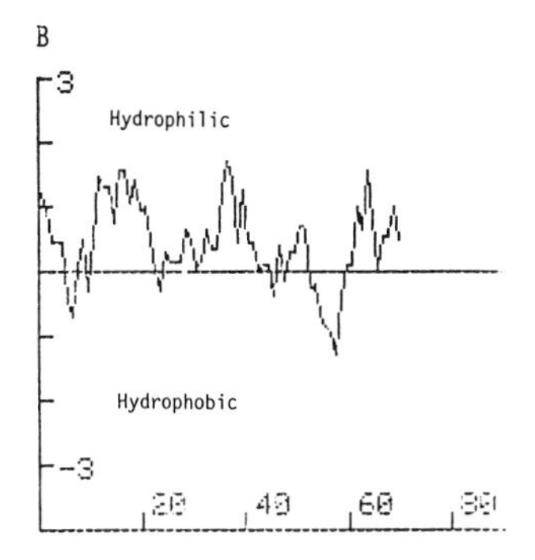

Fig. 5. Hydrophilicity profiles of beef heart (A) OSCP and (B) F_6, determined according to Hopp and Woods [16].

76 amino acids, with a molecular weight of 8960 Da. We find F_6 also to occur in a 75-amino-acid form, lacking the N-terminal asparagine. Independently, the structure of beef heart F_6 has been determined by Fang *et al.* [15]. Their structure is identical with our 76-amino-acid sequence, except residue 62, which is a threonine in our case and a phenylalanine in theirs.

The hydrophilicity profiles of OSCP and F_6 (Fig. 5) were determined according to Hopp and Woods [16]. These are similar to those independently reported by Dupuis *et al.* [17] for OSCP and by Fang *et al.* [15] for F_6. In the case of OSCP (Fig. 5A) the profile reveals an even distribution of hydrophobic and hydrophilic residues along the amino acid sequence. The hydrophilicity profile of F_6 (Fig. 5B) shows 4 hydrophilic segments and one hydrophobic, the latter between amino acid residues 50 and 60.

Fig. 6 shows the secondary structures of OSCP and F_6 as determined by the method of Chou and Fasman [18]. OSCP contains 48% α-helix (Fig. 6A), which again is in good agreement with results reported by Dupuis *et al.* [17]. Somewhat surprisingly, the more hydrophobic F_6 contains only 38% α-helix (Fig. 6B).

We have recently established the occurrence of repeating sequences within OSCP and F_6 (Fig. 7A, B), suggesting that both proteins have evolved by a process of gene duplication. In addition, OSCP and F_6 have been found to contain homologous structural elements which is especially noticeable when comparing the repeating C-terminal sequence of F_6 with the N-terminal sequence of OSCP (Fig. 7C). In the case of OSCP, the internal homology was substantiated by DIAGON computer analysis (Fig. 8A) [19].

Relationship of OSCP and F_6 to other F_0F_1-ATPase subunits and related proteins

It was first recognised by Walker *et al.* [21] that the structure of OSCP is homologous to that of the δ subunit of F_1 from *E. coli.* This observation was based on the amino acid sequences of the N- and C-terminal portions of OSCP [22], and was later substantiated [23] when the complete amino acid sequence of OSCP was established (Fig. 9). This homology could also be confirmed by DIAGON computer analysis (Fig. 8B) [24].

It has been shown that F_1 and OSCP from beef heart mitochondria form a stable complex. The complex can either be extracted from the membrane [25] or formed by incubating the two purified proteins together [5, 26]. It has also been shown [10] that mild trypsin treatment of F_1 from *E. coli*, leading to a limited digestion of the N-terminal segment of the α subunit, causes a loss of the ability of F_1 to retain its δ subunit. In the case of mitochondrial F_1 we found [27, 28] that the same treatment leads to a similar modification of the α subunit, and simultaneously to a loss of the ability of F_1 to bind OSCP. We found [29], moreover that the binding of OSCP to mitochondrial F_1 protects the enzyme from trypsin digestion of its α subunit, suggesting that the α subunit contains a binding site for OSCP. Recent evidence suggests that the β [30] and γ [31] subunits also may be involved in the binding of F_1 to OSCP.

Recently, we also established the occurrence of a certain homology between beef-heart OSCP and the b subunit of *E. coli* F_0 (Fig. 10) [19, 22, 31, 32] which is most pronounced between the repeating sequences of OSCP and the central part of the *b* subunit (residues 21–116). Thus OSCP appears to represent the product of a hybrid gene, resulting from a fusion of the genes for the *b* and δ subunits of the bacterial type of F_0F_1-ATPase. In fact, the two genes are adjacent in the *unc* operon (cf. ref. 4). The fusion of the two genes, with the resulting emergence of OSCP, may explain the relatively firm link between mitochondrial F_0 and F_1 as compared to the bacterial system.

Fig. 11 summarizes a comparison of the structures of OSCP and F_6 with those of some other subunits of the F_0F_1-ATPase and related proteins. As already reported briefly [19] an unexpected homology was found between the repeating regions of OSCP (residues 1–84) and the N-terminal sequence of the mitochondrial ADP/ATP carrier (residues 15–102). Certain homologous regions were also found between F_6 and the mitochondrial ATPase-inhibitor protein, as well as the amino acid sequence corresponding to the unidentified reading frame (URF) A6L of mitochondrial DNA. Furthermore, the N-terminal sequence of F_6 is homologous to the C-terminal region of the c subunit of *E. coli* F_0, including the DCCD-binding aspartyl residue [40]. However, treatment of the mitochondrial F_0F_1 complex with [^{14}C]DCCD, under conditions which resulted in a binding of DCCD to the c subunit (subunit 9), did not give rise to a labeling of F_6 [31].

Fig. 12 summarizes the above findings, together with earlier reported homologies between various subunits of the F_0F_1 complexes from *E. coli* and beef-heart (cf. refs 1 and 39). The subunits of the *E. coli* F_0F_1-ATPase are listed in the order in which they occur in the *unc* operon [33]. There is a well-established homology between the α, β and γ subunits of the bacterial and mitochondrial F_1. Likewise, there is a homology between the a and c subunits of bacterial F_0 and the corresponding subunits of mitochondrial F_0. The ϵ subunit of bacterial F_1 is homologous to the δ subunit of mitochondrial F_1. The OSCP of mitochondria is homologous

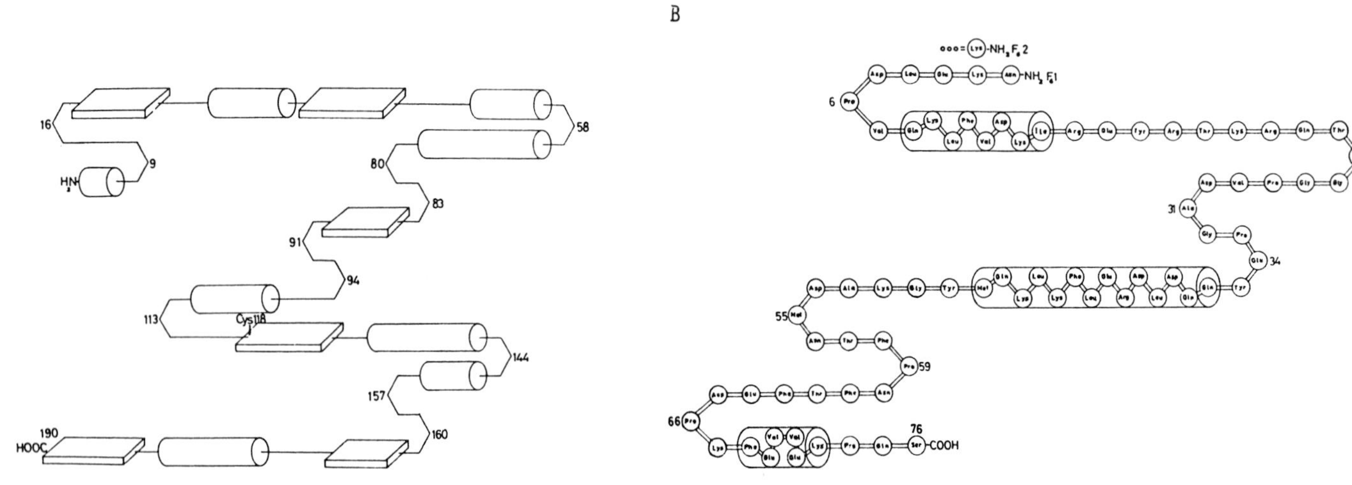

Fig. 6. Secondary structures of beef heart (A) OSCP and (B) F_6, determined by the method of Chou and Fasman [18].

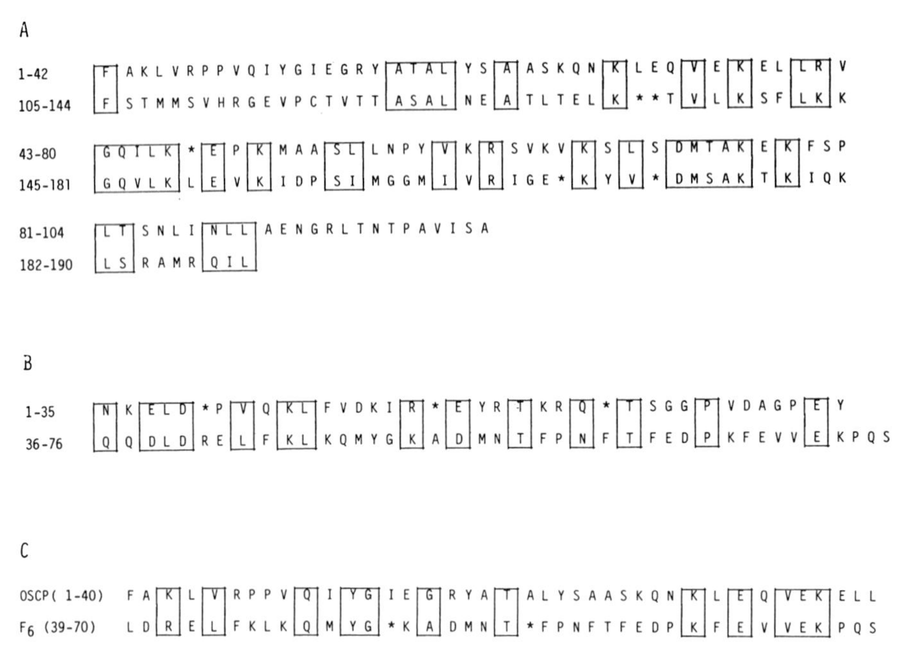

Fig. 7. Repeating sequences in beef-heart (A) OSCP and (B) F_6 (from refs 19, 24, 32, 40). (C) shows a comparison of the partial amino acid sequences of OSCP and F_6 (from ref. 40).

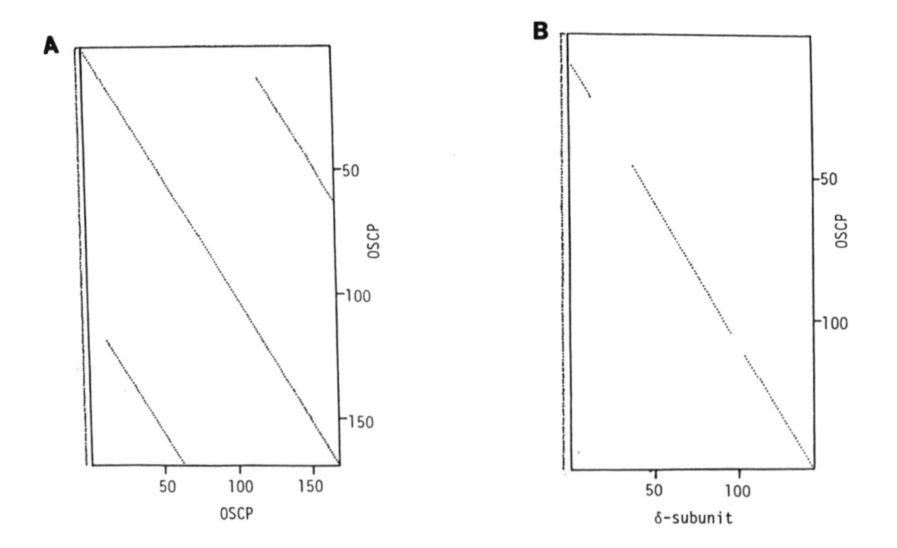

Fig. 8. Comparison matrix of the amino acid sequence of (A) beef heart OSCP with itself (from ref. 19) and (B) the δ subunit of *E. coli* F_1. The DIAGON computer model of McLachlan [20] was used in the calculation.

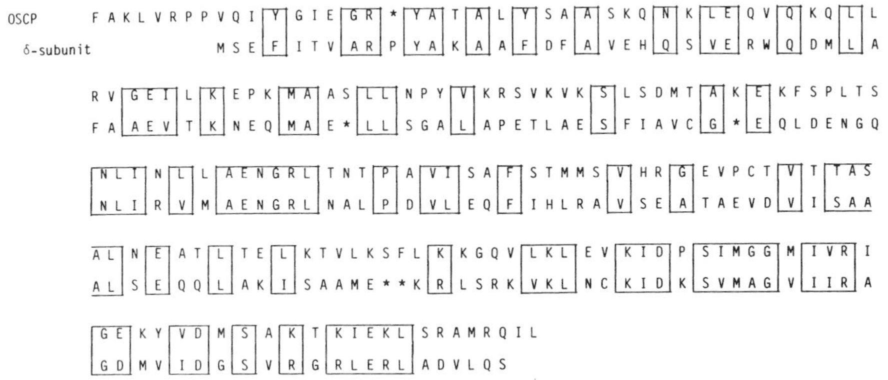

OSCP F A K L V R P P V Q I Y G I E G R * Y A T A L Y S A A S K Q N K L E Q V Q K Q L L
δ-subunit M S E F I T V A R P Y A K A A F D F A V E H Q S V E R W Q D M L A

R V G E I L K E P K M A A S L L N P Y V K R S V K V K S L S D M T A K E K F S P L T S
F A A E V T K N E Q M A E * L L S G A L A P E T L A E S F I A V C G * E Q L D E N G Q

N L I N L L A E N G R L T N T P A V I S A F S T M M S V H R G E V P C T V T T A S
N L I R V M A E N G R L N A L P D V L E Q F I H L R A V S E A T A E V D V I S A A

A L N E A T L T E L K T V L K S F L K K G Q V L K L E V K I D P S I M G G M I V R I
A L S E Q Q L A K I S A A M E * * K R L S R K V K L N C K I D K S V M A G V I I R A

G E K Y V D M S A K T K I E K L S R A M R Q I L
G D M V I D G S V R G R L E R L A D V L Q S

Fig. 9. Comparison of the amino acid sequences of OSCP and the δ subunit of *E. coli* F_1 (from ref. 23). The amino acid sequence of the δ subunit of *E. coli* F_1 is taken from ref 33. Identical residues and conserved substitutions are boxed, deletions are indicated by asterisks.

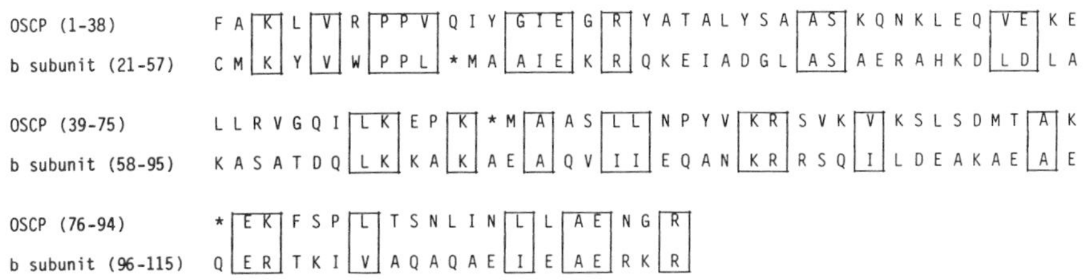

OSCP (1-38) F A K L V R P P V Q I Y G I E G R Y A T A L Y S A A S K Q N K L E Q V E K E
b subunit (21-57) C M K Y V W P P L * M A A I E K R Q K E I A D G L A S A E R A H K D L D L A

OSCP (39-75) L L R V G Q I L K E P K * M A A S L L N P Y V K R S V K V K S L S D M T A K
b subunit (58-95) K A S A T D Q L K K A K A E A Q V I I E Q A N K R R S Q I L D E A K A E A E

OSCP (76-94) * E K F S P L T S N L I N L L A E N G R
b subunit (96-115) Q E R T K I V A Q A Q A E I E A E R K R

Fig. 10. Comparison of the amino acid sequences of beef heart OSCP and the b subunit of *E. coli* F_0 (from refs 19, 24, 32, 33). The amino acid sequence of the b subunit is taken from refs 33–35.

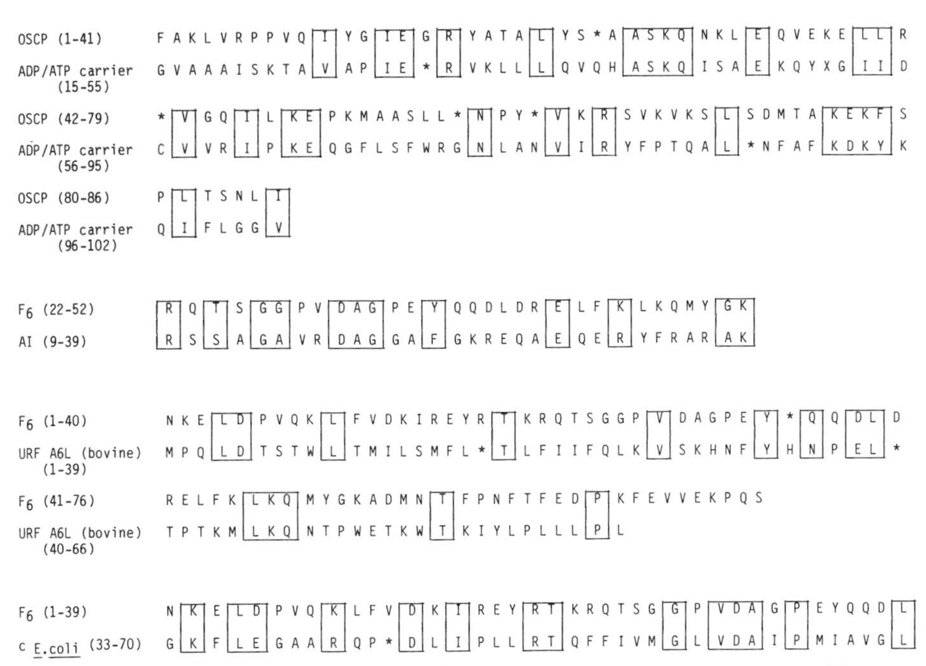

Fig. 11. Comparison of the amino acid sequences of beef heart OSCP and F_6 with some other proteins. The amino acid sequences of the various proteins are taken from the following references: ADP/ATP carrier, ref. 36; ATPase inhibitor (AI) ref. 37; URF A6L, ref. 38; c subunit of *E. coli* F_0, *ref.* 39.

to both the δ and b subunits of the *E. coli* ATPase, probably resulting from a fusion of two adjacent genes in the *unc* operon. The mitochondrial F_6 contains elements of the bacterial c subunit, and shows, in addition, some homology with OSCP. Both OSCP and F_6 contain repeating sequences, indicative of gene duplication. OSCP shows homology with the mitochondrial ADP/ATP carrier, and F_6 shows certain homology with both the mitochondrial ATPase-inhibitor protein and with the protein corresponding to the URF A6L of mitochondrial DNA.

Comparison between F_0F_1- and E_1E_2-ATPases

Proton-translocating F_0F_1-ATPases are generally considered to be radically different in their structure and reaction mechanism from so called E_1E_2-ATPases, i.e. transport ATPases [41] such as the eucaryotic plasma membrane Na^+, K^+-ATPase, the Ca^{2+}-ATPase of sarcoplasmic reticulum or the K^+-dependent ATPase of *E. coli.* The latter enzymes contain only one or two polypeptides and their reaction mechanism involves the formation of a phosphoenzyme inter-

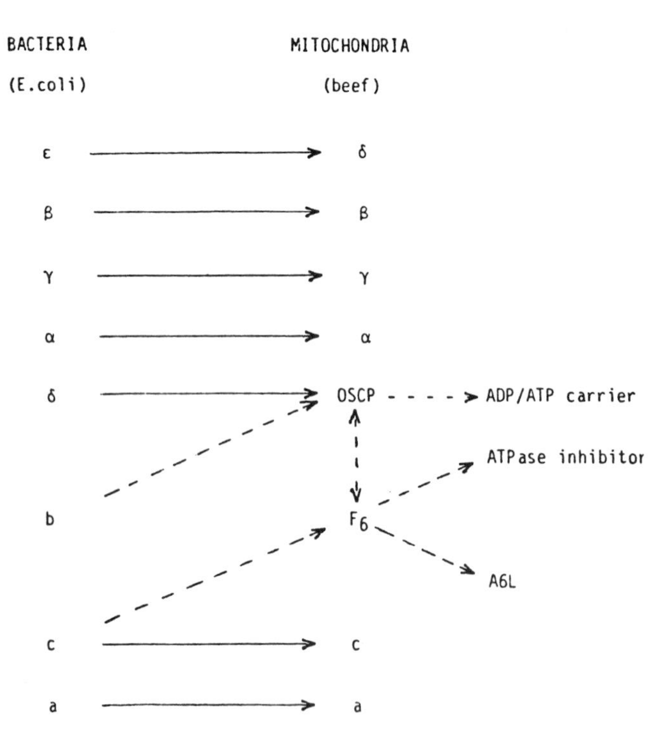

Fig. 12. Structural relationships between subunits of bacterial and mitochondrial F_0F_1-ATPases and some related proteins. The subunits of *E. coli* ATPase are listed in the order corresponding to the sequence of genes in the *unc* operon [4]. Solid arrows refer to earlier established homologies (see refs 1, 39). Dotted arrows indicate homologies suggested by the comparison reviewed in the present paper.

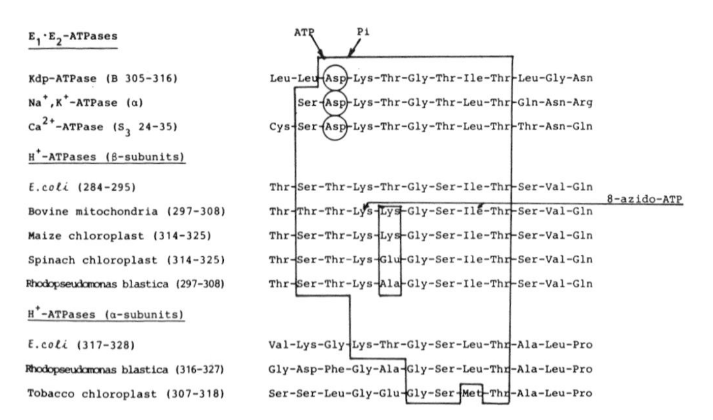

Fig. 13. Comparison of the amino acid sequences of the catalytic centers in F_0F_1- and E_1E_2 ATPases. From Modyanov *et. al.* [42].

mediate in which phosphate derived from the γ phosphoryl group of ATP is bound to an aspartyl residue of the enzyme.

Modyanov *et al.* [42] have recently detected a striking homology between the catalytic regions of various E_1E_2-ATPases and the β subunit of F_0F_1-ATPases from different sources (Fig. 13). The aspartyl residue of E_1E_2-ATPases involved in the phosphoenzyme formation is replaced in the case of the β subunit of F_0F_1-ATPases by threonine. Adjacent to this residue in both E_1E_2- and F_0F_1-ATPases is a lysine, which in the case of F_0F_1-ATPases is the binding site for 8-azido-ATP [43] and thus probably for ATP. In the α subunit of F_0F_1-ATPases the region of homology is considerably shorter.

Very recently the complete amino acid sequences of Na^+, K^+-ATPase from two sources [44, 45] and of sarcoplasmic reticulum Ca^{2+}-ATPase [46] have been reported. This opens the way to a further comparison of these structures with those of F_0F_1-ATPases.

Concluding remarks

From the data summarized in this paper it is evident that those parts of the F_0F_1-ATPase which are involved in the actual catalytic machinery of ATP synthesis, namely the α, β and γ subunits of F_1, with their well-documented role in the catalytic, regulatory, and gating functions of the enzyme, and the a and c subunits of F_0, which form the proton channel, are conserved to a high degree during evolution. Regarding the functional significance of other suggested homologies, one can only speculate at this point.

It is tempting to think that the fusion of the b and δ subunits of the bacterial ATPase into the mitochondrial OSCP has implied a stabilization of the binding between F_1 and F_0. In bacteria (and also in chloroplasts), F_1 can be removed from F_0 by treatment with EDTA (cf. ref. 2), indicating that the binding involves divalent cations. Such a binding of F_1 to F_0 would not be compatible with mitochondrial function, in view of the important role that mitochondria play in the regulation of intracellular Ca^{2+} homeostasis [47]. OSCP, which is firmly anchored in both F_1 and F_0, provides a stable link between the two moieties, independent of the prevailing intramitochondrial concentration of divalent cations. F_6, which contains elements of both bacterial c subunit and OSCP, probably serves as the anchorage for OSCP in F_0.

F_6 and OSCP seem also to have an evolutionary relationship to two specifically mitochondrial proteins with special functions, namely, the ATPase-inhibitor protein, the function of which is thought to be to regulate ATP synthesis in relation to other energy linked processes, e.g. Ca^{2+} transport [48], and the ADP/ATP carrier, a uniquely mitochondrial protein which exchanges cytoplasmic ADP for ATP synthesized inside the mitochondria.

Thus, the picture that emerges suggests a complex evolutionary process that led to the acquisition of those accessory functional features of the ATP-generating machinery which are specifically required for mitochondrial function and are not found in chloroplasts or bacteria.

Finally the data also reveal a striking structural homology between the catalytic centers of F_0F_1- and E_1E_2-ATPases, suggesting a common evolutionary origin of the enzymes involved in protonmotive ATP synthesis and ATP-driven cation transport.

References

1. Papa, S., Altendorf, K., Ernster, L. and Packer, L. (eds), *H⁺-ATPase: Structure, Function, Biogenesis.* ICSU Press and Adriatica Editrice, Bari (1984).
2. Senior, A. and Wise, J. G., *J. Membr. Biol.* **73**, 105 (1983).
3. Futai, M. and Kanazawa, H., *Microbiol. Rev.* **47**, 285 (1983).
4. Walker, J. E., Saraste, M. and Gay, N. J., *Biochim. Biophys. Acta* **768**, 164 (1984).
5. Vàdineanu, A., Berden, J. A. and Slater, E. C., *Biochim. Biophys. Acta* **449**, 468 (1976).
6. Russel, L. K., Kirkley, S. A., Kleyman, T. R. and Chan, S. H. P., *Biochem, Biophys. Res. Commun.* **73**, 434 (1976).
7. Norling, B., Glaser, E. and Ernster, L., in *Frontiers of Biological Energetics: From Electrons to Tissues* (ed. L. P. Dutton, J. S. Leight and A. Scarpa), pp. 504–515, Academic Press, New York (1978).
8. Sandri, G., Wojtczak, L. and Ernster, L., *Arch. Biochem. Biophys.* **239**, 595 (1985).

9. Loo, T. W., Stan-Lotter, H., MacKenzie, D., Molday, R. S. and Bragg, P. D., *Biochim. Biophys. Acta* **733**, 274 (1983).
10. Dunn, S. D., Heppel, L. A. and Fullmer, L. S., *J. Biol. Chem.* **255**, 6891 (1980).
11. Gautheron, D. C., Godinot, C., Deléage, G., Archinard, P. and Penin, F. in *Achievements and Perspectives of Mitochondrial Research* (ed. E. Quagliariello, E. C. Slater, F. Palmieri, C. Saccone and A. M. Kroon), vol. I, pp. 279–288, Elsevier, Amsterdam (1985).
12. Ernster, L., Hundal, T. and Sandri, G., *Meth. Enzymol.* **126**, 428 (1986).
13. Downie, J. A., Gibson, F. and Cox, G. B., *Annu. Rev. Biochem.* **48**, 103 (1979).
14. Walker, J. G., Tybulewicz, V. L. J., Falk, G., Gay, N. J. and Hampe, A., in *H^+-ATPase: Structure, Function, Biogenesis* (ed. S. Papa, K. Altendorf, L. Ernster and L. Packer), pp. 1–14. ICSU Press and Adriatica Editrice, Bari (1984).
15. Fang, J.-K., Jacobs, J. W., Kanner, B. I., Racker, E. and Bradshaw, R. A., *Proc. Natl. Acad. Sci. USA* **81**, 6603 (1984).
16. Hopp, T. P. and Woods, K. R., *Proc. Natl. Acad. Sci. USA* **78**, 3824 (1981).
17. Dupuis, A., Issartel, J.-P., Lunardi, J., Satre, M. and Vignais, P. V., in *Achievements and Perspectives of Mitochondrial Research* (ed. E. Quagliariello, E. C. Slater, F. Palmieri, C. Saccone and A. M. Kroon), vol. I, pp. 237–246, Elsevier, Amsterdam (1985).
18. Chou, P. Y. and Fasman, G. D., *Annu. Rev. Biochem* **47**, 251 (1978).
19. Ovchinnikov, Yu. A., Modyanov, N. N., Grinkevich, V. A., Aldanova, N. A., Kostetsky, P. V., Trubetskaya, O. E., Hundal, T. and Ernster, L., *FEBS Lett.* **175**, 109 (1984).
20. McLachlan, A. D., *J. Mol. Biol.* **61**, 409 (1971).
21. Walker, J. E., Runswick, M. J. and Saraste, M., *FEBS Lett.* **146**, 393 (1982).
22. Grinkevich, V. A., Modyanov, N. N., Ovchinnikov, Yu. A., Hundal, T. and Ernster, L., *EBEC Reports* **2**, 83 (1982).
23. Ovchinnikov, Yu. A., Modyanov, N. N., Grinkevich, V. A., Aldanova, N. A., Trubetskaya, O. E., Nazimov, I. V., Hundal, T. and Ernster, L., *FEBS Lett.* **166**, 19 (1984).
24. Grinkevich, V., Aldanova, N., Kostetsky, P., Trubetskaya, O., Modyanov, N., Hundal, T. and Ernster, L., in *H^+ATPase: Structure, Function, Biogenesis* (ed. S. Papa, K. Altendorf, L. Ernster and L. Packer), pp. 155–162. ICSU Press and Adriatica Editrice, Bari (1984).
25. Van de Stadt, R. J., Kraaipoel, R. J. and Van Dam, K., *Biochim. Biophys. Acta* **267**, 25 (1972).
26. Hundal, T. and Ernster, L., in *Membrane Bioenergetics* (ed. C. P. Lee, G. Schatz and L. Ernster), pp. 429–445. Addison-Wesley, Reading (1979).
27. Hundal, T. and Ernster, L., *FEBS Lett.* **133**, 115 (1981).
28. Hundal, T., Norling, B. and Ernster, L., *FEBS Lett.* **162**, 5 (1983).
29. Hundal, T., Norling, B. and Ernster, L., *J. Bioenerg. Biomembr.* **16**, 535 (1984).
30. Dupuis, A., Lunardi, J., Issartel, J.-P. and Vignais, P. V., *Biochemistry* **24**, 734 (1985).
31. Ovchinnikov, Yu. A., Modyanov, N. N., Grinkevich, V. A., Belogrudov, G. I., Hundal, T., Norling, B., Sandri, G., Wojtczak, L. and Ernster, L., in *Achievements and Perspectives of Mitochondrial Research* (ed. E. Quagliariello, E. C. Slater, F. Palmieri, C. Saccone and A. M. Kroon), vol. I, pp. 223–236, Elsevier, Amsterdam (1985).
32. Grinkevich, V. A., Aldanova, N. A., Kostetsky, P. V., Modyanov, N. N., Ovchinnikov, Yu. A., Hundal, T. and Ernster, L., in *Proc. 16th FEBS Meeting*, Part B, pp. 483–488 VNU Science Press, Utrecht (1985).
33. Gay, N. J. and Walker, J. E., *Nucleic Acid Res.* **9**, 3919 (1981).
34. Nielsen, J., Hansen, F. G., Hoppe, J., Friedl, P. and Von Meyenburg, K., *Mol. Gen. Genet.* **184**, 33 (1981).
35. Kanazawa, H., Mabuchi, K., Kayano, T., Noumi, T., Sekiya, T. and Futai, M., *Biochem. Biophys. Res. Commun.* **103**, 613 (1981).
36. Aquila, H., Misra, D., Eulitz, M. and Klingenberg, M., *Hoppe-Seyler's Z. Physiol. Chem.* **363**, 345 (1982).
37. Frangione, B., Rosenwasser, E., Penefsky, H. S. and Pullman, M. E., *Proc. Natl. Acad. Sci. USA* **78**, 7403 (1981).
38. Anderson, S., de Bruijn, M. H. L., Coulson, A. R., Eperon, I. C., Sanger, F. and Young, I. G., *J. Mol. Biol.* **156**, 683 (1982).
39. Walker, J. E., Fearnley, I. M., Gay, N. J., Gibson, B. W., Northrop, F. D., Powell, S. J., Runswick, M. J., Saraste, M., Tubulewicz, V. L. J., *J. Mol. Biol.* **184**, 677 (1985).
40. Grinkevich, V. A., Aldanova, N. A., Kostetsky, P. V., Modyanov, N. N., Hundal, T., Ovchinnikov, Yu. A. and Ernster, L., *EBEC Reports* **3**, 307 (1984).
41. Carafoli, E. and Scarpa, A. (eds.) *Transport ATPases. Ann. NY Acad. Sci.*, vol. 402 (1982).
42. Modyanov, N. N., Arzamazova, N. M., Arystarkhova, E. A., Gevondyan, N. M. and Ovchinnikov Yu. A., *Biologicheskie Membrany* **2**, 844 (1985).
43. Hollemans, M., Runswick, M. J., Fearnley, I. M. and Walker, J. E., *J. Biol. Chem.* **258**, 9307 (1983).
44. Shull, G. E., Schwartz, A. and Lingrel, J. B., *Nature (London)* **316**, 691 (1985).
45. Kawakami, K., Noguchi, S., Noda, M., Takahashi, H., Ohta, T., Kawamura, M., Nojima, H., Nagano, K., Hirose, T., Inayama, S. Hayashida, H., Miyata, T. and Numa, S., *Nature (London)* **316**, 733 (1985).
46. MacLennan, D. H., Brandt, C. J., Korczak, B. and Green, N. M., *Nature (London)* **316**, 696 (1985).
47. Carafoli, E. and Sottocasa, G., in *Bioenergetics* (ed. L. Ernster), *New Comprehensive Biochemistry*, vol. 9, pp. 268–289. Elsevier, Amsterdam (1984).
48. Gómez-Puyou, A., Tuena de Gómez-Puyou, M. and Ernster, L., *Biochim. Biophys. Acta* **547**, 252 (1979).
49. Ernster, L., *Curr. Top. Cell. Regul.* **24**, 313 (1984).
50. Kagawa, Y., in *Bioenergetics* (ed. L. Ernster), *New Comprehensive Biochemistry*, vol. 9, pp. 149–186. Elsevier, Amsterdam (1984).

Chemica Scripta 1986, **26B**, 281–284

Glutathione and the Evolution of Enzymes for Detoxication of Products of Oxygen Metabolism

Bengt Mannervik

Department of Biochemistry, University of Stockholm, Arrhenius Laboratory, S-106 91 Stockholm, Sweden

Paper presented at the Conference on 'Molecular Evolution of Life', Lidingö, Sweden, 8-12 September 1985

Introduction

Glutathione, the tripeptide γ-glutamylcysteinylglycine, has generally been regarded as the most abundant low-molecular-mass thiol in all types of cells [1]. The experimental basis for this generalization has been limited to analyses of tissues of a small number of organisms such as man, rat, and mouse. It has also been found that a higher homologue, homoglutathione, occurs in mung beans, *Phaseolus aureus* [2, 3]. However, more thorough analyses using improved analytical methods have recently demonstrated that although glutathione is the dominating thiol in all animal tissues examined, it is not always the most abundant sulfhydryl-containing compound in other organisms [4]. In fact, glutathione is not present in detectable concentrations in many classes of bacteria, and is in some cases less abundant than other thiols such as coenzyme A. No glutathione was detected in the anaerobic bacteria analyzed, which included the archaebacterium *Methanococcus voltae* and eubacteria such as the gram-positive *Clostridium pasteurianum* and the gram-negative *Bacteroides fragilis*. Several facultative or aerobic gram-negative bacteria, including *Escherichia coli*, contain high concentrations of glutathione, whereas others lack this compound. The latter group of bacteria in most cases have significant concentrations of H_2S or coenzyme A. The methanogen *M. voltae* has a high level of H_2S (> 0.2 mM) and lower levels (< 0.2 mM) of cysteine and coenzyme M (2-mercaptoethanesulfonate).

Thus, although glutathione is not essential to all prokaryotes, it is generally occurring in eukaryotes, the exception being homoglutathione present in certain leguminous plants [4]. Since homoglutathione reacts like glutathione with the glutathione-linked enzymes investigated [3], the presence of one or the other of these thiols in a particular organism would appear to lack importance for the evolution of this group of enzymes.

The results of the analysis of the occurrence of thiols in different organisms suggest that glutathione emerged as an important biomolecule at a time in evolution when oxygen had appeared in the atmosphere [5]. Other sulfhydryl-containing cofactors such as pantetheine and coenzyme A may have been generally occurring earlier. The higher concentration of glutathione as compared to other thiols in eukaryotic cells suggests a role for glutathione related to the metabolism of oxygen. The present paper summarized data supporting this tenet.

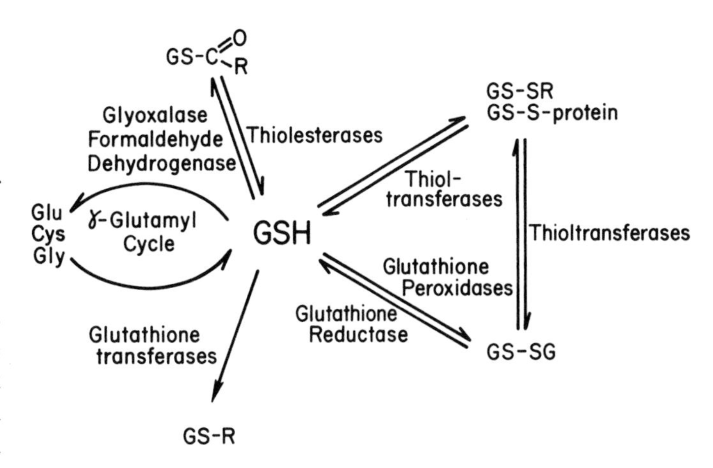

Fig. 1. Scheme of glutathione-linked enzymatic reactions related to biotransformation of products of oxidative metabolism. (After Fig. 1 in ref. 6.)

Glutathione-linked enzymes

A variety of different enzymes utilizing glutathione can be related to the further biotransformation of products of oxidative metabolism. The further discussion of these enzymes will be limited to those present in mammalian cells. Figure 1 shows some of the enzymes catalyzing reactions that have been well characterized. Oxygen-containing products such as H_2O_2 and organic hydroperoxides may be reduced by the action of glutathione peroxidase, which utilizes glutathione as a reductant [7]. Similarly, low-molecular-mass disulfides as well as protein disulfides can be reduced by means of glutathione in the presence of thioltransferase [8]. In both cases, the glutathione disulfide formed is reduced by the NADPH-dependent glutathione reductase [9].

Simple organic molecules that arise by biological oxidations include formaldehyde and 2-oxoaldehydes. These electrophilic compounds are detoxified by conversion to S-formylglutathione and S-(2-hydroxyacyl) glutathione; reactions catalyzed by formaldehyde dehydrogenase [10] and glyoxalase I, respectively [11]. These thiolesters are subsequently hydrolyzed, whereby glutathione is regenerated.

A wide variety of electrophilic organic compounds are metabolized to stable S-conjugates of glutathione that are precursors of excretion products. The latter include mercapturates, thioglucuronides and methylthio derivatives and arise by degradation of the glutathione moiety (see ref. 6). The conjugation reactions are catalyzed by glutathione transferase and are generally considered as major detoxication processes. Glutathione transferase occurs in multiple forms in the same organism and is present in high concentration in organs such as liver and kidney [12]. This group of isoenzymes probably has a prominent role in the metabolism of reactive intermediates. The finding that glutathione transferase is inducible by compounds inducing drug-metabolizing enzymes supports this view [13].

Glutathione transferase

Substrates arising from biological oxidations

Glutathione transferases as a group of enzymes catalyze several types of reaction that are relevant to detoxication of products of oxygen metabolism [12, 14].

Organic hydroperoxides (ROOH) are substrates not only for the selenium-dependent glutathione peroxidase, but also for most of the glutathione transferases:

$$ROOH + 2GSH \rightarrow ROH + H_2O + GSSG$$

where GSH and GSSG are glutathione and its corresponding disulfide, and ROH is an alcohol. Only the selenoprotein is active with H_2O_2 as the hydroperoxide substrate [7].

A second group of substrates is composed of activated alkenes. A variety of such electrophilic compounds are produced by lipid peroxidation and include the toxic 4-hydroxyalk-2-enals [15]. These reactive metabolites are detoxified by conjugation with glutathione, and it has been shown that they rank among the substrates giving the highest activity with the glutathione transferases [16].

$$\underset{\substack{|\\OH}}{R\text{—}CH}\text{—}CH\text{=}CH\text{—}CHO + GSH \rightarrow R\text{—}\underset{\substack{|\\OH}}{CH}\text{—}\underset{\substack{|\\SG}}{CH}\text{—}CH_2\text{—}CHO$$

The primary glutathione adduct is transformed to a cyclic hemimercaptal derivative [17]. The importance of this detoxication reaction is indicated by experiments in which the mutagenicity of the 4-hydroxyalkenals was to be tested. It was found that these compounds were too toxic for the *Salmonella* bacteria to survive in the assay system [18].

A third group of substrates include various epoxides. It has been found that purified cytosolic glutathione transferases catalyze the conjugation of glutathione with aliphatic epoxides, such as the biologically important leukotriene A_4 [19, 20]. Also other epoxide products of arachidonic acid have been found to serve as substrates [21].

Arene oxides represent a special type of epoxides generally recognized as mutagens and carcinogens. These compounds occur as products of monooxygenase reactions and are metabolically deactivated by reaction with glutathione [22].

$$\text{—}\overset{|}{C}\text{—}\overset{|}{C}\text{—} \ (\text{bridged by } O) + GSH \rightarrow \text{—}\underset{\substack{|\\HO}}{\overset{|}{C}}\text{—}\underset{\substack{|\\SG}}{\overset{|}{C}}\text{—}$$

Genetic differences in the occurrence of a particular form of glutathione transferase in human liver suggest that individuals of particular genotypes may be less well protected against certain genotoxic arene oxides [23]. Furthermore, it was recently found that the acidic human glutathione transferase π, in distinction from certain other forms of the enzyme, shows high stereoselectivity for the tumorigenic isomer of 7,8-dihydrodiol-9,10-oxy-benzo(a)pyrene, the *anti*(+)-isomer of R,S,S,R absolute stereochemical configuration (I. C. G. Robertson, C. Guthenberg, B. Mannervik, and B. Jernström, submitted for publication). This specificity may be considered to indicate that particular forms of glutathione transferase may have evolved to afford protection against mutagenic arene oxides of this stereochemistry.

The multiple forms of glutathione transferase

Glutathione transferase occurs in multiple forms in the same organism. At least a dozen cytosolic enzymes as well as a microsomal (membrane-bound) enzyme occur in rat tissues [12]. Also in mouse and human tissues, several forms of the enzyme have been isolated. The human enzymes were first divided into three discrete classes, called basic (or α-ϵ), near-neutral (or μ), and acidic (or π) [6, 23]. The three classes were originally distinguished by substrate specificities, physical properties, and reactions with antibodies [6]. The classification was further supported by inhibition studies [24] and amino-acid sequence analyses [25].

The classification of the cytosolic glutathione transferases was finally extended to rat and mouse enzymes, the properties of which could be related to those of the human enzymes [26]. The three classes of cytosolic glutathione transferase have been named *Alpha*, *Mu*, and *Pi*, and each of the three mammalian species investigated have been found to contain members of each enzyme class, even though the tissue distributions differ among the species. The microsomal enzyme is completely different from the cytosolic forms [27, 28].

Evolution of glutathione-linked enzymes

Glutathione transferases

The available information on amino acid sequences of glutathione transferases reveals extensive homologies within each of the three classes of the cytosolic enzymes, and clear, but more distant, relationships among the classes. Table I shows N-terminal primary structures for each class in which the known alternative residues are displayed for each position. The composite amino acid sequence for Class *Mu* is constructed from seven enzymes of rat, mouse, human and bovine origin. The composite structures of Class *Pi* and Class *Alpha* were derived from three (rat, mouse, and human) and two (rat) enzymes, respectively. It may be added that additional Class *Alpha* enzymes isolated from man and mouse, like one of the rat enzymes, appear to be N-terminally blocked [25, 26].

Comparison of the Class *Mu* and Class *Alpha* structures shows that six of the 40-odd residues aligned are identical in corresponding positions of each structure. An additional set of nine positions have residues that are identical in at least one protein from each class. Similarly, Class *Alpha* and Class *Pi* have nine identities in all structures and four identities between at least two proteins (one from each class). Finally, Class *Mu* and Class *Pi* have four identities in all structures and nine identities in at least one protein from each of the two classes. Two residues are identical in all twelve sequences, and four are common to eleven. Even if the comparison may somewhat exaggerate the number of identities, since some of the structures have only been assigned in the first 20 positions,

Table I. *Composite N-terminal amino acid sequences of three classes of cytosolic glutathione transferase*[a]

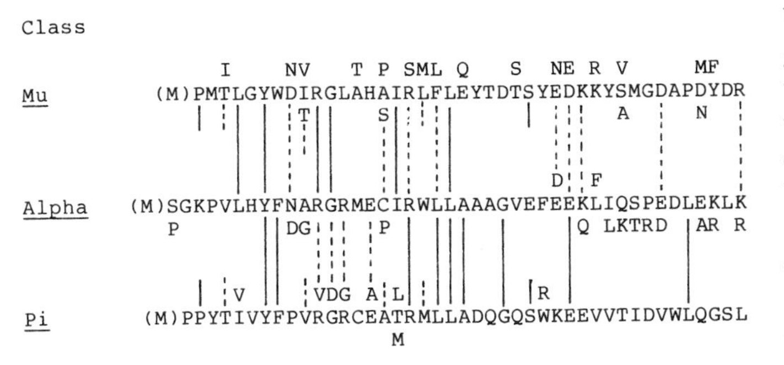

[a] The classes are defined in ref. 26. The sequences (in one-letter code) are composites of available N-terminal primary structures showing alternative amino acid residues in the positions indicated. The N-terminal methionine residues (M) deduced from complementary DNA structures are not present in the peptides analyzed. The Class *Mu* sequence is derived from rat 3-3 [26, 29, 30], rat 4-4 [26, 29, 31], mouse GT-8.7, and mouse GT-9.3 [32], mouse MIII [26], human [25] and a bovine glutathionetransferase [33]; the Class *Alpha* sequence from two variants of rat 1-1 [34–36] and from rat 2-2 [26, 37]; the Class *Pi* sequence from rat 7-7 [26, 28], human [25, 39] and mouse MII [26]. Long full lines mark complete identities between residues in the Class *Alpha* and the Class *Mu* and/or *Pi* sequences; long dashed lines indicate that certain individual, but not all, structures in the respective classes have identical residues. Corresponding short lines indicate the similarities between the Class *Mu* and Class *Pi* sequences.

the comparison leads to the conclusion that the three classes of glutathione transferase have evolved from a common ancestral protein. The divergence of the structures dates before the divergence of the mammalian species from which the enzymes have been derived, since the similarities within a class are significantly stronger than those between classes irrespective of the source of the enzyme. This conclusion is supported by comparisons of the few complete primary structures available.

It should be noted that the primary structure of the 'microsomal' glutathione transferase [28] shows no obvious similarity with any of the known structures of the cytosolic forms of the enzyme. The possible evolutionary relationship to the latter group of proteins is therefore obscure.

Possible relationships among glutathione-linked enzymes

A common denominator for most glutathione-linked enzymes is high specificity for the thiol substrate. In most cases glutathione is the only naturally occurring thiol that is active in the enzymatic reaction. This finding raises the possibility that these enzymes may display structural similarities in the form of a binding site for glutathione. The topography of the active site of human glyoxalase I has been probed by measurements of nuclear relaxation enchancement and been found to be complementary to an extended, Y-shaped glutathione molecule [40]. The computed structure of glutathione has been built into the active site of the bovine selenium-dependent glutathione peroxidase and been shown to fit the binding site for glutathione (R. Ladenstein and B. Mannervik, unpublished work). Thus, the question arises whether glutathione-binding sites have evolved through divergent or convergent processes. The appearance of protein binding sites for glutathione probably occurred with the change to an oxygen-containing atmosphere, since enzymes such as glutathione reductase do not seem to be present in strict anaerobic bacteria or archaebacteria [41].

It is possible that the limited sequence homologies noted between glutaredoxin, glutathione peroxidase, and cytosolic glutathione transferases [42] may indicate similarities related to glutathione binding. However, the primary structures would have to be related to the corresponding three-dimensional structures before any definitive conclusions can be drawn.

The structures of the genes for glutathione-dependent enzymes are unknown at present. The genomic sequences may bear on the hypothesis that, at least some, glutathione-linked enzymes may have evolved by combination of a gene coding for a glutathione-binding domain with genes coding for the catalytic functions of the various enzymes [12]. A binding function independent of catalysis is expressed by a receptor protein for glutathione [43].

Acknowledgement

This work was supported by the Swedish Natural Science Research Council.

References

1. Jocelyn, P. C., *Biochemistry of the SH Group*, pp. 1–14, Academic Press, London and New York (1972).
2. Price, C. A., *Nature* **180**, 148 (1957).
3. Carnegie, P. R., *Biochem. J.* **89**, 471 (1963).
4. Fahey, R. C. and Newton, G. L., in *Functions of Glutathione: Biochemical, Physiological, Toxicological, and Clinical Aspects* (ed. A. Larsson, S. Orrenius, A. Holmgren and B. Mannervik), pp. 251–260, Raven Press, New York (1983).
5. Fahey, R. C., *Adv. Exp. Med. Biol.* **86A**, 1 (1977).
6. Mannervik, B., Guthenberg, C., Jensson, H., Warholm, M. and Ålin, P., in *Functions of Glutathione: Biochemical, Physiological, Toxicological, and Clinical Aspects* (ed. A. Larsson, S. Orrenius, A. Holmgren and B. Mannervik), pp. 75–88, Raven Press, New York (1983).
7. Wendel, A., in *Enzymatic Basis of Detoxication* (ed. W. B. Jakoby), vol. 1, pp. 333–353, Academic Press, New York (1980).
8. Mannervik, B., in *Enzymatic Basis of Detoxication* (ed. W. B. Jakoby), vol. 2, pp. 229–244, Academic Press, New York (1980).
9. Mannervik, B., Boggaram, V., Carlberg, I. and Larson, K., in *Flavins and Flavoproteins* (ed. K. Yagi and T. Yamano), pp. 173–187, Japan Scientific Societies Press, Tokyo (1980).
10. Uotila, L. and Koivusalo, M., in *Functions of Glutathione: Biochemical, Physiological, Toxicological, and Clinical Aspects* (ed. A. Larsson, S. Orrenius, A. Holmgren and B. Mannervik), pp. 175–186, Raven Press, New York (1983).
11. Sellin, S., Aronsson, A.-C., Eriksson, L. E. G., Larsen, K., Tibbelin, G. and Mannervik, B., in *Functions of Glutathione: Biochemical, Physiological, Toxicological, and Clinical Aspects* (ed. A. Larsson, S. Orrenius, A. Holmgren and B. Mannervik), pp. 187–197, Raven Press, New York (1983).
12. Mannervik, B., *Adv. Enzymol. Relat. Areas Mol. Biol.* **57**, 357 (1985).
13. Guthenberg, C., Morgenstern, R., DePierre, J. W. and Mannervik, B., *Biochim. Biophys. Acta* **631**, 1 (1980).
14. Mannervik, B., Ålin, P., Guthenberg, C., Jensson, H. and Warholm, M., in *Microsomes and Drug Oxidations* (ed. A. R. Boobis, J. Caldwell, F. De Matteis and C. R. Elcombe), pp. 221–228, Taylor and Francis, London and Philadelphia (1985).
15. Esterbauer, H., in *Free Radicals, Lipid Peroxidation and Cancer* (ed. D. C. H. McBrien and T. F. Slater), pp. 101–122, Academic Press, London (1982).
16. Ålin, P., Danielson, U. H. and Mannervik, B., *FEBS Lett.* **179**, 267 (1985).
17. Esterbauer, H., Zollner, H. and Scholz, N., *Z. Naturforsch.* **30C**, 466 (1975).
18. Marnett, L. J., Hurd, H. K., Hollstein, M. C., Levin, D. E., Esterbauer, H. and Ames, B. N., *Mutation Res.* **148**, 25 (1985).

19. Mannervik, B., Jensson, H., Ålin, P., Örning, L. and Hammarström, S., *FEBS Lett.* **175**, 289 (1984).
20. Söderström, M., Mannervik, B., Örning, L. and Hammarström, S., *Biochem. Biophys. Res. Commun.* **128**, 265 (1985).
21. Spearman, M. E., Prough, R. A., Estabrook, R. W., Falck, J. R., Manna, S., Leibman, K. C., Murphy, R. C. and Capdevila, J., *Arch. Biochem. Biophys.* **242**, 225 (1985).
22. Chasseaud, L. F., *Adv. Cancer Res.* **29**, 175 (1979).
23. Warholm, M., Guthenberg, C. and Mannervik, B., *Biochemistry* **22**, 3610 (1983).
24. Tahir, M. K., Guthenberg, C. and Mannervik, B., *FEBS Lett.* **181**, 249 (1985).
25. Ålin, P., Mannervik, B. and Jörnvall, H., *FEBS Lett.* **182**, 319 (1985).
26. Mannervik, B., Ålin, P., Guthenberg, C., Jensson, H., Tahir, M. K., Warholm, M. and Jörnvall, H., *Proc. Natl. Acad. Sci. USA* **82**, 7202 (1985).
27. Morgenstern, R., Guthenberg, C. and DePierre, J. W., *Eur. J. Biochem.* **128**, 243 (1982).
28. Morgenstern, R., DePierre, J. W. and Jörnvall, H., *J. Biol. Chem.* **260**, 13976 (1985).
29. Frey, A. B., Friedberg, T., Oesch, F. and Kreibich, G., *J. Biol. Chem.* **258**, 11321 (1983).
30. Ding, G. J.-F., Lu, A. Y. H. and Pickett, C. B., *J. Biol. Chem.* **260**, 13268 (1985).
31. Ålin, P., Mannervik, B. and Jörnvall, H., *Eur. J. Biochem.* (in the press).
32. Pearson, W. R., Windle, J. J., Morrow, J. F., Benson, A. M. and Talalay, P., *J. Biol. Chem.* **258**, 2052 (1983).
33. Asaoka, K., *J. Biochem.* **95**, 685 (1984).
34. Pickett, C. B., Telakowski-Hopkins, C. A., Ding, G. J.-F., Argenbright, L. and Lu, A. Y. H., *J. Biol. Chem.* **259**, 5182 (1984).
35. Lai, H.-C. J., Li, N., Weiss, M. J., Reddy, C. C. and Tu, C.-P. D., *J. Biol. Chem.* **259**, 5536 (1984).
36. Taylor, J. B., Craig, R. K., Beale, D. and Ketterer, B., *Biochem. J.* **219**, 223 (1984).
37. Telakowski-Hopkins, C. A., Rodkey, J. A., Bennett, C. D., Lu, A. Y. H. and Pickett, C. B., *J. Biol. Chem.* **260**, 5820 (1985).
38. Suguoka, Y., Kano, T., Okuda, A., Sakai, M., Kitagawa, T. and Muramatsu, M., *Nucleic Acid Res.* **13**, 6049 (1985).
39. Dao, D. D., Partridge, C. A., Kurosky, A. and Awasthi, Y. C., *Biochem. J.*. **221**, 33 (1984).
40. Rosevear, P. R., Sellin, S., Mannervik, B., Kuntz, I. D. and Mildvan, A. S., *J. Biol. Chem.* **259**, 11436 (1984).
41. Ondarza, R. N., Rendón, J. L. and Ondarza, M., *J. Mol. Evol.* **19**, 371 (1983).
42. Jörnvall, H. and Persson, B., in *Thioredoxin and Glutaredoxin Systems: Structure and Function* (ed. A. Holmgren, C.-I. Brändén, H. Jörnvall and B.-M. Sjöberg), pp. 111–120, Raven Press, New York (1986).
43. Cobb. M. H., Heagy, W., Danner, J., Lenhoff, H. M., Marshall, G. R., *Mol. Pharmacol.* **21**, 629 (1982).

Chemica Scripta 1986, **26B**, 285–286

Evolutionary Relationship between Metal Centres in Cytochrome Oxidase and Blue Oxidases

Bo G. Malmström

Department of Biochemistry and Biophysics, University of Göteborg and Chalmers University of Technology, S-412 96 Göteborg, Sweden

Paper presented at the Conference on 'Molecular Evolution of Life', Lidingö, Sweden, 8–12 September 1985

Abstract

Comparisons are made of the amino acid sequences around metal-binding residues in cytochrome oxidase, a number of blue copper proteins and superoxide dismutase. These show that Cu_A in cytochrome oxidase is structurally related to type 1 Cu in blue oxidases as well as in simple blue proteins, like plastocyanin. In addition, there are homologies which indicate similarities in coordination of Cu_B and type 3 Cu. These metal sites are also related to the copper coordination in a simpler protein, superoxide dismutase, with a bimetallic site.

Introduction

Cytochrome *c* oxidase, the terminal electron-transport complex of the mitochondrial respiratory chain, couples electron transfer to the translocation of protons from the matrix to the cytosol side of the inner membrane [1]. The catalytic reaction, which provides the driving force for proton pumping, is the oxidation of four molecules of ferrocytochrome *c* by dioxygen, which is in this process reduced to two molecules of water. There are only a few other oxidases, notably the blue, copper-containing oxidases, laccase, ascorbate oxidase and ceruloplasmin, which also can utilize the full oxidizing capacity of dioxygen [2].

The catalytic centres of cytochrome oxidase and blue oxidases are similar in design, which suggests related catalytic mechanisms. The simplest blue oxidase, laccase, has four metal ions in its catalytic unit, like cytochrome oxidase. In the oxidized enzymes two of these ions are detectable by EPR, namely cytochrome a^{3+} and Cu_A^{2+} in cytochrome oxidase and types 1 and 2 Cu^{2+} in laccase. The EPR-detectable centres serve as the primary acceptors of electrons from the reducing substrates. From these metal centres the electrons are transferred intramolecularly to the EPR-nondetectable, binuclear centres, cytochrome a_3-Cu_B and the type 3 copper ions, respectively, which are the dioxygen-reducing sites of the enzymes.

Spectroscopic data [3] suggest that the coordination of Cu_A^{2+} is unique among all known copper (II) complexes. The spectroscopic properties of type 1 Cu^{2+} is also unusual [3], although similar among all blue oxidases as well as simple blue electron-transfer proteins, such as azurin, plastocyanin, stellacyanin and umecyanin. Amino-acid sequence data from several laboratories have now established [4] that not only are all blue proteins evolutionary related, but they also display homologies with subunit II of cytochrome oxidase.

Type 3 copper and Cu_B have been shown to be spectroscopically similar [5,6] and, in addition, related to the copper ion in the bimetallic zinc-copper site of superoxide dismutase (SOD) [5]. Thus, it is interesting to note that a short sequence around the ligand histidine residues in SOD is also found in human ceruloplasmin and cytochrome oxidase [7].

Sequence homologies among copper proteins

Type 1 copper and copper-A

The three-dimensional structure of two simple blue copper proteins, plastocyanin and azurin, is now known [8, 9]. In both cases the metal ligands of the type 1 copper are provided by two residues of histidine, one of cysteine and one of methionine. It is possible on the basis of the structure to explain [10] the unusual spectroscopic properties of type 1 Cu^{2+} in terms of a strain induced by a tetrahedral distortion forced on the metal by the protein tertiary structure and a ligand-to-metal charge transfer (LMCT) transition [$S \rightarrow Cu(II)$] involving the ligand cysteine.

The complete amino acid sequences have been determined for 9 azurins, 14 plastocyanins [4] and stellacyanin [11]; most of the sequence of umecyanin is also known [12]. All these small blue proteins show a large degree of homology [4], but interestingly enough stellacyanin lacks the ligand methionine. It has been suggested [10] that this is the reason that it has the lowest reduction potential of all blue proteins.

For the blue oxidases the available sequence information is more limited. The complete sequence of human ceruloplasmin has recently been reported [13], but partial sequences only have been determined for fungal [14] and tree laccase (Betty Malmström and L. Strid, unpublished work). These data suggest that the blue oxidases have copper-binding sites similar to plastocyanin and azurin, as shown in Fig. 1 by the comparison of sequences around the ligand cysteine. It is interesting that in all three multi-copper proteins, but not in the simple blue proteins, the ligand cysteine is flanked by two histidine residues. It has been argued [14, 15] that these are ligands to type 2 copper, which is, of course, absent in the simple proteins.

Bovine cytochrome oxidase contains 12 different polypeptides, which have all been sequenced [16, 17]. The three largest ones are coded for by mitochondrial DNA, and for these the primary structure has been deduced from the mitochondrial

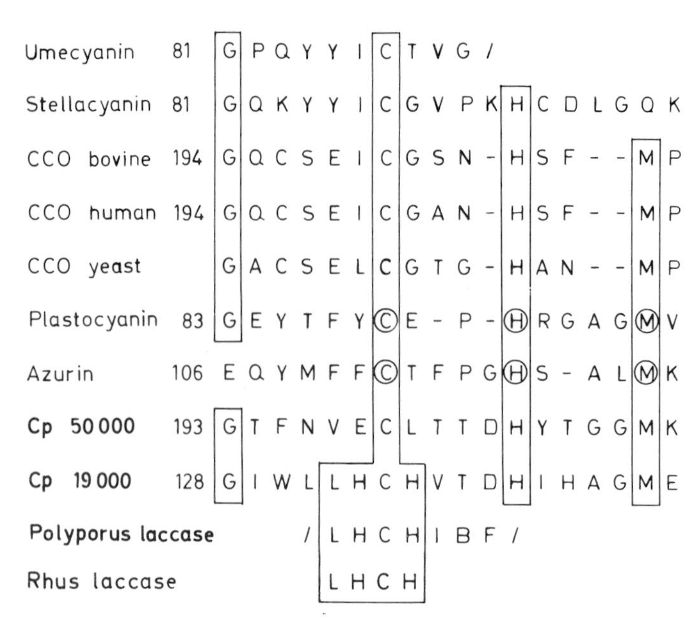

Fig. 1. Sequence homology around an invariant cysteine residue in some blue copper proteins and polypeptide II of cytochrome oxidase. Established metal ligands are circled. (Adapted from [7]).

DNA sequences of the genes for human oxidase [18]. A small part of the sequence around an invariant cysteine in polypeptide II is included in Fig. 1. It is seen that there is considerable homology with the sequences around the metal ligands in the blue proteins. In cytochrome oxidase there is, however, an additional invariant cysteine residue, and spectroscopic data suggest that it is also a ligand to Cu_A [19]. This could explain why Cu_A is unique despite its obvious relationship to type 1 copper.

A surprising finding is that human factor VIII is homologous with human ceruloplasmin [20]. This suggests that factor VIII also has a binding site for type 1 Cu.

Type 3 copper and copper-B

In Fig. 2 a comparison is made between specific sequences in ceruloplasmin, cytochrome oxidase and SOD. These proteins all contain bimetallic sites, which are, however, quite distinct: type 3 Cu-Cu in ceruloplasmin, cytochrome a_3-Cu_B in cytochrome oxidase and Cu-Zn in SOD. It is interesting that sequences homologous to those around the known copper ligands in SOD are found in the other two proteins. It has already been mentioned that spectroscopic data indicate similar coordination of the copper ion in all three proteins.

Concluding remarks

It is now well established that cytochrome oxidase and blue oxidases reduce dioxygen by very similar mechanisms [2], despite the fact that one is a haem-copper protein and the other ones pure copper proteins. The occurrence of similar functions in proteins with different structures is sometimes taken in evidence for convergent evolution. The amino acid sequence data have, however, shown that these enzymes are, in fact, also structurally related. Thus, Cu_A in cytochrome oxidase and type 1 Cu in blue oxidases show similar coordination, which is in turn the same as in simple blue proteins. Type 3 Cu and Cu_B are likewise kindred species, related to the copper coordination in SOD. It would appear that complex

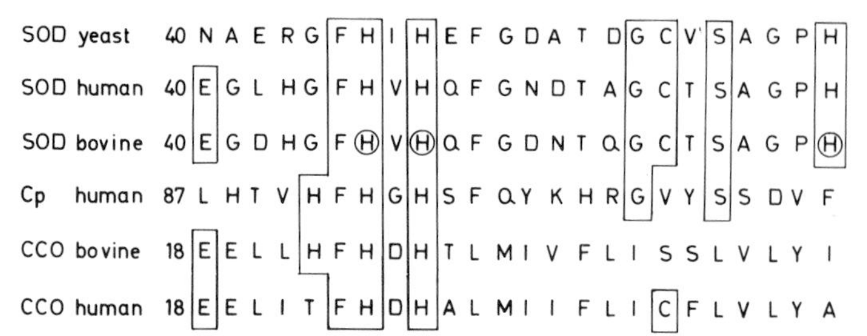

Fig. 2. Sequence homology in copper-containing oxidases and superoxide dismutase [7].

metalloenzymes, containing several metal-binding sites, have evolved by the combination of genes for simpler proteins, each one derived by divergent evolution of some ancestral metalloproteins.

Acknowledgements

Investigations in my own laboratory have been supported by the Swedish Natural Science Research Council. I wish to thank Dr Bengt Reinhammar for helpful discussions, and Mrs Betty Malmström and Mr Lars Strid for allowing me to include their unpublished results.

References

1. Malmström, B. G., *Biochem. Biophys. Acta* **811**, 1 (1985).
2. Malmström, B. G., *Annu. Rev. Biochem.* **51**, 21 (1982).
3. Vänngård, T., in *Biological Applications of Electron Spin Resonance* (ed. H. M. Swartz, J. R. Bolton and D. C. Borg), p. 411. Wiley, New York (1972).
4. Rydén, L., in *Copper Proteins* (ed. R. Lontie), vol. I, p. 157. CRC Press, Boca Raton (1984).
5. Reinhammar, B., Malkin, R., Jensen, P., Karlsson, V., Andréasson, L.-E., Aasa, R., Vänngård, T. and Malmström, B. G., *J. Biol. Chem.* **255**, 5000 (1980).
6. Cline, J., Reinhammar, B., Jensen, P., Venters, R. and Hoffman, B. M., *J. Biol. Chem.* **258**, 5124 (1983).
7. Reinhammar, B., in *Copper Proteins* (ed. R. Lontie), vol III, p. 1. CRC Press, Boca Raton (1984).
8. Colman, P. M., Freeman, H. C., Guss, J. M., Murata, M., Norris, V. A., Ramshaw, J. A. M. and Venkatappa, M. P., *Nature (London)* **272**, 319 (1978).
9. Adman, E. T., Stenkamp, R. E., Sieker, L. C. and Jensen, L. H., *J. Mol. Biol.* **123**, 35 (1978).
10. Gray, H. B. and Malmström, B. G., *Comments Inorg. Chem.* **2**, 203 (1983).
11. Bergman, C., Gandvik, E.-K., Nyman, P. O. and Strid, L., *Biochem. Biophys. Res. Commun.* **77**, 1052 (1977).
12. Bergman, C., Ph.D. thesis, University of Göteborg (1980).
13. Takahashi, N., Ortel, T. L. and Putnam, F. W., *Proc. Nat. Acad. Sci. USA* **81**, 390 (1984).
14. Briving, C., Gandvik, E.-K. and Nyman, P. O., *Biochem. Biophys. Res. Commun.* **93**, 454 (1980).
15. Malmström, B. G., in *The Evolution of Protein Structure and Function* (ed. D. S. Sigman and M. A. B. Brazier), p. 87. Academic Press, New York (1980).
16. Buse, G., in *Copper Proteins* (ed. R. Lontie), vol. III, p. 119. CRC Press, Boca Raton (1984).
17. Buse, G., Meinecke, L. and Bruch, B., *J. Inorg. Biochem.* **23**, 149 (1985).
18. Anderson, M., de Bruijon, M. H. L., Coulson, A. R., Eperon, I. C., Sanger, F. and Young, I. G., *J. Mol. Biol.* **156**, 683 (1982).
19. Blair, D. F., Martin, C. T., Gelles, J., Wang, H., Brudvig, G. W., Stevens, T. H. and Chan, S. I., *Chemica Scripta* **21**, 43 (1983).
20. Vehar, G. A., Keyt, B., Eaton, D., Rodriguez, H., O'Brien, D. P., Rotblat, F., Opperman, H., Keck, R., Wood, W. I., Harkins, R. N., Tuddenham, E. G. D., Lawn, R. M. and Copon, D. J., *Nature (London)* **312**, 337 (1984).

Complex Systems and Organization

Chemica Scripta 1986, **26B**, 289–297

Human T-Lymphotropic Retroviruses: Their Role in Malignancy and Immune Suppression

Robert C. Gallo

Laboratory of Tumor Cell Biology, National Cancer Institute, National Institutes of Health, Bethesda, Maryland 20892, USA

Paper presented at the Conference on 'Molecular Evolution of Life', Lidingö, Sweden, 8-12 September 1985

Abstract

Retroviruses probably originated from cellular genes through reverse transcription complexes (RNA molecules associated with particular DNA polymerases) associated with genes for membrane structures and transposon-like elements. These complexes may play some normal role in cell development. Through genomic change these particles may escape, infect cells inappropriately, and in some cases infect other animals including other species. As fully evolved infectious particles, retroviruses often cause fatal disease, usually infecting cells of the hematopoietic and nervous systems. Human retroviruses were first discovered in 1978–9. There are now 3 known categories; each one is tropic for the critical cell of the immune system, the T4 or helper T-cell. Because of this tropism they have been called human T-lymphotropic viruses I, II, and III. HTLV-I and II cause T-cell proliferation *in vitro* and T-cell leukemia in people. HTLV-III produces T-cell death *in vitro* and is the cause of acquired immune deficiency (AIDS) in people. All three are transmitted by blood, sexual contact, or congenital infection, and they probably entered humans from certain African monkeys, especially the African green monkey. The mechanisms of disease induction, the epidemiology, and ideas for their control are discussed.

Retroviruses probably originated from cellular nucleotide sequences. In time these sequences associated with a DNA polymerase (reverse transcriptase) capable of transcribing RNA molecules into DNA. Since this enzyme has little or no capacity to discriminate between various RNA molecules, compartmentalization of this enzyme with its RNA template must have been an early requirement. The basic (and minimum) genome of a *replicating* retrovirus consists of three genes. These are the gene for the virus core structural proteins (*gag* gene), the gene for reverse transcriptase (*pol* gene), and the gene for the envelope (*env* gene). These three genes are flanked at both 5′ and 3′ ends by repetitive and unique sequences which consist of several regulating elements including promoter sites, enhancer elements, CAP site, poly A site, and some additional functional sequences. These sequences are known as long terminal repeats or LTRs, and they bear structural similarities to transposon like elements found in the genome of cells. Sequences, *related* to some of the retroviral structural genes are also found in the DNA of normal cells, but even more striking is the presence of multiple copies of retroviral genomes in the DNA of most if not all vertebrates. These sequences are normally not expressed, but sometimes they can be activated by treatment of normal cells with chemicals like iododeoxyuridine. In these instances fully formed retrovirus particles may be formed. These are referred to as *endogenous* retroviruses, and must be viewed as normal genetic Mendelian elements transmitted in the germline. These retroviruses are rarely infectious for their own host, i.e. when expressed they do not re-infect cells of the host species. In some species, e.g. man, endogenous retroviruses have never been unequivocally demonstrated. However, by using molecular probes of endogenous retroviruses of other species, related sequences can always be found in other species even without ever demonstrating formation of a virus. The reason for failure to detect the endogenous retrovirus may be because of incomplete expression of the viral genome. Alternatively, many species may not contain the full complement of sequence needed for virus formation.

The expression of endogenous retroviral sequences in most species is tightly controlled, and those viruses have no known disease causing capacity in out-bred animals. Many people have proposed that endogenous viruses have a normal function, e.g. in gene amplification and gene transfer from one cell to another during differentiation. However, no specific function has been demonstrated and if one exists it is likely to be very transient.

There is evidence that during evolution some endogenous retroviruses escaped from their normal host and entered a new species. If it becomes an infectious agent in this species it is referred to as an exogenous retrovirus, and it is likely to cause disease in the new recipients.

The exogenous retroviruses may cause malignant disease, non-malignant disease (neurological abnormalities and chronic degenerative diseases) or both. Leukemias and lymphomas are the most frequently induced malignancies because the major target cell of a retrovirus is usually in the hematopoietic system.

Exogenous retroviruses are enveloped RNA viruses, which usually bud from cell membranes and contain an electron dense, central, core structure surrounding the viral RNA genome. The *sine qua non* of a retrovirus, however, is the presence of the reverse transcriptase (RT) which is complexed to the RNA genome in the viral core. When a retrovirus infects a cell the reverse transcriptase catalyzes the transcription of the RNA genome into a DNA form known as the provirus. The DNA form, as a double stranded circular

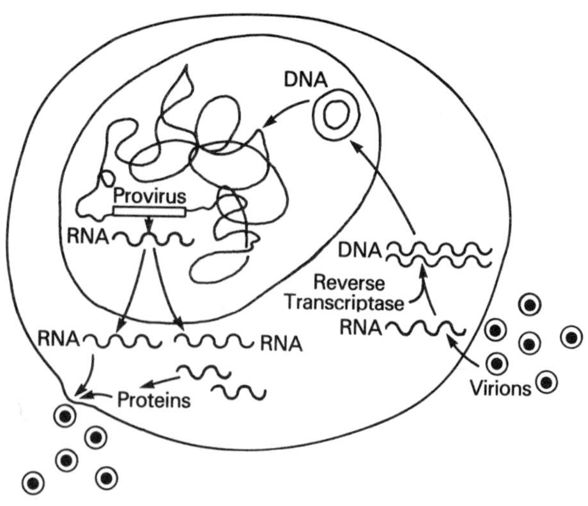

Fig. 1. Life cycle of a retrovirus. Intact virions are endocytosed via a specific cellular receptor. The uncoated viral single-stranded RNA is then transcribed into double-stranded DNA, enters the cell nucleus and integrates into the host genome. The DNA provirus in some conditions is unexpressed. In other cases, it is transcribed giving rise to viral RNA encoding viral proteins and to genome length viral RNA molecules which then reassemble with viral proteins to make complete virions. These progeny are released by budding from the cell membrane.

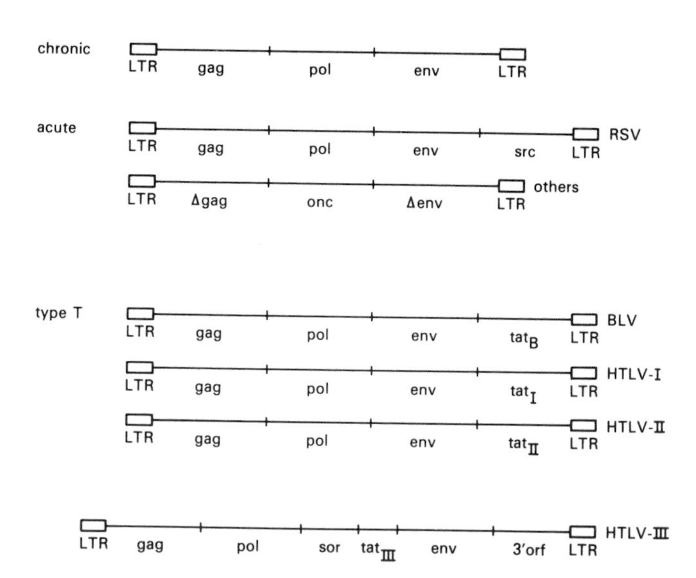

Fig. 2. Genetic structure of retroviruses. *gag*, core proteins; *pol*, reverse transcriptase; *env*, envelope; LTR, long terminal repeat; Δ*gag*, Δ*env*, incomplete genes; *src*, one of the *onc* genes; BLV, bovine leukemia virus; *tat*, transacting transcriptional activator gene; sor, short open reading frame; 3′ orf, 3′ open reading frame. The latter two are genes in HTLV-III of unknown function.

element, usually migrates from the cytoplasm to the nucleus and then integrates into the host cell DNA where the viral genes may remain for the lifetime of the cell (Fig. 1). Because the provirus is duplicated with the cell DNA during S phase of the cell cycle, the viral genes are passed to daughter cells. Therefore, infection of an organism is generally life long. The existence of an integrated DNA provirus derived from the viral RNA genome was predicted by Howard Temin during the 1960s in a moment of brilliant scientific insight. Temin and D. Baltimore, went on to prove this to be correct during the 1970s [1, 2]. The provirus can remain unexpressed, be partially expressed or fully expressed. In the latter case, the viral RNA and proteins will be found in the cell cytoplasm, and under the right circumstances will assemble at the cell membrane, where budding and release eventually occurs, thus completing the virus life cycle (see Figs 1 and 2). Because of recombinational events which may occur when the provirus integrates into the host cell DNA, on rare occasions the provirus may permanently acquire host cellular sequences. The subsequent formation of virus with these newly acquired nucleotide sequences, of course, will have some properties which differ from the original virus.

Major progress has come from studies of retrovirus genomes in recent years, and their properties can also be used in classifying these viruses. Structural features of their genome govern the basic type of molecular mechanisms by which these viruses alter a cell. The most common type of retroviral genome contains only the 3 essential genes for virus replication referred to above (Fig. 2), and retroviruses like this are important causes of animal disease.

As already noted, the viral genes are flanked at their ends by the LTRs. These sequences may influence the expression of the viral genes and sometimes of nearby cellular genes. In addition, the LTRs form the sites of integration into the host cell DNA, i.e. the cellular sequences are covalently linked to the LTRS. Viruses with this type of genome are usually called chronic leukemia viruses (Fig. 2). Examples are feline leukemia virus (FeLV), murine leukemia virus (MuLV), avian leukosis virus (ALV), and Gibbon ape leukemia virus (GaLV). These viruses replicate extensively in the host prior to their induction of leukemia, and there is evidence that they cause leukemia by integration within a certain region of a chromosome so that an LTR may promote extensive and continual expression of a cellular gene involved in growth. The best example of this *cis* mechanism is in chickens where the LTR of the ALV promotes expression of a cellular oncogene, which is believed to be the first step in the leukemic process induced by this virus. Since integration by retroviruses is random, the need to integrate in a specific region for induction of disease may explain the apparent need for viremia and extensive virus replication prior to the malignancy. High rate of replication should favor the chance of integration into regions sufficiently near the cellular oncogene to enable the LTR to activate this gene.

Sometimes retroviruses acquire a host cell gene, and when this gene gives the virus the property to rapidly transform cells and induce acute malignancies in an animal the virus is often called an acute leukemia or sarcoma virus and the gene a viral *onc* gene (Fig. 2). Viruses with *onc* genes are fortunately rare, have never been identified in humans, and when found in animals are of interest chiefly for studying mechanisms of neoplastic transformation but not as causes of naturally occurring cancer. Every cell infected by these virtues can be transmformed (polyclonal) because the product of the viral *onc* gene is directly transforming. Therefore, a common site of integration is not needed. The induction of disease then is by a *trans*-mechanism. The other categories of known retroviral genomes are described below.

General features of human retroviruses

A third category of retrovirus genomes was only recently discovered. It consists of all known human leukemia viruses: human T-lymphotropic (or leukemia) virus types I and II (HTLV-I and HTLV-II) (Fig. 2) and bovine leukemia virus

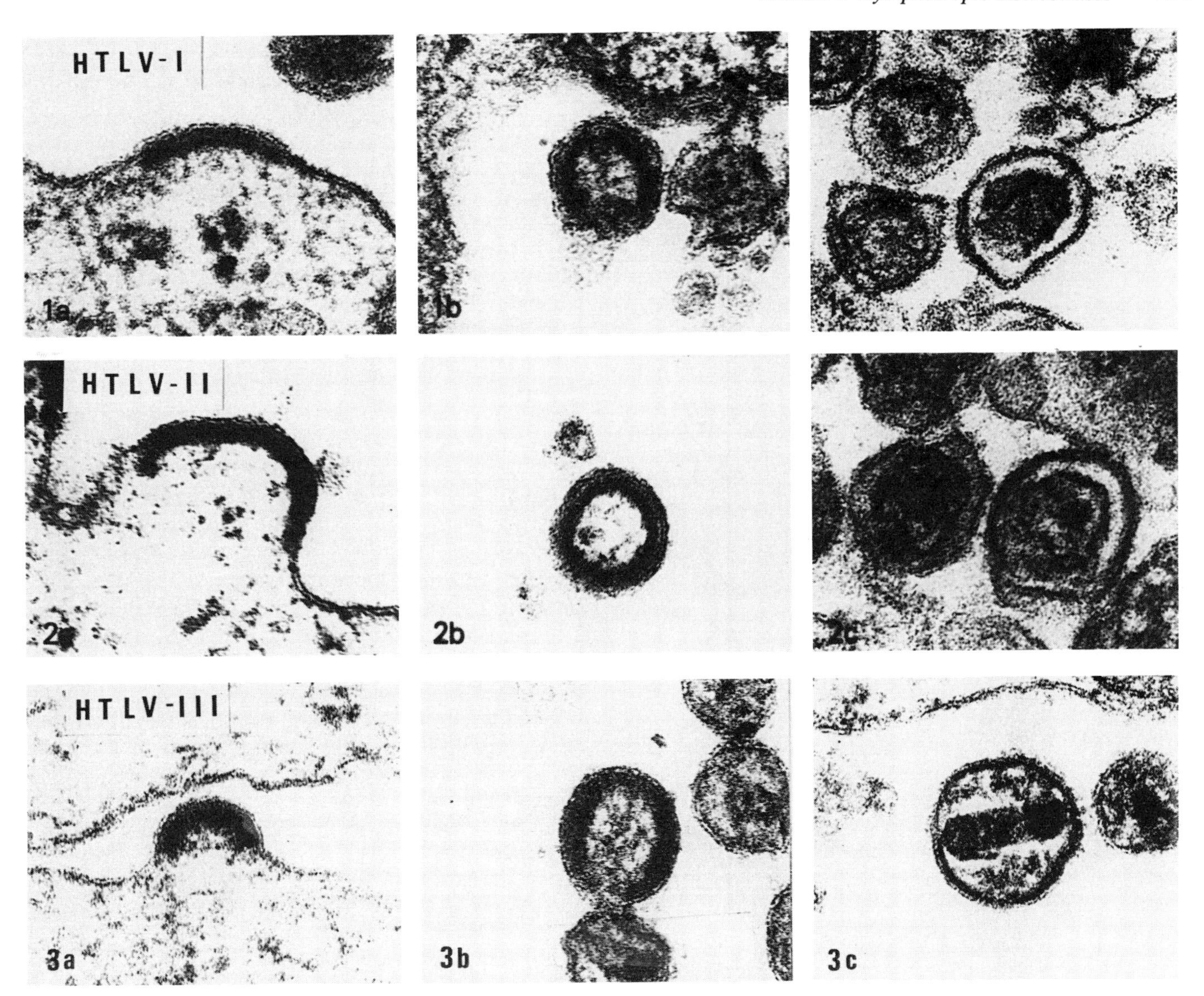

Fig. 3. Human retroviruses. Electron micrographs of maturation of HTLV-I, II, and III showing formulation, budding, and release of retrovirus. ×90000.

(BLV). The third known human retrovirus, human T-lymphotropic virus II (HTLV-III) forms a special sub-category (Figs 2 and 3), and although not yet known, it is likely that the lentiretroviruses will be similar to HTLV-III. The genomes of all these viruses have the following properties: (1) in addition to the viral replication genes (*gag*, *pol*, *env*) they contain one or more extra genes; (2) the extra gene(s) is not homologous to a mammalian cell gene, i.e. not an *onc* gene. Like the DNA tumor virus transforming genes, the origin of the extra gene(s) is unknown; [3] at least one of these extra genes codes for a protein which is involved in activating expression of other viral genes (likely by binding to the viral LTRs) and probably of certain cell genes (presumably by binding to regulatory enhancer elements in the cell which are similar to the viral LTRs). Recent results have indicated that major aspects of the biological effects of these viruses are mediated by this gene, called *trans*-acting transcriptional activator or *tat*, and involve this mechanism, which is referred to as transacting transcriptional activation. Since *tat* codes for a nuclear protein which can activate other genes, these viruses do not have to integrate in a special region to induce disease. This may be why extensive virus replication is not needed for them to cause neoplastic or non-neoplastic disease in man. A similar phenomenon is seen in lymphoma of cows induced by BLV. HTLV-III, also known as lymphadenopathy associated virus or LAV, not only contains the 3 genes for virus replication and a *tat* gene, it also contains at least 2 other genes of still undefined function.

Electron micrographs of the human retroviruses are shown in Fig. 3. HTLV-I and II are similar. HTLV-III differs in its mature form, showing a highly condensed cylindrical core structure. HTLV-I, the first human retrovirus, was first isolated in my laboratory in 1978 and reported in 1980 after its characterization. It was isolated from a young black man with an aggresive T-cell malignancy from the U.S. The methods we developed for isolation of human retroviruses depended upon the idea that a retrovirus might cause disease without an earlier viremia or without evident extensive replication in the tumor [3]. Therefore, sensitive means of virus detection and careful growth of the putative target cells were developed in my laboratory during the 1970s. The sensitive assays were based on the use of the enzyme, reverse transcriptase (RT), as a 'footprint' of a retrovirus since this assay can be made several orders of magnitude more sensitive than electron microscopy. A breakthrough in the growth of the target cells was achieved with the discovery of T-cell growth factor (interleukin 2 or Il-2) (see refs 4 and 5 for review). The combination of growing mature T-cells with Il-2 and the use of RT assays eventually led to the routine isolation of HTLV-I, and this is the same basic technique used

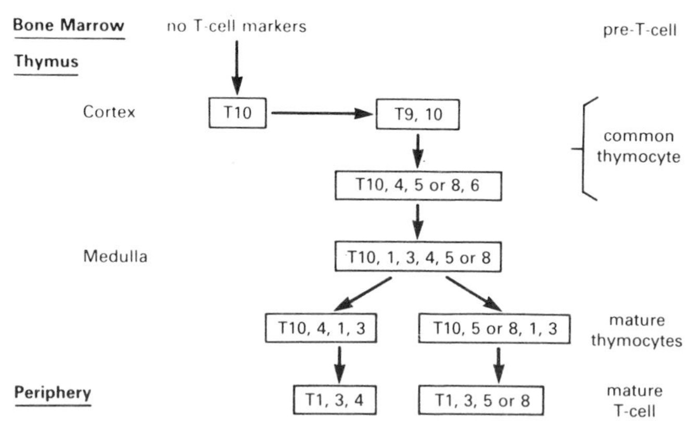

Fig. 4. The thymus is an epithelio-lymphoid organ with three distinct maturational layers: subcapsular cortex; cortex; and medulla. Lymphocyte maturation involves the induction and rearrangement of receptor genes and the expressions of antigen specific receptors and surface markers that define the mature peripheral T-cell subsets.

Table I. *The T4-helper cell plays a central role in the initiation, amplification, and regulation of the immune system*

Function of the T4-helper T cells
T-Cell dependent antibody responses
1. Antigen specific immunoglobin production
2. Class switching Igm → IgG
T-Cell induction of cellular immune response
1. Alloantigen response, graft rejection, immune surveillance
2. Antigen specific cytotoxic T-cells
3. Regulatory T-cells
4. Delayed hypersensitivity reactions
5. Activation of macrophages
6. Release of soluable factors (Il-2, IFNγ).
T-Cell regulation of other hematopoitic cells
1. Granulocytes
2. Erythrocytes
3. Megakaryocytes

for the isolation of the AIDS virus (HTLV-III or LAV) in various laboratories (see below).

One of the most remarkable features about the currently recognized human retroviruses is their tropism for the mature T4 helper lymphocyte (Fig. 4). Although other cells can be infected *in vitro* in a mixed cell population by all three types of known human retroviruses, and the primary diseases caused by them almost always involve this cell. Since the T4 cell regulates many of our immune functions and may even regulate some functions of non-lymphoid cells (Table I), it is not difficult to understand why these viruses induce such profound clinical disease (see below).

Another extraordinary feature of the human retroviruses is their ability to cause T4 cell alterations *in vitro*. Infection of T4 lymphocytes by HTLV-I or II leads to the immortalized growth of a small number of these cells, and the properties of the transformed cells are very similar to the primary HTLV-I positive leukemic cells from an ATL patient. Other T4 cells and other T cells infected by HTLV-I may not be transformed but exhibit impairment of one or more T-cell functions. *In vitro* infection of T4 cells by HTLV-III can lead to the premature death of these cells when they are activated and the viral genes are expressed, resembling what we assume to be the case *in vivo* in patients with AIDS. This affords us with a unique opportunity: the laboratory study of the molecular and cellular mechanisms of induction of two forms of fatal human disease utilizing the precise target cell and the known causative agents.

The disease associated with HTLV-I

As noted above, the vast majority of known HTLV-I induced leukemias or lymphomas involve the T4 cell. A major and apparently consistent difference between the HTLV-I positive leukemic T-cells and normal T-cells is the constitutive expression and increased number of receptors for Il-2 in the former (about 300000 receptors per cell). Receptors for the growth factor are present in normal T-cells only after they are activated (about 20000 receptors per cell). Then they are soon down regulated. Their continued expression and abundance in the leukemic cells may be central to the abnormal growth of these cells. Despite their T4 characteristics, the virus positive leukemic cells generally do not have detectable helper function, and some function as suppressor T-cells. Also, *in vitro* infection of cloned functional T-cells of different types leads to changes or loss of function. These findings parallel the immune defects and opportunistic infections that occur in these viral leukemias. Leukemias/lymphomas caused by HTLV-I usually fit a particular form of lymphoid malignancy known as adult T-cell leukemia/lymphoma (ATLL), clinically characterized chiefly by its aggressive course (median survival about 4 months), hypercalcemia (mechanism unknown), and several forms of skin abnormalities occur in over half the cases. The disease has many similarities to cutaneous T-cell lymphomas (mycosis fungoides and Sézary leukemias) except for the more aggressive course, and without the presence of the characteristic morphological changes in the T-cell or the positive virological studies for HTLV-I, it is doubtful whether one could confidently make the distinction. HTLV-I can also be involved in T4 cell leukemias and lymphomas that exhibit a more chronic course (15–20% of cases) with other features differing from ATLL. These may not be pathologically or clinically distinguishable from T-cell CLL, diffuse histiocytic lymphoma, large and mixed cell lymphomas, and typical mycosis fungoides or Sezary leukemias in individual cases. In the U.S. only a small percent of these disease are HTLV-I positive. In contrast, in typical ATLL close to 100% are virus positive. Rarely, cases of T8 leukemia have been found, similar to the rare induction of T8 transformed cell lines *in vitro* by HTLV-I or II. In areas of the world where HTLV-I is endemic some B-cell lymphoid malignancies, and certain other cancers have been associated with HTLV-I infection much more frequently than expected from the prevalence of the virus in the general population. For example, whereas studies of healthy Jamaicans suggest an incidence of HTLV-I infection of 2–3% of people, HTLV-I was found in close to 30% of B-cell chronic lymphocytic leukemias. In contrast to virus positive T-cell leukemias where the viral genes are integrated into the DNA of the leukemic cell, HTLV-I was not found in the B-cell tumor. Instead, the virus was found in the normal T-cells of the patients. Recent results indicate that the malignant B-cells of these patients make a single type of antibody directed against a protein of HTLV-I. Therefore, in contrast to the T-cell malignancies where the virus integrates into the DNA of the initial transformed cell, and viral sequences are, therefore, found in all T-cells of the tumor, the

B-cell tumors may arise in part by an indirect effect of HTLV-I, e.g. by a chronic stimulation combined with diminished T-cell surveillance, thereby increasing the chance for a neoplastic transformation in the expanding B-cell compartment. Therefore, although neither HTLV-I nor HTLV-II can cause AIDS *per se*, there is much *in vitro* evidence that both can impair T-cell function, and substantial *in vivo* evidence suggesting that HTLV-I infection increases the risk of opportunistic infections and certain malignancies.

Origin and spread of HTLV-I and epidemiology of adult T-cell leukemia (ATL)

Although the original discoveries of HTLV-I were in two sporadic cases of T-cell malignancies in U.S. blacks and the first clusters of this disease were found in Japanese and subsequently in Caribbean-born blacks, I proposed that the virus originated in Africa [6]. This was based on several observations. First of all, later studies of the geographic distribution of HTLV-I demonstrated that it is widely distributed throughout Africa, especially in central Africa among certain tribes. Moreover, the ATL in the Americas and in Europe is present chiefly in immigrant blacks and their descendants, although there are many exceptions. Finally, a highly related virus termed simian T-cell leukemia (lymphotropic) virus-I or STLV-I has been found in African green monkeys and in rhesus monkeys. In Japan HTLV-I is prevalent in the two small southwestern islands, Kyushu and Shikoku. We suspect that it was brought there by the sixteenth century arrival of European adventurers who came with African blacks and monkeys. Unlike ubiquitous viruses such as EBV and the papilloma viruses, the geographic distribution of HTLV-I has been very restricted. For example, far less than 1% of the white U.S. and European populations are infected, and it also very unusual in most of Asia. Therefore, establishing an epidemiological link of this virus to the disease it causes was not complicated.

The incidence of HTLV-I infection can vary dramatically over small distances. For instance, in some areas of Kyushu 15–20% of the population may be positive, and in nearby towns the incidence may be less than a few percent. This and several other results lead to the opinion that transmission is not casual but by intimate contact, by blood or blood products, and as some studies show by infection of the developing fetus *in utero*. Finally, it seems that HTLV-I is usually not transmitted as an extracellular virus but with the infected T4 cell. Because of the increase in travel, changes in sexual habits, drug abuse (blood contaminated needles), and wide use of blood and blood products in transfusion it is likely that problems with HTLV-I (and perhaps HTLV-II) will increase in the future.

Mechanism of HTLV-I T-cell transformation

Much has been learned about the mechanism by which HTLV-I initiates T-cell transformation. Upon infection of T-cells with HTLV-I (or II), a small number (oligo- or mono-clonal population) of T-cells rapidly become immortalized, losing their need for exogenous Il-2 to maintain growth. This phenomenon appears to be at least in part mediated by the *tat*-I gene product which is believed to 'activate' the HTLV-I LTR to promote virus expression and bind to regulatory elements on T-cells which in turn will activate the expression of gene(s) involved in T-cell proliferation (Fig. 5). One such gene is the Il-2 receptor which, as noted earlier, is constitutively expressed in these transformed cells. Since cells other than T4 cells can be infected, the usual limitation of *in vitro* transformation of these cells and the striking consistency of the leukemias as involving the T4 cells is puzzling. Perhaps the *tat* binding site in T4 cells differs from other cells in a way that facilitates binding of *tat* with greater efficiency or that its position in T4 cells is closer to genes involved in proliferation of that cell so that it can more efficiently induce their expression. We assume that the maintenance of malignancy may require additional genetic changes in the cell.

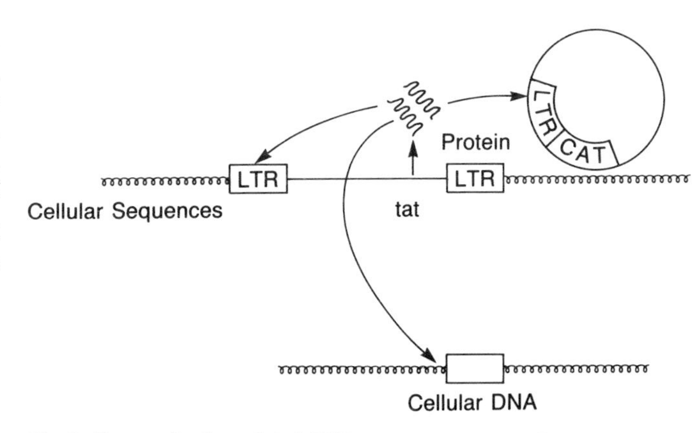

Fig. 5. Transactivation of viral LTR. *tat* gene messenger RNA is produced by a unique double splice mechanism in HTLV-I, -II, and -III. This m-RNA encodes a viral protein that may increase gene expression by the mechanism of transacting transcriptional activation (*tat* protein). The receptor site for this activity may be on a transfected DNA construct consisting of the viral LTR and a chloramphenicol acetyltransferase gene (CAT) integrated viral LTR or on cellular genes.

We isolated HTLV-II in 1981 from a cell line developed by A. Saxon and D. Golde from a young man with a T-cell variant of hairy cell leukemia. The disease was not aggressive. The virus was recently isolated on two other occasions, again from adult T-cell malignancies. All isolates of HTLV-II have been from Caucasians. Although considerable details exist on the nature of HTLV-II genome and on its *in vitro* effects (it is overall about 50% homologous to HTLV-I), little major biological differences have been found to warrant a separate detailed description. Although it is likely that this virus was involved in the cause of these few leukemias, as yet there is insufficient data to conclude that it is a cause of human leukemia. It is clear that HTLV-II is unusual in the U.S.–European population.

Concluding remarks on HTLV-I and II

The first known human retrovirus (HTLV-I, II) probably came into central African man from monkeys directly or via intermediaries in the distant past. A precise time for this is unknown but must be many centuries ago or earlier. HTLV-I has been linked to the cause of some T-cell malignancies in a way that probably provides the clearest evidence we have on the direct cause of a human cancer. This evidence includes: (1) epidemiological results clearly linking the two; (2) *in vitro* transformation of primary human T-cells (Fig. 6); (3) lymphoma induction in monkeys with the very closely related STLV-I; (4) molecular biological results showing a monoclonal distribution of the viral genes (provirus) in the tumor cell

DNA and no viral sequences in other tissues, providing evidence that the virus entered the patient before or at the time of the first T-cell neoplastic transformation and not later as a passenger virus; and (5) numerous animal models (chicken, mouse, rat, cat cow and Gibbon ape) showing that leukemias and lymphomas may be induced by retroviruses. HTLV-I appears also to play an indirect role in the cause of some other malignancies such as some B-cell tumors in HTLV-I endemic regions. HTLV-II has only been isolated a few times. It too probably causes some malignancies but more evidence is needed. (For more detailed reviews on HTLV-I and II see refs 7–9.)

HTLV-III and the origin of acquired immune deficiency syndrome (AIDS)

AIDS was first recognized in the U.S. in 1981 as a new disease associated with homosexual men. Subsequently, certain other risk groups, notably those exposed to blood either by infusion or by contaminated needles, and recent immigrants to the U.S. from Haiti were also found to be afflicted with this disease. It was recognized that a similar disease developed still earlier in parts of central Africa among promiscuous heterosexuals, including prostitutes. Now, of course, AIDS is recognized as a global problem beginning to involve every segment of society with enormous social, economic, ethical, legal, and medical consequences. The pathogenesis of the disease begins with alteration and eventual premature death of the T4 cell. This in turn leads to an overwhelming immune dysfunction and eventually to opportunistic infections, malignancies, and death. The cause of AIDS was proven to be a new retrovirus in the spring of 1984. This virus has been called Human T-Lymphotropic Virus-III (HTLV-III) because it was the third type of human T-lymphotropic virus to be discovered and Lymphodenopathy Associated Virus (LAV) because AIDS often is preceeded by lymphodenopathy.

There are lessons to be learned from the early ideas on the origin of AIDS. From 1981 to 1983 speculations outdistanced experiments and the major ideas included sperm (autoimmune disease with antibodies to sperm reactive also with T-cells); diverse reactions to HLA (exposure to various leukocytes from promiscuous sexual experiences) once again invoking some autoimmunity; amyl nitrite (used by some of the people in risk groups); a fungus; several types of viruses; and multiple and chronic antigenic stimulation, i.e. no specific cause. We proposed the idea that AIDS was caused by a new human T-lymphotropic retrovirus in 1982, based on the following considerations: (1) Epidemiological data strongly argued that the disease was new and spreading and in keeping chiefly with a new infection. (2) Blood transfusions gave rise to AIDS in non-risk groups, thereby eliminating causes other than a transmissible agent. (3) Factor VIII preparations induced AIDS in hemophiliacs; these preparations are filtered so as to remove bacteria, parasites, and fungi. Therefore a virus became the most logical candidate. (4) The disease seemed to specifically involve the T4 cell. Such specificity is more consistent with a virus than a bacterium or fungus. (5) A human retrovirus was thought to be the most logical candidate based on the earlier experiences with HTLV-I (T4 cell tropism; likely African origin; transmission by blood, sex and congenital infection; and known capacity to induce some T-cell functional changes) and by previous experiences with an animal retrovirus, the feline leukemia virus (FeLV). The FeLV can cause T-cell leukemia, but as noted by M. Essex and co-workers, more frequently an AIDS-like disease occurs possibly by a variant of the leukemia virus. Against the idea of a retrovirus was the prevailing feeling that most retroviruses are usually not very cytopathic.

We soon found retroviruses in AIDS patients which were not HTLV-I or II, but due to the extreme and unexpected cytopathogencity of what was later learned to be HTLV-III, the infected T-cells died, and the virus could not be characterized nor linked to the disease. However, Barré Sinoussi *et al.*, at the Pasteur Institute in Paris did report identification of this virus in 1983 in one man with lymphadanopathy [10]. Again, the virus could not be characterized nor linked to AIDS because of insufficient amounts due to rapid death of the primary infected T-cells.

A major step forward was the finding that certain clones of human leukemic T4 cell lines, free of any retroviruses, could be productively infected by HTLV-III [11]. This allowed for the first mass production of virus and accomplished the following: (1) The first specific reagents (antibodies, antigens, nucleic acid probes) so that all other isolates could be characterized. (2) Sufficient viral proteins for specific and wide scale sero-epidemiological studies. (3) A blood bank assay.

We were then able to describe 48 isolates of HTLV-III and clear epidemiologic evidence that this virus was the cause of AIDS [12–14] (see Table II). The blood test is based on detection of specific antibodies to HTLV-III envelope and core structural proteins in HTLV-III infected carriers. The assay approach is usually to screen by the ELISA method followed by a confirmatory test which is usually an immunoblot (Western blot) procedure or a specific competition assay using defined viral proteins to bind antibodies. These tests showed for the most part to be successful in eliminating blood transfusion associated AIDS in the U.S.

Using these specific viral reagents the distribution of HTLV-III in an infected individual has been determined. Other than T4 lymphocytes the virus has also been found in a small percent of B-cells, macrophage, and in the brain (probably in glial cells). The latter is particularly important in that neurological abnormalities have been noted in some infected people. Free extracellular virus has also been isolated from semen, saliva, plasma, tears, and cerebrospinal fluid (see refs 9 and 15 for reviews).

Table II. *Disorders caused by and/or associated with HTLV-III*

1. AIDS
2. LAS
3. Neurological
4. Teratogenic
5. ITP
6. Carcinomas
7. KS
8. Various lymphomas (mostly B-cell and Hodgkin's)
9. Interstitial pneumonitis
10. Acute malaria*
11. Tuberculosis*
12. Acute leukemia*
13. Others

* Reported in Zairian cases.

HTLV-III asscciated diseases

Other than AIDS and neurological abnormalities, infection with HTLV-III can lead to lymphoproliferative disorders such as generalized lymphodenopathy, interstitial pneumonitis, and an increased incidence of B-cell lymphonas, Hodgkin's Disease, some carcinomas, notably cloacogenic squamous cell carcinomas, head and neck tumors, and, of course, an extraordinary increase of Kaposi Sarcoma in the HTLV-III infected homosexual. The reason for the increase in these diseases is not completely understood. It is clear, however, that HTLV-III is not the direct cause, because viral sequences are not found in the DNA from the majority of tumor cells. For the B-cell lymphomas the mechanism may be identical to the indirect role described for HTLV-I in these diseases: namely, a chronic viral antigen stimulation of B-cell proliferation combined with diminished T4 cell function which in turn leads to a decrease in cytotoxic T-cell control of EBV. The result of these changes favor the change emergence of a neoplastic B-cell clone. The reason for the very high incidence of Kaposi Sarcoma is completely unknown, but we suspect another virus, highly efficient in transforming endothelial cells in a setting of T4 dysfunction. Finally, there are suggestions that HTLV-III infection may cause thrombocytopenia and congenital abnormalities, but more epidemiological information is needed.

The cytopathic effect of HTLV-III on T4 cells

Infection of T4 cells by HTLV-III leads to the premature death of these cells (Fig. 6). The mechanism is not by direct lytic effect. On the other hand, indirect effects, e.g. mediated by an autoimmune process, need not be postulated since one or more of the cloned genes of HTLV-III lead to T4 cell death upon transfection of the DNA provirus into these cells [16]. It appears that expression of one or more of the viral genes is necessary. In this regard, it is possible to infect T4 cells *in vitro* without cell killing until the T-cells are activated by exposure to a mitogen or antigen. The activated infected T-cells appear to go through the same process of cell gene expression as uninfected cells except that the viral genes are eventually expressed. When this occurs a higher percentage of the cells than normal will terminally differentiate, and the rate of terminal differentation is faster than in the activated uninfected T-cells. This process may involve the *tat* III gene (see Fig. 1). Expression of this gene may in turn activate an extremely high level of transcription of another viral gene or a set of cellular genes augmenting terminal differentation. If a viral gene is involved in this cell death, it may be the product of the SOR gene or 3′ ORF gene (see Fig. 1) whose functions are unknown.

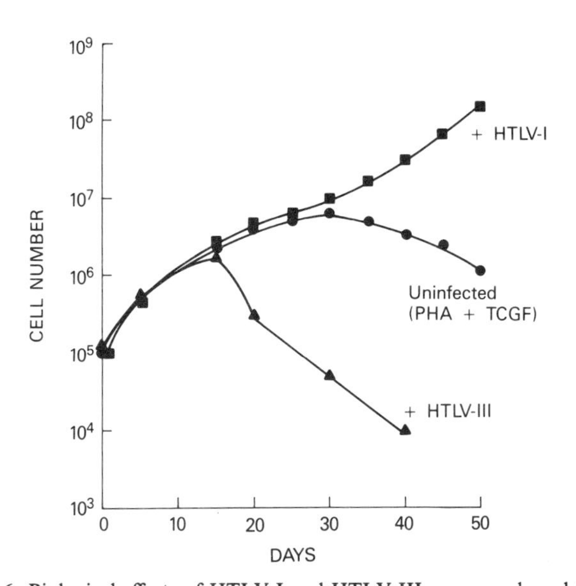

Fig. 6. Biological effects of HTLV-I and HTLV-III on normal cord blood T lymphocytes. Cumulative cell numbers were recorded after different days of infection. ●, Normal, uninfected, mitogen-stimulated T cells derived from the umbilical cord blood grown in the presence of 10% TCGF in long-term cultures. ■, HTLV-I-infected cord blood T cells cultured under the same *in vitro* conditions as uninfected cells. ▲, HTLV-III-infected cord blood T cells cultured under the same conditions.

Origin and evolution of HTLV-III genome

HTLV-III is clearly a new virus of man. An exception may be that small groups of rural central Africans may have been infected for a longer period of time. The time of first entry into man can not be approximated. However, there is serological evidence that HTLV-III or a related virus (? progenitor) was present in some central Africans at least as early as 1972, and other epidemiological results indicating presence of an AIDS like disease in the mid to late 1970s thus predating presence of virus or AIDS in the U.S. or Western Europe. Moreover, recently M. Essex and colleagues in Boston described a new virus in African green monkeys, called STLV-$III_{(AG)}$ more closely related to HTLV-III than any previously known virus. We propose, in agreement with the Boston group, that this may be the progenitor of HTLV-III which entered some central Africans and in this past decade began to spread to other regions of the world.

An early finding in the molecular analysis of various HTLV-III isolates was the variation of nucleotide sequences of certain parts of the genome, especially in the envelope gene (Fig. 7). Analysis of more than 30 isolates of the now over 200 isolates available in our laboratory indicates that there are not different strains of HTLV-III. Instead there is a continuum – from very closely related isolates (1–2% variation) to those that vary by more than 5%. The variation develops after successive infections and does not occur during prolonged tissue culture. This suggests that these changes occur during reverse transcription of the viral RNA genome to the DNA form and/or during the recombinational process

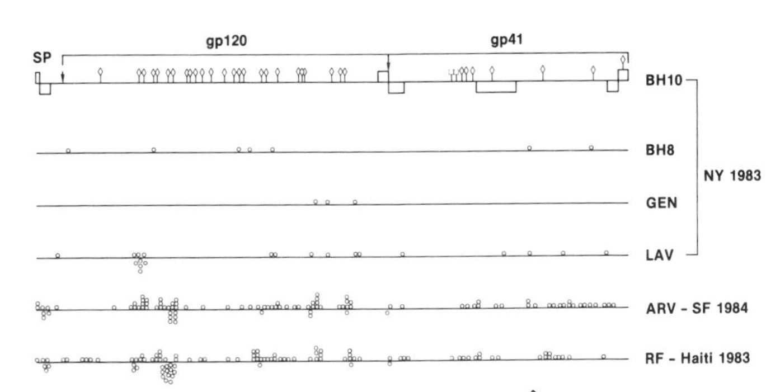

Fig. 7. Divergence of the env gene of HTL-III. ◇, Glycosylation sites; □, hydrophilic stretches; □, hydrophobic stretches; □, placement of cysteine residue; ○, non-conservative amino-acid change from the prototype BH10. Changes involving amino acids having similar side chains are considered conservative.

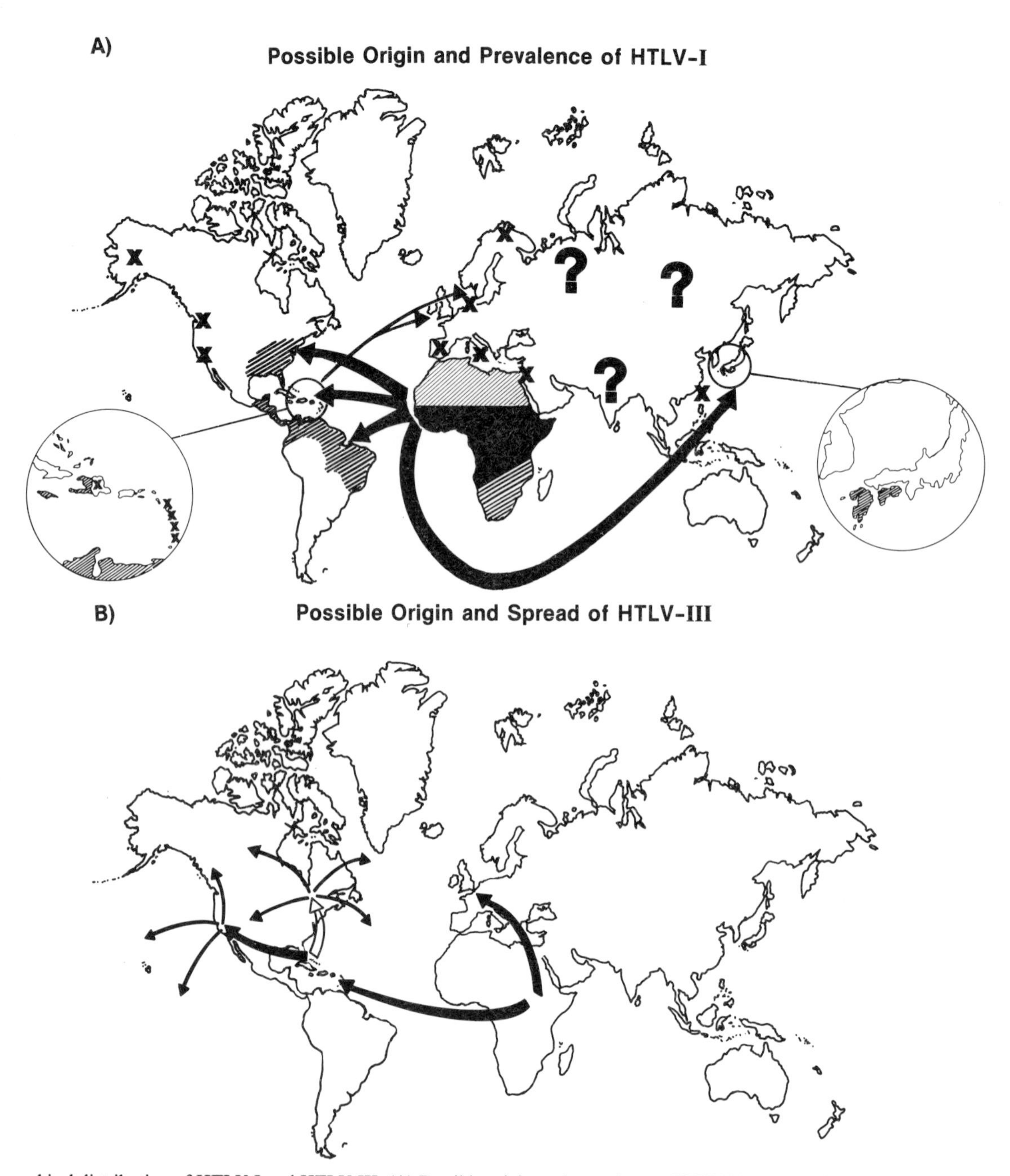

Fig. 8. Geographical distribution of HTLV-I and HTLV-III. (A) Possible origin and prevalence of HTLV-I; (B) Possible origin and spread of HTLV-III. (A) ■, Regions where the incidence of infection (seropositivity) are highly prevalent (>5%); ▨, moderate (1–5%). ×, Places where HTLV-I infection has been detected but extensive seroepidemiology has not been reported. ?, areas not studied. Insets are enlargements of two endemic areas, the Caribbean islands and Japan. (B) Thick arrows indicate primary spread; thin arrows, secondary spread.

when the DNA provirus integrates into the host cell DNA. Reverse transcriptases tend to be error prone. It is possible that the RT of HTLV-III is particularly error-prone. Since most of the genomic variation is limited to the envelope it may be that variation in other parts of the viral genome leads to non-infectious particles.

Summary and concluding remarks on HTLV-III

HTLV-III is a new infection of mankind with severe and often fatal consequences. Like HTLV-I (and probably also HTLV-II), it is likely that it entered African man from African green monkeys or related primates directly or through intermediary vectors and subsequently spread to other regions (Fig. 8). Also similar to HTLV-I is its mode of transmission, T4 tropism, *in vitro* mimicry of the disease, and presence of the *tat* gene. Unlike HTLV-I or II, the AIDS virus contains at least two additional genes, has strong cytopathic effects, greater structural similarities to the lenti-retroviruses of ungulates, and is generally more infectious. Despite the speed or progress on human retroviruses, diseases caused by them have outdistanced the science. To avoid a still more serious and global medical and economic problem from them, we will need even more rapid progress.

References

1. Temin, H. M. and Mizutani, S., *Nature* (*London*) **226** 1211 (1970).
2. Baltimore, D., *Nature* (*London*) **226**, 1209 (1970).
3. Poiesz, B. J., Ruscetti, F. W., Gazdar, A. F., Bunn, P. A., Minna, J. D. and Gallo, R. C., *Proc. Natl. Acad. Sci. USA* **77**, 7415 (1980).
4. Saxinger, W. C. and Gallo, R. C., in *Human Cancer Immunology* (ed.

B. Serrou and C. L. Rosenfeld). Amsterdam, North Holland Press (in the press).
5. Robert-Guroff, M., Sarngadharan, M. G. and Gallo, R. C., in *Growth and Maturation Factors* (ed. G. Guroff) vol. 2, pp. 267–308. New York, Wiley (1984).
6. Gallo, R. C., Sliski, A. and Wong-Staal, F., *Lancet* **2**, 962 (1983).
7. Gallo, R. C., in *Human T-Cell Leukemia/Lymphoma Virus* (ed. R. C. Gallo, M. Essex and L. Gross), pp. 1–8. New York, Cold Spring Harbor Laboratory (1984).
8. Wong-Staal, F. and Gallo, R. C., *Blood* **65**, 253 (1985).
9. Wong-Staal, F. and Gallo, R. C. *Nature (London)* **317**, 395 (1985).
10. Barré-Sinoussi, F., Chermann, J. -C., Rey, F., Nugeyre, M. T., Charmaret, S., Gruest, J., Dauguet, C., Axler-Shin, C., Vezinet-Brun, F., Rouzioux, C., Rosenbaum, W. and Montagnier, L., *Science* **220**, 868 (1983).
11. Popovic, M., Sarngadharan, M. G., Reach, E. and Gallo, R. C., *Science* **224**, 497 (1984).
12. Gallo, R. C., Salahuddin, S. Z., Popovic M., Shearer, G. M. Kaplan, M., Haynes, B. F., Palker, T. J., Redfield, R., Oleske, J., Safai, B., White, G., Foster, P. and Markham, P. D., *Science* **224**, 500 (1984).
13. Schupbach, J., Popovic, M., Gilden, R. V., Gonda, M. A., Sarngadharan, M. G. and Gallo, R. C., *Science* **224**, 503 (1984).
14. Sarngadharan, M. G., Popovic, M., Bruch, L., Schupbach, J. and Gallo, R. C., *Science* **224**, 506 (1984).
15. Broder, S. and Gallo, R. C., *N. Engl. J. Med.* **311**, 1292 (1984).
16. Fisher, A. G., Collatti, E., Ratner, L., Gallo, R. C. and Wong-Staal, F., *Nature (London)* **316**, 262 (1985).

Chemica Scripta 1986, **26B**, 299–303

Developmental Expression of Murine Homeo Box Sequences

A. M. Colberg-Poley, S. D. Voss and P. Gruss

Zentrum für Molekulare Biologie der Universität Heidelberg, Im Neuenheimer Feld 282, D-6900 Heidelberg, Federal Republic of Germany

Paper presented by P. Gruss at the Conference on 'Molecular Evolution of Life', Lidingö, Sweden, 8-12 September 1985

Abstract

The developmental expression of two previously cloned murine homeo box sequences, m6-12 and m5-4, was examined both *in vitro* in embryonal carcinoma cells and *in vivo* during mouse embryogenesis. Probes taken from a cluster of homeo boxes which spans 30 kbp of genomic DNA each contained unique murine DNA as well as 125 bp of homeo box sequence and hybridized to novel and distinct transcripts induced upon differentiation of embryonal carcinoma cells. Analysis of murine embryonic and extraembryonic tissues revealed transcripts which were similar to those observed *in vitro* and temporally expressed during embryogenesis. We also discuss the observation that homeo box-containing transcripts have been detected in various adult tissues, indicating the potential involvement of these genes in postnatal cellular and tissue development as well as during embryogenesis.

Introduction

The study of vertebrate embryology and the events which regulate developmental decisions in higher organisms has been hampered by the complexity of the vertebrate genome. Indeed, classical genetic analysis of the mouse, an enormous and painstaking effort, has been rewarded by the identification of only a few genes which possibly play a role in developmental processes; for review, see [1, 2]. Although unable to provide us with an understanding of the complexities of mammalian biology (e.g. immunogenicity, tumorigenesis), *Drosophila melanogaster*, the classic developmental genetic system, has recently provided us with a probe – the homeo box – lending us access to mammalian genes which may, in a manner analogous to the homeotic genes of *Drosophila*, be involved in regulatory decisions during development of the mouse. This manuscript reviews recent progress made toward the understanding of the role of these mammalian homeo boxes in development.

The *Drosophila* homeo boxes are confined to 180 bp of highly conserved DNA and were initially identified, by virtue of their cross homology, in two major clusters of *Drosophila* homeotic genes, the *Antennapedia* complex (ANT-C) and the bithorax complex (BX-C) [3, 4]. Subsequent use of these boxes as probes has allowed the identification of more than 10 additional homeo box containing regions in the fruit fly genome [5]. Furthermore, hybridization under reduced stringency showed the presence of these sequences in the genomes of an evolutionary gamut of organisms, including the frog, earthworm, mouse, and human [6]. Since the initial studies, several murine and human homeo box-containing sequences have been cloned and sequenced, corroborating the high degree of homology initially suggested by the hybridization studies [7–10].

Further detailed mapping and sequence analysis has shown the presence of additional murine homeo boxes in a homeo box cluster [10], an arrangement structurally analogous to that of the two *Drosophila* homeotic gene complexes. Such clustering had been previously found in human homeo box containing regions [8], and, based on homology to this region had also been proposed for the corresponding region of murine DNA as well [11, 12].

We have used two homeo box-containing probes from this cluster, m6 and m5, and examined the expression of transcripts containing these sequences during the differentation of F9 cells into parietal or visceral endoderm [9, 10, 13] as well as during the differentiation of P19 cells into myogenic cells [14]. Although embryonal carcinoma (EC) cells have served as useful models for events occurring during mammalian embryogenesis [15], it was necessary to demonstrate *directly* the expression and hence, the potential involvement of these homeo box-containing genes in development of the mouse embryo. Along these lines we examined mouse embryonic and extraembryonic tissues at various stages during embryonic growth. If one presumes homeo box genes play a role in establishing developmental programmes, one might expect, depending on the gene involved, a temporal pattern of expression during the first and second weeks of prenatal development. Moreover, we have examined the expression of these sequences in a variety of adult tissues [10] on the assumption that, since some cellular and tissue development occurs throughout the adult life of an organism, homeo box-containing genes might exert a regulatory influence on these postnatal events.

Materials and methods

The methods used to perform these experiments are as described previously [9, 10].

Results

Organization of the murine homeo boxes

A murine genomic library, constructed by Frischauf and co-workers in λEMBL3A [16], was screened using two *Drosophila* probes, *Antennapedia* (*Antp*, [17]) and *fushi tarazu* (*ftz*, [18]) under reduced stringency conditions. This allowed us to isolate 8 independent phage, each hybridizing to both probes, 6 of which contained overlapping inserts. These overlapping phage encompass approximately 30 kbp of murine genomic

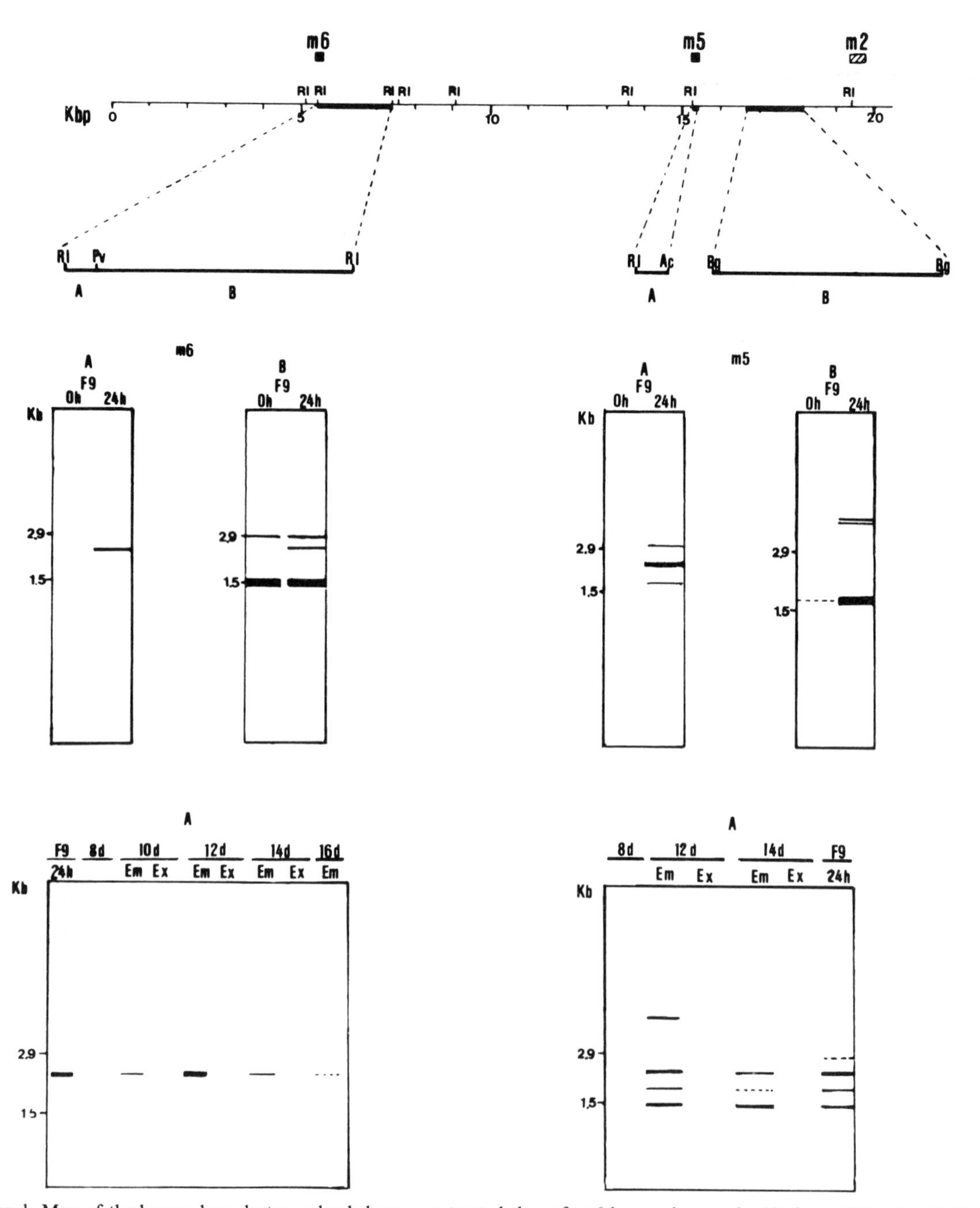

Fig. 1. Upper Panel: Map of the homeo box cluster and subclones constructed thereof. m6 homeo box probe (A) is a 187 bp *Eco* RI/*Pvu* II fragment; the m6 flanking DNA probe (B) is a unique 1.7 kb *Pvu* II/*Eco* RI fragment. m5 homeo box probe (A) is a 180 bp *Eco* RI/*Acc* I fragment and the m5 flanking DNA probe (B) is a 1.5 kb *Bgl* II/*Bgl* II fragment lying approximately 1.1 kb downstream from the m5 homeo box. All fragments were individually subcloned to avoid hybridizations due to contamination and co-purification of fragments during gel isolation.

Middle Panel: Poly A^+ RNAs prepared from F9 stem cells or from differentiated F9 cells (parietal endoderm) 24 h following treatment with RA and cAMP. A and B refer to the probe used, that is, homeo box or flanking sequence, respectively.

Lower Panel: Poly A^+ RNAs prepared from embryonic and extraembryonic tissues. 8 day RNA contains both embryonic (Em) and extraembryonic (Ex) tissue; 16 day extraembryonic tissue was not examined. Control was RNA from F9 cells differentiated for 24 h. The Northern blots represented here were hybridized only to the respective m5 and m6 homeo box probes (A).

DNA, within which a cluster of three homeo boxes is located. The homeo box sequences are in the same 5′ to 3′ orientation and contain significant homologies both to each other and to all other homeo boxes of known sequence [9, 10, and our unpublished data].

Expression of m5 and m6 homeo box-containing genes during EC cell differentiation

In order to establish a functional significance for these murine homeo box sequences, we first examined the expression of the respective homeo box genes and their flanking sequences during the differentiation of F9 EC cells. As represented in Fig. 1, the m6 homeo box probe (A) detects a single 2.4 kb transcript in poly A^+ RNA isolated 24 h following the induction of differentiation using retinoic acid (RA) and dibutyryl cAMP (cAMP). This transcript was not detectable in F9 stem cells and appeared to decrease in abundance during later stages of differentiation [9]. Interestingly, sequences just flanking the m6 homeo box (B) were transcribed both in EC stem cells and in their differentiated counterparts, suggesting that sequences from this m6 homeo box region may not be *directly* involved in the control of the differentiation event induced by RA and cAMP. Similar observations were obtained in F9 cells induced to differentiate into either visceral or parietal endoderm using either m6 homeo box or flanking sequence probes [13]. Furthermore, in PC13 cells, which can be differentiated into visceral endoderm-like cells [13] and in P19 cells differentiated into myogenic cells [14], induction of a 2.4 kb transcript following differentiation was also observed, as was the presence of the 2.9 and 1.5 kb transcripts in the respective stem and differentiated cell populations. These observations led us to conclude that the expression of the m6 homeo box information alone was neither tissue-specific nor an F9-associated phenomenon; rather, that its expression might have a more general impact and is not sufficient to direct a cell into a specific differentiation program.

Having examined m6 homeo box expression in some detail, we then examined the expression of the adjacent m5 homeo box and its adjoining sequences in the F9 cells. As shown in Fig. 1, the m5 homeo box probe (A) detected no transcripts in poly A^+ RNA prepared from stem cell populations, whereas 24 h following differentiation at least three transcripts were detected, one of which, 1.9 kb, was also detected by the unique flanking sequence probe (B). Whether these transcripts all contain m5 homeo box sequences or whether some were detected on the basis of strong homology to the m5 probe is not entirely clear. However, when sequences downstream of the m5 homeo box were used to probe the F9 cell RNAs, in addition to the 1.9 kb transcript detected, the appearance of a novel 5.5 kb doublet was observed following differentiation. The 1.9 kb transcript has occasionally been detected in stem cells and its induction upon differentiation appears to be in the threefold range [10].

Expression of m5 and m6 homeo box-containing genes during embryonic development

Finally, to corroborate our *in vitro* results, the expression of m5 and m6 homeo box genes was examined *in vivo* in poly A^+ RNAs isolated from various stage mouse embryonic and extraembryonic tissues. Both regions were expressed in embryonic tissues in a manner patterning that seen previously in F9 cells. In the case of m6, as shown in the bottom panel of Fig. 1, the homeo box probe hybridized exclusively to a 2.4 kb transcript similar to that seen in differentiated F9 cells. The expression of these sequences is both temporally regulated (as verified by hybridization to c-ras^{Ha}, ref. 10), and specific for, at least in these later developmental stages, embryonic tissue. Similar temporal regulation and tissue specificity of the 2.9, 2.4 and 1.5 kb transcripts detected by the m6 flanking sequences in differentiated F9 cells was also observed [10].

m5 homeo box sequences were also regulated in an embryonic tissue specific, and temporal manner, although the expression of m5 sequences would appear to be under different control mechanisms than m6. As indicated in Fig. 1, the m5 homeo box probe detects all the transcripts observed previously in F9 cells. Although the 5.5 kb doublet was not seen using the 180 bp m5 probe in differentiated F9 cells, this probe detects both the 1.9 and 5.5 kb transcripts in embryonic tissue. Important in this regard is the observation that these 5.5 kb transcripts are considerably more abundant in 12 day embryonic tissues than in differentiated F9 cells [10]. Moreover, this confirms that the 5.5 kb doublet contains homeo box sequences as well. Although the 1.9 kb transcript seen previously appears to be detected using the m5 homeo box probe in 12 day embryonic tissues as well as in differentiated F9 cells, other transcripts were observed in the 2 kb range. These transcripts are present in 14 day embryonic tissue as well. We have interpreted these results to suggest that the 5.5 kb and 1.9 kb transcripts contain m5 homeo box sequences, reasoning that the different abundances of these transcripts in F9 cells and embryonic tissues has resulted in our inability to detect these sequences using the smaller, lower specific activity homeo box probe. Nevertheless, the presence of other similarly sized transcripts makes unequivocal identification of the 1.9 kb transcript difficult. These other transcripts (one of which is in the familiar 2.4 kb m6 range) seen both in F9 cells and embryonic tissues are likely the result of cross hybridization of the m5 homeo box-containing transcripts. Why such cross hybridization has not been observed with the similarly sized (187 bp) highly homologous m6 probe remains unclear.

Discussion

The rapid advances being made in the identification and cloning of *Drosophila* genes whose expression controls pattern development, and the discovery that sequences common to these genes are present in the genomes of other organisms has provided us for the first time with *direct* access to mammalian genes possibly involved in the control of development. The excitement generated by this possibility has resulted in an increasing interest in mammalian homeo box-related research, as well as a certain degree of speculation and controversy regarding how closely the study of *Drosophila* pattern formation should be applied to concepts of mammalian embryology. In this manuscript we have chosen not to address this controversy directly, but rather to collect various pieces of published work and present them in summary form.

We have shown the presence of at least three homeo boxes clustered within 30 kbp of mouse DNA. These boxes were initially identified by their hybridization to two *Drosophila* homeo box probes, *Antp* and *ftz*, and confirmed as homeo boxes by subsequent sequence analysis. These boxes all lie in the same orientation relative to one another and are highly homologous, but not identical to other known homeo boxes [9, 10]. Putative homeo box translation products showed this homology to be even greater at the amino acid level, implying the conservation of a functional protein-coding domain essential to basic pattern formation in all multicellular animals. Of additional structural significance is the theory that in *Drosophila* duplication of homeotic genes occurred. These duplicated regions, following diversification, resulted in clusters of homeotic genes whose responsibility was the control of segment specific pattern formation [19]. Thus, similar clustering of homeo box containing genes in mammals [7–12] points to the conservation of a genetic structural organization within which the highly conserved homeo boxes are but a subset.

We were interested in first establishing the *in vitro* significance of the murine homeo boxes using EC cells as model systems. The differentiation of F9 cells into either visceral or parietal endoderm was accompanied by the expression of a novel 2.4 kb m6 homeo box-containing transcript. The additional presence of two non-homeo box-containing transcripts from the m6 region and the observation that they are transcribed from the same coding strand of DNA as the m6 homeo box led us to propose two alternative models for the transcription and/or splicing within this region [13]. Furthermore, the induction of P19 stem cells to differentiate into myogenic cells was also accompanied by the appearance of m6 homeo box-containing transcripts [14]. F9 cell differentiation was also shown to be accompanied by the induction of the 5.5 kb transcripts containing the m5 homeo box, suggesting the simultaneous involvement of numerous homeo box-containing genes in the *in vitro* differentiation process.

Having shown the involvement of homeo box sequences in *in vitro* developmental systems, we sought to establish their expression *in vivo* during mouse embryogenesis. The homeotic genes in *Drosophila* are expressed at various stages during embryogenesis and the regulated expression of these genes results in the establishment of segment number and segment identity in the embryo. We hoped, therefore, to detect temporal, stage specific expression of murine homeo box sequences in poly A^+ RNAs isolated at various stages during embryonic development. Poly A^+ RNA containing m6 homeo box sequences was seen to accumulate maximally in day 12 embryonic tissue; no m6 transcripts, however, were detected in extraembryonic tissues. m6 flanking sequences as well as m6 homeo box sequences were also detectable in 10, 14, and 16 day embryonic tissues, although the level of expression during these times was significantly lower than that observed at day 12.

In contrast to the increase and gradual decrease of m6 expression, m5 containing transcripts were seen to accumulate concomitantly with m6 transcripts in day 12 embryonic tissues, whereas no detectable m5 expression using unique DNA probes was detected at the other stages examined. Thus, the regulatory mechanisms controlling the expression of these closely linked homeo box-containing genes relative to one another appears to be different. These findings gain support from other mammalian systems in which homeo box expression during embryogenesis has been oberseved. In *Xenopus laevis* the AC1 gene, beginning at gastrulation, produces three classes of transcripts, 2.3, 1.6, and 1.2 kb, each of which – similar to m5 and m6 – appeared to be under temporal control during development [20]. Furthermore, differential expression during embryogenesis of an engrailed-like murine homeo box mapping to chromosome 1 has been documented [21], while the developmental expression of homeo box sequences on mouse chromosome 11 has also been noted [26].

Our data may at first appear somewhat paradoxical and self-contradictory in that we fail to detect homeo box transcripts in the extraembryonic tissues examined, despite observing abundant homeo box transcription upon differentiation of F9 cells into visceral or parietal endoderm, both extraembryonic tissues [9, 13]. Based on the P19 data showing homeo box expression during the differentiation of P19 stem cells into myogenic cells [14], we fully expected to detect expression, if not temporal expression, in embryonic tissue. The absence of homeo box expression in the late stage extraembryonic tissue examined is perhaps best explained by noting that visceral and parietal endoderm derive early during embryogenesis (day 6) from the primitive endoderm. It is thus unlikely that homeo box transcripts which are expressed only transiently during the differentiation of EC cells would be detected late in the development of the extraembryonic tissues. These observations also serve to confirm further F9 EC cells as a reliable *in vitro* model system in which visceral and parietal endoderm derives from an apparent stem cell population [15]. High resolution *in situ* analyses will be necessary to define better the involvement of homeo box containing genes in the establishment of visceral and parietal endoderm.

Our observation that homeo box-containing genes are also expressed in adult mouse tissues [10] is the first evidence to our knowledge showing that homeo boxes and/or their surrounding sequences are involved in postnatally occurring processes. Whether these processes are developmental in nature remains to be determined. Expression was most abundant in the testis of the adult mouse. Interestingly, only kidney and ovary cells show a pattern of homeo box transcription similar to that observed in embryonic tissue and EC cells. The variation in both size and abundance of transcripts stemming from this homeo box region suggests different regulatory mechanisms for the expression of these genes, reflecting perhaps the diverse requirements of the various tissues for homeo box gene products. Only direct analysis of cDNAs constructed from tissue specific mRNAs can hope to unravel the complicated patterns of expression observed. Similar to the data presented here, variation in the size of c-mos transcripts in the embryo, and adult testes and ovaries has also been observed recently [22].

As noted at the outset, the discovery of homeo box sequences in higher vertebrates has sparked a great deal of controversy and speculation as to the function of these sequences in mammals. Certainly one must agree that the process of segmentation in humans and other mammals is a fundamentally different process from that in *Drosophila* and thus one does not expect the involvement of mammalian homeo boxes in *directly* analogous pattern forming events. The process of segmentation in the mouse was recently reviewed by Hogan and co-workers [23]; they point out that although no *homeotic* transformations have been observed in the mouse, the overt differences in segmentation between *Drosophila* and vertebrates necessitate caution and perhaps re-direction of experimental approach in defining the role of these highly conserved homeo box genes.

Further along these lines, Stack [24] has recently presented a convincing set of arguments for the existence of homeotic-like transformations in man. He reasons that the well characterized *Drosophila* homeotic mutations have been selected for the laboratory, and that the extent of their penetrance and expressivity would not be likely to be naturally occurring. He presents evidence of localized changes in the phenotype of epithelial tissues – well known metaplasias and heteroplasias – which he correlates with stable changes in the epigenetic coding of epithelial cells. Although there is no evidence for such observations in the mouse, this does not rule out their existence, since the study of murine epithelial metaplasias is likely to be somewhat limited. We note, finally, that these recent theoretical observations can now be directly tested due to the isolation of mammalian homeo box genes. One might thus speculate the involvement of homeo box genes in a hierarchy of decisions during the development process [25], a process beginning with embryogenesis and continuing with ever-increasing complexity into the maintenance and regeneration of adult tissues.

Acknowledgements

We thank E. Vosshans-Bosbach for technical assistance and S. Mähler for photography. S. D. V. holds a Fulbright Fellowship. This research is supported by the Deutsche Forschungsgemeinschaft (DFG Ba 384/18-4).

Note added in proof:

During the time this manuscript was in press, two reports Jackson *et al. Nature* (*London*) **317**, 745 (1985); Awgulewitsch *et al. Nature* (*London*) **320**, 328 (1986)] documenting differential expression of other murine homeo box genes in adult tissues were published.

References

1. Gluecksohn-Waelsch, S., *Cold Spring Harb. Conf. Cell Prolif.* **19**, 3 (1983).
2. McLaren, A., *Ann. Rev. Genet.* **10**, 361 (1976).
3. Scott, M. P. and Weiner, A. J., *Proc. Natl. Acad. Sci. USA* **81**, 4115 (1984).
4. McGinnis, W., Levine, M. S., Hafen, E., Kuroiwa, A. and Gehring, W. J., *Nature* (*London*) **308**, 428 (1984).
5. Gehring, W. J., *Cell* **40**, 3 (1985).
6. McGinnis, W., Garber, R. L., Wirz, J., Kuroiwa, A. and Gehring, W. J., *Cell* **37**, 403 (1984).
7. McGinnis, W., Hart, C. P., Gehring, W. J. and Ruddle, F. H. *Cell* **38**, 675 (1984).
8. Levine, M., Rubin, G. M., and Tjian, R., *Cell* **38**, 667 (1984).
9. Colberg-Poley, A. M., Voss, S. D., Chowdhury, K., and Gruss, P., *Nature* **314**, 713 (1984).
10. Colberg-Poley, A. M., Voss, S. D., Chowdhury, K., Stewart, C. L., Wagner, E. F. and Gruss, P., *Cell* **43**, 39 (1985).
11. Joyner, A. L., Lebo, R. V., Kan, Y. W., Tjian, R., Cox, D. R. and Martin, G. R., *Nature* (*London*) **314**, 173 (1985).
12. Rabin, M., Hart, C. P., Ferguson-Smith, A., McGinnis, W., Levine, M. and Ruddle, F. H., *Nature* (*London*) **314**, 175 (1985).
13. Colberg-Poley, A. M., Voss, S. D. and Gruss, P., *Cold Spring Harb. Symp. Quant. Biol.* **50**, 285 (1985).
14. Dony, C., Colberg-Poley, A. M., Voss, S. D. and Gruss, P., Manuscript in preparation.
15. Martin, G. R., *Science* **209**, 768 (1980).
16. Frischauf, A.-M., Lehrach, H., Poustka, A. and Murray, N., *J. Mol. Biol.* **170**, 827 (1983).
17. Garber, R. L., Kuroiwa A. and Gehring, W. J., *EMBO J.* **2**, 2027 (1983).
18. Kuroiwa, A., Hafen, E. and Gehring, W. J., *Cell* **37**, 825 (1984).
19. Lewis, E. B. *Nature* (*London*) **276**, 565 (1978).
20. Müller, M. M., Carrasco, A. E. and De Robertis, E. M., *Cell* **39**, 157 (1984).
21. Joyner, A. L., Kornberg, T., Coleman, K. G., Cox, D. R. and Martin, G. R., *Cell* **43**, 29 (1985).
22. Propst, F. and Vande Woude, G. F., *Nature* (*London*) **315**, 516 (1985).
23. Hogan, B., Holland, P., Schofield, P., *Trends in Genet.* **1**, 67 (1985).
24. Slack, J. M. W., *J. Theor. Biol.* **114**, 463 (1985).
25. Slack, J. M. W., *From Egg to Embryo. Determinative Events in Early Development.* Cambridge University Press, Cambridge (1983).
26. Hart, C. P., Awgulewitsch, A., Fainsod, A., McGinnis, W. and Ruddle, F. H., *Cell* **43**, 9 (1985).

Chemica Scripta 1986, **26B**, 305–307

The Rapid Generation of Genomic Change as a Result of Over-replication of DNA

Robert T. Schimke and Steven W. Sherwood

Department of Biological Sciences, Stanford University, Stanford, CA 94305, USA

and Anna B. Hill

Department of Radiation Oncology, The University of Arizona, College of Medicine, Tucson, Arizona 85724, USA

Paper presented by Robert T. Schimke at the Conference on 'Molecular Evolution of Life', Lidingö, Sweden, 8–12 September 1985

Abstract

As the structure of the genomes of higher eukaryotes is being elucidated progressively, many changes of evolutionary significance, including the generation of multigene families, changes in genome size and complexity, rapid change in sequence copy number, maintenance of homogeneity of gene clusters, ploidy changes, and multiple chromosomal rearrangements, cannot be explained by point mutations within coding and/or regulatory regions of genes. In addition, the inactivation of genes by transposon mutagenesis cannot account for such changes. Our laboratory has been studying the process whereby cultured animal cells and human tumors become resistant to methotrexate as a result of amplification of the dihydrofolate reductase (DHFR) gene [1]. Gene amplification is a common phenomenon in biology [1, 2]. This paper describes our current understanding of the mechanism of gene amplification, and provides an explanation for the rapid generation of genome changes. We propose that any cell that is perturbed during the process of DNA replication can over-replicate DNA, generating a substratum of free DNA strands which can undergo different recombination events to generate a variety of chromosomal aberrations–rearrangements. We propose that this phenomenon occurs in germ-cell lineages during the rapid mitotic expansion phase, thereby occasionally fixing such chromosomal changes in gametes

On the mechanism of gene amplification

The frequency of a spontaneous doubling of the DHFR gene in CHO cells is 1×10^{-3} per cell generation [3] and the frequency of gene amplification can be increased 10-fold or more by pretreatment with such agents as hydroxyurea, aphidicolin, ultraviolet light and carcinogens [4, 5]. Mariani and Schimke [6] provided evidence that following the removal of hydroxyurea (which inhibits DNA synthesis) in CHO cells treated in the 2nd hour of S-phase, the DHFR gene is again replicated in the same cell cycle. In fact, they found that virtually all of the DNA replicated prior to the hydroxyurea block was re-replicated. That multiple initiations of DNA replication can occur in the same region of a chromosome, in particular when DNA replication patterns are perturbed, is consistent with the results of Prichard and Lark [7] with an *E. coli* auxotroph and of Billen [8] in *E. coli* following UV irradiation. Rice *et al.* [9] have demonstrated over-replication of DNA in CHO cells as a result of growing cells at a lower partial O_2 pressure. Evenson and Prescott [10] showed that brief heat treatment of the protozoa *Euploytes* results in the generation of multiple origins of DNA replication. Lastly, the developmental amplification of the chorion genes in *Drosophila* [11] is a result of multiple initiations at a single origin [12]

Treatment of cells with hydroxyurea [13] and aphidicolin [14] results in the generation of a wide variety of chromosomal aberrations–rearrangements observed in the first M-phase following inhibition of DNA synthesis. These aberrations include normal chromosomes with extrachomosomal DNA, increased frequencies of sister chromatid exchange, polyploidization, breakage–bridge fusions, and gapped–fragmented chromosomes. Most striking is the finding that virtually all of the chromosomal aberrations are found in that subset of hydroxyurea-treated cell population that contains greater than the normal content of DNA/cell as determined by flow cytometric techniques [13].

Recent studies from our laboratory suggest a possible mechanism for the over-replication phenomenon. Johnson *et al.* and Sherwood *et al.* (in preparation) have shown that the time of the inhibition of DNA synthesis by either aphidicolin or hydroxyurea is a critical parameter in determining the extent of over-replication. We have found that during the period of DNA synthesis inhibition, dihydrofolate reductase enzyme levels increase 10-fold, with a comparable increase in specific mRNA levels. By 2-D gel analysis we find that at least five other proteins increase as well. Following the release of DNA synthesis inhibition, there is a marked increase in DNA content/cell that is time-dependent, and the amount of this increase is dependent on the time that DNA synthesis has been inhibited. We propose that the synthesis of a number of proteins involved in initiation of DNA synthesis are regulated in a fashion similar to dihydrofolate reductase, and that inhibiting DNA synthesis allows them to accumulate. Thus following recovery of DNA synthesis (by washing out the drugs) there is an increased 'replication potential' such that over-replication of DNA in a single cell cycle occurs. Inasmuch as any number of drugs and other agents can result in inhibition of DNA synthesis and result in extensive chromosomal aberrations, we suggest that the inhibition of DNA synthesis, with the attendant increase in replication potential, is central to the generation of chromosomal aberrations.

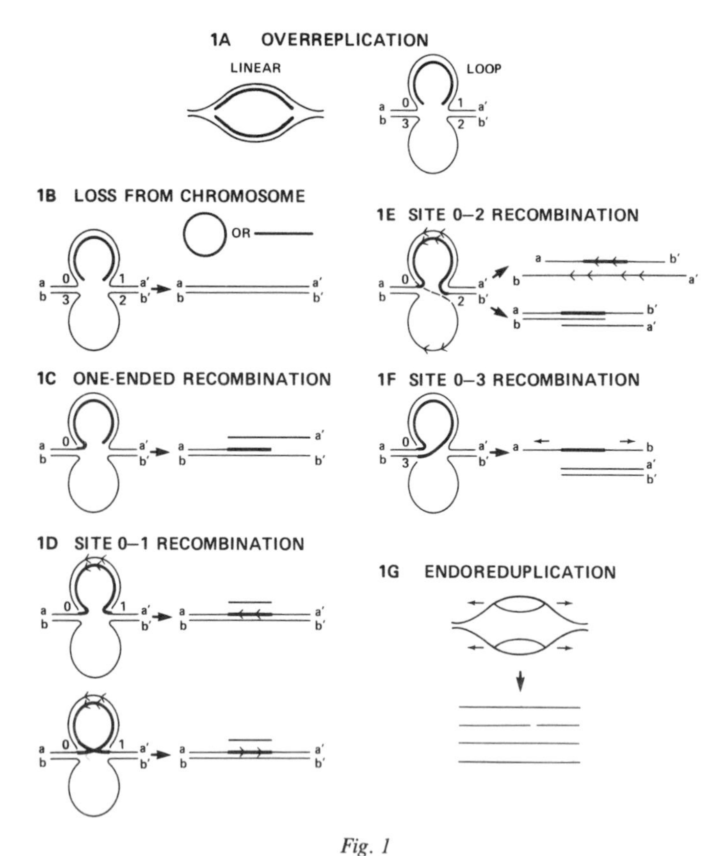

Fig. 1

Over-replication–recombination and generation of chromosomal rearrangements–aberrations

We have presented this model to account for gene amplification [1]; a model that proposes that the over-replicated strands can either recombine into the chromatid to generate expanded chromosomal regions containing amplified gene or circularize to generate extrachromosomal elements. Here we deal only with consequences of recombination.

Figure 1 A presents an over-replicated region of chromatids and employs the 'loop' configuration of over-replicated strands. For simplification, only a single over-replicated strand is depicted and the sister chromatids are completely replicated and ligated (i.e. these lines constitute double-stranded DNA).

Figure 1 B depicts the circularization of such strands to generate extrachromosomal structures. Non-circularized DNA segments, DNA circles without replication origins or without selective advantage, will be lost in dividing cells. In post-replicative cells they may simply be retained.

Figure 1 C shows a one-sided recombination (denoted site 0). The consequence is a broken chromatid. If this occurs at multiple replication complexes, the result is variable gapping–fragmentation. Note that maintenance of chromatid integrity would be better served if no recombination occurred.

Figure 1 D shows site 0–1 recombination. This can result in either a 'silent' event or in an inversion of a DNA segment. If the over-replicated strand were displaced and underwent recombination within a gene, the consequence would be a non-reciprocal 'gene conversion' event [15, 16]. Additional consequences include recombinational activation or inactivation of genes, generation of deletions and insertions, as well as recombination point mutations. At the HGPRT locus in humans Turner *et al*, [17] have reported that 57% of mutations involve major deletions–rearrangements, some of which are associated with amplification events.

Figure 1 E shows a 0–2 recombination, resulting in a sister chromatid exchange. Note that if the *b* to *a'* recombination does not occur, a broken chromatid is generated.

Figure 1 F shows a 0–3 recombination, which generates a breakage–bridge fusion chromosome.

Figure 1 G shows the consequence of over-replication in which multiple regions of the chromatids are re-replicated [7]. Such events will result in partial or complete endoreduplication of chromosomes, generating various forms of polyploidy, including so-called B chromosomes, as well as the generation of so-called mitotic recombination chromosomes.

One can readily envisage that translocations could result from recombination of chromatid fragments in the occasional cell in which chromosome (and gene) integrity is maintained to allow replicative viability of cells. Hill and Schimke [13] have described examples of virtually all of the above predicted chromosomal alterations, and find that they occur only in cells with greater than 2C DNA content following recovery from hydroxyurea. In addition, the presence of inversions and breakage–bridge fusion chromosomes have been described within regions containing amplified dihydrofoliate reductase genes in methotrexate-resistant cultured cells [18, 19].

Implications for rapid genomic change and speciation-evolution

As complex organisms have evolved with an increase in genome size, the requirement for retention of genetic material in a form that can be replicated faithfully and rapidly, and segregated properly into daughter cells, has required the evolution of chromosomes with multiple DNA replication initiation sites. We suggest that the problems attendant to replication of complex chromosomes and the potential for generation of chromosomal aberrations, in particular under circumstances where DNA replication patterns are perturbed, are critical to maintaining genomic constancy from generation to generation in both somatic cells and germ cells. We also suggest that a chain of causal relationships extends from perturbation of DNA replication to over-replication to the generation of various forms of genomic rearrangements and aberrations.

Chromosomal aberrations are a common phenomenon in cancer [20, 21] and in normal, but aging cells in mammalian species [22, 23]. For our purposes here we deal only with such aberrations in an evolutionary context. Among treatments of cells–organisms that result in over-replication are heat [10] and altered oxygen tension [9] and we suggest that other environmental perturbations may do likewise [24]. In many organisms the germ cells are potentially subject to direct environmental perturbation of DNA replication. Major genomic change often is an aspect of population divergence in plants [25, 26]. The rapid mitotic expansion of the germ-cell lineage in homeotherms is buffered against direct environmental influence. Yet metabolic perturbation of proliferating primordial germ cells can induce a high level of chromosomal abnormality in gametes of female mice [27]. In mammals, the vast majority of germ cells degenerate during primordial germ-cell mitosis and meiosis. This occurs in dividing rather than interphase cells, and such degenerating cells show chromosomal abnormalities [28, 29]. Germ-cell degeneration also occurs in birds [30] and cyclostomes [31].

As suggested by Bernstein [32], meiosis may have an important 'repair' function in which pairing 'edits' chromosomes either to rearrange chromosomal aberrations or ensure that such potential gametes do not progress to capacity for fertilization. Such a meiotic editing function does not appear to be foolproof, inasmuch as 60% of spontaneous human abortions have gross karyotypic abnormalities [33].

Thus, we suggest that the problems attendant upon replication of complex chromosomes, indeed, occur during rapid proliferation of germ-cell lineages. If over-replication–recombination occurs early in the proliferative process, a significant number of gametes from a single individual could carry the same and/or other chromosomal alterations. If such gametes are viable, chromosomal changes could be introduced rapidly into a population and fixed in a homozygous state by brother–sister mating.

References

1. Schimke, R. T., *Cell* **37**, 705 (1984).
2. Stark, G. R. and Wahl, G. M., *Annu. Rev. Biochem.* **53**, 477 (1984).
3. Johnston, R. N., Beverley, S. M., and Schimke, R. T., *Proc. Natl. Acad. Sci. USA* **80**, 3711 (1983).
4. Brown, P. C., Tlsty, T. D. and Schimke, R. T., *Mol. Cell. Biol.* **3**, 1097 (1983).
5. Tlsty, T. D., Brown, P. C. and Schimke, R. T., *Mol. Cell. Biol.* **4**, 1050 (1984).
6. Mariani, B. and Schimke, R. T., *J. Biol. Chem.* **259**, 1901 (1984).
7. Pritchard, R. H., and Lark, K. G., *J. Mol. Biol.* **9**, 288 (1964).
8. Billen, D. *J. Bact.* **97**, 1169 (1969).
9. Rice, G. C., Spiro, I. J. and Ling, C. C., *Int. J. Radiat. Oncol.* (in the press).
10. Evenson, D. P. and Prescott, D. M., *Exp. Cell Res.* **63**, 245 (1970).
11. Spradling, A. C. and Mahowald, A. P., *Proc. Natl. Acad. Sci. USA* **77**, 1096 (1980).
12. Osheim, Y. N. and Miller, O. L., *Cell* **33**, 543 (1983).
13. Hill, A. B. and Schimke, R. T., *Cancer Res.* **45**, 5050 (1985).
14. Rainaldi, G., Sessa, M. R. and Mariani, T., *Chromosoma* **90**, 46 (1984).
15. Holliday, R. *Gen. Res.* **5**, 282 (1964).
16. Jackson, J. A. and Fink, G. R., *Nature (London)* **292**, 306 (1981).
17. Turner, D. R., Morley, A. A., Haliandros, M., Kutlaca, R. and Sanderson, B. J., *Nature (London)* **315**, 343 (1985).
18. Biedler, J. L., Melera, P. W. and Spengler, B. A., *Cancer Genet. Cytogenet.* **2**, 47 (1980).
19. Fougere-Deschatrette, C., Schimke, R. T., Weil, D. and Weiss, M. C., *J. Cell Biology* **99**, 497 (1984).
20. Nowell, P. C. *Virchos. Arch.* B (*Cell. Path.*) **29**, 145 (1978).
21. Yunis, J. J. and Soreng, A. L., *Science* **226**, 1194 (1984).
22. Harnden, D. G., Benn, P. A., Oxford, J. M., Taylor, A. M. R. and Webb, T. P., *Somatic Cell Genet.* **2**, 55 (1976).
23. Martin, G. M., Smith, A. C., Ketterer, D. J., Ogburn, C. E. and Disteche, C. M., *Israel J. Med. Sci.* **21**, 296 (1985).
24. McClintock, B., *Science* **226**, 792 (1984).
25. Lewis, H., *Science* **152**, 162 (1966).
26. Stebbings, G. H., *Chromosomal Evolution in Higher Plants.* Addison-Wesley, New York (1971).
27. Hansmann, I., *Mutat. Res.* **22**, 175 (1974).
28. Beaumont, H. M. and Mandl, A. M., *Proc. R. Soc. London, Ser. B* **155**, 557 (1962).
29. Baker, T. G., and Franchi, L. L., *J. Cell Sci.* **2**, 213 (1967).
30. Hughes, G. C., *J. Embryol. Exp. Morph.* **11**, 513 (1963).
31. Hardisty, M. W., *The Biology of the Lampreys* (ed. M. W. Hardisty and J. C. Potter), vol. 2, p. 295. Academic Press, New York (1971)
32. Bernstein, H., *J. Ther. Biol.* **69**, 281 (1977).
33. Boue, A., and Boue, J., *Chromosomes Today* **7**, 281 (1981).

Chemica Scripta 1986, **26B**, 309–312

Evolution of Regulatory Signals of the Chromosomal β-lactamase Gene in Enterobacteria

Frederik Lindberg, Susanne Lindquist and Staffan Normark

Department of Microbiology, University of Umeå, S-901 87 Umeå, Sweden

Paper presented by Frederik Lindberg at the Conference on 'Molecular Evolution of Life', Lidingö, Sweden, 8–12 September 1985

Abstract

The chromosomal β-lactamase of Gram-negative enterobacteria is encoded by the *ampC* gene. The amino acid sequence of the β-lactamases from the different species are well conserved and more than 70% homologous. Enzyme expression is either constitutive, such as in the intestinal species *Escherichia coli* and *Shigella*, or inducible by β-lactams as in most soil bacteria, e.g. *Citrobacter*, *Enterobacter* and *Pseudomonas*. We discuss the regulatory mechanism governing β-lactamase expression in both groups, and argue that the constitutive subgroup has evolved from the inducible configuration through a deletion event, followed by several point mutations.

Introduction

β-lactams are a group of bacteriocidal substances that share the common feature of having the β-lactam ring. The β-lactams consist of several groups; among them the cephalosporins and the penicillins. Many of them are made as secondary metabolites by fungi. Discovered by Ernest Duchesne in 1896 and by Alexander Flemming in 1928, these compounds have been wildly exploited for medical use. Due to their high efficacy and low toxicity they are today the most commonly used drugs in the treatment of bacterial infections. The β-lactams act by inhibiting bacterial transpeptidases and carboxypeptidases (Fig. 1) [1]. Normally these enzymes have important functions in bacterial cell wall peptidoglycan synthesis and when they are irreversibly inhibited by the β-lactam they can no longer fulfill their function, resulting in bacterial death. The main bacterial counter-measure against this attack is the expression of β-lactamases which hydrolyse the β-lactam ring, releasing an inactive compound.

We have studied the regulation of the chromosomally encoded β-lactamase present in almost all enterobacteria and related species. It is encoded by the *ampC* gene, originally found and defined in *E. coli* by Boman and co-workers [2–3]. The enzymes of the different species are related both by chemical properties [4–6] and by sequence [7–10]. From the sequence information so far available, the enzymes from the different species have more than 70% of their amino acid sequence in common [10–11, unpublished data].

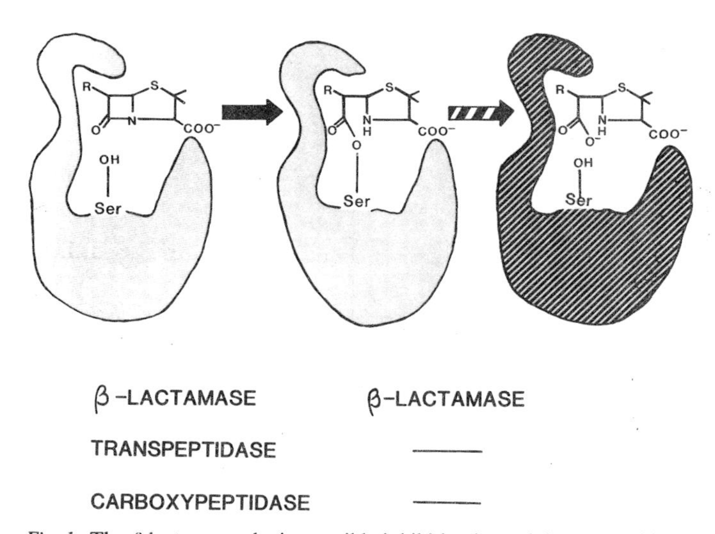

Fig. 1. The β-lactams act by irreversibly inhibiting bacterial transpeptidases and carboxypeptidases by acylating their active-site serines. Reaction with the β-lactamase is similar, but here the acyl-complex is only an intermediate, and the enzyme proceeds by hydrolyzing it, releasing a now inactivated compound and regenerating the enzyme.

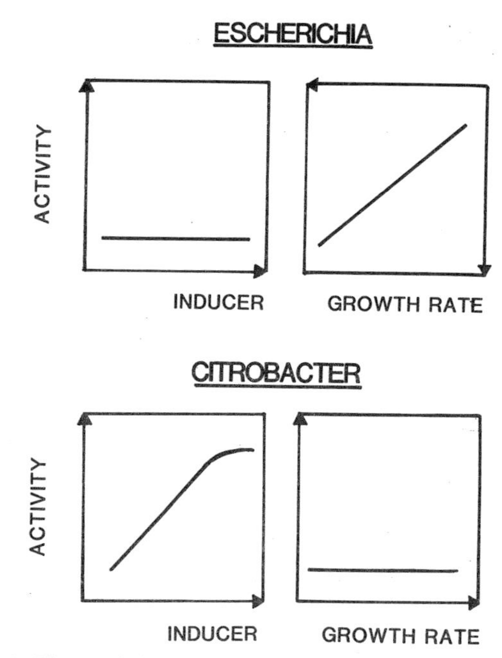

Fig. 2. The regulation of β-lactamase expression in enterobacteria and related species falls into two major groups – constitutive expression as in *E. coli* or inducibility of expression by β-lactams as in *C. freundii*. In *E. coli*, *ampC* transcription increases with increased growth rate, whereas this does not affect β-lactamase expression in *Citrobacter*. *Salmonella* is so far the sole representative of a third group of organisms which lack the *ampC* β-lactamase gene altogether.

The mode of regulation of the *ampC* gene divides the different species into three groups. Organisms such as *Citrobacter*, *Enterobacter* and *Pseudomonas* have an enzyme that can be induced by the addition of β-lactam to the medium (Fig. 2). In other genera, such as *Escherichia* and *Shigella*, the enzyme is produced at a constant level which is not affected by β-lactams [5–6]. Members of the *Salmonella* species do not produce β-lactamase at all as they lack the β-lactamase gene [12]. We will discuss the regulatory mechanisms governing gene expression in the two first groups, exemplifying the constitutive group with *Escherichia coli* and the inducibly regulated group with *Citrobacter freundii*. Experiments with *Shigella* [13] and *Enterobacter* [unpublished data] give results very similar to those that will be presented for *Escherichia* and *Citrobacter*, respectively. From these data we would like to hypothesize that inducible regulation, which is still present in soil bacteria, evolved first. We believe that the constitutive trait has occurred later by a single deletion event and subsequent adapting-point mutations. Possibly, constitutive β-lactamase expression together with the reduction in replication load is advantageous for species inhabiting the intestinal tract, where bacteria rarely encounter β-lactams (except in modern therapy).

Constitutive β-lactamase expression in *Escherichia coli*

The gene map of a small section of the *E. coli* chromosome is shown in the top part of Fig. 3. The four subunits of the fumarate reductase are encoded by the *frd* operon, consisting of the genes *frdA* through *frdD*. These enzymes are required by *E. coli* under some conditions of anaerobic growth [14–16]. The last gene in this operon, *frdD*, is immediately followed by the *ampC* β-lactamase gene. β-lactamase expression in *E. coli* cannot be induced by the addition of β-lactams to the medium, but enzyme expression increases with growth rate [17–18]. The molecular basis for this regulation is summarized in the bottom part of Fig. 3, where the β-lactamase promoter is shown to overlap the end of the *frdD* gene. Immediately following the promoter is a ribosome binding site and a hairpin structure with the features of a rho-independent terminator. Most of the transcripts initiated at the promoter will terminate at the hairpin and thus not contribute to β-lactamase expression. However, when a ribosome binds to the ribosome binding site it will prevent the formation of the stem, and thereby permit RNA polymerase to continue transcribing. Thus, as ribosome concentration increases as a function of growth rate, an increasing fraction of the initiated transcripts will read through and contribute to the expression of the *ampC* gene [19–20].

Inducible β-lactamase expression in *Citrobacter freundii*

The *ampC* gene is also closely linked to the genes of the fumarate reductase operon in *C. freundii*. However, here the end of the *frdD* gene is separated from *ampC* by 1100 base pairs that carry the regulatory gene *ampR* (Fig. 4, bottom). In *Citrobacter* the β-lactamase is inducibly expressed but is not regulated by growth rate [9, 21 and unpublished data]. A closer look at the end of the *frdD* gene (Fig. 4, bottom) reveals a typical terminator structure. However, all the features which in *E. coli* make this terminator into a growth-rate-dependent attenuator are absent in *C. freundii*. Following the terminator is not the β-lactamase gene, but the *ampR* gene which is transcribed from the opposite strand and ends here.

When the *ampR* gene is deleted or destroyed by frameshift mutations, *ampC* β-lactamase expression can no longer be induced and is made at a low level, suggesting a positive regulatory role for *ampR* [21]. Experiments with β-galactoside gene fusions show that this effect is at the transcriptional level [unpublished data]. Also, mutants can be selected that produce the β-lactamase at a high constitutive level, close to that obtained after maximal induction of the wild type. This over-expression is absolutely dependent on the presence of a functional *ampR* gene. When the *C. freundii ampR* and *ampC* genes are cloned into *E. coli*, *C. freundii* β-lactamase expression can still be induced without affecting expression of *E. coli's* own β-lactamase. Also in *E. coli*, mutants can be selected that constitutively overproduce the *C. freundii* β-lactamase. These mutants behave like those obtained in *Citrobacter*, i.e. they map outside of *ampR* and *ampC*, they occur at a high frequency, and they are absolutely dependent on *ampR* [21]. The mutated locus is tentatively called '*ampX*' and its function is currently being investigated in this laboratory.

From these data we have constructed a working hypothesis which is summarized in Fig. 5. AmpR increases *ampC*

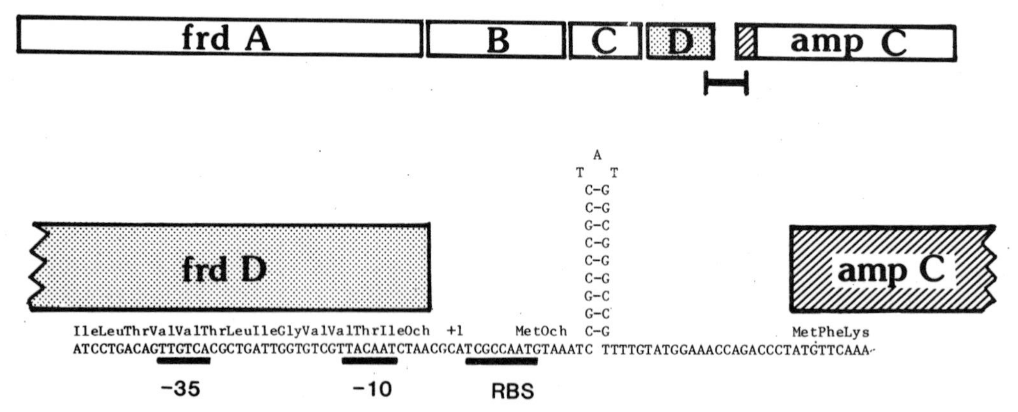

Fig. 3. Shown in the upper part of the figure is the organization of the *frd-ampC* region on the *E. coli* chromosome. The *ampC* regulatory sequences are shown in detail below. The potential hairpin structure is shown, although it will form at the mRNA level only. Transcripts are initiated at the promoter which overlaps the end of *frdD*, and most of them terminate at the hairpin structure (terminator) between *frdD* and *ampC*. These transcripts will not contribute to β-lactamase expression. When ribosomes bind to the ribosome binding site preceding the terminator, the stem does not form, permitting transcription to continue over *ampC*. Thus, as ribosome concentration increases as a function of growth rate the proportion of transcripts reaching *ampC* will increase, leading to a higher β-lactamase expression. Owing to its regulatory role, this terminator is often referred to as an attenuator.

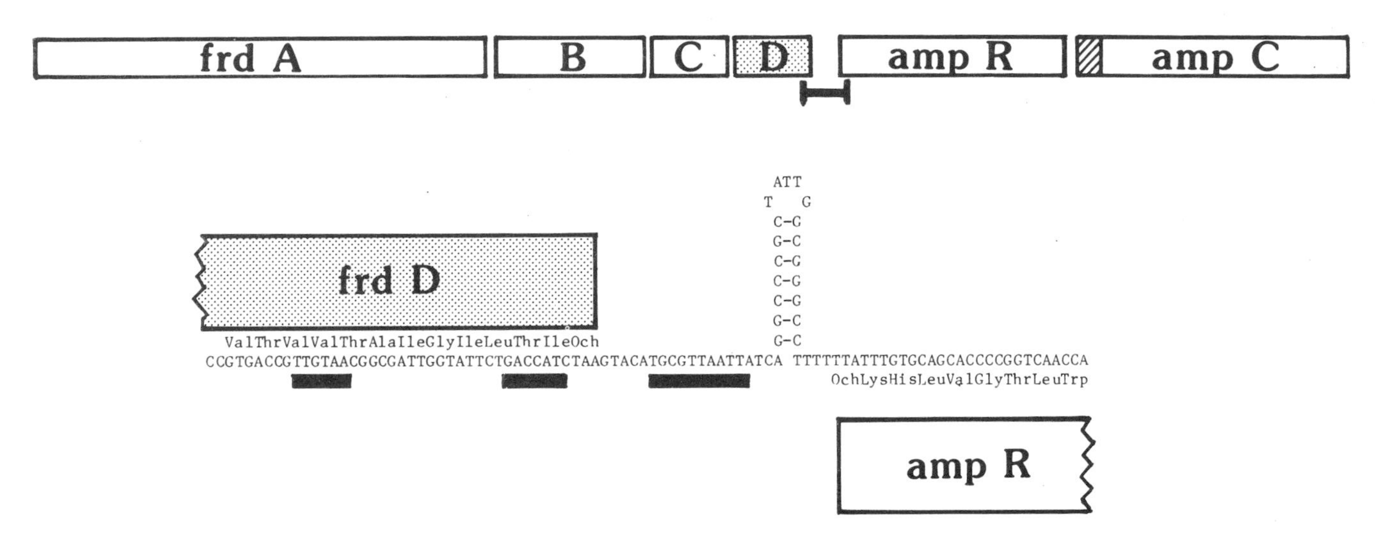

Fig. 4. In *C. freundii* the regulatory gene *ampR* is interposed between *frdD* and *ampC*. Also in this species, *frdD* is followed by a terminator, but here it lacks the special features of a growth-rate-dependent attenuator. No promoter can be found overlapping *frdD*, and instead *ampC* is transcribed from a separate promoter positioned between *ampR* and *ampC*. This promoter is directly or indirectly under positive control by the *ampR* gene product.

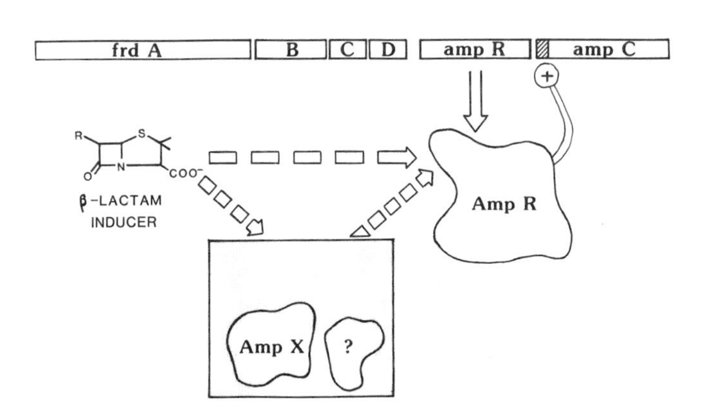

Fig. 5. In *C. freundii* transcription of the *β*-lactamase gene is greatly increased by inducing *β*-lactams if a functional *ampR* gene is present. Without *ampR* the rate of synthesis remains low and constant. Constitutive mutants, which make the *β*-lactamase at a level corresponding to maximal induction, are readily obtained. These mutants map outside of the *frd-amp* region. Similar mutations are obtained when the *Citrobacter ampR* and *ampC* genes are cloned in *E. coli*. In this case the locus of the mutation, '*ampX*', lies on the *E. coli* chromosome. This suggests that a process (boxed), involved in induction and including AmpX, is present in both *Citrobacter* and *E. coli*. This is indicated with a black box in the figure. However, *ampX* does not regulate the expression of the *E. coli ampC* gene. At present it is still unclear whether the inducer interacts directly with the AmpR protein or acts via the postulated 'black-box' process.

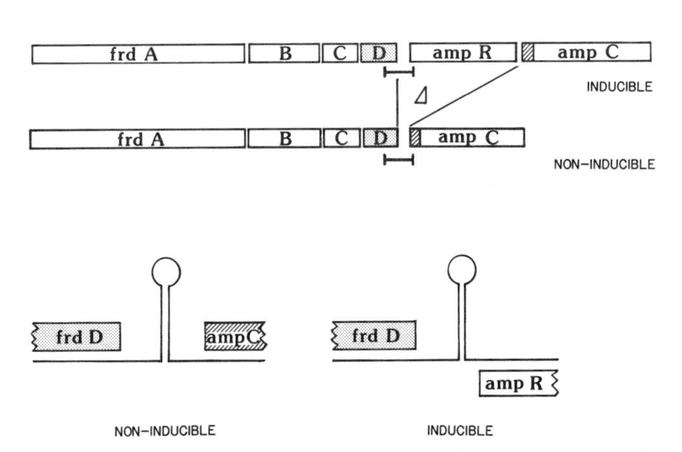

Fig. 6. Proposed evolution of *ampC* regulatory factors. The constitutive regulatory system could have evolved from the inducible configuration by a deletion of *ampR* between the end of *frdD* and the beginning of *ampC*. Expression of *β*-lactamase would then have been dependent on transcription from the *frd* operon. The subsequent accumulation of point mutations would have led to the evolution of the *ampC* promoter overlapping *frdD*, and to the development of the growth-rate dependent attenuator from a simple terminator structure.

β-lactamase transcription approximately 180-fold in the presence of either the inducer or the *ampX* mutation. It is not known whether the inducer interacts directly with *ampR*, or if it exerts its function via some intermediate factor. Since *ampX* is present in *E. coli*, which has a non-inducible *β*-lactamase, we believe that *ampX* takes part in some other function, e.g. cell wall synthesis. It is possible that the inducing *β*-lactam disturbs this process in a way similar to the *ampX* mutation. This change does not affect *β*-lactamase expression in *E. coli* but in inducible species like *C. freundii* it would be sensed by AmpR leading to an increased expression of the protecting *β*-lactamase. Future studies of the precise function of AmpR and *ampX* are needed to test the validity of the hypothesis.

Evolution from inducible to non-inducible *β*-lactamase expression

It is difficult to make convincing arguments about evolutionary patterns for two reasons, mainly. The first is the obvious fact that we are looking at a static situation – the result of evolution up to today, although we would like to discuss a dynamic process, such as the evolution from one species to another. The second difficulty is that we do not know the exact disadvantage or advantage of a certain genetical make-up versus another. Only very small differences would be needed to give one organism a slight selective advantage over its neighbors, so as to eventually form a new species. With this in mind we would, nevertheless, like to formulate a hypothesis on the evolution of *β*-lactamase regulation. We believe that the original state was the inducible one (Fig. 6, top), as present today in *Pseudomonas*, *Enterobacter*, *Citrobacter* and other soil bacteria. We believe that these species have to compete with *β*-lactam-producing fungi in some of their niches, and thus need a flexible response to β-lactams. The enteric organism, such as *Escherichia* and *Shigella*, almost never sees β-lactams. These bacteria have, through a deletion, lost the *ampR* gene and consequently β-lactamase inducibility. This results in a lower burden on replication which might give the slight selective advantage needed. The evolution of the β-lactamase attenuator (Fig. 6, bottom) could then have followed by selection of point mutants, allowing a slightly more flexible response without extra cost. This evolutionary pathway would imply that the control signals for the chromosomal β-lactamase in *E. coli* are of a much later date than the structural gene itself.

Salmonella lacks not only *ampR*, but also the *ampC* structural gene, whereas the genes following *ampC* in other species are present also in *Salmonella* [12]. Either this has occurred by a large deletion from the inducible configuration, or by an additional deletion from an *E. coli*-like organism. The high overall homology between *E. coli* and *Salmonella* would favor the latter hypothesis. If so, we would expect a closer look at the *Salmonella* sequence to reveal a fairly well conserved remnant of the *ampC* attenuator.

Acknowledgments

We are very grateful to all former students of this laboratory for their contribution to the effort of understanding β-lactamase regulation, and to Dr Christopher Korch for carefully reading the manuscript. The work discussed in this paper was supported by grants from the Swedish Board of Technical Development (Dnr 81–3384), the Swedish Medical Research Council (Dnr 5428) and the Swedish Natural Sciences Research Council (Dnr 3373).

Note added in proof:

The locus called *ampX* in this paper is now designated *ampD*.

References

1. Waxman, D. J. and Strominger, J. L., *Annu. Rev. Biochem.* **52**, 825 (1983).
2. Eriksson-Grennberg, K. G., Boman, H. G., Jansson, J. A. T. and Thorén, S., *J. Bact.* **90**, 54 (1965).
3. Burman, L. G., Park, J. T., Lindström, E. B. and Boman, H. G., *J. Bact.* **116**, 123 (1973).
4. Richmond, M. H. and Sykes, R. B., *Adv. Microbiol. Physiol.* **9**, 31 (1973).
5. Sykes, R. B. and Matthew, M., *J. Antimicrob. Chemother.* **2**, 115 (1976).
6. Sykes, R. B. and Smith, J. T., in *Beta-Lactamases* (ed. J. M. T. Hamilton-Miller and J. T. Smith), p. 369. Academic Press, London (1979).
7. Knott-Hunziker, V., Petursson, S., Jayatilake, G. S., Waley, S. G., Jaurin, B. and Grundström, T., *Biochem. J.* **201**, 621 (1982).
8. Bergström, S., Olsson, O. and Normark, S., *J. Bact.* **150**, 528 (1982).
9. Bergström, S., Lindberg, P., Olsson, O. and Normark, S., *J. Bact.* **155**, 1297 (1983).
10. Joris, B., De Meester, F., Galleni, M., Reckinger, G., Coyette, J., Frère, J. M. and Van Beeumen, J., *Biochem. J.* **228**, 241 (1985).
11. Jaurin, B. and Grundström, T., *Proc. Acad. Sci. USA 78*, 4897 (1981).
12. Olsson, O., Doctoral thesis, University of Umeå, Umeå, Sweden (1983).
13. Olsson, O., Bergström, S., Lindberg, P. and Normark, S., *Proc. Natl. Acad. Sci. USA* **80**, 7556 (1983).
14. Lemire, B. D., Robinson, J. J. and Weiner, J. H., *J. Bact.* **152**, 1126 (1982).
15. Cole, S. T., Grundström, T., Jaurin, B., Robinson, J. J. and Weiner, J. H., *Eur. J. Biochem.* **126**, 211 (1982).
16. Grundström, T., and Jaurin, B., *Proc. Natl. Acad. Sci. USA* **79**, 1111 (1982).
17. Normark, S., Edlund, T., Grundström, T., Bergström, S. and Wolf-Watz, H., *J. Bact.* **132**, 912 (1977).
18. Jaurin, B. and Normark, S., *J. Bact.* **138**, 896 (1979).
19. Jaurin, B., and Grundström, T., Edlund, T. and Normark, S., *Nature (London)* **290**, 221 (1981).
20. Grundström, T. and Normark, S., *Mol. Gen. Genet.* **198**, 411 (1985).
21. Lindberg, F., Westman, L. and Normark, S., *Proc. Natl. Acad. Sci. USA* **82**, 4620 (1985).

The Structure of a Human Common Cold Virus (Rhinovirus 14) and its Evolutionary Relations to Other Viruses

Michael G. Rossmann, Edward Arnold, John W. Erickson,* Elizabeth A. Frankenberger,† James P. Griffith, Hans-Jürgen Hecht,‡ John E. Johnson, Greg Kamer, Ming Luo, Gerrit Vriend

Department of Biological Sciences, Purdue University, W. Lafayette, Indiana 47907, USA

Anne G. Mosser, Ann C. Palmenberg, Roland R. Rueckert and Barbara Sherry

Biophysics Laboratory, University of Wisconsin, 1525 Linden Drive, Madison, Wisconsin 53706, USA

Paper presented by Edward Arnold at the Conference on 'Molecular Evolution of Life', Lidingö, Sweden, 8-12 September 1985

Abstract

We report here the first atomic resolution structure of any animal virus, namely that of human rhinovirus 14. The structure has been solved to 3.0 Å resolution using primarily a technique dependent on the viral symmetry rather than isomorphous replacement. The course of all four capsid polypeptides has been traced and correlated with the known amino acid sequences. The tertiary structures of the three larger proteins are each strikingly similar to those of the known icosahedral plant RNA viruses, as is also their quaternary organization in the virus coat. Four neutralizing immunogenic regions have been identified by sequencing mutants selected for their ability to survive in the presence of neutralizing antibodies. The altered amino acids, as well as corresponding antigenic sequences in the homologous polio and foot-and-mouth disease viruses, reside on protrusions. A large cleft, spanning the centre of each icosahedral face, is most probably the host cell receptor binding site. Evolutionary relationships among the three larger coat proteins of human rhinovirus 14 and the coat proteins of known icosahedral plant RNA viruses are assessed by the method of three-dimensional structure superposition.

Introduction

Picornaviruses are associated with serious diseases in humans and other animals, and they comprise one of the largest families of viral pathogens. For example, the common cold, poliomyelitis, foot-and-mouth disease and hepatitis can be caused by these viruses. They are among the smallest RNA-containing animal viruses [1–3]. Their molecular weight is around 8.5×10^6 and they contain about 30% by weight RNA. Their external diameter is roughly 300 Å and they form icosahedral shells. Picornaviridae have been subdivided into four genera on the basis of their buoyant density, pH stability and sedimentation coefficients: enterovirus (e.g. polio, hepatitis A and coxsackie viruses), cardiovirus (e.g. encephalomyocarditis and Mengo viruses), aphthovirus (e.g. foot-and-mouth disease virus) and rhinovirus. They differ also in the number of known serotypes. For instance, there are three known serotypes for polioviruses, seven for foot-and-mouth disease viruses (FMDV) and at least 89 for human rhinoviruses (HRV). Accordingly, it has been possible to produce effective vaccines for poliomyelitis and, with greater difficulty, for foot-and-mouth disease, but not for the common cold.

Picornavirions contain 60 protomers [4], each composed of four structural proteins, VP1, VP2, VP3 and VP4 (for nomenclature see ref. [2]). Their molecular weights in HRV14 are 32000, 29000, 26000 and 7000. The capsid protein VP0 is cleaved into its components VP4 and VP2 only in the final stages of assembly [5]. The cleavage occurs at an Asn-Ser peptide in HRV14, and hence is not effected by the viral protease whose specificity is for Gln-Gly peptides. The assembly of the capsid occurs in a series of steps culminating in the insertion of RNA into capsids to produce mature virions with the concomitant cleavage of VP0 [3, 5].

The virions contain a single, positive strand of RNA which is translated into a single polyprotein and then processed into its component proteins in a series of steps [6]. The gene order is essentially the same for HRV, FMDV, poliovirus and encephalomyocarditis virus (EMCV). The initial cleavage of capsid precursor protein from the polyprotein may be due to a host-cell protease, but subsequent cleavages are mostly dependent on the release of viral protease excised from the polypeptide [7].

Considerable effort has been devoted to mapping topological relationships among VP1, VP2, VP3 and VP4 within the capsid using chemical labelling of the surface of intact particles [8, 9], treatment with cross-linking reagents [10], reaction with specific antibodies and cross-linking with UV light [11]. The consensus is that VP1 is the most external and immunodominant protein, while VP4 is inaccessible from the outside but can be cross-linked with RNA on the interior. Heat treatment or mild denaturing agents cause a conformational change to the capsid, thus altering the response to antisera.

Present addresses:

* Department of Physical Biochemistry, AP-9A D-47E, Abbott Laboratories, Abbott Park, North Chicago, Illinois 60064, USA.

† Department of Agronomy, Purdue University, W. Lafayette, Indiana 47907, USA.

‡ F. G. Roentgenstrukturanalyse, Universitaet Wuerzburg, Zentralbau Chemie, Am Hubland, D-8700 Weurzburg, West Germany.

Surprisingly, the internal capsid protein VP4 can dissociate and escape from the capsid during antigenic conversion [3].

The RNA, and hence by inference the polyprotein, has been sequenced for all three strains of poliovirus, for various FMDV strains, for two strains of rhinovirus [12–14], for EMCV and for hepatitis A virus. Protein-to-protein comparisons between HRV, poliovirus, EMCV, FMDV and hepatitis A virus sequences show that HRV and poliovirus are closely related, whereas sequence homology of HRV to EMCV, FMDV or hepatitis A virus is not immediately obvious, particularly for the structural proteins.

X-ray diffraction studies of crystalline picornaviruses have been limited. Coxsackievirus crystals [15] and poliovirus crystals [16, 17] were reported a long time ago and rhinovirus strain 1A crystals a little later [18]. However, the technical problems involved in a complete structure determination were not solved until the elucidation of the small plant RNA viruses tomato bushy stunt virus (TBSV) [19], southern bean mosaic virus (SBMV) [20] and satellite tobacco necrosis virus (STNV) [21]. This encouraged the renewal of the crystallographic study of poliovirus [22], and stimulated work on rhinovirus [23] as well as Mengo virus. It was shown [23] that rhinovirus and poliovirus crystals can be roughly isomorphous, suggesting close similarities of the viral capsids, a result later supported by sequence homologies.

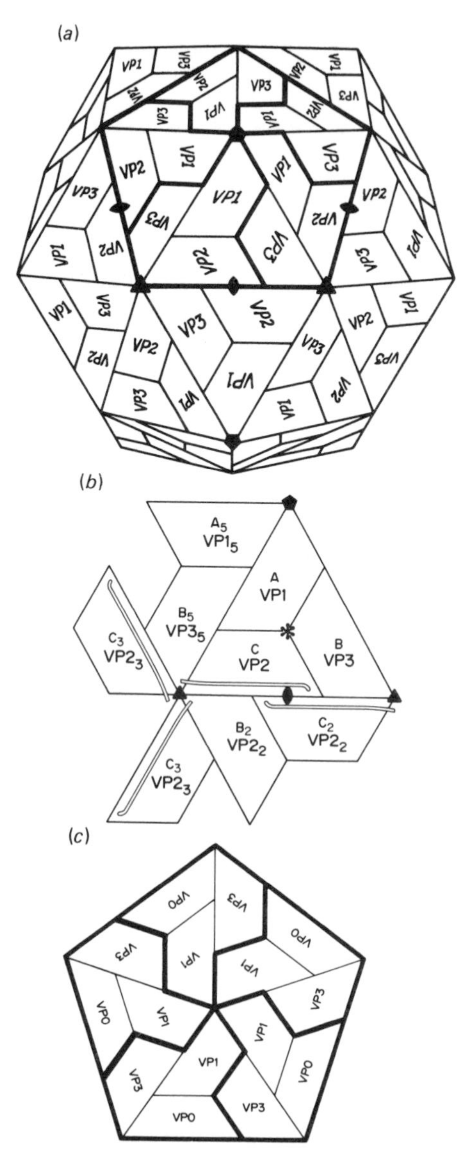

Fig. 1. Relation of the pseudo-equivalent VP1, VP2 and VP3 subunits to the quasi-equivalent subunits A, C and B in TBSV and SBMV. (*a*) Icosahedral capsid. (*b*) Icosahedral asymmetric unit. (*b*) shows the ordered amino-terminal arm βA present only in the C subunit of the plant viruses and VP2 of HRV14. The amino end of the arm interacts with two other VP2 arms across the threefold axis, while the carboxy end of the arm interacts with a VP2 across the twofold axis. Asterisks indicate the position of the quasi-threefold axis in SBMV and TBSV analogous to the pseudo-threefold axis in HRV14. Subscripts designate the symmetry operation required to obtain the given subunit from the basic triangle. (*c*) The thickly outlined VP1, VP3, VP0 unit corresponds to the 6S protomer, and the 15-mer cap to the 14S pentamer observed in assembly experiments.

The structure determination

HRV14 crystals were prepared as described by Arnold *et al.* [23]. The crystals are cubic with $a = 445.1$ Å belonging to space group $P2_13$. There are four particles per crystal cell with each virion situated on a crystallographic threefold axis along the body diagonal. Thus, one-third of the virus, or 20 icosahedral asymmetric units, is in the crystallographic asymmetric unit. The data used for the results given here were collected at CHESS (Cornell High Energy Synchrotron Source). A new crystal was used for every exposure. A total of 83, 0.3° oscillation, film packs were eventually included in the native data. An electron density map, based on two poor low-resolution isomorphous derivatives phased between 25 and 6 Å resolution and averaged over the 20 different non-crystallographic asymmetric units, showed clearly the viral protein envelope between about 100 and 150 Å radius. Phase extension beyond 5 Å, to eventually 3.0 Å resolution, was based only on the icosahedral symmetry and an assumption of an envelope within which was the protein shell of the virus.

When phases had been extended to 3.5 Å resolution, after 40 cycles of molecular replacement, a map was calculated of quite extraordinarily good quality. Most residues could be easily recognized. It was this map which was interpreted fully. Indeed, the polypeptide chains were traced and amino acids correlated to the sequences in a period of only two days. The final electron-density map was averaged and displayed in sections at 1 Å intervals perpendicular to an icosahedral twofold axis. A mini-map of the 3.5 Å resolution electron density was used for the original chain tracing and amino acid identification. An atomic model was then built into the final 3.0 Å resolution map using an Evans and Sutherland PS300 computer graphics system and the FRODO program [24]. Most of the carbonyl oxygens and some water solvent molecules were readily recognizable in this map.

Phase determination based on non-crystallographic symmetry was first suggested by Rossmann and Blow [25, 26]. It was used to compute very-low-resolution maps of SBMV [27] and polyoma virus [28] and also used for phase extension in the determination of STNV [29] and hemocyanin [30]. However, there was considerable trepidation that the molecular replacement method might fail in extending the phases all the way from low resolution (5 Å) to high resolution (3 Å). The high quality of the map reflects the power and exactness of the non-crystallographic constraints as opposed to the approximations required in an isomorphous replacement phase determination. The success of these techniques holds promise for a rapid series of structural determinations of viruses and other biological assemblies of high symmetries, by reducing the dependence upon finding satisfactory isomorphous heavy atom derivatives.

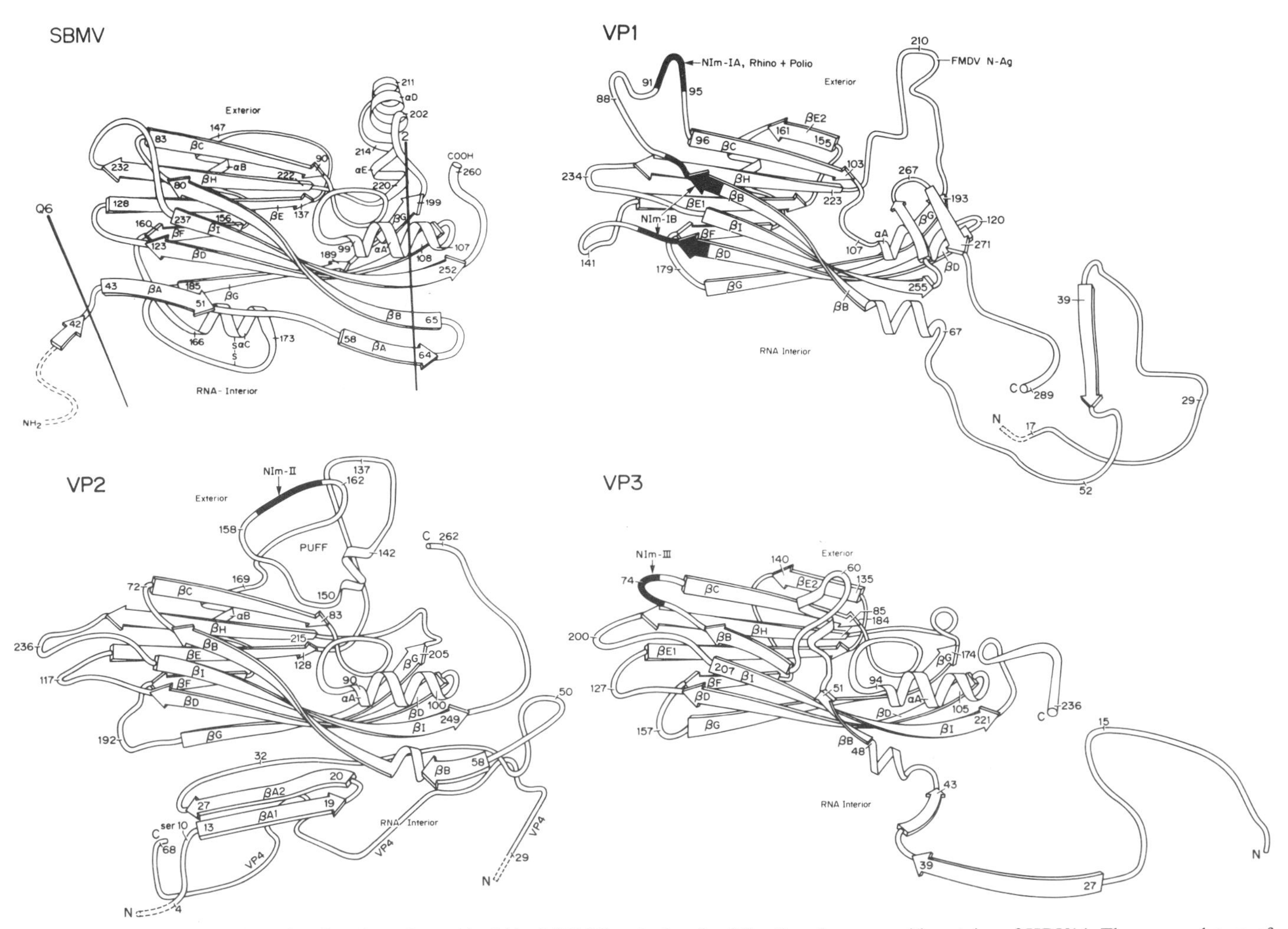

Fig. 2. Diagrammatic drawings showing the polypeptide fold of SBMV and of each of the three larger capsid proteins of HRV14. The nomenclature of the secondary structural elements is derived from that of SBMV [20]. Amino acid sequence numbers, appropriate for each protein, are also shown. (*a*) SBMV, (*b*) VPI of HRV14, (*c*) VP2 and VP4 of HRV14, (*d*) VP3 of HRV14. (Adapted from a drawing of SBMV by Jane Richardson.)

The structure

The particle consists of an icosahedral protein shell (Fig. 1*a*) surrounding an RNA core. The lack of visible structure in the central cavity results from the random orientation of the asymmetric RNA molecule. Both the tertiary fold of the VP1, VP2 and VP3 polypeptide chains and their quaternary organization within the HRV14 capsid are closely similar to the two published high-resolution structures of $T = 3$ (180 identical subunits per capsid) RNA plant viruses, TBSV [19] and SBMV [20]. Although the subunit organization within the icosahedron is somewhat different, a similar tertiary structure has also been found for $T = 1$ (60 subunits per capsid) STNV [21]. The radial position and orientation of structurally equivalent atoms of HRV14 and SBMV generally agree to better than 3 Å relative to the icosahedral symmetry axes. In the plant viruses, the three quasi-equivalent subunits A, B and C have the same amino acid sequence but cannot have identical geometrical environments. For SBMV there are 260 amino acids per subunit, but the first 63 residues are associated with the RNA and therefore do not have icosahedral symmetry. These are said to be in the 'random' domain. In the C subunit, unlike the A and B subunits, residues 38 to 63 are ordered with their ends forming a 'β-annulus' about the icosahedral threefold axes (Fig. 1*b*). Residues 64–260 of SBMV are referred to as the 'shell' domain and form an eight-stranded anti-parallel β-barrel (Fig. 2*a*). The β-barrels in HRV14, like those in SBMV, are wedge-shaped with the thin end (left in Fig. 2) pointing toward the five- or threefold (quasi-sixfold) axes. There are four excursions of the polypeptide chain toward the wedge-shaped end. Each excursion makes a sharp bend or 'corner'. The most exterior (top left in Fig. 2) corner is formed between the β-sheets βB and βC, the second corner down is formed between βH and βI, the third corner is between βD and βE1, while the most internal corner connects βF and βG. The βF-βG corner is the site of a 25-residue insertion in SBMV, including the α-helix αC, that is not present in any of the viral proteins of picornaviruses, nor in TBSV or STNV. The alignment of the VP1, VP2 and VP3 chains of several picornaviruses is shown relative to the capsid protein of SBMV in Fig. 3.

The three larger capsid proteins VP1, VP2 and VP3 in HRV14 are oriented and situated at essentially the same radius and position as the A, C and B subunits in SBMV, respectively. The capsid proteins of HRV14 are related by a pseudo-threefold axis, analogous to the quasi-threefold axis in SBMV and TBSV, at the asterisk in Fig. 1(*b*). However, VP1 and VP3 have additions at their amino and carboxy termini. The amino ends, as in plant viruses, are in contact with the RNA, but unlike plant viruses they are more acidic than basic and have only a very few amino-terminal residues 'disordered' (lacking icosahedral symmetry). The first 64 of the 73 amino-terminal residues of VP1 reside under VP3, while the first 42 of the 71 amino-terminal residues of VP3

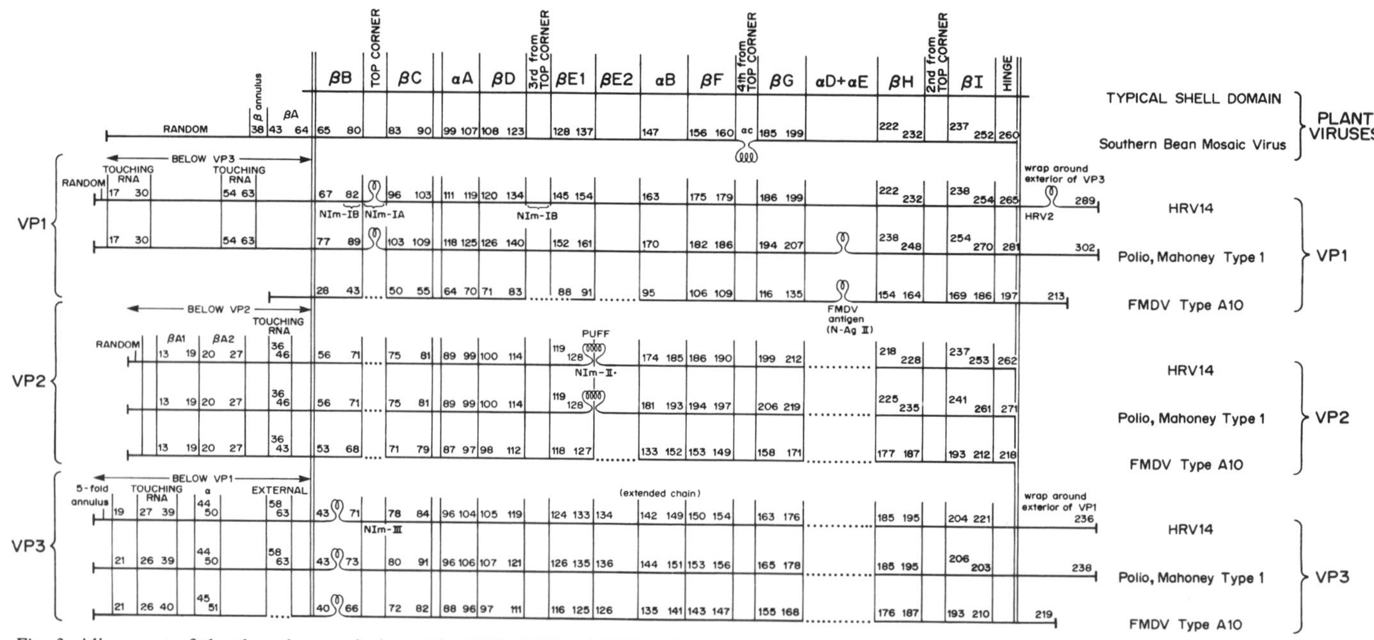

Fig. 3. Alignment of the three larger viral proteins VP1, VP2 and VP3 with a typical plant viral capsid protein as, for instance, the SBMV coat protein. The symbols (coil) and ... represent insertions and deletions with respect to the typical plant virus capsid protein. Immunogenic sites found in HRV14 are also shown. Three different picornaviruses are considered. The assignments of HRV14 are derived from the electron density map and the known sequence of its RNA. The alignment of polio type 1 Mahoney was deduced from the amino acid homology. The alignment of FMDV type A10, determined from amino acid homology with all available picorna sequences, is uncertain from residue 1 to 190 in VP1.

are under VP1. Thus, the predominant positions of VP1 and VP3 at the RNA–protein interface are exchanged relative to their positions at the exterior surface.

The first 25 of the 69 residues of the internal structural protein VP4 are not seen in the electron-density map, implying that they lack icosahedral symmetry. VP4 is positioned in part below VP1 and VP2, with its visible amino end surrounding the fivefold axis. The carboxy ends of VP1 and VP3 are external and function in part to associate proteins within a protomer (Fig 1*c*).

Large sequence insertions relative to the typical shell domain (Fig. 3) form protrusions on VP1, VP2 and VP3 and create a deep cleft or 'canyon wall' on the viral surface. The canyon separates the major part of five VP1 subunits (in the 'north') clustered about a pentamer axis from the surrounding VP2 and VP3 subunits (in the 'south'), thus forming a moat around the VP1 protrusions on the fivefold axis. The south canyon walls are lined with the carboxy-terminal ends of VP1 and a large sequence insertion in VP1 corresponding to helices αD and αE in the equivalent SBMV capsid protein. The north canyon wall is partially lined with the carboxy terminus of VP3. VP2 is hardly associated at all with the canyon, while VP1 is the major contributor to the residues lining the canyon. The canyon is 25 Å deep and 12–30 Å wide.

Due to the additional elaborations which VP1 has on the surface relative to VP2 and VP3, its overall shape is that of a kidney, with the depression forming a large part of the canyon. The first 16 residues of VP1 (Fig. 2*b*) are not seen in the electron-density map. The shell domain of VP1 in HRV14 starts at residue 74. The small sequence insertion between βB and βC in rhinovirus and poliovirus is not found in FMDV. This loop forms a major immunogen in HRV14 (NIm-IA* to be discussed below) and poliovirus. Five carbonyl oxygens of fivefold-related Asn 141 residues converge on the vertex to chelate a presumably cationic ligand on the fivefold axis. The residues in VP1 of HRV14, that are analogous to αD and αE helices in SBMV, protrude to the surface and form part of the south rim of the canyon, but do not form helices in HRV14. There is an 8-residue insertion in polio and FMDV relative to HRV14 (Fig. 3) at the most external portion of this segment. These additional residues contain the major antigenic site of FMDV and have been predicted to form an α-helix [32]. The carboxy-terminal 23 residues of VP1 line the south rim of the canyon in HRV14. There is little conservation of these residues among picornaviruses.

The first 3 residues of VP2 (Fig. 2*c*) are not seen in the electron density. The proximity of serine 10 in VP2 to the carboxy terminus of VP4 suggests that the VP0 cleavage, which occurs during the virus maturation step, may be autoproteolytic. There is no histidine in the immediate vicinity of Ser 10 and the carboxy end of VP4. However, nucleotide bases of the RNA might act as proton acceptors in the autocatalysis. Thus, the insertion of RNA into the growing capsids might trigger the cleavage VP0 into VP2 and VP4. Ser 10 is completely consistent for all sequences of HRV, polio, EMC, and coxsackie viruses but not in FMDV. However, FMDVs have a conserved pair of serines in sequences in position corresponding to 27–28 in HRV, also close to the carboxy terminus of VP4. The αD and αE helices are absent in VP2 (Fig. 2*c*). There is a large 43-residue insertion, in the VP2 position corresponding to βE2 of VP1 and VP3, forming an external mushroom-shaped 'puff'. This is positioned adjacent to the VP1 elaborations, associated with the major antigenic site in FMDV, which line the south canyon wall. The most external residues of this puff correspond to NIm-II of HRV14. In contrast to VP1 and VP3, the carboxy terminus of VP2 has no extension beyond the shell domain.

All residues of VP3 (Fig. 2*d*) can be seen in the electron density. The 26 amino-terminal residues form a fivefold

* The previously reported [31] N-AgI and N-AgII have been renamed NIm-IA and NIm-III to identify them as neutralizing immunogens and to indicate their major coat protein locations on VP1 and VP3, respectively.

β-barrel about the pentamer axis analogous to the β-annulus [19] about trimer axes in SBMV and TBSV. This fivefold annulus extends down into the RNA to a radius of 111 Å. The polypeptide emerges from the β-annulus, circles around the base of the VP1 shell domain while making extensive contact with the RNA, emerges on the viral surface near residue 61 and then enters the shell domain at residue 72. The top corner of the VP3 shell domain, between βB and βC, is the NIm-III site of HRV14, structurally equivalent to NIm-IA in VP1. The external helices αD and αE of SBMV are absent as in VP2. When the shell domains of VP3 and VP1 are superimposed, much of the amino- and carboxy-terminal arms are also structurally equivalent.

Immunogenic sites on HRV14

An animal's immunological response to a virus is one of the major defenses against disease. Antibodies can bind to viruses but they do not necessarily neutralize infectivity. In spite of the 60-fold equivalence of each potential binding site on the virus, as few as four neutralizing antibodies per virion can be sufficient to inhibit infectivity of poliovirus [33]. Neutralizing antibodies usually change the isoelectric point of the picornavirions [34, 35], indicating that a conformational change frequently accompanies neutralization. Antibodies may neutralize by interfering with cell attachment, membrane penetration or virus uncoating. Antibodies that bind to poliovirus may require bivalent attachment for neutralization of the virus [36, 33].

Amino acid residues within the major neutralization immunogens* of HRV14 have been identified by Sherry and Rueckert [31] and by Sherry *et al.* [37]. They isolated mouse hybridoma lines which secrete monoclonal antibodies that neutralize HRV14. Each was then used to select several viral mutants resistant to neutralization by that antibody. Finally, every monoclonal antibody was assayed for its ability to neutralize the mutants. The results revealed four major immunogenic neutralization sites. Each immunogenic site was composed of overlapping epitopes where a given mutant was resistant to many or all of the antibodies directed against the site. The amino acid residues defining the four immunogens are summarized in Table I. These results were obtained without knowledge of the three-dimensional structure. When, however, the electron-density map became available, it was immediately clear that the substitutions that could confer resistance to neutralization, regardless of their location in the amino acid sequences, were localized into four distinct areas (NIm-IA, NIm-IB, NIm-II and NIm-III) corresponding exactly to the proposed immunogens. Moreover, these residues invariably faced outward toward the viral exterior.

NIm-IA is an insertion between βB and βC at the top-most corner of the VP1 wedge. It is the most external portion of the complete virion being 160 Å from the centre. Resides 91 and 95 are on the extreme external portion of the loop. NIm-IB on VP1 is at the carboxy ends of βB and βD, which are situated on either side of the amino end of strand βI. No mutations conferring resistance have yet been found on the βI strand. The NIm-IB site is close to the fivefold axis, a little below the canyon rim. NIm-II on VP2 is at the extreme outside of the puff, 155 Å from the viral centre. This immunogenic puff is adjacent to the external loop formed by the sequence insertion in VP1 corresponding to αD and αE of

Table I. *Amino acids associated with each of the four major immunogenic sites in HRV14*

(Substitutions in mutants selected for resistance to neutralization by monoclonal antibodies.)

	Amino acid no.	Wild type	Observed mutations
NIm-IA			
VP1	91	D	A F G H N V Y
VP1	95	E	G K
NIm-IB			
VP1	83	Q	H
VP1	85	K	N
VP1	138	D	E G
VP1	139	S	P
NIm-II			
VP2	158	S	F
VP2	159	A	V
VP2	161	E	D V* D* K
VP2	162	V	M A*
VP1	210	E	D*
VP2	136	E	G
NIm-III			
VP3	72	N	I
VP3	75	R	G K M
VP3	78	E	K V
VP1	287	K	I
VP3	203	G	D

* A double mutation to V161 and A162 occurred in two cases; a double mutation to D161 and D210 occurred in a third case.

SBMV. One of the residues (Glu 210 on VP1) that is associated with NIm-II is on VP1 (Table I). NIm-III on VP3 is 149 Å from the viral centre, to some extent in the shadow of a larger protrusion of VP3 (residues 58–63) and the carboxy end of VP1 (282–286). Residue Lys 287 on VP1, associated with this site (Table I), points directly toward and is adjacent to the other residues on VP3 associated with NIm-III. The large protrusion near NIm-III, consisting of residues 282–286 in VP1 and 58–63 in VP3, has not been shown to be antigenic for HRV14, but in FMDV residues associated with the carboxy-terminal end of VP1 have shown antigenicity (Table II).

Relation to antigenic sites in polio and FMDV

Table II summarizes results obtained in mapping the immunogenic and antigenic surfaces of other picornaviruses. There is a dominant immunogen in poliovirus corresponding to NIm-IA of rhinovirus. Absence of a corresponding immunogen in FMDV is explained by deletion of this loop from its VP1.

The dominant antigenic site in FMDV resides in a region homologous to NIm-II in HRV14 (Table II). The VP1 contribution to this antigen has also been shown for HRV14 (E210 of NIm-II in Table I) and for poliovirus (residues 222–241, Table II). In FMDV, however, the immunogenic puff on VP2 is absent and 8 amino acids are inserted at the extreme surface of the αD-αE region of VP1. The resulting protrusion of NIm-II in FMDV would be unsupported by the puff. This protrusion may, therefore, occupy the space which in HRV14 is occupied by the puff. The inability of FMDV

Table II. *Evidence for antigenic regions in poliovirus and FMDV*

	Virus	Coat protein	Amino acid number	HRV14[a] amino acid equivalent	Sero-type[b]	Ref.	Method[c]	Position in HRV14 structure
1	Polio	VP1	11–17	Deleted	1M	35	3	Within RNA
2	Polio	VP1	24–40	14–30	1M	51	5	Faces RNA
3	Polio	VP1	70–75	60–65	1M	35	3,6	Faces RNA
4	Polio	VP1	61–80	51–70	1M	81	2	Faces RNA
5	Polio	VP1	86–103	77–97	1M	81	5	NIm-IB+NIm-IA
6	Polio	VP1	91–109	84–103	1M	81	2	NIm-IB+NIm-IA
7	Polio	VP1	100–109	93–103	1M	81	2	Part of NIm-IA
8	Polio	VP1	93–103	86–97	1M	35	2,3,6	NIm-IB(?)+NIm-IA[d]
9	Polio	VP1	93–104	86–98	1	54	4	NIm-IB(?)+NIm-IA[d]
10	Polio	VP1	93–100	86–93	3L	55, 56	1	NIm-IB(?)+part of NIm-IA[d]
11	Polio	VP1	141–147	135–141	1M,S	82	3	NIm-IB
12	Polio	VP1	161–181	154–174	1M	81	5	Partly exposed on canyon
13	Polio	VP1	182–201	175–193	1M	81	2	Buried on 4th corner down
14	Polio	VP1	222–241	212–225	1M	81	2	NIm-II
15	FMDV	VP1	141–160	210–228	O_1K	83	2	NIm-II
16	FMDV	VP1	144–159	211–227	O_1K	32	2	NIm-II
17	FMDV	VP1	145–168	211–236	A12	84	8	NIm-II
18	FMDV	VP1	146–154	211–223	O_1K	51	7	NIm-II
19	FMDV	VP1	169–179	237–247	A12	52, 84	8	NIm-IB central strand
20	Polio	VP1	270–287	254–272	1M	81	5	Partly exposed on canyon wall
21	FMDV	VP1	200–213	268–294	O_1K	83	2	NIm-III(?)
22	FMDV	VP1	200–213	268–294	O_1K	51	7	NIm-III(?)
23	Polio	VP2	162–173	161–170	1M	62	2,3	NIm-II
24	Polio	VP3	71–82	70–81	1M	62	6	NIm-III

[a] Aligned by eye and by fitting to the shell domain structure.
[b] 1M (type 1, Mahoney), S (Sabin), 3L (type 3, Leon), O_1K (type O_1, strain Kaufbeuern), A12 (type A, subtype 12).
[c] Method key: (1) Monoclonal antibodies raised against intact virus select for resistant mutations. (2) Synthetic peptides induce neturalizing antibodies. (3) Synthetic peptides prime for high titer neutralizing response. (4) Synthetic peptide competes with monoclonal antibody to inhibit neutralization. (5) Synthetic peptides induce antibodies which bind virus but neutralize poorly. (6) Neutralizing antisera or monoclonal antibodies bind peptide in ELISA. (7) Deduced from ability or inability of protein fragments to induce neutralizing antibodies. (8) Deduced from ability or inability of neutralizing monoclonal antibodies to bind protein fragments.
[d] Uncertainty due to somewhat arbitrary nature of computer alignments.

VP1 to elicit neutralizing antibodies [38] is consistent with the lack of structural support for this NIm-II protrusion in the absence of the neighboring VP2.

Poliovirus can also be neutralized by antibodies that bind to NIm-II or NIm-III (Table II). Thus there is an overall consistency in identifying at least three of the major HRV14 neutralization sites in poliovirus.

Many of the methods used to determine antibody binding sites depended upon the use of synthetic peptides as antigens (Table II). Peptides associated with neutralizing antigenic regions do, in general, elicit antibodies that can neutralize the intact virus. In a significant number of cases (lines 1–4, 8, 13 of Table II), however, the sequence in question lies far below the viral surface or even buried in the RNA. This suggests that some peptides can elicit antibodies which subsequently bind to totally unrelated portions of the native virus. Alternatively, some of the results might be accounted for by conformational changes occurring during the isoelectric transition of the virus [39].

The correspondence of sequence variability with antigenic sites has frequently been pointed out in the study of picornaviruses. This is certainly apparent on comparing the two available rhinovirus sequences of HRV14 [12, 13] and HRV2 [14], as well as in comparative studies of polio [40] and FMDV [41, 42] sequences. Nevertheless, the surface protrusion caused by HRV14 VP3 58–63 together with VP1 282–286 is equally variable, but has not yet been associated with an antigenic site in any picornaviruses.

The canyon as receptor binding site

In spite of the sequence and surface similarities of picornaviruses, they have different host and tissue specificity. Abraham and Colonno [49] have shown, using 24 rhinovirus serotypes in competition binding assays, that, while the majority recognize one receptor, a second smaller group recognizes a different receptor. HRV14 (sequenced by Callahan *et al.* [13] and by Stanway *et al.* [12]) belongs to the larger receptor group while HRV2 (sequenced by Skern *et al.* [14]) belongs to the smaller one. Krah and Crowell [45] have characterized some properties of HeLa cell receptors for group B coxsackieviruses. They found that concanavalin A and other lectins adsorbed to receptors and inhibited virus attachment, a finding similar to that of Lonberg-Holm [46].

The nature of the canyon 25 Å deep, circulating around each of the 12 pentamer vertices, suggests this to be the site for cell receptor binding. An antibody molecule, whose Fab fragment would have a diameter in the order of 35 Å, would have difficulty in reaching the canyon floor, its entrance being blocked by the canyon rim. Thus, the residues in the deeper recesses of the canyon would not be under immune selection and could remain constant, permitting the virus to retain its ability to seek out the same cell receptors.

While retention of the canyon structure for all picornaviruses is to be expected, variation in the residues lining the canyon should be anticipated between viruses that attach themselves to different host-cell receptors. That is, FMDV, polio, HRV14 and HRV2, all of which recognize different receptors, should exhibit some variation in the residues lining the canyon wall. It is thus noteworthy that those parts of the carboxy-terminal ends of VP1 and VP3 which line the canyon walls are some of the least conserved amino acids among picornaviruses.

Since the topology of the canyon should be retained, the highly conserved, structurally equivalent, sequences (MYVPPGAPNP starting at 151 of VP1 and AYTPPGARGP starting at 130 of VP3 in HRV14) in rhino, polio and FMD viruses situated near the floor of the canyon may be significant.

FMDV can be treated with trypsin causing cleavage at residues between 138 and 154 of VP1. This causes the virus to lose its ability to attach to cells and its ability to stimulate neutralizing antibody [50–52, 54]. The enzymic cleavages occur in NIm-II on the αD-αE protrusion of VP1, a large loop which also forms part of the presumed host receptor binding site.

Argos *et al.* [53] have shown that there is a similarity in the folding topology of concanavalin A and the shell domain of TBSV (and hence also to VP1 of HRV14), thus providing some rationalization of the competition experiments at the molecular level [45, 46]. The functional sugar binding site of concanavalin A corresponds roughly to the NIm-IB site in VP1. Comparison of the amino acid sequence between type 1 Mahoney (virulent) and type 1 Sabin (attenuated) shows that the differences are concentrated near the NIm-IB site. This raises the possibility that virulent and attenuated strains may have altered receptor specificity for carbohydrate recognition. Similarly, the deletion of HRV2 relative to HRV14 in βB near NIm-IB may, in part, be responsible for the different receptor specificity [49] of these two rhinoviruses. However, the non-coding region, also contained in the recombinant RNA segment, has been implicated as well [55].

Neutralization and serotypes

Since an antibody itself has a twofold axis, bivalent attachment could occur across icosahedral twofold axes. The distance between the nearest twofold-related immunogenic sites IA, IB, II and III is 120, 120, 50 and 60 Å, respectively. All other symmetrical counterparts of these antigens related by twofold axes are at least 170 Å apart. Lower and upper limits of the distance of the antibody binding sites on an immunoglubulin molecule are not well known but probably lie in the range 50–180 Å [33]. Hence, symmetrical bivalent attachment of antibodies to the known immunogens may be possible, although NIm-II and III are close to the likely limit. Since Fab arms of antibodies are capable of some rotation, it is conceivable that bivalent asymmetric attachment might also occur across three- or fivefold axes. The resultant torque may produce conformational changes sufficient to interfere with the normal process of infection.

Polyclonal antisera raised against virus are likely to contain antibodies against all immuogenic regions on the viral surface. Thus, different serotypes must express at least one mutation at each immunogenic site. If, say, four separate mutational events (one for each immunogenic site in HRV14) occurred in the corresponding limited region of the RNA, it is highly probable that mutations also will have occurred at other regions. Apparently, then, there could be some constraint on additional mutations accumulating over time which limits the known number of polio serotypes, while permitting numerous serotypes for rhinoviruses. Constraints limiting the diversity of viruses might be the cell receptor specificity (mediated perhaps by the nature of the carbohydrate attachment near NIm-IB), RNA structure providing hot spots for mutations, a reduced number of generations in particularly virulent pathogens, and the particular viral habitat.

Table III. *Structural Comparisons – Benchmarks and New Results*

Proteins		Number of equivalences	Percentage of equivalences		RMS (Å)
Molecule 1	Molecule 2		Molecule 1	Molecule 2	
Hbβ	Hbα	139	95	99	1.9
SBMV(C)	TBSV(C)	179	81	90	2.2
SBMV(A)	STNV	104	56	60	3.7
GAPDH(NAD)	LDH(NAD)	83	56	58	2.9
T4 lysozyme	HEW lysozyme	64	39	50	4.1
Viruses					
HRV14:					
VP1	VP3	168	62	71	3.8
VP2	VP3	152	60	65	2.9
VP1	VP2	124	45	49	3.2
VP3	SBMV(B)	136	58	69	2.6
VP2	SBMV(C)	134	52	62	2.6
VP1	SBMV(A)	123	45	57	3.4
VP1	STNV	90	33	49	3.0
VP1	ConA	61	22	26	3.0
VP3	ConA	87	38	37	3.5

Assembly

Assembly of picornaviruses [57–59] proceeds from 6S protomers of VP1, VP3 and VP0, via 14S pentamers of five 6S protomers, to mature virions. The final step involves inclusion of the RNA into empty capsids or partially assembled shells with simultaneous cleavage of VP0 into VP2 and VP4. Conversely, *in vitro* disassembly, produced by mild denaturation, proceeds via the expulsion of VP4 followed by the RNA [60].

Both the amino and carboxy ends of VP1 and $VP3_5$ are intertwined with each other as shown in Fig. 5. Furthermore, if VP4 and VP2 are considered as VP0, then VP0 is also intertwined with VP1 and $VP3_5$. This strongly suggests that the 6S protomer is as shown in Fig. 2(*c*). These protomers are themselves intertwined by virtue of the fivefold β-annulus formed by the amino ends of the VP3's and the proximity of the observed amino ends of VP4's to the fivefold axis. Thus, the 14S pentamers closely correlate with the observed structure, shown diagrammatically in Fig. 2(*c*). Such an assembly sequence matches that observed in plant viruses, in particular that of SBMV, where the building blocks are dimers corresponding to VP1 and $VP3_5$ and where the formation of intermediates with fivefold symmetry is considered to be a critical stage in the formation of $T = 1$ and $T = 3$ capsids [61].

The protein VP2, once cleaved from VP4, is globular and does not contact the other proteins extensively. There are large solvent accessible regions between VP2 and the surrounding proteins. This, as well as the extraordinarily internal heavy atom sites on VP2 (Table III), is consistent with the loose binding of VP2 to the capsid. (The ability of the heavy atoms to penetrate deeply into the shell suggests cautious interpretations of data based upon the use of small chemical labels for mapping viral surfaces.) Disruption of pentamer–pentamer contacts, mediated by a slight reorientation of VP2 or its complete removal, could provide a port by which the VP4 and RNA exit. Binding of a cell receptor to VP3 could facilitate this process, possibly accompanied by an isoelectric change.

The amino ends of the capsid proteins are invariably associated with the RNA in both plant and animal viruses. However, in TBSV, SBMV and STNV they are basic, whereas in HRV14 they are mildly acidic. These properties may be significant for the initial events of assembly.

Evolution

Conservation of three-dimensional structure is almost invariably greater than conservation of amino acid sequence homology [47, 48]. Thus structural comparison can be used to trace divergent evolution over longer time-spans than is possible by amino acid comparisons. Numerous structural comparisons have now been made and their probability for divergence assessed. These provide benchmarks to which other comparisons can be related. Thus, the considerable similarity of the VP1, VP2 and VP3 structures to those of the plant viruses leaves little doubt as to their divergence from a common ancestor. Possibly the corresponding gene for a protein such as concanavalin A provided a useful tool for the formation of biological assemblies that could attach themselves to animal cell receptors. Franssen *et al.* [63] and Argos *et al.* [64] also have shown that the virally coded polymerase, the genomic protein VPg and the protease of poliovirus are homologous in sequence and gene order to cowpea mosaic virus. These comparisons have been extended to alfalfa mosaic virus, brome mosaic virus, cucumber mosaic virus, tobacco mosaic virus, Sindbis virus and other viruses [65–68].

Three-dimensional superpositions of the Cα backbones of the β barrels of the available viral coat proteins and concanavalin A have been performed. The β barrels used were those of SBMV [20, 86], TBSV [19], STNV [29] and of VP1, VP2 and VP3 of HRV14 [this work] and of concanavalin A [69, 70]. The criteria used to assess similarity were [71, 72]:

(1) number and percentage of the residues in the barrel which could be equivalenced,

(2) RMS distance between equivalenced Cα coordinates and

(3) the minimum base change per codon for the structurally and topologically equivalenced amino acid residues.

The benchmarks provided by previous comparisons can be used to estimate the significance of the current equivalences. The striking degree of structural similarity between the α and β chains of hemoglobin was immediately suggestive [73] of evolutionary divergence of the corresponding genes from a common ancestor. The results of comparison [74] of the α and β chains of hemoglobin are shown as the first entry in Table III: 99% of the residues in the α chain have been successfully equivalenced with the β chain (90%), with an RMS distance of 1.9 Å between equivalenced Cα coordinates for the optimal superposition. Comparison [75] of the SBMV(C)

Table IV. *Structural comparisons – similarity matrix*

		SBMV	TBSV	HRV14 VP1	HRV14 VP2	HRV14 VP3	STNV	ConA
	SBMV	—	1.13	1.42	1.40	1.34	—	—
	TBSV	176	—	1.40	1.42	1.36	—	—
HRV14	VP1	123	—	—	1.27	1.34	1.48	1.43
HRV14	VP2	134	—	124	—	1.23	—	1.24
HRV14	VP3	136	—	168	152	—	—	1.33
	STNV	—	—	90	—	—		
	ConA	—	—	61	49	87		

Upper triangle of matrix gives minimum base change per codon. Bottom triangle gives number of equivalences.

subunit with the TBSV(C) subunit, also shown in Table III, indicates that 90% and 81% of the residues from SBMV and TBSV C subunits, respectively, have been successfully equivalenced, with an RMS distance of 2.2 Å between C_α in the equivalenced residues and a minimum base change/codon of 1.13 for the equivalenced amino acids. Thus the folding domains of the $T = 3$ plant viruses SBMV and TBSV exhibit a high degree of structural similarity, which correlates with the demands for successful packing of this domain in different quasiequivalent environments to form icosahedral shells. Indeed, the overall quaternary organization of the shell domains within the capsid is identical for these $T = 3$ plant viruses. The comparison [75] of the shell domain of the plant $T = 1$ RNA virus STNV with the SBMV A subunit indicates a significant, yet lesser degree of structural similarity: the percentage equivalences are 60% and 56% for STNV and SBMV A subunit, respectively, with an RMS Cα distance of 3.7 Å. Although the tertiary folds of the shell domains of SBMV and STNV are similar, the quaternary organization of the subunits in these $T = 3$ and $T = 1$ RNA plant viruses is different [75]. The NAD binding domains of GAPDH and LDH provide an example of structurally similar domains in different metabolic enzymes which provide a similar overall function, yet whose specificities are modulated by the rest of the enzyme as well as the particular amino acids which are essential for cofactor binding. The overall degree of similarity for the two NAD-binding domains is of the same order as for the STNV-SBMV comparison [47]. The final entry [76] in the table compares the structure of HEW lysozyme with phage T4 lysozyme, which show a low, yet still significant, degree of similarity.

The quantitative comparison results from the present study are shown in Table III and Table IV. Based on percentage of equivalenced residues and minimum base changes/codon, we see that

(1) the shell domains for TBSV and SBMV are the most similar subunits;

(2) VP1, VP2 and VP3 of HRV14 are almost as similar to each other as are the shell domains of SBMV and TBSV;

(3) VP1, VP2, and VP3 of HRV14 are slightly less similar to SBMV than to each other, but roughly half of the residues for each polypeptide chain can be successfully equivalenced;

(4) the coat protein of STNV and VP1 of HRV14 show the least degree of similarity for any pair of virus coat barrels compared;

(5) VP1 and VP3 of HRV14 show additional similarity in their N-termini; and

(6) portions of the barrel domain of concanavalin A can be successfully equivalenced with VP1, VP2, and VP3.

Although the animal virus HRV14, as well as other picornaviruses, has a nominal $T = 1$ icosahedral surface lattice, the quaternary arrangement and tertiary similarities of the major coat proteins VP1, VP2 and VP3 make the capsid structure resemble a $T = 3$ suface lattice, or a 'pseudo' $T = 3$ virus. The high degree of similarity between VP1, VP2 and VP3 of HRV14 and the coat proteins of the $T = 3$ plant viruses, SBMV and TBSV, can be viewed partially as satisfying the demands for packing in a $T = 3$ surface lattice. The picornavirus coat, with three distinct polypeptide chains, has more flexibility in terms of acceptable mutations (e.g. the wedge-shaped end of VP1 does not have to accommodate both icosahedral fivefold and quasi-sixfold environments). The differences in host characteristics for the small plant and animal RNA viruses are also reflected in the capsid structures of these viruses. Picornaviruses initiate the cycle of infection by binding to specific receptors embedded in the membranes of animal cells, and have the need to evade the immune system of the host. The elaborations on the surface of HRV14 relative to the $T = 3$ plant virus shells include protrusions which involve three of the four neutralizing immunogenic sites (NIm-IB is structurally analogous to a portion of the plant virus surface) and also provide the overall topography which allows for features such as the canyon. Most of the major differences among the HRV14 coat proteins map to insertions or deletions in loops connecting strands and helices in the β-barrel although several of the insertions occur in the middle of a β strand on the surface of the virion (VP2 puff and VP3 knob).

The coat proteins of the picornaviruses lack an analogue to the histone-like N-terminal arms of the $T = 3$ plant viruses and the inner protein surface facing the RNA is also not particularly basic. The RNA-binding and phosphate-neutralizing capability of the plant virus coat proteins has been functionally replaced by the spermidine in rhinoviruses, and Mg^{2+} in poliovirus [85]. The differences in the arrangement of the basic residues in plant and animal virus coat proteins may reflect differences in the modes and sequence of events in association of the coat proteins with RNA during assembly or disassembly or at the ribosome during translation of the polypeptide chain.

The similarity between the β-barrels of concanavalin A and the viral coat proteins may be significant in the context of receptor-binding. Concanavalin A, a plant lectin, has been shown [45, 46] to block attachment of picornaviruses to their cellular receptors. Whereas this ability of concanavalin A probably corresponds to a general sugar-binding function of the lectins, the evidence of the three-dimensional similarity of the lectin and virus barrels suggests that they may share similar strategies for binding cellular receptors, especially in the utilization of suger-binding as an initial guide for docking (cellular receptors for picornaviruses are glycosylated). In the light of the mitogenic effect that concanavalin A has been shown to exert on animal cells [77], it is also apparent that such a structure as concanavalin A is not only able to bind cellular receptors, but is also able to exert powerful regulatory effects on animal cells. By analogy, the sequence of events following picornavirus reception at the animal cell membrane results in powerful regulatory effects on the cell as exemplified by the nearly complete shutdown of host-cell protein synthesis [78]. Although the particular cascades which result from the mitogenic reception of concanavalin A and picornavirus reception are likely to be entirely different, the overall strategy of initiating the mitogenesis and infection may be similar. Thus not only do virus barrels and the concanavalin A barrel possess similar tertiary structures, but they may have similar functional capabilities. It is attractive to consider that the gene for the coat proteins of small RNA viruses and the gene for the concanavalin-A sugar-binding domain diverge from a common ancestor. This could explain the origin of viral coat proteins in that a gene for a protein able to bind the cell receptors could have been incorporated into the viral RNA. Such a borrowing could enable the viruses to enter through the 'front door' of the cell, not only to allow entry, but also the initiation of a regulatory cascade in the host cell.

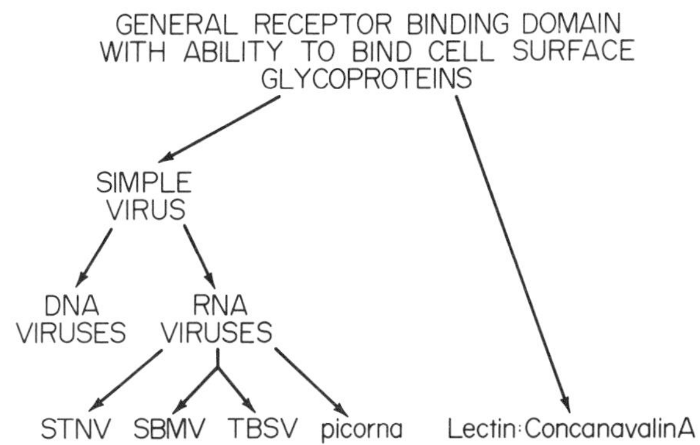

Fig. 4. Proposed evolutionary tree illustrating possible common ancestry of the coat proteins of small RNA viruses as well as receptor-binding proteins such as concanavalin A. Divergence from a common ancestor (although a detailed etiology is not being suggested) may explain some of the shared structural and functional characteristics of the gene products included in the scheme.

Figure 4 shows a possible scheme for the evolution of the coat proteins of plant and animal picornaviruses as well as concanavalin A from a common ancestor. Perhaps the gene for a primordial receptor-binding domain has been modified for the additional purpose of packaging RNA in virions. A series of gene duplications will have occurred on the pathway from $T = 3$ viruses to the 'pseudo' $T = 3$ picornaviruses in going from a single coat protein to three distinct viral coat proteins. In the light of the similarity between VP1 and VP3 N-termini, these two coat proteins may have diverged more recently from a situation in which only two distinct polypeptide chains were utilized. Alternatively, this feature may have arisen independently in either VP1 and VP3 and then be transferred to the other's coat protein gene by a picornaviral recombination process [79]. The possibility that the pseudo $T = 3$ picornaviruses are an ancestor of $T = 3$ viruses seems remote in that VP1, VP2 and VP3 have adapted to their particular packing environments and in doing so have probably relaxed their abilities to adapt to different quasi-symmetry environments.

As described previously, the proximity of Ser10 in VP2 to the N-terminal Asn69 of the VP4 chain is suggestive of an autoproteolytic cleavage of VP0 during the final stages of virion assembly. The activation of the serine to its nucleophilic alkoxide form (see Fig. 5) would conveniently coincide with the accessibility of a nucleotide base of the viral RNA – a perfect encapsidation signal. We would expect this apparent functionality to be conserved if this maturation mechanism applies to other picornaviruses in addition to HRV14; indeed,

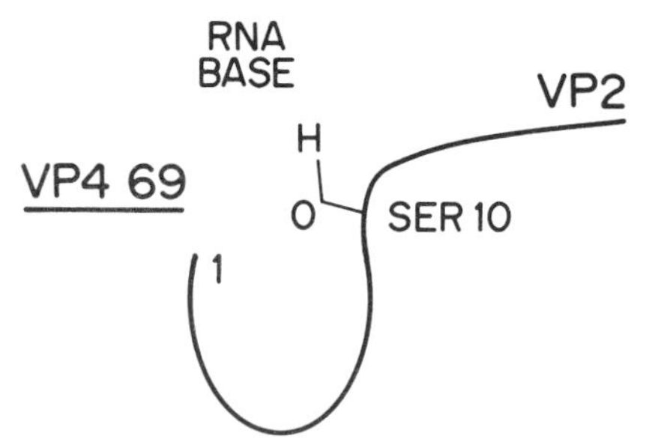

Fig. 5. Schematized representation of the proposed autoproteolytic cleavage of VP0 to produce VP2 and VP4. As the RNA encapsidation occurs during virion maturation, an RNA base could abstract a proton from ser10 of VP2 and initiate the proteolysis, which would share mechanistic similarities with the serine proteases.

serine is conserved in the homologous position on VP2 for HRV14, HRV2, poliovirus 1 (Mahoney), poliovirus 3 (Leon), poliovirus 2 (Sabin), coxsackie B3, and encephalomyocarditis virus, as shown in Table V. Although no serine is present in foot-and-mouth disease virus (FMDV) at the analogous bend in VP2, there is a set of two serines in all available FMDV sequences at positions 27 and 28 corresponding to a bend only one rung up in VP2 of HRV14. Thus this putative serine protease function is completely conserved among all picornaviruses for which sequence information is available, including the type member of each picornavirus class.

This hypothesis also provides a possible explanation for the observation [85] that mature picornavirions seem to contain at least a few uncleaved VP0 chains. As the virion maturation progresses, the RNA structure could be expected to become less flexible, and the likelihood of the proton abstraction from the serine becomes correspondingly less. When the vast majority of cleavages have already occurred, and the virion has nearly completely 'snapped shut', the nucleotide bases may not reach the remaining serines, and therefore a few VP0 chains may remain uncleaved. Alternatively, VPg could block access to a set of serines if the 3′-end of the RNA strand with its covalently attached VPg were close to the proposed site of VP0 cleavage (near the icosahedral threefold axis of symmetry).

Whereas this apparent conservation supports divergence of picornaviruses from a common ancestor which incorporated this function, it almost certainly represents convergent evolution to a catalytic mechanism which is similar in outline to other serine proteolytic mechanisms. Not only is the structural framework for this autodigestion completely different from other serine proteases, but even the identity of the activating base has changed from a histidine to a propitiously placed nucleotide base. Interestingly, the serine proteases have previously provided one of the clearest cases for convergent evolution. Subtilisin Novo and chymotrypsin were found to contain very similar active site geometry of the essential residues serine, histidine, and aspartic acid [80], yet the overall three-dimensional structures and folds of the two enzymes are totally different. Although the picornaviral autocatalytic mechanism is different from the serine protease enzyme mechanisms in detail, the use of the activated serine as the nucleophile is a recurrent theme, and reflects in part the limited number of selectively modulated nucleophiles among the twenty common amino acids. We may therefore expect that a number of successful proteases may have evolved independently in molecular and biological systems over time which capitalize on the use of a few amino acid functionalities which share only the outline of a catalytic mechanism. As our knowledge of enzyme structures expands, it is therefore anticipated that other active sites which have evolved independently, yet incorporating analogous mechanistic strategies, will be discovered. In this context the recognition by Kamer & Argos [65] of a conserved cysteine and histidine in many viral proteases including those of picornaviruses is especially relevant: they predict the proteolysis to share mechanistic, yet not structural, similarity with papain, a prototypical sulfhydryl protease. As the three-dimensional structures of viral proteases and other viral protein products becomes available, it will become possible to address such evolutionary questions more powerfully.

Table V. *Conservation of catalytic serine in VP2: portions of the aligned amino acid sequences for VP2 of representative picornaviruses*

	10
HRV14	C G Y S D R V L Q
HRV2	C G Y S D R I I Q
Polio 1 and 3	C G Y S D R V L Q
Polio 2s	C G Y S D R V M Q
Coxsackie B3	C G Y S D R V R S
EMC	E N L S D R V S Q
	28
FMDVA10	T T Q S S V G
FMDVA12	T T Q S S V G
FMDV01K	T T Q S S V G

Acknowledgements

This paper was adapted from one published in *Nature* [87]. We are grateful to Sharon Wilder, Kathy Shuster, Bill Boyle, Jun Tsao, Gale Rhodes, Diana Delatore and Tim Schmidt for help in data collection and presentation; to Keith Moffat, Wilfried Schildkamp, Robert Hunt, Don Bilderback, Aggie Sirrine and Boris Batterman at CHESS; to Hans Bartunik and Klaus Bartels at DESY; to Saul Rosen, John Steele, Tom Putnam and Paul Townsend at the Purdue University Computer Center; to Richard Colonno, Jeffrey Bolin, Abelardo Silva, Ignacio Fita, Celerino Abad-Zapatero, R. Usha, M. V. Hosur, Cynthia Stauffacher, Patrick Argos and J. K. Mohana Rao for helpful discussions. The work was supported by an NIH grant, an NSF grant for supercomputer time, a Purdue University Showalter Foundation grant and a contribution by Merck Sharp & Dohme Co. to M.G.R. and an ACS grant to R.R.R. CHESS is supported by an NSF grant and the macromolecular diffraction facility by an NIH grant. A postdoctoral Walter Winchell-Damon Ruyon fellowship supported E.A. and a predoctoral NIH training grant supported B.S.

References

1. Putnak, J. R. and Phillips, B. A., *Microbiol. Rev.* **45**, 287–315 (1981).
2. Sangar, D. V., *J. Gen. Virol.* **45**, 1 (1979).
3. Rueckert, R. R., *Comprehensive Virology* (ed. H. Fraenkel-Conrat, and R. R. Wagner), vol. 6, pp. 131. Plenum, New York (1976).
4. Rueckert, R. R., Dunker, A. K. and Stoltzfus, C. M., *Proc. Natl. Acad. Sci. USA* **62** 912 (1969).
5. Jacobson, M. F., Asso, J. and Baltimore, D., *J. Mol. Biol.* **49**, 657–669 (1970).
6. Pallansch, M. A., Kew, O. M., Semler, B. L., Omilianowki, D. R., Anderson, C. W., Wimmer, E. and Rueckert, R. R., *J. Virol.* **49**, 873 (1984).

7. Hanecak, R., Semler, B. L., Anderson, C. W. and Wimmer, E., *Proc. Natl. Acad. Sci. USA* **79**, 3973 (1982).
8. Longberg-Holm, K. and Butterworth, B. E., *Virology* **71**, 207 (1976).
9. Beneke, T. W., Habermehl, K. O., Diefenthal, W. and Buchholz, M., *J. Gen. Virol.* **34**, 387 (1977).
10. Hordern, J. S., Leonard, J. D. and Scraba, D. G., *Virology* **97**, 131 (1979).
11. Wetz, K. and Habermehl, K. O., *J. Gen. Virol.* **59**, 397 (1982).
12. Stanway, G., Hughes, P. J., Mountford, R. C., Minor, P. D. and Almond, J. W., *Nucl. Acids Res.* **12**, 7859 (1984).
13. Callahan, P. L., Mizutani, S. and Colonno, R. J., *Proc. Natl. Acad. Sci. USA* **82**, 732 (1985).
14. Skern, T., Sommergruber, W., Blaas, D., Gruendler, P., Fraundorfer, F., Pieler, C., Fogy, I. and Kuechler, E., *Nucl. Acids Res.* **13**, 2111 (1985).
15. Mattern, C. F. T. and duBuy, H. G., *Science* **123**, 1037 (1956)
16. Schaffer, F. L. and Schwerdt, C. E., *Proc. Natl. Acad. Sci. USA* **41**, 1020 (1955).
17. Finch, J. T. and Klug, A. *Nature (London)* **183**, 1709 (1959).
18. Korant, B. D. & Stasny, J. T. *Virology* **55**, 410 (1973).
19. Harrison, S. C., Olson, A. J., Schutt, C. E., Winkler, F. K. and Bricogne, G., *Nature (London)*, **276**, 368 (1978).
20. Abad-Zapatero, C., Abdel-Meguid, S. S., Johnson, J. E., Leslie, A. G. W., Rayment, I., Rossmann, M. G., Suck, D. and Tsukihara, T. *Nature (London)*, **286**, 33 (1980).
21. Liljas, L., Unge, T., Jones, T. A., Fridborg, K., Lövgren, S., Skoglund, U. and Strandberg, B., *J. Mol. Biol.* **159**, 93–108 (1982).
22. Hogle, J. M., *J. Mol. Biol.* **160**, 663 (1982).
23. Arnold, E., Erickson, J. W., Fout, G. S., Frankenberger, E. A., Hecht, H. J., Luo, M., Rossmann, M. G. and Rueckert, R. R., *J. Mol. Biol.* **177**, 417 (1984).
24. Jones, T. A., *J. Appl. Crystallogr.* **11**, 268 (1978).
25. Rossmann, M. G. & Blow, D. M., *Acta Crystallogr.* **15**, 24 (1962).
26. Rossmann, M. G. & Blow, D. M., *Acta Crystallogr.* **16**, 39 (1963).
27. Johnson, J. E., Akimoto, T., Suck, D., Rayment, I. and Rossmann, M. G., *Virology* **75**, 394 (1976).
28. Rayment, I., Baker, T. S., Caspar, D. L. D. and Murakami, W. T., *Nature (London)* **295**, 110 (1982).
29. Liljas, L., Unge, T., Jones, T. A., Fridborg, K. Lövgren, S., Skoglund, U. & Strandberg, B., *J. Mol. Biol.* **159**, 93–108 (1982).
30. Gaykema, W. P. J., Hol, W. G. J., Vereijken, J. M., Soeter, N. M., Bak, H. J. and Beintema, J. J., *Nature (London)* **309**, 23 (1984).
31. Sherry, B. and Rueckert, R., *J. Virol.* **53**, 137 (1985).
32. Pfaff, E., Mussgay, M., Böhm, H. O., Schulz, G. E. and Schaller, H., *EMBO J.* **1**, 869 (1982).
33. Icenogle, J., Shiwen, H., Duke, G., Gilbert, S., Rueckert, R. and Anderegg, J., *Virology* **127**, 412 (1983).
34. Mandel, B., *Virology* **69**, 500 (1976).
35. Emini, E. A., Jameson, B. A. and Wimmer, E., *Nature (London)* **304**, 699 (1983).
36. Emini, E. A., Ostapchuk, P. and Wimmer, E., *J. Virol.* **48**, 547 (1983).
37. Sherry, B., Mosser, A. G., Colonno, R. J. and Rueckert, R. R., *J. Virol.* **57**, 246 (1986).
38. Meloen, R. H., Briarie, J., Woortmeyer, R. J. and Van Zaane, D., *J. Gen. Virol.* **64**, 1193 (1983).
39. Mandel, B., *Comprehensive Virology* (ed H. Fraenkel-Conrat and R. R. Wagner), vol. 15, 37 (Plenum, New York, 1979).
40. Toyoda, H., Kohara, M., Kataoka, Y., Suganuma, T., Omata, T., Imura, N. and Nomoto, A., *J. Mol. Biol.* **174**, 561 (1984).
41. Makoff, A. J., Paynter, C. A., Rowlands, D. J. and Boothroyd, J. C., *Nucl. Acids Res.* **10**, 8285 (1982).
42. Beck, E., Feil, G. & Strohmaier, K., *EMBO J.* **2**, 555 (1983).
43. Rossmann, M. G., Abad-Zapatero, C., Hermodson, M. A. and Erickson, J. W., *J. Mol. Biol.* **166**, 37 (1983).
44. Erickson, J. W., Silva, A. M., Murthy, M. R. N., Fita, I. and Rossmann, M. G., *Science*, **229**, 625 (1985).
45. Krah, D. L. and Crowell, R. L., *J. Virol.* **53**, 867 (1985).
46. Lonberg-Holm, K., *J. Gen. Virol.* **28**, 313 (1975).
47. Rossmann, M. B., Liljas, A., Brändén, C. I. & Banaszak, L. J., *The Enzymes* (ed, P. D. Boyer), 3rd edn., Vol XI, 61. Academic Press, New York (1975).
48. Wierenga, R. K., De Maeyer, M. C. H. and Hol, W. G. J. *Biochemistry* **24**, 1346 (1985).
49. Abraham, G. and Colonno, R. J., *J. Virol.* **51**, 340 (1984).
50. Cavanagh, D., Sangar, D. V., Rowlands, D. J. and Brown, F., *J. Gen. Virol.* **35**, 149 (1977).
51. Strohmaier, K., Franze, R. and Adam, K. H., *J. Gen. Virol.* **59**, 295 (1982).
52. Baxt, B., Morgan, D. O., Robertson, B. H. and Timpone, C. A., *J. Virol.* **51**, 298 (1984).
53. Argos, P., Tsukihara, T. and Rossmann, M. G., *J. Mol. Evol.* **15**, 169 (1980).
54. Wychowski, C., van der Werf, S., Siffert, O., Crainic, R., Bruneau, P. and Girard, M., *EMBO J.* **2**, 2019 (1983).
55. Evans, D. M. A., Minor, P. D., Schild, G. S. and Almond, J. W., *Nature (London)* **304**, 459 (1983).
56. Minor, P. D., Schild, G. C., Bootman, J., Evans, D. M. A., Ferguson, M., Reeve, P., Spitz, M., Stanway, G., Cann, A. J., Hauptmann, R., Clarke, L. D., Mountford, R. C. and Almond, J. W. *Nature (London)* **301**, 674 (1983).
57. McGregor, S. & Rueckert, R. R., *J. Virol.* **21**, 548 (1977).
58. Fernandez-Tomas, C. B., Guttman, N. & Baltimore, D., *J. Virol.* **12**, 1181 (1973).
59. Jacobson, M. F. & Baltimore, D., *J. Mol. Biol.* **33**, 369 (1968).
60. Fernandez-Tomas, C. B. and Baltimore, D., *J. Virol.* **12**, 1122 (1973).
61. Rossmann, M. G., Abad-Zapatero, C., Hermodson, M. A. and Erickson, J. W., *J. Mol. Biol.* **166**, 37–83 (1983).
62. Emini, E. A., Jameson, B. A. and Wimmer, E., *Modern Approaches to Vaccines* (ed. R. M. Chanock and R. A., Lerner, p. 65. Cold Spring Harbor Laboratory, Cold Spring Harbor (1984).
63. Franssen, H., Leunissen, J., Goldbach, R., Lomonossoff, G. and Zimmern, D., *EMBO J.* **3**, 855 (1984).
64. Argos, P., Kamer, G., Nicklin, M. J. H. & Wimmer, E., *Nucl. Acids Res.* **12**, 7251 (1984).
65. Kamer, G. and Argos, P., *Nucl. Acids Res.* **12**, 7269 (1984).
66. Haseloff, J., Goelet, P., Zimmern, D., Ahlquist, P., Dasgupta, R. and Kaesberg, P., *Proc, Natl. Acad. Sci. USA* **81**, 4358 (1984).
67. Ahlquist, P., Strauss, E. G., Rice, C. M., Strauss, J. H., Haseloff, J. and Zimmern, D., *J. Virol.* **53**, 536 (1985).
68. Rezaian, M. A., Williams, R. H. V., Gordon, K. H. J., Gould, A. R. and Symons, R. H., *Eur. J. Biochem.* **143**, 277 (1984).
69. Hardman, K. D. and Ainsworth, C. F., *Biochemistry* **11**, 4910 (1972).
70. Reeke, G., Jr., Becker, J. W. and Edelman, G. M., *J. Biol. Chem.* **250**, 1525 (1975).
71. Rossmann, M. G. and Argos, P., *J. Mol. Biol* **109**, 99 (1977).
72. Rao, S. T. and Rossmann, M. G., *J. Mol. Biol.* **76**, 241 (1973).
73. Perutz, M. F., Rossmann, M. G., Cullis, A., Muirhead, H., Will, G. and North, A. C. T., *Nature (London)* **185**, 416 (1960).
74. Argos, P. and Rossmann, M. G., *Biochemistry* **18**, 4951 (1979).
75. Rossmann, M. G., Abad-Zapatero, C., Murthy, M. R. N., Liljas, L., Jones, T. A. and Strandberg, B., *J. Mol. Biol.* **165**, 711 (1983).
76. Rossmann, M. G. and Argos, P., *J. Mol. Biol.* **105**, 75 (1976).
77. Barondes, S. H., *Annu. Rev. Biochem.* **50**, 207 (1981).
78. Ehrenfeld, E., *Comprehensive Virology* (ed. H. Fraenkel-Conrat and R. R. Wagner), vol. 18, 177. Plenum, New York (1984).
79. Romanova, L. I., Tolskaya, E. A., Kolesnikova, M. S. and Agol, V. I., *FEBS Lett.* **118**, 109 (1980).
80. Kraut, J. Robertus, J. D., Birktoft, J. J., Alden, R. A., Wilcox, P. E., and Powers, J. C., *Cold Spring Harbor Symp. Quant. Biol.* **30**, 117 (1971).
81. Chow, M., Yabrov, R., Bittle, J., Hagle, J. & Baltimore, D., *Proc. Natl. Acad. Sci. USA* **82**, 910 (1985).
82. vanderWerf, S., Wychowski, C., Brunearu, P., Blondel, B., Crainic, R., Horodniceanu, F. and Girard, M., *Proc. Natl. Acad. Sci. USA* **80**, 5080 (1983).
83. Bittle, J. L., Houghten, R. A. Alexander, H., Shinnick, T. M., Sutcliffe, J. G, Lerner, R. A., Rowlands, D. J. and Brown, F., *Nature (London)* **298**, 30 (1982).
84. Robertson, B. H., Morgan, D. O. and Moore, D. M., *Virus Res.* **1**, 489 (1984).
85. Rueckert, R. R., *Virology* (ed. B. N. Fields *et al.*), p. 705. Raven Press, N.Y. (1985).
86. Silva, A. M. and Rossmann, M. G., *Acta Crystallogr.* **B41**, 147 (1985).
87. Rossmann, M. G., Arnold, E., Erickson, J. W., Frankenberger, E. A., Griffith, J. P., Hecht, H. J., Johnson, J. E., Kamer, G., Luo, M., Mosser, A. G., Rueckert, R. R., Sherry, B. and Vriend, G., *Nature (London)*, **317**, 145 (1985).

Chemica Scripta 1986, **26B**, 325–335

Genetic Variability and Evolution of Adenoviruses

Göran Wadell, Annica Allard, Magnus Evander, Li Quan-gen

Department of Virology, University of Umeå, S-901 85 Umeå, Sweden

Received November 1985

Paper presented by Göran Wadell at the Conference on 'Molecular Evolution of Life', Lidingö, Sweden, 8–12 September 1985

Abstract

The Adenovirus family consists of 2 genera – Aviadenovirus and Mastadenovirus.

Adenoviruses of each studied host species can be divided into subgenera – six among human adenoviruses – that show a low degree of DNA homology. They can therefore function as genetically closed canons. Within each subgenus frequent recombination events will generate the numerous serotypes and genome types that are characteristic for adenoviruses.

The subgenera may have evolved early. Consequently adenovirus types of subgenera of different host species could be related, i.e. the genetic homology between a given adenovirus type and an adenovirus type with a corresponding subgenus of a different host species should be higher than between adenovirus types belonging to different subgenera of the same host species.

One among all adenovirus types of each host species should be expected to display a broader genetic relatedness than the others and be expected to be related to the archetypes of adenoviruses. This serotype corresponds to Ad4 (Subgenus E) among the 41 human adenoviruses.

The common architecture and the common complex genome organization of adenoviruses of different host species are taken as evidence for a common origin of adenoviruses.

Introduction

Adenoviruses have been detected in all host species studied [1]. They are characterized by being strictly host-cell specific and furthermore by a wide expressed genetic variability. Different isolates (strains) of viruses in general are defined and registered as distinct serotypes. This is a useful concept but of limited information as regards genetic variability within the serotype and on the more or less remote relation to other adenoviruses. To introduce useful concepts the following taxonomic terminology has been suggested for adenoviruses:

Genus. Members of each genus share common epitopes on the hexons.

Subgenus. Each subgenus is defined by a DNA homology of more than 50% between members within a subgenus and less than 20% betweeen members of different subgenera.

Serotype. A quantitative neutralization with hyperimmune sera defines a serotype. The ratio of homologous to heterologous neutralization titre must be greater than 16.

Recombinant. This designation should be used only when the two parent genomes have been identified.

Evolutionary variant. This term is used when the genetic alteration was generated via insertion or intragenomic recombination in progeny of the same strain.

Genome type. A distinct viral entity within a serotype identified by DNA restriction site patterns when their generation as recombinants or evolutionary variants could not be ascertained.

Strain. The progeny of each wild type isolate is designated strain.

Morphological and biological properties of the virus capsid

Forty-one different serotypes of human adenoviruses have been identified [2]. They can be grouped into six different subgenera on the basis of a number of biophysical and biological characteristics [3–6] (Table I). All adenoviruses have a common architecture (Fig. 1). The virion is composed of 252 capsomers, arranged in icosahedral symmetry, that form a capsid of 73 nm in diameter. Of the capsomers 240 are arranged in a hexagonal symmetry. They are hence designated hexons. Twelve capsomers, penton bases, are placed in a pentagonal symmetry at each vertex. An antenna-like fibre protrudes from each of the 12 vertices and the tip of the fibre is responsible for binding of the virion to the host cell. The fibre of members of different subgenera differ distinctly in length from 10 nm for fibres of adenoviruses of subgenus B to 30 nm for fibres of subgenus C [7]. It is expected that the structure of the fibre is one of the dominant properties that determines the characteristic differences in tropisms between adenoviruses of different subgenera.

Haemagglutination by adenoviruses was first demonstrated by Rosen [8]. All adenovirus serotypes can agglutinate erythrocytes. Members of different subgenera can be distinguished by differential agglutination of erythrocytes from rat, monkey and other species. Haemagglutination can serve as model for studies of the fibre as an attachment unit.

Analysis of the externally localized structural proteins, hexons, penton bases and the fibres, revealed characteristics serotype specific epitopes on hexons and different serotype-specific epitopes on the tip of the fibre. An axis from the interior to the exterior of the adenovirus particle extending from a pronounced cross-reactivity between epitopes of different types to a distinctly serotype specific reactivity has been established. This implies that analyses of the cross-reactivity between adenovirus types as regards hexons, penton bases and the fibres can be used to divide adenoviruses into different subgenera [9, 10].

The variability expressed among structural proteins of the adenovirus particles

Internal structural proteins would be expected to be evolutionary conserved in analogy with histones. On the other hand externally localized structural proteins subjected to selective pressure by the immune response should be expected to vary.

Table I. *Properties of human adenovirus serotypes of subgenera A to F (modified from Wadell,* 1984)

Sub-genus	Serotype	DNA Homology (%) Intra-generic	Inter-generic	G+C (%)	Number of *Sma* I[a] fragments	Apparent molecular mass of the major internal polypeptides (kDa) V	VI	VII	Haema-glutination pattern[b]	Length of fibres (nm)	Oncogenicity in newborn hamsters	Tropism symptoms
A	12, 18, 31	48–69	8–20	48	4–5	51.0–51.5 46.5–48.5[c]	25.5–26.0	18	IV	28–31	High (tumours in most animals in 4 months)	Cryptic enteric infection
B:1 B:2	3, 7, 16, 21 14[d], 11, 34 35	89–94	9–20	51	8–10	53.5–54–5	24	18	I	9–11	Weak (tumours in few animals in 4–18 months)	Respiratory disease Persistent infections of the kidney
C	1, 2, 5, 6	99–100	10–16	58	10–12	48.5	24	18.5	III	23–31	nil	Respiratory disease persistent in lymphoid tiss
D	8, 9, 10, 13 15, 17, 19, 20, 22–30, 32, 33, 36, 37, 38, 39[e]	94–99	4–17	58	14–18	50.0–50.5[f]	23.2	18.2	II	12–13	nil	Kerator conjunctivitis
E	4	—	4–23	58	16–19	48	24.5	18	III	17	nil	Conjunctivitis respiratory disease
F	40, 41	62–69	15–22	52	9–12	46.0–48.5	25.5	17.5	IV	28–33	nil	Infantile diarrhoea

[a] The restricted DNA fragments were analyzed on 0.8–1.2% agarose slab gels. DNA fragments smaller than 400 pb were not resolved.
[b] *I*, Complete agglutination of monkey erythrocytes; II, complete agglutination of rat erythrocytes; III, partial agglutination of rat erythrocytes (fewer receptors); IV, agglutination of rat erythrocytes discernible only after addition of heterotypic antisera.
[c] Polypeptide V of Ad31 was a single band of 48 kDa.
[d] Members of subgenus B are divided into two clusters of DNA homology based on pronounced differences in DNA restriction sites.
[e] Only DNA restriction and polypeptide analysis have been performed with Ad32–39.
[f] Polypeptides V and VI of Ad8 showed apparent molecular mass of 45 and 22 kDa, respectively. Polypeptide V of Ad30 showed an apparent molecular mass of 48.5 kDa.

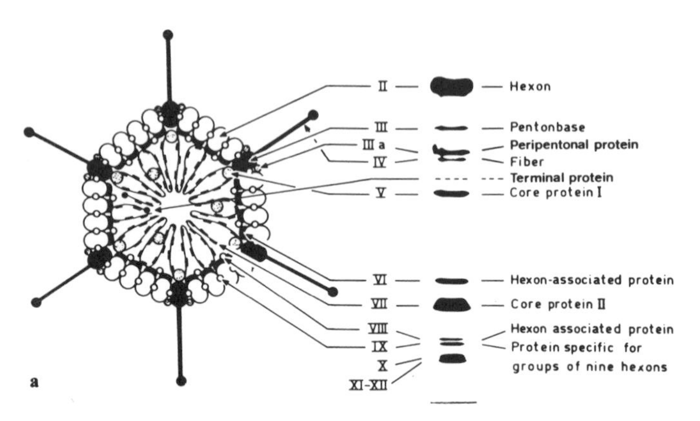

Fig. 1. Schematic presentation of the polypeptide composition of adenovirus 2 virions (reprinted from Philipson, L., *Curr. Top. Microbiol. Immunol.* **109**, 1, 1983).

An analysis of SDS polyacrylamide electrophoresis patterns of the 41 human adenovirus prototypes was therefore performed. It was noted that the size of the internal structural polypeptides V, VI, VII, VIII did not vary between serotypes being members of the same subgenus, whereas distinctly different sizes were noted in comparison between members of different subgenera (Fig. 2). It was consequently concluded that analyses of the polypeptide pattern could be used to classify the adenovirus serotypes into subgenera. Both human adenoviruses [5, 11] and simian adenoviruses [12] have been classified in this way.

Ultimate comparative analyses of viruses in general should be based on nucleotide sequence differences

The DNA homology of adenoviruses has been studied by filter hybridization [13] heteroduplex mapping [14] and liquid hybridization [4]. This classification was in agreement with data obtained by classification on the molecular weight of the internal structural proteins of the virion.

Human adenoviruses are also characterized by distinct differences in GC-content in their DNA: subgenus A (47–49%), subgenus B (49–52%), subgenus F (52%) and subgenus C, D and E (57–59%) [4, 15]. In order to exploit this obvious variation in nucleotide composition the DNA of all adenovirus prototypes was digested by restriction endonuclease *Sma* I which cleaves the DNA at 5′ CCC GGG (Fig. 3). Two principles were documented: (*a*) the limited range of number of *Sma* I restriction fragments of adenovirus DNA that was characteristic for each subgenus; (*b*) pairwise comparisons of the restriction patterns of DNA from two members of the same subgenus revealed that in general more than 50% of the fragments co-migrated [5]. This rule is not valid for subgenera A, B and F.

All adenoviruses of different subgenera studied display a

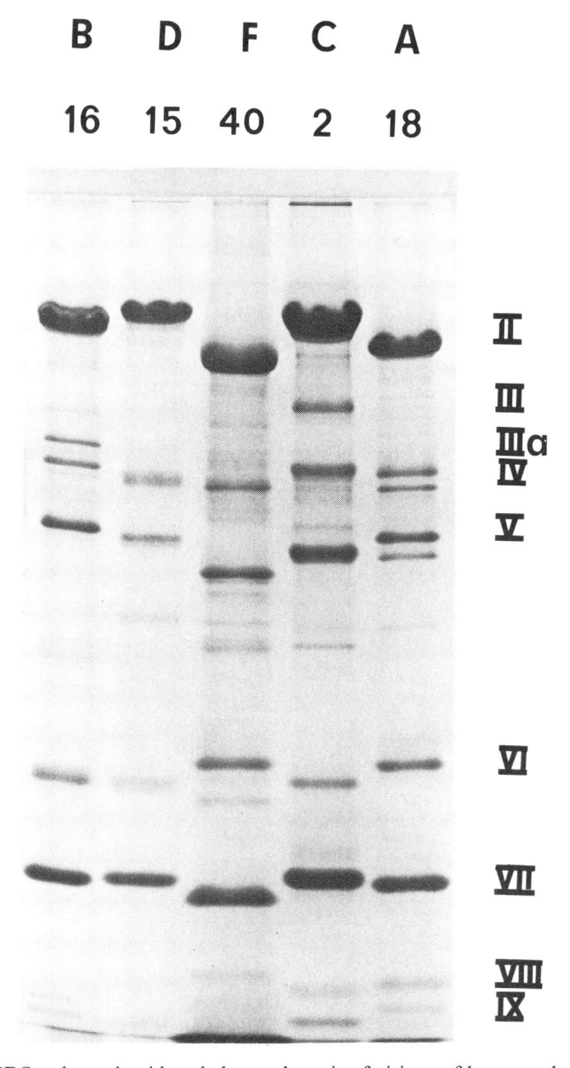

Fig. 2. SDS-polyacrylamide gel electrophoresis of virions of human adenoviruses Ad16 (Subgenus B), Ad15 (Subgenus D), Ad40 (Subgenus F), Ad2 (Subgenus C) and Ad18 (Subgenus A).

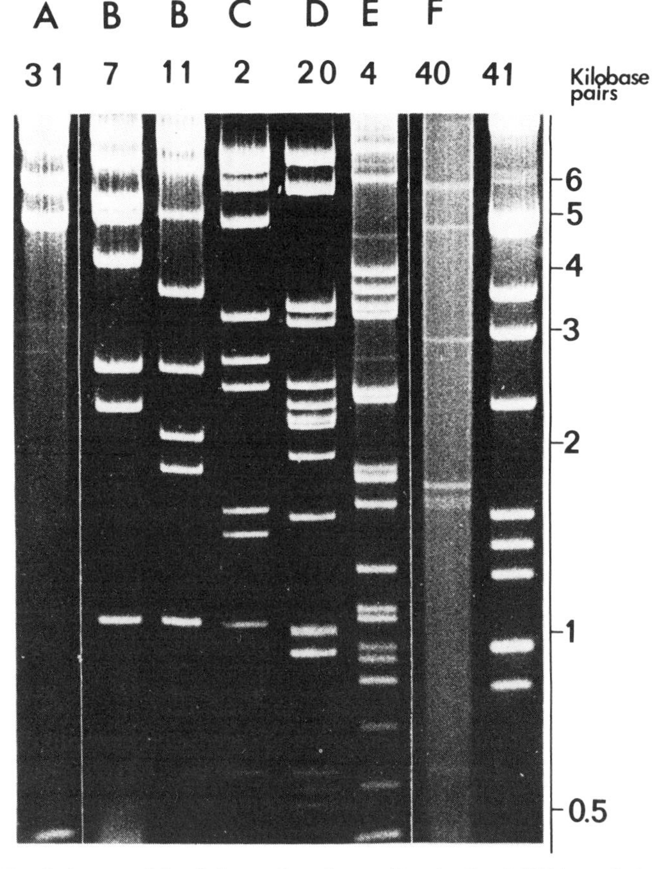

Fig. 3. Agarose slab gel electrophoresis revealing the *Sma* I DNA restriction patterns obtained after digestion of DNA from adenovirus serotypes representing subgenera A through F (reprinted from Wadell, G., *Curr. Top. Microbiol. Immunol.* **110**, 191, 1984).

similar organization of the viral genome. The DNA homology between serotypes of different subgenera is only 4–23%. The DNA homology between members of the same subgenera is above 48% [4]. This means that recombination between adenoviruses of different subgenera must be expected to be highly infrequent. Subgenera can thus be expected to serve as recombination barriers.

In view of the general similarity of the organization of the adenovirus genome between different subgenera, adenoviruses serve as an interesting model for oncogenicity in newborn hamsters. [16]. Adenoviruses of subgenus A induce tumours in most animals within 2 months. Adenoviruses of subgenera B and E induce tumours in few animals after 14–18 months. Adenoviruses of subgenera C, D and F do not induce tumours in newborn hamsters, whereas adenoviruses of subgenera C, D and F can transform rat cells *in vitro*. Ad40 and Ad41 of subgenus F cannot grow in established cell lines that support the growth of all other human adenoviruses [2, 6]. This property is a function of the organization of early region E1A and E1B. These regions code for enhancers and oncogene like proteins which may act in consort with proteins of early region E3 coding for membrane proteins. They have been demonstrated to associate with the heavy chain of MHC class 1 antigen *in lieu* of β-2 microglobulin. This results in impaired glycosidation of MHC class 1 antigen and reduced transportation to the cell membrane [17]. Furthermore, E4 gene products may together with E1B proteins shut off the expression of the host cell [18].

Subgenus A

Generally the adenovirus subgenera can be described as canons of genetically closely related members [19]. However, Ad12, Ad18 and Ad31 are distinctly different [Fig. 4]. The DNA homology is no more than 48–69% in pairwise comparison [4]. The structure of the tip of the fibre and the hexon of these three serotypes can be expected to be largely similar since haemagglutination–inhibition and neutralization epitopes are shared between the members of subgenus A.

Subgenus B

The DNA homology determined by liquid hybridization between the different members of subgenus B is above 89% [4]. However, within subgenus B two distinctly different clusters of adenovirus genomes can be discerned by using a more sensitive pairwise comparison of DNA restriction fragment co-migration (Fig. 4). The frequency of co-migration was noted to be 49–82% for members within the cluster and 11–22% in pairwise comparison of adenoviruses belonging to the two different DNA homology clusters (Table II). Distinctly different biological features are characteristic for the two DNA homology clusters.

B:1:Ad3, Ad7 and Ad21 cause outbreaks of respiratory disease of significant medical importance. These serotypes account for 33% of all adenovirus isolates typed and reported to WHO [20]. Severe manifestations in small children including

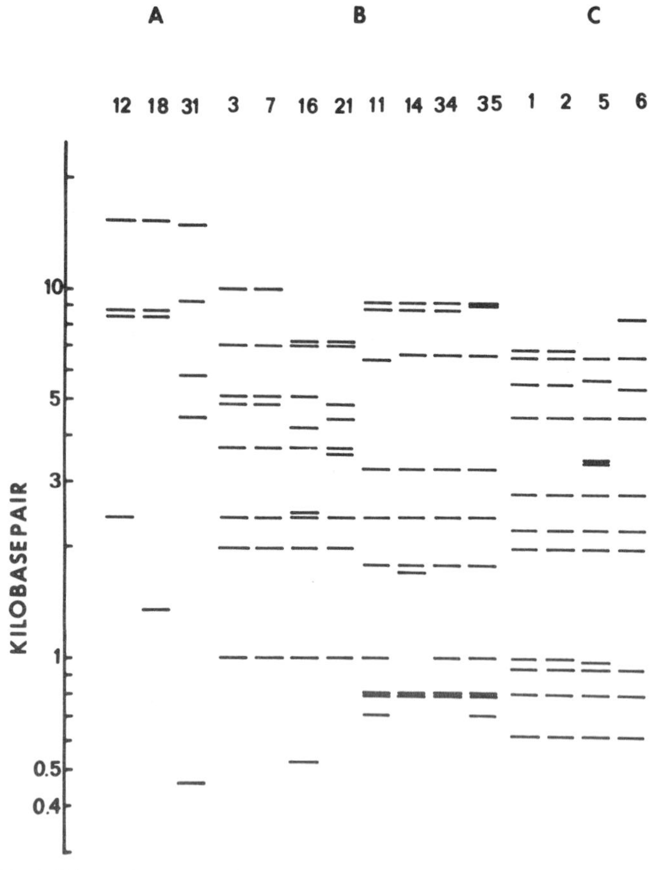

Fig. 4. Schematic presentation of the *Sma* I DNA restriction patterns of all adenovirus prototypes belonging to subgenera A, B and C.

Table II. *Pairwise comparison of the percentage of co-migrating DNA restriction fragments of human adenoviruses belonging to subgenus B*[a]

	3	7	16	21	11	14	34	35
3		75	65	55	15	12	18	17
7			55	49	18	17	22	20
16				56	17	11	15	15
21					22	22	21	21
11						57	74	80
14							62	65
34								82
35								

[a] This information is based on analysis by restriction endonucleases *Bam* HI, *Bgl* I, *Bgl* II, *Eco* RI, *Hind* III, *Hpa* I, *Sal* I, *Sma* I, *Xba* I and *Xho* I.

residual lung damage and bronchioectasis have been reported from follow-up studies.

B:2:AD11, Ad14, Ad34, Ad35 display a relative number of co-migrating DNA restriction fragments in pairwise comparison with members of B:1 being equal to or lower than 22%. The corresponding relation between Ad14 and the other B:2 members was 57–65% whereas pairwise comparisons between Ad11, Ad34 and Ad35 revealed a higher homology of 74–82%. Ad14 can cause respiratory outbreaks with the same clinical features as those caused by Ad7 and Ad21. Ad11, Ad34 and Ad35 are closely related on the genome level and are also characterized by their propensity to infect the urinary tract [21]. Furthermore, a kidney from an Ad11 seropositive donor has transmitted Ad11 infection to a seronegative kidney transplant recipient [22].

Ad11/34/35 are highly significantly over-represented in urine specimens from bone-marrow transplant recipients [23] and AIDS patients [24]. They can cause persistent infection. Ad11 was demonstrated to be shed to urine for 5 months in a healthy woman during the pregnancy and after delivery (Gardner, personal communication). All adenovirus serotypes of subgenus B can agglutinate monkey erythrocytes at +37 °C. Adenoviruses of DNA homology cluster B:1 are eluted from monkey erythrocytes at +4 °C, whereas Ad11 of DNA homology cluster B:2 still agglutinates monkey erythrocytes at this temperature indicating interaction with a distinctly different cell receptor [25].

The serotype and the genome type concept. The definition of a serotype relies on the distinct epitopes that are capable of inducing neutralizing antibodies. These are the mediators of host protection against reinfection. The gene product carrying these epitopes represent only a few per cent of the viral genome. The genetic variation within a serotype can be evaluated by analyses with DNA restriction endonucleases enabling identification of distinct viral entities designated genome types. Ad7 is particularly instructive in this context since this is the adenovirus type that has been most frequently associated with disease. The parallel use of DNA restriction endonucleases *Bam* HI and *Sma* I has enabled identification of 41 different adenovirus prototypes. These enzymes have been used in the analyses of 314 adenovirus 7 strains obtained from Africa, Asia, Australia, Europe, North and Latin America since 1958. By this approach seven Ad7 genome types (Ad7p, Ad7a, Ad7b, Ad7c, Ad7d, Ad7e and Ad7f) were identified [26] (Fig. 5).

An in-depth analysis of 26 additional adenovirus 7 strains by use of 12 different DNA restriction endonucleases revealed five additional adenovirus 7 genome types. All strains were recovered in Beijing, China, since 1958. Analysis of the pairwise DNA restriction co-migration based on 165-171 DNA fragments from the Ad7 isolates indicated a homology

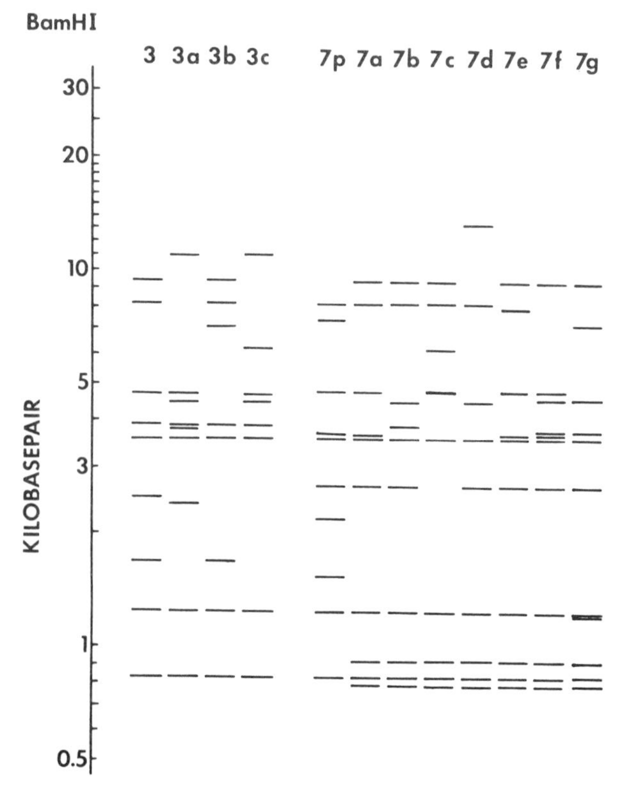

Fig. 5. Schematic presentation of *Bam* HI DNA restriction patterns of four genome types of Ad3 and seven genome types of Ad7.

Table III. *Pairwise comparison of the percentage[a] of co-migrating fragments of* 15 *genome types of Ad*7 *obtained after cleavage with* 12 *restriction endonucleases.*

	pl	a	a1	a2	a3	a4	a5	b	c	d	d1	e	f	g
p	93	78	74	71	72	74	72	72	73	74	68	75	72	73
pl		75	71	68	70	71	70	69	71	72	66	78	75	71
a			91	87	88	89	90	86	90	88	83	94	88	75
a1				94	97	99	98	95	98	94	89	91	88	78
a2					97	93	93	87	91	88	84	86	83	74
a3						96	95	92	95	91	86	88	85	76
a4							97	93	97	93	87	90	87	78
a5								93	96	92	90	89	86	76
b									94	96	92	87	92	80
c										93	88	91	88	77
d											95	89	88	79
d1												83	84	74
e													89	77
f														76

[a] 165–171 restriction fragments were compared in each pair of the genome types.

from 66 up to 99% (Table III). The corresponding values in pairwise comparison of Ad3, Ad7 and Ad16 prototypes was 55–75%. This indicates that the genetic variability within a serotype and between serotypes of the same subgenus can be of the same order.

The distribution of the different genome types in different parts of the world may shed light on the origin of the expressed genetic variability of adenoviruses (Fig. 6).

Africa. Ad7c dominated.

Australia. The distribution of Ad7 genome types was similar to that in Europe. However, the shift from Ad7c to Ad7b did not take place until 1975.

Brazil. Ad7e was the only genome type detected.

China. A detailed analysis revealed seven different genome types. The Ad7d and Ad7g genome types were unique to China. Ad7d dominated during the epidemic 1981–4. At the end of the epidemic a new genome type Ad7d1 that was closely related to Ad7d appeared. The Ad7a$_1$ and Ad7a$_4$ genome types isolated from autopsies are slightly different from the 7a genome type.

Europe. More than 90% of all isolates were genome typed as Ad7b or Ad7c. These types appear to be mutually exclusive – a shift from Ad7c to Ad7b occurred in 1969.

Japan. Ad7p was the only genome type detected.

USA. Ad7b predominated among the Ad7 strains recovered since 1962 [26].

Japan is one of the regions which has the highest detected prevalence of antibodies against Ad7. During the observation period 1966–79 Japan has had the lowest registered incidence of Ad7 isolates (Fig. 7). This is interpreted so that the Ad7p genome type(s) circulating in Japan can be of low virulence but widely spread and effectively immunizing the population.

Similar analysis has been performed with the closely related Ad3. Two genome types Ad3p and Ad3a predominated. The Ad3 prototype was identified in Australia, Brazil and Europe whereas the Ad3a genome type was found in China, Japan and the United States. No mutually exclusive shift in genome types was detected among adenovirus 3 strains analysed during a 20-year period in Europe [6].

Subgenus C

Ad1, Ad2, Ad5 and Ad6 display a DNA homology of 98% in pairwise comparisons [4]. They represent 59% of all adenovirus isolates reported to WHO [20]. However, they do not have a corresponding share of the adenovirus associated illness, since they persist for years in lymphoid tissue and can be intermittently shed into stools. This situation and the observation that one fifth of all progeny of reinfection with

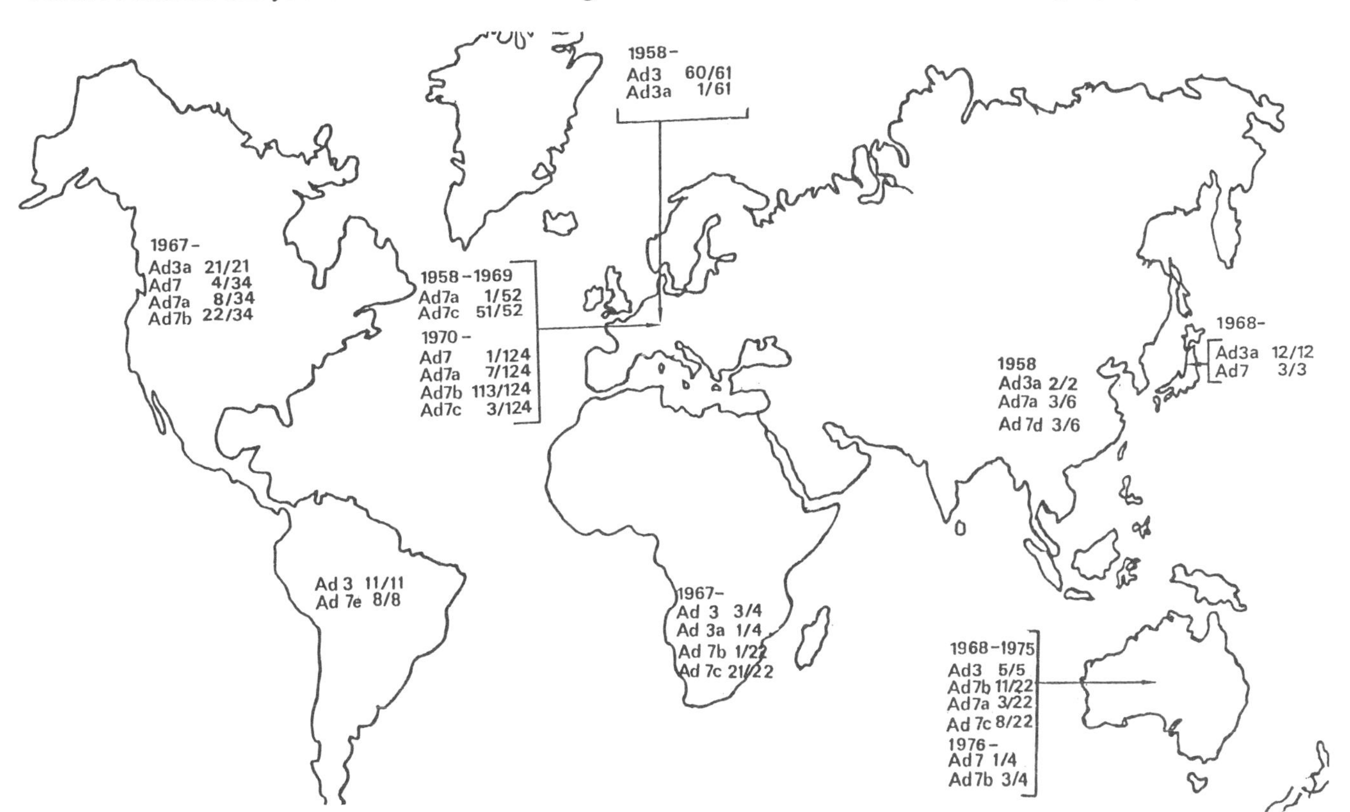

Fig. 6. Global distribution of genome types of adenovirus 3 and 7.

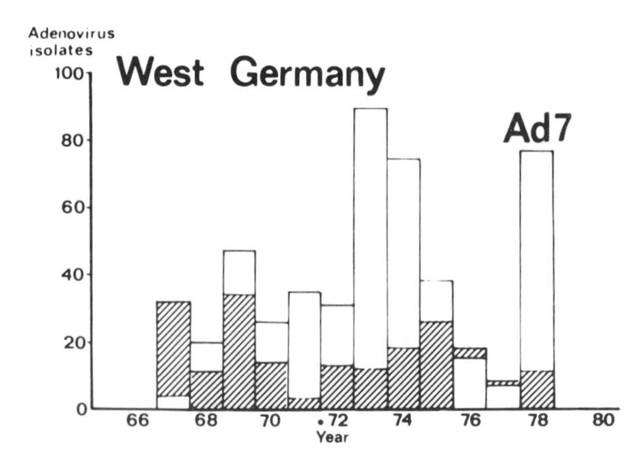

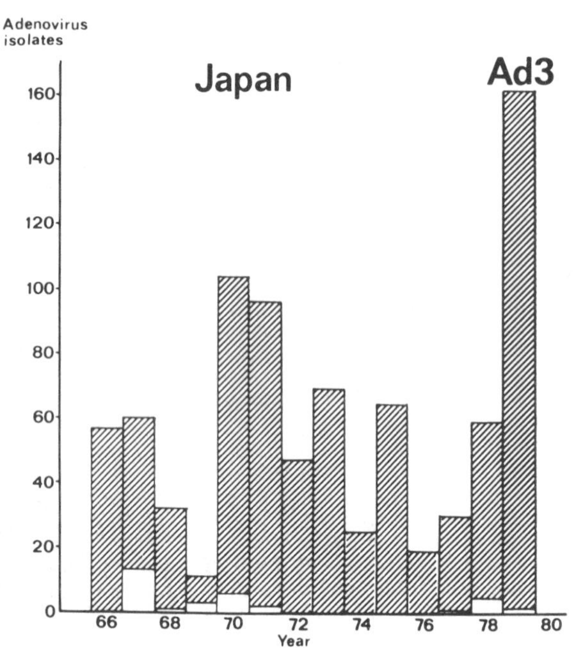

Fig. 7. Frequency of Ad3 (hatched columns) and Ad7 (open columns) strains isolated in West Germany/FRG) and Japan 1966 to 1979 (reprinted from Wadell, G., *Curr. Top. Microbiol. Immunol.* **110**, 191, 1984).

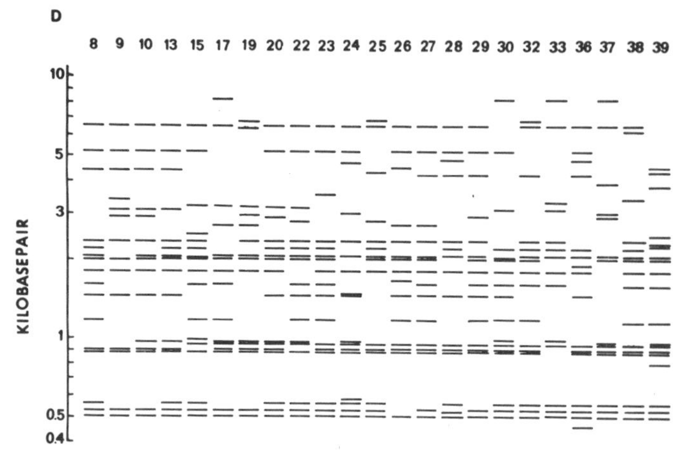

Fig. 8. Schematic presentation of the *Sma* I DNA restriction patterns of all adenovirus prototypes belonging to subgenus D.

two adenoviruses from this subgenus are recombinants between the two parent adenvovirus strains serve as an explanation for the numerous genome types detected within this subgenus [1, 19].

Subgenus D

Twenty-three serotypes and several intermediate strains have been identified within subgenus D. Several adenoviruses of this subgenus have a characteristic tropism for the eye. However numerous prototypes are not associated with overt disease. The DNA homology was determined by liquid hybridization to 94–99 [4]. Pairwise comparison of DNA restriction fragments can provide a more modulated information (Fig. 8). This issue can be illustrated by Ad8, Ad19a and Ad37 that cause outbreaks of epidemic keratoconjunctivitis. This disease was known already in the nineteenth century. It was designated shipyard conjunctivitis appearing in the naval shipyards of Pearl Harbor and on the west coast of the United States during the Second World War. The causative agent was later demonstrated to be Ad8 [27]. The Ad19 prototype was fortuitously isolated in 1955 during a trachoma survey in Saudi Arabia. Ad19 was then never isolated until 1973 when numerous outbreaks of epidemic keratoconjunctivitis caused by this virus was reported both in Europe and the United States. DNA restriction analyses of several Ad19 isolates demonstrated that a new virus Ad19a was associated with this clinical syndrome [28, 29]. In 1976 a new member of subgenus D emerged as a cause of epidemic keratoconjunctivitis. This virus was designated Ad37. DNA restriction analyses demonstrated some resemblance to Ad19a. The clinical features of infections caused by Ad8 and Ad37 were similar. However, comparative pairwise analyses of DNA restriction patterns between members of subgenus D revealed only 36% co-migration fragments common for Ad8 and Ad37 representing the most distant relationship within this subgenus [1].

Ad19 and Ad19a could not be distinguished by serum neutralization and haemagglutination inhibition assays used to type adenoviruses. Analyses of the differential agglutination of erythrocytes revealed that all members of subgenus D agglutinated rat erythrocytes. In addition Ad8, Ad19a and Ad37 agglutinated erythrocytes from dogs and guinea-pigs, whereas Ad19 prototype could not agglutinate these erythrocytes indicating a clear difference between the pathogenic and the apathogenic adenovirus strains as regards differential affinity for host-cell receptors [29].

Subgenus E

Ad4 is the only human adenovirus serotype classified into this subgenus (Fig. 9). Ad4 has been associated with respiratory disease that can be severe. A live Ad4 vaccine is used to protect against Ad4 associated pneumonias among conscripts in the United States Army. It was reported 1964 [30] that Ad4 can cause conjunctivitis without respiratory symptoms. DNA restriction analysis on various Ad4 revealed that the Ad4 prototype is a cause of the respiratory symptoms in the USA, whereas the Ad4a genome type is associated with acute haemorrhagic conjunctivitis which occur in epidemics. The Ad4a genome type has predominantly been isolated in Japan and has furthermore been demonstrated in recent oubreaks of eye disease in Europe and the United States [6]. The Ad4 prototype and the Ad4a genome type are strikingly different on the genome level (Fig. 10), only 45% of the DNA restriction fragments co-migrate [6]. No other adenovirus studied has displayed such a pronounced genetic variability within the serotype.

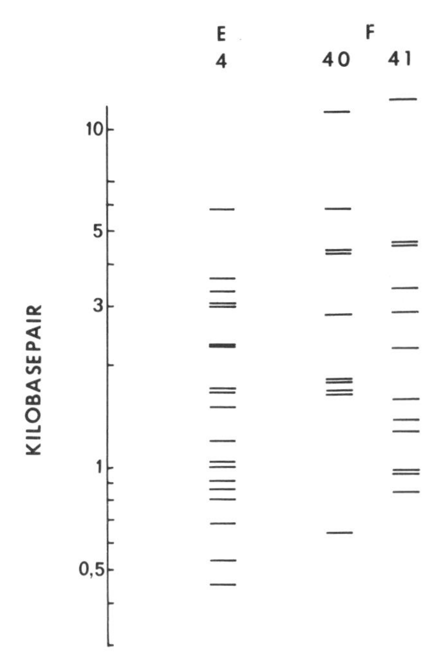

Fig. 9. Schematic presentation of the *Sma* I DNA restriction patterns of Ad4 (Subgenus E) and Ad40, Ad41 (Subgenus F).

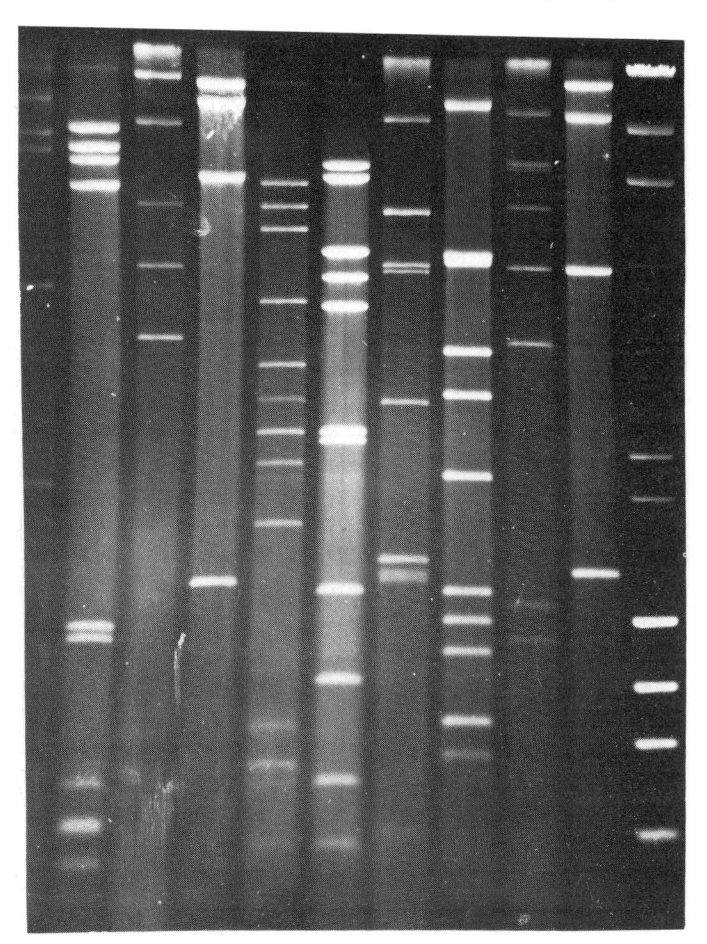

Fig. 11. DNA restriction patterns of the enteric adenoviruses Ad40 and Ad41 (Subgenus F) obtained with restriction endonucleases *Bam* HI, *Sal* I, *Pst* I, *Sma* I and *Nru* I. Lambda DNA and ϕX174 digested with *Hind* III and *Hinc* II, respectively, were used as size references.

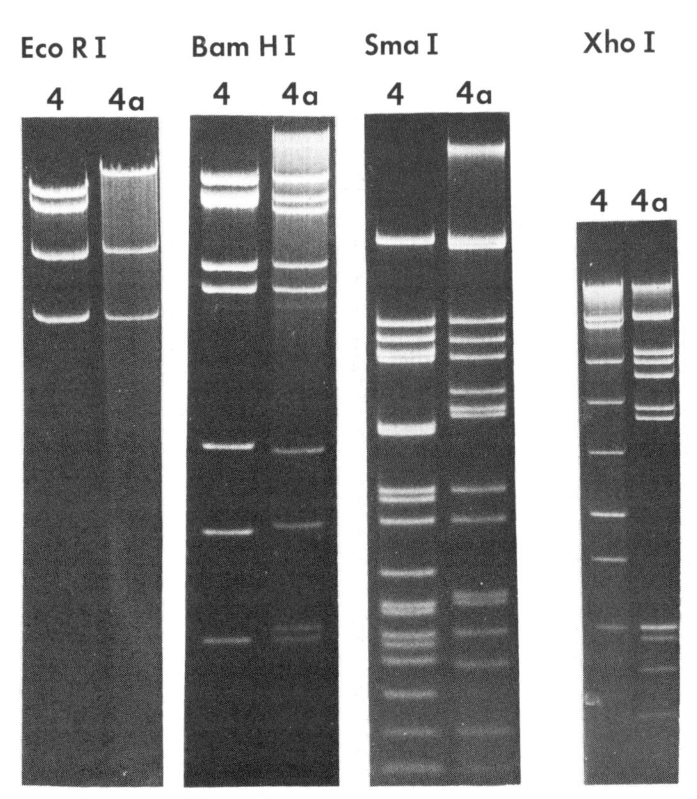

Fig. 10. DNA restriction patterns of the Ad4 prototype and the Ad4a genome type obtained with DNA restriction enzymes *Eco* RI, *Bam* HI, *Sma* I and *XhoI* I.

Subgenus F

This subgenus contains two serotypes Ad40 and Ad41 (Fig. 9). The DNA homology between them determined by liquid hybridization was 62–69% [15]. An extended analysis performed with *Bam* HI, *Bgl* I, *Eco* RI, *Eco* RV, *Hind* III, *Hpa* I, *Nru* I, *Nru* I, *Pst* I, *Pvu* I, *Sal* I and *Sma* I DNA restriction endonucleases indicated that only 18 out of 177 DNA restriction fragments co-migrated (Figs. 11, 12). This result reveals pronounced differences on the genome level between the two enteric adenoviruses. Both cause diarrhoea in children with a mean duration of 8.6 and 12.2 days for Ad40 and Ad41, respectively [31]. The enteric adenoviruses differ from all other human adenoviruses in their inability to grow in established human cell lines. However, a semi-permissive growth can be obtained in primary monkey kidney cells [2] and in 293 cell – a human kidney-cell line immortalized by transfection with Ad5, E1A and E1b regions. Again there is a clearcut difference between Ad40 and Ad41 since Ad41 replicates to significantly higher titres than Ad40 in 293 cells [32].

Limitations of studies in genetic variability of viruses

Studies of genetic variability of viruses circulating in the society are strongly biased towards viruses that cause symptoms of a sufficient degree to warrant medical care and towards viruses that are persistently shed in stools or urine. Furthermore fastidious tissue specific viruses will not be recovered in established cell lines and are consequently not available for analysis unless they occur in numbers being sufficiently high to allow direct biochemical identification.

All viruses of medical importance are isolated and registered as serotypes. Complete information on the variability on the

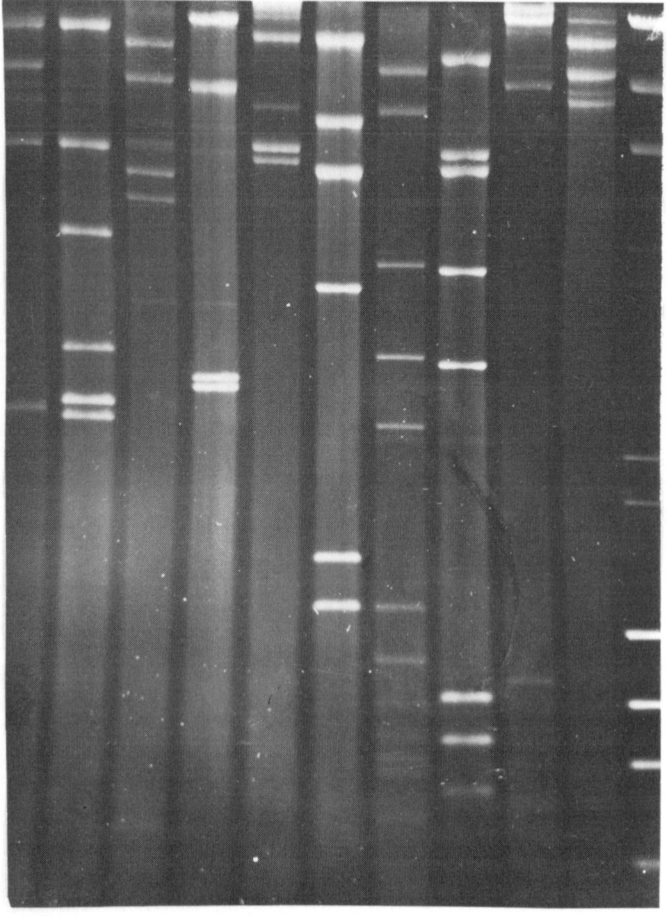

Fig. 12. DNA restriction patterns of the enteric adenovirus Ad40 and Ad41 (Subgenus F) obtained after digestion with *Eco* RI, *Eco* RV, *Pvu* I, *Hpa* I, *XhoI* I. The size reference is described under Fig. 11.

genome level can only be obtained by nucleotide sequence analysis. This is a cumbersome task when numerous strains have to be analysed. Consequently, DNA restriction endonucleases are indispensable as probes of the relations on the genome level between and within serotypes of viruses.

Origin of adenoviruses

The common architecture and the common complex genome organization of adenoviruses of different host species are taken as evidence for common origin of adenoviruses.

If we believe in a divergent evolution of adenoviruses one of the human adenoviruses should be expected to be more closely related to the archetypes of adenoviruses. This adenovirus should be expected to share properties with adenoviruses of all the different subgenera to greater extent than other adenoviruses. Ad4 is the serotype that has been suggested to be most closely related to common archetypes of human adenoviruses [5]. This suggestion is based on the following reasoning. Subgenus E contains only one member Ad4. This serotype has fibres with a length of 17 nm [33] intermediate between the fibres of other adenoviruses allowing for deletions and duplications of Ad4 fibre gene. Epitopes on the Ad4 fibre are preserved on all fibres that are 13–30 nm but not on fibres of subgenus B members which are only 10 nm [7]. Fibres of members of subgenera A, C, D, E and F all have affinity for receptors on rat erythrocytes [3, 34] whereas subgenus B members with the shortest fibres agglutinate monkey erythrocytes [25]. However, several other properties of Ad4 are rather characteristic of members of subgenus B. Ad4 displays immunological cross-reactivity on the hexon and capsomer level with members of subgenus B [9, 10]. In addition hyperimmune sera against Ad4 virions cross-react with epitopes on the surface of virions of Ad40 a member of subgenus F [35]. Ad4 form dodecons which are characteristic of subgenera B, D and E only. Hyperimmune sera against adenovirus serotypes of subgenera A to E all cross-react with early antigens of Ad4 whereas antigens of all other tested adenoviruses showed restricted cross-reactivity [36].

The forces of divergent evolution could have operated on the original adenovirus to account for the numerous adenoviruses now recognized in each host species studied according to the following model.

The organization of the adenovirus genome is complex. However, the late structural proteins are transcribed from one major late promoter. The fibre gene is localized with ancillary leaders in the fifth late region (L-5) at the most distant 3′ position. In contrast to other late structural regions the fibre is the only polypeptide encoded within region L-5. This region is separated from the other late regions by the interspersed recombination prone early region E3.

This organization may endow the archetype(s) with a mechanism for presentation of several alternatives to the prototype fibre region. The fibre serves as the ligand of the virus particle to the host cell receptor. A broad range of different conformations of the tip of the fibre will allow selection of host cells representing various organs. This can be regarded as a mechanism for appearance of different tropisms of adenoviruses.

Once a tropism has become manifest, infection with different adenovirus types can be considered as being compartmentalized, since the physical restriction impairs recombination between members of different subgenera. This situation will allow adenoviruses with different tropisms to undergo divergent evolution. On the other hand adenoviruses that have the same tropism, i.e. are infecting the same cells provide feasible conditions for recombination within genetically closed canons. Compartmentalization may be further enforced by tissue specific expression and function as a selective pressure on the adenovirus genome or genome products to adapt to host proteins. This mechanism could operate through host proteins, e.g. interacting within the region for initiation of viral DNA replication, with the E-1b gene products or the adenovirus specific DNA-binding protein.

Furthermore, experimental coinfection of adenovirus serotypes belonging to different subgenera have revealed a hierarchy of interserotypic subgenus-specific inhibition of viral DNA synthesis [37]. Once established the low level of DNA homology between adenovirus serotypes being members of different subgenera further reduces the probability of appearance of recombinants with parental genomes from two different subgenera. Thus, the subgenera in the present form can be regarded as genetically closed canons [19].

Within subgenera diversity can be expected to be continuously generated. In experimental coinfections of serotypes of subgenus C recombinant DNA was detected among the fifth of the progeny genomes [19]. Consequently the serotype members of a subgenus are expected to be subject to a divergent rather than a convergent evolution. Among all six human adenovirus subgenera the four menbers Ad1, Ad2, Ad5 and Ad6 of subgenus C are most closely related. The immune response may be regarded as an inefficient selective force of diversity of subgenus C members also mirrored by their inherent property to persist in the gut for several years. This situation allows frequent coinfection and recombination that may serve as a mechanism to even out the pool of subgenus C adenovirus genomes.

Subgenus D hosts 23 human adenovirus serotypes that have been suggested to be closely related since the DNA homology has been determined to be above 94% [4]. However, it was demonstrated using the more sensitive pairwise comparison of the DNA restriction fragments that the diversity between the members of this subgenus can be pronounced as expressed by the 36% relation between the first identified type Ad8 and the most recently identified Ad37 serotype. Furthermore Ad8 shows a pronounced difficulty to grow in established cell lines whereas the other junior serotypes have evolved a broader tissue specificity.

In the remaining subgenera, A, B, E and F a pronounced heterogeneity was noted among the members of each subgenus. Within subgenus A the DNA homology between two members is as low as 49%. In subgenus B two different clusters of DNA homology can be sorted out. Subgenus B:1 and subgenus B:2 show distinctly different tropism for the respiratory tract and the kidney, respectively. In subgenus F the two members Ad40 and Ad41 are clearly different both by DNA homology estimated by liquid hybridization to be 62–69% and by pairwise DNA restriction pattern analysis. Even in subgenus E with the single serotype Ad4 two distinctly different genome types were found. They are more different from each other than any other genome types within one serotype of a human adenovirus.

Serotypes of these four heterogenous subgenera display epitopes on the hexon that are capable of inducing neutralizing antibodies common to several members of each subgenus. Cross-reacting epitopes are also frequently detected on the tip of the fibre. This cross-reactivity can be pronounced in particular as regards the members of subgenus A, E and F, but also Ad7 and Ad11 of the two different DNA clusters of subgenus D have been demonstrated to cross-react immunologically. This can be taken as a sign that the neutralizing epitopes reside in regions of the hexon polypeptide that will hamper diversification due to requirements to retain a conserved structure.

The relation between the six subgenera of human adenoviruses is schematically expressed in Fig. 13. Each subgenus is presented as as genetically closed canon, i.e. a barrier against recombination between members of different subgenera.

The model of divergent evolution of adenoviruses implies that the subgenera of human adenoviruses should have their corollaries among adenoviruses of other host species.

Evolution of adenoviruses into different tropisms with a subsequent compartmentalization is suggested to have taken place early during the speciation of the host. The following observations favour this hypothesis.

In our early analysis of epitopes on fibres of adenoviruses belonging to different subgenera it was noted that antisera directed against fibres of Ad4 displayed a higher cross-

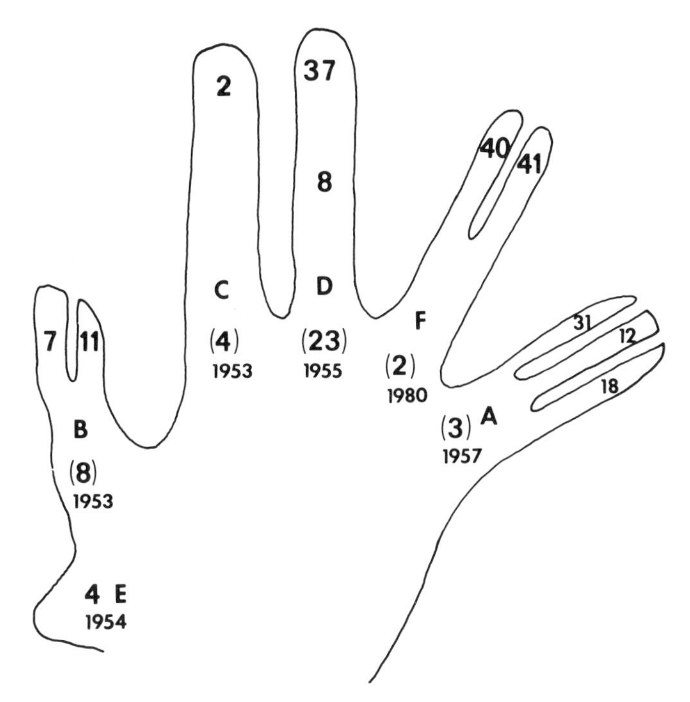

Fig. 13. A tentative schematic description of the relation between the six different subgenera of human adenoviruses.

reactivity with fibres of other subgenera than the reactivity of anti fibre sera of other adenovirus to Ad4 fibre. This is an additional observation supporting the notion that Ad4 could be a senior adenovirus (Fig. 14).

Furthermore, Ad4 fibres are distinctly different from fibres of Ad1, Ad2, Ad5 and Ad6 as regards the reactivity of subgenus specific epitopes. This was one of the criteria to distinguish Ad4 from Ad1, Ad2, Ad5 and Ad6 with whom Ad4 originally was classified on the basis of their common receptors on rat erythrocytes [34].

A corresponding analysis of epitopes on fibres of adenoviruses from different host species has been performed. It was noted that fibres of simian adenovirus 20 reacted distinctly better with antisera against human adenoviruses 2 and 6 of

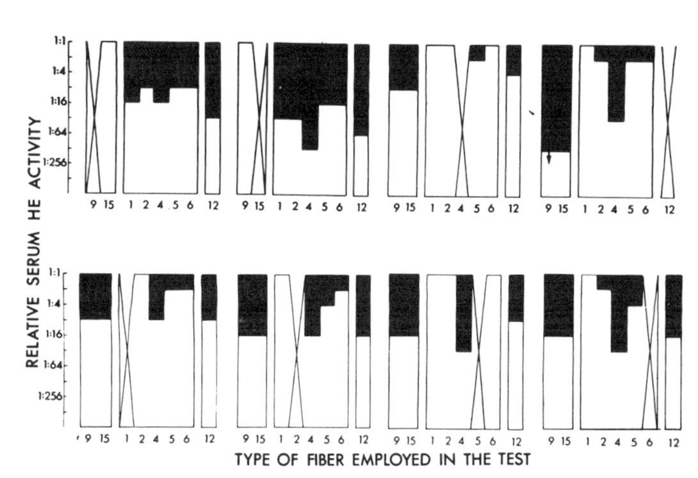

Fig. 14. Analysis of the relation between epitopes on fibres of human adenoviruses Ad9, Ad15 (subgenus D), Ad1, Ad2, Ad5, Ad6 (subgenus C), Ad4 (subgenus E) and Ad12 (subgenus A). The cross indicates the specificity of the anti-fibre serum evaluated by haemagglutination enhancement (HE) using fibres of the above mentioned serotypes. The HE titres of sera are presented in a relative fashion. The unfilled portion of a column gives the ratio of serum HE activity with the given fibre over the maximal titre obtained with any heterotypic fibre (reprinted from Wadell and Norrby, *J. Virol.* **4**, 671, 1969).

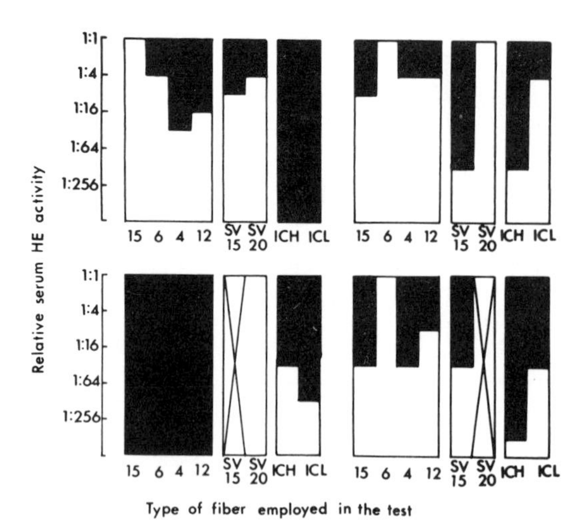

Fig. 15. Analysis of the relation between epitopes on fibres by haemagglutination enhancement (HE) using heterotypic fibres of human adenoviruses Ad15, Ad6, Ad4 and Ad12 (subgenus D, C, E and A respectively; SV15, SV20; infectious canine hepatitis (ICH) and infectious canine laryngo-tracheitis virus (ICL). The upper left and right part of the diagram represent sera of human adenoviruses 9 (subgenus D) and 2 (subgenus C) respectively. The lower porition of the diagram represent sera against fibres of simian adenovirus types SV15 and SV20 (reprinted from Norrby *et al., Can. J. Microbiol.* **17**, 1227, 1971).

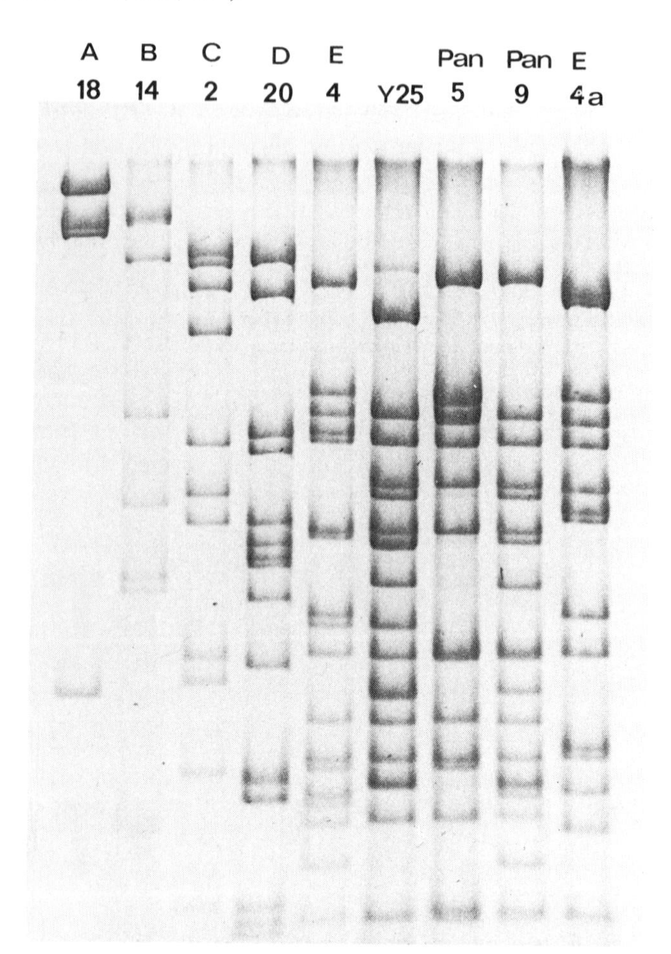

Fig. 16. Comparison of the *Sma* I DNA restriction patterns of chimpanzee adenoviruses Y25, Pan5, Pan9 with human adenovirus types representing subgenera A to E (reprinted from Wadell, *Curr. Top. Microbiol. Immunol.* **110**, 191, 1984).

subgenus C than with antisera against simian adenovirus 15. Furthermore, sera against human adenoviruses 2, 6 and simian adenovirus 20 reacted preferentially with fibres of canine laryngo-tracheitis virus in comparison with fibres of infectious canine hepatitis virus (Fig. 15). These two canine adenoviruses display cross-reacting epitopes on their hexons, but have been demonstrated to be distinctly different by DNA restriction analysis [38].

Furthermore, sera against pentons of simian adenovirus 15 reacted preferentially with vertex capsomers (the bases of pentons) of human adenoviruses belonging to subgenus B [39]. These data indicate that adenoviruses of a subgenus can be more closely related to adenoviruses of a subgenus of a different host species than to members of a different subgenus of the original host species.

More recently DNA restriction fragment co-migration analysis has been used in a comparison of the adenovirus genomes of human and chimpanzee origin. It was found that the relation between chimpanzee adenoviruses Y25, Pan5, Pan9 and human Ad4 and 4a was of the same order as was found within any of the subgenera of human adenoviruses (Fig. 16). In conclusion the Y25, Pan5 and Pan9 adenoviruses should be classified within the human adenovirus subgenus E. This statement is also corroborated by the apparent molecular mass of the internal virion polypeptides V, VI, VII and VIII of Y25, Pan5, Pan 9 and Ad4.

Analysis of other available chimpanzee adenoviruses indicated that C1 and Y15 were closely related and form one

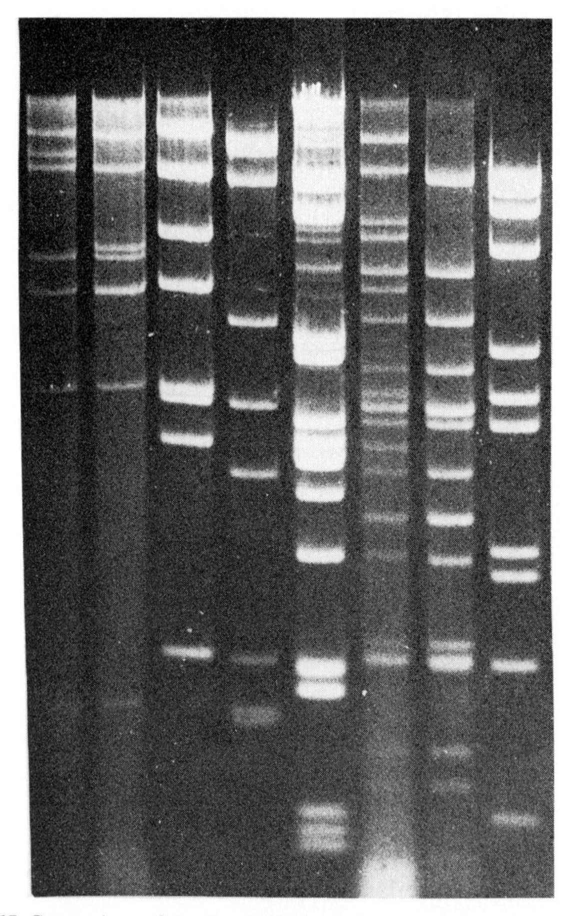

Fig. 17. Comparison of the *Sma* I DNA restriction patterns of chimpanzee adenoviruses C1, Y15; human adenoviruses 7 (subgenus B:1) 34 (subgenus B12), 20 (subgennus D; chimpanzee adenoviruses C2, Y41 and human adenovirus 2.

additional chimpanzee adenovirus subgenus with a remote relation to human adenovirus subgenus B. Chimpanzee adenoviruses C2 and Y41 form a third subgenus with a remote relation to human adenovirus subgenus C (Fig. 17).

Taken together these observations are compatible with the hypothesis that adenovirus subgenera have evolved early during the speciation of the host.

Acknowledgements

We are indebted to Katrine Isaksson and Lea Similä for excellent secretarial help and to Kristina Lindman and Gunnar Sundell for skilled assistance. The work in our laboratory has been supported by grants from The Swedish Medical Research Council, The Swedish Institute and The Swedish Medical Society.

References

1. Wigand, R. and Adrian, T., *Develop. Molecular Virology* **5**, (in the press).
2. deJong, J. C., Wigand, R., Kidd, A. H., Wadell, G., Kapsenberg, J. G., Muzerie, C. M. Wermenbol, A. G. and Firtzlaff, R. G., *J. Med. Virol.* **11**, 215 (1983).
3. Rosen, L., *Am. J. Hyg.* **71**, 120 (1960).
4. Green, M., Mackey, J. K., Wold, W. S. M. and Rigden, P., *Virology* **93**, 481 (1979).
5. Wadell, G., Hammarskjöld, M.-L., Winberg, G., Varsanyi, T. M., and Sundell, G., *Ann. N.Y. Acad. Sci.* **354**, 16 (1980).
6. Wadell, G., *Curr. Top. Microbiol. Immunol.* **110**, 191 (1984).
7. Norrby, E., *J. Gen, Virol.* **5**, 221 (1969).
8. Rosen, L., *Virology* **5**, 574 (1958).
9. Norrby, E. and Wadell, G., *J. Virol.* **4**, 663 (1969).
10. Wadell, G. and Norrby, E., *J. Virol.* **4**, 671 (1969).
11. Wadell, G., *Intervirology* **11**, 47 (1979).
12. Tikchonenko, T. I., *Curr. Top. Microbiol. Immunol.* **110**, 169 (1984).
13. Green, M., *Ann. Rev. Biochem.* **39**, 701 (1970).
14. Garon, C. F., Berry, K. W., Hierholzer, J. C. and Rose, J. A., *Virology* **54**, 414 (1973).
15. Van Loon, A. E., Rozijn, Th. H. deJong, J. C. and Sussenbach, J. S., *Virology* **140**, 197 (1985).
16. Bos, J. L. and van der Eb, A. J., *TIBS*, **310** (1985).
17. Burgert. H. G., Kvist, S., *Cell* **41**, 987 (1985).
18. Sarnow, P., Hearing, P., Anderson, C. W., Halbert, D. N., Shenk, T. and Levine, A. J., *J. Virol.* **49**, 692 1984.
19. Sambrook, J., Sleigh, M., Engler, J. A. and Broker, T. R., *Ann. N.Y. and Acad. Sci.* **354**, 426 (1980).
20. Schmitz H., Wigand, R. and Heinrich, W., *Am. J. Epidemiol.* **117**, 455 (1983).
21. Numazaki, Y., Shigeta, S., Kumasaka, T., Miyazawa, T., Yamanaka, M., Vano, N., Takai, S. and Ishida, N., *N. Engl. J. Med.* **278**, 700 (1968).
22. Harnet, G. B., Bucens, M. R., Clay, S. J. and Sakev, B. M., *Med. J. Australia* **1**, 565 (1982).
23. Shields, A. F., Hackman, R. C., Fife, K. H., Corey, L, Meyers, J. D., *New Engl. J. Med.* **312**, 529 (1985).
24. deJong, P. J., Valderrama, G., Spigland, I. and Horwitz, M. S., *Lancet* 1, 1293 (1983b).
25. Simon, M., *Acta Virologica* **6**, 302 (1962).
26. Wadell, G., Cooney, M. K., Linhares, C., Desilva, L., Kennett, M. L., Kono, R., Ren, G-f., Lindman, K., Nascimento, J. P., Schoub, B. D. and Smith, C. D. *J. Clin. Microbiol.* **21**, 403 (1985).
27. Jawetz, E., *Br. Med. J.*, **1**, 873 (1959).
28. Wadell, G. and deJong, J. C., *Infect. Immuni.* **27**, 292 (1980).
29. Kemp, M. C., Hierholzer, J. C., Cabbradilla, C. P. and Obijeski, J. F., *J. Infect. Dis.* **148**, 24 (1983).
30. Grayston, J. T., Yang, Y. F., Johnson, P. B. and Ko, L. S., *Am. J. Trop. Med. Hyp. Med.* **13**, 492 (1964).
31. Uhnoo, I., Wadell, G., Svensson, L., Johansson, M., *J. Clin. Microbiol.* **20**, 365 (1985).
32. Brown, M., *J. Clin Microbiology* **22**, 205 (1985).
33. Wadell, G., Norrby, E. and Schönning, U., *Arch. Ges. Virusforsch.* **21**, 234 (1967).
34. Wadell, G., *Proc. Soc. Exp. Biol. Med.* **132**, 413 (1969).
35. Svensson, L., Wadell, G., Uhnoo, I., Johansson, M. and v. Bonsdorff, C. H., *J. Gen. Virol.* **64**, 2517 (1983).
36. Gerna, G., Cattaneo, E., Grazia Revello, M. and Battaglia, M., *J. Infect. Dis.* **145**, 678 (1982).
37. Delsert, C., d'Halluin, J.-C., *Virus Research* **1**, 365 (1984).
38. Whetstone, C. A., *Intervirology* **23**, 116 (1985).
39. Norrby, E., Marusyk, R. G. and Wadell, G., *Can. J. Microbiol.* **17**, 1227 (1971).

Experiments on the Evolution of Bacteria with Novel Enzyme Activities

Patricia H. Clarke

Department of Chemical and Biochemical Engineering, University College London, London WC1E 7JE, United Kingdom

Paper presented at the Conference on 'Molecular Evolution of Life', Lidingö, Sweden, 8–12 September, 1985

Abstract

Bacteria have been shown to be able to evolve new metabolic activities under laboratory conditions. The genetic events concerned include mutations in enzyme structural genes resulting in altered enzymes, mutations in regulator genes producing higher enzyme levels, activation of cryptic genes, the acquisition of plasmids and the transfer of genes between species.

Introduction

One of the main problems of the early bacteriologists was to satisfy themselves, and even more their critics, that they were working with pure cultures. For this reason any variations in appearance or growth characteristics were suspect. Microorganisms were isolated from a variety of natural sources and every effort was made to obtain a single strain with stable and defined characteristics. These strains were studied in the laboratory and changes in their properties were usually, and often correctly, ascribed to casual contamination. If contamination of the cultures could be ruled out it was sometimes possible to recognise significant changes in the organism under study. In the medical field it might be found that a laboratory strain had lost some of its pathogenic properties, becoming attenuated or non-virulent. In other cases the nutritional characters might change. In the early years of this century Beijerinck and Winogradsky pioneered the use of enrichment culture to obtain microorganisms with the ability to utilize particular organic compounds as growth substrates [1]. Such strains were selected, and named, for their specialized biochemical activities and if these changed there could be problems of nomenclature to add to the uncertainty about whether a new organism had replaced the original culture. Although this made it difficult for taxonomists it was the first indication that bacteria might be used as model systems for experimental evolution in the laboratory.

Advances in microbial biochemistry made it possible to focus on individual metabolic pathways and on single enzymes and to examine the properties of bacterial cultures grown in different media. In an important series of experiments bacteria were 'trained' to do without amino acids that had appeared to be essential to the culture when it was first isolated and such observations led Stephenson in 1949 to the view that '...with bacteria constant evolutionary changes occur under our eyes and can be controlled and imitated in the laboratory' [2]. With rapid developments in techniques for studying microbial genetics these changes in enzyme activities could be explored in more detail. An essential contribution to such studies was to understand the control of gene expression. The operon model of Jacob and Monod in 1961 [3] described a mechanism for the phenotypic response of microbial cultures to changing chemical environments.

There are two threads running through this account of experiments on the evolution of bacterial enzymes. One is the assumption that studies in experimental evolution are applicable to evolution in nature and the other is the conviction that enzyme structures can be deliberately manipulated to obtain new and desirable properties. The first example to be considered is tryptophan synthetase and the effects of single site mutations in the structural gene. The classical experiments of microbial genetics depended on the isolation of defective mutants which could undergo recombination. Two auxotrophic mutants might give rise to a prototroph by genetic transfer involving conjugation, transduction or transformation. In any such experiment it was important to ensure that the prototrophs had been derived by genetic recombination rather than by reversion of an original mutation. However, a revertant or pseudorevertant may arise by a second mutation within the same gene, or in another gene, and this gives the opportunity to examine the effect of amino acid substitutions at specific sites on the catalytic properties of an enzyme.

The second approach is to study bacterial strains that give rise to mutants that are able to utilize compounds that cannot be used as growth substrates by the parental strains. In some cases these appear to arise by the activation of cryptic genes and experiments on these lines may reveal genes whose existence was previously unknown.

The third general method is to take a strain with a known and active enzyme and to select mutants that produce altered enzymes or have altered regulatory phenotypes. This method offers the opportunity to examine whether or not the same phenotype can arise from more than one type of mutation and whether successive mutations can result in stepwise changes in substrate specificities.

Concentration on single strains, single pathways and single enzymes has made it possible to examine the molecular events in the acquisition of novel phenotypes. However, in nature there are very few places where microorganisms exist in pure culture and it is now clear that novel microbial activities may involve interactions between different members of microbial communities. This will be the subject of the final section.

Tryptophan synthetase and single-site mutations

Conventional genetic methods established that the mutations of amino acid auxotrophs could result from single-site mutations or deletions. An important contribution to understanding gene–enzyme relationships was the demonstration by Yanofsky and his colleagues that single-site mutations in the tryptophan synthetase A gene were colinear with alterations in the amino acid sequence of the protein [4]. The *trpA* gene codes for the α subunit of tryptophan synthetase which, together with the β subunit, catalyses the synthesis of tryptophan from indoleglycerol phosphate and serine. Tryptophan auxotrophs were screened to find those that produced inactive proteins that cross-reacted with antiserum to the wild-type α subunit. These were missense mutants in which the wild-type amino acid had been replaced by another amino acid. Some showed partial activity in the presence of an extract containing a wild-type β subunit. Some of the missense mutant proteins were temperature sensitive.

The analysis of mutants A46 and A23 gave particularly interesting results. The wild type protein has glycine at position 211 which is replaced by arginine in mutant A23 and by glutamic acid in mutant A46. Both these mutants were capable of reversion to prototrophy but analysis of the products showed that functional proteins could exist with amino acids other than glycine in position 211 of the polypeptide chain. A23 reverted to give threonine or serine as alternatives to glycine and A46 to give alanine and valine. Only single base changes were required to get these amino acids from a single codon and the results provided valuable confirmation of the genetic code [5].

$$\begin{array}{ccc} \text{Arg [AGA]} & \leftarrow \text{Gly [GGA]} \rightarrow & \text{Glu[GAA]} \\ \downarrow & & \downarrow \\ \text{Thr [ACA] or Ser [AGU/C]} & & \text{Ala [GCA] or Val [GUA]} \end{array}$$

The revertants carrying Ala and Ser substitutions grew at the wild-type rate while those with Thr or Val grew more slowly and were classed as partial revertants. Later it became possible to extend the analysis of the effects of other substitutions at position 211 by using *in vivo* methods for site-specific mutagenesis. These methods included selection of missense revertants of amber mutations and suppression of non-functional substitutions by the insertion of another amino acid by a mutant transfer RNA [6]. Some were derived by more than one step and a total of 15 different substitutions at this site were reported in 1974 [5]. Non-functional proteins resulted from the substitution of Gly by Asp, Arg, Glu, Lys, Gln, Tyr and Trp, all of which involved charged or bulky groups. Other amino acids gave proteins which were functional but less active than the wild type. The efficiencies of the amino acid substituents at position 211 were Gly > Ala > Ser > Thr > Ile > Val > Asn.

Another class of revertants of A46 proved to have a secondary mutation at a second site within the *trpA* gene. The double mutant A46A446 produced an active enzyme which retained the Gly → Glu substitution at position 211 but also had a Tyr → Cys substitution at position 175. If only the latter substitution is present the enzyme is inactive [7]. Thus, although the activity of the wild type enzyme is dependent on the presence of glycine (or certain other amino acids) at position 211 there are alternative amino acid sequences that can lead to an appropriate tertiary structure for catalytic activity.

Some of the *trpA* revertants had suppressor mutations in unlinked genes which inserted the wild-type amino acid at the mutant codon [5]. Altered tRNA synthetases, produced by suppressor mutants, can insert the 'correct' amino acid at the site of nonsense or missense mutations. With the *trpA* missense mutants it was shown that some suppressor mutations were only partially effective and that strains carrying missense and suppressor mutations produced both wild-type and mutant proteins. Amber and ochre suppressors allow an amino acid to be inserted at the site of a nonsense mutation and this provides a general method for *in vitro* site-specific mutagenesis when the gene has been mapped and the sequence is known.

Site-specific mutagenesis can now be carried out *in vitro* using techniques developed for genetic engineering. It is possible to use synthetic oligonucleotides to prime DNA synthesis so as to produce DNA sequences that will result in amino acids changes at known sites in an enzyme protein. Studies on tyrosyl tRNA synthetase are providing fascinating information about the roles of the amino acids at the active site [8]. It has been suggested that the method is applicable only to enzymes whose tertiary structure and mechanism of action are known and this would seem to preclude most enzymes [9]. However, the catalytic activity of an enzyme and its physical properties are affected by mutations that are not necessarily concerned with the active site. If a fairly large number of strains are available with single-site mutations it may be worth making *in vitro* substitutions at those sites to test whether they might yield enzymes with useful novel properties.

Evolved β-galactosidase (Ebg) and cryptic genes

The *lacZ* gene of *E. coli* K12 codes for the enzyme β-galactosidase which enables lactose to be hydrolysed and used as a carbon source.

$$\text{Lactose} \xrightarrow[ebgA]{lacZ} \text{Glucose} + \text{Galactose}$$

A mutant carrying a *lacZ* deletion gave rise to lactose-utilizing strains which produced a novel β-galactosidase (Ebg). The Ebg enzyme differed from the classical β-galactosidase in K_M and V_{max}, had a different molecular weight and failed to give a cross-reaction with antiserum to the *lacZ* enzyme. The Ebg enzyme is coded by the *egbA* gene which maps in the opposite section of the chromosome [10]. The Ebg enzyme is induced by lactose (unlike the *lacZ* enzyme which is induced by allolactose) and although it can be induced in the wild-type strain the specific activity is very low. Mutants were first selected as coloured papillae growing out from the whitish Lac$^-$ colonies growing on nutrient agar plates containing lactose and a fermentation indicator. The production of acid by fermentative bacteria provides an excellent method for identifying strains that have acquired new catabolic activities. Hall [11] points out that it is possible to isolate 20 or so independent mutants from a single plate by this means.

Examination of the Ebg mutants revealed that two mutations were necessary for growth on lactose; a mutation in *ebgA* to improve enzyme activity and a mutation in the regulatory gene, *ebgR* to give either a higher level of induction or constitutive enzyme synthesis. From these results it could be concluded that the *ebg* genes did not contribute to the

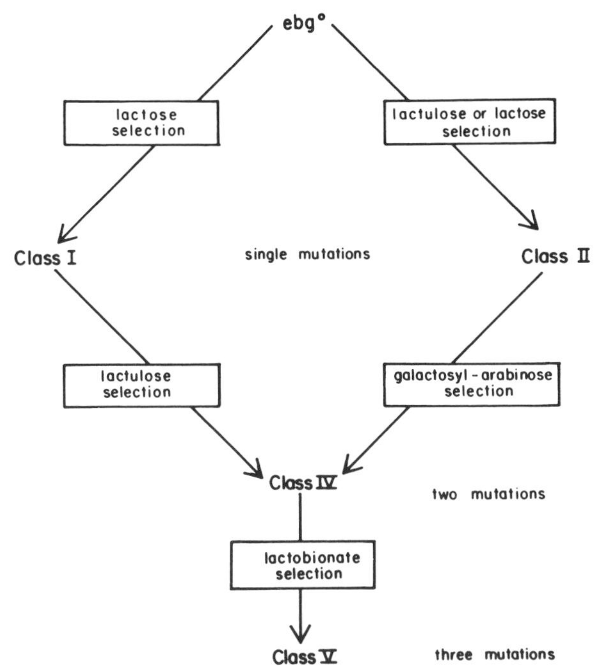

Fig. 1. Pathways for the experimental evolution of novel β-galactosidases in *Escherichia coli* K12. The Class V enzymes allow growth on lactobionate and these can be derived by successive single site mutations in the *ebgA* gene by either of the routes shown above.

utilization of lactose in the wild-type strain. Hall and colleagues found that, provided that appropriate *ebgR* mutations were present, it was sufficient to have single-site mutations in *ebgA*. These might occur at different sites and result in enzymes that differed from each other in K_M, V_{max} and substrate specificities [12]. With mutants carrying different *ebgR* and *ebgA* mutations it could be shown that the rate of growth was dependent on the *in vivo* lactose-hydrolyzing ability and that there was a threshold below which the rate was too low to support growth [13]. These experiments demonstrated the significance of co-evolution of regulatory and structural genes to obtain a novel catabolic activity.

Selection for growth on lactose produced some mutants that grew rapidly on lactose but not on lactulose (Class I); selection on lactulose produced mutants that grew on lactulose, less well on lactose and very slowly on galactosyl-arabinose (Class II). The Class I and Class II mutations occurred at different sites in the *ebgA* gene and mutants carrying both types of mutations (Class IV) were able to grow on galactosyl-arabinose. From the Class IV mutants it was possible to derive mutants that grew on lactobionate (Class V). Figure 1 shows the evolutionary pathways for the mutants with novel growth phenotypes [11].

Evidence for the presence of a cryptic gene in an organism is provided by mutation to an active state. Such silent genes will be undetected unless suitable selection methods have been devised. Cryptic genes may be fairly common in bacteria. For example, *E. coli* K12 does not normally utilize β-glucosides but mutants can be obtained that grow on arbutin and salicin (*bgl* mutants). Other mutants have been obtained that grow on cellobiose and it was shown that a different cryptic gene had been activated [14]. Genes can be defined as cryptic if they are phenotypically silent but capable of being activated in a few individuals of a population [15]. They may become cryptic by accumulation of deleterious mutations or by the insertion of IS elements or by any other genetic event that can be reversed to give a functional sequence. However, if a new activity can be related to the activation of a cryptic gene the mutant gene might not be exactly the same as its ancestor. The genetic origin of some novel activities, including enzymes acting on unnatural pentoses and pentitols, is still unknown [16].

Amidases with new substrate specificities

The amidase of *Pseudomonas aeruginosa* allows growth on the aliphatic amides, acetamide and propionamide.

$$\text{Acetamide} \xrightarrow{amiE\,amiR} \text{Acetate} + \text{Ammonia}$$

These two amides are good substrates and good inducers but no other aliphatic amides and only a few substituted amides, including glycollamide and lactamide, can be used for growth. The advantage of this system for experimental evolution is that an amide can be used as either a carbon or nitrogen source [17]. This made it possible to devise media for selecting for a series of mutant enzymes with altered regulatory or substrate specificities (Table I). Acetamide yields acetate, which can be utilized via the tricarboxylic acid and glyoxylate cycles, together with ammonium ions. Some of the selective media required to be supplemented. For example, phenylacetamide is a potential nitrogen source, but not a carbon source, since phenylacetate cannot be metabolized by *P. aeruginosa*. On the other hand, acetanilide (*N*-phenylacetamide) is a potential carbon source but not a nitrogen source, since aniline is not readily utilized. Amidase is subject to catabolite repression and this property provided another selection pressure for mutant isolation.

The inducer specificity is similar to, but not identical with, the substrate specificity of the amidase. Some of the mutants with new growth phenotypes carried mutations only in the regulatory gene *amiR*. Selection on succinate/formamide medium gave constitutive mutants or formamide-inducible mutants. The wild-type enzyme hydrolyses formamide at about 10% of the acetamide rate but the rate of induction by formamide, and thus the *in vivo* enzyme activity, is too low for growth to occur. Selection on butyramide medium gives constitutive mutants but not all constitutive mutants grow on butyramide. The reason for this is that butyramide not only blocks amidase induction in the wild-type but it also represses amidase synthesis by many constitutive mutants. Amidase is under positive regulation and while the *amiR* product of some constitutive strains is unaffected by butyramide there are others in which butyramide prevents activation of transcription. Although butyramide is hydrolyzed by the wild-type enzyme at only about 2% of the acetamide rate the amount produced by the butyramide-resistant mutants is sufficient to allow growth to occur although at the very early stages the growth rate is limited by the *in vivo* amidase activity [18].

It had been recognized by Mortlock and others that a regulatory mutation by itself could account for growth on certain pentoses and pentitols [16]. If the substrate specificity of an enzyme is such that it has at least some activity on a minor substrate then a constitutive mutant might be able to initiate a novel catabolic pathway. For example, mutants of *Klebsiella aerogenes* that are constitutive for L-fucose isomerase are able to grow on the unnatural pentose L-arabinose and mutants that are constitutive for ribitol dehydrogenase are

Table I. *Amide media used for selection of* Pseudomonas aeruginosa *amidase mutants*

Selection medium		
Carbon source	Nitrogen source	Mutant phenotypes
Succinate	Formamide	Constitutive Butyramide-sensitive or butyramide-resistant Formamide-inducible
Succinate	Lactamide	Catabolite-repression resistant Inducible or constitutive
Succinate	Butyramide	Constitutive and butyramide-resistant and catabolite-repression resistant Constitutive and altered enzyme
Butyramide	Butyramide	Constitutive and butyramide-resistant Constitutive and altered enzyme Butyramide-inducible and altered enzyme
Valeramide	Valeramide	Constitutive and altered enzyme
Succinate	Phenylacetamide	Constitutive and altered enzyme
Acetanilide	Ammonium sulphate	Constitutive and altered enzyme

The composition of the selective media depends on the known properties of amides as substrates and inducers. Some mutant classes can be selected directly from the wild type and others from particular mutant strains [17].

able to grow on xylitol. In the latter case the rate of growth was very low and selection for faster growth yielded predominantly mutants with gene duplications [19].

The families of mutant amidases were all derived from various classes of regulatory mutants (Fig. 2). A constitutive mutant that was unable to grow on butyramide, since it had retained sensitivity to butyramide repression, gave rise to a mutant producing an amidase with altered substrate specificity with about 10-fold higher activity on butyramide [17]. Additional mutations produced mutant amidases acting on valeramide and phenylacetamide (Fig. 2). All the B6 family of mutants had retained the original mutation in which a serine residue at position 7 had been replaced by phenylalanine but had a variety of additional mutations [20]. Thus, the substrate specificity of this enzyme could be progressively changed by successive mutations in the *amiE* gene. The selection method used in these experiments depended on growth on the novel substrate. Many other mutations must have led to inactive enzymes but these would not have been detected. The amidase of *P. aeruginosa* is very active on its optimal substrates but sufficient flexibility has been retained for single-site mutations to allow other amides to be accomodated at the active site. Events of this sort may occur when synthetic organic chemicals are degraded in the natural environment.

Many catabolic enzymes of bacteria are induced only in the presence of their substrates and it is a reasonable criticism that constitutive strains producing altered enzymes are laboratory artefacts. However, some Ebg mutants are inducible as are some of the mutants growing on the uncommon pentoses and pentitols [11, 16]. There are technical difficulties in direct selection of amidase mutants in which an altered enzyme is regulated by its new substrate but they have been obtained by indirect means. A recombinant strain was obtained which carried the wild-type *amiR* gene and the *amiE16* (butyramide-positive) mutation. This strain could not grow on butyramide because butyramide does not induce the enzyme. Selection for growth on butyramide gave mainly constitutive mutants but among them were some that were butyramide-inducible (Fig. 3). In this case mutation, recombination and further mutation contributed to the evolution of a strain in which an altered enzyme is induced by its novel substrate [21]. Inducible strains are generally thought to be at an advantage in nature since they thereby avoid synthesizing enzymes when they are not required for growth. However, stable mutants in which a single constitutive enzyme may be synthesized as 10% or more of the soluble protein may have useful industrial applications.

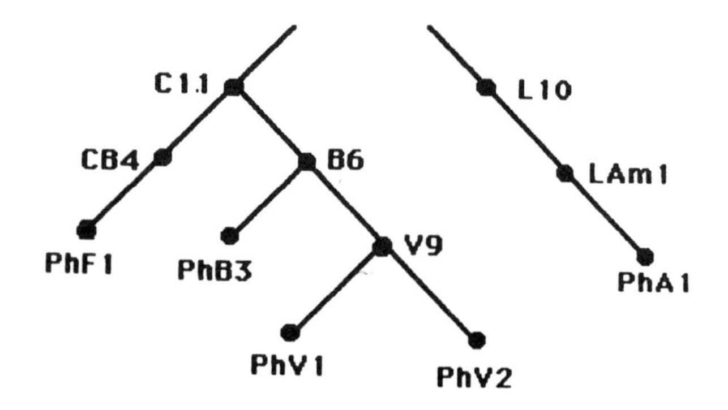

Fig. 2. Family tree of the experimental evolution of phenylacetamidases derived by successive single-site mutations in the *amiE* gene of *Pseudomonas aeruginosa* PAC1. The constitutive mutant C11 was derived by a single mutational step from PAC1 and gave rise to strain B6 producing a mutant butyramide-hydrolyzing enzyme. Additional mutations produced enzymes hydrolysing valeramide (V9) and phenylacetamide (PhB3, PhV1, PhV2). Strain CB4 is a constitutive butyramide-resistant mutant and PhF1 has a single mutation in the *amiE* gene. Strain L10 is a derepressed mutant derived from PAC1 and successive mutational steps gave the amidase-defective mutant LAm1 and the phenylacetamide-hydrolyzing mutant PhA1.

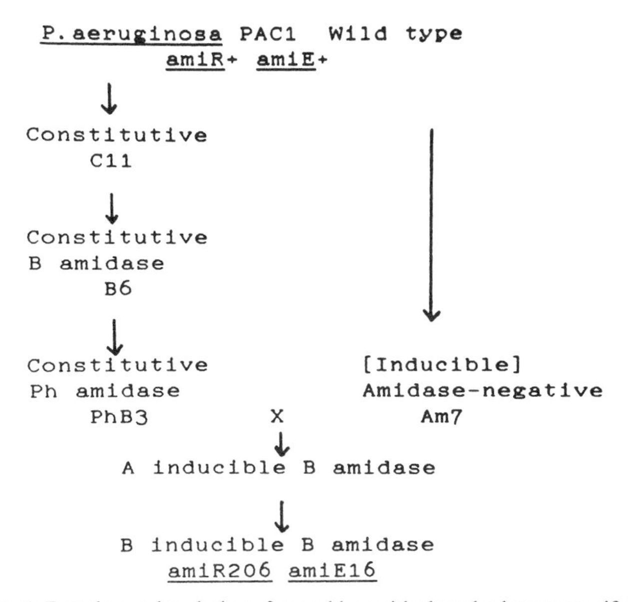

Fig. 3. Experimental evolution of an amidase with altered substrate specificity that can be induced by its novel substrate. A cross between the phenylacetamidase mutant PhB3 and the acetamide-negative mutant Am7 gave a recombinant that produced a butyramide-hydrolyzing enzyme when induced by acetamide. A mutation in the *amiR* gene resulted in a strain that was butyramide-inducible.

Degradation of halogenated compounds by microbial communities

Many halogenated compounds are released into the environment as pesticides and herbicides or as by-products of the chemical industry. It may be difficult to isolate single strains with the ability to degrade these compounds although they may be obtained by prolonged selection procedures [22]. The realization that synthetic organic chemicals may be cometabolized with related biologically produced compounds and that successive steps in degradation may be carried out by different microorganisms has led to a number of studies on microbial communities [23].

Dalapon (2,2′-dichloropropionate, 22DCPA) is a widely used herbicide known to be readily degraded in the soil. In 1976 it was reported that a microbial community had been established by enrichment culture that consisted of seven organisms. Of these there were three primary utilizers that were able to grow in pure culture with 22DCPA as carbon source and four others that were secondary utilizers. One of the secondary utilizers was readily lost but the other species were retained in a stable microbial community even when some of the growth parameters were changed [24]. The community had a higher rate of 22DCPA degradation than any single primary utilizer indicating that the community as a whole could have a significant growth advantage in nature.

After 3000 h of continuous growth a new primary utilizer, PP3, was detected derived from one of the original secondary utilizers, *Pseudomonas putida* PP1. This represented the acquisition of a novel catabolic activity by one member of a microbial population and could have been due to mutation in that strain or to transfer of genetic information from one of the other species of the microbial community. In this case the novel activity appears to be due to activation of cryptic genes and to be related to differences in the level of expression of dehalogenase activity rather than to mutations in structural genes. The primary utilizer, *P. putida* PP3 has dehalogenase activities that are 10–40 times greater than those of PP1. Later it was found that strain PP3 produces two dehalogenases with similar levels of activity for a range of halogenated aliphatic acids but more detailed investigations indicated that they had different reaction mechanisms [23]. The fraction I dehalogenase is severely inhibited by sulphydryl reagents while the fraction II dehalogenase is much less sensitive. Further, the products of dehalogenation of the enantiomers of DL-2-monochloropropionate by the fraction I dehalogenase are in the same configuration as the substrate while for the fraction II dehalogenase they are in the opposite configuration. Thus the novel dehalogenase activity of the new primary utilizer *P. putida* PP3 is due to a high level expression of two genes that are almost silent in the parent strain in which the enzyme activity never reaches the threshold necessary for growth.

The genes for many unusual catabolic activities of bacteria are carried on transmissible plasmids. Although catabolic plasmids may be present in a relatively small number of the bacterial population they provide a genetic reservoir for the utilization of a wide range of organic compounds in the natural environment. In certain cases, including octane and camphor degradation, plasmid genes code for the early enzymes of a pathway while chromosomal genes code for the later enzymes. The transfer of plasmids between strains may enable novel pathways to be established in which the enzymes coded by the plasmid provide the activity for a critical step. Strains that degrade 4-chlorobenzoate have been constructed in this way by transferring the TOL plasmid, which codes for a benzoate oxidase with broad substrate specificity, into *Pseudomonas* spp. B13 that carries the full complement of genes for the later steps of the pathway [22].

In some cases novel catabolic plasmids can be constructed by the transfer of chromosomal genes to transmissible plasmids carrying other markers such as drug resistance determinants. The dehalogenase I gene of *P. putida* PP3 was transferred to a broad host range drug-resistance plasmid to give an R prime dehalogenase I plasmid [25]. When this plasmid was transferred into *P. aeruginosa* PAC1 it conferred the ability to utilize 2-monochloropropionate. Mutants were selected for growth on plates containing 2-chloropropionamide as carbon source. These were found to synthesize the wild-type aliphatic amidase at a high level. The low rate of hydrolysis of 2-chloropropionamide is similar to its activity on butyramide and can be compensated for in the same way by a high enzyme level [26]. In this case the chromosomal amidase of *P. aeruginosa* initiates the hydrolysis of 2-chloropropionamide and the plasmid-coded dehalogenase converts 2-chloropropionate to lactate which is a normal growth substrate for this organism. A strain with a novel catabolic pathway has been constructed by capture of a chromosomal gene on a plasmid, transfer of the plasmid to another species followed by a mutation in a regulator gene.

Genetic events in experimental evolution

There are many ways in which bacteria can acquire novel enzyme activities. Selection pressures include growth in the absence of a hitherto essential nutrient, growth on organic compounds that cannot be utilized by parental strains, growth at faster rates in continuous culture, growth in the presence of toxic analogues and selection from mixed cultures. Mutagenic treatment may increase the numbers in cases where the spontaneous mutation rate is very low. Strains obtained in these ways can then be analyzed to discover what genetic events have taken place. More direct intervention in the evolutionary process includes the exchange of genetic information by recombination and the transfer of plasmids between strains. It is reasonable to suppose that most of the novel enzme activities described in this account could have evolved in a similar way in nature. These methods can all be used to select new strains producing enzymes with particular properties but with more detailed information at the molecular level we can now also use *in vitro* mutagenesis to obtain the novel enzymes.

Acknowledgement

The author is the holder of a Leverhulme Emeritus Fellowship.

References

1. Brock, T., in *Milestones in Microbiology*. American Society for Microbiology, Washington D.C. (1961).
2. Stephenson, M., in *Bacterial Metabolism*. Longmans, Green & Co., London (1949).
3. Jacob, F. and Monod, J., *J. Mol. Biol.* **3**, 318 (1961).
4. Yanofsky, C., Carlton, B. C., Guest, J. R., Helinski, D. R. and Henning, U., *Proc. Natl. Acad. Sci. USA* **51**, 266 (1964).
5. Murgola, E. J. and Yanofsky, C., *J. Mol. Biol.* **86**, 775 (1974).
6. Murgola, E. J. and Yanofsky, C., *J. Bacteriol.* **117**, 439 (1974).
7. Yanofsky, C., in *The Chemical Basis of Life* (ed. P. C. Hanawalt and R. H. Haynes), Freeman, San Francisco.
8. Fersht, A. R., Shi, J.-P., Knill-Jones, J., Low, D. M., Wilkinson, A. J., Blow, D. M., Brick, P., Carter, P., Waye, M. M. Y. and Winter, G., *Nature (London)* **314**, 235 (1985).
9. Winter, G. and Fersht, A. R., *Trends in Biotechnology* **2**, 115 (1984).
10. Campbell, J. J., Lengyel, J. and Langridge, J., *Proc. Natl. Acad. Sci. USA* **70**, 1841 (1973).
11. Hall, B. G., in *Microorganisms as Model Systems for Studying Evolution* (ed. R. P. Mortlock). Plenum Press New York (1984).
12. Hall, B. G., *Biochemistry* **20**, 4042 (1981).
13. Hall, B. G. and Clarke, N. D., *Genetics* **85**, 193 (1977).
14. Kricker, M. and Hall, B. G., *Mol. Biol. Evol.* **1**, 171 (1984).
15. Hall, B. G., Yokoyama, S. and Calhoun, D. H., *Mol. Biol. Evol.* **1**, 109 (1983).
16. Mortlock, R. P., in *Microorganisms as Model Systems for Studying Evolution* (ed. R. P. Mortlock). Plenum Press New York (1984).
17. Clarke, P. H., in *Microorganisms as Model Systems for Studying Evolution* (ed. R. P. Mortlock). Plenum Press New York (1984).
18. Brown, J. E. and Clarke, P. H., *J. Gen. Microbiol.* **64**, 329 (1970).
19. Hartley, B. S., in *Microorganisms as Model Systems for Studying Evolution* (ed. R. P. Mortlock). Plenum Press New York (1984).
20. Paterson, A, and Clarke, P. H., *J. Gen. Microbiol.* **114**, 75 (1979).
21. Turberville, C. and Clarke, P. H., *FEMS Microbiol. Lett.* **10**, 91 (1981).
22. Reinecke, W., in *Microbial Biodegradation of Organic Compounds* (ed. D. T. Gibson). Marcel Dekker, New York (1984).
23. Slater, J. H. and Lovatt, D., in *Microbial Biodegradation of Organic Compounds* (ed. D. T. Gibson). Marcel Dekker, New York (1984).
24. Senior, E., Bull, A. T. and Slater, J. H., *Nature (London)* **263**, 476 (1976).
25. Beeching, J. R., Weightman, A. J. and Slater, J. H., *J. Gen. Microbiol.* **129**, 2071 (1983).
26. Wyndham, R. C. and Slater, J. H. (in the press).

Chemica Scripta 1986, **26B**, 343–349

Evolutionary Aspects of Immunoglobulin-related Genes

Susumu Tonegawa and Haruo Saito*

Center for Cancer Research and Department of Biology, Massachusetts Institute of Technology, 77 Massachusetts Avenue, Cambridge, Massachusetts 02139, USA

Paper presented by Susumu Tonegawa at the Conference on 'Molecular Evolution of Life', Lindingö, Sweden, 8–12 September 1985

Abstract

In the vertebrates' immune system the recognition of nonself molecules is carried out by two sets of evolutionarily related glycoproteins, immunoglobulin (Ig) and T cell receptors (TCR). The most characteristic feature of these proteins is their enormous variability. Unlike most other known genes, an Ig or TCR gene is carried in the germline as gene segments (*V* and *J* or *V*, *D* and *J*) which are assembled at the DNA level specifically during lymphocyte differentiation. This somatic recombination generates much of the variability observed in the so called variable (V) regions of these polypeptide chains.

Ig and TCR polypeptide chains are composed of internally repeated common homology units which fold into similar globular domains: The V regions of all these chains, and the C (constant) regions of Ig light and TCR chains consist of a singled homology unit, while the C regions of Ig heavy chains contain three or four homology units. At the DNA level each of the C region homology units corresponds to an independent exon. As to the V homology units, the coding DNA is split into two (*V* and *J*) or three (*V*, *D*, and *J*) segments in the germline but composes a single exon in the B and T lymphocytes, for Ig and TCR genes, respectively. These results suggest that: (1) both Ig and TCR genes evolved from a common primordial gene of the size of one homology unit as a result of a series of duplication, sequence divergency, and fusion of the duplicated DNA copies by RNA splicing; (2) introns of some eukaryotic genes arose from the flanking non-coding sequences of ancient genes subsequent to gene duplication; (3) *V–J* (and possibly *V–D–J*) joining is a reversal of an ancient insertion of a transposition-like DNA element into one of the duplicated primordial genes.

Ig and TCR genes are two of an increasingly larger number of evolutionarily related genes, most of which encode integral membrane proteins, although somatic recombination seems to be restricted to the two gene families. These findings suggest that the Ig-like combining sites and/or Ig-like domains are widely used by cell surface proteins either for ligand binding or cell-cell interaction. The finding also suggests that many eukaryotic proteins may be classified into evolutionarily related but functionally distinct families.

Introduction

Vertebrates are endowed with a highly effective body defense mechanism referred to as the immune system. The critical event in mounting the immune response is the recognition of chemical markers present on the surface of molecules or a collection of molecules that are foreign to the body (antigens). This task is entrusted to two groups of proteins, namely immunoglobulins synthesized and secreted by B lymphocytes and the so called T cell receptors present on the surface of T lymphocytes. The most intriguing properities of these proteins is their variability of structure: A single organism can synthesize tens and hundreds of millions of structurally different antibodies and T cell receptors. During the past ten years the analysis of genes coding for these proteins uncovered not only the genetic origins of the enormous variability but also several intriguing features of the behaviour of these genes in evolution. Some of these findings are highly relevant to our understanding of the evolution of eukaryotic genes in general.

* Present address: Division of Tumor Immunology, Dana-Farber Cancer Institute and Harvard Medical School, Boston, MA 02115, USA.

Structure of antibody

The basic unit of an immunoglobulin molecule is a tetramer composed of two identical light chains and two identical heavy chains. Both chains consist of linear arrays of a homology unit of about 110 amino acid residues: a light chain is a dimer while a heavy chain is a tetramer or a pentamer. The homology units located at the amino terminal ends of both chains are diverse in the sequence (variable or V region) and are primarily responsible for the determination of antigen-binding specifications. The vast majority of antibody variability is in the V regions. The other homology units form the constant (C) regions which define the 'type' and 'class' of immunoglobulin molecules and carry polypeptides involved in specific effector functions of the immunoglobulin molecule. A given mammalian antibody molecule contains one of two types of light chains, κ, or λ, and one of five classes of heavy chains, μ, γ, α, etc. At the three dimensional level each homology unit folds into a relatively independent globular domain characterised by several antiparallel β strands. The domain is stabilized by a disulfide linkage provided by a pair of cysteins universally present in the immunoglobulin homology units. The antibody combining site can be viewed as a cleft formed between the light and heavy chain V domains that are held together by characteristic noncovalent bonds. Realization of the occurrence of internally repeated units in the polypeptide chains lead to the hypothesis that the immunoglobulin genes arose by a series of duplications followed by sequence divergence of a promodial gene corresponding in size to a single homology unit. As seen below this idea was born out by the direct structural analysis of the genes.

Structure of antibody genes and the origin of introns

Mammalian immunoglobulin chains are encoded in three unlinked gene families: λ light-chain genes, κ light-chain genes, and heavy-chain genes residing on separate chromosomes. The most intriguing feature of these genes is their dynamic behaviour in the life-span of an individual organism. In fact none of these genes is carried in the germ line in the

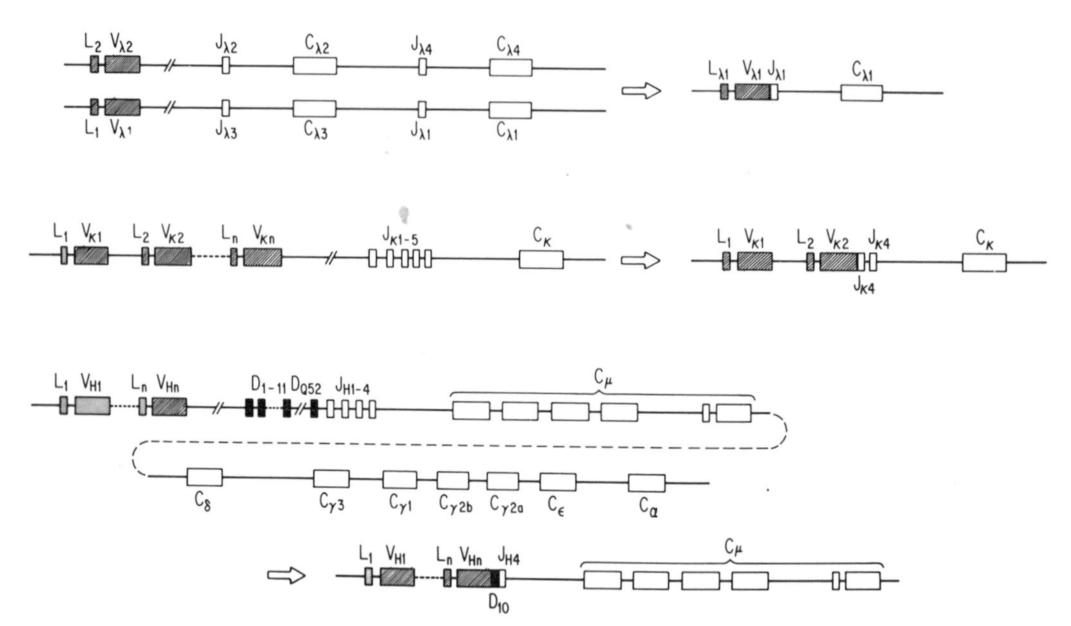

Fig. 1. Organization and somatic reorganization of immunoglobulin gene families. The diagram is for mouse Balb/c. The exon structures of heavy chain genes other than μ is abbreviated. Arrows indicate the change from germline configuration to the configuration observed in B lymphocytes.

complete form. Rather, they are transmitted through generations as gene segments which undergo a series of highly characteristic recombinational events which are strictly restricted to the differentiation of B lineage cells. More specifically, as shown in Fig. 1, the coding potential of the V regions of a light and heavy chains are split in the germline into two (*V* and *J*) and three (*V*, *D* and *J*) gene segments, respectively [1]. In B lineage cells which produce immunoglobulin chains, the corresponding gene segments are assembled into a continuous stretch of DNA as a consequence of *V–J* or *V–D–J* joining. Since a given immunoglobulin gene family contains multiple and different copies of each of the two (*V* and *J*) or three (*V*, *D* and *J*) types of gene segments and that *V–J* or *V–D–J* joinings can occur in a variety of combinations, a large number of different V regions can be generated starting with a relatively limited number of gene segments carried in the germ line. Enzymology of the recombinational events is poorly known. However, it is highly likely that B lymphoid cells contain a specific recombinase that recognizes specific sequences in or around the recombination sites. Indeed the nucleotide sequencing studies revealed that a characteristic heptamer and nonamer sequences are conserved adjacent to the recombination sites in the non-coding regions.

Another important finding made during the analysis of immunoglobulin genes is concerned with the positions of introns. It was shown that each homology unit is encoded by an independent exon [2]. Furthermore the signal peptide present at the amino terminal end of a nascent immunoglobulin chain and the so called hinge region of immunoglobulin γ chains are encoded by their own exons. On the basis of these findings we postulated that some eucaryotic genes may have been generated by shuffling of exons each of which encodes a polypeptide with a distinct function and/or a relatively independent three dimensional structure [3]. According to this hypothesis the origin of introns is non protein-coding sequences that flank mobile exons (see below). Interestingly such shuffling of exons at the DNA level does occur during the late stage of B cell development within the immunoglobulin

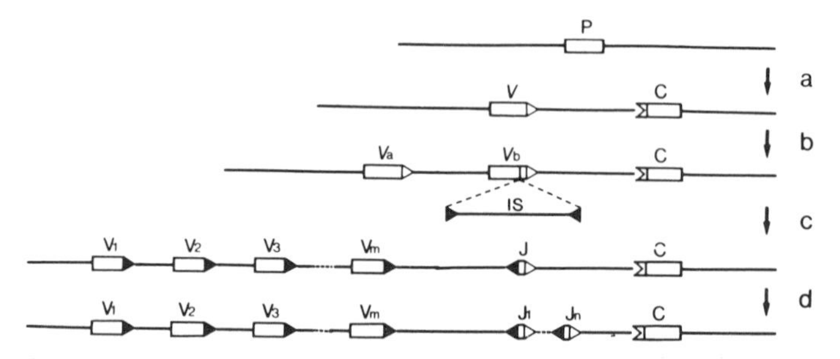

Fig. 2. Evolution of immunoglobulin and T cell receptor genes. 'IS' designates a transposition-like element thought to be inserted to one of the multiple DNA copies for a V region.

heavy chain gene family leading to a 'switch' of the expression of the heavy chain from one class (such as μ) to another (such as γ). While not all introns may have arisen in evolution by the proposed scheme, some recent molecular genetic studies on proteins with internal repetitions support the idea that some introns indeed are generated by gene duplications [4–6].

V–J joining may be a reversal of an ancient insertion event

It is reasonable to assume that the light chain gene containing two homology units was generated by duplication of a primordial gene of the size of a single homology unit followed by divergency and fusion of the two units. Let me consider these processes somewhat more in detail. Since the major source of gene duplication is an unequal crossing over between two homologous stretches of DNA it is very unlikely that the primordial gene corresponds exactly to the duplication unit. Rather, the flanking sequences are also duplicated, leaving a spacer between the two copies of the duplicated genes. Sakano *et al*, assumes that a 'dimeric gene' coding for two immunoglobulin domains was created from the duplicated primordial gene by exploitation of a pre-existing RNA splicing mechanism [7]. (Fig. 2) In this ancient *V–C* gene the *J* gene segment is thought to be an integral part of the ancestral *V* DNA segment because the J region is homologous to the caboxyl end of the C region. The ancestral *V* DNA,

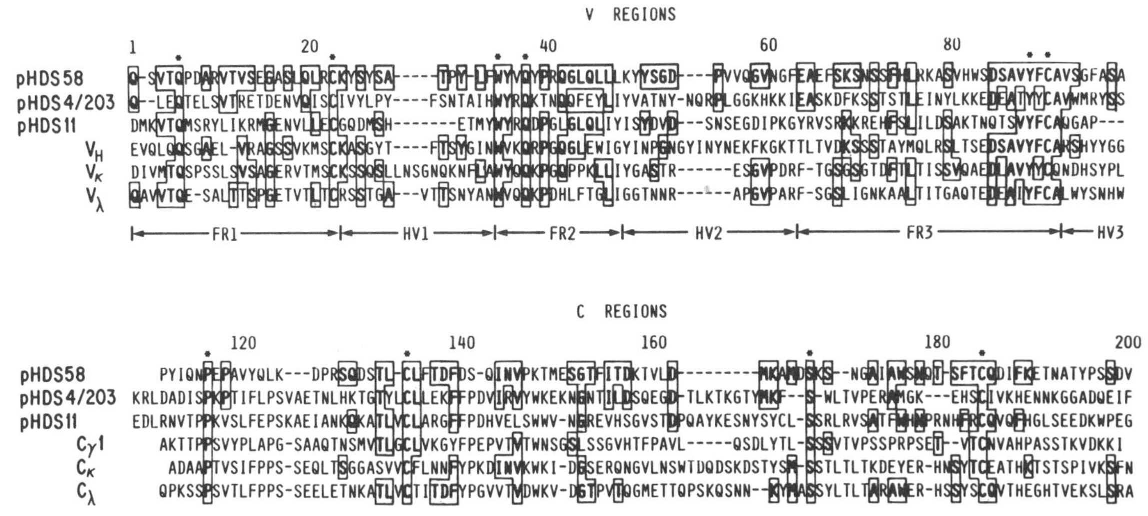

Fig. 3. Amino acid sequence homology among T cell receptor α (pHDS58), β (pHDS11), and γ (pHDS4/203) chains and immunoglobulin chains. Stars designate the residues conserved universally.

together with some flanking sequences, duplicated, triplicated, and so on under pressure to increase V-region diversity. As splicing occurs most efficiently as an intramolecular reaction, one requirement of such a 'polymeric gene' would be that the *V* and *C* DNA sequences are co-transcribed. In the long RNA transcript every V-coding sequence would have a splice donor site at its 3′ end, so that a series of mature mRNA containing different V-coding sequences and the same C-coding sequence could be generated. However this strategy of increasing variability seems to have severe limitations by at least two reasons. First, as the number of *V* DNA segments increases the transcription unit would become intolerably long. Second, as the gene grows it would become increasing difficult to restrict the expression to only one *V* gene segment, which would be required to fulfill the monospecificity of B lymphocytes.

Sakano *et al*, believe that these limitations were resolved by the invention of the somatic *V–J* joining event [7] (Fig. 2.) They propose that such a mechanism was initiated when an IS-like DNA element was accidentally inserted into one of the multiple *V* DNA copies of the ancestral polymeric gene, splitting it into two portions, one corresponding to the present day germ line *V* DNA segment and the other to the *J* DNA segment. While this insertion was fixed in the germ line genome, a mechanism to excise the inserted DNA and rejoin the split V-coding sequences was established in lymphocytes. As excision is a common and major manifestation of prokaryotic IS elements, the basic mechanism for such a process would probably have been available in vertebrates. Once an ancestral *V* DNA is split, the major body of the V-coding sequence and the 3′ flanking sequence containing one end of the IS-like DNA element can multiply independently of the *J* DNA segment. Becaused every copy of the duplicated *V* DNA segment would have the excision-recombination site at its 3′ end, it can rejoin with the *J* DNA segment which is kept in the vicinity of the C-coding DNA. Because the rearranged DNA can bring in its own transcription promoter, it is no longer necessary to confine the duplicated *V* DNA segments and the *C* DNA segment within the same transcription unit in the germ line genome. Furthermore this genetic strategy can provide a means to restrict the expression of immunoglobulin chains to that encoded by the rearranged *V* gene segment. RNA transcribed from unrearranged *V* DNA segments will not be spliced to that from the *C* DNA segment because the former RNA lacks the RNA splice signal.

T cell receptor genes

Unlike immunoglobulins which recognize and bind free antigens and are responsible for humoral immunity, T cell receptors recognize cell-bound antigens in the specific molecular context of self major histocompatibility complex (MHC) products and are responsible for cellular immunity [8–10]. The MHC-restriction appears to be largely acquired by a differentiating T cell population under the influence of MHC antigens expressed in the thymus, suggesting that precursor T cells are selected on the basis of their reactivity with MHC determinants expressed in the host thymus (thymic education) [11–13]. In order to understand the molecular basis of MHC restriction and thymic education it is of paramount importance to determine the structure of a T cell receptor. For many years T cell receptors were very elusive. However the first glimpse of these molecules was obtained in 1982 and 1983 through the development of effective antibodies [14–16]. These studies lead to the conclusion that the portion of the receptor determining specificity is a heterodimeric glycoprotein corresponding to a molecular weight (M_r) of about 90 K consisting of a 40–45 K α subunit and a 40–45 K β subunit. Moreover, peptide fingerprint analysis suggested that both the α and β subunits, like immunoglobulin heavy and light chains, are composed of variable (V) and constant (C) regions.

More recently the genes coding for these polypeptide chains were cloned [17–20]. The nucleotide sequences of these DNA clones permitted the deduction of the complete primary structure of the α and β subunits. When the predicted amino acid sequences of the α and β chains were compared with all other known protein sequences it was discovered that these chains are significantly homologous to immunoglobulin chains, particularly light chains (30–35% homology) (Fig. 3). The conserved residues include the two pairs of cysteines that are involved in the intradomain disulfide linkages in the V and C regions, the invariant aromatic rings Trp (34), Tyr (88), and

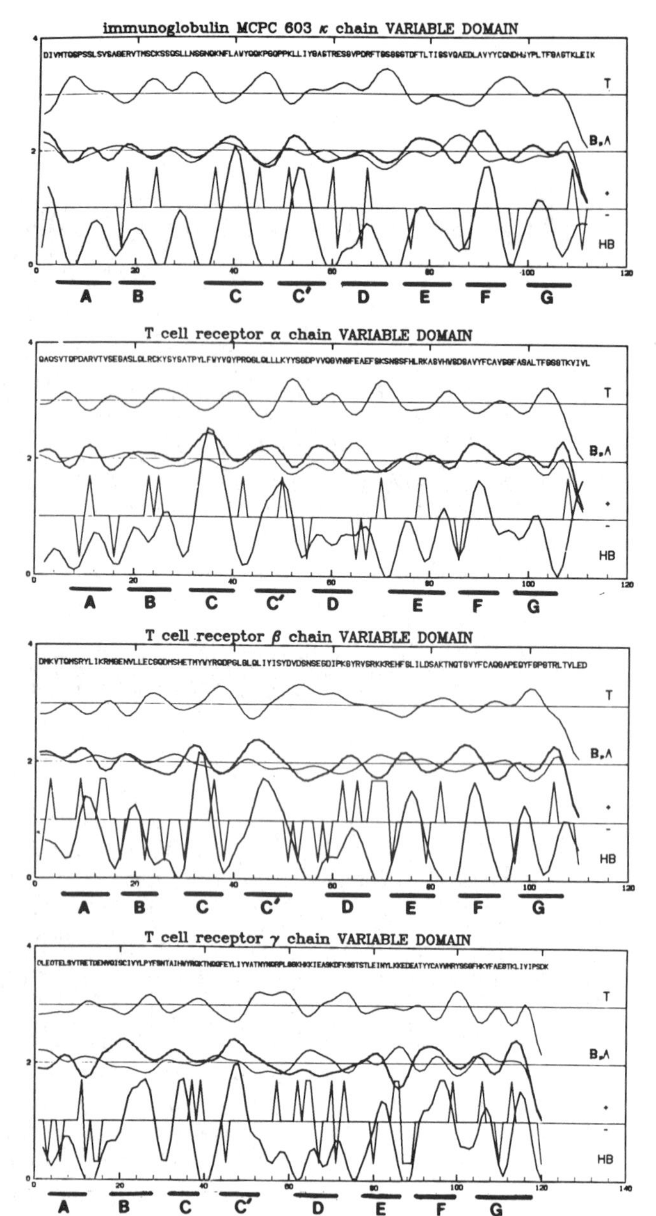

Fig. 4. 'Secondary Structure Profiles' of immunoglobulin and T cell receptor V regions. Each box shows from top to bottom, reverse turn propensity, T; α-helix (thin line A) and β sheet (heavy line B) propensities; profile of charged residues (+ −); and hydrophobicity. These were computed by J. Novotný of Massachusetts General Hospital in Boston as described [42]. Bars below the boxes indicate β strands identified from X-ray data (MOPC 603 V_L domain [43]) or deduced from the profiles (T cell receptor chains).

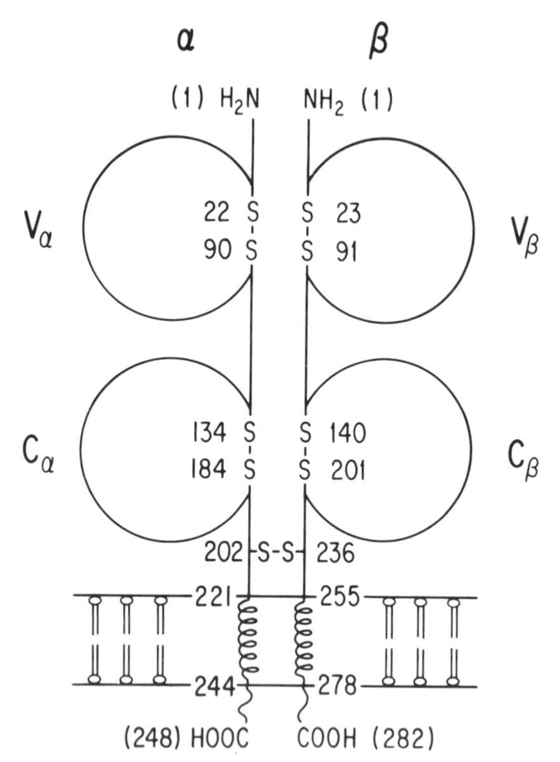

Fig. 5. Structure of the T cell receptor αβ heterodimer on the surface of a cloned functional CTL, 2C. Numbers represent the amino acid positions. 'S' represents the sulfur involved in intrachain or interchain disulfide linkages.

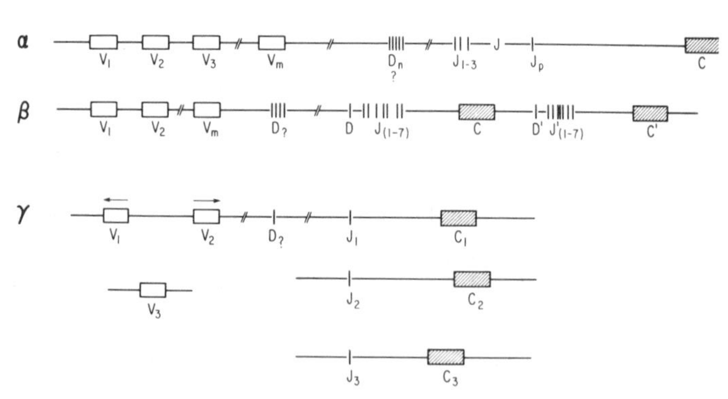

Fig. 6. Germline organization of T cell receptor genes α, β and γ.

Phe (101) at V–V interface, and the invariant Gln (37) that form a hydrogen-bond across the V–V interface. Furthermore the 'secondary structure profile' of the V_α and V_β regions (see Fig. 4) strongly suggest that these V regions are folded into the typical immunoglobulin domains characterized by anti-parallel β strands [21]. Figure 5 depicts the gross structure of the αβ heterodimer deduced from these analyses.

The striking structural similarity between immunoglobulin chains and T cell receptor subunits parallels with the equally striking resemblance of the organization and structure of the corresponding genes. As summarized in Fig. 6 the α and β chain genes are also split to *V*, (*D*), *J*, and *C* gene segments in the germline genome [22–25]. In T cells a selected set of a *V*, (*D*), *J* gene segments are assembled to form a continuous stretch of DNA coding for the V region. Furthermore the T cell receptor gene segments seem to share with immunoglobulin gene segments the same set of conserved heptamer and nonamer sequences as the signals for the V–J or V–D–J joining, although some interesting variations can be seen in the association of the spacer lengths and the gene segments.

In conclusion, immunoglobulins the T cell receptors show clear resemblance both at the level of the protein and DNA. It is inescapable to conclude that they have diverged from a common primordial gene.

T cell γ gene

During the search for the α and β chain gene we encountered the third gene (referred to as γ) expressed in the same cytotoxic T cell clone, 2C from which the α and β cDNA clones were isolated. [26] This gene has many properties in common with α and β genes which include [26–28]: (1) assembly from gene segments resembling the gene segments for immunoglobulin V, J, and C regions; (2) rearrangement and expression of these genes segments in T cells and not in B cells; (3) low but distinct sequence homology to immuno-

Table I. *Cell surface proteins with immunoglobulin-like domains*

	Polypeptide chains		Function	Distribution	DNA rearrangement
1	Immunoglobulin	κ light	Antigen recognition	B lymphocytes	
2		λ light			+
3		Heavy			
4	T cell receptor	α	Antigen and MHC recognition	T lymphocytes	
5		β			+
6		γ			
7	MHC Class I	Heavy	Restriction of	Many cell types	
8		β_2M	T cell response	but low in neuronal or glial cells	−
9	MHC Class II	α	Restriction of both	B cells, activated	
10		β	B and T cell responses	macrophages, dendritic cells, ephitherial cells	
11	T4 (L3T4)		Unknown	Most T helper cells	−
12	T8 (Ly2)		Unknown	Most T killer and T suppressor cells	−
13	Thy 1		Unknown	Thymocytes, neuronal cells and others	−
14	OX-2		Unknown	Thymocytes, brain cells and others	−
15	Poly Ig receptor		Transcellular transport of poly IgA and IgM	Glandular epithelial cells	−

globulin V, J, and C regions; (4) conservations of key residues involved in domain folding and interdomain interaction; (5) secondary structure profile; (6) sequences reminiscent of the transmembrane and intracytoplasmic regions of integral membrane proteins; (7) a cystein residue at the position expected for a disulfide bond linking two subunits of a dimeric membrane protein molecule. In spite of the long list of similarity between the γ gene and the α and β genes there seems to be some features unique to the γ gene which include [29–32]: (1) the diversity of the germline gene segments is significantly limited (see Fig. 6); (2) the expression seems to be dispensable for some helper T cells; (3) the expression is primarily in immature T cells while the β gene expression is constitutive throughout T cell differentiation and the α gene expression is primarily in mature T cells.

While no definitive function has been assigned to the putative γ chain we believe that the available data are best interpreted by assuming that the γ chain constitutes at least part of the receptor on immature T cells that undergo intrathymic selection (i.e. thymic education). Regardless whether this idea is correct it is intriguing that this gene and its putative gene product is so analogous to the α and β genes and their gene products.

Involved in Antigen Recognition

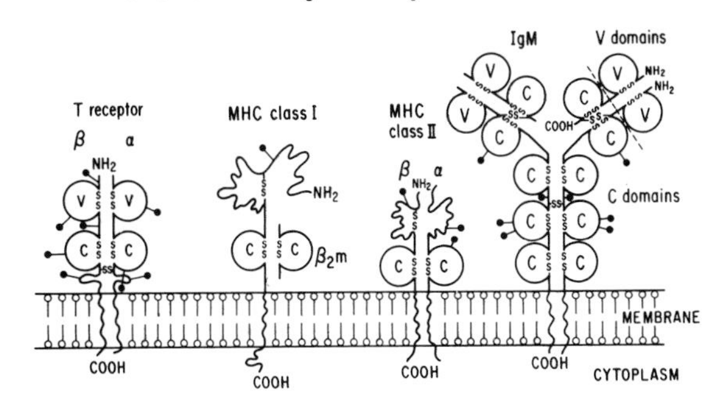

No Role in Antigen Recognition

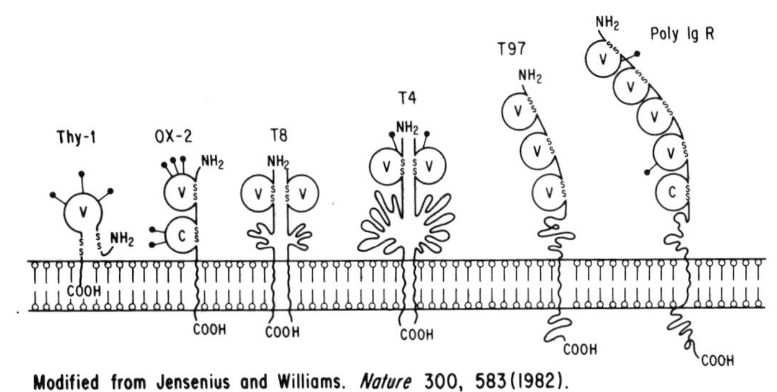

Modified from Jensenius and Williams. *Nature* 300, 583 (1982).

Fig. 7. Immunoglobulin superfamily.

Other immunoglobulin-like proteins and genes

Prior to the discovery of the T cell receptor α. β, and γ chains, several proteins were shown to have significant homology to immunoglobulin chains. These include the Class I and Class II MHC proteins [33, 34], the T cell surface protein Thy-1 [35], and the poly IgA and IgM receptor which is present on the surface of glandular epitherial cells and mediates transcellular transport of these immunoglobulins (Table I) [36]. Furthermore more recent studies are expanding the list of immunoglobulin-like proteins. For instance the T4 and T8 antigens present on the surface of most human helper T cells and cytotoxic (and suppressor) T cells, respectively, each contains

Table II. *Similarities between Ig V-like domains*

V_L	V_A	OX-2	T8	Thy-1	IgRc	T97-1	T97-2	T97-3	
31	33	30	25	23	22	20	22	24	V_H
	34	29	33	24	21	19	22	24	V_L
		27	28	22	31	21	23	20	V_A
			22	29	21	22	25	15	OX-2
				25	26	25	28	27	T8
					22	22	23	19	Thy-1
						22	14	17	T97-1
							16	23	T97-2
								18	T97-3

V_H, Human NEWM; V_L, mouse lambda, MOPC104E; V_A, mouse T cell receptor alpha, HDS58; Thy I, Rat Thy-1; IgRc, poly Ig receptor (4th domain).

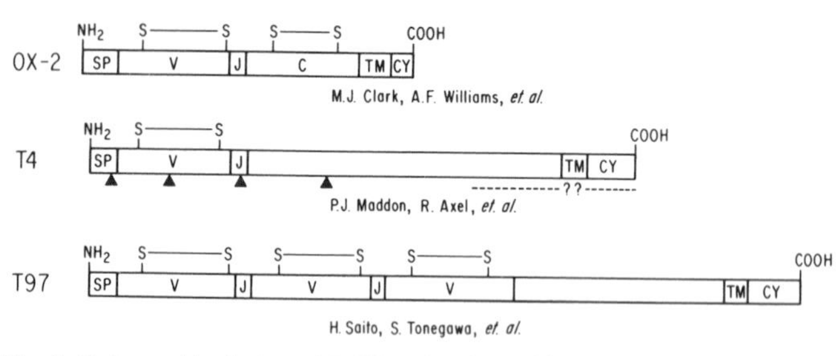

Fig. 8. Polypeptide chains with VJ region but without gene rearrangement.

an immunoglobulin V-like domain [37, 38, 39]. (Table I and Fig. 7). In addition OX-2 glycoprotein known to be present on the surface of rat thymocytes and some brain cells is composed to one V- and one C-like domain plus a transmembrane and intracytoplasmic peptides [40] (Table I and Fig. 7). Furthermore we have recently identified a cDNA clone (T97) in the library prepared from the cytotoxic T cell clone, 2C. The T97 gene seems to be expressed specifically in T cells and its putative gene product contains three immunoglobulin V-like domains, a long (about 160 residues) non immunoglobulin-like sequence, a transmembrane peptide and an intracytoplasmic peptide (Fig. 7).

The sequence homology among these immunoglobulin-like polypeptide chains can be as low as 14% (T97 domain 2 *vs* T97 domain 1) but in most cases it falls to a value between 20 and 35% (see Table II). In those cases where the sequence homology to an immunoglobulin domain is relatively low, some of the residues central to the immunoglobulin fold such as the invariant cysteins and tryptophane are missing (example, the N-terminal homology units of the Class II MHC products and Thy-1, etc). In these cases it is highly unlikely that the homology unit in question will fold into the immunoglobulin-like domain. On the other hand it is virtually certain that some other homology units such as those at the N-terminal of T4 and T8 fold into immunoglobulin-like domains (Figs. 4, 8).

Conclusions

A great deal has been learned by molecular genetic studies of immunoglobulins and T cell receptors. The genetic origins of antibodies, one of the most debated problems in immunology has now been resolved at least in the outline: somatic recombination among germline-carried gene segments and somatic hypermutation (not discussed in this article, see ref. 1) in the assembled immunoglobulin genes play key roles in the generation of the tremendous diversity seen in the population of antibodies synthesized by a single organism. Another problem, also heatedly debated in immunology, namely whether a T cell uses an immunoglobulin as its antigen receptor is now resolved beyond doubt: It does not. Instead, T cell receptors are encoded by two sets of genes that are clearly distinct from immunoglobulin genes but share many features in common with them including somatic rearrangement. At least one additional set of genes (γ), also immunoglobulin gene-like, may also be involved in the determination of T cell specificity in some stages of T cell differentiation. The

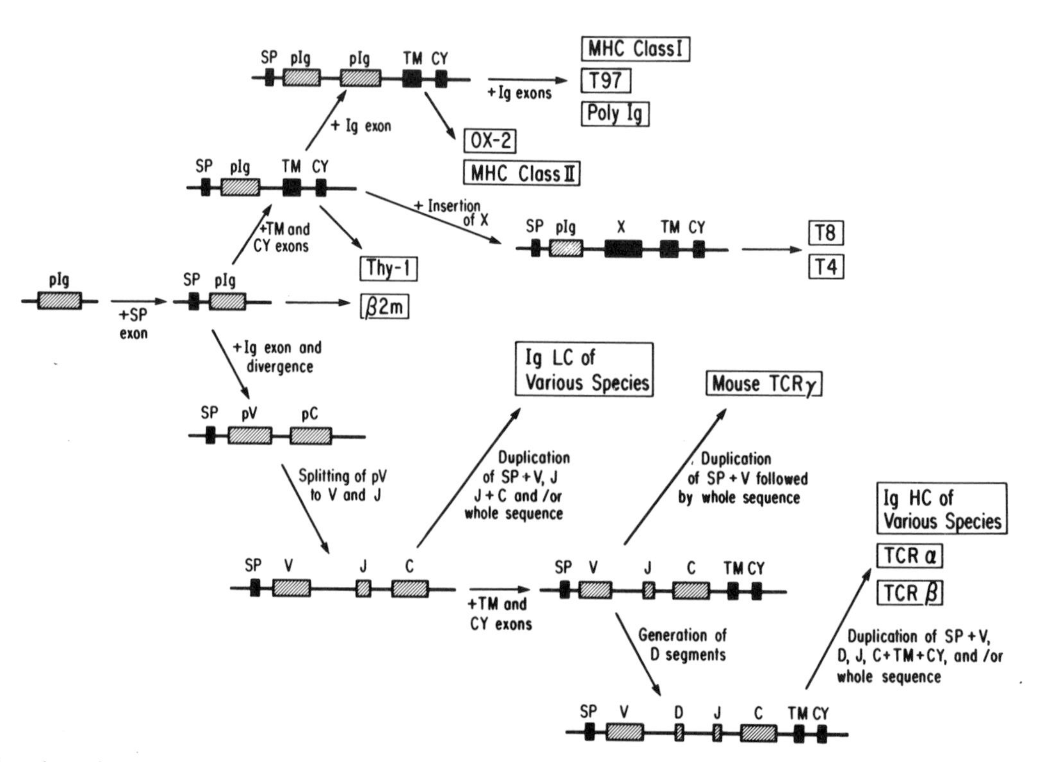

Fig. 9. A possible pathway for the evolution of immunoglobulin supergene family.

organization of the various immunoglobulin and T cell receptor gene segments and their nucleotide sequences shed light on the evolutionary origin of the somatic rearrangement: It seems that under the pressure to diversify the V region and to keep the monospecific nature of B or T cell clones, the ancestor of the vertebrates took advantage of the accidental insertion of a transposition-like element into an ancestral V copy. The present somatic *V–J* joining can be viewed as the reversal of this insertion event.

The detailed analysis of immunoglobulin and T cell receptor genes and discoveries of the ever increasing number of proteins containing immunoglobulin homology units and even immunoglobulin-like domains are contributing to our understanding of the behaviour of eukaryotic genes in general in evolution. First, the exon–intron structure of these genes clearly demonstrate the correlation of an exon and a functional and/or structural unit of the protein. Second, all immunoglobulin-like genes can be generated from a common primordial gene of one homology unit by duplication, subsequent sequence divergency of the duplicated exon, and an occasional addition of a new exon or a various combination of these events (Fig. 9). This dramatically illustrates the tremendous versatility of the gene creation mechanism based on exon shuffling. Third, at least some introns in these genes and by inference in some other genes are almost certainly derived from the spacer between the duplicated exons. Fourth, it is intriguing that most of the immunoglobulin-like proteins seem to be integral membrane proteins. This suggests that the immunoglobulin fold is repeatedly used in evolution for binding of a ligand. Alternatively, the domain–domain interaction akin to the $V_L - V_H$ or $C_L - C_{H2}$ may be widely used by cell surface receptors in order to accomplish interactions between cells.

Finally, the discovery of the structurally related and functionally distinct sets of proteins (sometimes referred to the immunoglobulin supergene family) make one wonder whether other eucaryotic genes may belong to their own supergene families. If so, the eucaryotic genome may be composed of no more than a few thousand supergene families.

Acknowledgements

We wish to thank many colleagues both at Basel Institute for Immunology and M.I.T. for their contribution. We also thank P. J. Maddon and R. Axel for communicating unpublished results. The work carried out at Basel Institute was supported by Hoffmann La-Roche Co. and that at M.I.T. by NIH grants CA28900 and AI17879.

References

1. Tonegawa, S., *Nature (London)* **302**, 575 (1983).
2. Sakano, H., Rogers, J. H., Hüppi, K., Brack, C., Traunecker, A., Maki, R., Wall, R. and Tonegawa, S., *Nature (London)* **277**, 627 (1979).
3. Tonegawa, S., Maxam, A. M., Tizard, R., Bernard, O. and Gilbert, W., *Proc. Natl. Acad. Sci. USA* **75**, 1485 (1978).
4. Jung, A., Sippel. A. E., Grez, M. and Schutz, G., *Proc. Natl. Acad. Sci. USA* **77**, 5759 (1980).
5. Yamada, Y., Avvedimento, V. E., Mudry, J. M., Ohkubo, H., Vogeli, G., Irani, M., Pastan, I and de Crombrugghe, B., *Cell* **22**, 887 (1980).
6. Odermatt, E., Tamkun, J. W. and Hynes, R. O., *Proc. Natl. Acad. Sci. USA* **82**, 6571 (1985).
7. Sakano, H., Hüppi, K., Heinrich, G. and Tonegawa, S., *Nature* **280**, 288 (1979).
8. Kindreed, B. and Shreffler, D. C., *J. Immun.* **109**, 940 (1972).
9. Katz, D. H., Hamaoka, T., Dorf, M. E., Maurer, P. H. and Benacerraf, B. J., *J. Exp. Med.* **138**, 734 (1973).
10. Zinkernagel, R. M. and Doherty, P. C., *Nature (London)* **248**, 701 (1974).
11. Zinkernagel, R. M., Callahan, G. N., Althage, A., Cooper, S., Klein, P. A. and Klein, J., *J. Exp. Med.* **147**, 882 (1978).
12. Bevan, M. J. and Fink, P. J., *Immun. Rev.* **42**, 3 (1978).
13. von Boehmer, H., Haas, W. and Jerne, N. K., *Proc. Natl. Acad. Sci. USA* **75**, 2439 (1978).
14. Allison, J. P., McIntyre, B. W. and Bloch, D., *J. Immunol.* **129**, 2293 (1982).
15. Haskins, K., Kubo, R., White, J., Pigeon, M., Kappler, J. and Marrack, P., *J. Exp. Med.* **157**, 1149 (1983).
16. Meuer, S. C., Fitzgerald, K. A., Hussey, R. E., Hogdon, J. C., Schlossman, S. F. and Reinherz, E. L., *J. Exp. Med.* **157**, 705 (1983).
17. Yanagi, Y., Yoshikai, Y., Leggett, K., Clark, S. P., Aleksander, I. and Mak, T. W., *Nature (London)* **308**, 145 (1984).
18. Hedrick, S. M., Nielsen, E. A., Kavaler, J., Cohen, D. I. and Davis, M. M., *Nature (London)* **308**, 153 (1984).
19. Saito, H., Kranz, D. M., Takagaki, Y., Hayday, A. C., Eisen, H. and Tonegawa, S., *Nature (London)* **312**, 36 (1984).
20. Chien, Y.-H., Becker, D. M., Lindsten, T., Okamura, M., Cohen, D. J. and Davis, M. M. *Nature (London)* **312**, 31 (1984).
21. Novotný, J., Tonegawa, S., Saito, H., Kranz, D. and Eisen, H., *Proc. Natl. Acad. Sci. USA* **83**, 742 (1986)
22. Gascoigne, N., Chien, Y.-H., Becker, D., Kavaler, J. and Davis, M. M., *Nature (London)* **310**, 387 (1984).
23. Mallissen, M., Minard, K., Mjolsness, S., Kronenberg, M., Goveriman, J., Hunkapiller, T., Prystowsky, M. B., Yoshikai, Y., Fitch, F., Mak, T. W. and Hood, L., *Cell* **37**, 1101 (1984).
24. Hayday, A. C., Diamond, D. J., Tanigawa, G., Heilig, J. S., Folsom, V., Saito, H. and Tonegawa, S., *Nature (London)* **316**, 828 (1985).
25. Winoto, A., Mjolsness, S. and Hood, L., *Nature* **316**, 832 (1985).
26. Saito, H., Kranz, D. M., Takagaki, Y., Hayday, A. C., Eisen, H. N. and Tonegawo, S., *Nature (London)* **309**, 757 (1984).
27. Hayday, A. C. Saito, H., Gillies, S. D., Kranz, D. M., Tanigawa, G., Eisen, H. N. and Tonegawa, S., *Cell* **40**, 259 (1985).
28. Kranz, D. M. *et al.*, *Science* **227**, 941 (1985).
29. Kranz, D. M., Saito, H., Heller, M., Takagaki, Y., Haas, W., Eisen, H. N. and Tonegawa, S., *Nature (London)* **313**, 752 (1985).
30. Heilig, J. S., Glimcher, L. H., Kranz, D. M., Clayton, L. K., Greenstein, J. L., Saito, H., Maxam, A. M., Burakoff, S. J., Eisen, H. N. and Tonegawa, S., *Nature (London)* **317**, 68 (1985).
31. Raulet, D. H., Garman, R. D., Saito, H. and Tonegawa, S., *Nature* **314**, 103 (1985).
32. Collins, M. K. L., Tanigawa, G., Kissonerghis, A.-M., Ritter, M., Tonegawa, S. and Owen, M. J., *Proc. Natl. Acad. Sci. USA* **82**, 4503 (1985).
33. Strominger, J. *et al.*, *Scand. J. Immunol.* **11**, 573–593 (1980).
34. Larhammar, D. *et al.*, *Cell* **30**, 153 (1982).
35. Williams, A. and Gagnon, J., *Science* **216**, 696 (1982).
36. Mostov, K. E. *et al.*, *Nature (London)* **308**, 37 (1984).
37. Kavathas, P., Sukhatme, V. P., Herzenberg, L. A. and Parnes, J. R., *Proc. Natl. Acad. Sci. USA* **81**, 7688 (1984).
38. Littman, D. R., Thomas, Y., Maddon, P. J., Chess, L. and Axel, R., *Cell* **40**, 237 (1985).
39. Maddon, P. J., Littman, D. R., Godfrey, M., Maddon, D. E., Chess, L. and Axel, R., *Cell* **42**, 93 (1985).
40. Clark, M. J., Gagnon, J., Williams, A. F. and Barclay, A. N., *EMBO J.*, **4**, 113 (1985).
41. Tonegawa, S., *Scientific American* **253**, 122 (1986).
42. Novotný, J. and Auffray, C., *Nucleic Acids Res.* **12**, 243 (1984).
43. Segal D., Padlan, E. A., Cohen, G. H., Rudikoff, S., Potter, M. and Davies, D., *Proc. Natl. Acad. Sci. USA* **71**, 4298 (1974).

Chemica Scripta 1986, **26B**, 351–355

The Intelligent Immune System

Hans Wigzell

Department of Immunology, Karolinska Institute, Box 60400, S-104 01 Stockholm, Sweden

Paper presented at the Conference on 'Molecular Evolution of Life', Lidingö, Sweden, 8–12 September, 1985

Abstract

In order to defend the multicellular society that our body is composed of, a sophisticated immune system has evolved. This system has the innate capacity to react against virtually everything above a certain molecular size but is undergoing a selective education and evolution within the individual to sense the difference between self and non-self. The present article gives some insight into how this is achieved whilst still leaving the system extremely sensitive to even minor perturbations of self structures.

Introduction

The meaning of the word 'immune' is to be exempt – that is, spared from something. In ancient Rome the word *immunis* may have meant the privilege of being exempt from paying tax, but 'immune' in the language of the layman soon became intimately linked to induced resistance to infectious diseases. People long ago realized that when a plague passed through a district leaving many diseased and dead, it rarely returned to cause disease in those previously infected. They were exempt, immune. This immunity was selective, applying only to that specific disease, and could be maintained for lengthy or sometimes life-long periods. The immune system thus has the unique capacity, besides the central nervous system, of being able to learn and remember, in a selective manner, previous experiences – that is, to become educated. Over the millions of years that passed until the creation of the present higher animals with their sophisticated immune systems, many steps of evolution occurred in the creation of those systems. Every multicellular organism has thus developed various means to defend itself against living pathogens. The means vary widely, as seen by comparing immune reactions in plants and animals. Although less plastic and variable in their defence mechanisms, plants do have a sizeable battery of defence mechanisms of an immune nature. They can even display a prolonged state of enhanced resistance somewhat reminiscent of a rather non-specific memory. However, only in the higher animals does one encounter truly 'intelligent' immune systems with an enormous capacity to cognize, recognize and remember. The present article will deal predominantly with the immune system as it is organized in higher animals.

The demands put to an intelligent immune system

A prime concern of a multicellular organism is to protect its territory from pathogens. A functioning immune system must be able to defend the tissues of the individual from invading organisms and their potentially toxic products in a rapid and efficient manner, despite the fact that the shapes of the molecules produced by the invader may occur in innumerable and sometimes entirely new forms. In order to save 'energy' the system must be selective and thus have an inherent capacity to produce a corresponding variability in the pathogens amongst the defender molecules, and allow a sharp increase in the relevant defence molecules when called upon. As time is of vital matter when it comes to combat, the ability to mount a rapid defence should be possible even by a virgin immune system and even more so would immune memory have become induced.

The creation of a highly flexible immune machinery with an innate ability to react against any possible invading organism also creates a danger – the possibility of autoimmune reactivity. Reactions against self constituents in an aggressive manner by the immune system must not be allowed, despite the fact that most 'self' molecules do not carry any general unique markers of a chemical nature. We now know that the immune system in fact has developed a mirror-image capacity for specific immunity, called immune tolerance, which largely allows the immune system to avoid causing direct autoimmune disease.

Our immune system has thus gradually evolved into a system with the ability to react against virtually anything of 'non-self' nature provided certain physicochemical requirements are fulfilled. At the same time damaging reactions against 'self' by the same system do not normally take place. Both kinds of reactions are specific and reached by a process of education of cells in a positive or negative manner. The basic requirements of education and the construction of a sensitive yet fairly safe immune defence machinery will serve as the main topic of this article.

The cellular society of the immune system

During evolution the immune system of the higher animals becomes increasingly complex with regard to the number of cell types and their corresponding products. The general trend during evolution is one of more specialized function, where each cell type or subset of cells is given a more and more restricted role to play in the immune orchestra. But in general it seems that, in the evolution of the immune system during phylogeny, ancient cell types are allowed to remain, although maybe in reduced numbers or with less importance, and in addition new and more sophisticated but also more restricted cell types are evolving. The more primitive cell types in our immune system have as a common feature a broad potential claviature – that is, a capacity to function in a variety of ways in a multispecific manner. The prototype for this cell type is the macrophage or monocyte, mostly known for its ability to ingest (macrophage = big eater) micro-organisms in order to destroy them via digestion. This group of cells is endowed with a certain talent to recognize many micro-organisms, notably bacteria, in a rather non-specific manner. They will function in a highly efficient manner if such organisms are coated by products from the more recently evolved and antigen-specific lymphocytes. During evolution also the monocyte/macrophage family has become diversified and representatives of this family can now be found in differentiated forms as more specialized cells, such as the Langerhans cells in epidermis, the Kuppfer cells in the liver and the glia cells in the CNS.

Typical for our immune system is the complex build-up with many different cell types acting together in concert to create a functional and highly intelligent immune defence system. Communication between the various cells and their products is here an essential part and it can be safely stated that a *single* cell from the immune system is largely incapable of performing well when confronted with a foreign structure or organism. This is particularly evident when it comes to the induction of an immune response, whereas the induced effector cells generated during this response may function at the single-cell level for a limited time-span. In the immune response in higher animals a key role is played by the antigen-specific cells of the immune system – that is, the lymphocytes. They are cells with an inherent capacity to react against a particular structure, antigen, and each lymphocyte when becoming immunocompetent will normally find itself equipped with antigen-binding receptors, largely unique for that particular lymphocyte and its subsequent descendants. The B lymphocytes have as their sole function to produce antibodies, immunoglobulins, whereas the thymus-derived T lymphocytes carry out their functions in a cellular, cell-bound fashion through catalysing, inhibitory or cytolytic reactions. The specific immune response normally occurs only when several different cell types react together against antigen, and a schematic picture of such a reaction is depicted in Fig. 1. Through this collaborative network of cells our immune system is able to cognize and recognize very low molar concentrations of foreign material amongst high molar concentration of self molecules and respond to these foreign molecules with a rapid, largely exponentially developing specific immune response.

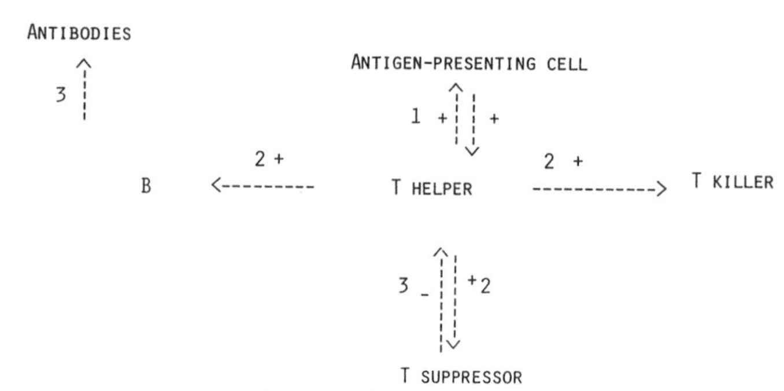

Fig. 1. Reactions initiated when antigen is recognized by T and B lymphocytes. + or − indicate consequence for target-cell function. 1, 2 or 3 indicate the sequence of events in the reaction.

Cognition of self versus recognition of non-self

A key element in understanding the evolution of our sophisticated immune system resides at the level of self versus non-self cognition. Whereas in the early evolution of eukaryotes the single cellular organisms could defend themselves against attack by, for instance, phagocytosis a request for restrictive aggressiveness toward self becomes self-evident when multicellular organisms are formed. As the immune defence reactions at this stage of evolution are largely dominated by cell-mediated reactions, cell surface molecules identifying cells as being of the 'right' type and not belonging to any invading army would be of value.

It is a general belief amongst immunologists that there exists a particular genetic system in each species which, among other things, may have this role to fulfill. The system is called the major histocompatibility complex, owing to the fact that the gene products of the system were first recognized by their ability to initiate violent cell-mediated immune reactions if introduced into individuals lacking the particular variants of MHC molecules present on the grafted cells. The MHC antigens are glycoproteins on the outer surface of virtually all nucleated cells in the body and they come in two major forms called class I and II [1]. In man the number of allelic forms of each MHC gene is quite high and the possible number of MHC gene constellations that a single individual may express supersedes by far the number of human beings on earth. The MHC system (in man called HLA) is thus in virtually any species analysed a most polymorphic system, where polymorphism has obvious advantages for the survival of the species. Part of the importance of the MHC genes resides at the level of their capacity to function as highly important molecules within our own bodies during immune reactions.

What is then the role of the MHC class I and II molecules in our immune system? These molecules play a key role as antigen-presenting molecules for the T lymphocytes in such a manner that antigenic fragments in association with particular MHC molecules will constitute the antigen against which a particular T cell with fitting antigen-binding receptors will react. To exemplify this one can look at Table I, where a virus infection in an individual has induced s.c. virus-specific killer T lymphocytes. An analysis of the specificity of such cytolytic T cells does reveal, however, that their specificity is not for the virus, nor for the MHC, but each single T cell reacts with a particular virus fragment in conjunction with a particular self-MHC molecule [2]. We now know that T lymphocytes normally cannot 'see' antigen in free solution but only at cell-surface levels, where MHC molecules and exogenous antigen may associate to create 'antigen'. This is

Table I. *The specificity of an 'influenza-specific' killer T cell scenario.*

(The killer T cell is isolated from a human being with the HLA-$A^FA^MB^FB^MC^FC^M$ and infected with influenza. Which target will be killed by this cytolytic cell?)

Example:	Target HLA						
A^F	A^M	B^F	B^M	C^F	C^M	Influenza?	Lysis?
+	+	+	+	+	+	+	+
+	+	+	+	+	+	−	−
+	−	−	−	−	−	+	−
−	+	−	−	−	−	+	−
−	−	+	−	−	−	+	−
−	−	−	+	−	−	+	+
−	−	−	−	+	−	+	−
−	−	−	−	−	+	+	−

Conclusion. The single killer T cell is specific for HLA-B^M plus 'Influenza' and for any target cell from any other individual which happens to have that *HLA*-antigen and is infected with influenza virus.

in contrast to B lymphocytes, whose antigen-binding receptors do not express a similar obsession for MHC structures and which can readily react with tertiary configurations on antigenic molecules in solution. There exist several experimental animal systems where the presence of a particular MHC gene constellation results in an ability or inability to respond by T cell immunity against a defined antigen [3]. The constitution of an individual as to MHC genes does accordingly contribute to the immune capacity of that individual to respond more or less vigorously against a particular micro-organism.

The genes coding for the antigen-binding receptors on T lymphocytes are not linked to those coding for the MHC genes of the same individual. The potential specificity-generating machinery of a T lymphocyte has thus no innate capacity to generate preferentially receptors which do not react strongly with self-MHC in a damaging manner or alternatively react efficiently with the self-MHC molecules if the latter associate with a relevant antigenic fragment. Our immune system is solving this via an evolutionary process in part taking place within the thymus during the differentiation of T lymphocytes within the individual.

From the bone marrow there is a constant flow of prothymocytes to the thymus. Upon entering the thymus these cells are immuno-incompetent, but within this tissue they will soon rearrange their genes coding for T cell antigen-binding receptors and express the gene products. This takes place in the cortex of the thymus in the presence of a high rate of cellular proliferation and also cell death. Eventually a minority of the cells will leave from the cortex to the medulla of the thymus in the form of immuno-competent mature T lymphocytes for further export into the peripheral tissues. Although the exact molecular mechanisms remain obscure there occurs within the cortex of the thymus a specific education or evolution according to Darwinian concepts. This leads to the following with regard to T-cell specificity in relation to self-MHC molecules, in particular of the class II type. *If* the antigen-binding receptors when they first become expressed on the cortex T cells happen to display a high binding affinity for any self-MHC molecule, this will lead to the elimination of that cell – that is, immune tolerance via clonal elimination. On the other hand, if the receptors would have a significantly lower but yet measurable affinity for self-MHC molecules, this will be of survival advantage to the T cells and they will be given the possibility to final maturation and export from the thymus. Within this T-cell population selected for weak self-MHC reactivity there are also present the cells which react against foreign MHC; for instance, in a kidney-graft situation, with vigorous rejection reactions. This can be easily understood as the very same receptors for antigen on a single T cell can (*a*) react weakly with self-MHC, (*b*) strongly with the same self-MHC molecule if a special peptide fragment is around as well, and also (*c*) avidly with a particular foreign MHC molecule if it happens to resemble the peptide + self-MHC complex. The cells in the thymus which are responsible for 'teaching' T cells this self-MHC preferential selection according to the above scenario are probably long-lived bone-marrow-derived macrophages acting in concert with locally produced thymus hormones with an impact on growth and differentiation of the thymocytes. As a result of this education within the thymus our T lymphocytes can be regarded as 'half' autoimmune with regard to a particular self-MHC molecule, which makes the cell extremely sensitive to further MHC perturbation by, for example, peptide fragments emanating from the interior of a virus-infected cell.

But our immune system has not been satisfied with these solutions for eliminating damaging self-reactive T cells whilst maintaining high T-cell sensitivity against modified self MHC molecules. During the stages of differentiation within the thymus the T cells in parallel in the capacity to display antigen-binding receptors, also start production of other cell-surface proteins. These proteins, which also serve as differentiation markers, have in several cases been identified as to functional properties. In man, two glycoproteins on mature T lymphocytes called T4 and T8 are of particular interest in this context. A single T cell in the peripheral tissues is either T4- or T8-positive. Most T4 cells have the capacity to function as helper cells for other cell types whereas most T8 cells are either suppressive or cytolytic. However, an even better correlation between presence of T4/T8 and function comes when the antigen-specificity of the T cells are considered. Virtually every T4-positive T cell does 'see' antigen together with class II MHC molecules, HLA-D, whereas the T8 cells, in contrast, see antigen in conjunction with class I MHC (that is, HLA-A, B or C molecules), and the same relationship occurs in the mouse [4, 5]. The functions of T4 and T8 respectively have been shown by several investigators to involve an invariant capacity to bind to HLA-D or HLA-A, B and C molecules respectively on other cells. In fact the binding of an antigen-specific T cell of $T4^+$ type towards its antigen (an $HLA\text{-}D^+$ antigen-presenting cell coated with the relevant antigen) is predominantly taking place via binding of T4 to HLA-D molecules and not via the antigen-specific receptors present on the same T lymphocytes. It has been known for a long time that T4-positive T lymphocytes 'like' to attach to any HLA-D-positive cell regardless of whether the latter has been in contact with a relevant antigen or not. This contact is limited in time and will leave no imprints on the two cells involved. *If*, however, during this time the antigen-binding receptors find a fitting antigen-MHC complex on the presenting cell as well, the contact will not only last longer but will also very likely result in T-cell activation into immune effector function. In this contact, T4 and T8 molecules accordingly play an important role in facilitating the possibilities of T-cell contact with potentially interesting target cells, but the actual signals for immune activation of the T cells would seem to require the binding of the antigen-receptor molecules to a relevant antigen. This construction not only reduces the binding strength required between antigen-binding receptors and antigen, which will significantly enhance the number of potentially antigen-responsive lymphocytes and thus enhance the force of the immune reaction. The situation with the T4 and T8 molecules acting as focusing entities of an invariant type enforcing cellular contact between T lymphocytes and other MHC-positive cells may even allow antigen-receptors on T cells to create possible antigen-MHC complexes on target cells from free antigenic fragments and MHC molecules. Support for such a possibility has in fact recently been coming from energy transfer studies using T-cell clones with defined specificity for a peptide fragment in conjunction with a class II MHC molecule [6]. This will, of course, act to enhance even further the ability of the T lymphocyte system to recognize foreign molecules in our body.

I have so far discussed MHC in a positive context – that

is, the presence of MHC as a Jekyll- and Hyde-like molecular system, where the MHC molecules may initially function like primitive antibody-like molecules for certain peptide fragments to create a molecular complex which is then recognized as antigen by specific T lymphocytes. T cells also, however, are largely unable to react against cells which for one reason or another may have lost their MHC expression. This could be considered to constitute a potentially dangerous hole in our immune defence shield. It is thus highly interesting to note that during the evolution of our immune system this problem also has at least in part, been taken care of by a cell type called NK or natural killer, discovered only some ten years ago [7]. NK cells have been known for some time to be able to kill tumour cells and certain normal cells, especially if the latter are of a primitive nature. Presence of MHC on the target cells has been known not to be required for NK lysis to occur. However, it has recently been found that class I MHC molecules in fact act in a directly opposite manner on target cells for NK cells from the way they do for killer T cells. Select depletion of class I MHC molecules from the surface of the target cells will thus, on a target cell with moderate sensitivity to NK cells, result in a dramatic increase in NK susceptibility, with a corresponding loss in susceptibility to cytotoxic T lymphocytes [8]. NK cells can thus function as naturally occurring cytotoxic cells, with an efficient capacity to kill MHC negative variant cells, which may occur within our body, for instance, during malignant transformation. The development of NK cells would also seem logical during evolution as a way to select positively for stability within a multicellular society with regard to self-markers.

Evolution of antigen-binding specificity

The ability of specific immune recognition by special lymphocytes producing cell-bound or humoral antigen-specific molecules is present already in the most primitive vertebrates, and evidence of specific recognition of antigen can be noticed in invertebrates as well, although the details in these latter species at present are poorly understood. I will here only briefly bring forward some of our knowledge about evolution of antigen-binding specificity as it occurs during ontogeny in vertebrates.

T and B lymphocytes use different gene families to produce and construct their respective antigen-binding receptors/antibodies but the genes are related and the general principle of construction in essence identical [9]. In order to save DNA when creating extreme variability, each receptor chain participating in the creation of an antigen-binding site is composed of several mini genes, which are brought together during the differentiation of the lymphocyte to create a unique combination. This will in itself allow a high number of combinations to become generated from a relatively low number of minigenes within families and is further expanded by the use of two different chains to create the antigen-binding site. The latter will result in a multiplication of the combinatorial possibilities for the respective polypeptide chain, and for both T and B lymphocytes the genetic germline situation as to minigenes is such that around 10^7 antigen-binding sites can be created in each lymphocyte population by somatic recombinations only. On top of this should be added the proven ability within the B-cell system to add further somatic mutations to increase variability. Mismatch possibilities when combining minigenes will further enhance the number of potential antigen binding sites that can be generated during immune evolution within the individual. This would then mean that the immune system in the form of antigen-binding receptors will create a molecular variability within the body which dominates *in toto* at least the variability determined by protein structure. Depending on which figure is accepted for the number of genes in a vertebrate, even a figure of 10^6 genes would make proteins made by the immune system constitute more than 99 out of 100 possible structural variations. This extreme variability is necessary in order for the immune system to be prepared via initial random generation of antigen-binding receptors to create a close-to-complete repertoire with regard to demand – that is, prepared to react against virtually anything foreign that could be introduced into the body. We can produce antibodies against virtually anything provided certain basic requirements as to size and solubility are fulfilled. This also means that we can produce antibodies against our own antibodies, in particular against the antigen-binding sites. This would also mean that we have molecules in the form of antigen-binding receptors which can mimic any antigenic determinant that can be introduced into our body, or in short that our immune system can make internal images of any antigen that can be introduced. This allows for an extremely important possibility, namely that there exists in the immune system a specific network where lymphocytes can communicate via such select receptor–anti-receptor interactions [10]. Evidence has accumulated that such a network does indeed exist and antibodies directed against antibodies against microorganisms have, for instance, in many diseases been shown to constitute powerful vaccines against these diseases [11]. As the embryo of the vertebrate will virtually always have its immunity maturing in the presence of maternal antibodies it has been speculated that maternal antibodies via such a molecular mimicry situation may in fact have an active impact on the specificity of the immune system of the foetus [12]. Data exists supporting such a possibility, and as the response pattern of the developing immune system is different from that of the mature system [in the embryo contact with an antigen leads to elimination of activity against that antigen (if antibodies are considered as antigen to more production of the same antibody)] whereas in the adult the opposite occurs (antibodies against the antigen or if antibodies are concerned anti-antibodies will be produced. It is quite possible that, for instance, antibodies against a malaria antigen in a given concentration from the mother when passing the placenta may lead to an active induction of immunity of a similar specificity. This is speculation, but experimental data supports this possibility. If true it would have obvious survival value for the individual if passively transferred antibodies in the foetal situation not only lead to passive immunity but also may have some active impact.

The discovery of the existence of the above-depicted network has also stimulated some new thought about the development of immune memory. Specific immune memory can exist for life, and it is unlikely that memory cells can last for such a long period, although in man T lymphocytes can indeed survive for several years. Immunity once induced may merely represent a shift in the balance between the lymphocytes that can make antibodies against the antigen (= the winners) and those that make antibodies which 'looked' like antigen (= the loosers). It would then be quite possible to consider a system where the memory consists of an expanded population of antigen-specific lymphocytes which, in the absence of antigen, can survive for a prolonged but limited time period. From the bone marrow, however, there will always be generated a certain number of virgin lymphocytes whose antigen-binding receptors, for reasons of chance, will look like the antigen. Such lymphocytes, when brought in contact with the memory cells, will be at a disadvantage and are expected to loose in a 'combat' situation. In contrast, the memory cells will be stimulated in a similar manner as would contact with proper antigen have done. Such a situation may then allow for a low level of 'inborn' stimulation by molecular mimicry of memory cells, once they are initiated, and this could then fully explain the longevity of immune memory.

Summary

The immune system is a most complex system in the higher animals. It is intelligent and undergoes evolution and education both during differentiation in the absence of exogenous antigen and during the selective processes involved at the cellular level when a specific immune response is initiated. As a final result, from close to unlimited chaos, order in a logical and efficient manner will be generated.

References

1. Klein, J., Immunology, in *The Science of Self–nonself Discrimination*, p. 270. John Wiley (1982).
2. Zinkernagel, R. and Doherty, P., *Nature* (*London*), **248**, 701 (1974).
3. McDevitt, H. O. and Chinitz, A., *Science*, **153**, 1207 (1969).
4. Swain, S. L., Dennert, G., Wormsley, S. and Dutton, R. W., *Eur. J. Immunol*, **3**, 175 (1981).
5. Krensky, A. M., Reiss, C. S., Mier, J. W., Strominger, J. L. and Burakoff, S. J., *Proc. Natl. Acad. Sci. USA* **79**, 2365 (1982).
6. Watts, T. H., Gaub, H. E. and McConnell, H. M., *Nature* (*London*) **320**, 179 (1986).
7. Kiessling, R., Klein, E., Pross, H. and Wigzell, H., *Eur. J. Immunol.* **5**, 118 (1975).
8. Kärre, K., Ljunggren, H. G., Piontek, G. and Kiessling, R., *Nature* (*London*) (in the press).
9. Patten, P., Yokota, T., Rothbard, J., Chien, Y., Arai, K. and Davis, M. M., *Nature* (*London*), **312**, 40 (1984).
10. Jerne, N. K., *Sci. Am.* **229**, 52 (1973).
11. Kennedy, R. C., Eichberg, J. W., Lanford, R. E. and Dreesman, G. D., *Science*, **232**, 220 (1986).
12. Wikler, M., Demeur, C., Dewasme, G. and Urbain, J., *J. Exp. Med.* **152**, 1024 (1980).

Chemica Scripta 1986, **26B**, 357–362

Polymorphism and Gene Duplication in the Human IFN-α and β Gene Family

Alexander von Gabain, Monica Ohlsson, Eleonor Lindström, Mona Lundström and Erik Lundgren
Institute for Applied Cell and Molecular Biology, University of Umeå, S-901 87 Umeå, Sweden

Paper presented by Alexander von Gabain at the Conference on 'Molecular Evolution of Life', Lidingö, Sweden, 8–12 September 1985

Abstract

A human cosmid clone carrying three interferon-α (IFN) genes was comparatively analyzed with chromosomal regions derived from another individual. The segment seemed to be an allele to corresponding chromosomal sections, with one of the genes showing discrete amino acid differences between the individuals. A similar variation was found when homologous cDNA clones encoding IFN-α2 types were compared. The complexity of the IFN gene family is further demonstrated by the finding of a duplicated IFN-β gene in some individuals. Calculations of divergency rates between different IFN-α genes is best compatible with concerted evolution within the species.

Introduction

Interferons (IFN) comprise a group of proteins and glycoproteins, which were originally defined by the antiviral state they induce in target cells; later on they were found to mediate other biological phenomena such as modulation of functions of immunocompetent cells or inhibition of cell growth. Classically, they are categorized into three groups by serological criteria: IFN-α, β and γ. In humans the comparison of the amino acid and nucleotide sequences discloses that IFN-α and IFN-β have a relatively similar structure with a homology of 45% at the nucleotide level [1], while IFN-γ shows more remote similarities with the other two [2].

On the genetic level three corresponding IFN gene clusters have been described: IFN-α and β genes are located on the short arm of chromosome 9, while the IFN-γ gene is located in the long arm of chromosome 12 [3]. Both IFN-α and β genes have no introns, a feature that they only share with histone genes. In humans, IFN-α is encoded by a multigene family, that can be subdivided in two groups: IFN-αI and αII. IFN-αII genes were not detected until very recently and seem to fall in between IFN-αI and IFN-β genes as regards their structural and evolutionary relationship [4]. The number of different IFN-αI gene sequences per haploid chromosome has been given as about 20 [5], while IFN-β exists as a single gene in the haploid genome of most individuals [6]. The different IFN-α species depart from one another in about 20 of the 166 amino acids of the primary structure [7]. Different IFN-αs and IFN-β have overlapping biological activities, but can in some respects also be distinguished [8]; in one extreme case the exchange of a single amino acid completely abrogated one biological trait, leaving the others untouched [9].

Human IFN genes have been isolated from cDNA libraries derived from induced hemopoietic cells or from chromosomal banks [10, 11]. At least 30 different IFN-αI genes have been identified in the different gene libraries [5]. Additionally, the analysis of restriction fragment length polymorphism (RFLP) in about 25 families has established markers in the IFN-α and β locus, proving the polymorphic nature of that entire gene locus and disclosing an extreme degree of heterozygosity [12].

Here we present a refined analysis of a chromosomal segment, carrying a cluster of IFN-α genes of the αI-type, that has been derived from a gene bank other than the Maniatis' bank employed for isolating the majority of chromosomal IFN genes and thus representing another individual. The structure of those genes confirms the diversification of the IFN-α gene pool; the evolutionary distance of those genes and some putative allelic variants was determined. The IFN-β locus was found to be amplified in the human gene pool with a low frequency. The results are discussed against the background of IFN functions.

Materials and methods

The human lymphoid cell line Namalwa, the cDNA library derived from it, bacterial strains, the human cosmid library derived from a placenta, plasmids and M13 cloning vehicles have previously been described [13, 14]. The designation of the plasmids, the cosmid pcosα and their derivatives containing the IFN genes follows the one introduced before [13, 14]. DNA from the λ-clone λHLeIF2 isolated from the Maniatis' library was kindly provided by Dr R. N. Lawn from Genentech Corp., San Francisco [15]. All techniques and methods employed to complete the DNA sequences and to analyse the cleavage products of restriction digests are contained in previous publications [13, 14]. We are grateful to Drs S. Josephsson and H. Hultberg from KabiGen, Stockholm, for synthesizing the oligonucleotides. The sequencing methods were according to Sanger or Maxam and Gilbert [16, 17]; the sequence of the oligonucleotides is contained in the Results' section. DNA sequence data were stored and computed using a program developed recently [18]. The calculation of the divergency between the aligned sequences was carried out as described by Miyata *et al.* [19] and the obtained corrected K_s values recalculated into million years of evolutionary distance assuming that the mammalian radiation occurred 75 million years ago.

Results

Two major cDNA clones encoding IFN-α were identified in a cDNA library derived from the human Burkitt's lymphoma

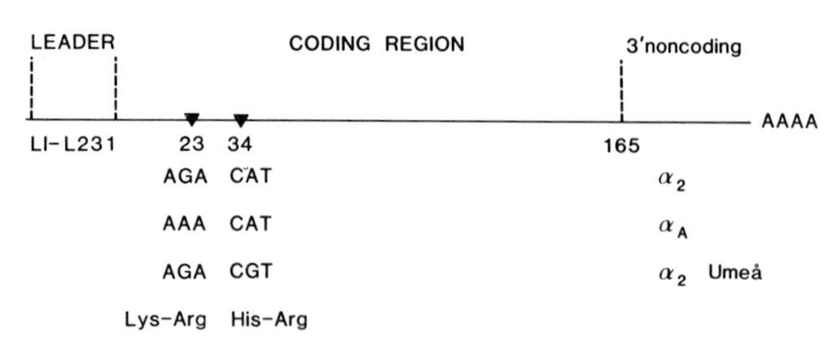

Fig. 1. The figure displays the three sequences representing the IFNα2 type that were isolated from different cDNA libraries. L1–L23 and 1–165 mark the amino acid positions of the leader peptide and the mature protein, respectively. The triangles indicate the codon positions found to be different in the respective IFN's, the nucleotide exchange at position 34 is documented in Fig 2A. α2 is derived from pooled buffy coats [10], αA from a myeloblastoid cell line (LeIF A) [11] and α2 Umeå from a substrain of the Namalwa cell line [13].

cell line Namalwa, designated 2337 and 88 [13]; the former has been characterized by partial sequencing and corresponds to the major type of IFN-α isolated before from other cDNA libraries as LeIF A or IFN-α2 [10, 11], while the latter, characterized by its complete nucleotide sequence, was found to be most similar to LeIF C [13]. The nucleotide sequence of 2337 was completed either by inserting the cDNA insert in the unique PstI site of the M13 vectors mp8 and mp9 and decoding the sequence according to Sanger [16] using synthetic primers (see Fig. 4), or by kinasing unique restriction sites (see Materials and Methods) and decoding the sequence following the protocol of Maxam and Gilbert [17]. Comparison of the sequence with those of LeIF A and IFN-α2 revealed a unique nucleotide exchange found in the second position of codon 34 of the frame encoding the mature protein (Figs 1 and 2), exchanging a histidine by an arginine, while at position 23 the sequence follows that of IFN-α2 (Fig. 1).

The heterogeneity of the major IFN-α2 type isolated from different individuals exemplifies the diversification of the IFN-α family. However, the degree and the nature of the diversification can only be explored by comparing identical loci of the chromosome. Two human genomic libraries were employed to identify chromosomal segments encoding IFN-α genes and gene-like sequences [5, 14]. By comparing restriction maps and sequence homologies, it was possible to combine the segments in seven non-continuous regions encoding altogether 20 IFN-α genes and gene-like sequences [5]. It seems likely, that the enumeration and cataloguing of the chromosomal loci of IFN-α genes is not complete, because a similar number was found in a study of IFN-α gene segregation employing heterozygotic RFLP markers [12]. Therefore, homologous regions derived from different individuals are likely to represent alleles. The only suggestive example so far is the previously isolated DNA segment, pcosα, containing three IFN-α genes (Fig. 3) [14]. The partial sequence of two of them, αN and αT, and the intergenic pattern of the respective restriction sites strongly indicated that it represented the same chromosomal region as λHLeIF2, previously

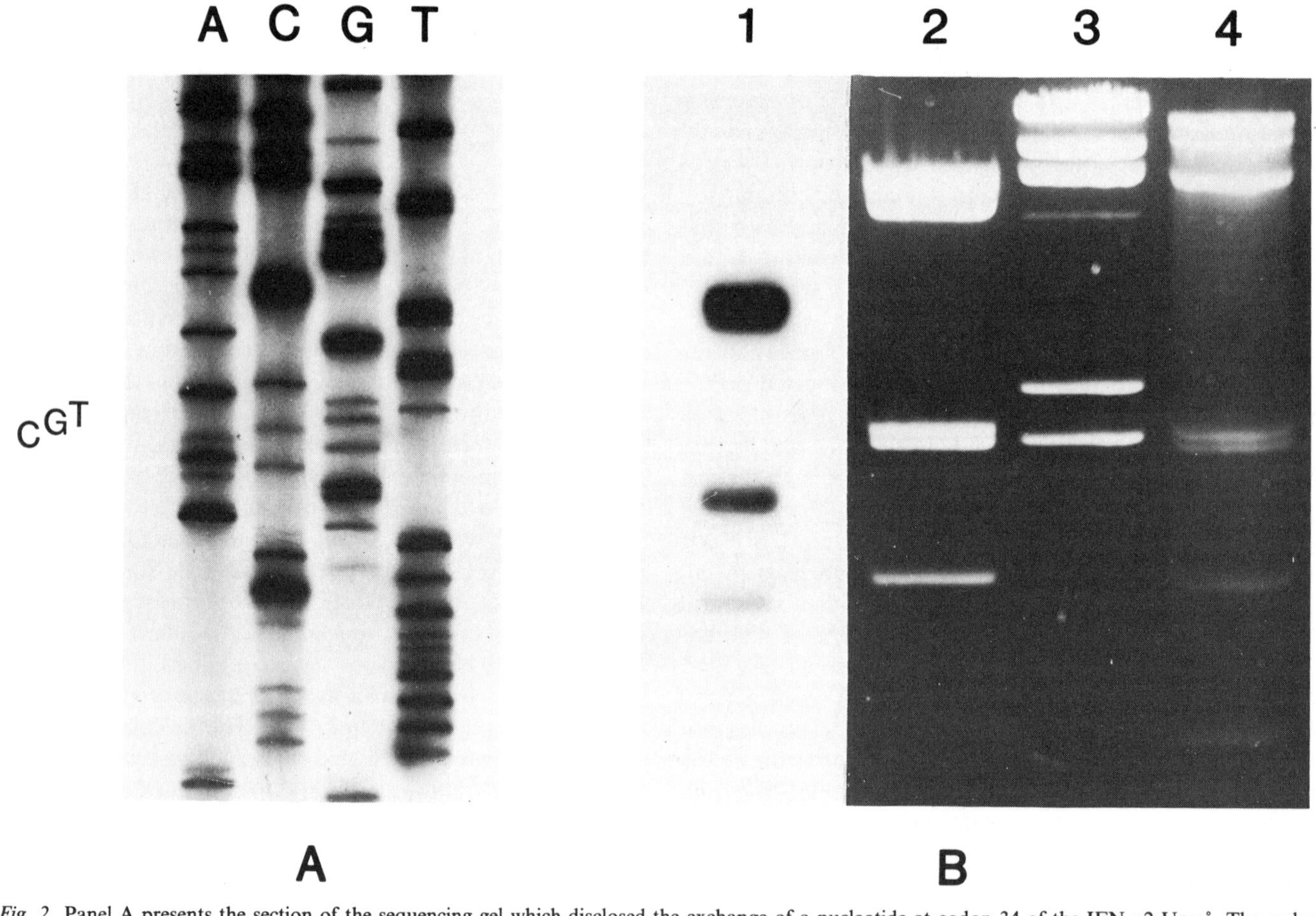

Fig. 2. Panel A presents the section of the sequencing gel which disclosed the exchange of a nucleotide at codon 34 of the IFN-α2 Umeå. The codon is CGT instead of CAT found for IFN-α2 and A (Fig. 1). Panel B shows the photo of the ethidiumbromide stained agarose gel after fractionation of the *Eco* RI fragments of pcosα (lane 2) and λHLeIF2 DNA (lane 4); the *Hin* dIII fragments of λDNA (lane 3) were used as size markers and obtained from New England Biolabs, Beverly, MA, USA. For the analysis of pcosα a subclone was used, subclone 23 [14], containing only the part corresponding to λHLeIF2 (Fig. 3). Lane 1 shows the autoradiogram of a nitrocellulose filter prepared and processed according to Southern (Materials and Methods), in which the *Eco* RI fragments of pcosα DNA were probed with nick-translated λHLeIF2 DNA.

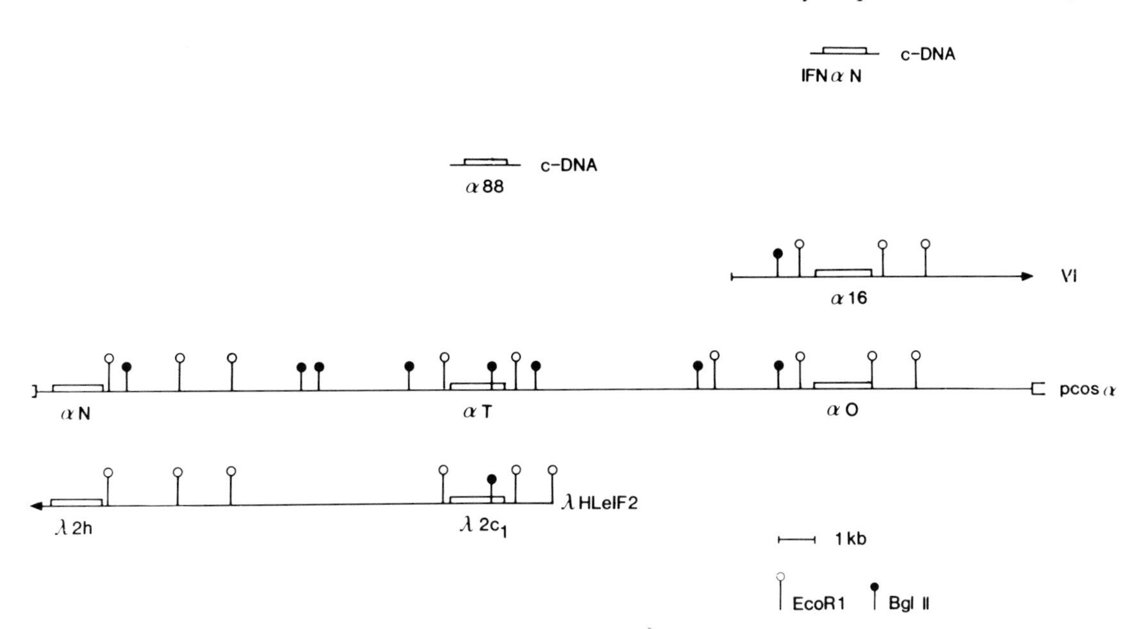

Fig. 3. The figure is showing the physical map of three chromosomal segments which overlap. VI, represents a linkage group described previously [5]; λHLeIF2, a chromosomal region isolated by Lawn *et al.* [15] and pcosα, a continuous segment derived from a human cosmid library [14]. IFN-α-N [20] and α88 [13] show cDNA clones exhibiting extreme homologies to the corresponding chromosomal loci.

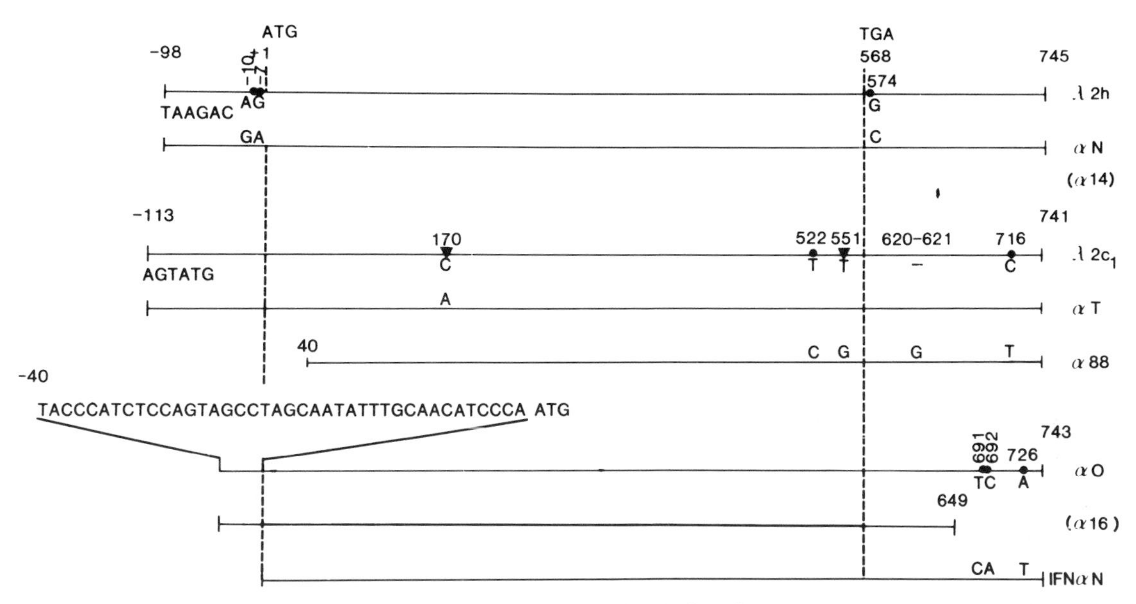

Fig. 4. The figure presents the sequences of the following IFNs: λ2h, αN, α14, λ2cl, αT, α88, αO, α16 and IFN-α-N. λ2H and λ2cl are taken from a previous publication [15], positions -98 and -113 correspond to the nucleotide positions in that paper; the first 6 nucleotides starting at the respective positions are indicated. The sequence of αO between the position 1 and 743 is taken from a homologous cDNA clone, IFN-α-N, published previously [20] and starts with the ATG; the position -40 to -1 was derived from the 5′ noncoding region of αO. Triangles, indicate nucleotide exchanges replacing an amino acid, and dots nucleotide exchanges; the positions of those exchanges are marked. In α88 the insertion of one nucleotide was found between position 620 and 621 of the homologous λ2cl. The synthetic oligonucleotides used as primers for establishing the sequences of αT, αN and αO were complementary to the sequence of λ2H at positions 140–157, 300–317, 485–499 and 750–767, or identical to the same sequence at positions 70–84 depending on whether the sense or the nonsense strand of the genes were contained in the single stranded form of the respective M13 F chimeras employed for sequencing. The commercially available M13 17-mer from New England BioLabs was used in some experiments.

isolated from the Maniatis' gene bank (Fig. 3) [15]. The partial sequence of the remaining IFN-α gene, αO, and the intergenic region of the other half of that segment was found to match another homologous region, linkage group VI, isolated from the same gene bank [5], as reported by Henco *et al.* (Fig. 3).

In order to confirm the previously concluded homology between the two chromosomal regions, pcosα and λHLeIF2, the DNA from them was hydrolysed with the restriction endonuclease *Eco* RI and *Bgl* II, analysed in parallel by agarose gel electrophoresis and further treated according to Southern by crossprobing the fragments with the entire DNA of λHLeIF2. In Fig. 2 *b* the result obtained for the *Eco* RI patterns is displayed as an example. All together, 11 recognition sites were tested and found to coincide in all positions of the two compared regions. The data show the identical locations of the respective cleavage sites and the close homology of the two chromosomal segments.

The three IFN-α genes, IFN-αN, IFN-αT and IFN-αO, clustered on the cosmid pcosα were previously subcloned in the phage vectors M13 mp8 and mp9 [14]. Relevant parts of the inserts were sequenced according to Sanger [16] by using

Table I. *Nucleotide differences between the three pcosα genes and the related genes α88, λ2h and λ2cl.*

		Silent changes				
		αT	α88	λ2cl	αN	αO
Replacement changes	αT	—	1	0	10	13
	α88	1	—	1	9	12
	λ2cl	1	2	—	10	13
	αN	27	26	28	—	14
	αO	25	24	26	25	—

α14, αN and λ2H are identical and αO and α16 are also identical, and therefore not listed in the table.

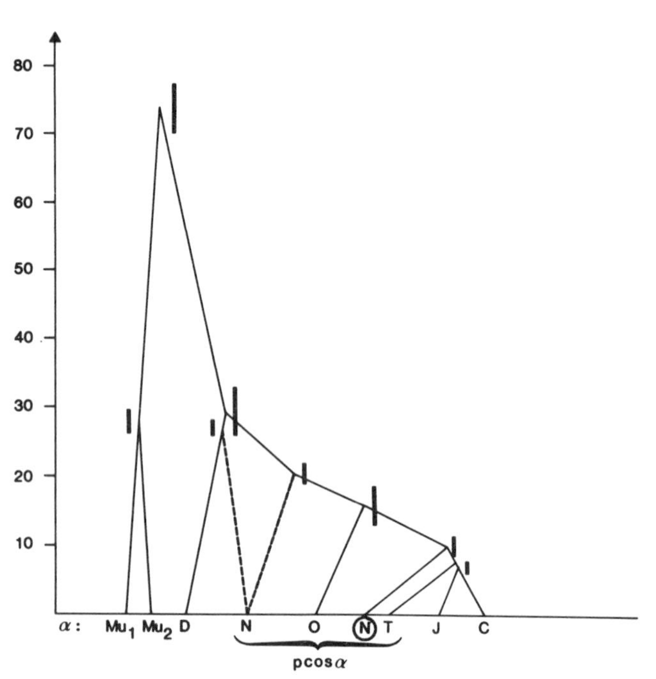

Fig. 5. Evolutionary relationship between the three IFN-α genes clustered on pcosα and some other IFN-sequences in comparison to LeIF C (C) and LeIF D (D), the most divergent members of the gene family [5] and two murine α-genes (Mu_1 and Mu_2 [25]). The calculations are made as described [19] and the evolutionary distance has been recalculated into million years based on the assumed mammalian radiation point of 75 million years ago. The bars represent standard errors. Ⓝ represents the 5′ part of αN (nucleotide positions 198–567). The calculations are based on the following sequences: LeIF D, G, ψL, J, F, C, αT, αN and αO.

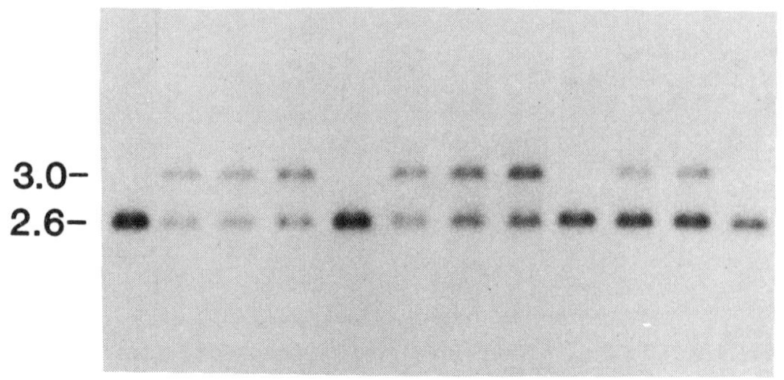

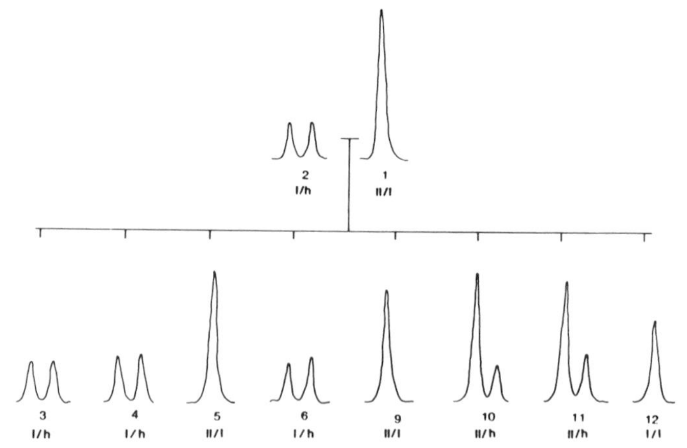

Fig. 6. Part A of the figure shows the autoradiogram of a hybridization according to Southern; *Ban* II digested DNAs from all members of a family were analysed with an IFN-β specific probe as described previously [12]. The sizes of the fragments are marked. 1 and 2 represent the parents and 3 to 12 the children. Part B displays the densitometric tracings of the signals seen in the autoradiogram for the respective individuals. l/h and ll/l are the genotypes of the parents, while l/h, ll/l, ll/h and l/l are the genotypes of the children.

synthetic primers as described in Materials and Methods. Thereafter the obtained sequences were aligned and compared to all known IFN-α sequences [5] and found to match each other as follows (Figs 3 and 4): αN and λ2h were identical, while λ2cl and αT differed in 2 nucleotides. Both genes were contained in λHLeIF2 at homologous positions. αO was identical to α16 [15] contained in the linkage group VI at homologous position. αT and αO also show only minor differences to clone 88 [13] and a sequence designated as IFN-α-N by Gren *et al.* [20] respectively, the two latter sequences being derived from cDNA clones. Extreme homology or identity was only found within the compiled groups, while comparing IFN-α genes in any other combination yields significantly less homology (Table I). Nevertheless, the number of amino acid and nucleotide exchanges among the compared genes proves the high degree of diversification of genes which are located on homologous segments of the chromosome.

The sequences offered the opportunity to analyse the evolutionary relationship of three IFN-α genes which are comparatively closely linked within a unitary DNA segment. The coding regions were compared by registering the silent and replacement positions among the relevant nucleotide sequences (Materials and Methods); the data were then further processed using an algorithm deviced to determine evolutionary distances [19]. The coding regions were compared by registering the silent nucleotide differences between the aligned sequences. After correction an evolutionary tree was constructed based on the K_s values [19]. The calculations are based on two murine α genes and nine human α genes, including αN, αT and αO. However, for simplification only these genes and the diverged sequences LeIF D and LeIF C are shown (Fig. 5). The analysis shows that αO is closest to the LeIF C cluster (comprising also LeIF F, LeIF J and ψLeIF L). αT comes between LeIF C and LeIF D, while αN is paradoxally close to both LeIF D ad LeIF C, as was previously shown for the synonymous sequence α14 [5]. However, the 5′ part (nucleotides 198–567, Fig. 4) is very close to LeIF C, differing only to a minor extent from αO.

IFN-α and β both exist as multigene families in cattle [21, 22] whereas in humans normally only one copy of IFN-β is found per haploid genome. In a previous study we undertook to investigate the diversification of the IFN-α and β locus with RFLP markers, and disclosed a duplication of the IFN-β locus in some individuals [12]. The duplication was noticed by departure of intensities from the expected 1:1 ratio in DNA fragments from individuals with heterozygous RFLP markers (Fig. 5).

This finding was confirmed by using two different restriction endonucleases [12] providing a heterozygote RFLP marker for the same individuals. Finally, the duplication was independently verified, when the children of those individuals were probed for the same markers (Fig. 5) and it was found that the duplicated IFN-β locus segregated according to Mendel's laws [12].

Conclusions

IFN-α and β genes are found as multigene families in rodents, cattle and man. However, in a variety of mammals, including man, only the IFN-α genes have been diversified, while IFN-β exists normally as a single gene. The separation of IFN-α and β has been calculated as dating back about 300 million years or even more [1, 19], whereas the further diversification of the two IFNs found in different species in all studies presented so far seems to have occurred after the mammalian radiation.

Human IFN genes were derived from only a small number of individuals [10, 11] or from buffy coat cells, pooled from several individuals. The IFN-α2 sequence we present differs in one or two amino acids respectively from these previously described. To our knowledge, identical IFN-α2 sequences have not been demonstrated from two individuals. Although this shows the complexity and diversification found, when IFN-α's are induced in different cell lines, it is difficult to conclude allelic variation from this, because the divergence could be due to different gene loci.

The comparison of αT, αN and αO with similar chromosomal sections in other individuals, namely λ2cl, λ2h and α16, is highly suggestive of an allelic diversification, if the compared chromosomal regions reflect identical loci. This seems very plausible, taking into consideration the extreme homology of the intergenic regions that was found, when pcosα and λLHeIF2 were compared by hybridization and restriction analysis. The fact that λ2h, αN and α14 are identical, as well as αO and α16, supports this conclusion. Furthermore, as outlined in Results, the numbers of IFN-αI genes and gene-like sequences identified in the Maniatis' gene bank are very likely to represent all human IFN-α loci. Therefore, we propose the pcosα is an allelic variant of its counterparts found in the Maniatis' gene bank [15]. α-88 [13] and IFN-α-N [20] derived from different cDNA banks are likely to present allelic variants of the genes with which they match best (Fig. 3). However, concerning their assignment to the respective loci the same caveat holds as was raised above for the IFN-α2 genes.

The differences identified within the homologous genes vary between one and three nucleotides, giving rise to one and two amino acid differences. The numbers seem to be small, when they are compared to the number of different amino acids found among major types of human IFN-αs, namely up to 20 exchanges. On the other hand, it should be remembered, that other human genes, e.g. the plasminogen activator [23] have been isolated from several gene libraries without finding any departure of amino acids. The result is also remarkable that human IFN-β was derived several times from different individuals and not more than one single silent nucleotide exchange was noticed [1, 13, 24]. Finally, the fact that the amino acid exchanges in homologous genes were found by comparing two individuals, allows us to expect more pronounced differences, when more individuals are analyzed.

When the evolutionary relationship between the different IFN-α genes identified in pcosα was compared with those published by others [5], it became evident that they fall within the rate of divergence displayed by the other members of the family. αT and αO are closely related to LeIF C, while αN has a more peculiar relationship to the other genes. It could be shown to be related both to LeIF D and the LeIF C cluster and, as Henco *et al.* have pointed out [5], the genes probably reflect a hybrid between two other IFN-α genes. However, the 5′ part (nucleotides 198–567, Fig. 4) is more closely related to the LeIF C cluster. It is also evident from the evolutionary relationship that the 5′ regions of αN and αT are more closely related to each other than the one of αO, which to some extent, reflects the position on the physical map. This parallelism between evolutionary distance and the physical map is compatible with gene duplication events giving rise to higher similarities between juxtaposed genes than to genes more separate on the chromosome.

The results displayed in Fig. 5 suggest that the IFN-α gene family in humans diverged after the mammalian radiation and this seems also to be the case for the two sequenced murine α genes. It is assumed that mammalian radiation took place 75 million years ago, and hence all diversification seems to be less than 30 million years old. On the other hand, the diversification could be alternatively explained by convergence, namely when genes are rectified within a species and therefore seem to have evolved after the branching of humans and rodents. We favour the latter idea of concerted gene evolution of the IFN family within the species, implying that the loci in fact duplicated before human/rodent separation.

Amplification and subsequent exchange of amino acids can be considered as potential tools driving forward the diversification of IFN genes. The duplication of the IFN-β locus found in some individuals reveals the imbalance of IFN genes in the human gene pool. The nature and the functions of the duplicated genes have not yet been elucidated. However, the detection of the amplification was found with an appreciable frequency [12], indicating that more and other types of amplifications might be identified, when large populations are screened. The result is especially intriguing as IFN-β genes seem to be maintaind as single genes in most mammals and as it is more difficult to explain the duplication of a single copy gene compared to multicopy genes. On the other hand, the existence of multiple IFN-β genes in some individuals might reflect the beginning of a diversification of IFN-β as found in cattle or the remnant of a former amplification, that is disappearing.

Polymorphism and gene amplification present evidence for the fast evolution of IFN-α and β genes. Two extreme conditions might explain diversification of a gene family: (i) the absence of any selection constraining the structure; (ii) the presence of a selection causing divergence. The first condition anticipates a response of target cells that are inert to structural diversifications. On the other hand, it conflicts with the finding that small structural differences change the apparent biological activities. The second condition maintains a paradox until the mechanism of the selection is identified, e.g. neoplastic or virus infected cells escaping the action of interferons.

Acknowledgements

This work was supported by grants from KabiGen AB and KabiVitrum AB, Stockholm. We thank Drs H. Hultberg and S. Josephsson for synthesizing the oligomers, Dr R. M. Lawn for supplying us with λHLeIF2 DNA and Dr C. Weissmann for making a preprint available to us. We are grateful to E. Ivarsson and M. von Gabain for skilful preparation of the manuscript and N. Shrimpton for brushing up the English text.

References

1. Taniguchi, T., Mantei, N., Schwarzstein, M., Nagata, S., Maramatsu, M. and Weissmann, C., *Nature (London)* **285**, 547 (1980).
2. Devos, R., Cheroutre, H., Toya, Y., Degrave, W., Van Heuverswyn, H. and Fiers, W., *Nucleic Acids Res.* **10**, 2487 (1982).
3. Trent, J. M., Olson, S. and Lawn, R. M., *Proc. Natl. Acad. Sci. USA* **79**, 7809 (1982).
4. Capon, D. J., Shepard, H. H. and Goeddel, D. V., *Mol. Cell. Biol.* **5**, 769 (1985).
5. Henco, K., Brosius, J., Fujisawa, A., Fujisawa J.-I., Haynes, J. R., Hochstadt, H., Kovacic, T., Pasek, M., Schamböck, A., Schmid, J., Todokoro, K., Wälchli, M., Nagata, S. and Weissmann, C., *J. Mol. Biol.* (in the press).
6. Houghton, M., Jackson I. J., Porter, A. G., Doel, S. M., Catlin, G. H., Barber, C. and Carey, N. H., *Nucleic Acids Res.* **9**, 247 (1981).
7. Tavernier, J., Derynck, R. and Fiers, W., *Nucleic Acids Res.* **9**, 461 (1981).
8. Alton, K., Stabinsky, Y., Richards, R., Ferguson, B., Goldstein, L., Altrock, B., Miller, L. and Stebbing, N., in *The Biology of the Interferon System 1983* (eds. E. De Maeyer and H. Schellekens) pp. 119–128. Elsevier, Amsterdam.
9. Liu, X.-Y., Bhatt, R. S., Pestka, S., *Antiviral Research*, Abstr. 1, No. *3*, p. 74 (1984).
10. Streuli, M., Nagata, S. and Weissmann, C., *Science* **209**, 1343 (1980).
11. Goeddel, D. V., Leung, D. W., Dull, T. J., Gross, M., Lawn, R. M., McCandliss, R., Seeburg, P. H., Ullrich, A., Yelverton E. and Gray, P. W., *Nature (London)* **290**, 20 (1981).
12. Ohlsson, M., Feder, J., Cavalli-Sforza, L. L. and von Gabain, A., *Proc. Natl. Acad. Sci. USA* **82**, 4473 (1985).
13. Lund, B., von Gabain, A., Edlund, T., Ny, T. and Lundgren, E., *J. of Interferon Research* **5**, 229 (1985).
14. Lund, B., Edlund, T., Lindenmaier, W., Ny, T., Collins, J., Lundgren, E. and von Gabain, A., *Proc. Natl. Acad. Sci. USA* **81**, 2435 (1984).
15. Lawn, R. N., Adelman, F., Dull, T., Gross, M., Goeddel, D. and Ullrich, A., *Science* **212**, 1159 (1981).
16. Sanger, F., Nicklen, S. and Coulson, A. R., *Proc. Natl. Acad. Sci. USA* **74**, 5463 (1977).
17. Maxam, A. M. and Gilbert, W., in *Methods Enzymol.* (eds. L. Grossman and K. Moldave), vol. 65, pp. 499–560. Academic Press, New York.
18. Harr, R., Fällman, P., Häggström, M., Wahlström, L. and Gustafsson, P., to be published in *Nucleic Acids Res.*, special issue, **14** (1) (1986).
19. Miyata, T., Hayashida, H., Kikuno, R., Toh, H. and Kawade, Y., *Interferon* **6**, 1 (1985).
20. Gren, E., Berzin, V., Jansone, I., Tsimanis, A., Vishnevsky, Y. and Apsalons, U., *J. IFN Res.* (in the press).
21. Wilson, V., Jeffreys, A. J., Barrie, P. A., Boseley, P. G., Slocombe, P. M., Easton, A. and Burke, D. C., *J. Miol. Biol.* **166**, 457 (1983).
22. Leung, D. W., Capon, D. J. and Goeddel, D. V., *Biotechnology* **2**, 458 (1984).
23. Ny, T. Elgh, F. and Lund, B., *Proc. Natl. Acad. Sci. USA* **81**, 5355 (1984).
24. Derynck, R., Content, J., De Clercq, E., Volckaert, G., Tavernier, J., Devos, R. and Fiers, W., *Nature (London)* **285**, 542 (1980).
25. Shaw, G. D., Boll, W., Taira, H., Mantei, N., Lengyel, P. and Weissmann, C., *Nucleic Acids Res.* **11**, 555 (1983).

Chemica Scripta 1986, **26B**, 363–375

Molecular Mechanisms of Morphologic Evolution

Gerald M. Edelman

The Rockefeller University, 1230 York Avenue, New York, New York 10021, USA

Paper presented at the Conference on 'Molecular Evolution of Life', Lidingö, Sweden, 8–12 September 1985

Abstract

A major basis of morphologic evolution is heterochrony, constituting changes during embryonic development in the relative rate of appearance of characters that are present in ancestors. It has been suggested that heterochrony results from mutations in regulatory genes that affect the timing of biochemical events, the integration of structural gene expression, and the fates of embryonic cells. One of the outstanding problems in modern evolutionary theory is to provide a molecular basis for heterochrony. In this paper, it is proposed that a special set of regulatory genes controlling the types and sequences of expression of cell adhesion molecules (CAMs) during animal development plays a central role in mediating heterochronic effects. Data on the structure, function, genes, sequences of expression, and regulation of cell adhesion molecules during development are reviewed in support of this hypothesis. The main conclusion is that the control of molecules such as the CAMs provides a link between the genetic code and the mechanochemical events that lead to the specification of three-dimensional form in the phenotype.

The central role of morphology and embryology in natural selection was clearly recognized by Darwin. Nonetheless, these fields played only peripheral roles in the modern synthesis [1], and there has been only sporadic recognition [2] that the main problem of developmental genetics must be solved before attempting to solve the outstanding problem of morphologic evolution [3, 4]. These two problems may be succinctly framed in the form of two questions: (1) How can a one-dimensional genetic code specify a three-dimensional organism? (The developmental genetic question). (2) How can the mechanism proposed to answer this question be reconciled with the large changes in form that can occur in relatively short evolutionary times? (The evolutionary question).

A satisfactory answer to the evolutionary question would shed great light on issues related to the evolution of phenotypic complexity and on the part played by the embryonic milieu in providing strong selection pressure during evolution. As implied in its statement, the evolutionary question is a contingent one – the developmental genetic question must be answered first. In other words, the mechanisms of morphologic evolution depend strongly upon embryonic mechanisms regulating morphogenesis.

A fruitful area in which such events may be studied is regulative animal development. In regulative development, cells of different history are brought together by morphogenetic movements to result in embryonic induction, or milieu-dependent differentiation. Embryonic induction can be reciprocal or asymmetric, and it can occur in a series of stages that are either stimulatory or inhibitory depending on the prior paths and local contacts of the induced and inducing cell populations [5]. Induction depends upon position, a fact which leads to an apparent paradox: the genome itself cannot contain specific information about the exact position of the participating cells in time and space and yet morphogenesis is under genetic control.

In order to resolve this paradox, the problem of morphogenesis must be viewed in terms of developmental genetic and mechanochemical factors acting jointly to yield a common phenotype within a taxon [6]. The requisite combination of molecular genetic control and macroscopic mechanism implies that regulation of pattern formation must range across a wide variety of levels of organization from the gene and functional gene products, to cells, to tissues and organs, and back to the gene. This multi-level control is exercised by regulatory primary processes of development (cell adhesion, differentiation) interacting with a number of other primary processes (cell division, cell motion, cell death) that serve as driving forces for the emergence of form.

Given this picture, how does form evolve? The current view is perhaps best represented by the idea that differences in morphology in related taxa can be explained by heterochrony: changes during development in the relative rate of appearance of characters already present in ancestors. Such changes are now interpreted as either local morphogenetic accelerations or retardations that lead, for example, to progenesis, neoteny or paedomorphosis [7]. The putative explanation for these effects is that they result from mutations in regulatory genes that change the timing of biochemical events, the fates of embryonic cells, or the integration of structural gene expression. The evidence [8] that mutations in regulatory genes are likely to be the major loci source of morphological differences within classes and that heterochrony provides a major means by which morphological change occurs during evolution raises the question of the kinds of regulatory genes and the mechanisms that are involved in morphologic evolution. In this paper, I shall consider the evidence supporting the hypothesis that genes regulating the expression of cell adhesion molecules (CAMs) are prime candidates for regulation of morphogenesis [9]. The main rationale underlying this hypothesis is that CAMs are structural gene products whose expression at the cell surface serves to link epithelia, condense mesenchyme and constrain movement. The CAMs serve as mechanochemical links between the genetic code and the organization of complex tissue structures.

If their role is relevant to the developmental genetic and

evolutionary questions, CAMs should possess a number of key properties: (1) They should show definite sequences of expression at cell surfaces during morphogenesis. (2) Their expression should either regulate or be correlated with cell movement and other primary processes of development at sites of embryonic induction. (3) Their molecular structure and binding properties should allow rapid switching on and off of this mechanochemical function. (4) Evolutionary similarities of developmental expression sequences of the CAMs in different species as well as evolutionary conservation of CAM structure and binding mechanisms should be observed. I shall review the evidence that CAMs share such properties and describe an hypothesis on their role in morphologic evolution. Before describing CAM properties, however, it is useful briefly to review the role of primary processes in regulative development and to describe two contrasting ideas related to the specificity of tissue formation.

Primary processes in regulative development

One may distinguish a number of processes at the cellular level which combine in varying degrees as a function of time directly to influence pattern formation and histogenesis. Besides the movement of cells and tissue sheets and cellular differentiation, these are cell division, cell adhesion, and cell death. Each of these primary processes is complex, consisting of many linked biochemical interactions. Form arises as a result of the differential effects of the driving forces provided by the processes of cell division, cell movement, and cell death. The key issue is concerned with how these forces are controlled [3]. The basic features of regulative development suggest that a central role is played by sequences of morphogenetic movements and by the places at which cells or collectives of cells interact as a result of these movements. The differentiation that occurs as a consequence of such interactions can change the shape of a cell, the products it synthesizes, and the movements themselves. In some sense, such inductive events create the environment necessary for triggering ensuing sequences, all of which occur in a historical fashion over time periods that are long as compared to those of intracellular processes.

What regulates the relative contribution of each of the primary processes to these sequences? Some aspects of the primary processes are clearly under genetic control. Others, however are epigenetic: certain sequences of events must occur in order for others to occur in certain regions of embryonic space at particular instants of time. But, inasmuch as the genes which ultimately regulate the primary processes cannot directly specify space-time coordinates, additional molecular mechanisms involving means other than genes must also regulate these epigenetic events. In searching for regulatory principles, it is well to keep in mind that the primary processes are partially dissociable, as Needham first pointed out [10]. This implies that to a certain extent they may be isolated experimentally from each other, facilitating tests of their relative contributions to the morphogenetic scheme.

Because genes cannot store the spatial or temporal coordinates of a local embryonic frame, and because cell motion cannot itself be genetically programmed so that each cell can reach a specific target, the existence of a system of individual cellular addresses expressed at the molecular level [11] in order to ensure appropriate cellular assembly seems highly unlikely [12]. Even if information for such an addressing system could be stored in the genes, another problem would obtrude: the need to induce the addressing molecule precisely on a given cell at a given place and time would be almost impossible to meet, given observations on the plasticity of development [13]. For example, in many embryonic sites at times before certain differentiation events have occurred, embryonic cells can be mixed and exchanged with equivalent morphogenetic results. It is difficult to see how individual cellular addresses specific at the molecular level could adapt to such maneuvers. It is the local environment, not the particular cell that is critical.

All of these considerations, which bear upon the central issue of positional specification of cells, make it highly unlikely that there is a prespecified, instructive, or precise informational mode by which cells know their place. The problem of position and regulation must be solved at the molecular level in another way. An analysis of the structure, function and expression of CAMs [14–18] provides important clues to the solution.

CAM properties

Cell adhesion molecules were first isolated from chick embryonic tissues [19, 20] by developing short-term adhesion assays for dissociated cells and then identifying their molecular-binding function by blockade with antibodies to the cell surface that contained specificities against a given CAM. The specific inhibiting antibody was then used to immunoprecipitate the particular CAM. CAMs have been identified in a large variety of vertebrate species. As shown in Table I, three different CAMs of different specificities, molecular weights, binding dependence on ions, and time and place of first

Table I. *Some characteristics of CAMs*

CAM	Type	Molecular weight (kD)	Carbohydrate characteristics	Binding mechanism	Ion dependence
N-CAM (neural CAM)	1°	180–250 (E form); 180, 140, 120 (A form)	Glycoprotein with unusual polysialic acid	Homophilic	None
L-CAM (liver CAM)	1°	124	Glycoprotein	Homophilic	Ca^{2+}
Ng-CAM (neuron-glia CAM)	2°	200 135 80	Glycoprotein	Heterophilic	None

appearance have been chemically characterized; others have since been found but less is known of their structure [21]. L-CAM (liver cell adhesion molecule [22]) and N-CAM (neural cell adhesion molecule [15, 16]) are called primary CAMs and appear early in embryogenesis upon derivatives of multiple germ layers. The third, Ng-CAM (neuron-glia CAM), is a secondary CAM that is not seen in early embryogenesis and that appears only on neuroectodermal derivatives, specifically on postmitotic neurons [23, 24]. CAMs are named according to the initials of the tissues from which they were first isolated (e.g. N-CAM, neural CAM) or additionally in terms of their heterotypic binding (e.g. Ng-CAM, neuron-glia CAM). This convention should not lead to the misconception that a named CAM is necessarily seen only in that tissue; nevertheless, the tissue is obviously representative of derivatives of one of the germ layers upon which the CAM first appears.

All of the well-characterized CAMs are glycoproteins synthesized by the cells on which they function. Structural comparisons (Fig. 1) show that they differ greatly. N-CAM and L-CAM are intrinsic membrane proteins; this appears also to be true of Ng-CAM but has not been as firmly established. N-CAM and probably L-CAM bind by homophilic mechanisms, i.e. CAM on one cell binds to another identical CAM on an apposing cell (Fig. 2A). While N-CAM binding is calcium independent, L-CAM depends on this ion both for its integrity and its binding; these two primary CAMs show no binding cross-specificity for each other. In contrast, the secondary Ng-CAM on neurons appears to bind by a heterophilic mechanism, i.e. Ng-CAM on a neuron to another CAM or a chemically different receptor on glia [23].

It has been suggested [14, 15] that CAMs act to regulate binding via a series of cell surface modulation mechanisms including changes in their prevalence at the cell surface, in their position or polarity, and in their chemistry of binding (Fig. 2B). All of these mechanisms have been shown to occur for one CAM or another at different developmental times [15]. The known case of chemical modulation is related to the presence [25] on Ng-CAM of α-2-8 linked [26] polysialic acid at 3 sites [27] present in the middle domain of the molecule (see Fig. 1). In the microheterogeneous embryonic (E) form of N-CAM, there are 30 grams per 100 grams of polypeptide, and in the discrete adult (A) forms this is reduced to 10 grams/100 grams polypeptide. Recent *in vitro* studies [28] suggest that N-CAM turns over at the cell surface and that the E form is replaced by newly synthesized A forms.

Although the carbohydrate of N-CAM is not directly involved in binding, kinetic studies [29] of CAM vesicle binding suggest that E-A conversion can result in a fourfold increase in binding rates (Table II). It seems likely that the charged polysialic acid either modulates the conformation of the neighboring CAM-binding region or directly competes with homophilic binding from cell to cell by charge repulsion [14]. Even more striking than the effects of E-A conversion is the dependency of homophilic binding on changes in CAM prevalence or surface density: a twofold increase in E forms leads to greater than 30-fold increases in binding rates [29]. This 5th order dependence is likely to result from an increase in valence by *cis* association at the cell surface of two or more CAM polypeptides. Similar rate studies (Table II) across a variety of vertebrate species [30] suggest that the N-CAM binding mechanism is conserved during evolution.

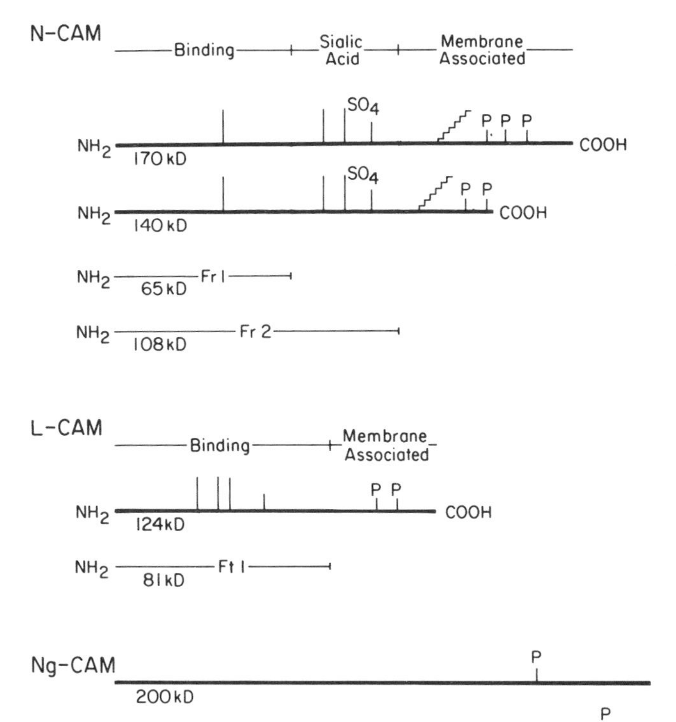

Fig. 1. Diagrams of the linear chain structure of two primary CAMs (N-CAM and L-CAM) and of the secondary Ng-CAM. N-CAM is seen as two closely related polypeptide chains, one shorter than the other. Below them are the fragments Fr1 and Fr2 derived by limited proteolysis. As indicated by vertical lines, most of the carbohydrate is covalently attached in the middle domain at three sites [27] and it is sulfated, although the exact sulfation site is unknown. Attached to these carbohydrates is polysialic acid. There are phosphorylation sites as well in the COOH terminal domains [55]. The diagonal staircases refer to covalent attachment of palmitate. L-CAM yields one major proteolytic fragment (Ft1) and has four attachment sites for carbohydrates (vertical lines) but lacks polysialic acid [22]. It is also phosphorylated in the COOH terminal region. Ng-CAM is shown without polarity but it is likely that the NH_2 terminus is to the left. There are two components (135 kD and 80 kD) that are probably derived from a post-translationally cleaved precursor. Each is related to the major 200 kD chain (which may be this precursor) and the smaller is arranged as shown on the basis of a known phosphorylation site.

The existence of such non-linear binding effects accompanying surface modulation is compatible with the idea that CAMs act as sensitive regulators of cell aggregation and cell motion. In order to understand their role as regulators, it is important to analyze CAM gene expression in relation to that of other products of differentiation. This entails a determination of the multiplicity of CAM genes, of the spatiotemporal sequences of CAM expression, and of the reciprocal relationship between CAM-binding function and morphology.

CAM genes

So far, cDNA clones for the two primary CAMs have been obtained [31, 32] and are being employed for these analyses. Two independent cDNA clones were derived from enriched mRNA coding for chick N-CAM that had been prepared by immunoprecipitation of polysomes with antibodies to N-CAM [31]. The plasmids hybridized to two discrete 6–7 kb long RNA species in poly(A)$^+$ mRNA from embryonic chick brain but not to comparably prepared RNA from liver. The two components detected by the cDNA probes for N-CAM appear to correspond separately to the two known polypeptide

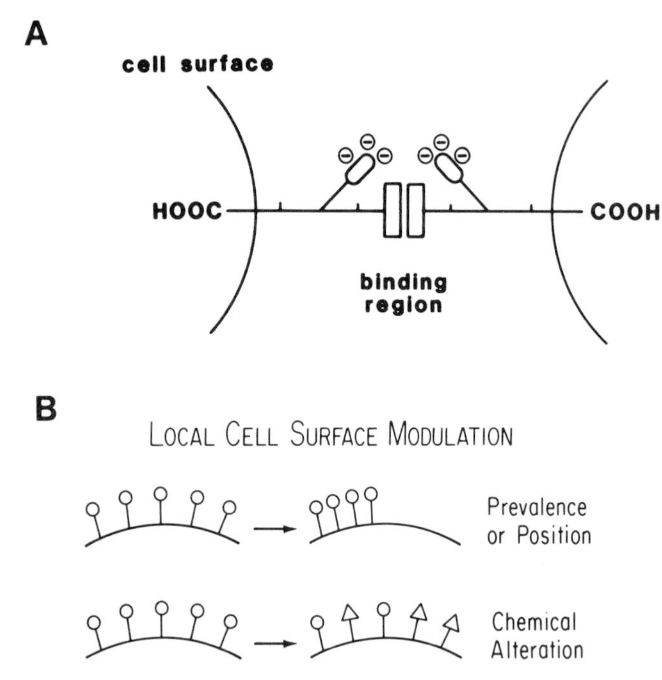

Fig. 2. CAM binding mode and cell surface modulation. (A) N-CAM binding is homophilic, i.e. to the binding domain of N-CAM present on an apposing cell. (B) Schematic representation of local cell-surface modulation. The various elements represent a specific glycoprotein (for example N-CAM) on the cell surface. The upper sequence shows modulation by alteration of both the prevalence of a particular molecule and its distribution on the cell surface. The lower sequence shows modulation by chemical modification resulting in the appearance of new or related forms (triangles) of the molecule with altered activities. Local modulation is distinct from global modulation, which refers to alterations in the whole membrane that affect a variety of different receptors independent of other specificity (see refs 14–16).

Table II. *Aggregation of reconstituted and native membrane vesicles containing N-CAM (see ref. 29)*

(A) Reconstituted vesicle aggregation using purified chick N-CAM

Vesicles (form of N-CAM)	N-CAM/ lipid (g/mg)	Kagg (units)[a]
E	17	3.5
A	17	12.5
E	14	1.5
E	19	6.4
E	28	54.0

(B) Co-Aggregation of native brain membrane vesicles

Vesicle Pairing	Efficacy of Co-Aggregation (%)[b]
Chick Brain (E)–Chick Brain (E)	96
Chick Brain (E)–Chick Brain (A)	98
Chick Brain (E)–Mouse Brain (E)	100
Chick Brain (E)–Mouse Brain (A)	91
Chick Brain (E)–Mouse Brain (A)	97
Chick Brain (E)–Tadpole Brain	98
Mouse Brain (E)–Tadpole Brain	96
Chick Brain (E)–Chick Liver (E)	10

[a] One unit is the rate of aggregation of a sample (measured in nl of superthreshold particles) divided by the square of the concentration of vesicles in the sample (measured in mg of lipid per ml).

[b] The rate of aggregation shows a 2nd-order dependence on vesicle concentration. The efficacy of co-aggregation is a measure of the extent that the rate of aggregation of a mixture of vesicle preparations exceeds the sum of the rates of aggregation of each preparation alone. 100% = full co-aggregation; 0% = no co-aggregation.

chains existing for N-CAM; recently, a probe has been found (Murray, Edelman and Cunningham, unpublished observations) which detects only one of the two mRNA species. A cDNA clone for L-CAM has been obtained from poly-(A)$^+$RNA of embryonic liver using the λgtll expression vector [32]. The clone was complementary to a single 4 kb mRNA which was found in tissues from all organs expressing L-CAM but not from those that lacked this CAM.

All of these cDNA probes have been used to explore the multiplicity of genes coding for primary CAMs (Fig. 3). In neither case of the two primary CAMs was there evidence of a large family of genes. Southern blotting analysis with one of the probes for N-CAM detected only one fragment in chicken genomic DNA digested with several restriction enzymes [31] suggesting that sequences corresponding to those of the probe are present at most a few times and possibly only once in the chicken genome. Southern blotting analysis of the cDNA probe for L-CAM [2] showed components consistent with the presence of from one to three L-CAM genes (Fig. 3).

These results support the notion that the complexity of CAM expression during development (to be discussed in the next section) is not a reflection of structural gene complexity. Moreover, there does not appear to be a complex processing pathway for messages during synthesis of the CAMs. Instead, it appears that initial control of CAM expression is seated in early steps of control of expression of the respective structural genes and possibly of control of the rate of mRNA turnover. A system of perturbation of CAM differentiation using temperature sensitive RSV mutants to transform rat cerebellar cell lines [33, 34] has been used to analyze the level of the control of expression of N-CAM. The data (Brackenbury and Edelman, unpublished observations) are consistent with the notion that a major element of control occurs at the transcriptional level.

CAM expression sequences

In this section, I will review recent evidence indicating (i) that CAMs have definite sequences of expression that correlate strongly with major morphogenetic events related to epithelial–mesenchymal transformations [35–37], and (ii) that modulation of their expression at sites of embryonic induction occurs in two distinct modes: in mesenchyme, N-CAM diminishes at the surface and then reappears; in epithelia, both N-CAM and L-CAM appear together and one or the other subsequently disappears [38].

The two characterized primary CAMs, N-CAM and L-CAM, are both present at low levels on the chick blastoderm before gastrulation (Fig. 4A, B). As gastrulation occurs in the chick, epiblast cells express both N-CAM and L-CAM (Fig. 4C, D); as they ingress through the primitive streak, the amount of detectable CAMs decreases [35–38], presumably reflecting the fact that they have been down-regulated or masked. As the ingressing cells condense into the mesoblast, they re-express N-CAM. Thus, epiblast cells lose both CAMs as they undergo epithelial–mesenchymal transformation and move into the middle layer. Some of the ingressing cells become the chordamesoderm which also stains for N-CAM and subsequently takes part in neural induction.

Following gastrulation and coincident with neural induction, there is a marked change in the distribution of the two

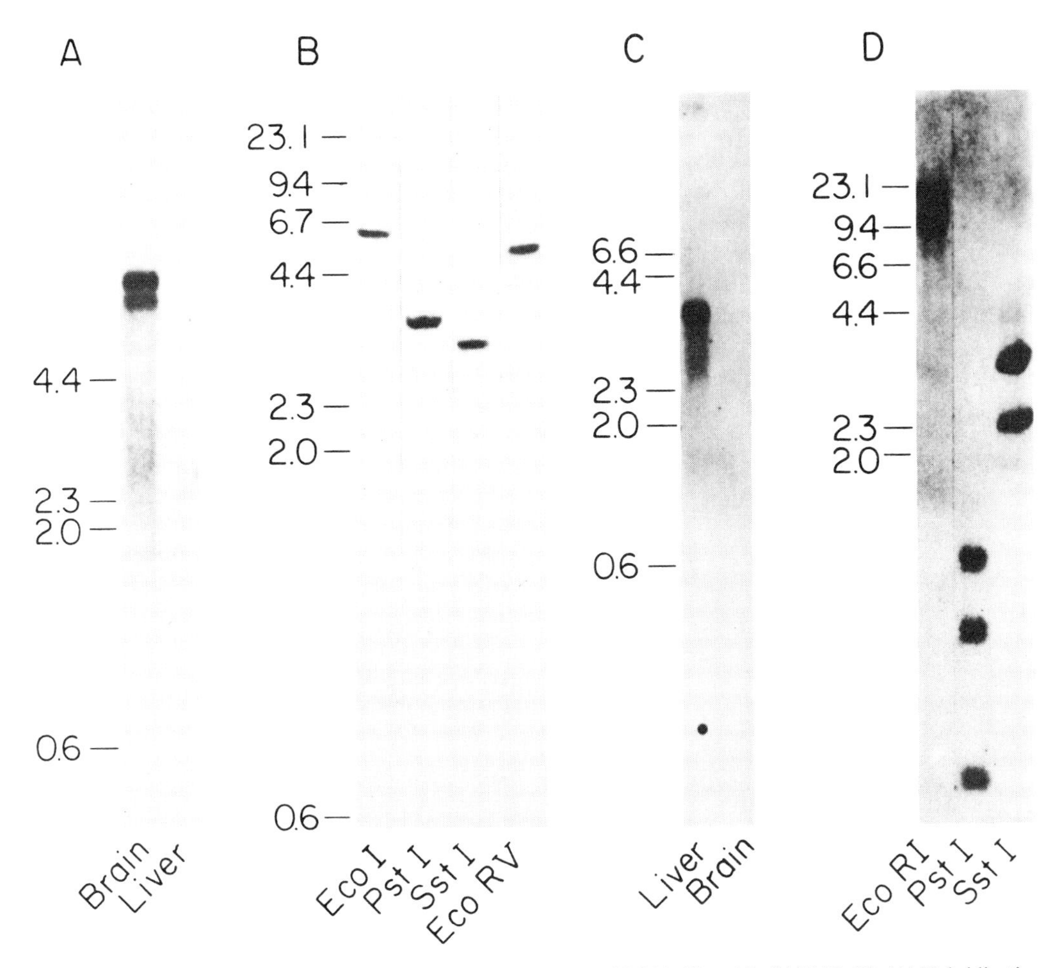

Fig. 3. Hybridization analyses of mRNA species and genomic sequences for chick N-CAM (A, B) and L-CAM (C, D). (A) Hybridization analysis ('Northern blot') of poly(A)$^+$ RNA from a 9-day embryonic chicken brain or 14-day embryonic chicken liver probed with the N-CAM cDNA clone pECOO1. (B) Hybridization analysis ('Southern blot') of adult chicken liver DNA digested with the indicated restriction endonucleases and probed with nick-translated pECOO1. (C) Hybridization analysis of L-CAM mRNA. Poly(A)$^+$ RNA of liver and brain from 11-day embryos were hybridized with the PEC301 L-CAM cDNA clone. (D) Embryonic chicken liver DNA, digested with the indicated restriction endonucleases and probed with pEC301. Hybridized ^{32}P-labeled probes were detected by autoradiography. Sizes of marker fragments (in kilobase pairs) are shown to the left of each group of lanes.

primary CAMs. An increase in immunofluorescent N-CAM staining appears in the region of the neural plate and groove as L-CAM staining disappears (Fig. 4E, F). In conjugate fashion, L-CAM staining is enhanced in the surrounding somatic ectoderm as N-CAM staining slowly diminishes. Migration of neural crest cells is also accompanied by down-regulation or masking of cell surface N-CAM [35]. Somewhat later, after neurulation, all sites of secondary induction show changes in cell surface prevalence of N-CAM, L-CAM or both (see Fig. 5 for some examples).

At sites of secondary induction, patterns of CAM expression correlate with the apposition of epithelia and mesenchyme prior to their further differentiation. As shown in Table III (see ref. 38 for further details), expression sequences of CAMs observed on cells in epithelia and mesenchyme follow two general modes. In mode I, expression of N-CAM (or of both CAMs, as seen in the epiblast) in mesenchyme decreases to low amounts at the cell surface and then N-CAM is re-expressed. In mode II, one or the other CAM disappears from epithelia initially expressing both CAMs, as seen earliest in neural induction. As a result of the primary processes of development, collectives of cells linked by N-CAM and undergoing modulation mode I are brought into the proximity of collectives of cells linked by L-CAM plus N-CAM or by L-CAM undergoing modulation mode II.

Such adjoining cell collectives or 'CAM couples' have been found at all sites of embryonic induction examined [38]. In the pharynx, for example, the early thyroid epithelium expresses both N-CAM and L-CAM as it is induced by N-CAM positive mesoderm (Fig. 5A, B). Similarly, strongly N-CAM positive mesoderm induces the lung buds from the laryngo-tracheal epithelium which expresses both N-CAM and L-CAM (Fig. 5C, D). More posteriorly, the pancreas is induced from the open anterior intestinal portal by an N-CAM containing mesenchyme (Fig. 5E, F). Later in development, the epithelia of the thyroid and pancreas lose N-CAM and express only L-CAM [37] whereas, in the lung, a population of epithelial cells retains both N-CAM and L-CAM into adulthood [38].

Similar expression sequences have been observed at other sites of secondary induction. The placodes that will give rise to neural structures first express both CAMs but eventually lose L-CAM. The apical ectodermal ridge of the limb bud expresses both CAMs as the limb is induced [35, 38]. Thus, from the time of primary induction, epithelia expressing both N-CAM and L-CAM are induced by N-CAM positive mesodermal tissues. This pattern is seen at sites of secondary induction in the lung and gut derivatives, in the skin (feather), and in the limb bud [38–40]. In the kidney, however, the direction appears to be reversed: the L-CAM and N-CAM positive Wolffian duct is the inductor for the N-CAM positive mesonephric mesenchyme [36, 38].

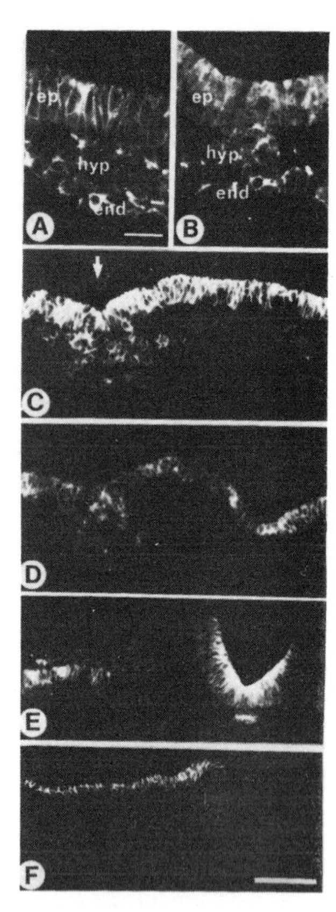

Fig. 4. Expression of CAMs in the early chick embryo. Transverse sections of full primitive streak stage (A, B), head process stage (C, D) or 5 somite chicken embryos (E, F) stained with anti-N-CAM (A, C, E) or anti-L-CAM (B, D, F). Arrow points to the primitive streak. Ep, epiblast, hyp, hypoblast, end, endophyll.

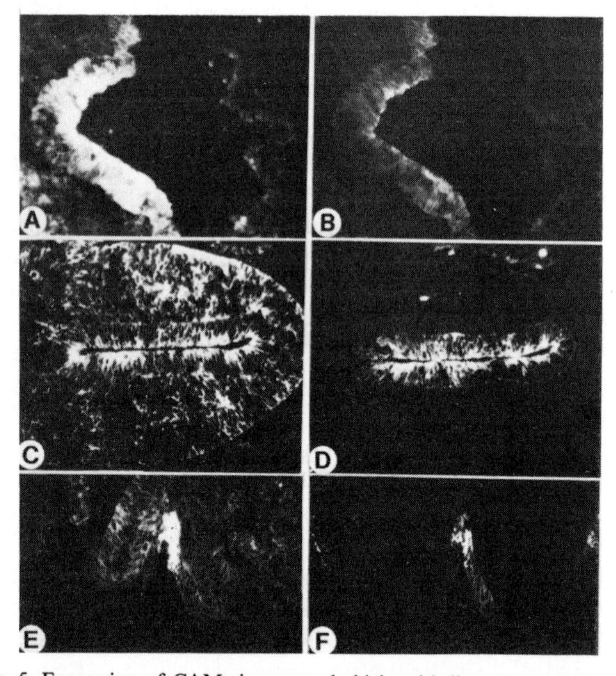

Fig. 5. Expression of CAMs in apposed chick epithelia and mesenchyme. Epithelia of the thyroid rudiment (stage 15) (A, B), laryngotracheal groove (stage 18) (C, D) and pancreatic rudiment (stage 15) (E, F) stain for both N-CAM (A, C, E) and L-CAM (B, D, F) and are adjacent to N-CAM containing mesenchyme (A, C, E).

The sequences reviewed above indicate that at certain times, one cell can express two CAMs, as first seen for the primary CAMs in blastoderm. In the induction of pharyngeal and gut appendages and in feather formation [39, 40], epithelial cells

Table III. *Modulation modes of CAM expression during chicken embryogenesis*[a]

Mode I: Mesenchymal conversions[b]	Mode II: epithelia[c]
Ectodermal	Ectodermal
N→O→N	NL→N (→*)
neural crest	neural plate
	NL→L (→*)
	somatic ectoderm
Mesodermal	Mesodermal
N→O→N	N→NL→L
dermis	kidney tubules
N→O→N→*	
chondrocytes	Endodermal
	NL→L
	thyroid
	pancreas

[a] Only one example is given in each case; for further examples, ref. 38.
[b] Mode I shows cyclic changes in N-CAM or disappearance. Some of these transitions occur with movement. O represents low levels of CAM; in some cases (marked *), CAM can be replaced by a differentiation product.
[c] Mode II shows replacement of one CAM by another or disappearance.

from ectoderm or endoderm can also express both L-CAM and N-CAM simultaneously; this is also true of kidney mesoderm [38]. In general, the expressions in epithelia are dynamic and change so that one or the other CAM disappears during maturation to the adult state (mode II, Table III).

Collections of cells linked by L-CAM (or cells expressing both L-CAM and N-CAM, mode II) have been found adjacent to collectives of cells linked by N-CAM (mode I) at all sites of secondary embryonic induction so far examined [38]. The potential significance of these so-called primary CAM couples in morphogenesis is not yet fully understood, but examination of periodic, hierarchically arranged structures such as the feather shows unequivocally that their CAM expression sequences are correlated with other cytodifferentiation events. The feather is a particularly compelling example of CAM couple formation occurring periodically in space and time.

The induction of the feather occurs by means of periodic accumulations in the skin of dermal condensations of mesodermal mesenchyme [41]. Each of these condensations acts upon ectodermal cells to induce placodes. After placode formation, a dermal papilla is formed involving another couple containing mesodermal (N-CAM) and ectodermal (L-CAM) elements [39, 40]. Afterwards, L-CAM positive papillar ectoderm produces collar cells that express both CAMs. These cells will provide the basis for the formation of barb ridges and barbule plates through alternating CAM couples ultimately yielding three hierarchical levels of branching: rachis, ramus, and barbules (Fig. 6).

In the formation of this hierarchy there is an extraordinary sequence [39, 40] of CAM couple expression linked first to cell movement, then to cell division, and finally to differentiation and cell death: (1) Initially, L-linked ectodermal cells are approached by CAM negative mesenchyme cells moving into the vicinity. Just beneath the ectoderm, the mesenchyme cells become N-CAM positive (mode I) and accumulate in lens-shaped aggregates that induce placode formation in the L-CAM linked ectodermal cells (Fig. 6A, B). Later, the L-CAM positive placode cells transiently express N-CAM

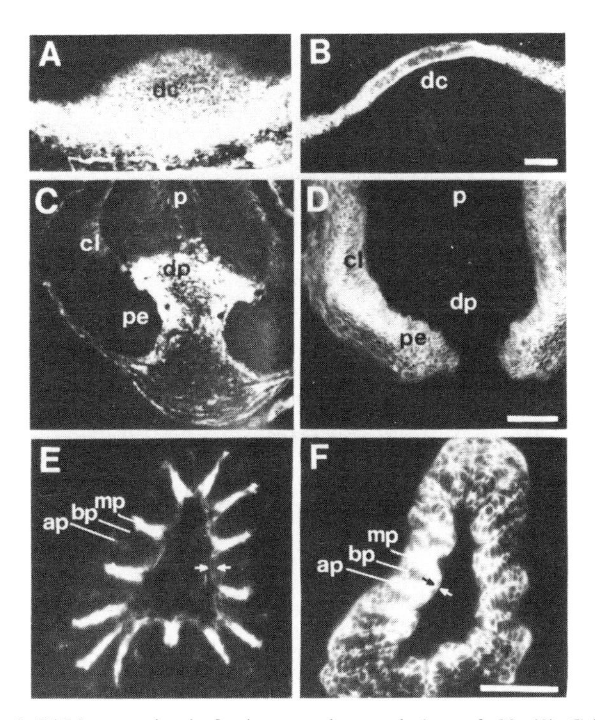

Fig. 6. CAM expression in feather morphogenesis (see refs 39, 40). Cells of the ectodermal placode (A, B) stain for both N-CAM (A) and L-CAM (B) and are underlain by strongly N-CAM positive (A) dermal condensations (dc). In the dermal papilla (dp), (C, D), N-CAM positive cells (C) are surrounded by collar epithelial cells (cl) expressing both N-CAM (C) and L-CAM (D). The epithelial papillar ectoderm (pe) expresses only L-CAM (D). P, pulp. As barbules are formed, cells of the marginal plate (MP) and axial plate (ap) express both N-CAM (E) and L-CAM (F) and are separated by L-CAM positive cells of the barbule plate (BP).

(mode II, Table III). (2) In the formation of the dermal papilla, N-CAM positive mesodermal cells adjoin L-CAM positive ectodermal cells (Fig. 6C, D). At this stage in the highly proliferative collar epithelium, these ectodermal cells express both L-CAM and N-CAM. (3) Derivatives of these cells lose N-CAM while retaining L-CAM as they form barb ridges by division. In the valleys between the ridges, single or small numbers of basilar cells then express N-CAM while losing L-CAM. This mode II process extends cell by cell up each ridge resulting in the formation of the N-CAM positive marginal plate. The net result is alternating barb ridges (L-CAM linked) and marginal plates (N-CAM linked) (Fig. 6E, F). (4) As ridge cells organize into barbule plates linked by L-CAM, a similar process recurs – N-CAM is expressed in cells lying between each of the future barbules resulting in yet another level of periodically expressed CAM couples. The net result is a series of cellular patterns in which cell collectives expressing L-CAM alternate with those expressing N-CAM at both the secondary barb level and the tertiary barbule level. (5) Finally, after further growth of these structures and extension of the barb ridges into rami, the L-CAM positive cells keratinize and the N-CAM positive cells die without keratinization. This deposits extracellular N-CAM, which appears to be reabsorbed or dispersed, leaving alternate spaces between rami and between barbules and yielding the characteristic feather morphology.

A key feature of this histogenetic CAM expression sequence is periodic CAM modulation in a cycle on particular cells, such as those of the inducing mesenchyme as it moves into sites near the ectoderm. Another important feature is the periodic and successive formation of adjacent N-linked and L-linked cell collectives. Finally, there is a striking association of particular cytodifferentiation events with one or the other member of a couple. This association clearly illustrates some means by which the primary processes of development constituting the driving forces for morphogenesis may be linked to the regulatory processes of adhesion. Morphogenetic movement is coupled to expression of a CAM cycle in mesenchymal cells in the original induction. Cell division is associated with the formation of papillar ectoderm and L-positive barb ridge formation. Cell death is linked to the existence of N-linked collectives in barb and barbule formation and it comprises the terminal stages of hierarchical pattern formation in the feather; the conjugate process, cell differentiation, as marked by keratin expression, is linked to prior morphoregulatory CAM differentiation events in L-positive cells. In areas of induction, an epithelial collective of cells linked by L-CAM plus N-CAM (or by L-CAM only) is adjoined by a collective of cells linked by N-CAM alone. Such CAM couples arise either from movement of mesenchymal cells to adjoin epithelia or from differential gene expression and cell division in cells of the same lineage, as seen in the feather.

These observations reveal that the primary CAMs are ubiquitous and that the uniform patterns of a small number of CAM expression sequences are repeated in many locales. The evidence suggests that the genes affecting CAM expression (morphoregulatory genes) act independently of and prior to those controlling tissue-specific differentiation (historegulatory genes) inasmuch as CAM expression in most induced areas does initially precede the expression of most cytodifferentiation products. This is consistent with the observation that the expression of each primary CAM in tissues overlaps several different tissue types of different germ layer origin, as seen in a classical fate map [36]. N-CAM is found in neural regions and muscles and L-CAM is seen in somatic ectoderm and endoderm. Such a map is presented in Fig. 7 which also summarizes in a diagram the expression sequence of the two primary CAMs discussed here.

As shown in this diagram for Ng-CAM, during histogenesis one observes the appearance of such secondary CAMs; one observes the appearance of certain specialized substrate adhesion molecules (SAMs) which also show a defined or restricted spatiotemporal sequence of appearance. Some of these events may be illustrated by considering neural histogenesis.

Neural histogenesis

Within the overall expression sequence diagrammed in Fig. 7, a set of microsequences can be discerned as histogenetic events characterized by cellular differentiation occur. Following neural induction and neurulation, N-CAM is found distributed throughout the nervous system. At E 3.5 in the chick, the secondary Ng-CAM appears on neurons already displaying N-CAM [42, 43]. In the CNS, it is seen mainly on extending neurites and is seen only faintly or not at all on cell bodies. This so-called polarity modulation is not seen on cells known to be undergoing migration along guide glia, for example, in the cerebellum (Fig. 8) or spinal cord. Instead, in regions containing such cells, Ng-CAM is found both on leading processes of the cell bodies and on neurites. A contrasting picture is seen in the peripheral nervous system:

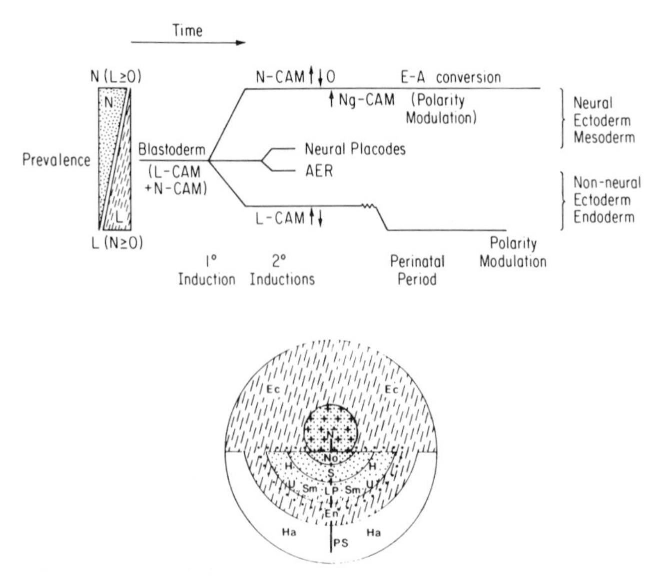

Fig. 7. Major CAM expression sequence and composite CAM fate map in the chick. (A) Schematic diagram showing the temporal sequence of expression of CAMs during embryogenesis. Vertical wedges at the left refer roughly to relative amounts of each CAM in the different parts of the embryo, i.e. the line referring to blastoderm has relatively large amounts of each CAM whereas that for neural ectoderm has major amounts of N-CAM but little or no L-CAM. After an initial differentiation event, N-CAM and L-CAM diverge in cellular distribution and are then modulated in prevalence (↑↓) within various regions of inductions or actually decrease greatly (○) when mesenchyme appears or cell migration occurs (see Table III). Note that placodes which have both CAMs echo the events seen for neural induction. Just before the appearance of glia, a secondary CAM (Ng-CAM) emerges; unlike the other two CAMs, this CAM would not be found in the fate map in B before 3.5 days. In the perinatal period, a series of epigenetic modulations occurs: E-A conversion for N-CAM and polar redistribution for L-CAM. (B) Composite CAM fate map in the chick. The distribution of N-CAM (stippled), L-CAM (slashed) and Ng-CAM (crossed) on tissues of 5- to 14-day embryos is mapped back onto the tissue percursor cells in the blastoderm. Additional regions of transient N-CAM staining (see Table III) in the early embryo (5 days) are shown by larger dots. In the early embryo, the borders of CAM expression overlap the borders of the germ layers, i.e. derivatives of all three germ layers express both CAMs. At later times, overlap is more restricted: N-CAM disappears from somatic ectoderm and from endoderm, except for a population of cells in the lung. L-CAM is expressed on all ectodermal and endodermal epithelia but remains restricted in the mesoderm to epithelial derivatives of the urogenital system. The vertical bar represents the primitive streak (PS); EC, intraembryonic and extraembryonic ectoderm; En, endoderm; N, nervous system; No, prechordal and chordamesoderm; S, somite, Sm, smooth muscle; Ha, hemangioblastic area.

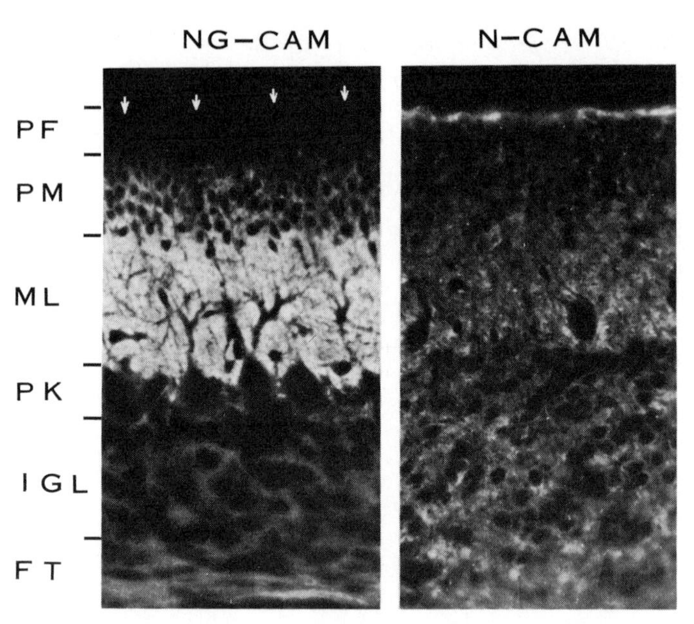

Fig. 8. Localization of Ng-CAM and N-CAM on 17-day embryonic chicken cerebellum (see refs 24, 43). Frozen sections were allowed to react sequentially with rabbit anti-Ng-CAM or anti-N-CAM IgG and fluorescein-conjugated goat anti-rabbit IgG. Flourescence micrographs of comparable fields are shown; treatment with preimmune sera gave no staining. Note that staining for N-CAM but not for Ng-CAM (arrows indicate the cerebellar surface) was visualized in the proliferative zone (PF) of the external granular layer. In contrast, in the molecular layer (ML), staining for Ng-CAM is more intense than for N-CAM. This reflects the expression of Ng-CAM on both neurites and cell bodies. In contrast, in regions in which cell migration is not occurring, Ng-CAM is seen only on neurites, i.e. it displays polarity modulation. PM, premigratory zone; PK, Purkinje cell; IGL, internal granular layer; FT, fiber tract. Bar = 10 μm.

after its appearance, Ng-CAM is found at all times on neurites and cell bodies [42, 43].

Ng-CAM shows a detailed microsequence (Table IV) of appearances that follows known sequences of neurite extension and cell migration [42, 43]. The order of appearance shown in Table IV is reproducible from animal to animal. This suggests that local signals related to cellular maturation and possibly to growth factors produced by glial precursors are responsible for both the appearance and the remarkable prevalence modulation of Ng-CAM at the cell surface.

Over the considerable period of time in which these Ng-CAM appearances occur, alterations of a lesser degree and with longer time courses can be seen in the prevalence of N-CAM on different cell surfaces. At a time in the sequence when many neural tracts have been established and myelination is to begin, Ng-CAM as detected by immunohistochemical methods is down-regulated at the cell surface in all CNS tracts that are to become white matter (Table IV). No such down-regulation is seen in the peripheral nervous system. At roughly similar times, N-CAM undergoes E-A conversion [42, 43] which as discussed above, leads to increases in binding rates *in vitro*.

The net result of the various types of cell surface modulation occurring developmentally for the two neuronal CAMs is a striking change in their relative distributions in most areas of the CNS (Fig. 8, Table IV). But in areas of the CNS remaining capable of forming new connections into adult life as well as in the peripheral nervous system, the relative distribution of the two CAMs does not change with time. A set of maps has been constructed [43] that display the changes and constancies in the two neuronal CAMs for many brain areas in the chick. At this point, the data strongly suggest that surface modulation events and CAM expressions can occur in relatively small cell populations in a defined order. This conclusion is reinforced by the data on sequences of CAM couple expression in feather histogenesis, findings which also suggest that the successive signals for CAM expression are sharply localized in time and space during the emergence of form.

Recently, it has been found that such localized expression can also occur for extracellular matrix proteins. A new matrix protein, cytotactin [45], has been found in a variety of embryonic sites in a defined sequence restricted to only certain tissues. Cytotactin is made by glia and mediates interaction between neurons and glia. Its initial distribution is in early basal laminae; later it is seen in smooth muscle as well as in the nervous system. In contrast to fibronectin and laminin, it appears in a sharply defined pattern during embryogenesis and histogenesis, suggesting that certain SAMs, like CAMs, may play a role in defining morphogenetic pathways and

Table IV. *Expression sequence of two neuronal CAMs in the developing chick nervous system (see ref.* 43)[a]

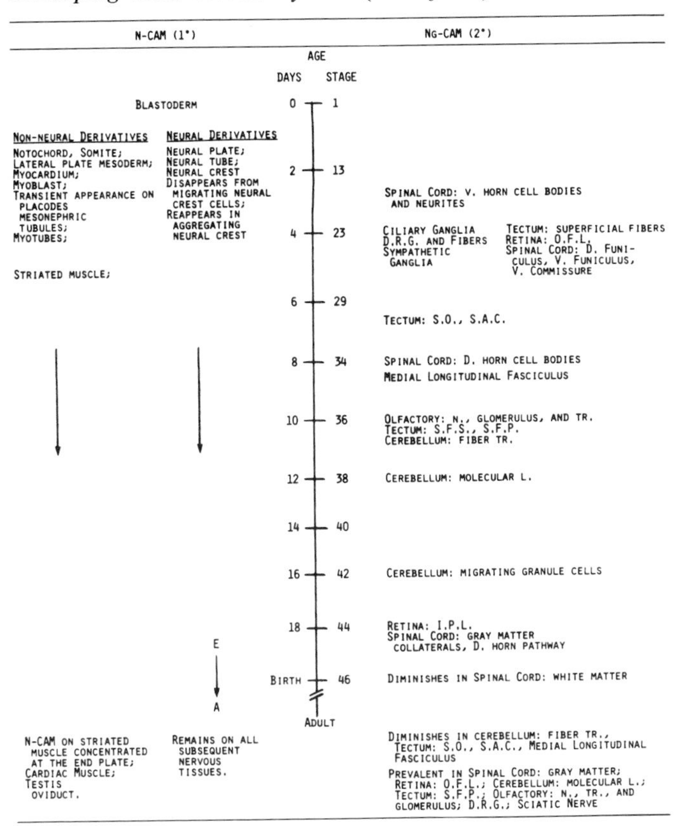

[a] D, dorsal; E to A, embryonic to adult conversion of N-CAM; IPL, inner plexiform layer; L, layer; OFL, optic fiber layer; SAC, *stratum album centrale;* SFP, *stratum fibrosum superficiale;* SO, *stratum opticum;* TR, tract; V, ventral.

regulating morphogenetic movements. The notion that cytotactin may play such a role is supported by perturbation experiments using antibodies to cytotactin, an approach first used with the CAMs: anti-cytotactin blocks or slows migration of external granule cells on Bergmann glia cells in cerebellar tissue slices *in vitro* (Chuong and Edelman, unpublished observations).

Perturbation of morphology and CAM expression

Expression sequences of the type described above clearly reveal correlations between CAM expression, cell surface modulation and key morphogenetic events, but they do not provide any direct indication of the causal role of the CAMs. Considered *a priori*, CAM expression could be a cause or an effect; indeed, if CAM expression occurs in cycles and in parallel with other processes, particularly those involving morphogenetic movements, it could be both cause and effect. By means of perturbation experiments, and by further characterizing CAM genes, one may provide appropriate bases to guide a search for signals initiating CAM expression and to analyze the causal sequences of cellular controls.

If the phenomenological descriptions of the expression sequences of CAMs are to be related confidently to molecular function, then one should be able to show that perturbation of their function alters movement and form. Form-producing movements in the embryo are various [4, 13, 46, 47]. They include the movement of cells following epithelial to mesenchymal conversion (as mentioned above in mesoblast formation and neural crest cell movements), the movement of tissue sheets in tension upon each other (as seen in neurulation, the folding up of the neural tube driven by the extension of the notochord), and also the exquisite tract and map-forming movements of neurons on each other and glia following neural process extension. By means of perturbation experiments, CAMs have been shown to have direct or indirect roles in these events.

Experiments on perturbation of CAM function or alterations of CAM expression have been carried out in a number of systems [14–16]. Addition of anti-L-CAM antibodies leads to failure *in vitro* of histotypic aggregate formation by liver cells [48]. Anti-N-CAM antibodies disrupt neural fasciculation *in vitro* and greatly disrupt layer formation in the chick retina during organ culture [49]. Implanted anti-N-CAM antibodies disrupt retinotectal map formation *in vivo* in the frog [50]. When neural cells were transformed by a temperature-sensitive mutant of RSV [33, 34], they retained normal morphology, adult N-CAM levels, and normal aggregation behavior at the non-permissive temperature. At the permissive temperature, the cells transformed and within hours, they down-regulated their surface N-CAM and became more mobile.

If, as implied by these data, CAMs alter morphology by their differential expression in time and space, that expression must depend upon the morphologic integrity of collectives of interacting cells. Perturbation of normal cell-cell interactions *in vivo* has in fact been shown to lead to alteration of CAM expression and distribution. For example, N-CAM is present at the end plate of striated muscles [51] but is absent from the rest of the surface of the myofibril. After cutting the sciatic nerve, the N-CAM disappears from the end plate, the anti-N-CAM staining is increased in the cytoplasm and the molecule appears diffusely at the cell surface (Fig. 9). Thus, perturbation of morphology can be accompanied by altered CAM modulation. Alteration of CAM modulations have also been seen in certain genetic defects. In the mouse mutant *staggerer*, which, in the double recessive, shows connectional defects in the cerebellum between parallel fibers and Purkinje cells accompanied by extensive granule cell death, E-A conversion of N-CAM is greatly delayed in the cerebellum [52].

In all of the above examples, morphogenetic movements play a key role. Both primary and secondary CAMs appear to have major roles in the regulation of movement [15, 16, 24]. These roles are various: (i) permissive (down-regulation of cell surface CAM to allow cells linked in an epithelium to convert to a mobile, loosely linked mesenchyme); (ii) formative (mediating interaction of two epithelia under tension and plastic flow); and (iii) initially requisite (needed for movement of cell bodies on cell processes already laid down). Perturbation experiments [24] have shown that anti-Ng-CAM antibodies in cerebellar tissue slices *in vitro* can inhibit the movement of external granule cells on guide glia (Bergmann glia). These glial cells also produce cytotactin, an extracellular matrix protein, and, as mentioned previously, anti-cytotactin also inhibits granule cell migration [45].

The examples given in this brief review indicate that perturbations in CAM binding can lead to altered morphogenesis and that altered morphogenesis can lead to changes in CAM modulation patterns. In addition, perturbation of tissue-restricted SAMs such as cytotactin can also affect morphogenetic movement. The cumulative evidence supports

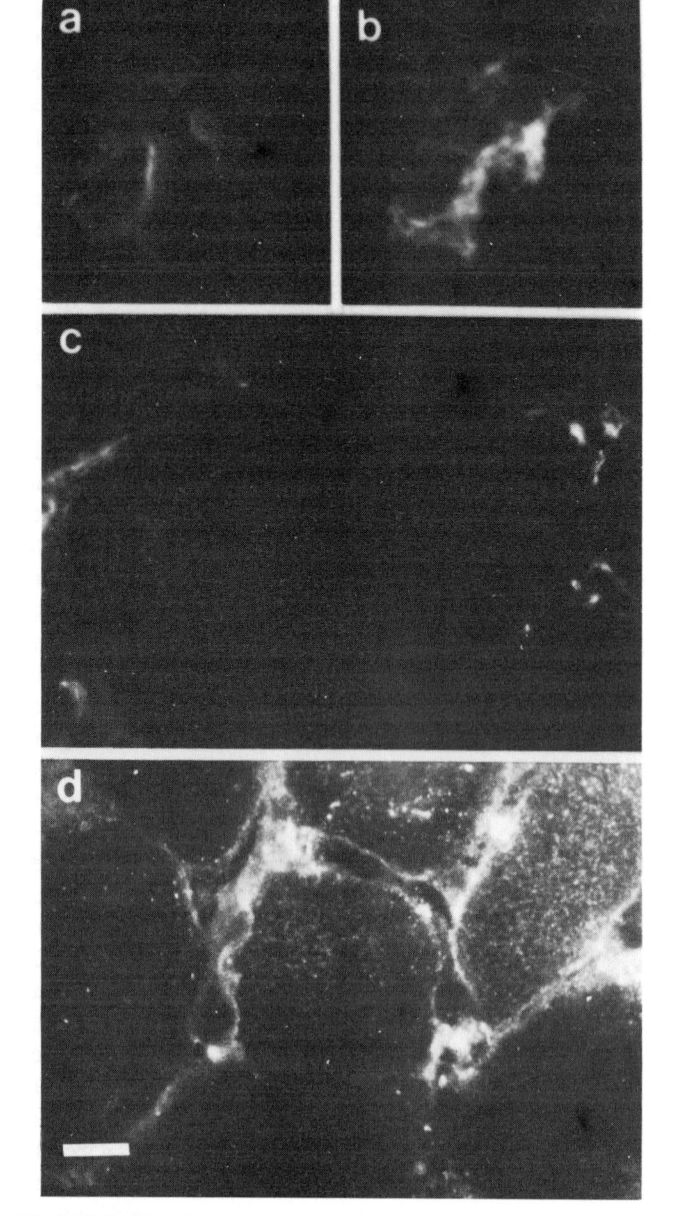

Fig. 9. N-CAM at the motor end plates and changes in N-CAM prevalence in muscle after denervation (see ref. 51). Staining of motor end plate with fluoresceinated-bungarotoxin (a) and rhodamine-labeled anti-N-CAM (b). Two weeks after section of the sciatic nerve in one thigh of the mouse, the gastrocnemius muscles from the unperturbed side (c) and the denervated side (d) were dissected, frozen, and cut in cross-section. The sections were stained with fluorescein-labeled anti-N-CAM IgG. Bar, 10 μm; × 800.

the notion that control of CAM expression and binding functions plays a key role in the morphogenetic movements leading to form and histogenesis. Although the functions of CAMs vary in different contexts of movement, cell type, and tissue production, there can be little doubt that, as a family of molecules subserving adhesion, CAMs and SAMs are major regulators of movement. This brings us back to the main issue: how can such regulation be related to the expression of CAM genes on the one hand and to epigenetic sequences on the other?

Epigenesis: mechanochemistry linked to developmental genetics

The foregoing evidence on the structure, function and sequences of CAM expression and the results of perturbation experiments suggest that these molecules play a major regulatory role in morphogenesis. This conclusion provides the basis for the regulator hypothesis [9] which states that, by means of cell surface modulation, CAMs are key regulators of morphogenetic motion, epithelial integrity, and mesenchymal condensation. According to the hypothesis, the genes affecting CAM expression (morphoregulatory genes) act independently of and prior to those controlling tissue specific differentiation (historegulatory genes). Consistent with this assumption, CAM expression in most induced areas does initially precede the expression of most cytodifferentiation products. This is consistent in turn with the observation that the expression of CAM types in tissues overlaps different tissue types as seen in the classical fate map (Fig. 7). Inasmuch as cell adhesion molecules are likely to be proteins specified by particular genes, their function provides a candidate mechanism for how the one-dimensional genetic code might regulate three-dimensional form. Obviously, this mechanism, if true, would serve as a major mechanism of morphologic evolution.

A main action of CAMs is to attach cells in collectives and to regulate movement and thus a major part of their function is mechanochemical in nature. Although CAMs exist in a relatively small number of specificities, they are capable of a large number of alterations in their binding properties which are graded and non-linear. To relate this mechanochemical CAM function to CAM expression sequences and to the expression of other gene products concerned with cytodifferentiation, it must be assumed that chemical or mechanochemical signals act *locally* on cells in collectives. Altered pressure, tension, or flow in the vicinity of a collective of cells held together by one kind of CAM in the neighborhood of another collective held together by a CAM of different specificity, together with chemical factors released by such collectives could alter the expression or modulation of the CAMs at cell surfaces. A varying temporal sequence of such expressions would alter morphogenetic movements and lead to changes in specific form and pattern. We must examine how such signals could effect morphoregulatory genes for CAMs and SAMs and historegulatory genes for tissues.

At the level of a given kind of cell and its descendants, CAM expression may be viewed as occurring in a cycle (Fig. 10). Traversals of the outer loop of this cycle affect morphoregulatory genes which can lead either to the switching on of one or another of the CAM genes or to their switching off. Switching on and off of the same genes is suggested in the case of mesenchymal cells contributing to dermal condensations as well as in the case of neural crest cells (mode I, see Table III). Switching to a different CAM gene is suggested in epithelia (mode II). The subsequent action of historegulatory genes (Fig. 10, inner loop) is pictured to be the result of inductive signals arising in the new milieux that occur through CAM-dependent cell aggregation, motion, and tissue folding. If the expression of certain historegulatory genes led to altered cell motion or shape or altered post-translational events, this would alter the effects on morphogenesis of subsequent traversals of the outer loop. One key example of a traversal of the inner loop, directly affecting the cells containing N-CAM, concerns the historegulatory genes specifying the enzyme responsible for E to A conversion, an event which leads to changes in N-CAM binding rates. Combination of the outer and inner loop of the cycle and the linkage of two

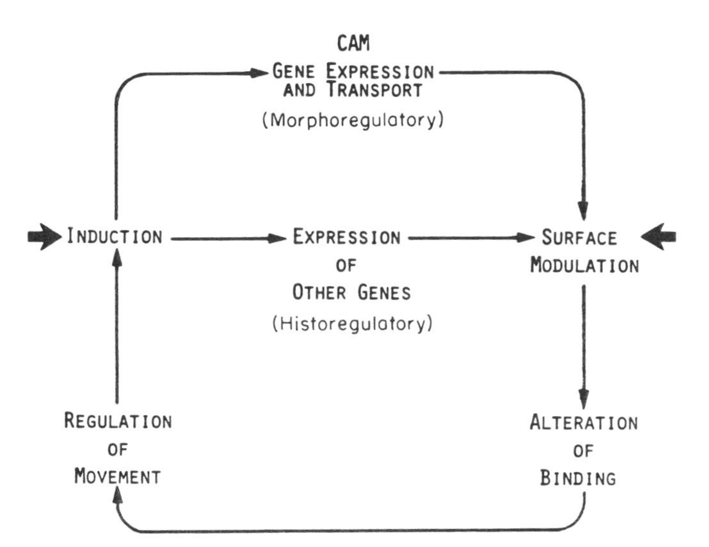

Fig. 10. The regulator hypothesis (see ref. 9) as exemplified in a CAM regulatory cycle. Early induction signals (heavy arrow at left) lead to CAM gene expression. Surface modulation (by prevalence changes, polar redistribution on the cell or chemical changes such as E to A conversion, see Fig. 2) alters the binding rates of cells. This regulates morphogenetic movements which in turn affect embryonic induction or milieu-dependent differentiation. The inductive changes can again affect CAM gene expression as well as the expression of other genes for specific tissues. The heavy arrows at left and right refer to candidate signals for initiation of induction which are still unknown. These signals could result from global surface modulation as a result of CAM binding (right) or from release of morphogens affecting induction (left) or both; in any case, a mechanochemical link is provided between gene expression and morphogenesis.

such cycles by the formation of CAM couples (see Table III) could lead to a rich set of effects altering the path of morphogenesis.

In considering how two such cycles (each related to a different CAM) might interact, we are compelled to consider the nature of the signals that activate morphoregulatory and historegulatory genes during induction. It remains unknown whether these signals are morphogens released by cells linked by a particular CAM (Fig. 10, left large arrow) or whether they are derived from mechanical alterations of the cell surface or cytoskeleton through global cell surface modulation (Fig. 10, right large arrow).

On the assumption that both types of signals are involved, a concrete testable model that accounts for periodicity in initial feather induction (see Fig. 6) can be formulated. The assumption of this model is that CAM morphoregulatory genes act prior to historegulatory genes and that, at least early in feather formation, the two kinds of genes are independent. Signals to express N-CAM in mesodermal mesenchyme cells near the ectoderm are assumed to be triggered by the release of signals from the L-CAM linked cell collectives in the ectoderm. Reciprocal signals arising from the resultant N-CAM linked mesodermal collectives then *locally* turn off the signals from L-linked cells while inducing placode formation in these cells. These reciprocal signals would reach threshold levels only when the N-linked mesodermal collective attained a sufficient size. Events embodied in this model would lead to the emergence of periodic feather induction sites and the model prompts the prediction that the blockade of cell linkage by anti-L-CAM in the ectoderm together with the blockade of linkage by anti-N-CAM in the dermal condensation would tend to alter or block periodic placode formation.

It is useful at this point to discriminate among the various levels of control in embryonic events leading to form. The regulator hypothesis assumes that the cell is the unit of control, that the cell surface is the nexus of control events, that cell adhesion and differentiation order the driving forces of the processes of cell movement and cell division, and that adhesion acts by generating *local* signals affecting collectives in CAM couples during epigenesis. These signals are assumed to consist of both morphogens *and* mechanochemical factors such as cross linkage of CAMs affecting global cell surface modulation. While the cell is considered to be the unit of control, the unit of induction is considered to be a cell collective of sufficient size linked by a particular primary CAM or combination of CAMs.

CAMs are thus hypothesized to be molecules that provide the linkage between the genes and the mechanochemical requirements of epigenetic sequences. Linked CAM cycles occurring in various contexts provide a potential solution to the problem of mechanochemical control of pattern ranging over the various levels from gene through organ and back to gene during regulative development. According to these views, 'place' in the embryo is not recognized by a microscopically assigned marker; it is established as a consequence of modulation and control. The expression of CAMs as gene products acting through their role as regulators of cellular mechanochemical events links the one-dimensional genetic code to the higher order assemblages of cells and tissues.

According to this view, the control of CAM binding makes the necessary rapprochement between developmental genetics and mechanochemistry at cell and tissue levels. CAMs are proposed as a bridge between 'entwicklungsmechanik' [53] and developmental genetics [2–4]. It is clear that while the CAMs are necessary as key regulators of pattern, other molecules such as SAMs (e.g. cytotactin) and other tissue-specific gene products must also be present to ensure cellular migration and specialization in histogenesis. In this dynamic model, CAMs themselves are clearly insufficient to generate a rich diversity of form; the regulatory cycle (Fig. 10) and the primary processes of development are the fundamental generators. Nonetheless, the picture of CAM regulation that emerges provides a basic mechanism for heterochrony and thus allows a connection between the developmental genetic and evolutionary questions with which this paper began.

Morphologic evolution: CAM regulation as a basis for heterochrony

As a potential answer to the developmental question, the regulator hypothesis is consistent with the demands of the evolutionary question. The hypothesis asserts that CAM regulatory genes are the *minimal* set likely to be instrumental in morphologic evolution. Other genes affecting the primary processes would obviously be equally important. It is likely that, during evolution, independent but *covariant* alterations occur in morphogenetic cell movements and timing of CAM regulatory gene expression [9]. Any mutational variation of either that fails to lead to successful embryonic induction could be selected against. On the other hand, any evolutionary variation in morphogenetic motion that is consistent with a pattern-producing epigenetic sequence or with a variation in the timing of gene expression that results from a mutation in

CAM regulatory genes may be selected for. Whether that selection takes place or not depends upon the phenotypic expression of the altered form and upon its adaptive advantage. A strong prediction of the regulator hypothesis related to this idea of covariance is that the CAM fate maps of the chick (see Fig. 7) and the frog will be highly similar despite the large differences in early morphogenetic movements seen in the two species of embryo. These assumptions are genetically parsimonious: small changes in timing or influence of CAM regulatory genes could have large non-linear effects upon form and pattern in short periods of evolutionary time. The parallel nature of the loops of the CAM cycle (Fig. 10) would allow florid variations within the same general morphological scheme. It is not difficult to incorporate heterochronic mechanisms leading to progenesis, neoteny and paedomorphosis [7] into this scheme: small changes in the timing of expression of morphoregulatory genes for CAMs in the cycle could lead to large changes in form.

The emphasis of the regulator hypothesis on local control ranging from the gene to tissue and back to the gene provides a basis for understanding how complexity arises epigenetically during development. Aside from their plasticity, the most striking feature of developmental processes is their parallel nature [3, 4, 10]. Independent parallel events can interact in such systems in a non-linear and complex fashion. In the scheme reflected by the CAM cycle (Fig. 10), this parallelism is subserved by the shunt pathway with the heading 'other genes'. But some of these genes, of course, specify and regulate key primary processes such as cell division and cell death, both of which are major driving forces in morphogenesis. Under certain circumstances, and at certain times, such processes can temporarily wrest control of morphogenesis away from the regulator loop, a phenomenon seen clearly in the terminal portions of feather histogenesis [39, 40].

The ideas elaborated here for CAMs can be extended to other morphoregulatory molecules. Specific molecular examples are provided by the functions of substrate adhesion molecules (SAMs) which are important for cell migration and of cell junctional molecules (found in gap junctions, tight junctions, etc.) which specify cell linkages and communications. The action of both these classes of molecules is required for orderly morphological development [54]. The example of intermodulation of N-CAM and fibronectin observed in the movement of neural crest cells [35] is pertinent: Fibronectin is required for crest cell migration but is not produced by crest cells. Down-regulation of N-CAM at the cell surface is a permissive condition for motion; concomitant expression of fibronectin is a necessary condition. The recent discovery of cytotactin [45], which is a site-specific SAM or extracellular matrix molecule that is concerned with regulation of cell movement, also raises the possibility that considerations similar to those in the regulator hypothesis may apply to certain SAMs. Should this turn out to be the case, the basic idea that there are morphoregulatory genes under prior and separate control from the historegulatory genes that lead to specific cytodifferentiation would be extended but not contradicted. It is a challenging task for the molecular embryologist to determine how many morphoregulatory genes will be required to determine the morphology of a complex appendage or organ. If the regulator hypothesis is correct, the number need not be large, and population genetic analyses of morphologically significant genes may not be unduly complex.

References

1. Hamburger, V., in *The Evolutionary Synthesis* (eds. E. Mayr, W. B. Provine), p. 96. Harvard University Press, Cambridge, MA (1980).
2. Waddington, C. H., *Principles of Development and Differentiation*. Macmillan Publishing Co., New York (1966).
3. Bonner, J. F., *Life Sci. Res. Report*, vol. 22. Springer Verlag, Berlin, Heidelberg and New York (1982).
4. Raff, R. A. and Kaufman, T. C., *Embryos Genes and Evolution, The Developmental Genetic Basis of Evolutionary Change*. Macmillan Publishing Co., New York (1983).
5. Jacobson, A., *Science* **152**, 25 (1966).
6. Edelman, G. M., *Cold Spring Harbor Symp. Quant. Biol. L*, 877–889 (1985).
7. Gould, S. J., *Ontogeny and Phylogeny*. The Belknap Press of Harvard University Press, Cambridge, MA, and London (1977).
8. King, M. C. and Wilson, A. C., *Science* **188**, 107 (1975).
9. Edelman, G. M., *Proc. Natl. Acad. Sci. USA* **81**, 1460 (1984).
10. Needham. J., *Biol. Rev.* **8**, 180 (1933).
11. Sperry, R. W., *Proc. Natl. Acad. Sci. USA* **50**, 703 (1963).
12. Edelman, G. M., *Trends Neurosci.* **7**, 78 (1984).
13. Weiss, P., *Principles of Development, a Text in Experimental Embryology*. Hafner Publishing Co., New York (1969).
14. Edelman, G. M., *Science* **219**, 450 (1983).
15. Edelman, G. M., *Annu. Rev. Neurosci.* **7**, 339 (1984).
16. Edelman, G. M., *Annu. Rev. Biochem.* **54**, 135 (1985).
17. Edelman, G. M., in *17th Stadler Genetics Symposium on Genetics, Development and Evolution*. Plenum Publishing Corporation, New York (1985).
18. Edelman, G. M., in *Molecular Bases of Neural Development*. John Wiley & Sons, New York (1985).
19. Brackenbury, R., Thiery, J.-P., Rutishauser, U. and Edelman, G. M., *J. Biol. Chem.* **252**, 6835 (1977).
20. Thiery, J.-P., Brackenbury, R., Rutishauser, U. and Edelman, G. M., *J. Biol. Chem.* **252**, 6841 (1977).
21. Hatta, K., Okada, T. S. and Takeichi, M., *Proc. Natl. Acad. Sci. USA* **82**, 2789 (1985).
22. Cunningham, B. A., Leutzinger, Y., Gallin, W. J., Sorkin, B. C. and Edelman, G. M., *Proc. Natl. Acad. Sci. USA* **81**, 5787 (1984).
23. Grumet, M. and Edelman, G. M., *J. Cell Biol.* **98**, 1746 (1984).
24. Grumet, M., Hoffman, S., Chuong, C.-M. and Edelman, G. M., *Proc. Natl. Acad. Sci. USA* **81**, 7989 (1984).
25. Hoffman, S., Sorkin, B. C., White, P. C., Brackenbury, R., Mailhammer, R., Rutishauser, U., Cunningham, B. A. and Edelman, G. M., *J. Biol. Chem.* **257**, 7720 (1982).
26. Finne, J., Finne, U., Bazen-Deagostini, H. and Goridis, C., *Biochem. Biophys. Res. Comm.* **112**, 482 (1983).
27. Crossin, K. L., Edelman, G. M. and Cunningham, B. A., *J. Cell Biol.* **99**, 1848 (1984).
28. Friedlander, D. R., Brackenbury, R. and Edelman, G. M., *J. Cell Biol.* **101**, 412 (1985).
29. Hoffman, S. and Edelman, G. M., *Proc. Natl. Acad. Sci. USA* **80**, 5762 (1983).
30. Hoffman, S., Chuong, C.-M. and Edelman, G. M., *Proc. Natl. Acad. Sci. USA* **81**, 6881 (1984).
31. Murray, B. A., Hemperly, J. J., Gallin, W. J., MacGregor, J. S., Edelman, G. M. and Cunningham, B. A., *Proc. Natl. Acad. Sci. USA* **81**, 5584 (1984).
32. Gallin, W. J., Prediger, E. A., Edelman, G. M. and Cunningham, B. A., *Proc. Natl. Acad. Sci. USA* **82**, 2809 (1985).
33. Greenberg, M. E., Brackenbury, R. and Edelman, G. M., *Proc. Natl. Acad. Sci. USA* **81**, 969 (1984).
34. Brackenbury, R., Greenberg, M. E. and Edelman, G. M., *J. Cell Biol.* **99**, 1944 (1984).
35. Thiery, J.-P., Duband, J.-L., Rutishauser, U. and Edelman, G. M., *Proc. Natl. Acad. Sci. USA* **79**, 6737 (1982).
36. Edelman, G. M., Gallin, W. J., Delouvée, A., Cunningham, B. A. and Thiery, J.-P., *Proc. Natl. Acad. Sci. USA* **80**, 4384 (1983).
37. Thiery, J.-P., Delouvée, A., Gallin, W. J., Cunningham, B. A. and Edelman, G. M., *Dev. Biol.* **102**, 61 (1984).
38. Crossin, K. L., Chuong, C.-M. and Edelman, G. M., *Proc. Natl. Acad. Sci. USA* **82**, 6942 (1985).
39. Chuong, C.-M. and Edelman, G. M., *J. Cell Biol.* **101**, 1009 (1985).
40. Chuong, C.-M. and Edelman, G. M., *J. Cell Biol.* **101**, 1027 (1985).

41. Sengel, P., *Morphogenesis of Skin*. Cambridge University Press, New York (1976).
42. Thiery, J.-P., Delouvée, A., Grumet, M. and Edelman, G. M., *J. Cell Biol.* **100**, 442 (1985).
43. Daniloff, J. D., Chuong, C.-M., Levi, G. and Edelman, G. M., *J. Neurosci.* **6**, 739 (1985).
44. Chuong, C.-M. and Edelman, G. M., *J. Neurosci.* **4**, 2354 (1984).
45. Grumet, M., Hoffman, S., Crossin, K. L. and Edelman, G. M., *Proc. Natl. Acad. Sci. USA* **82**, 8075 (1985).
46. Slack, J. M. W., *From Egg to Embryo*. Cambridge University Press, New York (1983).
47. Trinkhaus, J. P., *Cells Into Organs. The Forces that Shape the Embryo*, 2nd edition. Prentice Hall, Englewood Cliffs, N.J. (1984).
48. Bertolotti, R., Rutishauser, U. and Edelman, G. M., *Proc. Natl. Acad. Sci. USA* **77** 4381 (1980).
49. Buskirk, D., Thiery, J.-P., Rutishauser, U. and Edelman, G. M., *Nature (London)* **285**, 488 (1980).
50. Fraser, S. E., Murray, B. A., Chuong, C.-M. and Edelman, G. M., *Proc. Natl. Acad. Sci. USA* **81**, 4222 (1984).
51. Rieger, F., Grumet, M. and Edelman, G. M., *J. Cell Biol.* **101**, 285 (1985).
52. Edelman, G. M. and Chuong, C.-M., *Proc. Natl. Acad. Sci. USA* **79**, 7036 (1982).
53. Roux, W., *Die Entwicklungsmechanik der Organismen*, p. 283. W. Engelmann, Leipzig (1905).
54. Edelman, G. M., in *The Cell in Contact: Adhesions and Junctions as Morphogenetic Determinants* (eds. G. M. Edelman, J.-P. Thiery). John Wiley & Sons, New York (1985).
55. Sorkin, B. C., Hoffman, S., Edelman, G. M. and Cunningham, B. A., *Science* **225**, 1476 (1984).